Introduction to Modern Physics

Introduction

F. K. Richtmyer

Late Professor of Physics
Cornell University

E. H. Kennard

Late Professor of Physics
Cornell University

John N. Cooper

Professor of Physics
Naval Postgraduate School

to *Modern Physics*

SIXTH EDITION

TATA McGRAW-HILL PUBLISHING COMPANY LTD.

New Delhi

Introduction to Modern Physics

T M H Edition 1976

Reprinted	1978
Reprinted	1980
Reprinted	1981
Reprinted	1982

Reprinted in India by arrangement with McGraw-Hill, Inc. New York.

This edition can be exported from India only by the publishers,
Tata McGraw-Hill Publishing Company Ltd.

Published by Tata McGraw-Hill Publishing Company Limited and
Printed by Mohan Makhijani at Rekha Printers Pvt. Ltd., New Delhi-110020

Preface

Although the spectacular growth of physics in the past forty years has led to an extensive recasting of the material in Richtmyer's first edition and to a myriad of additions, subsequent revisions have retained so much of his goals and spirit that excerpts from his original preface are still pertinent. The present edition reflects the labors of several authors but Richtmyer's words, as appropriate for the sixth edition as they were for the first, speak for all. In 1928 he wrote:

"The purpose of this book is, frankly, pedagogical. The author has attempted to present such a discussion of the origin, development, and present status of *some* of the more important concepts of physics, *classical as well as modern*, as will give to the student a correct perspective of the growth and present trend of physics as a whole. Such a perspective is a necessary basis—so the author, at least, believes—for a more intensive study of any of the various subdivisions of the subject. While for the student whose interests are cultural, or who is to enter any of the professions directly or indirectly related to physics, such as engineering, chemistry, astronomy, or mathematics, an account of modern physics which gives the *origin* of current theories is likely to be quite as interesting and valuable as in a categorical statement of the theories themselves. Indeed, in *all* branches of human knowledge the 'why' is an absolutely indispensable accompaniment to the 'what.' 'Why?' is the proverbial question of childhood. 'Why?' inquires the *thoughtful* (!) student in classroom or lecture hall. 'Why?' demands the venerable scientist when listening to an exposition of views held by a colleague. Accordingly, if this book seems to lay somewhat greater emphasis on matters which are frequently regarded as historical, or, if here and there a classical experiment is described in greater detail than is customary, it is with a desire to recognize the importance of 'why.'

"If one were to attempt to answer all of the 'why's' raised by an intelligent auditor in answer to a categorical statement, such as, 'The atom of oxygen is composed of eight electrons surrounding a nucleus containing four alpha particles,' one would have to expound a large part of physical science from Copernicus to Rutherford and Bohr. To attempt a statement of even the more important concepts, hypotheses, and laws of modern physics and of their origin and development would be an encyclopedic task which, at least in so far as concerns the aim of this book, would fall of its own weight. Accordingly, it has been necessary to select those parts of the subject which best serve our purpose. This selection, as well as the method of presentation, has been based upon the experience gained in giving the above-mentioned lectures to numerous groups of students. Many very important developments, particularly the more recent ones, either have been omitted entirely or have been given only a passing comment. And even in those parts of the subject which have been discussed, there has been no attempt to bring the discussion strictly up to date. Indeed, with the present rapid growth of physics, it would be quite impossible for any *book*, even a special treatise, to be strictly up to date. Happily, for our purpose, up-to-date-ness is not an imperative requisite, since it is assumed that the student who wishes the *latest* knowledge will consult the current periodicals.

"In this connection, it should be emphasized that this book is an *introduction* to modern physical theories and is intended neither as a compendium of information nor as a critical account of any of the subjects discussed. In preparing the manuscript, the author has consulted freely the many very excellent texts which deal with the various special topics. Save for here and there a very minor item, or an occasional novelty in presentation, the book makes no claim to originality, except, perhaps, as regards the viewpoint from which some parts have been written.

"It is assumed that the student is familiar with the elementary principles of calculus, for no account of modern physics can dispense with at least a limited amount of mathematical discussion, if for no other reason than to emphasize the fact that, in the progress of physics, *theory* and *experiment* have gone hand in hand. Partly, however, for the sake of brevity and partly in the attempt always to keep the underlying physical principles in the foreground, numerous 'short cuts' and simplifications, some of them perhaps rather questionable from a precise standpoint, have been introduced. These elisions should cause no confusion. . . .

"There is no more fascinating story than an account of the development of physical science as a whole. (*Any* scientist would probably make the same statement about *his own* science!) Such a study leads to certain broad generalizations which are of outstanding importance in evaluating current theories and concepts. For example, one finds that, taken by and large, the evolution of physics has been characterized by *continuity*. That is to say: With a few exceptions, the ideas, concepts, and laws of physics have evolved *gradually;* only here and there do we find *outstanding* discontinuities. The discovery of photoelectricity, of x-rays, and of radioactivity represent such discontinuities and are correctly designated 'discoveries.' But we must use 'discover' in a quite different sense when we say that J. J. Thomson 'discovered' the electron. The history of the electron goes back at least to Faraday. Thomson's experiments are all the more brilliant because he succeeded in demonstrating, by direct experiment, the existence of something, evidence concerning which had been previously *indirect*. Then, there are the respective roles played by qualitative and by quantitative work. Numerous important discoveries have been made 'by investigating the next decimal place.' Witness the discovery of argon. And ever since Kepler proved that the orbits of the planets are ellipses, relations expressible in *quantitative* form have carried greater weight than those which could be stated only qualitatively. For example, Rumford's experiments on the production of heat by mechanical means were suggestive. But Joule's measurement of the mechanical equivalent of heat was *convincing*. If, directly, or indirectly by inference, the author has succeeded here and there in the text in pointing out such generalizations as these, one more object which he has had in mind will have been accomplished."

With the objective of making the sixth edition a more useful intermediate textbook, the material has been organized into shorter chapters to which problem sets have been added. New chapters have been prepared on relativity, wave mechanics, and solid-state physics. To cover the entire book would require a full year's course, but appropriate selection of chapters will make the book a suitable text for courses as short as one quarter or one semester. In many chapters topics which are often omitted in shorter courses are set off on a gray background. To keep the size of the book within reasonable bounds, many subjects treated in the fifth edition are omitted or discussed more briefly; the largest deletion is virtually the entire chapter on cosmic rays.

The general development has been carried out using rationalized MKSA units, but there has been no hesitancy to use units such as the gram, the electron volt, or the Angstrom unit whenever they seemed more appropriate (e.g., the tradional use of cm^{-1} for wavenumbers is retained). In general, symbols and abbreviations follow the system proposed by the commissions of the International Union of Pure and Applied Physics.

After the death of Professor Richtmyer in 1939, Professor E. H. Kennard undertook the preparation of the third and fourth editions with Professor L. G. Parratt collaborating on the chapter on x-rays. Professor T. Lauritsen joined Kennard in bringing out the fifth edition. When time came for a sixth edition, Dr. Kennard's health was failing (his death came in January, 1968) and Professor Lauritsen declined the invitation to assume responsibility for the revision. Although he read critically many chapters of the present manuscript and proposed numerous improvements, Professor Lauritsen elected to restrict his participation in the present edition and, at his own request, he is not listed as an author.

In addition to my indebtedness to Professor Lauritsen, I acknowledge help from many persons. First of all, my wife Elaine typed and helped with the physical preparation of the entire manuscript. Students who studied selected chapters in preliminary form have made valuable suggestions and caught many errors. Several colleagues have been very helpful in reviewing material in the areas of their specialties; in particular, I thank Professor R. L. Kelly for his aid on the chapters on atomic spectroscopy, Professors K. E. Woehler and W. B. Zeleny for assistance on relativity, and Professor G. E. Schacher for constructive comments on the chapters on the solid state.

In spite of great effort to avoid errors it is inevitable that some will creep into any book. I will be grateful to anyone who finds and reports mistakes and deficiencies in this book.

John N. Cooper

Contents

chapter one

The Heritage of Modern Physics

As a prelude to understanding modern physics, one should be familiar with the main events in the evolution of natural philosophy over the centuries. Physicists owe it to their science to possess sufficient knowledge of its history for a correct perspective of its development and present-day importance. That very history, the lives of the men whose labors formed our subject, and the way physics has molded human thought and contributed to modern civilization claim our first attention.

1.1   Classical vs. Modern Physics

Taken literally, *modern physics* means, of course, the *sum total* of knowledge included under the head of present-day physics.   In this sense, the physics of 1890 is still modern; very few statements made in a good physics text of that year would have to be changed today as untrue.

On the other hand, since 1890, there have been enormous advances in physics, some of which have necessitated radical modifications in certain theories that had seemed to be strongly supported by experimental evidence. In particular, the metaphysical assumptions of the nineteenth century proved to be an inadequate foundation for the growing structure of natural philosophy. New ideas were required, differing sharply with the older ones in special areas where the latter failed but still yielding the same predictions where the older ideas had been successful. The extensive physics of 1890, which includes analytical mechanics, thermodynamics, and maxwellian electrodynamics, is often called *classical physics*. Then "modern physics" implies the quantitative description of phenomena which could not be understood in terms of classical assumptions.

Of course, there is no sharp demarcation between classical and modern physics; the latter grows out of the former and is nourished by it. For example, few physicists in 1890 questioned the wave theory of light. Its triumph over the corpuscular theory seemed to be complete, particularly after the brilliant experiments of Hertz in 1887, which demonstrated the fundamental soundness of Maxwell's electromagnetic theory of light. And yet, by an irony of fate which makes the story of modern physics full of dramatic situations, these very experiments brought to light a new phenomenon, the photoelectric effect, which played an important role in establishing the quantum theory. In many of its aspects the quantum theory seems diametrically opposed to the wave theory of light; indeed, the reconciliation of these two theories, each based on incontrovertible evidence, was one of the great problems of the first quarter of the twentieth century.

For the brief historical sketch which follows, the evolution of premodern physics is arbitrarily divided into three periods.

The first period extends from the earliest times up to about A.D. 1550, this date marking approximately the beginning of the experimental method. During this long era, there was some advance in the accumulation of the *facts* of physics as a result of the observation of natural phenomena, but the development of extensive physical *theories* was rendered impossible, partly by the speculative, metaphysical nature of the reasoning employed, but more particularly by the almost complete absence of experiment to test the correctness of such theories as were proposed. The main characteristic of this period, therefore, is *the absence of systematic experiment*.

The second period extends from A.D. 1550 to 1800. Numerous basic advances were made during this period—by such men as Gilbert, Galileo, Newton, Huygens, Boyle—but its most important characteristic is *the development and the firm establishment of the experimental method* as a recognized and accepted means of scientific inquiry. This period was inaugurated by the classic work of Galileo (1564–1642); but it took nearly two centuries more to bring universal recognition of the following basic principle:

Science can advance only so far us theories, themselves based upon experiment, are accepted or rejected according as they either agree with, or are contrary to, other experiments devised to check the theory.

The third period, 1800 to 1890, is characterized by the development of what is now called classical physics in contrast with the quantum physics of the present century. The experiments of Count Rumford and Joule led to our present kinetic theory of heat. The observations of Thomas Young (1802) and his proposal of the principle of interference (of two beams of light) resulted ultimately in the triumph of Huygens' wave theory of light. The researches of Faraday and others gave Maxwell the material for the crowning achievement of this period, the electromagnetic theory.

First Period: Earliest Times to A.D. 1550

1.2   The Greeks

Relatively speaking, the contributions made by the Greeks to the natural sciences were far less than their contributions to mathematics, literature, art, and metaphysics. Nevertheless, in spite of their vague and misty philosophizing concerning natural phenomena and in spite of their general failure to test theory by experiment, the Greeks gave to the world much of the physics that was known up to A.D. 1400. In their writings, one finds the germ of such fundamental concepts as the conservation of matter, inertia, atomic theory, the finite velocity of light, and the like.

1.3   Thales of Miletus (640–546 B.C.[1])

According to Aristotle, Thales was acquainted with the attractive power of magnets and of rubbed amber. He is often said to have discovered the inclination of the ecliptic and the spherical shape of the earth, but Aristotle credited him with the doctrine that the earth was cylindrical in shape and rested on water.

[1] Few of these ancient dates are known with certainty, and authorities differ.

1.4   *Pythagoras* (580–500 B.C.)

One of the greatest of the .rly Greek philosophers and the founder of
the Pythagorean school, he held that the earth is spherical, although
the basis of this belief is not known; it may have been based on the idea
that the sphere is the most perfect of all figures.   Pythagoras believed
that the entire universe was spherical in shape with the earth at i's
center and that the sun, stars, and planets moved in independent circles
around the earth as a center.

1.5   *Anaxagoras* (500–428 B.C.) *and Empedocles* (484–424 B.C.)

According to Plato, Anaxagoras neglected his possessions in order to
devote himself to science.   He is credited with the view that the moon
does not shine by its own light but that "the sun places the brightness
in the moon" and "the moon is eclipsed through the interposition of the
earth."   Also, "The sun is eclipsed at new moon through the interposition of the moon."   Anaxagoras was accused of impiety, however,
because he taught that the sun was a red-hot stone and that the moon
was simply earth, and for holding this doctrine he was banished from
Athens.

To Anaxagoras is due the germ of the atomic hypothesis of Democritus, who lived in the next generation.   Anaxagoras denied the conten-
tion of the earlier Greeks regarding the creation or destruction of matter.
He taught that changes in matter are due to combinations or separations of small, invisible particles (*spermata*).   These particles themselves were conceived to be unchangeable and imperishable but different
from each other in form, color, and taste.   This doctrine foreshadowed
the law of the conservation of matter.

Empedocles, on the other hand, reduced the elements to four—
earth, water, air, and fire—through combinations and separations of
which the All exists.   He also held that light is due to the emission by
the luminous or visible body of small particles that enter the eye and
are then returned from the eye to the body, the two "streams" giving
rise to the sense of form, color, etc.

According to Aristotle, Empedocles believed that light "takes time
to travel from one point to another."   This idea was rejected by Aristotle,
who stated that "though a movement of light might elude our observation within a short distance, that it should do so all the way from east to
west is too much to assume."

1.6 Democritus (460–370 B.C.)

He gave more definite form to the atomic hypothesis of Anaxagoras by postulating that the universe consists of empty space and an (almost) infinite number of indivisible and invisible particles which differ from each other in form, position, and arrangement. In support of this hypothesis, Democritus argued that the creation of matter is impossible, since *nothing* can come from *nothing* and, further, nothing which is can cease to exist. Aristotle put the argument in this form: "If, then, some one of the things which are is constantly disappearing, why has not the whole of what is been used up long ago and vanished away?" But Aristotle rejected the atomic hypothesis.

1.7 Aristotle (384–332 B.C.)

A pupil of the philosopher Plato, Aristotle contributed so much to all branches of knowledge—logic, rhetoric, ethics, metaphysics, psychology, natural science—that it is difficult to sift out that which is germane to a brief history of physics. Perhaps the most important single fact is the tremendous influence which, as a result of his intellectual brilliance and achievements in *many* branches of learning, he exerted for many succeeding centuries in *all* branches, physics included. Viewed from our twentieth-century vantage point, however, not a little of his reasoning concerning the physical universe sounds like piffle.

Yet Aristotle frequently calls in observed facts to substantiate his speculations. For example, in "De Caelo" (Book II, Cap. XIV), after proving, by a more or less abstract argument, that the earth is spherical, he says:

> The evidence of all the senses further corroborates this. How else would eclipses of the moon show segments as we see them? . . . since it is the interposition of the earth that makes the eclipse, the form of this line [i.e., the earth's shadow on the moon] will be caused by the form of the earth's surface, which is therefore spherical.

He also points to the apparent change in altitude of the stars as one travels north or south and concludes that "not only is the earth circular, but it is a circle of no great size."

Indeed, in theory if not in his own practice, Aristotle places great emphasis on the importance of facts in connection with scientific development. In a paragraph in "De Generatione et Corruptione" (Book I, Cap. II), he says:

Lack of experience diminishes our power of taking a comprehensive view of the admitted facts. Hence, those who dwell in intimate association with nature and its phenomena grow more and more able to formulate, as the foundation of their theories, principles such as to admit of a wide and coherent development; while those whom devotion to abstract discussions has rendered unobservant of the facts are too ready to dogmatize on the basis of a few observations.

This is surely good doctrine even for twentieth-century scientists!

An attempt to summarize Aristotle's views on physics is beyond the scope of this book, but reference may be made to two of his doctrines because of their bearing upon subsequent history.

The first is his supposed views on falling bodies. The statement is commonly made that Aristotle held that a heavy body would fall from a given height with greater velocity than a light body. It is difficult to be sure from Aristotle's extant writings just what he actually held in regard to this point. The passages that seem to refer to it occur in the course of his arguments against the possibility of the existence of a void. For example, he states:[1]

We see the same weight or body moving faster than another for two reasons, either because there is a difference in what it moves through, as between water, air, and earth, or because, other things being equal, the moving body differs from the other owing to excess of weight or lightness. . . . And always, by so much as the medium is more incorporeal and less resistant and more easily divided, the faster will be the movement.

Here, as elsewhere, Aristotle speaks always of movement of a body through a medium. He may have meant *terminal* velocity, such as the constant velocity of rain drops as they approach the ground. The *heavier* drops *do* fall faster. It seems probable that Aristotle, believing that a medium of some sort must always be present, was unaware of such distinctions as that between terminal and nonterminal velocity and actually did believe that in all stages of its motion the heavier body falls faster.

The second doctrine referred to is that of the motion of the earth, sun, and planets. In his "De Caelo" (Book II, Cap. XIV), after a series of abstract arguments, in the course of which he states that "heavy bodies forcibly thrown quite straight upward return to the point from which they started even if they be thrown to an infinite [!] distance," Aristotle concludes "that the earth does not move and does not lie elsewhere than at the center." He supposed that the sun, planets, and stars are carried by a series of concentric spheres which revolve around the earth as a center.

[1] This and the following quotation from Aristotle are taken from "The Works of Aristotle Translated into English," vol. II, Clarendon Press, Oxford, 1930.

1.8   Aristarchus (310–230 B.C.)

He enunciated a cosmogony identical with that proposed by Copernicus nearly 2000 years later.   No mention of this hypothesis is made in his only extant work, "On the Sizes and Distances of the Sun and Moon," but Archimedes tells us, in a book called "The Sand-reckoner," that "Aristarchus of Samos brought out a book" containing the hypothesis "that the fixed stars and the sun remained unmoved; that the earth revolves around the sun in the circumference of a circle, the sun lying in the middle of the orbit"; and that "the sphere of the fixed stars" is very great compared with the circle in which the earth revolves. The prestige of Aristotle was too great, however, and the geocentric hypothesis that he supported was so completely satisfactory to the ancient mind that Aristarchus' theory was practically lost for nearly 2000 years.

1.9   Archimedes (287–212 B.C.)

Probably the most noted physicist of antiquity, Archimedes was a man of great ability in mathematics as well as a practical engineer.   He invented the endless screw, the water screw, a pulley block, and the burning mirror.   He was a founder of statics, and in his book "On Floating Bodies" he laid the foundations of hydrostatics, including his famous principle, "A solid heavier than a fluid will, if placed in it, descend to the bottom of the fluid, and the solid will, when weighed in the fluid, be lighter than its true weight by the weight of the fluid displaced."

1.10   From the Greeks to Copernicus

To give but a passing comment to the 17 centuries between Archimedes and Copernicus would seem to give the reader the false impression that no developments of moment occurred during that long period.   During ancient times Ptolemy of Alexandria (A.D. 70–147) collected the optical knowledge of his time in a book in which he discussed, among other things, reflection from mirrors—plane, convex, concave—and, particularly, refraction, which Ptolemy evidently studied experimentally.   He gave, in degrees, relative values of angles of incidence and of refraction for air-water, air-glass, and water-glass surfaces and described an apparatus by which he determined these quantities; he stated that for a given interface these two angles are proportional.   He also mentioned atmospheric refraction as affecting the apparent position of stars. He invented a complicated theory of the motions of the planets in their orbits about the earth in order to explain their apparent motions among the stars.

From Ptolemy to the Arabian Alhazen is a span of nine centuries—twice the total lapse of time from the discovery of America to the present—during which there was stagnation in almost all intellectual pursuits. But about the eighth century the Arabs began to cultivate chemistry, mathematics, and astronomy, in part by translating into Arabic the works of the Greeks but also by making original contributions. About the year 1000, Alhazen produced a work on optics in seven books. This treatise sets forth a clear description of the optical system of the eye, discusses the laws of concave and convex mirrors, both spherical and cylindrical, and carries the subject of refraction a little further by recognizing that the proportionality between the angles of incidence and refraction holds only for small angles.

During the next 500 years, relatively few advances in physics were made. Roger Bacon (1214–1294), English philosopher, scientist, and Franciscan monk, taught that in order to learn the secrets of nature *we must first observe.* He believed in mathematics and in deductive science, but he clearly realized that only as scientific conclusions are based on observed phenomena and tested by experiment can useful knowledge result.

About the same time Petrus Peregrinus recognized that magnetic poles are of two kinds, like poles repelling and unlike attracting each other.

Then there was Leonardo da Vinci (1452–1519), Italian painter, architect, sculptor, engineer, and philosopher, whose greatness as a scientist has come to be appreciated only in recent years. Since his works were left in manuscript form and were probably not widely known among his contemporaries, his influence on early science is comparatively insignificant. His belief in the value of experiment is worthy of the twentieth century: "Before making this case a general rule, test it by experiment two or three times and see if the experiment produces the same effect." Although expressed in the vague language of his time, some of his ideas concerning what we now refer to as force, inertia, acceleration, the laws of motion, etc., were qualitatively correct. Concerning perpetual motion, he wrote: "Oh, speculators on perpetual motion, how many vain projects of the like character you have created! Go and be the companions of the searchers after gold." Rejecting the Ptolemaic theory, he held that "the sun does not move." That he was not persecuted or even burned at the stake, as Bruno was a century later, for holding such revolutionary and therefore (!) heretical views is probably due to the fact that his doctrines were given little publicity; for, holding no academic position, he did not teach, and he published nothing.

Finally, in the sixteenth century, the full force of that period of intense intellectual activity known as the Renaissance began to be felt in the field of physics. Then came such men as Copernicus, Tycho, Kepler, Galileo, and Newton, who with their contemporaries and colleagues, in a space of hardly more than a century, broke the spell of Aristotle and made possible the beginnings of modern experimental science. Insofar as the heliocentric theory completely revolutionized man's conception of the universe and his place in it, it is quite correct to regard the work of Copernicus as beginning a new era in scientific thought. But had it not been for other discoveries coming immediately after Copernicus, such as the telescope, Kepler's laws, Galileo's famous experiments on falling bodies, and many others, it is quite possible that the theory of Copernicus would have had the same fate as that of Aristarchus centuries earlier. It is, therefore, fitting to regard the birth of the Copernican theory as *closing* the first period in the history of physics.

1.11 The Copernican System

Copernicus (1473–1543), a younger contemporary of Columbus, spent most of his life as one of the leading canons in the monastery at Frauenburg, near the mouth of the Vistula. His theory of the universe is set forth in his "De Revolutionibus Orbium Coelestium," published near the close of his life.

Copernicus perceived that, by assuming that the earth is a planet like the others and that all the planets move in circles around the sun, a great simplification, both philosophical and mathematical, could be made with regard to the world system. He could thus easily account for the seasons and for the apparent retrograde motion, at times, of the planets. The rotation of the earth on its axis causes the *apparent* daily motion of the sun, moon, and stars; and he pointed out that, probably, the stars are too far away for any motion of the earth to effect their apparent positions. He gave the correct order of the planets from the sun outward.

Whatever the system as proposed by Copernicus lacked quantitatively, it was correct, in its main outline, qualitatively. Its reasonableness set a few men thinking and did much to usher in a new era in science, an era that could come only when truth could have the opportunity of standing alone, *unaided* or *unhindered* by the "authority" of 2000 years.

Second Period (A.D. 1550–1800): Rise of the Experimental Method

1.12 *Galileo Galilei* (1564–1642)

Galileo is widely regarded as the father of experimental physics. To be sure, physics has grandfathers and still more remote ancestors, but none of them gave more than he. Galileo was descended from a noble family, and it is probable that he inherited from his father the spirit of free inquiry which characterized his life; for, in the writings of the elder Galileo, who was well educated and an accomplished musician, is the statement: "It appears to me that they who in proof of any assertion rely simply on the weight of authority, without adducing any argument in support of it, act very absurdly."

As a student in the monastery of Vallombrosa, near Florence, the young Galileo excelled in the classics and literature and was something of a poet, musician, and art critic. He also showed an aptitude for science and exhibited considerable mechanical inventiveness. At the age of seventeen he was sent to the University of Pisa to study medicine. It was here that he made his first discovery and invention. One day, in 1581, he noticed the regular oscillations of the great hanging lamp in the cathedral at Pisa. Although the amplitude of these oscillations became less and less, they were all performed in the same time, as he determined by counting his pulse. Turning the process around, he invented a pulsometer, a ball-and-string, i.e., simple pendulum, device, whose length, when adjusted to synchronism with the pulse, was a measure of its frequency.

But the urge toward mathematics and science overcame the pecuniary advantages of a medical career; at twenty-six, Galileo became professor of mathematics at Pisa. Here he began a systematic investigation of the mechanical doctrines of Aristotle. He soon demonstrated by experiment that Aristotle was in error in many of his assertions. Aristotle was commonly understood to teach that a heavy body falls faster than a light one. This doctrine had been questioned, on the basis of actual test, by various writers, e.g., by Philoponus in the fifth century and by Benedetto Varchi in the generation before Galileo. Nevertheless the authority of Aristotle had continued to be accepted. To test the point, Galileo is rumored to have tried the famous experiment of dropping bodies of unequal weight from the leaning tower of Pisa and finding that they all fell with practically equal velocities. While we know nothing of the details of the experiment, it is certain that Galileo publicized Aristotle's error.

Fig. 1.1 Galileo Galilei.

Then began a persecution which led Galileo to quit Pisa. In 1592 he became professor of mathematics at the University of Padua, where he remained 18 years, enjoying comparative liberty of thought and teaching. His fame as a teacher spread all over Europe, and his lectures were crowded.

In 1608, a Dutch optician, Lippershey, as a result of a chance observation of an apprentice, had succeeded in "making distant objects appear nearer but inverted" by looking through a tube in which were mounted two spectacle lenses. News of this invention reached Galileo in June, 1609. Grasping the principle involved, he made a telescope and exhibited it in Venice "for more than a month, to the astonishment of the chiefs of the republic." By January, 1610, Galileo had made a telescope with a power of 30 diameters,[1] with which he made a number of fundamental discoveries. He saw that the number of fixed stars was vastly greater than could be seen by the unaided eye, and thus he explained the agelong puzzle of the Milky Way. He saw that the

[1] Galileo's telescopes were similar to the modern opera glass—a double-convex (or plano-convex) object glass and a double-concave eyepiece. Thus, they had an erect image.

planets appeared as luminous disks while the stars still remained points of light, and he discovered the satellites of Jupiter. These discoveries made Galileo famous. He soon accepted an invitation to return to Pisa as "First Mathematician and Philosopher," at a substantial increase in salary, but at a sacrifice of his academic freedom in Padua. Continuing his astronomical investigations, he discovered the crescent phases of Venus, sunspots and the rotation of the sun, the faculae of the solar atmosphere, and the libration of the moon. In 1612, he published his "Discourse on Floating Bodies."

At first, it seemed as if his fame had silenced all opposition from the church. But the support that his discoveries gave to the hated Copernican theory and his vigorous attacks on Aristotelian philosophy roused his enemies to fury, with the result that in 1615 he was hauled before the Pope and, under threat of imprisonment and torture, was "enjoined . . . to relinquish altogether the said opinion that the sun is the center of the world and immovable . . . nor henceforth to hold, teach, or defend it in any way . . . ." Simultaneously, it was decreed that the works of Copernicus "be suspended until they be corrected." Galileo acquiesced to these decrees and was allowed to return to Pisa, where he continued his researches.

In 1623, one of Galileo's friends, Barberini, became Pope Urban VIII, from whom Galileo received assurances of "pontifical good will." Thereupon he began the writing of his great book, "Dialogues on the Ptolemaic and Copernican Systems." This was published in 1632, under formal permission from the censor. The form of these dialogues is ingeniously contrived to abide by the *letter* of the decree of 1615. Three "characters" carry on the discussion: Salviati, a Copernican; Simplicio, an Aristotelian; and Sagredo, a witty, impartial, good-natured chairman. The dialogues cover four "Days," during which the arguments for and against each system are set forth with apparent impartiality and without reaching any *stated* conclusion. Nevertheless, the general effect of the book was a powerful argument for Copernicanism.

At the instigation of his enemies Galileo was presently called before the Inquisition. He was sixty-seven years old and impaired in health and spirit. Bowing to the inevitable, he indicated his "free and unbiased" willingness to recant, to "abjure, curse, and detest the said heresies and errors and every other error and sect contrary to the Holy Church," and he agreed "never more in future to say or assert anything, verbally or in writing, which may give rise to a similar suspicion." Thereafter Galileo was a prisoner under suspended sentence, first at Rome, then at his home in Arcetri. Here, during the last years of his life, he prepared and in 1636 published his "Dialogues on Two New Sciences," i.e., cohesion and motion.

The dialogues on "Motion" sum up Galileo's earlier experiments and his more mature deliberations. He states that "if the resistance of the media be taken away, all matter would descend with equal velocity." He deduces the formulas of uniformly accelerated motion. He shows that the path of a projectile is parabolic under suitable limiting conditions and states that if all resistance were removed, a body projected along a horizontal plane would continue to move forever. His work on mechanics paved the way for the enunciation by Newton of the famous three laws of motion.

1.13 Tycho Brahe (1546–1601) and Kepler (1571–1630)

The work of Tycho and Kepler is particularly interesting not only because of its direct bearing on the development of physics but more particularly because of the mutual dependence of the work of each one upon that of the other, a relation common in present-day science. Tycho was the experimentalist who supplied the accurate data upon which Kepler built a new theory of planetary motion. Without a Kepler, Tycho's observations would have attracted hardly more than passing notice. Kepler, in turn, might have theorized to his heart's content but without the accurate data of a Tycho, those theories might ultimately have shared the fate of Aristotle's. Sometimes theory precedes, sometimes experiment. But neither can get far without the other.

Tycho Brahe, born of a noble family in Sweden, was educated for a career as a statesman, but he developed a consuming interest in astronomy. By observations of his own, he found that the current astronomical tables were incorrect. In 1575, he was put in charge of the observatory of Uraniborg by King Frederick II of Denmark, one of his duties being to make *astrological* calculations for the royal family. Here he spent 20 years making systematic observations of the planets, constructing a star catalog, and accumulating other astronomical data, always with the highest accuracy that could be attained without a telescope. In 1599, he undertook to establish a new observatory at Prague for the German emperor, Rudolph II, but in the midst of this work he suddenly died.

Among Tycho Brahe's assistants at Prague was a brilliant young mathematician, Johann Kepler. He succeeded Tycho as principal mathematician to the emperor and undertook to carry to completion the new set of astronomical tables based on the elaborate observations of his predecessor. Kepler remained at Prague until 1612; from then until his death, in 1630, he held a professorship at Linz.

Tycho Brahe had rejected the Copernican system for a geocentric system of his own. It is one of the ironies of science that his own data on planetary motions, taken in support of his own theory, became, in the hands of Kepler, the clinching argument for the Copernican system. Using Tycho Brahe's observations, Kepler made a special study of the motion of Mars. He tried to reconcile the recorded positions of the planet by assuming circular orbits for Mars and for the earth, trying various positions of these orbits relative to the sun. None worked. Resorting to the Ptolemaic notion of epicycles and deferents led to some improvement, but still the observed positions differed from the computed, in some cases by as much as 8 minutes of arc. Kepler knew that the observations could not be in error by that amount. Some new concept regarding planetary motion was necessary.

Then Kepler gave up *uniform* circular motion and assumed that the speed varied inversely as the planet's distance from the sun. This assumption is his famous "second law," that *the radius vector from the sun to the planet describes equal areas in equal times.* It worked approximately, but still there were systematic errors which exceeded the possible errors of observation. Finally, he cast aside the last traditions of the Ptolemaic system and tried orbits of other forms, first, an oval path and then an ellipse with the sun at one focus. At last, his years of computation bore fruit. The path *was* an ellipse. Theory and observation agreed! And one of the most important and far-reaching laws in all science had been discovered, all because of a discrepancy of 8 minutes of arc between observation and theory! One of the striking things in the growth of science is the fact that many fundamental advances have come about because of just such discrepancies, frequently very small ones, between observation and theory.

Eventually, Kepler handed down to posterity the three laws of planetary motion, which, sweeping away all remnants of the Ptolemaic system, paved the way for modern astronomy:

1. The planets move around the sun in orbits which are ellipses with the sun at one focus.
2. The radius vector (from sun to planet) sweeps over equal areas in equal times.
3. The squares of the periods of revolution of the planets around the sun are proportional to the cubes of the semimajor axes of their respective orbits.

But what makes the planets move? *Why* do the outer ones go more slowly? Is there "one moving intelligence in the sun, the common center, forcing them all around, but those most violently which are nearest?" Kepler speculated long on this question and arrived at the idea of an attraction acting between any two material bodies. This *qualitative* idea of Kepler's was later developed by Newton into his *quantitative* theory of universal gravitation. Kepler himself, however, seems to have had no idea that it is this very attractive force which keeps the planets in their orbits.

Kepler also made substantial contributions in the field of optics. He understood the principle of total reflection and how to determine what we now call the *critical angle*.   He studied atmospheric refraction as affecting the apparent position of the heavenly bodies and worked out an approximate formula to allow for this error for all positions, from zenith to horizon.   He was the first to propose the meniscus type of lens.   And he invented the Keplerian telescope, in which a *real* image is formed, thus making possible accurate measurements by means of cross hairs in the focal plane of the objective.

1.14   The Experimental Method Spreads

The impetus given to science by Galileo, Tycho, and Kepler resulted in an ever increasing number of investigators in the generations that followed.   We mention only a few of them.   Of great significance, too, is the formation in Europe at about this time of learned societies which brought together, for argument and discussion, men of kindred interests. The Lincean Society was founded in Italy, in 1603; the Royal Academy of Sciences, in France, in 1666; and the Royal Society, in England, in 1662.   Continued improvement of the art of printing facilitated the diffusion of scientific knowledge.

In 1600, William Gilbert, an English physician, published his famous work "De Magnete," based largely upon his own experiments, in which he showed the fallacy of such popular fancies as the belief that lodestones lost their magnetic power when rubbed with garlic and regained it again when smeared with goat's blood.   He was the first to recognize that the earth is a great magnet, and he actually magnetized a small sphere of iron and showed that it produced a magnetic field similar to that of the earth.

Among other workers in magnetism may be mentioned Kircher (1601–1680), who, by measuring the force required to pull a piece of iron from either pole of a magnet, demonstrated the equality of the two poles; Cabeo (1585–1650), who showed that an *unmagnetized* iron needle floated freely on water would place itself along the earth's magnetic meridian; and Gellibrand (1597–1637), who discovered the secular variation of the magnetic declination.

In optics, Scheiner (1575–1650) studied the optics of the eye; Snell (1591–1626) discovered the true law of refraction; and Cavalieri (1598–1647) gave the correct formula for the focal length of a thin glass lens in terms of the radii of curvature of the two sides. In acoustics, Mersenne (1588–1648), after having investigated the laws of vibrating strings, determined, in absolute measure, the frequency of a tone. He also measured the velocity of sound by observing the time interval between the flash of a gun and the arrival of the report.

In fluid mechanics, Torricelli (1608–1647), studied the flow of liquids from orifices, discovered the principle of the barometer, and observed variation in barometric height with altitude. Working independently, Guericke (1602–1686) invented the air pump. Pascal (1623–1662) measured the difference in barometric height between the base and the top of a mountain, correctly explaining the reason for the difference, and later announced the famous principle of hydrostatics that bears his name.

Not only was physics, as a subject, beginning to assume definite form, but even the subdivisions such as mechanics, light, sound, etc., were beginning to crystallize. Then came a man

> . . . towering head and shoulders above all his contemporaries, a veritable giant among the giants, a man whose intellect and whose contributions to knowledge are incomparably greater than those of any other scientist of the past, that prince of philosophers, Sir Isaac Newton.[1]

The other "giants" referred to, contemporaries of Newton, are such men as Boyle, Huygens, and Hooke.

1.15   Sir Isaac Newton (1642–1727)

Newton was born in the little hamlet of Woolsthorpe, England, in 1642, less than a year after the death of Galileo. In the public school at Grantham, he showed at first no exceptional aptitude for his studies, but ultimately he rose to the highest place in the class. Then, at the age of fifteen, he was removed from school to assist his widowed mother in running the family estate at Woolsthorpe. But he had little taste for farming. Rather, he was interested in studying and in devising various mechanisms. He made a water clock, waterwheels, sundials, and a working model of a windmill. One morning his uncle found him under a hedge studying mathematics when he should have been farming. Thereupon Newton's mother wisely decided that an educational career was more suitable for her son, and he was sent back to school and in 1661 to Cambridge. Here his creative genius began to appear. While still a student, he discovered the binomial theorem, developed the

[1] I. B. Hart, "Makers of Science," Oxford, 1923.

methods of infinite series, and discovered fluxions, or the differential calculus.

Soon thereafter, an outbreak of the plague closed the university for some months, during which time Newton, at the family estate at Woolsthorpe, began his speculations on the subject of gravity, which later led to his enunciation of the inverse-square law. It was here that the much-quoted falling-apple episode is said to have occurred, which is supposed to have given Newton the basic idea of *universal* gravitation. But Newton himself makes no mention of the incident, and it seems far more probable that at Cambridge he had read Kepler's qualitative proposal of a general principle of gravitation. Certainly, Newton was familiar with Kepler's laws of planetary motion.

In 1667, Newton returned to Cambridge as Fellow of Trinity. At the age of twenty-six, he was appointed Lucasian Professor of Mathematics, a chair which he held for nearly 30 years. In 1703, he resigned his professorship to devote himself to duties as Master of the Mint, to the scientific work of his contemporaries, and to defending his own work against the attacks of jealous rivals. In this same year, he was elected President of the Royal Society, an office which he held until his death. In 1705, he was knighted by Queen Anne.

Most of Newton's important scientific work was done before he vacated the professorship, although he remained thereafter "a power of the first magnitude in the world of science." In his later years, he devoted much time to theological studies. Throughout his life he shunned publicity, retaining a modesty and simplicity which are indicated by a sentiment uttered shortly before his death:

> I do not know what I may appear to the world, but to myself I seem to have been only like a boy playing on the seashore, and diverting myself in now and then finding a smoother pebble or a prettier shell than ordinary, whilst the great ocean of truth lay all undiscovered before me.

Any brief account of Newton's work must inevitably give a very inadequate impression of his contributions to science. We refer here only to a few of his researches on optics and on mechanics. Newton's work on *optics* arose out of an attempt to improve lenses. The inability of a lens with spherical surfaces to bring parallel rays to a point focus was early recognized. In 1629, Descartes had shown that lenses with hyperbolic or, under certain conditions, parabolic surfaces should be free from the defect which we now call *spherical aberration*. Newton found, however, that such lenses produced only a very slight improvement in the image, and he conjectured that perhaps the trouble lay not in the lens but in the light itself.

Fig. 1.2   Sir Isaac Newton.

Accordingly, he procured a prism of glass and, placing it over a hole $\frac{1}{4}$ in. in diameter through which sunlight was shining into a darkened room, he observed the "very vivid and intense colors" produced on a wall some 20 ft distant.   Newton was surprised to find that this *spectrum*, as we now call it, was so much longer than it was wide ($13\frac{1}{4}$ by $2\frac{5}{8}$ in.).   The *width* subtended at the hole an angle corresponding *exactly* to the sun's angular diameter, but the length could not be so explained. He made various surmises as to the origin of the colors, such as the varying thickness of the prism from apex to base, the unevenness of the glass, a curvilinear motion of the light after leaving the prism, etc. One by one, experiment proved these hypotheses wrong.   Finally, he isolated one ray, or "color," by suitable screens and caused it to pass through a second prism.   In this way, he could measure the refrangibility of a single ray.   He found that the refrangibility increased from red to violet; that, therefore, the first prism simply "sorted out" the colors, which, in combination, made "white" light.   In other words, he discovered that white light is made up of the spectral colors, an elementary concept to us but new and of far-reaching importance in 1666.

Newton at once saw that this dispersion of light was the cause of his failure to effect any substantial improvement in telescopes by use of paraboloidal lenses. Furthermore, he concluded, on the basis of a hurried experiment, that in different media dispersion was always proportional to refracting power. If this were so, no combination of lenses of different materials could eliminate chromatic aberration. This singular error of Newton's retarded the development of refracting telescopes for years. In 1730, Hall made several achromatic combinations of crown and flint glasses, but he published no account of his work, so that when Dolland, about 1757, rediscovered the method of making such combinations, he was able to secure a patent on it—an invention that had been within the grasp of Newton three-quarters of a century before.

Newton's theories concerning the nature of light are of historical interest. Much has been written concerning the extent to which he may have retarded the development of optics by espousing the corpuscular theory as against the wave theory of his contemporaries Huygens (1629–1695) and Hooke (1635–1703). Newton was by no means dogmatic in his support of the corpuscular theory. In a 1675 communication to the Royal Society concerning "An Hypothesis Explaining the Properties of Light," Newton states:

> I have here thought fit to send you a description . . . of this hypothesis . . . though I shall not assume either this or any other hypothesis, not thinking it necessary to concern myself whether the properties of light discovered by men be explained by this or any other hypothesis capable of explaining them; yet while I am describing this, I shall sometimes, to avoid circumlocutation . . . speak of it as if I assumed it.

He then proceeds to describe "an aetherial medium, much of the same constitution with air but far rarer, subtiler and more strongly elastic" and supposes that

> . . . light is neither aether, nor its vibrating motion, but something of a different kind propagated from lucid bodies. They that will may suppose it an aggregate of various peripatetic qualities. Others may suppose it multitudes of unimaginable small and swift corpuscles of various sizes springing from shining bodies . . . and continually urged forward by a principle of motion which in the beginning accelerates them, till the resistance of the aetherial medium equals the force of that principle much after the manner that bodies let fall in water are accelerated till the resistance of the water equals the force of gravity.

In 1704 Newton published his optical researches in his well-known book "Opticks," of which the third edition appeared in 1724. His first sentence reads, "My Design in this Book is not to explain the

Properties of Light by Hypotheses, but to propose and prove them by Reason and Experiment." He describes his researches on refraction, reflection, colors of thin plates, etc., and he concludes by "proposing only some queries in order to further search to be made by others." One of these queries expresses his objections to the wave theory of light:

> 28. Are not all hypotheses erroneous in which light is supposed to consist in pression or motion propagated through a fluid medium? If light consists only in pression propagated without actual motion, it would not be able to agitate and heat the bodies which refract and reflect it, and . . . it would bend into the shadow. For pression or motion cannot be propagated in a fluid·in right lines beyond an obstacle . . . but will bend and spread every way into the quiescent medium which lies beyond the obstacle.

To account for the colors of thin films, which are now regarded as strong evidence for the wave properties of light, he supposes that

> every ray of light in its passage through any refracting surface is put into a certain transient constitution or state, which in the progress of the ray returns at equal intervals and disposes the ray at every return to be easily refracted through the next refracting surface and between the returns to be easily reflected by it.

He even suggests that the effect might be due to vibrations excited by the "rays" in the material medium, vibrations which

> move faster than the rays so as to overtake them; and that when any ray is in that part of the vibration which conspires with its motion, it easily breaks through a refracting surface, but when it is in the contrary part of the vibration which impedes its motion, it is easily reflected. . . . But whether this hypothesis be true or false, I do not here consider.

Newton regarded his corpuscular theory as tentative and subject to confirmation on the basis of further experiments. If his theory did retard progress in optics, the fault lay rather with those who attached too great weight to his opinions.

Newton's speculations also serve as an example to illustrate the rule that even the greatest intellect works on the basis of the knowledge and viewpoints of its age. Had Newton lived a century later, he might have been one of the first believers in the wave theory. The fact that great scientists share the limitations of their age is an added reason for treating their speculative opinions chiefly as sources for further experiment.

Newton's researches on optics alone would have given him a high rank among the scientists of his time. But of still greater value was his work in *mechanics*. In announcing that "every particle of matter in the universe attracts every other particle with a force inversely proportional to the square of the distance between the two particles" and in showing that this one universal law governs the motions of the planets around the sun and of the satellites round their planets, he gave to the world a truth which exercised an enormous influence upon thought. This achievement of Newton's played a large part in bringing about the general conviction that the physical universe in its entirety is governed by law, not by caprice.

Newton himself has told us how he came to discover the law of gravitation. First he attacked the problem of finding a law of attraction between two bodies, such as the sun and a planet, which would result in Kepler's third law. He found that a gravitational attraction varying as the inverse square of the distance gives this law of planetary motion. Then he saw that a test of this inverse-square law could easily be made by comparing the acceleration of the moon toward the earth with the acceleration of falling bodies at the surface of the earth. It was known that the distance between the moon and the earth's center is about 60 earth radii. By the inverse-square law, therefore, in 1 s the moon should "drop" toward the earth $1/60^2$ as far as a body at the surface of the earth drops in 1 s. The latter distance being, from observations on falling bodies, 16 ft, the former should be $16/60^2$ ft, or 16 ft in 1 min. But the acceleration of the moon could be determined directly by applying to the moon the expression he had used for the motion of the planets in their orbits, viz.,

$$a = \frac{v^2}{r} = 4\pi^2 \frac{r}{T^2}$$

where v = velocity of moon in its orbit
  T = period of moon's motion around earth
  r = radius of orbit

Now r is equal to 60 times the earth's radius, which was then taken as 3436 miles on the then common assumption that a degree of latitude is 60 miles. On this basis, the moon "drops" 13.9 ft toward the earth in a minute, instead of 16 ft, as predicted by the inverse-square law.

Newton was twenty-three years old at the time, and he laid this calculation aside, not mentioning it to anyone. Some years later, however, he learned of a more accurate determination of the length subtended by a degree. Picard found it to be more nearly 70 miles. Meantime Newton had succeeded in proving that a homogeneous sphere attracts an external body as if all its mass were concentrated at its center, thereby removing one uncertainty in the calculation. On the basis of Picard's value Newton revised his computations on the moon's acceleration and, to his great joy, found that it falls toward the earth 16 ft in a minute, just as predicted by the inverse-square law. At last, he had discovered the law of gravitation on the basis of which he could derive all three of Kepler's laws.

These results, together with some propositions on the motion of the planets, were communicated in 1683 to the Royal Society, which requested permission to publish Newton's complete researches on the subject of motion and gravitation. In 1687, there appeared the first edition of his celebrated "Philosophiae Naturalis Principia Mathematica" (Mathematical Principles of Natural Philosophy). The treatise is divided into three books, the subject matter of each being presented by propositions, theorems, and corollaries. The first two books deal with general theorems concerning the motions of bodies, and the third contains applications to the solar system. The entire treatise is characterized by the exposition of the principle of universal gravitation and its ramifications, but, as the author carefully points out, without attempting any hypothesis as to the cause of gravitation.

The treatise assumes Newton's famous three laws of motion as axioms and makes other important contributions to mechanics. It introduces the concept of mass and describes a series of experiments showing that the period of a pendulum is independent of the material of which it is made, from which Newton concluded that for different bodies, mass and weight are proportional to each other. He also gave precision to the idea of force and formulated in general terms the principle of the parallelogram of forces.

Of Newton's invention of the method of fluxions, i.e., the calculus, of his very interesting miscellaneous writings, of the many controversies with his contemporaries into which he was unwillingly drawn in defense of his scientific work, we cannot take space to write. We can only urge our readers to make further study of the life and works of this renowned physicist.

1.16   Newton's Contemporaries

The true productive period of Newton's life ended about 1700. His biography is so full of interest and inspiration that it is a temptation to discuss similarly the work of his contemporaries, themselves eminent scientists—Robert Boyle (1627–1691), the discoverer of Boyle's law; Christian Huygens (1629–1695), whose wave theory of light was to triumph a century and a half later, and whose other contributions included important work on conic sections, the theory of probability, the pendulum clock, mechanics, and the improvement of the telescope and microscope; Robert Hooke (1635–1703), proponent of the undulatory theory and originator of Hooke's law of elasticity; Leibnitz (1646–1716), whose calculus ultimately replaced Newton's fluxions. But, remembering that the main business of this book is modern physics, we pass on to a rapid review of the developments of physics during the *eiahteenth* century.

1.17　Mechanics during the Eighteenth Century

The subsequent history of mechanics has consisted of the derivation, from Newton's three laws, of various secondary principles which are convenient for special purposes and of the solution of all sorts of problems. Among the prominent workers during the eighteenth century, we find such names as Daniel Bernoulli (1700–1782), who worked on hydrodynamics, the kinetic theory of gases, and the transverse vibrations of rods; Euler (1707–1783), who shares with Bernoulli the honor of discovering the general law of the conservation of angular momentum (1746); and Lagrange (1736–1813), who gave, in the equations that bear his name, a general method of attacking mechanics problems, using any sort of coordinates that may be convenient.

1.18　Heat during the Eighteenth Century

Galileo had invented an air thermometer in 1597; the first mercury thermometer was used by Kircher in 1643. About 1724, Fahrenheit proposed the temperature scale now known by his name; this was followed by the Réaumur scale and, in 1742, by the Celsius scale. James Black (1728–1799), a professor of chemistry at Glasgow and Edinburgh, made measurements of the heat of fusion and of vaporization of water, which led to modern calorimetry and gave definite form to the previously hazy distinctions between temperature and heat.

　　With regard to theories of heat, however, there was retrogression during the eighteenth century. From his writings, it is clear that Newton regarded heat as intimately connected with the motion of the small particles of which bodies are composed, and this view seems to have been shared by his contemporaries. But early in the eighteenth century, there was a return to the *caloric theory*, which held heat to be a fluid that could be extracted from, or added to, a body, thereby causing its changes of temperature. This heat fluid was indestructible, its particles were self-repellent but were attracted by ordinary matter, and it was all-pervading. The expansion of bodies when heated was the natural result of "swelling" due to forcing caloric into matter. The production of heat by percussion was due to the releasing or "pounding loose" of some of the caloric naturally condensed in or absorbed by the body, thereby increasing the amount of free caloric. Black explained latent heats and specific heats on the basis of this theory. Indeed, by the end of the eighteenth century, the caloric theory of heat was widely accepted.

1.19 Light during the Eighteenth Century

An event of special importance in the history of science was the discovery of the aberration of light by Bradley in 1728. The absence of any measurable stellar parallax had been one of the stumbling blocks in the way of the Copernican system and was, therefore, one of the outstanding problems of astronomy. Tycho Brahe had recognized that, viewed from opposite sides of the earth's orbit, the stars should show a perspective displacement, but his careful observations convinced him that no such displacement so great as 1 minute of arc existed. Later observers likewise sought such an effect in vain.

In hopes of being able to measure stellar distances, Bradley began, in 1725, systematic observations on the position of a zenith star, γ Draconis. If stellar parallax existed, this star should be farthest south in December and should then move north, reaching its maximum northerly position 6 months later. The position of the star was found to change but not in the manner expected. It reached farthest south in *March* and farthest north in *September*, the angular distance between the two positions being about 40 seconds of arc. Bradley continued his observations on other stars, and in 1728 he came to the conclusion that the observed displacement was not due to parallax at all but to an apparent shift in the star's position due to a combination of the velocity of light with that of the earth in its orbit. He was thus enabled to deduce a value for the velocity of light. The value so found was in substantial agreement with that determined by Römer half a century earlier, from a study of the motion of Jupiter's moons, which constituted the first determination of the velocity of light. This discovery of Bradley's was the first in the series that formed the basis for the modern theory of relativity.

Theories about the nature of light made no material progress during the eighteenth century because of the lack of any crucial experiment, as was also true of theories of heat. Science has seldom progressed on the basis of speculation only.

1.20 Electricity during the Eighteenth Century

Electricity received great attention during the eighteenth century, but research was principally in electrostatics. Stephen Gray (1670–1736) distinguished between conductors and nonconductors and proved that conducting bodies can be electrified provided they are insulated. Du Fay (1698–1739) showed that flames exercise a discharging power and that there are two kinds of electricity, which he called *vitreous* and *resinous*. He was thus led to propose the two-fluid theory of electricity. During the first half of the eighteenth century, the electroscope was invented by Hauksbee in 1706; frictional electric machines were developed; the Leyden jar was introduced in 1745; and there was considerable popular interest in electrical phenomena. During the latter half of the century, three names are preeminent: Benjamin Franklin (1706–1790), Henry Cavendish (1731–1810), and Charles A. de Coulomb (1736–1806).

Franklin's experiments began about 1745. One of his first observations was the effect of points "in drawing off and throwing off the electrical fire." He proposed the one-fluid theory of electricity, somewhat similar to the caloric theory of heat. This theory supposed that all bodies contained naturally a certain amount of the electric fluid. When a body had an excess of the fluid, it exhibited one of the two kinds of electrification, which Franklin chose to call for this reason *positive;* when it had a deficit, it exhibited the other kind, which he called *negative*. Certain features in the appearance of electric sparks led him to identify his positive electrification with that which had been called vitreous.

About 1750, Franklin began to speculate on the identity of electric discharges and lightning, pointing out many similarities and proposing, by means of a pointed iron rod, to "draw off the fire from a cloud." In 1752 Dalibard tried the experiment in Paris, confirming Franklin's prediction. A short time later, Franklin performed the famous kite experiment, well known to every schoolboy; this led to his study of atmospheric electricity and to his invention of the lightning rod. Franklin's researches occupied but a small portion of his long and busy life, but they were sufficient to give him a high standing among the scientists of the world.

Quantitative researches in electricity began with Cavendish and Coulomb. Cavendish is known not only for his work in electrostatics but also for his researches in chemistry and for the well-known Cavendish experiment, in 1798, in which he determined the constant of gravitation. His electrical researches were extensive, but most of his work remained unknown, for he published only one paper of importance. He left behind a wealth of manuscript notes, which were edited and published in 1879 by Maxwell. In these experiments, Cavendish tested the inverse-square law of electrostatic force; measured capacitance, recognized the principle of the condenser, and measured the specific inductive capacity of several substances; arrived at a reasonably clear idea of the quantity which we now call *potential;* and anticipated Ohm's law by 50 years.

Had these important measurements been communicated to his scientific contemporaries, the history of electricity might have been substantially modified.   As it is, the credit of *discovery* seems fairly to belong to others, for a discovery is of no importance to any one else if it is kept secret.

Coulomb's work in electricity grew out of his development of the torsion balance, originally used for studying the torsional elasticity of wires.   In the period from 1785 to 1789 he published seven papers on electricity and magnetism in the *Mémoires de l'Académie Royale des Sciences* in which he showed that electrostatic forces obey the inverse-square law; that, on conducting bodies, the charge exists only on the surface; and that the capacitance of a body is independent of the nature of the material of which it is composed.

1.21   Close of the Second Period

The end of the eighteenth century found rival theories contending in three divisions of physics: the caloric vs. the kinetic theory in heat; the corpuscular vs. the undulatory theory in light; and the one-fluid vs. the two-fluid theory in electricity.   The very fact that these issues were raised in clean-cut fashion is an indication of the tremendous strides that had been taken since Galileo.   But most important of all, men had learned the value of experiment and the fallacy of blindly following "authority."

Third Period (A.D. 1800–1890): the Rise of Classical Physics

1.22   The Nineteenth Century in Physics

So much was added to physical knowledge during the nineteenth century that an adequate history of this period would almost constitute a text-book of physics.   We comment only briefly upon the principal lines of advance and a few important discoveries, selecting especially those that form the background for the characteristic advances of the present century.

In mechanics, there was Hamilton, who discovered in the hamilto-nian equations a form of the equations of motion which is particularly valuable for attacking problems in quantum mechanics.   The theory of motion of rigid bodies, including the gyroscope, was worked out, as well as the mathematical theory of elasticity.   Hydrodynamics, dealing with the motion of fluid of all sorts, was developed. In the flow of viscous fluids, however, only simple problems could be solved; extensive study of such fluids, by half-empirical methods, was not made until the present century, after the invention of the airplane.

Work in other fields was more striking. The most significant discoveries and advances were the establishment of the kinetic theory of heat and the development of the kinetic theory of gases; the victory(?) of the wave theory of light over the corpuscular theory; the formulation of the general law of the conservation of energy; the discovery of the second law of thermodynamics; and above all the discovery, by Faraday and others, of the whole range of electromagnetic phenomena, culminating in Maxwell's theory of the electromagnetic field and of light.

Of these lines of advance we select three for brief discussion, choosing those which bear directly upon modern developments in physics; and to exemplify the great scientists of the period, we discuss Faraday and Maxwell.

1.23   Heat and Energy

The law of the conservation of energy is one of the most fundamental and far-reaching of all physical laws; yet it is of comparatively recent origin, for it was not announced until the middle of the nineteenth century. As exemplified in mechanics, it had been recognized during the eighteenth century, in the theory of the *vis viva;* but its announcement as a law of universal application awaited experimental work demonstrating the definite equivalence of heat and mechanical work.

An early *qualitative* experiment bearing on the nature of heat was performed in 1798 by Count Rumford, an American who had fled to England in 1775 and eventually became a sort of military engineer to the Bavarian government. Impressed by the large amount of heat that was developed in boring cannon, he performed experiments indicating that this heat was too great to be accounted for plausibly by the caloric theory. He could find no loss of weight when the chips made by boring were included and showed that the specific heat of the chips was the same as that of the block from which they had come. He concluded that heat "cannot possibly be a material substance" such as caloric but must be a form of "motion."

A more difficult experiment for the caloric theory to explain was one performed by Sir Humphry Davy, director of the Royal Institution, which had been founded by Count Rumford. Davy rubbed together two pieces of ice in a vacuum surrounded by walls kept below the freezing point and melted the ice. Here the mechanical work of rubbing accomplished exactly the same effect that could have been produced by the addition of heat from outside, yet there was no way in which caloric could have entered the ice. The majority of the supporters of the caloric theory were, however, unconvinced. Even Carnot (1796–1832), a pioneer in thermodynamics, based his reasoning on the caloric theory when he proposed the now famous cycle in 1824.

In 1842, R. J. Mayer (1814–1878) published a paper in which, partly on philosophical grounds, he announced the equivalence of heat and energy, and from data on the specific heats of a gas he deduced a value for the mechanical equivalent of heat. Meanwhile, Joule (1818–1889), in England, unacquainted with Mayer's work, was carrying on experiments in which he converted the mechanical energy of a falling weight into heat by a paddle wheel revolving in water and thus determined that 778 ft-lb of work would raise the temperature of 1 lb of water 1°F. Joule announced his results at a meeting of the British Association for the Advancement of Science in 1847. The paper would have passed almost unnoticed had it not been for William Thomson, later Lord Kelvin, who, grasping the real significance of the proposed theory, made the paper the event of the meeting.

Independently of the work of Mayer and of Joule, Helmholtz (1821–1894) in 1847 read a paper before the Physical Society in Berlin on "Die Erhaltung der Kraft," in which, on the basis of the impossibility of perpetual motion machines, he announced the law of the conservation of energy. The paper, rejected by the editor of the *Annalen der Physik*, was published in pamphlet form.

The caloric theory could not withstand these attacks, and by 1850 the mechanical theory of heat and the doctrine of the conservation of energy were generally accepted.

The second law of thermodynamics was announced by Clausius (1850) and, in another form, by Kelvin (1851), and in 1854 Kelvin proposed the thermodynamic scale of temperature. Thus was developed the highly successful *classical theory* of heat. We shall discuss later the facts that this theory, in turn, was unable to explain and how these facts led to the development of the quantum theory.

1.24 Light

The revival of the wave theory of light, begun by Thomas Young (1773–1829), is one of the important features of the history of the nineteenth century. Young pointed out that the division of a beam of light into a refracted ray at the interface between two media was to be expected from the wave theory but had not been satisfactorily explained by the corpuscular theory.   In 1801, he presented to the Royal Society a paper "On the Theory of Light and Colors," in which he proposed the principle of the interference of two wave trains as an explanation of Newton's rings and the colors of thin plates. From Newton's measurements of the thickness of the air layers necessary to produce the several colors, Young was enabled to compute wavelengths. In subsequent papers, he described the interference fringes which he had observed by placing hairs or silk threads in front of a narrow slit illuminated from the rear; he announced the change of phase on reflection; he explained diffraction bands by the principle of interference, showing that since the spacing of these bands gave values of the wavelength agreeing with those obtained from Newton's rings, both phenomena must therefore be due to a common cause.   Again, *quantitative* measurements became an indispensable link in the chain of reasoning.

But the dogmatic spirit in regard to scientific matters was not yet dead.   Young's paper aroused a storm of protest, even of derision and abuse.   His chief assailant was Henry Brougham, afterward Lord Chancellor of England, who "reviewed" Young's papers in the *Edinburgh Review* writing:

> We wish to raise our feeble voice against innovations that can have
> no other effect than to check the progress of science and renew all
> those wild phantoms of the imagination which Bacon and Newton
> put to flight from her temple.   We wish to recall philosophers to the
> strict and severe methods of investigation . . . .

In 1815, Fresnel (1788–1827) rediscovered the phenomenon of interference, performing the famous experiment with the two mirrors. A few years later he developed a mathematical theory of such phenomena (1818–1819). He also explained the polarization of light by assuming that the light vibrations in the ether are *transverse* to the direction of propagation of the light rather than longitudinal. (This suggestion had already been made by Young in a letter to Arago written in 1817.)   Fresnel supported the explanation by showing experimentally that two plane-polarized beams of light cannot interfere at all if their planes of polarization are perpendicular to each other. Phenomena of polarization had been known to Newton, the polarization of light by Iceland spar being discovered by Bartholinus in 1669. Newton had tried to fit these phenomena into the corpuscular theory by assuming a sort of structure in the corpuscles, but the explanation was not convincing, and polarization had remained an enigma to both theories of light. Fresnel's explanation assumed that light was a mechanical-wave motion through a medium known as the *luminiferous ether* which had the properties of rigidity and density needed to transmit transverse waves.

(And yet the planets move through this ether with no measurable drag!) On the basis of these mechanical waves Fresnel derived *correct* expressions for reflection and transmission coefficients as a function of incidence angle and polarization when light passes from one medium to another. It may seem surprising that correct equations could come from such a model, but both Fresnel's mechanical waves and electromagnetic waves involve equivalent wave equations, and analogous boundary conditions are applied. Thus the precise nature of the waves plays no significant role in determining reflection and transmission coefficients.

Experimental evidence for the wave theory continued to accumulate. In 1850 Foucault performed a crucial experiment, showing with his revolving-mirror apparatus that light travels more slowly in water than in air, as predicted by the wave theory.

From 1850 until the end of the third period (1890), the wave theory held the field undisputed. The frequent assertions that the corpuscular theory was finally disposed of certainly seemed justified, particularly after the development of Maxwell's electromagnetic theory of light and its experimental verification. Yet the corpuscular theory was not dead; it was only sleeping.

Some important discoveries in light from 1800 to 1890 not previously mentioned are:

	Discoverer
Dark (Fraunhofer) lines in the solar spectrum (1801)	Wollaston
Three-color theory of vision (1807)	Young
Rotary polarization of quartz (1811)	Arago
Polarization of scattered light (1813)	Arago
Rotary polarization by liquids (1815)	Biot
Light sensitivity of silver bromide (1826)	Balard
Change of conductivity of selenium on illumination (1837)	Knox
Doppler effect (1842)	Doppler
Foundation of spectral analysis (1859)	Kirchhoff and Bunsen

1.25   Electricity and Magnetism

The history of electricity during the nineteenth century is so extensive that even a sketchy outline would fill a small volume. We discuss little besides the fundamental discoveries of the opening decades and then the work of Faraday and Maxwell, which are closely related to each other and to recent developments in physics.

While the mathematical theory of electrostatics and of magnetism was being elaborated by Laplace, Green, and Poisson, fundamental discoveries were made in regard to electric *currents*. In 1786 Galvani, as a result of a chance observation that a frog's leg kicked convulsively when connected with the terminal of an electric machine, was led to an extensive study of "animal electricity." If the frog's leg was so suspended that the exposed nerves touched a metal plate, say silver, then a contraction of the muscle occurred whenever the foot touched another metal, say iron. He even observed slight muscular contraction when both plates were of the same kind of metal. This led him to believe that the nerve was the source of electricity and that the metal served simply as conductor. Volta later found that potentials could also be produced using inorganic materials, and in 1800 he described the first battery for producing an electric current—the historically famous *voltaic pile*, consisting of alternate zinc and copper plates separated by blotting paper moistened with brine. He also described a battery consisting of cups containing brine or dilute acid connected by copper and zinc strips joined together. Volta ascribed the effect to the contact of two dissimilar metals. We now know, however, that the electromotive force is due to chemical action at the contacts of the metals with the electrolyte.

This new source of electricity was received with great interest. A few weeks after hearing of Volta's work, Nicholson and Carlisle discovered the decomposition of water by the electric current. Thinking to secure better contact between two wires forming part of the circuit, they had joined the ends of the wires by a drop of water. At once they observed the formation of a gas, which they recognized as hydrogen. This was the beginning of the study of electrolysis. During this same period the heating effect of the current and the arc light were discovered.

It was early suspected that there was some relation between electricity and magnetism, but the first significant discovery was made in 1820 by Oersted, who found that a magnetic needle tends to set itself at right angles to a wire through which an electric current is flowing. Soon after, Biot and Savart discovered the law for the field of a long straight current, and toward the end of 1820 Biot proposed the formula

$$d\mathbf{B} = \varkappa I \, \frac{d\mathbf{s} \times \mathbf{r}}{r^3}$$

for the contribution to the magnetic induction $\mathbf{B}$ due to a current element $I \, d\mathbf{s}$; this relation is often miscalled "Ampère's formula." Soon after, the brilliant French physicist Ampère (1775–1836) showed that a closed current is equivalent in its magnetic effects to a magnetic shell. Then, reversing his line of thought, he suggested that magnetism itself might be due to currents circulating in the molecule. He also discovered the action of a magnetic field on a current. Within 5 years of the first discovery, the foundations of electromagnetism had been laid.

1.26   Michael Faraday

a. Biographical Sketch   Michael Faraday was born in 1791 in a small village near London.   To help his mother provide for the family, he started in 1804 as errand boy to a bookseller and stationer, and in the following year he was formally apprenticed to his employer to learn the art of bookbinding.   Faraday made good use of his spare time by reading some of the books that passed through the shop.   He was particularly interested in works on science, and in connection with his reading he showed one of the important characteristics of the great investigator-to-be by performing such of the simple experiments described "as could be defrayed in their expense by a few pence per week."

Aside from his own reading, *Faraday's scientific education consisted* *in a dozen lectures on natural philosophy by a Mr. Tatum and four lectures* *on chemistry by Sir Humphry Davy, in the winter of* 1812.   Submitting the very careful and neatly written notes which he made of these lectures "as proof of his earnestness," he made bold to apply to Sir Humphry Davy for a position, however menial, at the Royal Institution, of which Davy was director.   Davy was so pleased with the letter and the notes that in March, 1813, Faraday was engaged as apparatus and lecture assistant at 25 shillings per week.   In October, 1813, he accompanied Sir Humphry and Lady Davy on a trip to the Continent, which took them to many of the important scientific centers of Europe.   Assistant though he was, Faraday impressed others because of his modesty, amiability, and intelligence; said one writer, "We admired Davy, we *loved* Faraday."

On returning to England, under Davy's encouragement, Faraday began original investigations, initially in chemistry.   From 1816 to 1819 he published 37 papers concerned with such subjects as the escape of gases through capillary tubes, the production of sound in tubes by flames, the combustion of the diamond, and the separation of manganese from iron.   About 1820, he began his electrical researches, which continued for nearly 40 years.

Almost his entire scientific life was spent at the Royal Institution. In 1825, he was made Director of the Laboratory.   Declining offers of positions elsewhere which might have made him wealthy, he gave to his science and to the institution he served a devotion seldom equaled. The secret of his success, which brought him honors from all over the scientific world, is perhaps to be found in some excerpts from his many notes:

> Aim at high things, but not presumptuously.
> Endeavor to succeed—expect not to succeed.
> It puzzles me greatly to know what makes the successful philoso-

Fig. 1.3   Michael Faraday.

pher.   Is it industry and perseverance with a moderate proportion
of good sense and intelligence?   Is not a modest assurance or earnestness a requisite?   Do not many fail because they look rather to
the renown to be acquired than to the pure acquisition of knowledge . . . ?   I am sure I have seen many who would have been good
and successful pursuers of science, and have gained themselves a high
name, but that it was the name and the reward they were always
looking forward to—the reward of the world's praise.   In such there
is always a shade of envy or regret over their minds and I cannot
imagine a man making discoveries in science under these feelings.

b. The Principle of the Motor   Faraday had been interested in electromagnetism since April, 1821, when Wollaston attempted, at the Royal
Institution, to make a wire carrying an electric current revolve around
its own axis when the pole of a magnet was brought near.   The experiment was unsuccessful, but the phenomenon excited Faraday's interest,
and he determined to make a study of it.   First, he read what had been
done by others and repeated many of their experiments.   In the course
of these experiments, he observed that, when the magnetic pole was
brought near the wire, "the effort of the wire is always to pass off at
right angles from the pole, indeed to go in a circle around it. . . . "

On Sept. 4, 1821 he wrote in his laboratory notebook:

Apparatus for revolution of wire and magnet.   A deep basin with a bit of wax at bottom and then filled with mercury.   A magnet stuck upright in wax so that pole [is] just above surface of mercury. Then piece of wire, floated by cork, at lower end dipping into mercury and above into silver cup.

When a current passed through the wire, it revolved *continuously* around the magnet.   This was the first electric motor!

c. Electromagnetic Induction   Oersted's experiment and subsequent developments had clearly shown how "to produce magnetism by electricity."   Faraday seems to have held it as one of the tenets of his scientific philosophy that every physical relation (of cause and effect) has its converse.   If electricity can produce magnetism, then magnetism should produce electricity.   His repeated attempts to accomplish this failed. In 1825, he tried what seemed to be the obvious converse by looking for an electric current in a helix of wire coiled around a magnet.   Later, he tried to find a current in a wire placed near another wire carrying current.   Other scientists were looking for similar effects but without success.   They were all looking for the production of a *steady* current.

But several times investigators were very near to the discovery of induced currents.   In 1824, Arago observed the damping of the vibrations of a magnetic needle suspended over a copper plate.   This observation was extended by causing the needle to revolve by revolving the copper plate underneath it.   It was shown that this "dragging" effect was greater, the greater the electrical conductivity of the spinning plate. Even the effect of radial slits in the copper disk, in reducing the dragging action on the magnet, was observed.   Suggestive as these experiments were, the true explanation remained undiscovered.

In the summer of 1831 Faraday attacked the problem a fifth time. Instead of placing a *permanent* magnet inside a helix, he procured a soft iron ring, 6 in. in external diameter, on which he wound two coils of copper, *A* and *B*, "separated by twine and calico."   To detect a possible current in coil *B*, he "connected its extremities by a copper wire passing to a distance and just over a magnetic needle."   When coil *A* was connected to a battery, there was "a sensible effect on the needle.   It *oscillated* and settled at last in *original position*.   On breaking connection of side *A* with battery, again a disturbance of the needle."   Slight as these momentary effects were, Faraday recognized their importance, although he had been looking for a *continuous* effect.   On Aug. 30, he writes, "May not these transient effects be connected with causes of difference between power of metals at rest and in motion in Arago's experiments?"

From this slender clue, Faraday proceeded rapidly to the discovery of the real effect. On the third day of his experiments, he wound a coil of wire around an iron cylinder and placed the cylinder so as to join the north pole of one permanent magnet with the south pole of another. The coil was connected to a galvanometer:

> Every time the magnetic contact at N or S was made or broken there was a magnetic action at the indicating helix [i.e., galvanometer]—the effect being, as in former cases, not permanent but a mere momentary push or pull.

On the fourth day, he showed that the presence of iron was not necessary: the effect could be produced by the action of one helix on another. On the fifth day:

> A cylindrical bar magnet . . . had one end just inserted into the end of the helix cylinder, then it was quickly thrust in the whole length and the galvanometer needle moved; then pulled out and again the needle moved, but in the opposite direction. The effect was repeated every time the magnet was put in or out, and therefore a wave of electricity was so produced from mere *approximation* of a magnet and not from its formation *in situ*.

At last he had "converted magnetism into electricity." The essential requisite was *relative motion*, or a *change* of condition. On the ninth day, he produced a continuous current by turning a copper disk between the poles of a powerful electromagnet, the periphery of the disk being connected to its axis through an indicating galvanometer. This was the now well-known Faraday disk dynamo, the *first* dynamoelectric machine.

Thus, after only a few days' work in his laboratory, following years of patient and persistent experiment, Faraday had discovered a phenomenon for which the greatest scientists of his time had sought in vain—electromagnetic induction.

Following this discovery, Faraday devised and tried various electric machines to test and extend his newly discovered principle. One of these machines, consisting of a rotating rectangle of wire *with a commutator attached*, is the prototype of the modern dynamo. But his interest was always in pure science, for he writes:

> I have rather, however, been desirous of discovering new facts and relations dependent on magnetoelectric induction, than of exalting the force of those already obtained; being assured that the latter would find their full development hereafter.

Being unacquainted with mathematical symbols and methods, Faraday always sought to explain his discoveries and to extend his

researches by purely physical reasoning. To the mathematician, the
law of attraction between magnetic poles,

$$F = x\,\frac{m_1 m_2}{r^2}$$

may be a sufficient explanation of the phenomenon. To Faraday, this
gave a statement only of the *magnitude* of the magnetic forces; *it left
the phenomenon itself quite unexplained.* He insisted that two magnetic
poles, or two electric charges, could act on each other *only if the medium
between the two played some important part in the phenomenon.* This
insistence on the importance of the medium ultimately led him to the
concept of lines of force and of the "cutting" of these lines as essential
to electromagnetic induction. At first qualitative, this concept was
developed by Faraday into an essentially quantitative form, although
it was first stated in mathematical language by F. Neumann in 1845.
Commenting on Faraday's laws of electromagnetic induction, Maxwell
wrote:

> After nearly a half-century . . . , we may say that, though the
> practical applications of Faraday's discoveries have increased and are
> increasing in number and value every year, no exception to the state-
> ment of these laws as given by Faraday has been discovered, no new
> law has been added to them, and Faraday's original statement
> remains to this day the only one which asserts no more than can be
> verified by experiment, and the only one by which the theory of the
> phenomena can be expressed in a manner which is exactly and numer-
> ically accurate, and at the same time within the range of elementary
> methods of exposition.

d. The Laws of Electrolysis   Faraday next turned his attention to
proving that "Electricity, whatever may be its source, is identical in its
nature." He found, for example, that electricity from a friction machine
would deflect a galvanometer and would cause chemical decomposition
just as electricity produced by chemical action would. This led him into
the field of electrolysis. He found that many substances, e.g., certain
chlorides and sulfates, are nonconductors when solid but good conductors
when melted and that in the molten state they are decomposed by the
passage of current. This showed that water was *not* essential to electrolysis. To clarify description of his experiments, he introduced the
terms electrode, anode, cathode, ion, anion, cation, electrolyte, electro-
chemical equivalent, etc. A quantitative study of the phenomena
resulted in his discovery of the laws of electrolysis that bear his name and
which are the basis of all present-day work in that field.

Faraday recognized that a definite quantity of electricity is associated with each atom or ion in electrolysis. Had he been able to determine the number of atoms in unit mass of any substance, he would have anticipated, by 60 years, the determination of the fundamental charge *e*. For he says:

> Equivalent weights of bodies are simply those quantities of them which contain equal quantities of electricity; . . . it being the *electricity* which determines the combining force. Or, if we adopt the atomic theory or phraseology, then the atoms of bodies which are equivalent to each other in their ordinary chemical action, have equal quantities of electricity naturally associated with them.

e. The Conservation of Energy   In connection with a proof of the fact that the electricity from the voltaic pile results from chemical action and not from mere contact of one substance with another, Faraday stated clearly the doctrine of the conservation of energy several years before the statement of Helmholtz. In 1840, he wrote:

> The contact theory assumes that a force which is able to overcome a powerful resistance . . . can arise out of nothing. . . . This would indeed be a creation of power, and is like no other force in nature. We have many processes by which the form of the power is so changed that an apparent conversion of one into the other takes place. . . . But in no case is there a pure creation or a production of power *without a corresponding exhaustion of something to supply it.*

f. The Faraday Effect   Reference has already been made to Faraday's abhorrence of the doctrine of action at a distance. He believed that if two electric charges attract each other, the medium between the two plays some important role. Presumably, therefore, the medium between two such charges is in a different state than it would be if the charges were not present; and if so, such an altered state should be detectable by observing the alteration in some physical property of the medium. As early as 1822, Faraday experimented with a beam of polarized light passing through a transparent solution carrying a current to see whether the current caused any depolarizing action. Although he repeated the experiment several times in subsequent years, the results were all negative. In 1845, he returned to the problem, but still with negative results. He then tried solid dielectrics between plates of metal foil connected to a powerful electric machine to see whether, under electric strains, they would show any optical effects. No results! In 1875, this effect was found by Kerr.

Faraday then substituted a magnetic field for the electrostatic field to see whether the former would cause any depolarizing action on the beam of light. Various substances were tried but still with negative results. Finally, he placed in the magnetic field a piece of dense lead glass.   When the magnetic lines were parallel to the direction of the beam of polarized light, he observed that the plane of polarization was rotated.   At last, he had found a relation between magnetism and light. This magnetic rotation is now known as the *Faraday effect*.   Again, his persistent search, maintained during 20 years of repeated failures, was rewarded by the discovery of an effect in the existence of which he had had sublime confidence.

g. Miscellaneous   Among Faraday's other researches may be mentioned numerous investigations in chemistry; the liquefaction of several gases formerly thought "permanent"; the diffusion of gases through solids; self-induction; certain fundamental properties of dielectrics; diamagnetism; distinction between anode and cathode in the electric discharge through gases at low pressure; vibration of plates; regelation of ice; alloys of steel; and optical glass.

Well may this simple, modest, self-taught philosopher be given a conspicuous place among the great benefactors of mankind.

1.27   Joseph Henry (1799–1878)

Any account of Faraday's work, however brief, should be accompanied by at least a mention of the researches of the American physicist Joseph Henry, whose memory is honored by the name of the unit of inductance, the henry, which bears to electrokinetics a relation identical with that of the farad to electrostatics.   Had Henry been able to experiment continuously, and with more resources, instead of only during a summer vacation of 1 month while teaching mathematics at Albany Academy, and then only with such apparatus as he could make with his own hands, he would undoubtedly have anticipated Faraday in the discovery of electromagnetic induction, including the phenomena of self-induction. In all his work, furthermore, he was greatly hampered by his isolation from the scientific atmosphere of Europe.

Henry was interested especially in the design and use of electromagnets.   He constructed the first electric motor operating by an electromagnet, which rocked back and forth between two permanent magnets. He found that for maximum tractive effect, the cells of the battery and also the spools of the electromagnet should be connected in series if the magnet were a long distance from the battery, but they should be connected in parallel if the wires joining the magnet to the battery were short.   His work on electromagnets led *directly* to the commercial development of the telegraph.

1.28   James Clerk Maxwell (1831–1879)

It would be difficult to pick out two eminent scientists whose beginnings differed from each other more than did Maxwell's and Faraday's. Faraday came of very humble parentage; Maxwell, from a long line of distinguished ancestors. Faraday's early life was lived almost in poverty; Maxwell's family had abundant means. Faraday received only the most rudimentary education; Maxwell was given every advantage of school and university. They differed also in their aptitude for scientific work. Faraday was one of the greatest exponents of experimental science, whereas Maxwell, although an able experimenter, is one of the great figures in theoretical physics. And yet both made indispensable and mutually supplementary contributions to the classical theory of electromagnetics.

Maxwell early showed extraordinary interest in both theoretical and experimental research in physics. He invented a means of drawing certain types of oval curves, and a few years later he published a paper on "The Theory of Rolling Curves" and another on "The Equilibrium of Elastic Solids"—all before he was nineteen years old! During these same years he was also busy with experiments of many sorts, especially in his little laboratory in a garret on the family estate at Glenlair, where he spent his vacations.

After 3 years at the University of Edinburgh, Maxwell entered Trinity College, Cambridge, from which he graduated in 1854 with high honors. He spent 4 years at Aberdeen and was professor for 5 years at King's College, London (1860–1865); from here some of his most important papers were published, such as "Physical Lines of Force" (1862) and his greatest paper, "A Dynamical Theory of the Electromagnetic Field." After a retirement of several years, he was elected in 1870 to the newly founded professorship of experimental physics at Cambridge. In this capacity, he superintended the planning and equipping of the now famous Cavendish Laboratory, of which he was director until his untimely death in 1879.

A large proportion of Maxwell's papers, over 100 in number, may be grouped under three headings: color vision, molecular theory, and electromagnetic theory.

The work on color vision was undertaken to make a quantitative study of the physical facts pertinent to the theory of color sensations proposed by Thomas Young, according to which any luminous sensation is the result of exciting in the eye three primary sensations, red, green, and violet. For this purpose Maxwell invented a "color box," by means of which he could mix spectral colors.

Fig. 1.4 James Clerk Maxwell.

Maxwell's work on molecular physics is extensive. He discovered and, in part, established theoretically the law of the distribution of velocities among the molecules of a gas (Maxwell's law). He showed that when two gases are at the same temperature, the mean kinetic energy of translatory motion of their individual molecules is the same in both gases. From the kinetic theory of viscosity, he drew the surprising conclusion that the viscosity of a gas is independent of density so long as the mean free path is not too large, and he verified this conclusion by experiment. He brought to bear upon the whole subject mathematical methods "far in advance of anything previously attempted on the subject"; indeed, he is the cofounder with Clausius (1822–1888) of the kinetic theory of matter.

In electromagnetic theory, Maxwell's great contributions were the "displacement currents" and the formulation of the general equations of the electromagnetic field, which led to the electromagnetic theory of light. In the preface to his treatise "Electricity and Magnetism," he remarks, "Before I began the study of electricity I resolved to read no mathematics on the subject till I had first read through Faraday's 'Experimental Researches on Electricity.'" He became convinced that

Faraday was right in regarding the dielectric as the true seat of electrical phenomena and in supposing that it acted by becoming electrically polarized, the positive ends of its molecules pointing on the whole with the field and the negative ends in the opposite direction. (The term dielectric, as used here, must be understood to include a tenuous medium, or ether, filling all space, even in what we call a vacuum.) He drew the conclusion that when the polarization changes, this change must involve a displacement of electricity, and so there must exist in the dielectric, while the change is going on, a current having the same magnetic properties as the current in a conductor.

This assumption of displacement currents opened the way for the deduction of Maxwell's famous equations of the electromagnetic field. It is interesting, however, that he was first led to these equations through a mechanical analogy, i.e., in studying the behavior of a *mechanical system* filling all space, which would be capable of causing the observed electrical and magnetic phenomena. He showed (1862) that his hypothetical medium would be capable of transmitting transverse vibrations with a speed equal to the ratio of the electromagnetic to the electrostatic unit of charge. Although he did not take his model too seriously, he nevertheless remarks that the ratio of the units

> . . . agrees so exactly with the velocity of light calculated from the optical experiments of M. Fizeau, that we can scarcely avoid the inference that *light consists in the transverse undulations of the same medium, which is the cause of electric and magnetic phenomena.*

The theory was restated, without reference to any particular model, in his great paper of 1864, in which he says:

> The theory which I propose may therefore be called a theory of the *Electromagnetic Field*, because it has to do with the space in the neighborhood of the electric or magnetic bodies, and it may be called a *Dynamical Theory*, because it assumes that in that space there is matter in motion by which the observed phenomena are produced.

In 1873, Maxwell published his "Treatise on Electricity and Magnetism," one of the most important books in all science.

The *physical ideas* underlying Maxwell's new theory were left none too clear by him. In his treatise, we find the assumption that all space is full of incompressible "electricity"; in a conductor this electricity can move freely (except for ohmic resistance), thus constituting an electric current, but in a dielectric "there is a force which we have called electric elasticity which acts against the electric displacement and forces the electricity back when the electromotive force is removed." This is clear enough. But what is the origin of this electromotive force in empty space? And in what does electrification consist? Maxwell appears to have supposed that when a conductor is charged by allowing a current to flow onto it through a wire, what really happens is that electricity flows into the conductor along the wire and displaces some of the electricity out of the conductor into the space surrounding it. Thus a charged conductor would contain neither more nor less electricity than an uncharged one.   Such a hypothesis concerning electrification became improbable when in 1876 the American physicist Rowland showed that a moving charged conductor is surrounded by a magnetic field, the moving electrification evidently constituting a current.

The *mathematical theory*, on the other hand, was slowly developed by others, especially by H. A. Lorentz, and was shown to give a good account of electric and magnetic phenomena and of the principal properties of light.   In Germany, stimulated by Helmholtz, Hertz set out to search experimentally for the magnetic effects of Maxwell's displacement currents, and in 1887 he produced electromagnetic waves from electric circuits.   Later it was shown that the speed of propagation of these waves is the same as that of light.   Speculation as to the nature of the displacement currents in a vacuum gradually died out, until today it is usual to speak only of electric and magnetic *fields* governed by the Maxwell-Lorentz equations.   This development in electromagnetic theory illustrates a tendency, notable during the last century but regretted by many, for physics to become highly abstract and mathematical.

1.29   Clouds over Classical Physics

Before the end of the nineteenth century physics had evolved to the point at which mechanics could cope with highly complex problems involving macroscopic situations, thermodynamics and kinetic theory were well established, geometrical and physical optics could be understood in terms of electromagnetic waves, the atomic foundation of chemistry had been laid, the conservation laws for energy and momentum (and mass) were widely accepted, and classical physics had reached a proud maturity.   So profound were these and other developments that not a few physicists of note believed that all the important laws of physics had been discovered and that, henceforth, research would be concerned with clearing up minor problems and, particularly, with improvements of methods of measurement.   At that time few could foresee that the world of physics was on the eve of a series of epoch-making discoveries destined on the one hand to stimulate research as never before and on the other to usher in an era of the application of physics to industry on a scale previously unknown.

Late in the century x-rays and radioactivity were discovered.   No one had yet been able to explain quantitatively photoelectric phenomena, blackbody radiation, and the origin of spectral lines. Nevertheless, many believed that some future genius would find the answers in classical physics. As late as 1899 Michelson said:

> The more important fundamental laws and facts of physical science have all been discovered, and these are so firmly established that the possibility of their ever being supplanted in consequence of new discoveries is exceedingly remote. . . . Our future discoveries must be looked for in the sixth place of decimals.

But others were not so sure. In a lecture delivered in 1900 Lord Kelvin began with the words, "The beauty and clearness of the dynamical theory, which asserts light and heat to be modes of motion, is at present obscured by two clouds." The first of these clouds was the question of "how the earth [can] move through an elastic solid such as essentially is the luminiferous ether" (in which electromagnetic waves were assumed to propagate). The second was the failure of the "Maxwell-Boltzmann doctrine regarding the equipartition of energy" to predict results consistent with experiment in all cases.

Neither of these clouds were destined to be dissipated by classical approaches; new ideas were required. The first evaporated quickly when Einstein advanced the theory of relativity, introduced in the next chapter. The second evaporated more slowly, requiring not only Planck's introduction of the energy quantum $h\nu$ but also the evolution of a new mechanics.

chapter two

Introduction to Relativity

The theory of relativity put forward by Einstein in 1905 represented a vast revolution in physical thought. Relativity touches all branches of modern physics and plays a major role in many.

2.1 Galilean-Newtonian Relativity

Motion has been a subject of speculation since ancient times. It was early recognized that all motion involves displacement relative to something or other, but ideas have varied in regard to the entity relative to which the displacement occurs. In his treatise on mechanics, Newton says that "absolute motion is the translation of a body from one absolute place to another absolute place," but he does not explain what he means by "absolute place." He states explicitly that translatory motion can be detected only in the form of motion relative to other material bodies.

Motion involves, also, the passage of time. Until Einstein, time was regarded as something entirely distinct from space or from the behavior of material bodies. Newton says, "Absolute, true, and mathematical time, of itself, and by its own nature, flows uniformly on, without regard to anything external." Thus there was supposed to be a single time scale valid everywhere.

Consider an *event* such as the emission of an electron from a surface or the decay of a radioactive nucleus. Classically, to specify the location of an event we make use of a reference frame in which distances can be measured. To define a time, we must have available some reference process, e.g., the rotation of the earth, in terms of which times can be specified. The material means of fixing positions and times, together with the methods adopted for using them, are said to constitute a space-time *frame of reference*. When Newton's laws of motion are satisfied in any particular frame of reference, it is said to be an *inertial frame*. Einstein called such frames galilean reference systems because the galilean law of inertia holds in them. (Some physicists have argued that *newtonian frames* would be a more appropriate designation, since a frame is inertial only if Newton's laws of motion apply.)

Now suppose we have two frames of reference each in *uniform translatory motion* relative to the other. Let us call the two frames S and S' and let the velocity of S' relative to S be $\mathbf{V}$. Let coordinates and times of any event obtained when the frame S is used be denoted by x, y, z, t and those obtained for the same event when S' is used by x', y', z', t'. To make the relation between these variables as simple as possible, let us choose our axes so that the x and x' axes are parallel to $\mathbf{V}$ and thus slide along each other; let the y' and z' axes be parallel to y and z, respectively (Fig. 2.1). Let us also count time from the instant at which the origins of the coordinates O and O' coincide. Then the

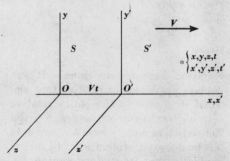

Fig. 2.1   Inertial frame S' moving with constant velocity $\mathbf{V}$ along x axis of inertial frame S.

coordinates of O' as measured in frame S are $x = Vt$, $y = 0$, $z = 0$. If an event occurs at a position and time specified by x, y, z, and t in frame S, according to classical physics the coordinates of the event in frame S' are x', y', z', t', where

$$\begin{aligned} x' &= x - Vt & z' &= z & &\textit{galilean} \\ y' &= y & t' &= t & &\textit{transformation} \end{aligned} \qquad (2.1)$$

These equations are the *space-time* transformation of classical mechanics. They enable us to pass from space-time coordinates of events in one reference frame to the space-time coordinates referred to another frame moving with constant velocity V relative to the first. Under the galilean transformation the laws of classical mechanics are the same in the two frames S and S', since mechanical forces and accelerations are the same in both frames. As we shall see, the classical laws are approximate only.

2.2   Galilean Relativity and Electricity

The laws of classical electromagnetism are not compatible with the galilean transformation. For example, imagine a charge Q a distance a from an infinite wire bearing a charge ρ_l per unit length, with both Q and the wire at rest in inertial frame S (Fig. 2.2). The charge Q experiences an electric field $\rho_l/2\pi\epsilon_0 a$ and no magnetic field in S, where all charges are at rest.

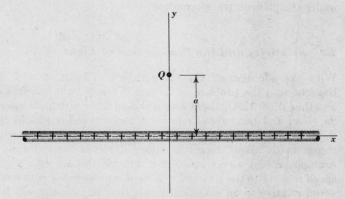

Fig. 2.2   A point charge Q a distance a from an infinite wire charged with ρ_l C/m.

If S' is moving to the right relative to S with speed V, all charges have a speed V to the left in S', and Q is then moving in a magnetic field given by

$$B'_z = -\frac{\mu_0 I}{2\pi a} = -\frac{\mu_0 \rho_l V}{2\pi a} \qquad \begin{array}{l} galilean \\ transformation \end{array}$$

According to the Lorentz equation, $\mathbf{F} = Q(\mathbf{E} + \mathbf{v} \times \mathbf{B})$, so that the force on Q in frame S' is the resultant of that due to the electric intensity (which is the same in S' as in S) and that due to the magnetic field above. Hence the charge Q experiences a net force:

in frame S'

$$F'_y = \frac{Q\rho_l}{2\pi\epsilon_0 a} - \frac{Q\mu_0\rho_l V^2}{2\pi a} \qquad galilean$$

while *in frame S*

$$F_y = \frac{Q\rho_l}{2\pi\epsilon_0 a} \qquad galilean$$

Clearly, classical electromagnetic theory, together with galilean relativity, predicts different forces on Q in the two inertial frames; on the other hand, Eqs. (2.1) predict the same acceleration. Thus something is not right—either in the galilean transformation or in classical electromagnetism or in newtonian mechanics. We say that the laws of classical electromagnetism, together with newtonian mechanics, are *not invariant* under the galilean transformation.

2.3   Relativity and the Propagation of Light

With the adoption of the wave theory of light, a new element was brought into the problem of motion. For if light consists of waves in an ether (Sec. 1.24), these waves should have a definite speed *relative to the ether*, and their speed *relative to material bodies* should change when the motion of these bodies through the ether varies. Analogous statements made about waves in material media are certainly true. For example, sound waves propagate through the air with a characteristic speed relative to the air itself. If light is propagated with a characteristic speed relative to an ether, certain optical phenomena will be influenced by a motion of optical apparatus through the ether. Of course, the velocity that can be given to a material body in the laboratory is extremely small compared with the velocity of light; however, the speed of the earth in its orbital motion about the sun is about one ten-thousandth of the speed c of light in free space.

An interesting case to consider is the formation of images by the object lens of a telescope. Suppose a light wave from a star enters the telescope sketched in Fig. 2.3. When the telescope is at rest in the ether, let the wave come to a focus so as to form a star image on the cross hairs at P. When the telescope is moving toward the star, the wave might be expected to focus on the same point P *in the ether* as before. While the wave is passing from the lens to this point, however, the telescope moves forward, carrying the cross hair to some other point P'. The eyepiece would therefore have to be drawn out farther in order to focus on the image of the star. Thus the focal length of the telescope would appear to be increased. Similarly, if the telescope were moving in the same direction as the light, its effective focal length would be shortened. Thus, as an astronomical telescope is pointed at stars in different directions, its apparent focal length might be expected to vary slightly because of the earth's orbital motion. This effect was looked for long ago by Arago but in vain.

This and other negative results led Fresnel in 1818 to propose that moving transparent bodies may partially drag the light waves along with them. In the case just considered, if the lens L were to drag the light with it in its motion (toward the right in Fig. 2.3), the part of the wave that goes through the center of the lens would spend a longer time in the lens, and hence would be retarded more than it would be if there were no motion. Consequently the wave would emerge from the lens more concave in shape and would focus on a point nearer the lens. If the amount of the drag were just right, there would be no effect at all on the apparent focal length, the star image falling on the cross hair however the telescope might be moving. Fresnel showed that all effects on phenomena of refraction would be prevented if it were a law of optics that any moving transparent medium of refractive index n changed the

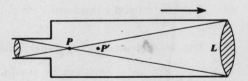

Fig. 2.3 Effect of the motion of a telescope through the ether on the position of the image formed by the objective lens relative to the cross hairs.

velocity of light in such a way as to add vectorially to its velocit. n the stationary medium the fraction

$$1 - \frac{1}{n^2}$$

of the velocity of the medium.   That is, in a medium moving with vector velocity **u**, the vector velocity of light in any given direction is the sum of a vector of magnitude $(1 - 1/n^2)$**u** and a vector of magnitude c/n.

In 1851 Fizeau showed, by an interference method, that a moving column of water *does* drag the light waves with it to the extent required by Fresnel's hypothesis!   The cause of the drag was assumed by Fresnel to lie in an actual partial dragging of the ether itself along with the moving medium.   When Lorentz worked out his electromagnetic theory, however, about 1895, on the assumption of a stationary ether, he found that the theory led automatically to Fresnel's formula for the light drag.

No effect of the earth's orbital velocity on terrestrial optical phenomena has been observed.   It is well known that annual *changes* in the direction of the earth's orbital motion cause a variation in the apparent positions of the stars, called *stellar aberration* (Sec. 1.19), but that is another story.   It may be doubted whether the *early* experiments were sufficiently precise to detect a possible effect due to the earth's motion, but at least one test of adequate delicacy was made in 1871 by Airy, who filled a telescope tube with water and observed that star images were formed in the same position with or without the water, in spite of the longer time required for the light to reach the focal plane when traveling through water instead of air.

Fresnel and Lorentz showed that there should never be any effect on optical phenomena that is of the *first order* in the velocity of the apparatus through the ether.   There might, however, be second-order effects.   Since the square of the earth's orbital velocity is only $10^{-8}c^2$, such effects would be difficult to detect.   In seeking a sufficiently delicate means of observation, Michelson invented his interferometer, with which in 1887 he and Morley performed a famous experiment which could have detected a second-order effect if it had been present.

2.4   *The Michelson-Morley Experiment*

In the interferometer arrangement used in this experiment (Fig. 2.4) a beam of light from a lamp L falls upon a half-silvered glass plate P, placed at 45° to the beam, which divides each wave into two parts. One partial wave, reflected from P, travels off sideways to a mirror M_1, by which it is reflected back to P; part of it is then transmitted through

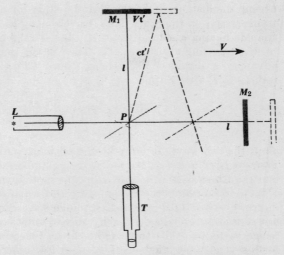

Fig. 2.4 The Michelson-Morley experiment.

P and enters the telescope T. The other part of the original wave, transmitted at once through P, travels ahead and is reflected back by a second mirror M_2; upon returning to P, it is partially reflected into the telescope on top of the first part of the wave, with which it forms an interference pattern.

Both mirrors are at the same distance from the plate P If the apparatus is at rest in the ether, the two waves take the same time to return to P and meet in phase both there and in the telescope. Suppose, however, that the apparatus is moving with speed V through the ether in the direction of the initial beam of light. Then, if the initial wave strikes the plate P when it has the position shown in the figure, the paths of the waves and the subsequent positions of reflection from mirrors and plate will be as shown by the dotted lines. The necessary change in the direction of the transverse beam is produced automatically, as an aberration effect, through the operation of Huygens' principle. But now the times taken by the two waves on their journeys are no longer equal. The wave moving longitudinally toward M_2 has a velocity *relative to the apparatus* of $c - V$ on the outgoing trip, c being the speed of light through the ether, and $c + V$ on the return trip; hence the time required by this wave is

$$t_2 = \frac{l}{c - V} + \frac{l}{c + V} = \frac{2lc}{c^2 - V^2} = \frac{2l}{c}\left(1 + \frac{V^2}{c^2} + \cdots\right)$$

l being the distance from the plate to either mirror. The wave moving transversely, on the other hand, travels along the hypotenuse of a right triangle having a side of length l. Let it take a time t' to go from P to M_1, traveling a distance ct'. Then in the same time the mirror M advances a distance Vt'. Hence, since $c^2t'^2 = V^2t'^2 + l^2$, this wave returns to P after a time

$$t_1 = 2t' = \frac{2l}{(c^2 - V^2)^{\frac{1}{2}}} = \frac{2l}{c}\left(1 + \frac{V^2}{2c^2} + \cdots\right)$$

Thus the two waves interfere in the telescope with a phase difference of $2\pi c(t_2 - t_1)/\lambda \approx 2\pi l V^2/c^2\lambda$ rad (λ being the wavelength of the light), and the fringe pattern is shifted by the motion through $V^2 l/c^2\lambda$ fringes.

In performing the experiment, the whole apparatus, floated on mercury, was rotated repeatedly through 90°. Since the two light paths are caused to exchange roles by such a rotation, it should cause the fringe pattern to shift twice as much; by reflecting the beam back and forth several times, the effective length l was increased to 11 m. Even then, with a wavelength of about 5.9×10^{-5} cm, if we insert for V the whole orbital velocity of the earth so that $V/c = 10^{-4}$, we find a shift of only

$$N = 2 \times 10^{-8}\frac{11 \times 10^2}{5.9 \times 10^{-5}} = 0.37 \text{ fringe}$$

Michelson and Morley were sure that they could detect a shift of a hundredth of a fringe. *Such shifts as were observed amounted only to a small fraction of the theoretical value and were not consistent.* Thus the result of the experiment was negative; the expected effects of motion through the ether did not appear.

It might happen, to be sure, that at a given moment the earth by accident had no resultant component of velocity parallel to the surface of the earth; for upon its orbital motion there would probably be superposed a motion of the entire solar system through the ether. But, if this happened at a certain time, then 6 months later the earth's orbital velocity about the sun would be reversed, and its velocity through the ether should be twice its orbital velocity.

Observations made by the Michelson method at various times of day and at different seasons of the year by physicists in different parts of the world always yielded the same negative result. In 1930 Joos in Germany found no evidence of motion through an ether with an interferometer capable of detecting an ether drift one-twentieth of the orbital velocity of the earth. In 1958 Cedarholm and Townes devised a new technique using masers, with which an ether drift of one one-thousandth the orbital velocity could have been observed, but no evidence of any drift was found.

The failure to detect the anticipated motion of the earth through the ether was an important experimental result, very hard to bring into harmony with existing theories of light and matter. The theoretical calculation rests on a peculiarly simple foundation, for the only properties of light that are made use of are the constancy of its velocity in space and Huygens' principle. Classically, three possible explanations of the negative result seem to offer themselves.

1. Perhaps the earth drags the ether with it, much as a moving baseball carries along the air next to it. On this assumption there would never be any motion of the earth through the ether at all, and no difficulties could arise. One objection to this explanation is that the ether next to the earth would then be in motion relative to the ether farther away; and this relative motion between different parts of the ether would cause a deflection of the light rays coming from the stars, just as wind is observed to deflect sound waves. This deflection would alter the amount of the stellar aberration. It was difficult to devise a plausible motion for the ether which would give a value of the aberration agreeing with observation and yet preserve the negative result of the Michelson-Morley experiment. A second objection arises from the fact that, as has already been stated, a transparent object of laboratory size does not drag the light waves with the full velocity of the moving matter, as it necessarily would do if it *completely* dragged the ether along with it; and the observed *partial* drag is fully accounted for by current electromagnetic theory.

2. As an alternative, we could assume that light projected from a moving source has added to its own natural velocity the velocity of the source, just as the velocity of a projectile fired from a moving ship is equal to the vector sum of its velocity of projection from the gun and the velocity of the ship. If this were true, the negative result obtained by Michelson and Morley would at once be explained, for the light from their lamp would have always the same constant velocity relative to the lamp and to the interferometer. Such an assumption, however, is in gross conflict with the wave theory of light. It is of the essence of waves in a medium that they have a definite velocity *relative to the medium*, just as sound waves have a definite velocity relative to the air. Furthermore, there is strong *experimental evidence against* the assumption in question, which we have not space to describe.

3. The third possible explanation put forward independently by FitzGerald and by Lorentz, is that motion through the ether might in some manner cause the material composing the interferometer to *shorten*

in a direction parallel to the motion. Such a contraction in the ratio $\sqrt{1 - V^2/c^2}$ would serve to equalize the light paths and thus to prevent a displacement of the fringes. This explanation was shown to be inadequate in 1932 by Kennedy and Thorndike, who found no shift even when the two arms of the interferometer had grossly different lengths.

2.5 The New Relativity of Einstein

From the situation just described one easily gains the impression that there exists something like a conspiracy in nature to prevent us from detecting motion through the ether. A similar situation can be shown to exist in the field of electromagnetism as well as in optics. A number of electrical or magnetic experiments can be invented which, at first sight, offer promise of revealing motion through the ether; but always there occurs some other effect which just cancels the effect sought.

In reflecting upon this extraordinary situation, Einstein arrived in 1905 at a radically new idea. He proposed that motion through an ether filling empty space is a *meaningless concept;* only motion *relative to material bodies* has physical significance. He then considered how this assumption could be made to harmonize with the known laws of optics. Possibilities of conflict arise in any argument involving the velocity of light. Consider, for example, a frame of reference S' (say on the earth) moving relative to another frame S (say on the sun), and suppose that S' carries a source of light. Then light from this source must move with the same velocity relative to S as light from a source on S itself, since, as explained above, we cannot suppose that the velocity of light is influenced by motion of its source. But this light must also move *with the same velocity relative to S';* otherwise the laws of nature would not be the same on S' as on S, and no reason could be assigned for their being different. Yet it seems quite impossible that light should move with the same velocity relative to *each* of *two* frames that are *moving relative to each other!*

Einstein accordingly put the laws of the propagation of light in the forefront of the discussion. He based his new theory, which is known as the *special* or *restricted theory of relativity*, upon two postulates which may be stated as follows:

1. *The laws of physical phenomena are the same when stated in terms of either of two inertial frames of reference (and can involve no reference to motion through an ether); no experiment can be performed which can establish that any frame is at absolute rest.*

2. *The speed of light is independent of the motion of its source and is the same for observers in any inertial frame.*

Of these two postulates, the second is believed to represent an experimental fact, whereas the first is a generalization from a wide range of physical experience. The first postulate is in no way self-evident; like the assumptions made in all physical theories, it is a hypothesis to be tested by experiment. Further, some of the apparently well-established laws of classical physics, such as the law of conservation of mass, fail to meet the requirement of the first postulate. Indeed, the usual form of Newton's second law of motion proves to be incompatible (see Sec. 3.3).

2.6 Simultaneity and Time Order

Einstein found the key by which these two postulates could be brought into harmony by giving up the newtonian conception of absolute time. Newton had supposed that it is always possible to say unambiguously which of two events precedes the other or that they occur simultaneously. But how can this be done if the two events occur in widely separated locations?

In practice, times at two different places are compared by reference to a clock at each place. It is necessary to synchronize these clocks, which in modern practice is done by means of electromagnetic signals. For precision, however, it is necessary to correct for the time required for light to travel from one place to the other; and making this correction requires a knowledge of the velocity of light *in one direction*. Ordinary measurements of the velocity of light furnish only its *average velocity in two opposite directions*. We could measure its velocity in one direction if we had our clocks already synchronized, of course, but this leads into a vicious circle.

As an alternative, we might carry a chronometer from one place to the other and set both clocks by comparison with the chronometer. But how could we prove experimentally that the chronometer runs at a constant rate while moving in various directions? Every method that can be devised to prove this or to measure the velocity of light in one direction turns out to rest upon some fresh assumption that cannot be tested in advance. It follows that a general time scale can be set up only on the basis of an artificial convention of some sort. The simplest procedure is the usual one of *assuming* the velocity of light to be the same in any one direction as in the opposite direction. We shall see, however, that a time scale set up on this basis varies somewhat according to the frame of reference that is used in measuring velocities; even the time order of two events may be different in some cases.

Before developing the mathematical theory further, the following additional remarks may be of interest.   Consider only events that happen at various points along a straight track, so that the situation can be diagrammed in two dimensions.   On a plot (Fig. 2.5) let horizontal distances represent displacements along the track and vertical distances lapses of time.   Consider an observer at the point P_0 at time t_0.   Through P_0 draw two lines representing the progress of two light signals traveling in opposite directions along the track which pass P_0 at time t_0.   Then points on the plot in the area above these two lines will represent possible events which, at time t_0, certainly lie in the *future* for *any* observer at P_0.   These events occur late enough so that the observer still has an opportunity to influence them causally, perhaps by sending a radio signal.   Below both lines, on the other hand, lies the observer's definite *past*.   This area consists of events on the track which may have had a causal influence on events in his neighborhood at time t_0 and of which he may have already acquired knowledge at that time by means of radio signals.   Between the lines, however, lies also a third region, representing what might be called the observer's *physical present*.   At time t_0 it is too late for him to influence events plotted in this region, but also too early for him to have any observational knowledge of them.   For events located at a given point of the track, the observer's physical present covers a range of time, such as QR, that increases with increasing distance from P_0.   If the observer is in New York, this interval is about $\frac{1}{30}$ s for events in San Francisco (the time required for light to make the double journey); it is about 16 min for events on the sun.   Here we

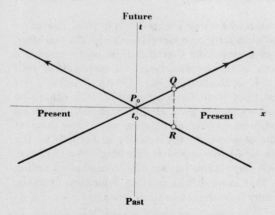

Fig. 2.5   The classification of space-time relations.

have an essential difference between Einstein's relativity and Newton's; for in the older theory the physical present was reduced to a single line.

2.7 The Lorentz Transformation

Like all physical theories, relativity must stand or fall according to whether or not the deductions drawn from it agree with experimental facts. To make deductions from relativity, we compare descriptions of some phenomenon in terms of two inertial frames which are in relative motion. We need to know, first of all, how measurements of space and time compare. (In discussing these, we speak of an *observer* who makes measurements referred to a particular frame. This does not imply that relativity has any closer connection with human psychology than the rest of physics does.)

Einstein's second postulate requires that the speed c of light in vacuo be the same in all directions and for observers in all inertial frames; thus c becomes a universal constant.

Consider again the inertial frames S and S' of Sec. 2.1. Let the two origins coincide at $t = 0$, $t' = 0$, at which instant a light pulse is emitted from the common origin. Imagine that observers in S and S' have arranged apparatus which enable them to follow the pulse as it moves outward from the source. By Einstein's second postulate, observer O in frame S and observer O' in S' find the locus of the wavefront to be given respectively by

$$x^2 + y^2 + z^2 - c^2 t^2 = 0 \tag{2.2a}$$
$$x'^2 + y'^2 + z'^2 - c^2 t'^2 = 0 \tag{2.2b}$$

Thus each observer finds the wavefront to be a *sphere centered at his own origin*, even though the origins of the two systems no longer coincide!

Direct substitution of the galilean transformation into Eq. (2.2b) yields $x^2 - 2xVt + V^2 t^2 + y^2 + z^2 - c^2 t^2 = 0$, which is clearly incompatible with Eq. (2.2a). The equations $y = y'$ and $z = z'$ are not the source of the incompatibility, and we accept them without proof; powerful arguments based on the isotropy of space can be advanced for their validity. Therefore we seek compatible relations between x', x, t', and t. As the origins pass, $x = x'$, and we choose $t = 0 = t'$. Because of the homogeneity of space and of the uniformity of natural laws in time we assume the relationships are linear and try

$$x' = \alpha x + \eta t \tag{2.3a}$$
$$t' = \epsilon x + \gamma t \tag{2.3b}$$

where α, η, ϵ, and γ are constants to be determined. At the origin of S', $x' = 0$ and $x = Vt$, so that by Eq. $(2.3a)$ $0 = \alpha Vt + \eta t$. Thus $\eta = -\alpha V$, and so $x' = \alpha(x - Vt)$. Inserting this value of x' and Eq. $(2.3b)$ into Eq. $(2.2b)$ yields

$$\alpha^2 x^2 - 2\alpha^2 Vxt + \alpha^2 V^2 t^2 + y^2 + z^2 - c^2 \epsilon^2 x^2 - 2c^2 \epsilon \gamma xt - c^2 \gamma^2 t^2 = 0$$

This result is compatible with Eq. $(2.2a)$ only if

$$\alpha^2 - c^2 \epsilon^2 = 1 \qquad \alpha^2 V + c^2 \epsilon \gamma = 0 \qquad c^2 \gamma^2 - \alpha^2 V^2 = c^2$$

These three equations can be solved for the three unknowns α, γ, and ϵ in terms of V and c and give

$$\alpha = \gamma = \frac{1}{\sqrt{1 - V^2/c^2}} \quad \text{and} \quad \epsilon = -\frac{\gamma V}{c^2} = -\frac{V/c^2}{\sqrt{1 - V^2/c^2}}$$

All constants are now determined, and we have

$$x' = \frac{x - Vt}{\sqrt{1 - V^2/c^2}} = \gamma(x - Vt)$$

$$y' = y \qquad z' = z$$

$$t' = \frac{t - Vx/c^2}{\sqrt{1 - V^2/c^2}} = \gamma(t - Vx/c^2)$$

Lorentz transformation (2.4)

These equations, named the *Lorentz transformation* by Poincaré, were discovered by Voigt in 1887 and independently derived by Lorentz in 1904 in the course of his study of matter moving in an electromagnetic field. Lorentz assumed one frame of reference to be at rest in the ether and attached an immediate physical meaning only to measurements made with this frame. The new principle of relativity implies that all inertial frames are to be treated on an equal footing.

By solving Eqs. (2.4) for x, y, z, t, we obtain the inverse transformation:

$$x = \gamma(x' + Vt') \qquad y = y'$$

$$t = \gamma\left(t' + \frac{Vx'}{c^2}\right) \qquad z = z'$$

$(2.4a)$

From these equations, we can easily confirm the statement that events which happen at the same place at different times, as viewed from one frame, may be seen from another frame to happen at different places as well. Similarly, a difference in spatial position with respect to one frame may correspond to a difference in both space and time with respect to another. Thus a space difference can be converted partly into a time difference, or vice versa, merely by changing the frame of reference that is being used. For this reason it has become customary to speak of space and time as aspects of a four-dimensional continuum known as *space-time*.

2.8 Space Contraction and Time Dilation

In two famous cases the Lorentz transformation leads immediately to results of special interest.

Consider a body which, when at rest in S, has a length L_0 in the direction of the x axis. Let it be set moving relative to S at such speed that it is at rest in S'. Its length as measured in S' will now be L_0; for its length is determined by certain natural laws and hence must have a certain fixed value in any frame in which the body is at rest. Let us see how the length now measures in S, relative to which the body is moving with speed V. To do this, we must first consider what we mean by the *length* of a *moving* object. It seems reasonable to define the length as the distance between two points fixed in S which are occupied by the ends of the object simultaneously, i.e., at the same time t. If the coordinates of these points are x_1 and x_2, the length is then $L = x_2 - x_1$. By the same definition, since the body is at rest in S', its ends have fixed coordinates x_1', x_2' such that $L_0 = x_2' - x_1'$. If we substitute values of x_2' and x_1' from Eq. (2.4) for a given value of t, we obtain

$$L_0 = \gamma(x_2 - x_1) = \gamma L = \frac{L}{\sqrt{1 - V^2/c^2}} \qquad \text{or}$$

$$L = L_0 \sqrt{\frac{1 - V^2}{c^2}} \tag{2.5}$$

From this equation we may draw two distinct conclusions. First, any body measures shorter in an inertial frame relative to which it is moving with speed V than it does as measured in a frame relative to which it is at rest, the ratio of shortening being $\sqrt{1 - V^2/c^2}$. This is a relation between measurements referred to different frames. Second, relative to a single frame, any physical body set into motion with speed V shortens in the direction of its motion, as was postulated by Fitz-Gerald and Lorentz, in the ratio $\sqrt{1 - V^2/c^2}$. In one sense, the contraction is perhaps not a "real" one, since in a frame in which it is at rest the body measures the same as before; but as far as effects on surrounding bodies are concerned, the contraction[1] is as real as if it were due to a drop in temperature.

Next consider a vibrator, such as a good crystal clock or a radiating atom, whose frequency is determined by natural laws and hence is the same in any inertial frame in which the vibrator is at rest. Let the vibrator be stationary in S'. Then the time T for one vibration as measured in S can be written $t_2 - t_1$, where, using Eq. (2.4a) for t_2 and t_1 with x' held constant,

$$T = t_2 - t_1 = \gamma(t_2' - t_1') = \frac{T'}{\sqrt{1 - V^2/c^2}} \qquad (2.6)$$

T' representing the period as measured in S'. This result has again a meaning beyond a mere relation between measurements. For T' is also the period that the vibrator would have in S if it were stationary in that frame. Thus, relative to S, setting it in motion has lengthened its period of vibration in the ratio γ. More generally, when system A is in uniform translational motion at speed V relative to system B at rest in an inertial frame, all natural processes in A as observed from B are slowed down in the ratio $(1 - V^2/c^2)^{\frac{1}{2}}$; similarly, processes in B as observed from A are slowed down in the same ratio. For example, spectral lines from moving atoms show a relativistic doppler shift which includes a slight displacement toward the red due to time dilation.

Muons are unstable particles with a rest mass 207 times that of the electron and a charge $q \pm 1.6 \times 10^{-19}$ C. Muons at rest decay into electrons or positrons with an average lifetime of about 2 μs. However, for muons moving with a speed of 0.98c, the measured life is 5 times longer. Experimental confirmation of the time dilation of Eq. (2.6) comes from determination of the lifetimes of several unstable particles. Time dilation effects are as "real" as those due to a slowing down of any other sort. As a further example, suppose a good crystal clock were subjected to steady translation for some time, first in one direction and then in the opposite, until it returned to the starting point. As judged by clocks that did not share its variable motion, it would lose time, and

---

[1] It was once a widely held misconception that an object moving with a speed approaching c would appear to an eye or a camera to be contracted in the direction of motion in accord with Eq. (2.5). However, in 1959 Terrell pointed out that what any optical instrument "sees" depends on the light reaching the controlling aperture at a particular instant. Light from different parts of the same object will have left the object at different times, that from greater distances having left earlier than that from nearer points. This leads to a distorted "picture" in which the object appears rotated. [See *Phys. Today*, 18(9): 24 (1960).]

after its return it would be behind a similar clock that had remained at rest.

2.9 Velocity Transformations

Poincare used the Lorentz transformation to derive important formulas for the transformations of *velocities* from frame S to frame S'. By definition,

$$v_x = \frac{dx}{dt} \qquad v_y = \frac{dy}{dt} \qquad v_z = \frac{dz}{dt}$$

$$v'_x = \frac{dx'}{dt'} \qquad v'_y = \frac{dy'}{dt'} \qquad v'_z = \frac{dz'}{dt'}$$

On the other hand, from Eq. (2.4), $dy' = dy$, $dz' = dz$, and

$$dx' = \gamma(dx - V\,dt) \qquad dt' = \gamma[dt - (V\,dx)/c^2]$$

where $\gamma = 1/\sqrt{1 - V^2/c^2}$. Then we have

$$v'_x = \frac{dx'}{dt'} = \frac{\gamma(dx - V\,dt)}{\gamma[dt - (V\,dx)/c^2]} = \frac{dx/dt - V}{1 - (V\,dx)/(c^2\,dt)} = \frac{v_x - V}{1 - Vv_x/c^2}$$

and in general,

$$v'_x = \frac{v_x - V}{1 - Vv_x/c^2} \qquad v'_y = \frac{v_y}{\gamma(1 - Vv_x/c^2)} \qquad v'_z = \frac{v_z}{\gamma(1 - Vv_x/c^2)} \tag{2.7}$$

As a numerical example, consider two electrons ejected from a radioactive source at rest in S, one toward $-x$ with $v_x = -0.8c$ and the other toward $+x$ with $v_x = 0.9c$. Then the difference in their velocities, measured in S, is $1.7c$. But if we make $V = -0.8c$ so that the S' frame keeps up with the electron going toward $-x$, the velocity of the second electron relative to the first (measured now in S') is

$$v'_x = \frac{0.9c - (-0.8c)}{1 - (-0.8c)(0.9c)/c^2} = \frac{1.7c}{1.72} = 0.99c$$

When v_x and V are small compared with the speed of light, Eqs. (2.7) reduce to the classical velocity transformations: $v'_x = v_x - V$; $v'_y = v_y$; $v'_z = v_z$.

2.10   The Variation of Mass

The laws of mechanics as they left the hands of Newton are not in harmony with the new theory of relativity.   One approach for correcting them was discovered originally in studying the motion of charged particles in electromagnetic fields, but they can also be deduced from a study of ordinary mechanical phenomena.

For this purpose, we select for study an experiment the outcome of which can be inferred from considerations of symmetry.   Consider two identical balls A and B; let A be projected in the $+y$ direction with a speed u in frame S and B be projected in the $-y'$ direction of frame S' with speed u'.   Assume that $u' = u$, that both are small compared with V (the speed of S' relative to S), and that the balls collide elastically with each other in such a way that the velocity of A is reversed in S while that of B is reversed in S' (Fig. 2.6).   As seen in S, the y component of the velocity of ball B is, by the inverse of Eq. (2.7),

$$v_{yB} = \frac{v_y' \sqrt{1 - V^2/c^2}}{1 + Vv_x'/c^2} = u' \sqrt{1 - \frac{V^2}{c^2}}$$

since $v_x' = 0$.   If energy is conserved, the y components of the velocities of the spheres are simply reversed, while the x components are unchanged. Conservation of momentum requires that the y component of the momentum in frame S before the collision be equal to that after the collision, and so

$$m_A u - m_B u' \sqrt{1 - \frac{V^2}{c^2}} = m_B u' \sqrt{1 - \frac{V^2}{c^2}} - m_A u \tag{2.8}$$

Since $u' = u$ by hypothesis, as viewed from S, m_B is equal not to m_A but to $m_A/\sqrt{1 - V^2/c^2}$.   So long as u is arbitrarily small, we may

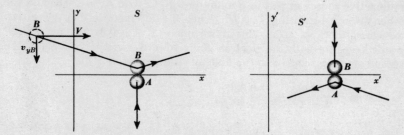

Fig. 2.6 Elastic collision of identical spheres viewed from inertial frames S and S' in relative motion.

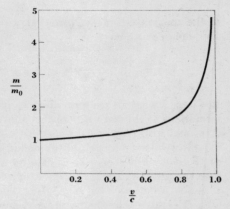

Fig. 2.7 Ratio of the mass of a body to its rest mass as a function of v/c.

replace m_A by m_0, where m_0 is the mass of one of the spheres at rest relative to S. Then clearly m_B is a function of V, which we write $m(V)$. From Eq. (2.8) we then have

$$m(V) = \frac{m_0}{\sqrt{1 - V^2/c^2}} \tag{2.9}$$

Here m_0 is called the *rest mass*.

The formula for m has been derived here for a special case; if we assume that it holds generally, we arrive at a relativistic mechanics which is compatible with Einstein's basic postulates. The momentum **p** of a body moving with velocity **v** is then

$$\mathbf{p} = m\mathbf{v} = \frac{m_0\mathbf{v}}{(1 - v^2/c^2)^{\frac{1}{2}}} \tag{2.10}$$

As v approaches c, the effective mass increases without limit (Fig. 2.7). No body with finite rest mass m_0 can quite achieve the speed of light; an infinite momentum is unknown.

2.11 Force and Kinetic Energy

All other mechanical terms and principles must now be scrutinized to see whether or not they require changes.

In newtonian mechanics the force **F** is equal to the time derivative of the momentum $m\mathbf{v}$. Now that we recognize that the mass m is a function of v, we have

$$\mathbf{F} = m\frac{d\mathbf{v}}{dt} + \mathbf{v}\frac{dm}{dv}\frac{dv}{dt} = m\mathbf{a} + \mathbf{v}\frac{dm}{dt} \tag{2.11}$$

In Chap. 3 we shall see that when we patiently apply the Lorentz transformations to the right side of this relation, forces too are different when measured with respect to co-moving inertial systems.

In relativity we retain the definitions of the work done by a force as $\int \mathbf{F} \cdot d\mathbf{s}$ and of energy as the ability or capacity to do work. We can then compute the kinetic energy K of a body of rest mass m_0 and velocity $\mathbf{v}$ in the usual way by calculating the work done by a constant net force $\mathbf{F}$ acting on the body, initially at rest, until it attains the velocity $\mathbf{v}$. This gives

$$K = \int_{v=0}^{v=v} F\, ds = \int_{v=0}^{v=v} \frac{d}{dt}mv\,\frac{ds}{dt}\,dt = \int v\frac{d}{dt}mv\,dt = \int v\,d(mv)$$

since $ds/dt = v$, the instantaneous velocity. Inserting the value of the mass m from Eq. (2.9), we have then

$$K = \int v\,d(mv) = \int v\,d\,\frac{m_0v}{(1 - v^2/c^2)^{\frac{1}{2}}}$$

$$= m_0 \int v\left[\frac{1}{(1 - v^2/c^2)^{\frac{1}{2}}} + \frac{v^2/c^2}{(1 - v^2/c^2)^{\frac{3}{2}}}\right]dv$$

$$= m_0 \int_0^v \frac{v\,dv}{(1 - v^2/c^2)^{\frac{3}{2}}} = m_0c^2\,\frac{1}{(1 - v^2/c^2)^{\frac{1}{2}}}\bigg|_0^v$$

Thus we find, for the *kinetic energy* of a body of which the rest mass is m_0, when moving with speed v,

$$K = m_0c^2\left(\frac{1}{\sqrt{1 - v^2/c^2}} - 1\right) \tag{2.12}$$

We can also expand in powers of v, obtaining

$$\left(1 - \frac{v^2}{c^2}\right)^{-\frac{1}{2}} = 1 + \frac{1}{2}\frac{v^2}{c^2} + \frac{3}{8}\frac{v^4}{c^4} + \frac{5}{16}\frac{v^6}{c^6} + \cdots$$

so that

$$K = \tfrac{1}{2}m_0v^2\left(1 + \frac{3}{4}\frac{v^2}{c^2} + \frac{5}{8}\frac{v^4}{c^4} + \cdots\right)$$

Thus, if $v \ll c$, K reduces approximately to the ordinary value, $\frac{1}{2}m_0v^2$. Under the same circumstances, the momentum can also be written, as usual, m_0v. In general, newtonian mechanics constitutes an approximate form of mechanical theory valid for any motion which is slow as compared with the speed of light.

2.12 Mass and Energy

Combining Eqs. (2.9) and (2.12), we write

$$K = (m - m_0)c^2 \tag{2.12a}$$

Thus the kinetic energy of a moving body equals c^2 times its gain in mass due to the motion. This relation suggests that we may think of the increase in energy as the actual cause of the increase in mass. It is then an attractive hypothesis to suppose that even the rest mass m_0 is due to the presence of an internal store of energy of amount m_0c^2. This may be called the *rest energy* of the body.

The total energy of a moving body would then be $m_0c^2 + K$, or

$$E = mc^2 = \frac{m_0c^2}{(1 - v^2/c^2)^{\frac{1}{2}}} \tag{2.13}$$

and we can write for its inertial mass m and momentum p

$$m = \frac{E}{c^2} \tag{2.14a}$$

$$p = mv = \frac{Ev}{c^2} \tag{2.14b}$$

From (2.13) and (2.14b) it also follows that

$$E^2 = m_0^2c^4 + p^2c^2 \tag{2.15}$$

The foregoing relations suggest that inertial mass may be a property of *energy* rather than of matter as such, each joule of energy possessing, or having associated with it, $1/c^2$ kg of mass. This idea has proved very useful in dealing with nuclear phenomena. For instance, a γ-ray photon impinging upon a nucleus can be converted into an electron and positron. Here the energy of the photon reappears in part as the kinetic energy of the particles, in part as their rest-mass energy $2m_0c^2$. A photon with frequency ν and wavelength λ has energy $E = h\nu$ and momentum $p = E/c = h\nu/c = h/\lambda$, where h is Planck's constant.

The electron, the least massive of charged particles, was the first for which the relativistic change of mass with speed was investigated. The experimental results are in complete accord with relativity theory. Since high-energy charged particles are usually produced by accelerating them through potential differences, it is convenient to measure their energies—kinetic, rest, and total—in electron volts.[1] For example, the rest mass of an electron corresponds to 0.511 MeV. An electron given a kinetic energy of 1 GeV in an accelerator has a total energy mc^2 of 1000.511 MeV; its mass m is almost 2000 times the rest mass.

Actually the energies equivalent to the masses of elementary particles, such as the electron and proton, are known to higher precision in electron volts than the actual masses in kilograms. For this reason it is common to find the "mass" of a particle such as the proton quoted as 938 MeV. The masses of many elementary particles are listed in this way inside the back cover.

The association of mass and energy is not limited to kinetic energy or to rest mass. Relativity requires mass to be associated with potential energy as well. Lorentz and others explored the possibility that the rest mass of the electron might be potential energy arising from the charge distribution. The potential energy of a charge Q distributed uniformly through a sphere of radius r is $\frac{3}{5}Q^2/4\pi\epsilon_0 r$; for other spherical distributions the $\frac{3}{5}$ was replaced by other numerical constants, but the $Q^2/4\pi\epsilon_0 r$ appeared repeatedly. Eventually it became customary to write for the electron, $mc^2 = e^2/4\pi\epsilon_0 r_0$, where

$$r_0 = \frac{e^2}{4\pi\epsilon_0 mc^2} = 2.81784 \times 10^{-15} \text{ m} \tag{2.16}$$

is known as the *classical radius* of the electron. Experiments on the scattering of electrons by electrons at high energies have shown that the interaction remains coulomb repulsion down to separations of less than 2×10^{-16} m, so that clearly the classical radius is several times too large to be consistent with electron-electron interactions. On the other hand, for scattering x-rays the effective radius of the electron is of the order of r_0.

2.13[2] Mass and Potential Energy

To see that mass is associated with potential as well as kinetic energy, suppose, for example, that two equal masses moving with equal and

---

[1] One electron volt is the *energy gained by an electron in falling through a potential difference of one volt;* 1 eV = 1.602 × 10⁻¹⁹ J.

[2] Text covered by gray shading indicates supplementary material often omitted in shorter courses.

opposite velocities along the x axis as seen by an S observer collide with each other, a spring acting as a buffer between them. Just as they come to rest, let a lock snap shut and hold them combined into a single mass. The initial kinetic energy of the masses is converted by the collision into potential energy of the spring. Let us view this collision from a second frame of reference S' that is moving with velocity V parallel to x. Then, if the velocities of the two bodies before the collision, as measured in S, are $v_1 = v_{x1} = v$, $v_2 = v_{x2} = -v$, the same velocities as measured in S' are, by Eq. (2.7),

$$v_1' = v_{x1}' = \frac{v - V}{1 - Vv/c^2} = -V + \left(1 - \frac{V^2}{c^2}\right)\frac{v}{1 - Vv/c^2}$$

$$v_2' = v_{x2}' = \frac{-v - V}{1 + Vv/c^2} = -V - \left(1 - \frac{V^2}{c^2}\right)\frac{v}{1 + Vv/c^2}$$

Therefore,

$$\frac{v_1'}{(1 - v_1'^2/c^2)^{\frac{1}{2}}} = -\frac{V}{(1 - v_1'^2/c^2)^{\frac{1}{2}}} + \left(1 - \frac{V^2}{c^2}\right)\frac{v}{[1 - (V^2 + v^2)/c^2 + V^2v^2/c^4]^{\frac{1}{2}}}$$

the last term being obtained in this form after inserting for v_1' under the radical the first expression given for v_1' above. Similarly,

$$\frac{v_2'}{(1 - v_2'^2/c^2)^{\frac{1}{2}}} = -\frac{V}{(1 - v_2'^2/c^2)^{\frac{1}{2}}} - \left(1 - \frac{V^2}{c^2}\right)\frac{v}{[1 - (V^2 + v^2)/c^2 + V^2v^2/c^4]^{\frac{1}{2}}}$$

The total momentum *before* collision is, therefore,

$$\frac{m_0v_1'}{\sqrt{1 - v_1'^2/c^2}} + \frac{m_0v_2'}{\sqrt{1 - v_2'^2/c^2}}$$
$$= -m_0V\left(\frac{1}{\sqrt{1 - v_1'^2/c^2}} + \frac{1}{\sqrt{1 - v_2'^2/c^2}}\right) \quad (2.17)$$

On the other hand, if the rest mass of the combined body after collision were merely the sum of the rest masses of the separate bodies, or $2m_0$, the total momentum *after* collision would be

$$\frac{-2m_0V}{\sqrt{1 - V^2/c^2}}$$

by Eq. (2.10), since the velocity of the combined body is then $-V$. This is *not equal* to the momentum *before* collision, as given by (2.17); this can be seen very easily when v_1' and v_2' are both either greater or less than V.

Thus conservation of momentum fails if only the rest masses are taken into account. But suppose we include in the mass of the combined body a mass associated with the potential energy of the spring. Then the total mass is proportional to the total energy and is a constant. The total mass after the collision is, therefore, the same as it was before the collision, or

$$\frac{m_0}{\sqrt{1 - v_1'^2/c^2}} + \frac{m_0}{\sqrt{1 - v_2'^2/c^2}}$$

If we multiply this value of the total mass by the common velocity of the bodies after the collision, which is $-V$, we obtain for the total momentum *after* collision exactly the same expression as that given in Eq. (2.17) for the total momentum *before* collision. The principle of the conservation of momentum thus holds here if and only if we assume that the potential energy in the spring makes its contribution to the mass and momentum of the system.

Problems

1. An electron is moving at a speed of 1.8×10^8 m/s. Find the ratio of its mass to its rest mass. What is its kinetic energy? Its total energy?

Ans: 1.25; 128 keV; 0.639 MeV

2. A proton is accelerated in a synchrotron until its kinetic energy is just equal to its rest mass (938 MeV). Find the ratio v/c for this proton.

3. An electron is accelerated to a kinetic energy of 1.00×10^9 eV (1 GeV, or as it is sometimes known in the United States, 1 BeV). Find for this electron the ratios m/m_0 and v/c.

Ans: 1.96×10^3; 0.99999987

4. Through what potential difference must protons be accelerated to achieve a speed of $0.600c$? What is the momentum of such a proton? The total energy?

Ans: 235 MeV; 3.76×10^{-19} kg-m/s; 1173 MeV

5. Find the work which must be done on an electron to increase its speed from $0.5c$ to $0.9c$.

Ans: 0.582 MeV

6. How many electron volts of energy must an electron gain to bring its mass (a) to $1.05m_0$ and (b) to $2m_0$? In each case what is the speed of the electron?

   Ans: (a) 25,500, 9.2×10^7 m/s; (b) 511,000, 2.6×10^8 m/s

7. Carry out the operations required to obtain the velocity transformations for v_y' given in Eq. (2.7). Starting with Eqs. (2.4a), derive the transformations for v_x and v_y in terms of v_x' and v_y' for a particle with arbitrary velocity **v**.

8. An α particle moving east with a speed of $0.5c$ is passed by an electron moving west with a speed of $0.95c$. Find the speed of the electron relative to the α particle.

   Ans: $0.98c$

9. An electron moving to the right with a speed of 2.5×10^8 m/s passes an electron moving to the left with a speed of 2.8×10^8 m/s. Find the speed of one electron relative to the other.

   Ans: 2.98×10^8 m/s

10. Show that if two events occur simultaneously at different points (x_1,y_1,z_1) and (x_2,y_2,z_2) in the S frame, they occur at different times in S' with $t_2' - t_1' = \gamma V(x_1 - x_2)/c^2$.

11. Observer A is at rest in a frame S' which moves horizontally past an inertial frame S at a speed of $0.6c$. A boy in the latter frame drops a ball which according to observer A's clock falls for 1.5 s. How long would the ball fall for an observer at rest in the S frame?

   Ans: 1.2 s

12. The mean life of a muon at rest is about 2×10^{-6} s. If muons in cosmic rays have an average speed of $0.998c$, find the mean life of these muons and the approximate distance they move before decaying.

   Ans: 3.2×10^{-5} s; 9.5 km

13. The observed half-life of pions at rest is 1.8×10^{-9} s. A beam of pions has a speed of $0.95c$. In the laboratory system how long will it take for half the pions to decay? Through what distance will they travel in that time?

   Ans: 5.8 ns; 1.7 m

14. A constant force F acts on a body initially at rest for a time t. Show that its speed $v(t)$ is given by $c/[1 + (m_0c/Ft)^2]^{\frac{1}{2}}$. Show further that if $Ft \ll m_0c$, $v(t) \approx Ft/m_0$, while if $Ft \gg m_0c$, $v(t) \approx c$.

15. A particle has a rest energy m_0c^2 and a total energy E. Show that the velocity of this particle is $c\sqrt{1 - (m_0c^2/E)^2}$.

16. Fizeau showed that the speed of light in a liquid streaming toward a light source with velocity u was given by $v' = c/n - u(1 - 1/n^2)$, where n is the refractive index of the liquid. Show that this result agrees with the relativistic law, Eq. (2.7), provided that terms in u/c higher than the first power are neglected.

17. Show that if an observer in S determines the length of a meter stick in S' by determining the time required for the meter stick to pass a fixed point x_1 and multiplying this time by V, he obtains the same contraction found in Eq. (2.5).

18. A particle has momentum p, kinetic energy K, velocity v, and rest mass m_0. Show that

$$\frac{pv}{K} = \frac{K/m_0c^2 + 2}{K/m_0c^2 + 1}$$

Prove that this equation gives the classical $K = \frac{1}{2}mv^2$ for $K \ll m_0c^2$ and $K = mv^2$ for $K \gg m_0c^2$.

19. Compute the daily loss of mass of the sun associated with the isotropic emission of electromagnetic radiation, given that the average solar constant at the earth is 1400 W/m^2 and that the average radius of the earth's orbit is 1.5×10^{11} m.

Ans: 3.8×10^{14} kg/day

20. An electron moving perpendicular to a uniform magnetic field B is bent into a circular path. Show that $Br = \sqrt{K^2 + 1.02K}/300$ if B is in webers per square meter, r in meters, and the kinetic energy K is in MeV.

21. (a) A stick of length L_0 is at rest in S, making an angle θ with the x axis. Show that an observer in the S' frame finds the length to be $L' = L_0(\cos^2 \theta/\gamma^2 + \sin^2 \theta)^{\frac{1}{2}}$ and the angle θ' with his x axis to be given by $\tan \theta' = \gamma \tan \theta$.

(b) A particle moves at speed v an angle θ with the x axis in the S frame. Find its speed and direction relative to the x' axis of the S' frame.

$$Ans:\ v' = \frac{\{v^2 - 2Vv \cos \theta + V^2 - [(Vv \sin \theta)/c]^2\}^{\frac{1}{2}}}{1 - (Vv \cos \theta)/c^2}$$

$$\tan \theta' = \frac{\sqrt{1 - v^2/c^2}\, v \sin \theta}{v \cos \theta - V}$$

22. Show that for a particle having acceleration components a_x and a_y in the S frame, the corresponding accelerations in the S' frame are

$$a'_x = \frac{(1 - V^2/c^2)^{\frac{3}{2}}a_x}{(1 - Vv_x/c^2)^3}$$

and

$$a'_y = \frac{1 - V^2/c^2}{(1 - Vv_x/c^2)^2} \left(a_y + \frac{Vv_y/c^2}{1 - Vv_x/c^2} a_x \right)$$

23. In 1906 Einstein proposed a gedanken (thought) experiment to show that radiant energy has associated with it a mass equivalent equal to the energy E/c^2. He supposed that a burst of photons with total energy E and momentum E/c left one end of a cylinder of length L and mass M. While the photons traverse the length of the cylinder, the latter recoils in the opposite direction until the burst is absorbed at the other end. Show that the photons must have transferred a mass E/c^2 if the center of mass of the isolated system remains fixed.

24. Show that if the force between two particles is a function of only their separation, Newton's second law $F = ma$ is invariant under a galilean transformation.

25. Show that the Lorentz-FitzGerald contraction leads to a nonzero fringe shift if the interferometer paths differ in length and if the velocity of the interferometer relative to the ether varies.

chapter three

Relativity and Four-vectors

In relativity, space and time are no longer separate concepts but are interwoven in a way that makes the introduction of a four-dimensional space-time continuum very useful. In such a four-dimensional space many of the laws of classical physics can be expressed in a particularly elegant form which greatly simplifies the shift from one inertial system to another.

3.1 The Interval between Events

As we saw in Chap. 2, observers O and O' in inertial frames S and S' moving relative to one another ordinarily assign different time and space coordinates to an event which both witness. By an *event* we mean some physical occurrence detected by an observer; e.g., a cosmic-ray electron interacts with a nucleus in the atmosphere to produce a photon. To O the event is specified in terms of three spatial coordinates and the time, hence by the four coordinates (x_1, y_1, z_1, t_1).

Next consider a second event, perhaps the decay of a pion, to which O assigns the coordinates (x_2, y_2, z_2, t_2). We shall define the *interval* between the two events to be the quantity s_{12}, where

$$s_{12}{}^2 = c^2(t_2 - t_1)^2 - (x_2 - x_1)^2 - (y_2 - y_1)^2 - (z_2 - z_1)^2 \quad (3.1)$$

If observer O' detected the same events, he would specify them with coordinates (x_1', y_1', z_1', t_1') and (x_2', y_2', z_2', t_2'). Even though O and O' might find significantly different values of the time elapsed between the two events and of their spatial separation, *the interval is the same for both;* it is an *invariant* common to all observers in inertial frames.

When $s_{12}{}^2$ is positive, the interval s_{12} is a real number and is said to be *timelike*. For a timelike interval it is always possible to find an inertial frame in which the two events occur at the same place (values of x, y, z). However, it is impossible to find an inertial frame in which the two events occur simultaneously ($t_2 = t_1$) since in such, a frame $s_{12}{}^2$ would be negative, contrary to the first condition of this paragraph.

If $s_{12}{}^2$ is negative, the interval is imaginary and is referred to as *spacelike*. For any spacelike interval, it is always possible to find some inertial frame in which the two events are simultaneous, but it is impossible to find a frame in which the events occur at the same place. Thus the spacelike or timelike character of an interval is independent of the inertial frame of reference. In the discussions of Sec. 2.6 and in Fig. 2.5 all points in the regions marked "future" and "past" are separated from the origin by timelike intervals, while those in the regions marked "present" are separated from the origin by *spacelike* intervals.

3.2 Four-vectors

Let us introduce $w = ict$ ($i = \sqrt{-1}$) and rewrite Eq. (3.1) in the form

$$-s_{12}{}^2 = (x_2 - x_1)^2 + (y_2 - y_1)^2 + (z_2 - z_1)^2 + (w_2 - w_1)^2 \quad (3.1a)$$

Here we may interpret x, y, z, and w as cartesian coordinates in a *four-dimensional euclidean* "space" in which $\sqrt{-s_{12}{}^2}$ may be regarded as the "distance" between points (x_1, y_1, z_1, w_1) and (x_2, y_2, z_2, w_2). The corresponding "displacement" can be treated as a vector with components $x_2 - x_1$, $y_2 - y_1$, $z_2 - z_1$, and $w_2 - w_1$ in a four-dimensional space. Such a quantity is often called a *four-vector* or a *world vector*.

If x, y, z, w are the coordinates of an event in frame S, we may consider these quantities as the components of a *four-dimensional radius vector* for this event. Thus, if 1_x, 1_y, 1_z, and 1_w represent unit vectors in the x, y, z, and w directions,

$$\underline{r} = 1_x x + 1_y y + 1_z z + 1_w w$$

and

$$r^2 = \underline{r} \cdot \underline{r} = x^2 + y^2 + \tau^2 + w^2$$

since $1_x \cdot 1_x = 1$, $1_x \cdot 1_y = 0$, etc., just as for ordinary three-dimensional vectors. The magnitude of the four-vector $\underline{r}$ is given by

$$r = \sqrt{x^2 + y^2 + z^2 + w^2}$$

If we wish to find the components of the vector $\underline{r}$ in frame S' moving in the x direction with speed V and with times measured from the instant the origins of the two frames coincide (Secs. 2.1 and 2.7), we apply the Lorentz transformation, Eqs. (2.4), which can be written in matrix form

$$\begin{bmatrix} x' \\ y' \\ z' \\ w' \end{bmatrix} = \begin{bmatrix} \gamma_T & 0 & 0 & i\beta_T\gamma_T \\ 0 & 1 & 0 & 0 \\ 0 & 0 & 1 & 0 \\ -i\beta_T\gamma_T & 0 & 0 & \gamma_T \end{bmatrix} \begin{bmatrix} x \\ y \\ z \\ w \end{bmatrix} \tag{3.2}$$

where $\gamma_T \equiv 1/\sqrt{1 - V^2/c^2}$ and $\beta_T \equiv V/c$. It is left as an exercise (Prob. 1) to show that Eq. (3.2) is equivalent to Eq. (2.4).

Minkowski discovered another way of looking at the Lorentz transformation; Eqs. (2.4) are mathematically the same as the equations for the transformation of a four-vector under a rotation of the x and w axes in the four-dimensional x, y, z, w space through an imaginary angle $\phi = \tan^{-1} i\beta_T$. For any point P with coordinates (x,w), the corresponding coordinates (x',w') are, from Fig. 3.1 and analytical geometry,

$$x' = x \cos \phi + w \sin \phi \qquad w' = w \cos \phi - x \sin \phi$$

while the Lorentz transformation, Eqs. (2.4) with $w = ict$, gives

$$x' = \gamma_T x + i\beta_T\gamma_T w \qquad w' = \gamma_T w - i\beta_T\gamma_T x$$

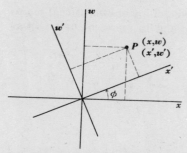

Fig. 3.1 Rotation of the x and w coordinate axes through angle ϕ.

Comparing these two pairs of equations shows that

$$\cos \phi = \gamma_T = \frac{1}{\sqrt{1 - V^2/c^2}} \quad \text{and} \quad \sin \phi = i\beta_T\gamma_T$$

$$= \frac{iV}{c\sqrt{1 - V^2/c^2}}$$

In addition to the radius four-vector many other useful four-vectors appear in relativistic physics. Here a four-vector is defined as a quantity whose components transform in the same way as (x,y,z,w) under a Lorentz transformation. Thus, if $\underline{A}$ is a four-vector, it must transform to $\underline{A}'$ in going from S to S' in the same way as $\underline{r}$, viz.,

$$\begin{bmatrix} A_x' \\ A_y' \\ A_z' \\ A_w' \end{bmatrix} = \begin{bmatrix} \gamma_T & 0 & 0 & i\beta_T\gamma_T \\ 0 & 1 & 0 & 0 \\ 0 & 0 & 1 & 0 \\ -i\beta_T\gamma_T & 0 & 0 & \gamma_T \end{bmatrix} \begin{bmatrix} A_x \\ A_y \\ A_z \\ A_w \end{bmatrix} \tag{3.2a}$$

The magnitude of a four-vector $\underline{A}$ is a scalar invariant, i.e., the same for all inertial frames.

If $\underline{A}$ and $\underline{B}$ represent four-vectors, $\underline{A} = 1_x A_x + 1_y A_y + 1_z A_z + 1_w A_w$ and $\underline{B} = 1_x B_x + 1_y B_y + 1_z B_z + 1_w B_w$, the scalar product $\underline{A} \cdot \underline{B} \equiv A_x B_x + A_y B_y + A_z B_z + A_w B_w$. However, the cross product $\underline{A} \times \underline{B}$ cannot be represented by a vector; rather it is defined as the antisymmetric world tensor Φ, the components of which are the determinants of the matrix

$$\begin{bmatrix} A_x & A_y & A_z & A_w \\ B_x & B_y & B_z & B_w \end{bmatrix}$$

or

$$\underline{A} \times \underline{B} \equiv \Phi \equiv \begin{bmatrix} 0 & \Phi_{xy} & \Phi_{xz} & \Phi_{xw} \\ \Phi_{yx} & 0 & \Phi_{yz} & \Phi_{yw} \\ \Phi_{zx} & \Phi_{zy} & 0 & \Phi_{zw} \\ \Phi_{wx} & \Phi_{wy} & \Phi_{wz} & 0 \end{bmatrix} \tag{3.3}$$

where $\Phi_{xy} = -\Phi_{yx} = A_x B_y - A_y B_x$, etc.

3.3 Proper Time and the Four-momentum

One of the postulates of special relativity is that properly formulated laws of nature are the same for any two inertial frames; thus they may not include anything which refers to the translational velocity of the reference frame. If we let **f** stand for the classical resultant force on a body, Newton's second law of motion in the form $\mathbf{f} = d\mathbf{p}/dt$ does not satisfy this postulate. Even when the relativistic change of mass with velocity is introduced, the law does not maintain its form under a Lorentz transformation. Applying Newton's second law with **p** given by Eq. (2.10) yields

$$\mathbf{f} = \frac{d\mathbf{p}}{dt} = \frac{m_0}{\sqrt{1 - v^2/c^2}} \left(\frac{d\mathbf{v}}{dt} + \frac{v \, dv/dt}{c^2 - v^2} \mathbf{v} \right) \tag{3.4}$$

When the direction of the force differs from that of the velocity, the acceleration is not in the direction of the force unless the speed is constant (as it is in the very important case of a particle moving perpendicular to a uniform magnetic induction **B**). Further, when the force is in the direction of the velocity, the ratio of force to acceleration is given by $m_0/(1 - v^2/c^2)^{\frac{3}{2}}$, which is sometimes called the *longitudinal mass*. When the force is perpendicular to the velocity, the force/acceleration ratio is $m_0/(1 - v^2/c^2)^{\frac{1}{2}}$, sometimes called the *transverse mass*. In the clumsy formalism using three space coordinates plus time, we have the fiction that mass, an intrinsic property, depends on reference frame, and worse, that force has complicated transformation properties. In four-vector formalism it is possible to remove these difficulties and arrive at a logical, self-consistent system.

In Sec. 3.4 we shall describe one way to put Newton's second law into a valid four-vector form compatible with the Einstein postulates, but first we must introduce the *proper time* τ, which is time measured in an inertial reference frame in which the body is instantaneously at rest. For this special frame S' the spatial coordinates x', y', z' do not change in an infinitesimal time interval dt', so that by the invariance of intervals

[Eq. (3.1)] we have

$$c^2\,d\tau^2 = c^2\,dt'^2 = c^2\,dt^2 - (dx)^2 - (dy)^2 - (dz)^2$$

or

$$d\tau = \sqrt{(dt)^2 - \frac{[(dx)^2 + (dy)^2 + (d\tau)^2]}{c^2}} = dt\sqrt{1 - \frac{v^2}{c^2}} \qquad (3.5)$$

where x, y, z, t are measured in the inertial reference frame S of observer O.   Since $c\,d\tau$ is an interval, the proper-time element $d\tau$ is the same for observers in all inertial frames.   As in ordinary three-dimensional space, a quantity which has the same value in all inertial frames is called a *scalar;* thus we say that $d\tau$ is a *Lorentz scalar.*   When the elapsed time in S is $t_2 - t_1$, the corresponding proper-time duration is

$$\tau_2 - \tau_1 = \int_{t_1}^{t_2} dt\sqrt{1 - \frac{v^2}{c^2}} \qquad (3.5a)$$

where v is a function of time t for the accelerated object.   From Eq. (3.5a) it can be seen that the proper-time interim for an accelerated body is always less than the corresponding interim in any inertial frame from which the motion is observed.

We now define the four-velocity $\underline{u}$ of the body as the derivative of its four-vector radius $\underline{r}$ with respect to proper time τ:

$$\underline{u} = \frac{d\underline{r}}{d\tau} = \frac{d\underline{r}}{dt}\frac{dt}{d\tau}$$

$$= \left(1_x\frac{dx}{dt} + 1_y\frac{dy}{dt} + 1_z\frac{dz}{dt} + 1_w\frac{dw}{d\tau}\right)\frac{1}{\sqrt{1 - v^2/c^2}} \qquad (3.6)$$

where v^2 is the square of the ordinary velocity.   Thus $u_x = \gamma v_x$; in general the spatial components of the four-velocity are γ times the components of the ordinary velocity, while the fourth component is $i\gamma c$.

The four-momentum $\underline{P}$ of any particle is the product of the four-velocity $\underline{u}$ and the proper mass m_0.   Note that the adjective *proper* means *relative to an inertial frame in which the body is instantaneously at rest.*   The proper mass m_0, the proper length L_0 of an object, and the proper time τ are all Lorentz scalars, whence comes their usefulness.   From the definition

$$\underline{P} = m_0\underline{u} = m_0\gamma(1_x v_x + 1_y v_y + 1_z v_z + 1_w ic) \qquad (3.7)$$

we observe that the spatial components are the ordinary momentum components. The fourth component is $im_0 \gamma c = i(mc^2)/c = iE/c$, where E is the total energy [Eq. (2.13)] of the body. Thus *the time component of the momentum four-vector is the energy multiplied by* $\sqrt{-1}/c$. When one transforms from one's rest frame S to a moving frame S', the four-momentum transforms as a four-vector, so that, by Eq. (3.2a),

$$p'_x = \gamma_T \left(p_x - \frac{VE}{c^2} \right) \qquad p'_y = p_y \qquad p'_z = p_z$$
$$E' = \gamma_T (E - V p_x) \tag{3.8}$$

The four-momentum has a magnitude $(\underline{\mathbf{P}} \cdot \underline{\mathbf{P}})^{\frac{1}{2}} = im_0 c$. Indeed, *the dot product of any two four-vectors is a scalar invariant.* Further $\underline{\mathbf{P}} \cdot \underline{\mathbf{P}} = p^2 - E^2/c^2 = -m_0^2 c^2$, which is Eq. (2.15) in a different form. While the rest mass is the same in every inertial frame, the momentum $\mathbf{p}$ and the total energy E are different.

The four-momentum concept can be applied to find the equations for the electromagnetic doppler effect. Consider a photon of energy $h\nu$ and momentum $\mathbf{1}(h\nu/c)$, where $\mathbf{1}$ is a unit vector in the propagation direction in frame S. Let θ be the angle between $\mathbf{1}$ and the x axis and let $\mathbf{1}$ lie in the xy plane. In frame S', moving with velocity V along the x axis of S, the components of the four-momentum of the photon are given by

$$
\begin{bmatrix} p'_x \\ p'_y \\ p'_z \\ \dfrac{iE'}{c} \end{bmatrix}
=
\begin{bmatrix} h\nu' \dfrac{\cos \theta'}{c} \\ h\nu' \dfrac{\sin \theta'}{c} \\ 0 \\ \dfrac{ih\nu'}{c} \end{bmatrix}
=
\begin{bmatrix} \gamma_T & 0 & 0 & i\beta_T \gamma_T \\ 0 & 1 & 0 & 0 \\ 0 & 0 & 1 & 0 \\ -i\beta_T \gamma_T & 0 & 0 & \gamma_T \end{bmatrix}
\begin{bmatrix} h\nu \dfrac{\cos \theta}{c} \\ h\nu \dfrac{\sin \theta}{c} \\ 0 \\ \dfrac{ih\nu}{c} \end{bmatrix}
\tag{3.9}
$$

where $\gamma_T = 1/\sqrt{1 - V^2/c^2}$ and $\beta_T = V/c$, from which

$$\nu' \cos \theta' = \gamma_T \nu \left(\cos \theta - \frac{V}{c} \right) \tag{3.10a}$$

$$\nu' \sin \theta' = \nu \sin \theta \tag{3.10b}$$

and

$$\nu' = \gamma_T \nu \left(1 - V \frac{\cos \theta}{c} \right) \tag{3.11a}$$

$$\tan \theta' = \frac{\sin \theta}{\gamma_T (\cos \theta - V/c)} \tag{3.11b}$$

In general, $\theta' \neq \theta$ and, unlike the case of acoustic waves, there is a doppler shift for $\theta = 90°$. This transverse doppler shift arises from the different rates at which clocks run in S' and S as seen by an observer in either frame.

3.4   Relativistic Mechanics

In classical mechanics the conservation of energy and the conservation of momentum are key laws. In relativistic mechanics the energy of a particle may be looked upon as the time component of the momentum four-vector multiplied by $c/\sqrt{-1}$. Conservation of momentum and conservation of energy can be regarded as the spatial and temporal aspects of a single conservation of four-momentum. In the absence of an external resultant force on a particle, p_x, p_y, p_z, and E are all constants of the motion. When N particles interact or decay in an isolated system, the laws of conservation of both momentum and energy are implicit in the relation

$$\Big(\sum_{i=1}^{N_i} \underline{P}_i \Big)_{\text{initial}} = \Big(\sum_{i=1}^{N_f} \underline{P}_i \Big)_{\text{final}} \tag{3.12}$$

where $\underline{P}_i$ is the four-momentum for the ith particle.

The rest mass M_0 of a system of particles, such as an atom, can be defined as the total energy of the system in the center-of-mass reference frame divided by c^2. Then, for a system of noninteracting particles, i.e., freely moving with no potential energy of interaction, the rest mass M_0 is greater than the sum of the rest masses of the particles by $1/c^2$ times the kinetic energy of the particles in the center-of-mass frame. However, when there are attractive forces between the particles, the negative potential energy arising from the interaction may bind the particles into a composite body with $M_0 < \sum_i (m_0)_i$. For an atom the difference $\Delta M = M_0 - \sum_i (m_0)_i$ is called the *mass defect*, and $\Delta M c^2$ is the *binding energy* of the atom. When ΔM is negative for every possible way in which an atom can be broken up, the atom is stable. When ΔM is positive for some form of breakup, it is energetically possible for the atom to decay spontaneously. This criterion is useful in atomic and nuclear physics.

While the statement that "mass is a form of energy" is frequently made and successfully applied in many problems, it is ambiguous in some cases. For example, when a number of uranium 236 atoms fission in a nuclear explosion, the *total* rest mass of the system (fission products, kinetic energies of all particles including photons and neutrinos, etc.) is unchanged, but the sum of the rest masses of the individual constituents is less by about 0.1 percent than the original rest mass. An unambiguous statement of the relationship between rest mass and energy is available from the four-momentum vector for the system, which is the resultant of the four-momenta of the constituents. Here the energy E is proportional to the time component of the momentum four-vector (this component is iE/c), while the rest mass of the system is a measure of the magnitude of the four-vector (the magnitude is iM_0c). We may summarize as follows.

In any reaction between two or more bodies, the resultant *momentum-energy four-vector of the system is conserved*, and so the ordinary laws of momentum and energy conservation coalesce into a single relativistic law. Further, for any particle the invariant magnitude of $\underline{\mathbf{P}}$ is im_0c, where m_0 is the rest mass of the particle. For particles with zero rest mass, such as photons or neutrinos, $p = E/c$. The relativistic conservation of momentum-energy is valid for both elastic and inelastic collisions.

As we have seen in Sec. 3.3, certain ambiguities arise when one considers Newton's second law in moving inertial frames. Minkowski devised a formulation of this law which is consistent with the postulates of special relativity by equating what we call the *Minkowski force four-vector* $\underline{\mathbf{F}}$ to the derivative of the four-momentum with respect to proper time

$$
\begin{aligned}
\underline{\mathbf{F}} = \frac{d}{d\tau} m_0\underline{\mathbf{u}} &= 1_x \frac{dp_x}{d\tau} + 1_y \frac{dp_y}{d\tau} + 1_z \frac{dp_z}{d\tau} + 1_w \frac{dp_w}{d\tau} \\
&= 1_x\gamma \frac{dp_x}{dt} + 1_y\gamma \frac{dp_y}{dt} + 1_z\gamma \frac{dp_z}{dt} + 1_w\gamma \cdot \frac{i}{c}\frac{dE}{dt} \\
&= \gamma \frac{d\mathbf{p}}{dt} + 1_w\gamma \frac{i}{c}\frac{dE}{dt}
\end{aligned}
\tag{3.13}
$$

The spatial part of the Minkowski force is equal to $\gamma\, dp/dt$, where $\mathbf{p}$ is the ordinary momentum. By the classical definition the ordinary force is $\mathbf{f} = dp/dt$, so that the spatial part of the Minkowski force is greater than the ordinary force by the factor γ and its spatial components are

$$
F_x = \frac{f_x}{\sqrt{1 - v^2/c^2}} \qquad F_y = \frac{f_y}{\sqrt{1 - v^2/c^2}}
$$

$$
F_z = \frac{f_z}{\sqrt{1 - v^2/c^2}}
\tag{3.14}
$$

The time component of the Minkowski force is

$$\frac{i}{c}\frac{dE}{dt}\frac{1}{\sqrt{1 - v^2/c^2}}$$

The time rate of change of energy of the particle is equal to the rate at which the force **f** does work on the particle, and this is the power **f** · **v** delivered by the force. Thus the time component of $\underline{F}$ becomes

$$F_w = \frac{i}{c}\frac{dE}{dt}\frac{1}{\sqrt{1 - v^2/c^2}} = \frac{i\mathbf{f}\cdot\mathbf{v}}{c\sqrt{1 - v^2/c^2}} \qquad (3.15)$$

where **f** and **v** are the classical forces and velocity, respectively, and the Minkowski force may be written

$$\underline{F} = \gamma\mathbf{f} + 1_w i\gamma\mathbf{f}\cdot\mathbf{v}/c$$

3.5   *Relativity and Electromagnetism*

Contrary to the situation in mechanics, a review of the laws of the electromagnetic field shows them to be in harmony with relativity as they stand. This might perhaps have been expected in view of the fact that the theory of relativity actually developed out of experiments in that part of the field of electromagnetism which is called optics. Nevertheless there are many aspects of electromagnetic theory to which it is fruitful to apply relativity theory. The advantages of using four-space notation turn out to be great.

Let us postulate that the charge q of a particle is a scalar invariant. Consider a uniform charge distribution at rest in frame S. The charge density in a small cube $\Delta x\,\Delta y\,\Delta z$ is $\rho_0 = \Sigma q_i/(\Delta x\,\Delta y\,\Delta z)$, where the sum is carried out over all charges in the cube. An observer O' in a moving frame S' agrees on the charge enclosed in the cube, but by Eq. (2.5), $\Delta x' = \Delta x\sqrt{1 - V^2/c^2}$, although $\Delta y' = \Delta y$ and $\Delta z' = \Delta z$. Thus O' finds the charge density to be $\rho' = \rho_0/\sqrt{1 - V^2/c^2} = \gamma_T\rho_0$; because of the Lorentz-FitzGerald contraction of the volume element, ρ' exceeds ρ.

Next consider a uniform charge distribution of density ρ moving with velocity **v** relative to frame S. The current density is then $\mathbf{J} = \Sigma q_i\mathbf{v}/(\Delta x\,\Delta y\,\Delta z) = \rho\mathbf{v} = \gamma\rho_0\mathbf{v}$, where $\gamma = 1/\sqrt{1 - v^2/c^2}$. Clearly, in S' the current density J' is different, both because v' differs from v and because of space contraction.

Although neither ρ nor **J** is invariant under Lorentz transformation, we may introduce a current-density four-vector in the following way.

Let $\underline{u}$ be the four-velocity of the charges in the S frame and ρ_0 be the *proper* charge density, viz., that in a frame in which the charges are at rest. Then the four-vector $\underline{J} = \rho_0\underline{u}$ is, by Eq. (3.6),

$$\underline{J} = 1_x\gamma\rho_0v_x + 1_y\gamma\rho_0v_y + 1_z\gamma\rho_0v_z + 1_w i\gamma c\rho_0$$
$$= 1_xJ_x + 1_yJ_y + 1_zJ_z + 1_wJ_w \tag{3.16}$$

where $\gamma = 1/\sqrt{1 - v^2/c^2}$. The spatial components of $\underline{J}$ correspond to the components of the current density, while the fourth component J_w is $ic\gamma\rho_0 = ic\rho$.

The equation of continuity in electromagnetism, which follows from the conservation of charge, is

$$\nabla \cdot \mathbf{J} + \frac{\partial\rho}{\partial t} = 0 \tag{3.17}$$

where the vector operator del is $\nabla = 1_x\partial/\partial x + 1_y\partial/\partial y + 1_z\partial/\partial z$ in cartesian coordinates. In relativity it is convenient to introduce the four-vector operator $\square$ defined by

$$\square = 1_x\frac{\partial}{dx} + 1_y\frac{\partial}{dy} + 1_z\frac{\partial}{dz} + 1_w\frac{\partial}{dw} \tag{3.18}$$

n this language the equation of continuity becomes simply

$$\square \cdot \underline{J} = 0 \tag{3.17a}$$

3.6 Maxwell's Equations and the Four-potential[1]

The ordinary force $\mathbf{f}$ acting on a charge q moving with velocity $\mathbf{v}$ is given by the Lorentz relation

$$\mathbf{f} = q[\mathbf{E} + (\mathbf{v} \times \mathbf{B})] \tag{3.19}$$

where the electric intensity $\mathbf{E}$ is the force per unit charge when the charge is at rest relative to the reference frame. The difference $\mathbf{f} - q\mathbf{E} = q(\mathbf{v} \times \mathbf{B})$ is the force due to the magnetic induction $\mathbf{B}$. The electric intensity $\mathbf{E}$ and the magnetic intensity $\mathbf{B}$ are thus defined for a particular reference frame; the distinction between electric and magnetic fields is in part a relative one, depending on the frame of

[1] Familiarity with Maxwell's equations is assumed in Secs. 3.6 and 3.7.

reference. For example, as we saw in Sec. 2.2, charges at rest relative to the S frame produce only an electrostatic field, but to an observer in the S' frame, these same charges constitute current elements with an associated magnetic field.

The electric displacement **D** is defined by the relation $D = \epsilon_0 E + P$, where **P** is the polarization (the electric dipole moment per unit volume) and ϵ_0 is the permittivity of free space. Similarly, the magnetizing field **H** is defined by $H = B/\mu_0 + M$, where μ_0 is the permeability of free space and **M** is the magnetization (magnetic moment per unit volume). If **J** represents current density and ρ charge density, these quantities are related through Maxwell's equations:

$$\nabla \times E = - \frac{\partial B}{\partial t} \tag{3.20a}$$

$$\nabla \cdot D = \rho \tag{3.20b}$$

$$\nabla \times H = J + \frac{\partial D}{\partial t} \tag{3.20c}$$

$$\nabla \cdot B = 0 \tag{3.20d}$$

In view of Eq. (3.20d), **B** can be expressed as the curl of a vector potential **A**;

$$B = \nabla \times A \tag{3.21}$$

When Eq. (3.21) is substituted into Eq. (3.20a) to find **E**, one obtains

$$E = - \frac{\partial A}{\partial t} - \nabla \phi \tag{3.22}$$

where the constant of integration is written as minus the gradient of a scalar electrostatic potential ϕ, determined by the distribution of charges in space. To reduce the complications we shall confine this discussion to free space, in which case we can find the electrostatic potential at any point P by making this point the origin of an S-coordinate system and evaluating the integral

$$\phi(t) = \int_{\substack{all \\ space}} \frac{1}{4\pi\epsilon_0} \frac{\rho(r, \, t - r/c)}{r} \, d\upsilon \tag{3.23}$$

where $d\upsilon$ is a volume element. To obtain ϕ at time t we must evaluate ρ at time $t - r/c$.

Similarly, the sources of the vector potential **A** are currents, and **A** is given by

$$A(t) = \int_{\substack{\text{all} \\ \text{space}}} \frac{\mu_0}{4\pi} \frac{J(r, \, t - r/c)}{r} \, d\mathcal{U} \tag{3.24}$$

Since in free space $c^2 = 1/\mu_0\epsilon_0$, Eq. (3.23) can be put in the form

$$\frac{i\phi}{c} = \int_{\substack{\text{all} \\ \text{space}}} \frac{\mu_0}{4\pi} \frac{ic\rho(r, \, t - r/c)}{r} \, d\mathcal{U}$$

$$= \int_{\substack{\text{all} \\ \text{space}}} \frac{\mu_0}{4\pi} \frac{J_w(r, \, t - r/c)}{r} \, d\mathcal{U}$$

which represents the time component A_w of a four-potential

$$\underline{A} = 1_x A_x + 1_y A_y + 1_z A_z + 1_w A_w = A + 1_w \frac{i\phi}{c} \tag{3.25}$$

In free space **P** and **M** are zero, and so $D = \epsilon_0 E$ and $B = \mu_0 H$. Then, taking the divergence of Eq. (3.22) yields, with the aid of Eq. (3.20b),

$$-\nabla \cdot E = \nabla^2 \phi + \nabla \cdot \frac{\partial A}{\partial t} = -\frac{\rho}{\epsilon_0} \tag{3.26}$$

while Eqs. (3.20) to (3.22) can be combined to give

$$\nabla^2 A - \nabla(\nabla \cdot A) - \mu_0\epsilon_0 \frac{\partial^2 A}{\partial t^2} - \mu_0\epsilon_0 \frac{\partial \nabla \phi}{\partial t} = -\mu_0 J \tag{3.27}$$

Now **A** is not completely defined by Eq. (3.21), since an infinity of vectors may have the same curl. We are further free to require that its divergence have some specified value; in this case we follow Lorentz in imposing the condition

$$\nabla \cdot A + \mu_0\epsilon_0 \frac{\partial \phi}{\partial t} = 0 \tag{3.28}$$

Since both **A** and ϕ are continuous functions of x, y, z, and t, we may exchange the order of differentiation. This, plus imposition of the Lorentz condition and use of $c^2 = 1/\mu_0\epsilon_0$, permits us to write Eqs. (3.27) and (3.26) in the form

$$\nabla^2 A - \frac{1}{c^2} \frac{\partial^2 A}{\partial t^2} = -\mu_0 J \tag{3.27a}$$

$$\nabla^2 \phi - \frac{1}{c^2}\frac{\partial^2 \phi}{\partial t^2} = -\frac{\rho}{\epsilon_0} = -\mu_0 c^2 \rho \tag{3.26a}$$

$$\text{With } \square^2 = \frac{\partial^2}{\partial x^2} + \frac{\partial^2}{\partial y^2} + \frac{\partial^2}{\partial z^2} + \frac{\partial^2}{\partial w^2} = \nabla^2 - \frac{1}{c^2}\frac{\partial^2}{\partial t^2}$$

these equations become

$$\square^2 \mathbf{A} = -\mu_0 \mathbf{J} \qquad \square^2 \phi = -\frac{\rho}{\epsilon_0} = -\mu_0 c^2 \rho$$

or

$$\square^2 \underline{\mathbf{A}} = -\mu_0 \underline{\mathbf{J}} \tag{3.29}$$

in four-vector notation.   In terms of the four-potential $\underline{\mathbf{A}}$ the Lorentz condition (3.28) is simply $\square \cdot \underline{\mathbf{A}} = 0$.

3.7   The Field Tensor

If one knows the components of the four-potential in one frame, one can make a Lorentz transformation to a moving frame and, from the potentials, find the electric and magnetic intensities.   However, often one knows the fields in one frame and would like to find the fields in another frame without going through the potentials.   This can be done through the *field tensor* f_{ik}, given by

$$f_{ik} = \square \times \underline{\mathbf{A}} = \begin{bmatrix} 0 & \dfrac{\partial A_y}{\partial x} - \dfrac{\partial A_x}{\partial y} & \dfrac{\partial A_z}{\partial x} - \dfrac{\partial A_x}{\partial z} & \dfrac{\partial A_w}{\partial x} - \dfrac{\partial A_x}{\partial w} \\ \dfrac{\partial A_x}{\partial y} - \dfrac{\partial A_y}{\partial x} & 0 & \dfrac{\partial A_z}{\partial y} - \dfrac{\partial A_y}{\partial z} & \dfrac{\partial A_w}{\partial y} - \dfrac{\partial A_y}{\partial w} \\ \dfrac{\partial A_x}{\partial z} - \dfrac{\partial A_z}{\partial x} & \dfrac{\partial A_y}{\partial z} - \dfrac{\partial A_z}{\partial y} & 0 & \dfrac{\partial A_w}{\partial z} - \dfrac{\partial A_z}{\partial w} \\ \dfrac{\partial A_x}{\partial w} - \dfrac{\partial A_w}{\partial x} & \dfrac{\partial A_y}{\partial w} - \dfrac{\partial A_w}{\partial y} & \dfrac{\partial A_z}{\partial w} - \dfrac{\partial A_w}{\partial z} & 0 \end{bmatrix}$$

$$= \begin{array}{c} \quad k \rightarrow \\ i \downarrow \begin{bmatrix} 0 & B_z & -B_y & -\dfrac{iE_x}{c} \\ -B_z & 0 & B_x & -\dfrac{iE_y}{c} \\ B_y & -B_x & 0 & -\dfrac{iE_z}{c} \\ \dfrac{iE_x}{c} & \dfrac{iE_y}{c} & \dfrac{iE_z}{c} & 0 \end{bmatrix} \end{array} \tag{3.30}$$

The field tensor is an antisymmetric second-rank tensor.   To transform such a tensor from frame S to frame S' is somewhat more involved than to transform a vector.   A four-vector, for example, transforms according to

$$V'_i = \sum_{k=1}^{4} C_{ik} V_k$$

where C_{ik} is the direction cosine between the i axis in the S' frame and the k axis in the S frame and V_k is respectively the x, y, z, or w component of V for $k = 1, 2, 3,$ or 4. The terms of a second-rank tensor have two indices and transform according to the equation

$$T'_{ik} = \sum_{j=1}^{4} \sum_{m=1}^{4} C_{ij} C_{km} T_{jm} \tag{3.31}$$

In matrix notation,

$$
\begin{bmatrix}
T'_{11} & T'_{12} & T'_{13} & T'_{14} \\
T'_{21} & T'_{22} & T'_{23} & T'_{24} \\
T'_{31} & T'_{32} & T'_{33} & T'_{34} \\
T'_{41} & T'_{42} & T'_{43} & T'_{44}
\end{bmatrix}
=
\begin{bmatrix}
C_{11} & C_{12} & C_{13} & C_{14} \\
C_{21} & C_{22} & C_{23} & C_{24} \\
C_{31} & C_{32} & C_{33} & C_{34} \\
C_{41} & C_{42} & C_{43} & C_{44}
\end{bmatrix}
$$
$$
\begin{bmatrix}
T_{11} & T_{12} & T_{13} & T_{14} \\
T_{21} & T_{22} & T_{23} & T_{24} \\
T_{31} & T_{32} & T_{33} & T_{34} \\
T_{41} & T_{42} & T_{43} & T_{44}
\end{bmatrix}
\begin{bmatrix}
C_{11} & C_{21} & C_{31} & C_{41} \\
C_{12} & C_{22} & C_{32} & C_{42} \\
C_{13} & C_{23} & C_{33} & C_{43} \\
C_{14} & C_{24} & C_{34} & C_{44}
\end{bmatrix}
\tag{3.32}
$$

as can be verified by multiplying out the matrices and comparing the result with the expansion of Eq. (3.31). Note that the third matrix on the right-hand side of the equation is the *transpose* of the first, i.e., the matrix formed by interchanging rows and columns so the element in the ith row and kth column of the original matrix appears in the kth row and ith column of the transpose matrix.

The field tensor in the S' frame is therefore

$$
f'_{ik} =
\begin{bmatrix}
\gamma_T & 0 & 0 & i\beta_T\gamma_T \\
0 & 1 & 0 & 0 \\
0 & 0 & 1 & 0 \\
-i\beta_T\gamma_T & 0 & 0 & \gamma_T
\end{bmatrix}
\begin{bmatrix}
0 & B_z & -B_y & -\dfrac{iE_x}{c} \\
-B_z & 0 & B_x & -\dfrac{iE_y}{c} \\
B_y & -B_x & 0 & -\dfrac{iE_z}{c} \\
\dfrac{iE_x}{c} & \dfrac{iE_y}{c} & \dfrac{iE_z}{c} & 0
\end{bmatrix}
$$
$$
\begin{bmatrix}
\gamma_T & 0 & 0 & -i\beta_T\gamma_T \\
0 & 1 & 0 & 0 \\
0 & 0 & 1 & 0 \\
i\beta_T\gamma_T & 0 & 0 & \gamma_T
\end{bmatrix}
\tag{3.33}
$$

from which we obtain

$$B'_x = B_x \qquad B'_y = \gamma_T \left(B_y + \frac{VE_z}{c^2} \right)$$

$$B'_z = \gamma_T \left(B_z - \frac{VE_y}{c^2} \right) \quad (3.34)$$

$$E'_x = E_x \qquad E'_y = \gamma_T(E_y - VB_z) \quad E'_z = \gamma_T(E_z + VB_y)$$

These equations can be used to ascertain the effect of uniform motion upon the field of a point charge. Let charge q be moving with speed v. We take the x axis through the charge and in the direction of its motion. Consider any point P not on the x axis and let this point and **v** establish the xy plane (Fig. 3.2). We choose an S' frame in which q is at rest; in this frame the field is purely electrostatic, and its components are

$$E'_x = \frac{qx'}{4\pi\epsilon_0 r'^3} \qquad E'_y = \frac{qy'}{4\pi\epsilon_0 r'^3} \qquad E'_z = 0 \quad (3.35)$$

Since the S frame is moving with a velocity $\mathbf{V} = -\mathbf{v}$ relative to S', the transformation equations to the S frame corresponding to Eq. (3.34) are

$$B_x = B'_x \qquad B_y = \gamma \left(B'_y - \frac{vE'_z}{c^2} \right) \qquad B_z = \gamma \left(B'_z + \frac{vE'_y}{c^2} \right) \quad (3.36)$$

$$E_x = E'_x \qquad E_y = \gamma(E'_y + vB'_z) \qquad E_z = \gamma(E'_z - vB'_y)$$

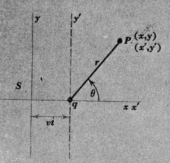

Fig. 3.2 Charge q at rest in frame S' moving with speed v along the x axis of frame S.

Now $\mathbf{B}' = 0$ in S', and so $B'_x = 0 = B'_y = B'_z$. By Eq. (2.4), $x' = \gamma(x - vt)$ and $y' = y$. Further $x - vt = r \cos\theta$, $y = r \sin\theta$, and

$$r'^2 = x'^2 + y'^2 = \gamma^2(x - vt)^2 + y^2$$
$$= \gamma^2 r^2 \cos^2\theta + r^2 \sin^2\theta = \gamma^2 r^2[\cos^2\theta + (\sin^2\theta)/\gamma^2]$$
$$= \gamma^2 r^2[1 - v^2(\sin^2\theta)/c^2]$$

Upon making these substitutions in Eqs. (3.36), we obtain

$$E_x = \frac{q(1 - v^2/c^2)\cos\theta}{4\pi\epsilon_0 r^2[1 - v^2(\sin^2\theta)/c^2]^{\frac{3}{2}}}$$

$$E_y = \frac{q(1 - v^2/c^2)\sin\theta}{4\pi\epsilon_0 r^2[1 - v^2(\sin^2\theta)/c^2]^{\frac{3}{2}}} \tag{3.37}$$

$$E_z = 0 = B_x = B_y \qquad B_z = \frac{qv(1 - v^2/c^2)\sin\theta}{4\pi\epsilon_0 r^2[1 - v^2(\sin^2\theta)/c^2]^{\frac{3}{2}}}$$

The fact that $E_y/E_x = \tan\theta$ shows that the electric field points radially outward from the instantaneous location of q. In spherical polar coordinates $\mathbf{E}$ is in the $\mathbf{r}$ direction and $\mathbf{B}$ in the φ direction in any inertial frame with its origin instantaneously at q. However, the field of a moving charge is concentrated into the equatorial plane (Fig. 3.3).

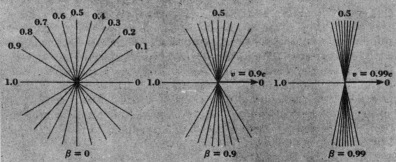

Fig. 3.3 The concentration of the electric flux of a moving charge into the equatorial plane; the decimals give the fraction of the total flux lying within a cone with axis **v**. The inverse-square law applies at all speeds. (*Modified from R. B. Leighton,* "*Principles of Modern Physics.*" *Copyright* 1959. *McGraw-Hill Book Company. Used by permission.*)

Many other conclusions can be drawn by transforming fields and potentials from one frame to another; e.g., one can easily show that $\mathbf{B} \cdot \mathbf{H} - \mathbf{D} \cdot \mathbf{E}$, $\mathbf{A} \cdot \mathbf{J} - \rho\phi$, and $\mathbf{E} \cdot \mathbf{B}$ are invariants having the same values in both S and S' frames. For plane waves in free space $\mu_0 H^2 = \epsilon_0 E^2$, so that $\mathbf{B} \cdot \mathbf{H} - \mathbf{D} \cdot \mathbf{E} = 0$ in any inertial frame; thus plane waves in one frame transform to plane waves in another frame, although both direction of propagation and frequency are altered, as we saw in Sec. 3.3.

3.8   General Theory of Relativity

In considering the bearing of the special theory of relativity upon physical laws, we have said nothing about *gravitation*. After publishing the special theory, Einstein took up the problem of harmonizing the laws of gravitation with the requirements of that theory. Since no physical effect can be transmitted from one place to another with a velocity exceeding that of light, it may be assumed that gravitational effects are propagated with a finite velocity. What, then, is the law of this propagation?

At the same time, another idea was active in Einstein's mind. In the special theory, only *unaccelerated* frames of reference had been compared. Why this limitation? Could not the principle of relativity be generalized somehow so that *frames of all sorts* would stand on an equal footing?

In studying these questions, Einstein was impressed by the fact that gravitational acceleration is exactly the *same for all bodies*, however much they differ in density or in other properties. In this respect, gravitational acceleration resembles the relative acceleration which appears when a frame of reference is itself subjected to acceleration, a matter of common experience. Everyone knows how, when riding on an elevator, he seems momentarily lighter when the elevator is accelerated downward. This effect simulates closely an actual change in the force of gravity. By no *mechanical* experiment, indeed, can an apparent gravitational field thus produced be distinguished from a true field due to gravitational attraction.

Eventually Einstein came to the conclusion that in the neighborhood of any given point, there should be *no difference* between a gravitational field due to attracting matter and the "apparent" field due to acceleration of a frame of reference. This proposition he adopted as a *postulate* called the *strong principle of equivalence*. If the principle is accepted, it leads to the prediction of a number of physical effects hitherto unobserved.

Light, for example, had not commonly been supposed to be subject to gravitational action. But suppose the earth's gravitational field were abolished within a laboratory by allowing the whole laboratory to fall freely. Then, relative to the laboratory, there would be no gravitational attraction; a ball thrown horizontally would travel in a straight line relative to the laboratory, not in a parabola. By the principle of equivalence, therefore, a ray of light projected horizontally would also appear to travel in a straight line; for conditions relative to the laboratory are the same as they would be out in space far from all attracting masses, and there is no doubt that in such locations rays of light are straight. Relative to the earth, however, the path of the ray of light would be slightly curved.

The principle of equivalence implies that a *uniform* gravitational field can be transformed away *in its entirety* by a proper choice of the frame of reference. Any field can be transformed away in the neighborhood of a single point, but, in general, the choice of frame that does this varies from point to point. For example, relative to a frame falling freely in New York there is no gravitational field in New York but one of double strength in Australia.

There remains the problem as to the law according to which gravitating matter determines which frames have the inertial property. The law must be such that its consequences agree with those derived from Newton's law of gravitation as a first approximation, since this law describes the motions of the solar system with high accuracy; and it must also be in harmony with the special theory of relativity. Einstein surmised that the law could probably be stated most simply in terms of a formulation that would permit the use not only of any frame of reference in the ordinary sense but of any sort of generalized coordinates. With the aid of the mathematician Grossmann he found out how to write physical laws in a form that is valid *for any choice of space-time coordinates whatever*. The method involves the use of general tensor analysis. Suffice it to say that Einstein found that among all possible guesses as to the correct law of gravitation one stood out in contrast to all others as the simplest in mathematical form. He adopted this law as a tentative hypothesis and then proceeded to look for predictions based on it which could be tested by experiment.

From the new law of gravitation thus obtained, Einstein deduced three novel effects that might be accessible to observation.

1. The motion of the planets should be very slightly altered. In particular, the perihelion of the orbit of Mercury should be caused

to precess about the sun at the rate of 43 seconds of arc per century. Perturbations by the other planets cause an advance of the perihelion by 5556 seconds of arc per century, but the observed advance is 5600 seconds of arc per century. Einstein's theory seemed to have removed this annoying discrepancy in astronomical theory. However, in 1961 Dicke and Brans introduced a modified relativity theory which predicts a 10 percent smaller relativistic precession, and in 1967 Dicke and Goldenberg found evidence of a solar quadrupole moment large enough to bring the calculated and observed precessional rates into reasonable agreement.

2. Rays of light passing close to a heavy body should be bent toward it. In the case of the sun, the deflection should be inversely proportional to the distance of closest approach of the ray to the sun's center and for a ray just grazing the sun's surface should amount to 1.75 seconds of arc. Stars seen adjacent to the sun during an eclipse appear to be displaced outward by about 2 seconds of arc (with considerable scatter in the data).

3. Physical processes in a region of low gravitational potential, when compared with similar processes at a point of high potential, should be found to take place more slowly. Consequently, atomic vibrations on the sun are slowed down, and spectral lines observed in the spectrum of sunlight are shifted toward the red as compared with lines emitted or absorbed by the same elements on the earth by an amount $\Delta\lambda/\lambda \approx 2.12 \times 10^{-6}$. In 1960 Pound and Rebka showed that 14.4-keV γ rays from excited Fe^{57} nuclei arising from the β decay of Co^{57} underwent 0.97 ± 0.04 times the predicted frequency shift in moving through a vertical distance of 22 m at the surface of the earth.

Einstein's theory of space, time, and gravitation, thus arrived at, is known as the *general theory of relativity*. According to this theory, the spatial behavior of matter is not quite euclidean. If a triangle of astronomical size near a heavy body like the sun were surveyed by means of rigid rods, with or without the help of light signals, the angles would not quite add up to 180°; and so on. Einstein's great intellectual achievement has been followed by various proposed modifications, some of which lead to predicted phenomena so close to those of Einstein that present observational data are not adequate to decide between theories.

Problems

1. Prove that the matrix equation (3.2) is equivalent to the Lorentz space-time transformation, Eq. (2.4).

2. Prove that the magnitude of a four-vector $\underline{A}$ is a scalar invariant for two inertial frames by showing that

$$A_x{}^2 + A_y{}^2 + A_z{}^2 + A_w{}^2 = A'_x{}^2 + A'_y{}^2 + A'_z{}^2 + A'_w{}^2$$

3. Show that the magnitude of the momentum four-vector is im_0c.

4. Show that the four-velocity has the dimensions of speed. Prove that it transforms as a four-vector by showing its magnitude is a constant independent of the inertial frame in which it is measured.

5. Let frame S' be moving with constant velocity V along the y axis of frame S. If corresponding axes of the two frames are parallel, write the Lorentz transformation for the radius four-vector.

6. Make a Lorentz transformation of the four-velocity from a frame S to a frame S' moving with speed V along the x axis of S. Show that the results are compatible with the Poincaré velocity transformation of Eq. (2.7).

7. Show that for any system of N freely moving particles, such as the molecules of gas in a box, the total rest mass is given by

$$M_0 = \sum_{i=1}^{N} \left[(m_0)_i + \frac{K_i}{c^2} \right]$$

where $(m_0)_i$ and K_i are the rest mass and kinetic energy of the ith particle respectively.

8. A proton of rest mass m is accelerated to a speed $v \approx c$ in a synchrotron. If it undergoes a perfectly inelastic collision with a nucleus at rest of mass M, find the rest mass and the speed of the resulting compound nucleus.

Ans: $\sqrt{m^2 + M^2 + 2\gamma m M}$; $\gamma m v/(\gamma m + M)$

9. (a) In frame S two identical balls A and B of rest mass m_0 have equal and opposite velocities of magnitude v. They collide and coalesce to form a single particle at rest in S. Apply the conservation of momentum-energy to the collision and find the rest mass of the system after the collision.

(b) Repeat the problem in frame S', in which ball A is at rest, and show that the rest mass of the system after the collision is the same as in part (a).

10. Transform the four-vector $\underline{F}$ in S to $\underline{F}'$ in S' by use of the Lorentz transformation. Use the result to show that the transformation equations for ordinary force are

$$f'_x = f_x - \frac{Vv_y f_y}{c^2 - Vv_x} - \frac{Vu_z f_z}{c^2 - Vv_x}$$

$$f'_y = \frac{\sqrt{1 - v'^2/c^2}}{\sqrt{1 - v^2/c^2}} f_y \qquad f'_z = \frac{\sqrt{1 - v'^2/c^2}}{\sqrt{1 - v^2/c^2}} f_z$$

11. Show that $\square \cdot \underline{\mathbf{A}} = 0$ is the Lorentz condition, Eq. (3.28), in four-vector notation.

12. A frame S_1 moves with velocity $V_1 = \beta_1 c$ with respect to a frame S. A second frame S_2 moves with velocity $V_2 = \beta_2 c$ relative to S_1. Find the velocity of S_2 relative to S and show that

$$\beta = \frac{\beta_1 + \beta_2}{1 + \beta_1 \beta_2}$$

13. Show that if a photon of energy $h\nu$ is absorbed by a stationary nucleus (or other particle) of rest mass M_0, the excited particle has a mass $M_0 + h\nu/c^2$ and a speed $h\nu c/(h\nu + M_0 c^2)$.

14. An electron of rest energy $m_0 c^2$ and total energy E strikes an electron at rest. Show that if $E \gg m_0 c^2$, the maximum energy available in the zero-momentum frame is $\sqrt{2m_0 c^2 E}$.

15. A particle of mass M, initially at rest, decays into two particles with rest masses m_1 and m_2. Show that the total energy of m_1 is $c^2(M^2 + m_1{}^2 - m_2{}^2)/2M$.

16. An isolated excited nucleus of mass M emits a γ ray in a transition to the ground state. If the excited state lies an energy ΔE above the ground state, show that the energy of the photon is $\Delta E(1 - \Delta E/2Mc^2)$. (Note that this is a special case of the preceding problem.)

17. Given an arbitrary velocity in frame S such that $v_x = v \cos \theta$. Show that in S' (in which $v'_x = v' \cos \theta'$)

$$\sqrt{1 - \frac{v'^2}{c^2}} = \frac{\sqrt{1 - V^2/c^2}\,\sqrt{1 - v^2/c^2}}{1 - vV(\cos \theta)/c}$$

and

$$\tan \theta' = \frac{\sqrt{1 - V^2/c^2}\,v \sin \theta}{v \cos \theta - V}$$

Show further that if $v = c$,

$$\tan \theta' = \frac{\sqrt{1 - V^2/c^2}\,\sin \theta}{\cos \theta - V/c}$$

which is the relativistic equation for the aberration of light derived by Einstein in his first paper.

18. A positive pion can be produced through the reaction $p + p \rightarrow p + n + \pi^+$ by bombarding protons at rest with high-energy protons. If the rest energies for p, n, and π^+ are respectively 938, 939.5, and 135 MeV, find the minimum kinetic energy for the incident protons for this reaction. *Hint:* Work first in a frame in which the center of mass is at rest, the zero-momentum frame.

Ans: 280 MeV

19. Antiprotons are produced by bombarding protons at rest with high-energy protons through the reaction $p + p \rightarrow p + p + (p + \bar{p}^-)$. In the center-of-mass frame find the minimum kinetic energy of each proton required for the reaction. In this frame and for this energy what are the values of β and γ? Transform to the laboratory frame and find the minimum kinetic energy of the incident proton for the reaction.

Ans: $m_0 c^2$; $\sqrt{3}/2$, 2; $6 m_0 c^2$

20. Show that $\mathbf{B} \cdot \mathbf{H} - \mathbf{D} \cdot \mathbf{E}$, $\mathbf{A} \cdot \mathbf{J} - \rho \phi$, and $\mathbf{E} \cdot \mathbf{B}$ are invariants under Lorentz transformation. (They are therefore valid world tensors.)

21. A long straight wire of radius r_0 in a evacuated tube bears a current I along the axis of a hollow cylindrical plate of radius R which is at a potential V relative to the wire. If the central wire is hot enough to emit electrons of negligible energy, show that electrons can reach the cylinder only if I is less than

$$2\pi \frac{\sqrt{(Ve/c)^2 + 2m_0 Ve}}{\mu_0 e \ln (R/r_0)}$$

22. In a crossed-field velocity selector particles of the desired velocity pass through undeflected because the forces due to the electric (qE_y) and magnetic ($qv_x B_z$) fields are equal and opposite. Show that the force on a particle of the desired velocity is zero in the frame in which the particle is at rest.

23. Stark showed that the spectral lines of elements are split into several components by an electric field. Wien showed that Stark patterns are observed in the spectrum of hydrogen when a beam of protons moves at high speed v perpendicular to a magnetic field B_z in an otherwise field-free space. Show that the effective electric field in which the proton finds itself in its own rest frame has magnitude $\gamma v B$ by finding the field tensor for this frame.

24. A long straight wire is electrically neutral in the S frame in which it bears a current density $j = -nev_D$, where n is the number of conduction electrons per unit volume and v_D is the drift velocity of the electrons. Show that an observer moving with speed V parallel to the wire sees it charged with a net charge density

$$\rho_{\text{net}} = \rho'_+ + \rho'_- = \frac{\gamma n e V v_D}{c^2}$$

25.   Show that the Lorentz force per unit volume $\rho(\mathbf{E} + \mathbf{v} \times \mathbf{B})$ is the space part of a four-vector.   Determine the fourth component and show that it is a measure of the power expended by the electric field per unit volume.

chapter four

Atoms and Molecules

The question whether matter is continuous or atomic in nature was fairly well settled during the nineteenth century. _A major part of modern physics is concerned with the structure and properties of atoms and of combinations of atoms as gaseous molecules and crystalline solids._ _Although classical physics did not achieve a thorough understanding of these topics, it left a vast heritage of enduring material, some part of which is the subject of this chapter._

4.1 Chemical Evidence for Atoms

Early in the nineteenth century chemists placed the atomic hypothesis on a quantitative basis. It was then known that there are many elements—substances which cannot be resolved into simpler components by chemical means—and a much larger number of compounds which can be broken down into elements.

Whenever a particular compound is formed, the mass of one element required to combine with a fixed mass of another is always the same.

This great generalization is the *law of definite proportions* or the *law of definite composition*. In 1803 Dalton proposed that each element consists of atoms, which are very small, indivisible particles, each atom of a given kind being identical. When different kinds of atoms combine, a compound is formed. Dalton's theory implies the law of definite proportions. Indeed, Dalton predicted that two elements A and B might form more than one compound and thus presaged the *law of multiple proportions:*

When the same two elements form more than one compound, the different quantities of elements A which combine with the same quantity of element B are in the ratio of small integers.

In 1808 Gay-Lussac showed that when two gases interact to form a gaseous compound, the ratios of the volumes of the reacting gases and of the resulting gas are represented by small whole numbers. Three years later Avogadro made the distinction between atoms and molecules and advanced his celebrated hypothesis:

Under identical conditions of temperature and pressure equal volumes of all gases contain equal numbers of molecules.

While an atom is the smallest unit of an element which can participate in a chemical reaction, a molecule is the smallest unit of an element or compound which exists stably by itself in the gaseous form. Molecules are said to be monatomic if they are single atoms, e.g., helium, neon, diatomic if they consist of two atoms, e.g., hydrogen, oxygen, and nitrogen, triatomic if they are composed of three atoms, e.g., water, carbon dioxide, and so forth.

A major contribution of Dalton's hypothesis was the concept of atomic weight. In his day it was impossible to determine the absolute mass of any kind of atom, but the relative weights of different atoms could be inferred by determining the weight of one element which combines with a given weight of another. For many years the chemical atomic weight system was based on assigning the number 16.0000 to natural oxygen. On this basis the atomic weight of the lightest element, hydrogen, was somewhat greater than unity. In 1960 to 1961 the International Union of Physics and Chemistry agreed to base atomic weights on the isotope carbon 12, which is taken as 12.000000. If A represents the atomic weight of an element, A g of the element is called *one gram atomic weight* and A kg is 1 kg at wt. It follows from the correctness of Avogadro's hypothesis that 1 kg at wt samples of all elements have the same number of atoms, now known to be 6.0225 × 10^{26}. We call it *Avogadro's number* and represent it by N_A; N_A molecules of a gaseous compound constitute 1 kmole of the substance.

4.2 The Ideal Gas

In 1738 Daniel Bernoulli explained gas pressure in terms of the collisions of molecules with the walls of the container and thus laid the quantitative foundation for the kinetic theory of gases, which received further treatment in the hands of Herapath (1821), Waterson (1843), Kronig (1856), and Clausius (1857). A highly unsophisticated model of a real gas is an *ideal gas*, consisting of a huge number of identical monatomic molecules, each occupying a negligible volume. These molecules are in constant motion in random directions with a broad range of speeds. When a molecule collides with a wall or another molecule, we shall assume for the moment that the collisions are elastic. (This restriction may be relaxed; see Sec. 4.3). In the molecular chaos postulated, all positions in the container and all directions of the velocities are equally probable.

Consider a single molecule of mass m and speed v in an otherwise empty cubical box with sides of length L, one corner of which serves as origin for cartesian coordinates with axes along edges of the box. Let v_x be the x component of the particle velocity. When this molecule collides with a face parallel to the yz plane, the x component of its momentum is changed by $2mv_x$. A time L/v_x later the molecule strikes the opposite wall, and in time $2L/v_x$ it returns to the first face. Hence the number of collisions per unit time with this face is $v_x/2L$, and the change in momentum per second at this face for one molecule is

$$(2mv_x)\left(\frac{v_x}{2L}\right) = \left(\frac{mv_x^2}{L}\right)$$

Let us now add identical molecules until there are Avogadro's number N_A molecules in the box. It can be shown that the resultant change in momentum per second at the face is just what it would be if the molecules did not collide with each other; hence

$$\text{Change in momentum/second} = \frac{m}{L}\sum_{i=1}^{N_A}(v_x)_i^2 = N_A\frac{m}{L}(v_x^2)_{\text{av}}$$

where $(v_x{}^2)_{av}$ is the average value of the square of v_x for all N_A molecules. By Newton's second law the change in momentum per second is equal to the average force on the face, which in turn is the product of the pressure p and the area L^2 of the face.

$$F = pL^2 = \frac{N_A m (v_x{}^2)_{av}}{L} \tag{4.1}$$

Now L^3 is the volume of the box and the volume occupied by 1 kmole. We call it V_m and rewrite Eq. (4.1)

$$pV_m = N_A m (v_x{}^2)_{av} \tag{4.1a}$$

Similar arguments for faces parallel to the xy and xz planes lead to relations $pV_m = N_A m (v_y{}^2)_{av}$ and $pV_m = N_A m (v_z{}^2)_{av}$, where p is known experimentally to be the same at all faces. Therefore, $(v_x{}^2)_{av} = (v_y{}^2)_{av} = (v_z{}^2)_{av}$, and since $(v^2)_{av} = (v_x{}^2)_{av} + (v_y{}^2)_{av} + (v_z{}^2)_{av}$,

$$pV_m = \frac{2N_A}{3}\left[\frac{1}{2} m(v^2)_{av}\right] \tag{4.2}$$

But, by the general gas law

$$pV_m = RT \tag{4.3}$$

where T is the absolute temperature and R is the universal gas constant, 8314 J/kmole-°K. Comparison of Eqs. (4.2) and (4.3) leads to

$$\tfrac{1}{2}m(v^2)_{av} = \frac{3}{2}\frac{R}{N_A} T = \tfrac{3}{2}kT \tag{4.4}$$

from which we see that the average kinetic energy of translation for an ideal gas molecule is directly proportional to absolute temperature. The ratio k of the universal gas constant to Avogadro's number is known as the *Boltzmann constant* and has the value 1.38×10^{-23} J/molecule-°K.

The energy kT, which will appear frequently in subsequent sections, has the value 4×10^{-21} J or $\frac{1}{40}$ eV at 290°K, just below room temperature. Less than 0.02 percent of the molecules in a gas have an energy as great as $10kT$.

Equation (4.4), derived above for an ideal gas, is applicable to any real gas for conditions under which the general gas law, Eq. (4.3), is obeyed. At room temperature molecules of hydrogen, nitrogen, oxygen, and carbon dioxide in the atmosphere all have the same average kinetic energies of translation, so that the *root-mean-square* speed $v_{rms} = [(v^2)_{av}]^{\frac{1}{2}}$ of each type is inversely proportional to the mass of the molecule. As the temperature is reduced, attractive forces between molecules become important and the gas liquefies. (Extension of classical ideas to liquids and solids led to the conclusion that at 0°K all atomic and molecular motion would cease, but this extrapolation is not valid. Quantum mechanics predicts a zero-point motion at 0°K.)

4.3 Degrees of Freedom and the Equipartition of Energy

Any mechanical system has a certain number of degrees of freedom, defined as the minimum number of independent coordinates in terms of which the positions of all the masses composing the system can be specified. For example, a block sliding in a groove has one degree of freedom; if it is sliding on ice with no possibility of being set into rotation, it has two degrees of freedom; if it can also rotate about a vertical axis, it has three. A monatomic gas molecule behaving as if it had only energy of translation has three degrees of freedom, corresponding to the three coordinates x, y, z (or r, θ, φ), that specify its position in space. An ideal diatomic molecule formed by two mass-point atoms has six degrees of freedom, which might be chosen as the coordinates of the two atoms or, alternatively, as the three coordinates of the center of mass, two angular coordinates to locate the axis of the molecule, and a sixth to locate the distance of one atom from the center of mass. If the distance between the atoms is fixed, only five coordinates are needed, and so there are only five degrees of freedom. In general, a triatomic molecule has nine degrees of freedom, but this number is reduced to six if the separations of the atoms are fixed in a plane and to only five if they are fixed along a line.

Degrees of freedom are additive: the total number of degrees of freedom of a system of bodies is the sum of the number of degrees of freedom possessed by all the bodies which make up the system. To each degree of freedom there corresponds an independent term in the expansion for the kinetic energy of the body in question. For example, monatomic molecules behave as though they had only translational motion. The average kinetic energy $K_{av} = \frac{3}{2}kT$ by Eq. (4.4), and there are three degrees of freedom per molecule; there is on the average $\frac{1}{2}kT$ of energy associated with each degree of freedom. The classical *principle of equipartition of energy* states:

> *For a system in thermal equilibrium the average kinetic energy associated with each degree of freedom has the same value, $\frac{1}{2}kT$.*

This principle, proposed by Clausius in 1857 and used by Maxwell in 1860, was useful in explaining the specific heats of monatomic gases and some solids at room temperature, but it failed for other gases and for all solids at low temperature. Not until the development of quantum mechanics were these specific-heat anomalies resolved. For an ideal monatomic gas the energy E_m per kilomole is the product of N_A and the average kinetic energy per molecule. Hence, by Eq. (4.4), $E_m = N_A(3RT/2N_A) = \frac{3}{2}RT$. The heat capacity per kilomole at constant volume is given by $(c_V)_{kmole} = (\partial E_m/\partial T)_V = \frac{3}{2}R$. This result is in excellent agreement with the measured values for monatomic gases; for example, $c_V/R = 1.52$ for helium and 1.51 for argon at room temperature.

For a rigid diatomic molecule with five degrees of freedom, equipartition predicts that $E_m = \frac{5}{2}RT$, so that c_V/R should be $\frac{5}{2}$; experimental values are 2.44 for hydrogen, 2.45 for nitrogen, 2.50 for oxygen. However, the ratio of 3.02 for chlorine presented a problem; if the two atoms were rigidly connected, the ratio should be 2.5. If the atoms could vibrate, it should be 3.5; the addition of this simple harmonic motion would give six degrees of freedom, and associated with the average kinetic energy of simple harmonic motion there is an equal average potential energy which adds another $\frac{1}{2}kT$ per molecule. Similar problems arose in connection with HOH and SO_2, for which c_V/R has the values 3.3 and 3.79, respectively, at room temperature.

If energy is exchanged between rotational (or vibrational) kinetic energy and translational energy in collisions of polyatomic molecules, the assumption of Sec. 4.2 that the collisions are elastic is not justified. Actually, it is not necessary that collisions be elastic but only that $(v^2)_{av}$ in Eq. (4.2) remain constant. This condition is satisfied when the system is in equilibrium, since equilibrium means that such macroscopic quantities as $(v^2)_{av} = v_{rms}^2$ are stationary in time.

In 1819 Dulong and Petit found that "the product of the atomic weight and the specific heat is the same for all elementary (solid) substances." This product is called the *atomic heat;* it is numerically equal to the heat capacity per kilomole. At ordinary temperatures this *law of Dulong and Petit* holds roughly for many substances, the heat capacities per kilomole for 58 elements ranging from $2.7R$ to $3.5R$ with an average of $3.1R$.

A few striking exceptions occur, however, notably boron, with heat capacity $1.7R$; beryllium, $1.9R$; diamond (carbon), $0.73R$; and silicon, $2.5R$. These are all light elements with high melting points. Extended research showed that the heat capacities of these elements increase rapidly with temperature; e.g., in 1872 Weber observed that the specific heat of diamond triples between 0 and 200°C. Later it was found that the heat capacities of all substances decrease rapidly if the temperature

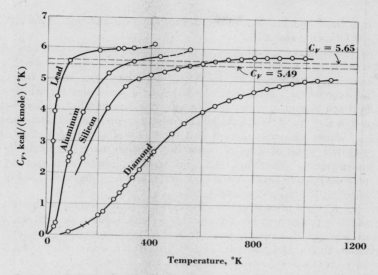

Fig. 4.1 Variation of the heat capacity per kilomole at constant volume with temperature.

is lowered sufficiently, tending toward zero at 0°K. Typical curves (Fig. 4.1) for elementary substances are all of the same form; i.e., they can be brought (almost) into coincidence by making a suitable choice of the temperature scale for each. Such a striking regularity must have its source in a simple general principle (treated in Sec. 20.7). Classical theory suggested that in a solid the atoms vibrate about certain mean positions, behaving like harmonic oscillators with three degrees of freedom. The heat capacity per kilomole should be the sum of $\frac{3}{2}R$ for kinetic energy plus $\frac{3}{2}R$ for the associated potential energy, a total of $3R$ in approximate agreement with observations. However, classical theory offers no explanation for the rapid decrease in the heat capacity at low temperature.

4.4   The Maxwellian Distribution

From the earliest days of kinetic theory it was recognized that the speeds of the gas molecules would vary over a wide range at any instant and that the speed of any one molecule would vary over a similar range in time as a result of collisions with other molecules. In 1859 Maxwell in a brief and brilliant paper deduced the distribution of velocities by the following reasoning.

Consider a gas of N identical molecules, each of which has velocity components v_x, v_y, v_z. Let the number of molecules for which v_x lies between v_x and $v_x + dv_x$ be $Nf(v_x)\,dv_x$, where $f(v_x)$ is the function to be determined. Similarly, the number of molecules for which v_y lies between v_y and $v_y + dv_y$ is $Nf(v_y)\,dv_y$ and the number with v_z between v_z and $v_z + dv_z$ is $Nf(v_z)\,dv_z$, where f always represents the same function.

Next Maxwell asserted that the value of v_x does not affect in any way the value of v_y or v_z, since they are mutually perpendicular and independent. Therefore, the number of molecules with velocity components lying between v_x and $v_x + dv_x$, v_y and $v_y + dv_y$, and v_z and $v_z + dv_z$ is $Nf(v_x)f(v_y)f(v_z)\,dv_x\,dv_y\,dv_z$.

If we represent each velocity by a scaled vector with its tail at the origin, the number terminating in any volume element $dv_x\,dv_y\,dv_z$ in this *velocity space* is just the same $Nf(v_x)f(v_y)f(v_z)\,dv_x\,dv_y\,dv_z$. But since the x, y, and z axes are perfectly arbitrary, this number must depend only on the distance from the origin, and so $f(v_x)f(v_y)f(v_z)$ must be some function of the speed alone (or of its square). Therefore,

$$f(v_x)f(v_y)f(v_z) = \phi(v_x{}^2 + v_y{}^2 + v_z{}^2) = \phi(v^2)$$

Solving this functional equation leads to the results that

$$f(v_x) = Ce^{-A^2v_x{}^2} \tag{4.5a}$$

and

$$\phi(v^2) = De^{-A^2v^2} \tag{4.5b}$$

where C, D, and A are positive constants. The negative sign accompanies A since a positive sign would lead to the number of particles in a given volume of velocity space increasing monotonically with speed.

In Appendix 4A the constants A, C, and D are evaluated by use of the conditions that the total number of particles is N and the total translational kinetic energy is $\frac{3}{2}NkT$. Substitution of these values in Eq. (4.5) leads to the result that the number of molecules having v_x between v_x and $v_x + dv_x$ is given by

$$Nf(v_x)\,dv_x = N\left(\frac{m}{2\pi kT}\right)^{\frac{1}{2}} e^{-mv_x{}^2/2kT}\,dv_x \tag{4.6}$$

while the number of molecules having a velocity v with components between v_x and $v_x + dv_x$, v_y and $v_y + dv_y$, and v_z and $v_z + dv_z$ is

$$N\phi(v^2)\,dv_x\,dv_y\,dv_z = N\left(\frac{m}{2\pi kT}\right)^{\frac{3}{2}} e^{-mv^2/2kT}\,dv_x\,dv_y\,dv_z \tag{4.7}$$

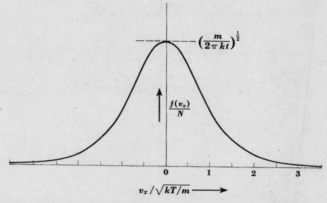

Fig. 4.2 Maxwellian distribution of the x component of molecular velocity.

To find the number of molecules having speed between v and $v + dv$, we replace the volume element $dv_x \, dv_y \, dv_z$ in velocity space by the volume element $4\pi v^2 \, dv$ and obtain

$$n(v) \, dv = N4\pi v^2 \phi(v^2) \, dv = N \left(\frac{m}{2\pi kT} \right)^{\frac{3}{2}} e^{-mv^2/2kT} \, 4\pi v^2 \, dv \qquad (4.8)$$

Figures 4.2 and 4.3 show the maxwellian distribution for v_x and v, respectively.

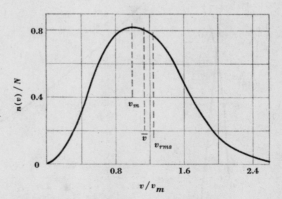

Fig. 4.3 Maxwellian distribution of molecular speeds in units of $v_m = \sqrt{2kT/m}$.

The average molecular speed $\bar{v}$ is then

$$\bar{v} = \int_0^\infty v \left(\frac{m}{2\pi kT}\right)^{\frac{3}{2}} \frac{N4\pi v^2}{N} e^{-mv^2/2kT} \, dv = \sqrt{\frac{8kT}{\pi m}} \tag{4.9}$$

while the most probable speed v_m is given by

$$v_m = \sqrt{\frac{2kT}{m}} \tag{4.9a}$$

Thus at a given temperature

$$v_m : \bar{v} : v_{\text{rms}} = 1 : 1.1284 : 1.2248$$

The maxwellian distribution of translational kinetic energies is readily found from Eq. (4.8) by making use of the fact that the kinetic energy K of a molecule is $\frac{1}{2}mv^2$ and $dK = mv \, dv$. Then the number of molecules $n(K) \, dK$ with kinetic energies between K and $K + dK$ is (Fig. 4.4)

$$n(K) \, dK = \frac{2N \sqrt{K} \, e^{-K/kT} \, dK}{\sqrt{\pi} \, (kT)^{\frac{3}{2}}} \tag{4.10}$$

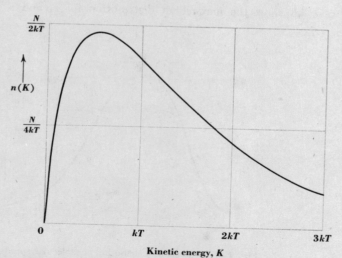

Fig. 4.4   Maxwellian distribution of molecular kinetic energies.

4.5 The Boltzmann Distribution

Consider the dynamic equilibrium of an isothermal atmosphere (Boltzmann, 1876) in which the molecules of the gas are subject to a constant gravitational field, which results in a decrease of pressure and density with height. Let n represent the number of molecules per unit volume. If p represents pressure at height z, the pressure at a height $z + dz$ is less than p by the weight of the air in a volume of unit area and height dz, so that

$$-dp = nmg \, dz \tag{4.11}$$

By Eq. (4.3), $pV_m = RT = N_A kT$, so that

$$p = \frac{N_A}{V_m} kT = nkT \tag{4.12a}$$

$$dp = kT \, dn \tag{4.12b}$$

Substitution of this result in Eq. (4.11) gives

$$-kT \, dn = nmg \, dz$$

and

$$n = n_0 e^{-mgz/kT} \tag{4.13}$$

where n_0 is the number of particles per unit volume at $z = 0$. The choice of the particular height at which $z = 0$ is arbitrary; it need not be at the bottom of the atmosphere. Equation (4.13) is equally valid for negative values of z in the isothermal atmosphere.

If Eq. (4.12a) is used to eliminate n from Eq. (4.13), one finds that, in the isothermal atmosphere, p also varies exponentially with height:

$$p = p_0 e^{-mgz/kT} \tag{4.14}$$

Since mgz is the potential energy P of a molecule relative to $z = 0$, we may write the Boltzmann distribution for the number of molecules per unit volume as a function of height [Eq. (4.13)] in the form

$$\frac{n}{n_0} = e^{-mgz/kT} = e^{-P/kT} \tag{4.13a}$$

Hence, the ratio of the number n_1 per unit volume at height z_1 to the number n_2 per unit volume at height z_2 is given by

$$\frac{n_1}{n_2} = \frac{e^{-mgz_1/kT}}{e^{-mgz_2/kT}} = \frac{e^{-P_1/kT}}{e^{-P_2/kT}} = e^{-(P_1-P_2)/kT} = e^{-\Delta P/kT} \qquad (4.15)$$

where $\Delta P = P_1 - P_2$.

4.6 Phase Space and the Maxwell-Boltzmann Distribution

In the preceding section we have derived an important result involving potential energy in an isothermal atmosphere. We now seek more general relations involving kinetic as well as potential energy. Following Maxwell, in Sec. 4.4 we began our discussion of gases in terms of a velocity space. However, there are advantages in changing to a momentum space in anticipation of quantum mechanics. This can be done in the nonrelativistic case by replacing v by p/m, where p is the *momentum*. (Note that up to this point in this chapter p has stood for pressure.)

Classically the state of a system of noninteracting particles is specified at any time when the position (x,y,z) and the momentum (p_x,p_y,p_z) of each particle are known. If we introduce the concept of a six-dimensional *phase space*, we can locate each particle in terms of six coordinates (x,y,z,p_x,p_y,p_z). Thus, phase space consists of coordinate space and momentum space taken together.

For an ideal isothermal gas, not subject to any external force field, the probability that the phase point for any one particle falls into a volume of phase space $dp_x \, dp_y \, dp_z \, dx \, dy \, dz$ is the product of the probability that its momentum lies in the element $dp_x \, dp_y \, dp_z$ of momentum space and the probability that it is in a volume $dx \, dy \, dz$ of coordinate space. The latter is $(dx \, dy \, dz)/V$, where V is the volume of the gas, while the former is $\phi(p^2) \, dp_x \, dp_y \, dp_z$, which comes from Eq. (4.7). Therefore,

$$\mathcal{P}(x,y,z,p_x,p_y,p_z) \, dp_x \, dp_y \, dp_z \, dx \, dy \, dz$$
$$= \frac{e^{-(p_x{}^2+p_y{}^2+p_z{}^2)/2mkT} \, dp_x \, dp_y \, dp_z}{(2\pi mkT)^{\frac{3}{2}}} \frac{dx \, dy \, dz}{V} \qquad (4.16)$$

The exponential term is equivalent to $e^{-K/kT}$, where $K = p^2/2m$ is the kinetic energy. If we consider any two cells in phase space, the ratio of the number of molecules per unit volume n_1 in cell 1 to the number

per unit volume n_2 in cell 2 is

$$\frac{n_1}{n_2} = \frac{\mathcal{P}_1}{\mathcal{P}_2} = \frac{e^{-K_1/kT}}{e^{-K_2/kT}} = e^{-(K_1-K_2)/kT} \tag{4.17}$$

a result to be compared with that of Eq. (4.15). In each case the ratio is $e^{-\Delta E/kT}$, where ΔE is the appropriate difference in energy.

In an isothermal atmosphere subject to a uniform gravitation field, the probability that a given molecule is a volume element $dx\ dy\ dz$ of real space is no longer $(dx\ dy\ dz)/V$, since the particle is less likely to be in a region of high potential energy than in one of low. The probability is $e^{-P/kT}\ (dx\ dy\ dz)/CV$, where C is the constant such that $\iiint_{\text{space}} e^{-P/kT}\ dx\ dy\ dz = CV$. Under these circumstances the probability that any one molecule lies in a volume of phase space $dp_x\ dp_y\ dp_z\ dx\ dy\ dz$ is

$$\begin{aligned}
\mathcal{P}&(x,y,z,p_x,p_y,p_z)\ dp_x\ dp_y\ dp_z\ dx\ dy\ dz \\
&= \frac{e^{-K/kT}\ dp_x\ dp_y\ dp_z}{(2\pi mkT)^{\frac{3}{2}}} \frac{e^{-P/kT}\ dx\ dy\ dz}{CV} \\
&= \frac{e^{-(K+P)/kT}\ dp_x\ dp_y\ dp_z\ dx\ dy\ dz}{(2\pi mkT)^{\frac{3}{2}}CV} \tag{4.18}
\end{aligned}$$

The sum $K + P$ is the total energy E of the ideal gas molecule. If Eq. (4.18) is multiplied by the total number of molecules N, the resulting term is the probable number of molecules dN in the infinitesimal volume element $dp_x\ dp_y\ dp_z\ dx\ dy\ dz$ in phase space, viz.,

$$dN = N \frac{e^{-E/kT}}{(2\pi mkT)^{\frac{3}{2}}CV}\ dp_x\ dp_y\ dp_z\ dx\ dy\ dz \tag{4.19}$$

The ratio of the number of particles n_1 expected in a small volume element $g_1 = \Delta p_x\ \Delta p_y\ \Delta p_z\ \Delta x\ \Delta y\ \Delta z$ in phase space to the number n_2 in another small volume element g_2 is

$$\frac{n_1}{n_2} = \frac{g_1 e^{-E_1/kT}}{g_2 e^{-E_2/kT}} = \frac{g_1}{g_2} e^{-(E_1-E_2)/kT} = \frac{g_1}{g_2} e^{-\Delta E/kT} \tag{4.20}$$

In Chap. 20 we shall derive the Maxwell-Boltzmann distribution function from elementary statistical mechanics and show that Eq. (4.20) is the ratio of two such functions. Although the Maxwell-Boltzmann relations derived above have many uses, we shall find that they are not generally applicable to elementary particles such as electrons.

4.7   The Sizes of Atoms

We can readily estimate the approximate radii of atoms, even though an atom has no well-defined boundary. Since most liquids and solids are difficult to compress, it is not unreasonable as a first approximation to think of them as assemblages of hard-sphere atoms in contact with one another. As an example, consider liquid nitrogen, which has a density ρ roughly equal to that of water, 1000 kg/m³. The atomic weight A of nitrogen is 14, and therefore 14 kg involves $N_A = 6.02 \times 10^{26}$ atoms, so that the number of atoms per cubic meter is $N_A \rho / A = 4.3 \times 10^{28}$ and the average volume per atom is 23×10^{-30} m³. A little cube of this volume would have a side of length $\sqrt[3]{23 \times 10^{-30}}$ m $= 2.8$ Å. By considering various ways in which the nitrogen atoms might be packed, we could vary the diameter somewhat, but not by an order of magnitude.

Similar estimates for other atomic species lead to the conclusion that atoms have diameters varying roughly from 1 to 4 Å, a remarkably small range when one remembers that they differ in relative mass from 1 to 250 and in number of electrons from 1 to 100. Nor are the heavy atoms necessarily among the larger ones. An atom of lithium occupies a larger volume than one of platinum. In general, the alkali atoms have the largest radii, while carbon in the diamond lattice involves an unusually small volume per atom.

Confirming evidence that atomic diameters are of the order of 10^{-10} m comes from measurements of collision probabilities and mean free paths in gases. A particle of radius r moving along a straight line with speed v through an ideal gas of molecules with radius R would collide in 1 s with every molecule in a cylinder of radius $r + R$ and length v. If n represents the number of molecules per unit volume, the number of collisions per second would be $\pi(r + R)^2 v n$, and the mean free path (average distance between collisions) λ is given in this crude approximation by

$$\lambda = \frac{1}{\pi(r + R)^2 n} \tag{4.21}$$

When a beam of particles from an oven or other source passes through a small thickness Δx of gas, those particles which collide with gas molecules are scattered out of the beam. The reduction in beam current $-\Delta I$ is proportional to the product of the incident current I and to the effective area $\pi(r + R)^2 n\ \Delta x$ subtended by the molecules, so that

$$-\Delta I = I\pi(r + R)^2 n\ \Delta x \tag{4.22}$$

from which

$$I = I_0 e^{-n\pi(r+R)^2 x} = I_0 e^{-x/\lambda} \tag{4.23}$$

Measurements of beam attenuation in gases at reduced pressure thus permit determination of mean free paths. By determining λ's for appropriate combinations of beam particles and scattering molecules, one can obtain a reasonably self-consistent set of values of molecular radii.

Kinetic theory predicts that the viscosity of a gas is proportional to the mean free path of the molecules. Viscosity values are also consistent with the conclusion that atomic radii are of the order of 10^{-10} m.

Since mean free paths depend on the number of particles per unit volume, they are proportional to the pressure so long as the temperature is held constant. At room temperature and pressure typical mean free paths for oxygen and nitrogen molecules are about 10^{-7} m. Since the rms speeds are roughly 500 m/s, there are of the order of 10^9 collisions per molecule per second.

Appendix 4A - Integrals Arising in the Kinetic Theory of Gases

From Eq. (4.5) we infer that the number of molecules with v_x in the range from v_x to $v_x + dv_x$ is $NCe^{-A^2 v_x^2} dv_x$. If we integrate from $v_x = -\infty$ to $v_x = \infty$, we must obtain the total number of molecules N, and so

$$N = \int_{-\infty}^{\infty} NCe^{-A^2 v_x^2} dv_x = NC \frac{\sqrt{\pi}}{A} \tag{4A.1}$$

hence $C = A/\sqrt{\pi}$. To evaluate the integral above we note that the indefinite integral $\int e^{-x^2} dx$ cannot be written in terms of elementary functions. However, a simple trick permits the evaluation of the definite integral. Let

$$I = \int_{-\infty}^{\infty} e^{-x^2} dx = \int_{-\infty}^{\infty} e^{-y^2} dy$$

Then

$$I^2 = \int_{-\infty}^{\infty} \int_{-\infty}^{\infty} e^{-(x^2+y^2)} dx \, dy$$

If we convert this integral over the xy plane to one in plane polar coordinates r, θ, we obtain

$$I^2 = \int_0^\infty \int_0^{2\pi} d\theta \; e^{-r^2} r \; dr = 2\pi \int_0^\infty re^{-r^2} \; dr = 2\pi \left[-\frac{e^{-r^2}}{2} \right]_0^\infty = \pi$$

so that

$$I = \int_{-\infty}^\infty e^{-x^2} \; dx = \sqrt{\pi} \qquad \text{and} \qquad \int_0^\infty e^{-A^2 x^2} \; dx = \frac{\sqrt{\pi}}{2A} \qquad (4A.2)$$

Since $f(v_x)f(v_y)f(v_z) = \phi(v^2)$, we have

$$C^3 e^{-A^2 v_x^2} e^{-A^2 v_y^2} e^{-A^2 v_z^2} = De^{-A^2 v^2}$$

and so

$$D = C^3 = A^3 \pi^{-\frac{3}{2}}$$

To evaluate A we find the total kinetic energy of translation and equate it to $\frac{3}{2}NkT$, since we know that the average kinetic energy per molecule is $\frac{3}{2}kT$. We have just found above that the number of molecules with speeds between v and $v + dv$ is $NA^3 \pi^{-\frac{3}{2}} 4\pi v^2 e^{-A^2 v^2} \; dv$. If we multiply this by the kinetic energy of translation $\frac{1}{2}mv^2$ and integrate over all possible speeds, we have

$$\int_0^\infty N(\tfrac{1}{2}mv^2) A^3 \pi^{-\frac{3}{2}} 4\pi v^2 e^{-A^2 v^2} \; dv = 2\pi Nm A^3 \pi^{-\frac{3}{2}} \frac{\frac{3}{8}\sqrt{\pi}}{A^5} \qquad (4A.3)$$
$$= \tfrac{3}{2}NkT$$

from which $A = \sqrt{m/2kT}$.

In kinetic theory integrals of the form $I_n = \int_0^\infty v^n e^{-A^2 v^2} \; dv$ appear frequently. As Eq. (4A.2) above shows,

$$I_0 = \int_0^\infty e^{-A^2 v^2} \; dv = \frac{\frac{1}{2}\sqrt{\pi}}{A}$$

which is Gauss' probability integral. Similarly,

$$I_1 = \int_0^\infty v e^{-A^2 v^2} \; dv = -\frac{1}{2A^2}[e^{-A^2 v^2}]_0^\infty = \frac{1}{2A^2}$$

Further,

$$I_2 = -\frac{\partial I_0}{\partial(A^2)} = \frac{1}{4}\frac{\sqrt{\pi}}{A^3} \qquad I_3 = -\frac{\partial I_1}{\partial(A^2)} = \frac{1}{2}A^{-4}$$

$$I_4 = \frac{\partial^2 I_0}{\partial(A^2)^2} = \frac{3}{8}\frac{\sqrt{\pi}}{A^5} \qquad I_5 = \frac{\partial^2 I_1}{\partial(A^2)^2} = \frac{1}{A^6}$$

and so forth.

Although $\int e^{-x^2}\,dx$ cannot be written in terms of elementary functions, the indefinite integral $\int_0^w e^{-x^2}\,dx$ appears often in probability theory. As we saw above, $\int_0^\infty e^{-x^2}\,dx = \sqrt{\pi}/2$. It is customary to define the *error function* or the *probability integral* as

$$\operatorname{erf} w = \frac{2}{\sqrt{\pi}}\int_0^w e^{-x^2}\,dx$$

This integral has been evaluated numerically for various values of w and is widely tabulated.

Problems

1. Show that the most probable velocity of a gas molecule is $(2kT/m)^{\frac{1}{2}}$. If v_m, $\bar{v}$, and v_{rms} stand respectively for the most probable, average, and root-mean-square velocities, show that for the maxwellian distribution $v_{rms}:\bar{v}:v_m = \sqrt{1.5}:\sqrt{4/\pi}:1$.

2. Find the rms speed, the average speed, and the most probable speed of N_2 molecules in the air at 27°C.

Ans: $v_{rms} = 517$ m/s

3. Derive Eq. (4.10) from Eq. (4.8).

4. Carry out the mathematical steps required to evaluate $\bar{v}$ in Eq. (4.9).

5. The statistical weight for the first excited state of sodium ($3^2P_{\frac{1}{2}}$) is the same as that of the ground state ($3^2S_{\frac{1}{2}}$), and the energy difference corresponding to the yellow line ($\lambda = 5896$ Å) is 2.1 eV. If the surface temperature of the sun is 6000°K, find the ratio of the number of Na atoms in this excited state to the number in the ground state.

Ans: 0.018

6. The standard atmosphere of the International Civil Aviation Organization is roughly that corresponding to average annual values of pressure,

temperature, and density at 40°N in the United States. At mean sea level it has $p_0 = 1013$ millibars (mb) and density 1.225 kg/m³. At 10 km the ICAO values are $p = 264$ mb, $\rho = 0.413$ kg/m³. Find the corresponding 10-km values for the isothermal atmosphere.

7. Show that the most probable kinetic energy of a gas molecule is $kT/2$.

8. Given that the average speed of O_2 molecules in a container is 450 m/s, the effective radius is 1.7 Å, and the number of molecules per cubic meter is 3×10^{25}. Find the average time between collisions and the mean free path.

Ans: 2×10^{-10} s; 9×10^{-8} m

9. The speed of sound in a gas is given by $V = \sqrt{\gamma p/\rho}$, where γ is the ratio of the specific heat at constant pressure to that at constant volume. Find the ratio of V to the average speed of a gas molecule. Obtain numerical values of the ratio for monatomic gases ($\gamma = \frac{5}{3}$) and a diatomic gas for which $\gamma = \frac{7}{5}$.

Ans: 0.804; 0.741

10. If v denotes the speed of a molecule, calculate the average value of $1/v$ for an ideal gas in thermal equilibrium. How does this value compare with the value of $\bar{v}$?

11. Find the average value of v^3 for molecules of mass m in a gas at temperature T.

12. The viscosity η of oxygen at 20°C and 1 atm pressure is 2.03×10^{-5} N-s/m². In 1860 Maxwell derived the relation $\eta = \frac{1}{3}\rho\bar{v}\lambda$ for a gas, where ρ is the density, $\bar{v}$ the average speed, and λ the mean free path. Calculate the mean free path λ and the "diameter" of an oxygen molecule.

Ans: 1×10^{-7} m; 3.3×10^{-10} m

13. Compute the rms values of the velocity for molecules of hydrogen, oxygen, and carbon dioxide at 300°K.

14. (*a*) At 100°C and 1 atm pressure 1 kg of water vapor occupies a volume of 1.67 m³. Compute the volume available per HOH molecule on the average.

(*b*) Given that the densities of ice and water are respectively 900 and 1000 kg/m³, compute the average volume per HOH unit in each phase.

15. Find the fraction of the molecules which have the y component of the velocity between 0 and $v_m = \sqrt{2kT/m}$.

Ans: 0.422

16. What fraction of the molecules of a gas have x components of velocity greater than the average speed $\bar{v}$? Greater than the rms velocity? Greater than the most probable speed v_m?

17. (a) Show that the fraction of molecules with speeds between 0 and v' is given by

$$\operatorname{erf}(Av') - \frac{2Av'e^{-A^2v'^2}}{\sqrt{\pi}}$$

where $A = \sqrt{m/2kT}$.

(b) Use this result to find the fraction of molecules with kinetic energy $K > 10kT$.

Ans: 1.8×10^{-4}

18. Use the result of the preceding problem to find the fraction of the molecules of a gas which have speed (a) less than the average speed $\bar{v}$; (b) less than the rms speed; (c) greater than $2\bar{v}$.

Ans: (a) 0.53; (b) 0.60

chapter five

The Origin of the Quantum Theory

Relativity and quantum theory represent the two great conceptual remodelings of classical physics which were required to bring physics into the modern era. The quantum theory, first proposed by Planck in 1900, arose out of the inability of classical physics to explain the experimentally observed distribution of energy in the spectrum of a blackbody. By assuming that electromagnetic radiation was emitted by harmonic oscillators, the energy of which could change only by discrete energy jumps of amount hν, Planck was able to bring theory and experiment into accord. In the following 25 years striking confirmations of the quantum hypothesis came from several areas of physics.

5.1 Thermal Radiation

A hot body emits radiant energy, the *quantity* and *quality* of which depend on the temperature. Thus, the rate at which an incandescent filament emits radiation increases rapidly with increase in temperature, and the emitted light becomes whiter.   If this light is dispersed by a prism, a *continuous* spectrum is formed.   Such radiation is ordinarily emitted in an appreciable degree only by objects which are dense enough or thick enough to be opaque.

Radiation falling upon bodies is at least partially absorbed by them. At all temperatures bodies are emitting and absorbing thermal radiation; if they are neither rising nor falling in temperature, this is because as much radiant energy is absorbed each second as is emitted.   In order to deal with the simplest possible case of *thermal equilibrium*, consider a cavity whose walls and contents are all at a common temperature.   An *isothermal enclosure* of this sort is of special interest because the field of radiation inside it possesses some remarkably simple properties.

In an isothermal enclosure the stream of radiation in any given direction must be the same as in any other direction; it must be the same at every point inside the enclosure; and it must be the same in all enclosures at a given temperature, irrespective of the materials composing them.   Furthermore, all these statements hold for each spectral component of the radiation taken separately.

Proof of these statements proceeds by showing that if any one of the statements were not true, a device could be constructed that would violate the second law of thermodynamics.   For example, if the stream of radiation traveling west were greater than that traveling north, we could introduce two similar absorbers, one facing east and the other south.   One of these absorbers would then become hotter than the other by absorbing radiant energy from the stronger stream.   We could, therefore, operate a Carnot engine, using the two absorbers as source and sink, and so could convert heat continuously into work without causing other changes in the system, in violation of the second law. Radiation of a particular wavelength can be tested similarly by using selective absorbers.

The radiation field in an enclosure has important relations with the energy emitted by the walls or other physical bodies.   An *ideal black* surface has the property that it absorbs completely all radiation falling upon it.   Radiation leaving a black surface consists entirely of radiation emitted by it.   Hence *the stream of radiation emitted by any black surface or body in any direction is the same as the stream of radiation that travels in one direction in an isothermal enclosure at the same temperature.*   The total *energy density* produced just in front of a black surface by radiation emitted by that surface alone is just half as great as the density of energy in an enclosure at the same temperature, since the radiation emitted by the body is confined to a hemisphere of directions, whereas in the enclosure radiation is traveling in all directions.

Blackbody radiation is a phenomenon of **great** interest from the theoretical standpoint, because its properties have a universal character, being independent of the properties of any particular material substance. Several questions press at once for an answer. How does the energy density in blackbody radiation vary with the temperature? And what is the spectral distribution of the radiation? Furthermore, we wish to understand how this particular distribution is brought into existence by the atomic processes.

Concerning the first two questions, it was found possible, during the last century, to obtain further information from thermodynamics *without making any assumption as to the atomic process*. The method consisted in considering the effect of expanding or contracting an isothermal enclosure and taking account of the work done on the walls by the radiation in consequence of *radiation pressure*. Appendix 5A summarizes many important classical developments in radiation theory.

5.2 Early Radiation Laws

In 1879 Stefan found empirically that the power emitted per unit area by a blackbody is proportional to the fourth power of the absolute temperature, or

$$R_B = \sigma T^4 \tag{5.1}$$

where R_B = blackbody radiant emittance = power radiated per unit area
σ = Stefan's constant = 0.56686×10^{-7} W/m^2 — °K^4
T = absolute temperature

Five years later Boltzmann deduced the *Stefan-Boltzmann law*, Eq. (5.1), from thermodynamical considerations (Sec. 5A.2).

Not only the quantity but also the quality of radiation emitted by a blackbody changes as the temperature increases. The distribution of energy in the blackbody spectrum is shown in Fig. 5.1. We define the *monochromatic emissive power* e_λ thus: the radiant power emitted per unit area in the spectral range λ to $\lambda + d\lambda$ is $e_\lambda \, d\lambda$; clearly, the radiant emittance $R = \int_0^\infty e_\lambda \, d\lambda$. From Fig. 5.1 one sees that e_λ is a function of both wavelength and temperature. In 1893 Wien showed (Sec. 5A.5) that for any blackbody

$$e_\lambda = T^5 f(\lambda T) = \lambda^{-5} F(\lambda T) \tag{5.2}$$

where $F(\lambda T) = (\lambda T)^5 f(\lambda T)$. Thus e_λ / T^5 is the same for all blackbodies, as is shown in Fig. 5.2, where the data of Fig. 5.1, obtained in 1899 by Lummer and Pringsheim, are presented in a different form. A single curve serves to represent blackbody radiation at all temperatures. It follows that if λ_m is the wavelength at which e_λ is maximum,

$$\lambda_m T = \text{const} = 2.898 \times 10^{-3} \text{ m-}°\text{K} \tag{5.3}$$

for all temperatures—a special case of *Wien's displacement law*.

Thus, by reasoning based on thermodynamics, the problem of blackbody radiation is reduced to the determination of a *single unknown function*, $f(\lambda T)$ or $F(\lambda T)$. *All attempts to obtain the correct form for this function from classical theory failed.*

Two of the formulas proposed on a classical basis nevertheless

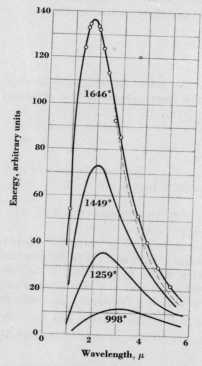

Fig. 5.1 Distribution of energy in constant wavelength interval $\Delta\lambda$ for the blackbody spectrum at four Kelvin temperatures.

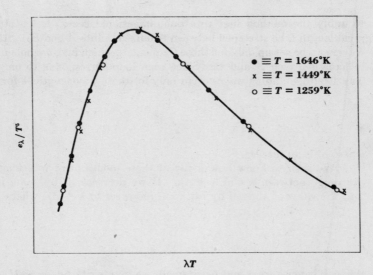

Fig. 5.2 Experimental verification of the Wien displacement law for blackbody radiation.

deserve mention. Wien derived a formula based on special assumptions regarding emission and absorption of radiation; he found that the energy density $U_\lambda \, d\lambda$ for wavelengths between λ and $\lambda + d\lambda$ is

$$U_\lambda \, d\lambda = \tfrac{1}{4}ce_\lambda \, d\lambda = C_1\lambda^{-5}e^{-C_2/\lambda T} \, d\lambda \tag{5.4}$$

where c is the speed of light in free space. By proper choice of the undetermined constants C_1 and C_2 Wien's radiation law can be made to fit the data of Fig. 5.2 except at higher values of λT, where it predicts too low a value. In 1900 Rayleigh applied the principle of equipartition of energy to blackbody radiation. This led, with a contribution from Jeans, to a formula for e_λ which fitted the high-λT part of the e_λ curve but rose to infinity with decreasing λT. The reasoning employed by Rayleigh and Jeans, discussed below, has applications in modern physics.

5.3 Degrees of Freedom in an Enclosure

A typical vibrating system, such as a violin string or organ pipe, is capable of many modes of vibration. We are interested in the number of modes within a given frequency range for a cavity, but it is instructive to apply the method first to a one-dimensional case. Let a string of great length L be stretched between two fixed points A and B. Standing waves can be set up only for those frequencies which give a whole number of loops between A and B. Since each loop corresponds to one-half a wavelength, standing waves occur only for those wavelengths λ for which

$$\frac{L}{\lambda/2} = n_x = \frac{2L}{\lambda} \tag{5.5}$$

where n_x is an integer.

We wish to know how many of these modes have wavelengths in the range between λ and $\lambda + \Delta\lambda$. If we decrease n_x by some integer Δn_x, the value of λ given by Eq. (5.5) increases to $\lambda + \Delta\lambda$, where

$$\frac{2L}{(\lambda + \Delta\lambda)} = n_x - \Delta n_x$$

When L is very great and $\Delta\lambda$ is small, we may write $\Delta n_x = -2L\,\Delta\lambda/\lambda^2$, where Δn_x now represents the number of modes of vibration with wavelengths in the range $\Delta\lambda$.

Now, associated with each mode there are two degrees of freedom, since the vibrations are transverse and any point on the string is free to move in a plane at right angles to the string. The total number of degrees of freedom per unit length of string in the wavelength range between λ and $\lambda + \Delta\lambda$ is therefore

$$\Delta n_l = \frac{4\Delta\lambda}{\lambda^2} \tag{5.6}$$

In two dimensions the situation is more complex but involves identical principles. Let us first discuss the system of waves in a square of side L (Fig. 5.3a). A set of waves moving initially in direction OM_1 will after reflection move in the direction M_1M_2. After reflection at M_2, the direction of propagation is M_2M_3, which is parallel but opposite to OM_1, etc. For this group of waves only four directions of motion are possible, $\pm OM_1$ and $\pm M_1M_2$.

The four wave trains thus formed combine to produce a set of standing waves only if in both the x and y directions there is a whole number of half waves. Consider a cube of side L with perfectly reflecting walls in which there are electromagnetic waves polarized with the electric intensity in the z direction. Since the z component E_z of the electric intensity must be zero at the bounding planes when $x = 0$ or L and

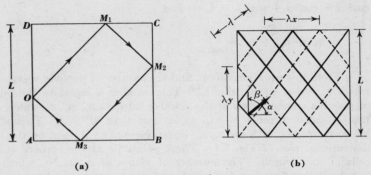

Fig. 5.3 (a) Reflection of waves inside a square. (b) Standing waves in the xy plane of a cube.

$y = 0$ or L, the allowed solutions of the wave equation are of the form

$$E_z = f(z,L) \sin \frac{n_x\pi}{L} x \sin \frac{n_y\pi}{L} y$$

where n_x and n_y are integers; this is the equation for a standing wave. The condition that there be nodes at the walls of Fig. 5.3b puts a limitation on the allowed wavelengths; specifically, the projection of any side along the propagation direction must be an integral number of half wavelengths. In Fig. 5.3b the heavy arrow represents one of the propagation directions, which makes angle $\alpha = 37°$ with the x axis; for this particular case there are four half wavelengths associated with the x direction and three with y.

From the figure it is clear that $\lambda_x = \lambda/\cos \alpha$ and

$$\lambda_y = \frac{\lambda}{\cos \beta} = \frac{\lambda}{\sin \alpha}$$

Since $\sin (n_x\pi x/L) = \sin (2\pi x/\lambda_x)$, we have $n_x\pi/L = (2\pi \cos \alpha)/\lambda$; similarly $n_y\pi/L = (2\pi \cos \beta)/\lambda$. If we solve for n_x and n_y, square each, and add, we have

$$n_x{}^2 + n_y{}^2 = \frac{4L^2}{\lambda^2} (\cos^2 \alpha + \cos^2 \beta) = \frac{4L^2}{\lambda^2} (\cos^2 \alpha + \sin^2 \alpha) = \frac{4L^2}{\lambda^2}$$

This method of constructing standing waves can be extended to a three-dimensional cube, for which eight sets of waves are involved. The condition $n_x\pi/L = (2\pi \cos \gamma)/\lambda$ must be applied, and since

$\cos^2 \alpha + \cos^2 \beta + \cos^2 \lambda = 1$, we find

$$n_x{}^2 + n_y{}^2 + n_z{}^2 = \frac{4L^2}{\lambda^2} \tag{5.7}$$

From this equation we can find the number of possible wavelengths which exceed a given value λ_m. This number is equal to the number of possible combinations of the positive integers n_x, n_y, n_z which make the left-hand member of Eq. (5.7a) less than $4L^2/\lambda_m{}^2$. To find this number, imagine each set n_x, n_y, n_z represented by a point on the three-dimensional plot of Fig. 5.4. These points lie at the corners of cubic cells of unit length. The number of values of n_x, n_y, n_z which give $\lambda > \lambda_m$ is therefore just the number of points inside a sphere of radius $2L/\lambda_m$ divided by 8 (since we are interested only in the all-positive octant), which is

$$n = \left(\frac{4\pi}{3} \frac{8L^3}{\lambda_m{}^3}\right) \frac{1}{8} = \frac{4\pi}{3\lambda^3} L^3$$

Differentiation of this equation gives the number of wavelengths in a range from λ to $\lambda + d\lambda$; thus

$$dn = \frac{4\pi \, d\lambda}{\lambda^4} \text{ (vol of cube)} \tag{5.8}$$

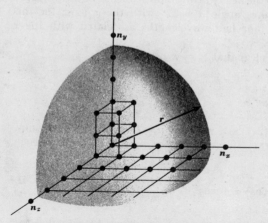

Fig. 5.4 The Rayleigh scheme for counting the number of allowed values of n_x, n_y, n_z for which $n_x{}^2 + n_y{}^2 + n_z{}^2$ is less than r^2.

where both dn and $d\lambda$ are taken as positive for convenience. For electromagnetic waves there are two polarizations per allowed wavelength, and so we finally obtain for the number of degrees of freedom per unit volume for standing waves in the enclosure with wavelength from λ to $\lambda + d\lambda$ or frequency from ν to $\nu + d\nu$

$$g(\lambda)\, d\lambda = \frac{8\pi\, d\lambda}{\lambda^4} \qquad g(\nu)\, d\nu = \frac{8\pi\nu^2\, d\nu}{c^3} \qquad (5.9)$$

5.4 The Rayleigh-Jeans Radiation Law

In an isothermal enclosure, the radiation is constant both in quantity and in spectral characteristics, so that any energy absorbed is reradiated. The result is the same as if all the energy were reflected at the walls. Rayleigh and Jeans assumed that oscillators in the wall absorbed and emitted radiation constantly, with each oscillator having its own characteristic frequency. For continuous operation of any given oscillator, standing waves must be set up in the enclosure. However for any enclosure of reasonable size the differences between neighboring frequencies are so small that the radiation appears to be continuous.

The principle of equipartition of energy requires assigning to these oscillators $\frac{1}{2}kT$ of kinetic energy per degree of freedom plus another $\frac{1}{2}kT$ for potential energy. Assignment of an average energy of kT to each mode of vibration leads to an energy density $U_\lambda\, d\lambda$ for waves with wavelength between λ and $\lambda + d\lambda$ given by

$$U_\lambda\, d\lambda = kTg(\lambda)\, d\lambda = 8\pi kT\lambda^{-4}\, d\lambda \qquad (5.10)$$

Equation (5.10) is the Rayleigh-Jeans formula for blackbody radiation. It will be noted that it contains *no new constants*. At *long wavelengths this formula agrees with observation*, but near the maximum in the spectrum and at short wavelengths it gives much too large values. Furthermore, it assigns no maximum at all *to* U_λ. In fact, the energy density U in the enclosure would be infinite; for there is no lower limit to the possible wavelengths, and $U = \int_0^\infty U_\lambda\, d\lambda \to \infty$. In reality, the radiation density in an isothermal enclosure is ordinarily much less than the energy density due to thermal agitation of the molecules of a solid body; e.g., at 1000°K U is about 7.5×10^{-4} J/m³, as compared with a total density on the order of 10^9 J/m³ in iron. At temperatures above $10^{6°}$K the comparison would be reversed. Infinite energy in the electromagnetic field would make the specific heats of all material bodies infinite.

5.5   Planck's Investigation of Blackbody Radiation

The correct blackbody formula was discovered in 1900 by Max Planck. By introducing a radical innovation quite at variance with previous concepts, he found a formula in complete agreement with experiment. *This was the birth of the quantum theory.*

The experiments of Hertz on electromagnetic waves had seemed to give final confirmation to the electromagnetic theory of light, and this convinced Planck that the key to the blackbody spectrum would be found in the laws governing the absorption and emission of radiation by electric oscillators. We may imagine that the walls of an isothermal enclosure contain oscillators of all frequencies, essentially similar to the hertzian oscillator, and that the emission and absorption of radiation by the walls are caused by these oscillators.

Investigation by means of electromagnetic theory led Planck to the conclusion that an oscillator, in the long run, would affect only radiation of the *same frequency* as that of the oscillator itself. In the state of equilibrium, there would be a definite ratio between the density of radiation of any frequency ν and the average energy of the oscillators of that frequency. The problem of the blackbody spectrum was thus reduced to the problem of the average energy of an oscillator at a given temperature. If he then assumed the classical value kT for this average energy, as derived from the equipartition of energy, he was led to the Rayleigh-Jeans formula.

Planck, however, did not accept the principle of the equipartition of energy for oscillators. On the basis of a different assumption he was led at first to Wien's formula for the radiation density, Eq. (5.4). In 1900 new measurements of the blackbody spectrum by Lummer and Pringsheim and by Rubens and Kurlbaum showed definitely that Wien's formula is not correct. Planck discovered an *empirical* modification of Wien's formula that fitted the observations. Then he sought to modify the statistical theory of the distribution of energy among a set of oscillators so that the theory would lead to his new formula. He succeeded in doing so only after making a new assumption that broke drastically with classical principles. To facilitate understanding his assumption, a brief description of the statistical theory will be given.

5.6   Distribution of Energy among Oscillators in Thermal Equilibrium

Consider a linear oscillator of mass m that vibrates in simple harmonic motion; let its displacement be x and its momentum $p = m\,dx/dt$. The potential energy of the oscillator can be written in the form $\frac{1}{2}bx^2$, b denoting the force constant; its kinetic energy is $\frac{1}{2}m(dx/dt)^2 = p^2/2m$. Thus its total energy is

$$\epsilon = \frac{p^2}{2m} + \frac{1}{2}bx^2 \tag{5.11}$$

and by elementary theory its frequency of vibration ν has the value

$$\nu = \frac{1}{2\pi}\sqrt{\frac{b}{m}} \tag{5.12}$$

When a large number of such oscillators are in thermal equilibrium, the energy of an individual oscillator varies widely but the energies of the entire group are distributed in energy according to the Maxwell-Boltzmann distribution (Sec. 4.6). By a contraction of Eq. (4.19) to two phase-space dimensions we find the number of oscillators dN with coordinates between x and $x + dx$ and momentum between p and $p + dp$ to be

$$dN = NAe^{-\epsilon/kT}\, dx\, dp \tag{5.13}$$

where A is a proportionality constant such that $\int dN = N$.

Since we are interested only in the energies of the oscillators, it is convenient to throw the formula into a different form. Let us construct a plot on which x and p are taken as cartesian coordinates (Fig. 5.5). Each oscillator is represented by a point on this plot which moves about as the oscillator vibrates. So long as the oscillator is free from disturbing forces, x and p change in such a way that the energy remains constant; hence the representative point moves along an ellipse given by Eq. (5.11) with a fixed value of ϵ. Let two such ellipses be drawn for slightly different energies ϵ and $\epsilon + \Delta\epsilon$. Throughout the elliptical ring between these ellipses, if $\Delta\epsilon$ is very small, the quantity $e^{-\epsilon/kT}$ is sensibly constant; hence, in a state of thermal equilibrium, the number of oscillators represented by points in this ring is, by Eq. (5.13),

$$\Delta N = NAe^{-\epsilon/kT}\iint dx\, dp \tag{5.14}$$

The integral $\iint dx\, dp$ is the area of the elliptical ring, and its value is easily found in terms of $\Delta\epsilon$. We first note that the area inside an ellipse whose semiaxes are x_m and p_m is

$$S = \pi p_m x_m \tag{5.15}$$

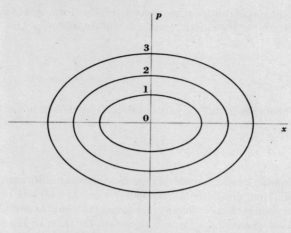

Fig. 5.5 The momentum-coordinate plane for a harmonic oscillator.

Here x_m and p_m also represent the maximum values of x and of p, respectively, during a vibration of an oscillator having an energy ϵ; substituting in (5.11) first $x = x_m$ and $p = 0$ and then $x = 0$ and $p = p_m$, we find

$$x_m = \sqrt{\frac{2\epsilon}{b}} \qquad p_m = \sqrt{2m\epsilon} \qquad S = 2\pi\epsilon\sqrt{\frac{m}{b}} = \frac{\epsilon}{\nu} \qquad (5.16)$$

by (5.12). The area of a ring corresponding to an increment $\Delta\epsilon$ is thus $\iint dx\,dp = \Delta S = \Delta\epsilon/\nu$. Hence (5.14) can be written in the form

$$\Delta N = N A_1 e^{-\epsilon/kT}\,\Delta\epsilon \qquad (5.17)$$

where $A_1 = A/\nu$.

The average energy of the oscillators can now be found by summing their individual energies and dividing by N. Ordinarily this would be done by means of integrals, but for our present purpose we employ discrete sums. Let ellipses centered at the origin be drawn so as to divide the xp plane into rings of equal area h; and let these rings, of which the innermost is actually an elliptical area, be numbered 0, 1, 2, . . . from the center outward, as in Fig. 5.5. Then the total area *inside* ring number τ is $S = \tau h$; and, by Eq. (5.16), the energy of an oscillator represented by a point on the inner boundary of ring number τ is

$$\epsilon = S\nu = \tau h\nu$$

The number of oscillators N_r on ring r can thus be written, from (5.17),

$$N_r = N_0 e^{-rh\nu/kT}$$

where N_0 replaces $NA_1 \Delta\epsilon$. The total energy E of all oscillators is then, approximately,

$$E = \sum_0^\infty rh\nu N_r = N_0 h\nu e^{-h\nu/kT}(1 + 2e^{-h\nu/kT} + 3e^{-2h\nu/kT} + \cdots)$$

$$= N_0 h\nu e^{-h\nu/kT}(1 - e^{-h\nu/kT})^{-2}$$

since by the binomial theorem the last series is of the form

$$1 + 2x + 3x^2 + \cdots = (1 - x)^{-2}$$

In a similar way it is found that

$$N = \Sigma N_r = N_0(1 + e^{-h\nu/kT} + \cdots) = N_0(1 - e^{-h\nu/kT})^{-1}$$

For the *average energy* per oscillator we have then, finally.

$$\bar\epsilon = \frac{E}{N} = \frac{h\nu}{e^{h\nu/kT} - 1} \tag{5.18}$$

This expression represents also the energy of a particular oscillator averaged over any length of time that is not too short. For, being similar, the oscillators will all have the same average energy in the long run, and this average must be E/N.

In classical theory it is now necessary to let $h \to 0$. The approximations made then disappear. Using the series $e^x = 1 + x + x^2/2 + \cdots$, we find that, to the first order in h,

$$e^{h\nu/kT} - 1 = \frac{h\nu}{kT}$$

Hence, in the limit as $h \to 0$, $\epsilon = kT$. This is the same value for the average energy of a harmonic oscillator that was deduced from the equipartition of energy in Sec. 5.4, and it leads to the incorrect Rayleigh-Jeans formula.

5.7   *Planck's Quantum Hypothesis*

The new assumption introduced by Planck was equivalent to *keeping h finite* in the preceding formulas. In the first formulation of the new theory, Planck assumed that the oscillators associated with a given ring all have the energy proper to the inner boundary of that ring. Then Eq. (5.18) for ϵ holds exactly. According to this assumption, the energy of an oscillator cannot vary continuously but must take on one of the discrete set of values $0, h\nu, 2h\nu \ldots , rh\nu, \ldots$. The actual original form of Planck's assumption was that the energy of the oscillator must always be an integral multiple of a certain quantity ϵ_0, but he then showed that for oscillators of different frequencies, ϵ_0 must be proportional to ν if the radiation law is to harmonize with the Wien displacement law. Thus he assumed that $\epsilon_0 = h\nu$, where h is a constant of proportionality. The connection between h and areas on the xp plane for the oscillator, as described above, was recognized by Planck later.

It must be emphasized that the assumption of a discrete set of possible energy values, or energy levels, for an oscillator was completely at variance with classical ideas. According to this assumption, if the energies of a large number of oscillators were measured, some might be found to have zero energy, some $h\nu$ each, others $2h\nu$, and so on. But not a single oscillator would be found which had energy, say, $1.73h\nu$. When the energy of a given oscillator changes, therefore, it must change suddenly and discontinuously. According to the older conceptions, the interchange of energy between two systems, e.g., between one gas molecule and another or between radiation and oscillators, is a continuous process, and the energy of an oscillator would likewise vary continuously. Such continuity of energy values is demanded by classical physics. For example, the electric and magnetic vectors in a light wave may have any values whatsoever, *from zero up;* and, accordingly, the wave may have any intensity, *from zero up.* The emission and absorption of this energy by the walls of an enclosure should, likewise, be a perfectly continuous process.

The problem of the absorption and emission of radiation presented serious difficulties for the new theory. If the energy of an oscillator can vary only discontinuously, the *absorption and emission of radiation* must be *discontinuous processes.* As long as the oscillator remains in one of its *quantum states,* as we now call them, with its energy equal to one of the allowed discrete values, it cannot be emitting or absorbing radiation according to the laws of classical physics, for then the conservation of energy would be violated. This is contrary to classical electromagnetic theory, which requires an isolated, accelerated electric charge to radiate energy.

According to Planck's new theory, *emission* of radiation occurs only when the oscillator "jumps" from one energy level to another; if it jumps down to the next lower energy level, the energy $h\nu$ that it loses is emitted in the form of a short pulse of radiation. *Absorption* was also assumed at first to be discontinuous. An oscillator, Planck assumed, can absorb a quantum $h\nu$ of radiant energy and jump up to its next higher energy level. This assumption met with special difficulties, however. For the quantum of radiant energy emitted by an oscillator, according to the classical wave theory, would spread out over an ever expanding wavefront, and it is hard to see how another oscillator could ever gather this energy together again so as to absorb it all and thereby acquire the energy for an upward quantum jump. Absorption ought, therefore, on Planck's theory, to be impossible.

To avoid this difficulty, Planck later modified his theory so as to allow the oscillators to absorb in a continuous manner, only the process of emission being discontinuous. The energy of an oscillator could then take on all values, as in classical theory; but every time the energy passed one of the critical values $\tau h\nu$, there was assumed to exist a certain chance that the oscillator would jump down to a lower energy level, emitting its excess energy as a quantum of radiation. This came to be known as the *second form* of Planck's quantum theory. It can be shown that in this form of the theory, the oscillators are evenly distributed over each ring on the px plane, instead of being all on the inner boundary, and the mean energy of all oscillators is, instead of the value given by Eq. (5.18),

$$\bar{\epsilon} = \frac{h\nu}{e^{h\nu/kT} - 1} + \frac{1}{2} h\nu \tag{5.19}$$

Having read thus far, the student may perhaps have reached a state of confusion as to *what were* the essential assumptions of Planck's quantum theory! This confusion can be no worse than that which existed in the minds of most physicists in the year, say, 1911. The situation was made still more puzzling by the success of Einstein's theory of the photoelectric effect (Sec. 6.7); for Einstein assumed not only that radiation came in quantized spurts but that each spurt was closely concentrated in space, contrary to the wave theory. Confusion usually reigns while important physical advances are being made; it is only afterward that a clear-cut logical path can be laid down leading straight to the goal.

One of the aims of this book will be to show how the theory gradually became clarified. It will be found that the following two new ideas introduced by Planck have been retained permanently and form a part of modern wave mechanics.

1. *An oscillator or any similar physical system has a discrete set of possible energy values or levels; energies intermediate between these allowed values never occur.*

2. *The emission and absorption of radiation are associated with transitions, or jumps, between two of these levels, the energy thereby lost or gained by the oscillator being emitted or absorbed, respectively, as a quantum of radiant energy, of magnitude $h\nu$, ν being the frequency of the radiation.*

The difficulties with the theory of electromagnetic waves were overcome when the theory of wave mechanics was applied to the electromagnetic field itself.

It should be emphasized that Planck's revolutionary assumptions were not based upon an extension of the ordinary lines of reasoning of classical physics. Quite the contrary; they represented an *empirical modification* of classical ideas made in order to bring the theoretical deductions into harmony with experiment. Had the magnitude of the quantum of energy turned out to be not $h\nu$ but something *independent of the frequency*, the new theory might well have taken the form of a simple atomicity of energy, similar to the atomicity of electricity represented by the electronic charge. Such is not the case, however. Rather, it is the new universal constant h that represents the essentially new element introduced into physics by the quantum theory. We shall find h playing an important part in a wide variety of atomic phenomena.

5.8 Planck's Radiation Law

Planck derived his new radiation formula by considering the interaction between the radiation inside an isothermal enclosure and electric oscillators which he imagined to exist in the walls of the enclosure. A more direct and equally satisfactory procedure is to combine Planck's new expression for the mean energy of an oscillator with the analysis of the electromagnetic field by the method of Rayleigh and Jeans, in which the various modes of oscillation of the field inside an enclosure are treated as if they were oscillators. In Sec. 5.3, Eq. (5.9), we found that there would be $8\pi \, d\lambda/\lambda^4$ such modes of oscillation or degrees of freedom per unit volume in the wavelength range λ to $\lambda + d\lambda$. If we multiply this number by $\bar{\epsilon}$ as given by Eq. (5.18), we obtain

$$U_\lambda \, d\lambda = 8\pi \frac{d\lambda}{\lambda^4} \frac{h\nu}{e^{h\nu/kT} - 1}$$

Let us substitute here $\nu = c/\lambda$, c being the speed of light in vacuum. Thus we obtain, as Planck's new radiation law,

$$U_\lambda = \frac{8\pi ch}{\lambda^5} \frac{1}{e^{ch/\lambda kT} - 1} \tag{5.20}$$

[Strictly speaking, we should have used for $\bar{\epsilon}$ the value given by Eq. (5.19), which agrees with the value obtained from wave mechanics. The effect of this change would be to add in U_λ a term independent of temperature. Since only *changes* in U are perceptible, this term would be without physical effect.]

Planck's formula reduces to Wien's formula near one end of the spectrum and to the Rayleigh-Jeans formula near the other end. In Fig. 5.6 is shown a comparison between the several spectral-energy distribution formulas and the experimental data. The circles show observations by Coblentz on the energy distribution in the spectrum of a blackbody at 1600°K. The reason that the curve for Planck's formula drops below the Rayleigh-Jeans formula is the failure of the classical principle of the equipartition of energy. The high-frequency modes of oscillation of the electromagnetic field in the enclosure, which should all have mean energy kT according to classical theory, remain almost entirely in their lowest quantum states and so contribute little to the observable density of radiant energy.

Planck's formula has the form that we found to be required by the principles of thermodynamics (Sec. 5.2); for Planck's formula can be

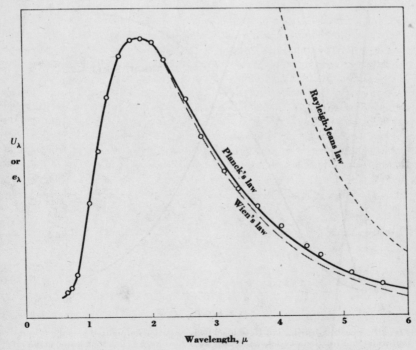

Fig. 5.6 Comparison of the three radiation laws with experiment at 1600°K.

written

$$U_\lambda = T^5 f(\lambda T) \qquad f(\lambda T) = \frac{8\pi ch}{(\lambda T)^5} \frac{1}{e^{ch/k(\lambda T)} - 1} \tag{5.20a}$$

From the formula in this form, both the Stefan-Boltzmann formula and the Wien displacement law follow as mathematical consequences.

The value of the new constant h can be determined by comparing the observed values of U_λ with the formula, in which the values of all other quantities are known.   In his original paper (1901) Planck obtained in this way the value $h = 6.55 \times 10^{-34}$ The accuracy of thermal data is relatively low, however, and better methods yield the value

$$h = 6.6256 \times 10^{-34} \text{ J-s}$$

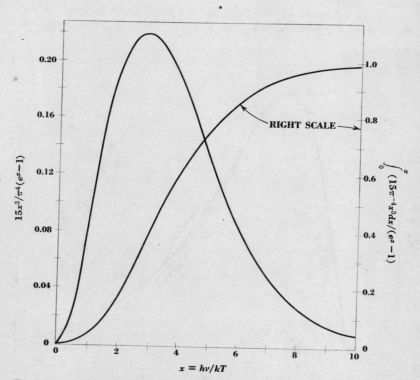

Fig. 5.7 Spectral distribution U_ν of blackbody radiation normalized to make U unity (*left scale*) and fraction of energy radiated with x ($= h\nu/kT$) less than value given on abscissa (*right scale*).

It is often desirable to write the Planck law in terms of frequency rather than λ. In this case the energy density U_ν in the range between ν and $\nu + d\nu$ is

$$U_\nu \, d\nu = \frac{8\pi h\nu^2 \, d\nu}{c^3(e^{h\nu/kT} - 1)} \tag{5.21}$$

The total energy density U is

$$U = \int_0^\infty U_\lambda \, d\lambda = \frac{8\pi k^4 T^4}{c^3 h^3} \int_0^\infty \frac{x^3 \, dx}{e^x - 1} = \frac{8\pi^5}{15} \frac{k^4 T^4}{c^3 h^3} \tag{5.22}$$

where $x = ch/\lambda kT$ and the integral has the value $\pi^4/15$. The normalized spectral distribution of a blackbody is shown in Fig. 5.7 along with a curve showing the fraction of the energy radiated at x less than a prescribed value.

As is shown in Sec. 5A.1, Eqs. (5A.4) and (5A.5), the power radiated per unit area, total, or in a given λ or ν interval, is just $c/4$ times the appropriate energy density. Thus the power radiated per unit area is $R_B = cU/4$, and we have recovered Eq. (5.1) with the Stefan constant expressed in terms of π, k, c, and h.

Appendix 5A - Classical Radiation Theory

5A.1   *Pressure and Energy Flux Due to Isotropic Radiation*

Suppose a stream of radiation in a vacuum falls normally on the surface of a body. Then, if w is the mean energy density in the oncoming waves, they carry also w/c units of momentum per unit volume (Sec. 2.12). Thus the waves bring up to each unit area of the surface, along with cw J of energy, w units of momentum per second, the momentum as a vector being directed normally toward the surface. If the waves are absorbed by the surface, it receives this momentum and experiences, therefore, a pressure equal to w.

Suppose, next, that the radiation is incident at an angle θ. Then the energy that crosses a unit area drawn perpendicular to the rays (*PQ* in Fig. 5A.1) is received by a larger area of magnitude $1/\cos\theta$ on the surface *PR*. Furthermore, the component of the momentum normal to the surface is less than in the case of normal incidence in proportion to $\cos\theta$. Thus the momentum in the direction of the normal that is delivered to unit area of the surface per second is decreased by the

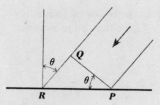

Fig. 5A.1

obliqueness of incidence in the ratio $\cos^2 \theta$, and the resulting pressure, if the radiation is entirely absorbed, is

$$p = w \cos^2 \theta$$

The same expression holds for the pressure caused by the *emission* of a beam at an angle θ or for the additional pressure caused by the occurrence of a reflected beam. If an incident beam is specularly reflected from a surface at the angle of incidence θ, the total pressure on the surface is $2w \cos^2 \theta$.

Finally, let radiation be streaming toward a surface and also away from it with equal intensities in all directions, as in an isothermal enclosure. Such a distribution of radiation is equivalent to a large number of beams of plane waves, all of equal intensity, with their directions of propagation distributed equally in direction. Let there be N beams in all, and let the energy density due to any one of them be w. Then the total energy density U just in front of the surface and the pressure p on it are, respectively,

$$U = Nw \tag{5A.1a}$$

$$p = \Sigma w \cos^2 \theta = w\Sigma \cos^2 \theta \tag{5A.1b}$$

To find this latter sum, imagine lines drawn outward from a point O on the surface to represent the various directions of the beams, whether moving toward the surface or away from it, and then about O as center draw a hemispherical surface of unit radius with its base on the surface (see Fig. 5A.2, where only two of the lines are shown). From the hemisphere cut out a ring-shaped element of area QS by means of two cones of semiangle θ and $\theta + d\theta$, drawn from O as apex and with the normal OP as axis. The edge of this element is a circle of perimeter $2\pi \sin \theta$, and its width is $d\theta$; hence its area is $2\pi \sin \theta \, d\theta$, whereas the area of the whole hemisphere is 2π. Now the lines of approach of the N beams of radiation, if drawn through O, will cut the hemisphere in points equally distributed over its surface. Hence, if we let dN denote the number of these lines that pass through the ring-shaped element, dN will be to N in the ratio of the area of the ring to the area of the hemisphere, so that

$$\frac{dN}{N} = \frac{2\pi \sin \theta \, d\theta}{2\pi} = \sin \theta \, d\theta$$

The value of $\cos^2 \theta$ is the same for all the dN beams. Hence their contribution to $\Sigma \cos^2 \theta$ is $\cos^2 \theta \, dN$ or, from the last equation, $N \cos^2 \theta \sin \theta \, d\theta$. Thus

$$\sum \cos^2 \theta = \int \cos^2 \theta \, dN = N \int_0^{\pi/2} \cos^2 \theta \sin \theta \, d\theta = \tfrac{1}{3}N \quad (5A.2)$$

(The limit is $\pi/2$ because directions all around the normal OP are included in the ring.)   For the pressure we thus obtain, from Eq. (5A.1), $p = \tfrac{1}{3}wN$, or

$$p = \tfrac{1}{3}U \tag{5A.3}$$

Thus the pressure on the walls of an isothermal enclosure equals one-third of the radiant-energy density at any interior point.

It will be useful to calculate also the total *energy* brought up to the surface. Since half of the N waves are moving toward the surface and each wave delivers energy $cw \cos \theta$, the total energy brought up to unit area per second is

$$\tfrac{1}{2}N \int_0^{\pi/2} cw \cos \theta \sin \theta \, d\theta = \tfrac{1}{4}cwN = \tfrac{1}{4}cU$$

by (5A.1a). In an isothermal enclosure an equal amount of energy is carried away from the surface. It can be shown in the same way that in an isothermal enclosure in which the energy density is U, energy $\tfrac{1}{4}cU$ per second passes in each direction across unit area of any imaginary surface drawn inside the enclosure.

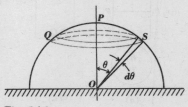

Fig. 5A.2

Furthermore, the energy emitted per second by unit area of any black surface, which is its radiant emittance R, is related to the energy density U in an isothermal enclosure having the same temperature by the equation

$$R = \tfrac{1}{4}cU \tag{5A.4}$$

The same relation holds for each wavelength separately. Let the energy emitted from the blackbody with a range of wavelengths $d\lambda$ be denoted by $e_\lambda \, d\lambda \left(\text{so that } R = \int_0^\infty e_\lambda \, d\lambda \right)$ and let the density of radiant energy in the enclosure within the same range of wavelengths be similarly represented by $U_\lambda \, d\lambda \left(\text{so that } U = \int_0^\infty U_\lambda \, d\lambda \right)$. Then

$$e_\lambda = \tfrac{1}{4}cU_\lambda \tag{5A.5}$$

5A.2   The Stefan-Boltzmann Law

In 1884, Boltzmann deduced a theoretical law for the variation of the total intensity of blackbody radiation with temperature. For this purpose, he applied the laws of the Carnot cycle to an engine in which the radiation played the part of the working substance.

The ideal Carnot engine consists of an evacuated cylinder with walls impervious to heat, a piston likewise impervious to heat and moving without friction, and a base through which heat may enter or leave. Let the walls, piston, and base be perfectly reflecting except for a small opening O in the base, which can be covered at will by a perfectly reflecting cover. Let this cylinder be placed with the opening O uncovered and opposite an opening in an evacuated isothermal enclosure B_1, which is maintained at temperature T_1 (Fig. 5A.3). Then the cylinder will fill

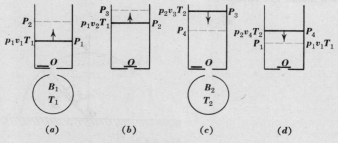

Fig. 5.1.3   Boltzmann's radiation engine.

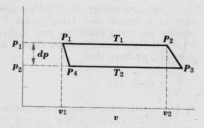

Fig. 5A.4 The p-v diagram for the Carnot cycle of the radiation engine.

up with radiation entering it through O from B_1 until there is the same density U_1 of radiation in the cylinder as there is in B_1, at which time radiation will be passing at the same rate from O to B_1 as from B_1 to O.

We may now consider the following cycle of events:

1. Starting with the piston in the initial position P_1 (Fig. 5A.3a), the initial volume of the cylinder being v_1 and the initial pressure due to the radiation $p_1 = \frac{1}{3}U_1$, we cause the piston to move upward, *slowly*, until position P_2 is reached, the volume increasing to v_2. During this process the radiation density within the cylinder remains constant at U_1. To keep it constant, additional radiation must enter the opening O from the enclosure B_1, for two reasons:

a. Work W_e is done by the radiation on the piston. If T_1 remains constant, so do U_1 and p_1 and

$$W_e = p_1(v_2 - v_1) = \tfrac{1}{3}U_1(v_2 - v_1)$$

b. The volume of the cylinder has *increased* by $v_2 - v_1$, which requires an additional influx of energy equal to $U_1(v_2 - v_1)$. Thus the total influx H_1 of radiation from B_1 must be

$$H_1 = \tfrac{4}{3}U_1(v_2 - v_1) \tag{5A.6}$$

This isothermal process is represented on the p-v diagram (Fig. 5A.4) by the horizontal line P_1P_2. The energy H_1 is equivalent to heat supplied to the space within the cylinder, just as in an ordinary Carnot cycle the first isothermal expansion is accompanied by an absorption of heat. An amount of heat equal to H_1 must also be supplied from external sources to B_1 in order to keep the temperature of B_1 constant.

2. When the piston has reached P_2, the perfectly reflecting cover is placed over the opening O (Fig. 5A.3b), thereby effecting complete thermal isolation of the interior of the cylinder, and a further expansion to position P_3 is made. External work is done, as before, on the piston, the energy required for this external work being supplied by the radiation. Partly because of this work and partly because of the increase in volume, *the energy density of the radiation within the cylinder must decrease* from U_1 to some smaller value U_2. The pressure, likewise, has decreased. This is obviously an adiabatic process. It is represented in Fig. 5A.4 by the line P_2P_3.

The new energy density U_2 is now equal to the energy density in an enclosure at a certain new temperature T_2. If the expansion during this second process was very small, we may represent the change in temperature $T_1 - T_2$ by dT and the corresponding change in energy density $U_1 - U_2$ by dU. Since $p = \frac{1}{3}U$, we have then

$$dp = \tfrac{1}{3}dU \tag{5A.7}$$

dp representing the change in radiation pressure.

3. The engine is now placed opposite a second isothermal enclosure B_2 (Fig. 5A.3c) at temperature T_2, the slide is removed from the opening O, and the piston is moved, by the application of suitable external force, from P_3 to P_4. On account of this compression, there is a tendency for the density of radiation within the cylinder to rise and for radiation to pass through O into B_2. The compression is supposed to take place so slowly, however, that the radiation density remains constant at a value only infinitesimally in excess of U_2. During this second isothermal process, radiant energy in amount H_2 leaves the engine.

4. The piston having reached a suitable point P_4, the opening O is closed, and the radiation is then compressed adiabatically until the initial position P_1 is reached.

The net external work done during this cycle is represented by the area $P_1P_2P_3P_4$ of Fig. 5A.4. If we assume the change of pressure to have been very small, this area equals $(v_2 - v_1)\,dp$. Calling the net external work dW, we have, therefore,

$$dW = (v_2 - v_1)\,dp = \tfrac{1}{3}(v_2 - v_1)\,dU$$

by (5A.7). Hence, by the usual rule for a Carnot cycle,

$$\frac{dW}{H_1} = \frac{T_1 - T_2}{T_1} = \frac{dT}{T_1}$$

and, using the value found for dW and also Eq. (5A.6), we have

$$\frac{dU}{U_1} = \frac{4dT}{T_1}$$

Thus, dropping the subscript,

$$\frac{dU}{U} = 4\,\frac{dT}{T}$$

Integrated, this equation gives $\log U = 4 \log T + \text{const}$ or

$$U = aT^4 \tag{5A.8}$$

where a is a constant, not yet known. From Eq. (5A.4) we have then also, for the emissive power or radiant emittance

$$R = \sigma T^4 \tag{5A.9a}$$
$$\sigma = \tfrac{1}{4}ca \tag{5A.9b}$$

Thus *both the energy density of the radiation within an isothermal enclosure and the total emissive power of a blackbody are proportional to the fourth power of the absolute temperature T.* This is the Stefan-Boltzmann law.

5A.3 Reflection from a Moving Mirror

In the preceding discussion no attention was paid to the spectral distribution of the radiation. The question presents itself, however, whether the same law can be applied also to the separate wavelengths. In order to investigate this question, we need to know what happens to the spectral distribution of a beam of radiation when it is reflected from a mirror that is in motion, such as the piston in the ideal apparatus that was described in the last section.

Let us consider first the effect of such motion upon a monochromatic beam. For this purpose we employ Huygens' principle. In Fig. 5A.5a, MM represents a mirror moving with a component of velocity V perpendicular to its plane. AB and DE represent parts of two incident waves which are one wavelength, or a distance λ, apart and are falling on the

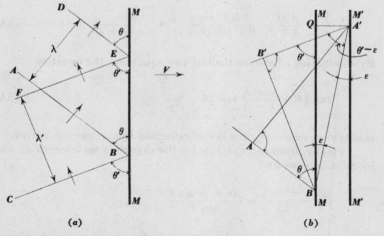

Fig. 5A.5 Reflection from a moving mirror.

mirror at an angle of incidence θ; CB and FE are parts of the same waves which have been reflected and are now leaving the mirror at an angle of reflection θ', a distance λ' apart. It is obvious from the figure that $\lambda = BE \sin \theta$, $\lambda' = BE \sin \theta'$, whence

$$\frac{\lambda'}{\lambda} = \frac{\sin \theta'}{\sin \theta} \tag{5A.10}$$

To obtain a second relation between λ' and θ', consider two successive positions of the same wave. In Fig. 5A.5b, the part AB of a wave is just beginning to fall on the mirror, whose instantaneous position is MM. The same portion of the wave at a later instant, after it has been reflected at the angle θ', is shown by $A'B'$, the point A being now in contact at A' with the mirror, which is in the new position $M'M'$. While A traveled along the ray AA', B traversed BB'; hence $AA' = BB'$. The angles BAA' and $BB'A'$, being angles between ray and wave, are right angles. It follows, therefore, by similar triangles, that angle

$$B'A'B = ABA'$$

or

$$\theta' - \epsilon = \theta + \epsilon \tag{5A.11}$$

where ϵ is the angle QBA'. Furthermore, while A went from A to A' at the speed c of light, the mirror moved from MM to $M'M'$ at the speed u. Hence, if $A'Q$ is a perpendicular dropped from A' onto MM,

$$\frac{V}{c} = \frac{A'Q}{AA'} = \frac{BA' \sin \epsilon}{BA' \sin (\theta + \epsilon)} = \frac{\sin \epsilon}{\sin (\theta + \epsilon)} \tag{5A.12}$$

By eliminating ϵ between the last two equations, the equation

$$\tan \tfrac{1}{2}\theta' = \frac{c + V}{c - V} \tan \tfrac{1}{2}\theta \tag{5A.13}$$

is obtained, expressing the law of reflection from a moving mirror. Finally, from Eq. (5A.10), for the change of wavelength $\Delta\lambda$ caused by reflection, we find

$$\frac{\Delta\lambda}{\lambda} = \frac{\lambda' - \lambda}{\lambda} = \frac{\sin \theta' - \sin \theta}{\sin \theta}$$

We need the value of $\Delta\lambda$, however, only for an infinitesimal value of V

and hence of $\theta' - \theta$.   For such a value

$$\sin \theta' - \sin \theta = (\theta' - \theta) \frac{d}{d\theta} \sin \theta = (\theta' - \theta) \cos \theta$$

Also, from Eqs. (5A.10) and (5A.11), in which ϵ is an infinitesimal,

$$\theta' - \theta = 2\epsilon = 2 \sin \epsilon = 2 \frac{V}{c} \sin (\theta + \epsilon)$$

to the first order; and here $\sin (\theta + \epsilon)$ may be replaced by $\sin \theta$.   We thus find, to the first order in V,

$$\Delta\lambda = 2 \frac{V}{c} \lambda \cos \theta \tag{5A.14}$$

5A.4 Effect of an Adiabatic Expansion upon Blackbody Radiation

Returning now to the sequence of operations described in Sec. 5A.2, let us consider the effect of the adiabatic process (step 2) upon the spectral distribution of the radiation.   In this process, blackbody radiation initially at temperature T_1 imprisoned in a cylinder with perfectly reflecting walls is slowly expanded from an initial energy density U_1 to a new energy density U_2.   Let the former restriction to a small expansion be dropped.   The change in direction of the rays that is produced by the moving piston, according to Eq. (5A.13), will tend to make the radiation no longer isotropic.   We can obviate this inconvenient effect, however, by letting part of the walls of the cylinder reflect perfectly *but diffusely*. A surface of magnesium oxide does this very well.   Then, if the expansion is made very slowly, because all rays (except a negligible few) strike the diffusing surface repeatedly, the radiation will be kept effectively isotropic; and the pressure on the piston, according to Eq. (5A.3), will be at all times equal to $\frac{1}{3}U$.

Let the cylinder have a cross section A and a (variable) length l. Then, if U is the energy density at any moment, when the piston moves outward a distance dl, work $p\, dv = \frac{1}{3}UA\, dl$ is done on it by the force due to radiation pressure.   This work is done at the expense of the enclosed energy, the total amount of which is lAU.   Hence

$$\frac{1}{3}UA\, dl = -d(lAU) = -AU\, dl - Al\, dU$$
$$\frac{dU}{U} = -\frac{4}{3}\frac{dl}{l}$$

and, after integration,

$$\log U = - \log l^{\frac{1}{3}} + \text{const} \tag{5.15a}$$
$$U = Cl^{-\frac{1}{3}} \tag{5.15b}$$

C denoting a constant.

Now it can be shown by thermodynamic reasoning that an expansion of the type considered here cannot destroy the blackbody property of the radiation. For, at a certain instant, suppose that the expansion has reduced the total energy density to U_2, and let T_2 be the temperature of an enclosure in which the density has this same value. Suppose that in the cylinder there were more radiation per unit volume of wavelengths near some value λ' than at the same wavelengths in the enclosure and less radiation near some other wavelength λ''. It would then be possible to cause a little radiation to pass from the cylinder into a second enclosure at a temperature T_2' slightly above T_2 by covering the opening in the base of the cylinder with a plate transmitting wavelengths near λ' but reflecting all others and putting the cylinder into communication with the second enclosure through this opening. In a similar way, enough radiation near λ'' could be passed *into* the cylinder from an enclosure at a slightly *lower* temperature T_2'' to restore the total energy to U_2. Then the radiation could be compressed back to U_1, the changes in U and l and the amount of the work done being just the reverse of these quantities during the expansion. Finally, putting the cylinder again into communication with the enclosure at T_1, we could allow the spectral distribution to be restored to that proper to a blackbody at T_1 but without any net transfer of energy between cylinder and enclosure, since the total energy density has already been restored to that corresponding to T_1. Thus we should have performed a cyclic operation, the *only* effect of which is to transfer heat energy from an enclosure at T_2'' to one at a higher temperature T_1''. But this is inconsistent with the second law of thermodynamics.

Thus black radiation must remain black during any slow adiabatic expansion or compression. Its density and temperature, however, decrease. In the case under discussion, the last equation, in combination with Eq. (5A.8), gives

$$T \propto \frac{1}{l^{\frac{1}{3}}} \tag{5A.16}$$

To determine the effect on the spectral distribution, we must find the average rate at which wavelengths are increased. Suppose, first, that the walls of the cylinder and the piston reflect specularly. Then

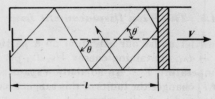

Fig. 5A.6

any ray preserves its angle of inclination to the axis of the cylinder θ in spite of repeated reflections (see Fig. 5A.6) and has, therefore, a constant component of velocity $c \cos \theta$ perpendicular to the piston. The ray strikes the piston $(c \cos \theta)/2l$ times a second, and its wavelength, according to (5A.14), increases each time by $(2V\lambda \cos \theta)/c$, V being the speed of the piston. Thus its wavelength increases at a rate

$$\frac{d\lambda}{dt} = \frac{c \cos \theta}{2l} \frac{2V\lambda \cos \theta}{c} = \frac{V\lambda}{l} \cos^2 \theta$$

If rays of wavelength λ are equally distributed in direction, the average value of $\cos^2 \theta$ for all of them is $\frac{1}{3}$, according to Eq. (5A.2); hence the average value of $d\lambda/dt$ for these rays is

$$\frac{d\lambda}{dt} = \frac{V\lambda}{3l} = \frac{\lambda}{3l} \frac{dl}{dt}$$

since $V = dl/dt$.

To simplify the calculation, suppose now that there is a diffusely reflecting spot on the walls. Then all the rays will take turns moving in the various directions, and all rays of wavelength λ will undergo the average change of wavelength just calculated. (This conclusion can be confirmed by a more complete analysis for which we have no space.) The last equation can be integrated thus:

$$\frac{d\lambda}{\lambda} = \frac{1}{3} \frac{dl}{l} \tag{5A.17a}$$

$$\log \lambda = \log l^{\frac{1}{3}} + \text{const} \tag{5A.17b}$$

$$\lambda \propto l^{\frac{1}{3}} \propto \frac{1}{T} \tag{5A.17c}$$

by (5A.16). Thus we have the conclusion that each spectral component of blackbody radiation must change in wavelength in such a way that $\lambda \propto 1/T$.

5A.5   *The Wien Displacement Law*

Now let us fix our attention on a particular spectral range from λ_1 to $\lambda_1 + d\lambda_1$, containing energy $U_{\lambda 1} d\lambda_1$ per unit volume in an enclosure at temperature T_1.   An adiabatic expansion which lowers the temperature to T_2 changes the limits of this range, according to the result just reached, to λ_2 and $\lambda_2 + d\lambda_2$, where

$$\frac{\lambda_2}{\lambda_1} = \frac{\lambda_2 + d\lambda_2}{\lambda_1 + d\lambda_1} = \frac{T_1}{T_2} \tag{5A.18a}$$

Therefore

$$\frac{d\lambda_2}{d\lambda_1} = \frac{T_1}{T_2} \tag{5A.18b}$$

The energy in $d\lambda_1$ is at the same time decreased, and in the same ratio as the total energy is; for in the argument that led up to Eq. (5A.15), we might have started with only the radiation in $d\lambda_1$ present in the cylinder. Hence, $U_{\lambda 2}$ being the new value of U_λ,

$$\frac{U_{\lambda 2} \, d\lambda_2}{U_{\lambda 1} \, d\lambda_1} = \frac{U_2}{U_1} = \frac{T_2{}^4}{T_1{}^4}$$

and using (5A.18b),

$$\frac{U_{\lambda 2}}{U_{\lambda 1}} = \frac{e_{\lambda 2}}{e_{\lambda 1}} = \frac{T_2{}^5}{T_1{}^5} \tag{5A.19}$$

Here e_λ is the emissive power of a blackbody, given in terms of U_λ by Eq. (5A.5).

Thus, if the values of U_λ or e_λ at two different temperatures are compared not at the same wavelength but at wavelengths inversely proportional to T, then their values will be found to be proportional to the *fifth power* of the absolute temperature.   This conclusion is known as the *Wien displacement law*.   It can also be expressed by saying that at a wavelength varied so that the product λT is held constant, U_λ/T^5 or e_λ/T^5 has the same value at all temperatures.   Thus if either of these ratios is plotted against λT as abscissa, a single curve valid at all temperatures will be obtained (Fig. 5.2).

Problems

1.   (a) Estimate the surface temperature of the sun, assuming it is essentially a blackbody of radius 7×10^8 m which is 1.5×10^{11} m from the earth, from the fact that the solar constant (power per unit area on a surface normal to the sun's rays) is 1400 W/m^2.

(b) Estimate the surface temperature of the sun from the fact that $\lambda_m = 4900$ Å.

2.   Show that Planck's radiation law, Eq. (5.20), reduces to Wien's law,

Eq. (5.4), when $h\nu \gg kT$ and to the Rayleigh-Jeans law, Eq. (5.10), when $h\nu \ll kT$.

3. If the fireball of a nuclear weapon can be approximated at some instant to be a blackbody of 0.5 m radius with a surface temperature of 10^7°K, find (a) the total electromagnetic power radiated, (b) the wavelength at which maximum energy is radiated, and (c) the value of kT in electron volts.

  Ans: (a) 1.8×10^{21} W; (b) 2.9 Å; (c) 0.86 keV

4. From Fig. 5.7, assuming that the sun is a blackbody with $\lambda_m = 4900$ Å, estimate the fractions of the solar radiation in the visible (3500 to 7000 Å), the infrared, and the ultraviolet parts of the spectrum.

  Ans: 0.43, 0.50, 0.07

5. Show that if λ_m represents the wavelength at which maximum energy is radiated in wavelength interval $d\lambda$, then $\lambda_m T = ch/4.965k$, which is equivalent to Eq. (5.3).

6. A very bright star of spectral class B0 has an effective surface temperature of 20,000°K. Find the wavelength at which e_λ is maximum, assuming the radiation is approximately blackbody. How does e_λ at 5000 Å for this star compare with that of the sun at this wavelength if the surface temperature of the sun is 6000°K?

7. The intensity of radiation per unit wavelength interval from a fireball sphere of radius 2 m has been measured. When the resulting curve of intensity as a function of wavelength is plotted, it has its maximum at 289.8 Å and is closely approximated by the blackbody curve. Find the temperature and the total power radiated by the sphere at the instant in question.

8. Show that the maximum of the U_ν curve (Fig. 5.7) in terms of $x = h\nu/kT$ occurs for $\nu = 2.82kT/h$.

9. Find the value of the Stefan constant σ in terms of k, c, and h by the use of Eq. (5.22).

  Ans: $2\pi^5 k^4/15c^2 h^3$

10. Planck originally computed his value of h and the value of the Boltzmann constant k by using experimental results of Rubens and Kurlbaum, which gave $U = (7.6 \times 10^{-16} \text{ J/m}^3\text{-}°\text{K}^4)T^4$ and

$$\lambda_m T = \frac{ch}{4.956k} = 2.9 \times 10^{-3} \text{ m-}°\text{K}$$

Find h and k from these data.

11. Estimate from Fig. 5.7 the fraction of the energy radiated by a blackbody which is in the visible region (3500 to 7000 Å) for temperatures of 1000, 2000, and 3000°K.

12. Show with the aid of Eq. (5A.5) that both the Stefan-Boltzmann formula and Wien's displacement law in the form of Eq. (5.3) follow as mathematical consequences of Planck's formula, Eq. (5.20a).

13. If in a blackbody cavity the number of photons per unit volume with wavelength between λ and $\lambda + d\lambda$ is $N_\lambda \, d\lambda$, find the wavelength at which N_λ is maximum in terms of h, k, and T.

14. Show that in a blackbody cavity the number of photons per unit volume with frequency between ν and $\nu + d\nu$ is

$$N_\nu \, d\nu = \frac{8\pi\nu^2 \, d\nu}{c^3(e^{h\nu/kT} - 1)}$$

and verify that N_ν has its maximum for $\nu = 1.59kT/h$.

15. Using the result of the preceding problem, show that the number of photons per unit volume in a blackbody cavity is

$$N = \frac{8\pi k^4 T^4}{c^3 h^3} \int_0^\infty \frac{x^2 \, dx}{e^x - 1} = \frac{8\pi k^4 T^4}{c^3 h^3} \times 2.404$$

where $x = h\nu/kT$. Verify that the integral has the value $2\zeta(3)$, where $\zeta(n) = 1 + 1/2^n + 1/3^n + \cdots$ [$\zeta(3) = 1.202$].

16. Using the results of the preceding problem, verify that the total number of photons emitted per second per unit area by a blackbody is

$$2.404 \frac{2\pi}{c^2} \left(\frac{kT}{h}\right)^3$$

and that the average energy per photon is $2.70kT$.

17. The *kinetic temperature* kT of a hot plasma or a nuclear weapon is sometimes used in preference to the Kelvin temperature. For a 2-keV weapon estimate (a) the Kelvin temperature, (b) the wavelength at which e_λ is maximum and the energy of photons of this wavelength, (c) the average energy per photon (see Prob. 16), (d) the photon energy for which U_ν is maximum (see Prob. 8), and (e) the photon energy for which the number of photons per unit frequency range is maximum (see Prob. 14).

Ans: (a) 23,200,000°K; (b) 1.25 Å, 9920 eV; (c) 5400 eV; (d) 5640 eV; (e) 3180 eV

18. Derive Eq. (5.22) from (5.21), showing that

$$\int_0^\infty \frac{x^3 \, dx}{e^x - 1} = 6\zeta(4)$$

where $\zeta(n) = 1 + 1/2^n + 1/3^n + \cdots$
[Accept the fact that $\zeta(4) = \pi^4/90$.]

chapter six

Electrons and the Photoelectric Effect

During the nineteenth century the atomic hypothesis was strongly buttressed, and the electromagnetic nature of light was established. Concurrent with these theoretical strides came great advances in experimental chemistry and electricity. The development of induction coils, electromagnets, and vacuum pumps paved the way for experimental work in electrical discharge through rarefied gases, which culminated in the discovery of the electron, of x-rays, and of radioactivity in the concluding decade. Suddenly it became clear that atoms have structure, even though the word "atom" originally meant "uncuttable."

6.1 Electricity in Matter

The atomicity of electricity was suggested by the laws of electrolysis. Faraday found that when the same quantity of electricity deposits univalent ions of different kinds, the amounts deposited as measured by weight are proportional to the ionic weights in the chemical sense. By the ionic weight is meant here the sum of the atomic weights of atoms, one or more, composing the ion. It follows that the same quantity F of electricity suffices to deposit a gram ion of any kind of univalent ion, i.e., a mass of the ions equal in number of grams to the ionic weight. The value of this quantity F, called the *faraday*, is 96,487 C.

The number of ions in a gram ion or of atoms in a gram atom is in all cases the same. It follows that all univalent ions carry the same unit of charge, usually denoted by e. For this natural unit Stoney proposed the name *electron* in 1891.

An independent line of thought which likewise suggested the presence of definite electrical charges in matter was the treatment of dispersion in developing the electromagnetic theory of light. L. Lorenz suggested in 1880 that refractive media may contain small charged particles which can vibrate with a natural frequency ν_0 about a fixed equilibrium position. By assuming one or more sets of such particles, with different natural frequencies, it was possible to account completely for the phenomena of dispersion. But the data on dispersion could not be made to furnish any clue to the magnitude of the charges on these particles or even to their sign.

In the middle of the eighteenth century it had been found that air in the neighborhood of hot bodies is an electrical conductor. In a systematic investigation begun in 1880 Elster and Geitel showed that at a red heat charged bodies lose their electrification. In 1883, in the course of his development of the incandescent lamp, Edison worked with some evacuated bulbs containing three electrodes connected to the outside. After a carbon filament was attached to two of these electrodes and brought to incandescence, Edison observed that when the third electrode was made positive relative to the filament, there was a current in the circuit connected to the third electrode. When the latter was made negative, no current was established. Thus Edison discovered thermionic emission several years before the existence of the electron was established. (Thermionic emission is discussed further in Secs. 6.8 and 21.8.)

6.2 Discovery of Photoelectricity

Still another suggestion that atoms might be divisible came from the work of Heinrich Hertz, who first produced from laboratory circuits the electromagnetic waves predicted by Maxwell. Hertz generated the waves with an oscillating circuit containing a spark gap P and detected them by means of sparks at another gap S in a suitably tuned circuit. In 1887, to facilitate observation, he enclosed S in a black box and found that S must be made shorter to allow sparks to pass.   Even a plate of glass interposed between P and S would cause this effect.   Hertz concluded that ultraviolet light coming from spark P facilitated the passage of sparks at S when no obstacle was interposed.   He found that light from another spark was equally effective but that in all cases the light must fall on the terminals themselves.   This discovery of the *photoelectric effect* was accidental in the sense that it was unplanned.   While investigating one physical phenomenon, Hertz discovered quite a different one whose existence he had not even suspected.

Hertz's discovery at once attracted numerous investigators.   Hallwachs found that a freshly polished zinc plate, insulated and connected to an electroscope, when charged *negatively* and illuminated by ultraviolet light, would lose its charge, but there was no effect if the charge was *positive*.   He observed, by using an electrometer instead of an electroscope, that a *neutral* insulated plate, when illuminated, acquired a small *positive* potential, i.e., lost a *negative* charge.   Stoletow devised an arrangement, shown diagrammatically in Fig. 6.1, for producing a continuous *photoelectric current*. P is a photoelectrically sensitive plate, say a polished zinc plate, connected to the negative terminal of a battery B of several cells. S is a wire grating or gauze connected to the positive terminal of the battery through a very sensitive galvanometer G.   When ultraviolet light falls upon P, a continuous current is observed in G, indicating that a negative charge passes from P to S.   No current exists if the battery is reversed.

Elster and Geitel showed that there is a close relation between the

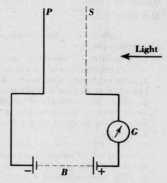

Fig. 6.1 Stoletow's arrangement.

contact-potential series of metals and the photoelectric effect: the more electropositive the metal, the longer the wavelength to which it responds photoelectrically.   The alkali metals, sodium, potassium, and rubidium, were found to be sensitive even to light of the visible spectrum.   That the charge is carried by negatively electrified particles was clearly indicated by experiments of Elster and Geitel, who showed that a transverse magnetic field diminishes the photoelectric current if the phenomenon takes place in a vacuum.   It was early found that the effect persisted even to the highest attainable vacuum and was quite independent of the "degree" of the vacuum after a certain low pressure had been reached. This seemed to indicate that the gas molecules themselves in the region between P and S could not be acting as carriers of the charge.   The suggestion was made that, perhaps, under the influence of light, negatively charged ions of the cathode became detached and moved to the anode.   This suggestion was rendered untenable by an experiment by Lenard, in which a clean platinum wire acted as anode and a sodium amalgam as cathode, both being in an atmosphere of hydrogen.   The photoelectric current was allowed to flow until about 3×10^{-6} C had passed through the circuit.   If the carriers of the charge were ions of sodium, each atom could hardly be expected to carry a larger charge than it carries in electrolysis.   There should, therefore, have been deposited on the platinum wire at least 0.7×10^{-6} mg of sodium, a quantity sufficient to be detectable by the well-known flame test.   On removing the wire from the bulb, however, no trace of sodium could be detected.

If the photoelectric current is carried neither by the gas surrounding the cathode nor by ions from the cathode, what are the carriers? The answer to this question came through the convergence[1] of a number of different lines of evidence, which finally culminated in the discovery of the electron.

[1] Collectively these developments illustrate many characteristics of the growth of modern physics:

1. A number of seemingly unrelated lines of research frequently converge to provide an explanation of, or a theory for, a group of phenomena not hitherto understood.
2. The explanation or theory thus evolved is then found to bear directly on other branches of physics and often on other sciences.
3. Thus, the methods of physics are both synthetic and analytic—a fortunate circumstance, which makes it possible for the physicist to comprehend physics as a whole in spite of the vast increase, particularly in recent years, of factual knowledge.
4. These discoveries, of both fact and theory, are the *sine qua non* of applied physics and of much of industry—witness, as a single example, the wide use of the various kinds of photo- and thermionic tubes.
5. This application of science to industry very frequently reacts to provide the research man with improved tools.   Electron tubes, manufactured primarily for industrial purposes, are a boon to the physicist in his research laboratory.

6.3   The Zeeman Effect

In 1862 Faraday, looking for a possible effect of a magnetic field upon a light source, placed a sodium flame between the poles of a strong electromagnet and examined the D lines by a spectroscope.   He was unable to detect any change in the appearance of the lines.

Faraday's failure to observe the effect that he expected was due to the inadequate resolving power of his apparatus.   For, in 1896, Zeeman, repeating Faraday's experiment with the improved techniques then available, discovered that spectral lines are split up into components when the source emitting the lines is placed in a very strong magnetic field; furthermore, following a suggestion by H. A. Lorentz, he found that these components are polarized.

The simplest case, shown in Fig. 6.2, involves what is known as the *normal Zeeman effect*.   A single line of wavelength λ_0 in the absence of a magnetic field is split into three components with wavelength $\lambda_0 - \Delta\lambda$, λ_0, and $\lambda_0 + \Delta\lambda$ when viewed at right angles to the magnetic induction $\mathbf{B}$; the lines of wavelengths $\lambda_0 \pm \Delta\lambda$ are plane-polarized with the electric vector perpendicular to $\mathbf{B}$, while the center line has the same wavelength as the original and is plane-polarized with the electric vector parallel to $\mathbf{B}$.   If one pole piece is drilled longitudinally so that one can view the source in a direction parallel to $\mathbf{B}$, only the two components with wavelengths $\lambda_0 \pm \Delta\lambda$ are seen; they are circularly polarized, as shown in Fig. 6.2.

A short description will be given of the classical Zeeman-Lorentz theory in spite of the fact that classical theory is not applicable to such

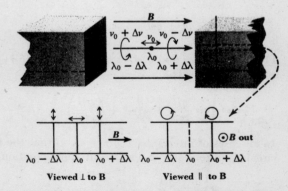

Fig. 6.2   The normal Zeeman effect; below are shown the lines and polarizations as viewed perpendicular to and along $\mathbf{B}$.

atomic phenomena.   Nevertheless it gives a simple, concrete picture which assists the memory and yields several correct and significant predictions.   Classically, a charge performing simple harmonic motion radiates light.   Assume, then, that there is within an atom a particle of charge q and mass m bound to its equilibrium position by a force $\mathbf{F}$ proportional to its displacement; $\mathbf{F} = -b\mathbf{r}$.   The cartesian components of $\mathbf{F}$ are $F_x = -bx$, $F_y = -by$, $F_z = -bz$.   Hence in the absence of a magnetic field each component of the displacement varies harmonically with the same frequency

$$\nu_0 = \frac{1}{2\pi} \sqrt{\frac{b}{m}} \tag{6.1}$$

although perhaps with different amplitudes and phases.

Now let a magnetic field be applied.   The component of the motion of the particle parallel to the field is unaffected.   This gives a line of the original wavelength polarized with the electric vector parallel to the applied $\mathbf{B}$ when observed at right angles to $\mathbf{B}$ but no line at all when observed along $\mathbf{B}$—precisely as observed.

In general any simple harmonic vibration can be regarded as the sum of two superposed circular motions executed with the same frequency but in opposite directions.   Application of a magnetic field perpendicular to the original vibration causes the frequencies of these two circular motions to be different, since the magnetic field introduces an additional force $q(\mathbf{v} \times \mathbf{B})$ on the particle.   Viewed perpendicular to $\mathbf{B}$, radiation from the circular motion would be plane-polarized with the electric vector perpendicular to $\mathbf{B}$.   For steady circular motion

$$m \frac{v^2}{r} = br \pm qvB \tag{6.2}$$

Putting $v = 2\pi r\nu$ and $b = 4\pi^2 m\nu_0{}^2$ from Eq. (6.1) into this equation and solving for ν yields

$$\nu = \pm \frac{qB}{4\pi m} + \sqrt{\nu_0{}^2 + \left(\frac{qB}{4\pi m}\right)^2}$$

In all cases of interest $\nu_0 \gg qB/4\pi m$, so that the frequencies of the two circular motions are, to a sufficient approximation,

$$\nu = \nu_0 \pm \frac{qB}{4\pi m} \tag{6.3}$$

For a positive particle the slower rotation (longer λ) occurs in a clockwise direction as seen by an observer looking at the source through the hole in the pole piece, whereas Zeeman found counterclockwise rotation for the line of lower frequency. Therefore the charge on the radiating particle is *negative*. The ratio of charge to mass for the particle can be computed from the separation between the outer lines, which is given by $\Delta\nu = Bq/2\pi m$, and so

$$\frac{q}{m} = \frac{2\pi}{B} \Delta\nu \tag{6.4}$$

Zeeman's first observation indicated that q/m was of the order of 10^{11} C/kg. Later, working with higher resolving power, he found $q/m = 1.6 \times 10^{11}$ C/kg, compared with the modern value of 1.759×10^{11} C/kg for the electron.

Later experiments have shown that the Zeeman effect is much more complicated than this picture suggests. Nevertheless the work of Zeeman and Lorentz provided the first evidence that spectral lines were radiated by *negative* charges, yielded a reasonably good value for e/m (the charge on the electron is $-e$), and stimulated the accumulation of data from which a better understanding could be achieved. Very often more than three components are observed, and the theory above fails completely to explain these more complicated patterns; indeed, it does not even explain why only vibrations parallel and perpendicular to **B** should be considered. (The Zeeman effect is discussed further in Chap. 18.)

6.4 Cathode Rays and the Electron

Before 1897, many studies had been made of that beautiful phenomenon, the discharge of electricity through rarefied gases. Let the discharge from an induction coil or an electrostatic machine pass between the negative terminal C (cathode) and the positive terminal A (anode) sealed into a glass tube (Fig. 6.3) which is being exhausted through the side

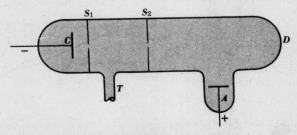

Fig. 6.3 Tube for studying electrical discharge at low pressure.

tube T.   At a very low pressure, there appears around the cathode a dark space, known as the *Crookes dark space*, which, with further decrease in pressure, grows longer, i.e., extends farther toward D, until finally it reaches the glass walls of the tube, which then have a fluorescent glow. If screens pierced with holes S_1 and S_2 are introduced, the glow is confined to a spot on the end of the tube at D, thus indicating that some things called *cathode rays* are proceeding from the cathode and causing the glass to fluoresce.   By suitably curving the cathode, these rays can be focused on a piece of platinum foil within the tube, heating it to incandescence. That cathode rays bear negative charge was inferred from their deflections in electric and magnetic fields; in 1895 Perrin caught the rays in an insulated chamber connected to an electroscope and confirmed that they carry negative charge.

In 1897 J. J. Thomson measured the ratio of charge to mass for cathode rays with the apparatus of Fig. 6.4.   A highly evacuated tube contained a cathode C and an anode A, having a small slit through which the cathode rays passed.   A second slit A_2 electrically connected to A_1 helped define a sharp beam of cathode rays, which produced a small fluorescent patch at O in the absence of fields beyond A_2.   But when a

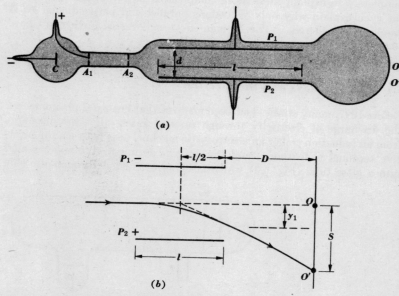

Fig. 6.4   (a) Thomson's apparatus for determining e/m for electrons.   (b) Deflection of an electron by the electrostatic field (not to scale).

potential difference V was applied between plates P_1 and P_2, the beam was deflected to point O'. By the use of a pair of Helmholtz coils, not shown, a uniform magnetic field out of the plane of the paper was produced throughout the volume between P_1 and P_2. Two observations were then made. With a given electric intensity E between the plates the magnetic induction $\mathbf{B}$ was adjusted until the beam struck at the original position O. Then the downward force $e\mathsf{E}$ due to the electric field was balanced by the force $ev \times \mathbf{B}$ due to the magnetic field. Hence $e\mathsf{E} = evB$ and

$$v = \frac{\mathsf{E}}{B} \tag{6.5}$$

Thus the first observation serves to measure the velocity v of the particles; the crossed electric and magnetic fields serve as a *velocity selector*.

The magnetic field is then removed, and the deflection OO' caused by the electrostatic field alone is measured. From this the deflection y_1 of the particles as they pass over the distance l between the plates can be determined. This deflection results from a uniform acceleration $a_y = e\mathsf{E}/m$ during a time l/v; hence,

$$y_1 = \tfrac{1}{2}\mathsf{E}\,\frac{e}{m}\left(\frac{l}{v}\right)^2 \tag{6.6}$$

from which e/m can be computed.

Thomson found that e/m determined in this way was independent of the kind of gas in the tube (air, H_2, or CO_2) and likewise independent of the electrodes (Al, Fe, or Pt). His final determination was $e/m = 1.7 \times 10^{11}$ C/kg, a value *almost identical with the value found by Zeeman for the particles taking part in light emission*. It is very much greater than e/m for hydrogen atoms in electrolysis.

6.5 The Electronic Charge

Shortly before Thomson's work the study of the charges on gaseous ions such as are produced in a gas by x-rays had begun. To measure their charges, Townsend, working in Thomson's laboratory, utilized the clouds that form about the ions in saturated air. By observing the rate of fall of the cloud and applying Stokes's law for the free fall of spheres through a viscous medium, he was able to determine the size of the droplets; from a measurement of the total amount of water in the cloud, he could then calculate the number of droplets it had contained. He assumed that each droplet contained just one ion; hence, having measured also the total charge on the cloud, he was able to calculate the charge on a single ion. For this charge he obtained a value of about 3×10^{-10} electrostatic unit. A repetition of the measurements with some modifications by Thomson gave the value 6.5×10^{-10} esu (2.2×10^{-19} C).

Thomson assumed that the charge on the gaseous ions was the same as that on his cathode particles. It followed, then, that the cathode particles must be previously unknown particles of extremely small mass, which he called *corpuscles* or *primordial atoms*. For many years English writers stuck to the name corpuscle for these particles, using the word electron in Stoney's original sense to denote the amount of charge carried by a corpuscle or a univalent ion; but others, including Lorentz, called the corpuscles themselves electrons, and this usage ultimately became well established. It was generally assumed that electrons are a constituent part of all atoms and are responsible for the emission of light by them, thus accounting for the fact that the ratio e/m had been found to be the same for the vibrating particles causing the Zeeman effect as for the cathode rays. Thus was made possible the explanation of a number of more or less diverse phenomena on the basis of a single concept.

A more reliable method of measuring ionic charges was developed by Millikan in 1909. He found that tiny droplets of oil in ionized air, viewed under a microscope, would frequently pick up charges and could then be held suspended, or accelerated upward or downward, by applying a suitable electric field. When uncharged, the droplets fall at a slow, uniform rate, their weight being balanced by the drag due to the viscosity of the surrounding air. By observing their rate of fall, Millikan was able to determine the size and weight of the droplets. According to Stokes's law, a sphere of radius a moving at a steady slow speed v_0 through a fluid whose coefficient of viscosity is η experiences a resisting force $F = 6\pi\eta a v_0$. If we equate this force to the weight of a droplet, which is $\frac{4}{3}\pi a^3 \rho g$ in terms of the density ρ of the oil, we have

$$\frac{4}{3}\pi a^3 \rho g = 6\pi\eta a v_0 \tag{6.7}$$

Now suppose the drop picks up a charge q and that a vertical electric field E is present. Then a force qE is added to the weight, and we have

$$\frac{4}{3}\pi a^3 \rho g + qE = 6\pi\eta a v_1 \tag{6.8}$$

v_1 being the new velocity of steady fall. From these equations we find

$$qE = 6\pi\eta a(v_1 - v_0)$$

and inserting in this equation the value of a given by (6.7) and solving for q, we find

$$q = 6\pi\eta^{\frac{3}{2}}(v_1 - v_0)\left(\frac{9v_0}{2\rho g}\right)^{\frac{1}{2}}\frac{1}{E} \tag{6.9}$$

All quantities in this last expression being known, q can be calculated.

Millikan found that the charges so calculated from his observations were all multiples of a smallest charge, assumed to be the charge on the electron. The value of this smallest charge he reported to be $-e$, where $e = 4.774 \times 10^{-10}$ esu $= 1.591 \times 10^{-19}$ C.

This value of e was accepted for the next 20 years. In 1928, however, Bäcklin pointed out that x-ray wavelengths as measured by means of a ruled grating could be made to agree with the same wavelengths measured by means of a calcite crystal provided Millikan's value of e was assumed to be 0.4 percent too small. The discrepancy was later traced to Millikan's having used too low a value of the viscosity of air, which appears as η in Eq. (6.9). Improved measurements gave a value of η nearly 0.5 percent greater than Millikan's, resulting in the value 1.602×10^{-19} C for e as calculated from the oil-drop observations.

It is now recognized, however, that what is obtained directly from the x-ray measurements is actually a value of N_A, the number of atoms in 1 kg atom; e is then calculated as $10^3 F/N_A$, where F is the value of the faraday (Sec. 6.1). The accuracy of the value of e obtained in this manner is greater than that of the oil-drop observations. Millikan's work nevertheless remains of fundamental importance. He demonstrated that electric charges in ionized gases occur in multiples of a single fundamental unit, and the close agreement of his value of e with that required by the x-ray data constitutes an important confirmation of the principles underlying the calculation of e from the x-ray data themselves.

6.6 Photoelectrons

The discovery of the electron suggested the hypothesis that the photoelectric effect is due to the liberation of electrons. This hypothesis was confirmed in 1900 by Lenard. By measuring the deflection of the "photoelectric rays" in a known magnetic field, he found a value of e/m in qualitative agreement with Thomson's value for electrons. Lenard's method of determining e/m for photoelectrons involves basic principles which, with ever increasing refinement, have been widely applied in charged-particle physics. His apparatus is shown diagrammatically in

Fig. 6.5a. A glass tube, evacuated through the side tube T, contained an aluminum cathode C, which could be illuminated by ultraviolet light from a spark S, the light passing through the quartz plate Q. A screen A, with a small hole at its center and connected to earth, served as anode. P_1 and P_2 were small metal electrodes connected to electrometers. When C was illuminated and charged to a negative potential of several volts, photoelectrons were liberated and accelerated toward the anode A. A few electrons passed through the hole in A and proceeded thereafter at uniform velocity to the electrode P_1, their reception there being indicated by electrometer 1. But if, by means of a pair of Helmholtz coils (represented by the dotted circle), a magnetic field directed toward the reader was produced in the region between A and P_1, the electrons were deflected upward in a circular path and, with a sufficient field strength, struck the electrode P_2.

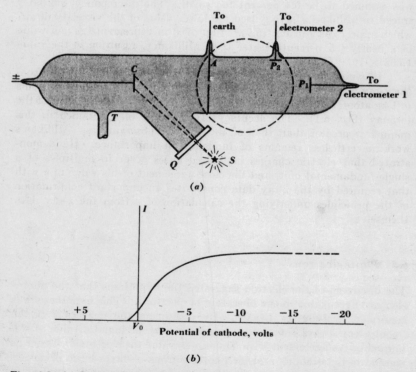

(a)

(b)

Fig. 6.5 (a) Lenard's apparatus for determining e/m for photoelectrons. (b) Variation of photoelectric current with cathode potential.

Lenard first investigated the relation between the current reaching the anode and the potential V applied to C. There was no photoelectric current when V was several volts positive. But, when V was dropped to about 2 V *positive*, a small current was observed. This indicated that the photoelectrons were not simply *freed* from the cathode but that *some were ejected with sufficient velocity to enable them to overcome the retarding potential* of 2 V. The current increased when V was reduced to zero and as V was made negative but attained a saturation value after V had reached some 15 or 20 V negative. These data are shown diagrammatically in Fig. 6.5b. V_0 is the positive potential that was required to prevent the escape of electrons.

The determination of e/m was made essentially as follows. Let a negative potential V, large compared with V_0, be applied to the cathode C. The photoelectron, on reaching the anode, will have a kinetic energy given approximately by

$$eV = \tfrac{1}{2}mv^2$$

where v is its velocity on reaching A (Fig. 6.5a). Assuming that after leaving A the electron moves in a uniform magnetic field, the circular path it follows is determined by the equation

$$evB = \frac{mv^2}{R}$$

where B is the magnetic field intensity just necessary to cause the electron to reach P_2, and R is the radius of the corresponding circular path, determined from the geometry of the apparatus. From the above two equations we have

$$\frac{e}{m} = \frac{2V}{B^2R^2} \tag{6.10}$$

6.7 What Is the Photoelectric Mechanism?

Efforts to imagine a mechanism for photoelectric emission on the basis of the classical electromagnetic theory of light encountered great difficulties. It was early assumed that metals contain a number of *free electrons* in the spaces between the atoms and that electric currents are due to a drift of these free electrons caused by the electric field. Electromagnetic waves entering a metal would agitate the free electrons, and it could readily be imagined that occasionally an electron near the surface of the metal would be given enough energy to eject it entirely.

If photoele trons originated in this way, however, it would be expected that they would emerge with smaller kinetic energies when the incident light is faint than when it is strong, whereas the experimental fact is that their kinetic energies are *independent of the intensity* of the light. On the other hand, their energies *do* depend on the *frequency* of the light.

The latter two experimental facts taken together led at one time to the proposal of an alternative hypothesis, viz., that the photoelectrons come from inside the atoms, and that their release is of the nature of a resonance phenomenon. The light might function as a trigger releasing the electron from inside an atom in which it already possessed energy of considerable magnitude. Light of a given frequency would then release only those electrons which were tuned to that frequency, and the atomic mechanism might conceivably be such that the energy of the expelled electron would be proportional to the frequency. It was hard to believe, however, that the atoms contained resonating systems of all frequencies for which photoelectric emission was observed. Furthermore, the fact remained to be explained that the number of the photoelectrons is proportional to the intensity of the light.

The experimental facts thus seem to demand the conclusion that the photoelectric energy comes directly from the energy of the incident light. Another difficulty is then encountered, however. It is hard to understand how so much energy can be absorbed by a single electron. According to the electromagnetic theory of light, the energy of light waves is distributed equally over the wavefront. Let it be assumed, for purposes of calculation, that the electron can absorb as much of this energy as falls on the area occupied by one atom lying on the surface of the metal. Then it can be calculated that the time required for a photoelectron to absorb the maximum energy of emission, $\frac{1}{2}mv_m^2$, from faint light of sufficient intensity to cause an easily measurable photoelectric emission from sodium would be more than 100 days. The situation is improved if the electron is assumed to vibrate inside the atom in exact resonance with monochromatic light, since it can be shown that the electron can then manage to absorb as much of the incident energy as falls upon a considerable fraction of a square wavelength. Even so, however, the calculated time exceeds 1 min. Thus, if the electron obtains its energy by an ordinary process of absorption, there should be an appreciable lag between the beginning of illumination and the start of the photoelectric current. Precise measurements showed, however, that, if such a lag exists, it is less than 3×10^{-9} s.

A satisfactory explanation of the photoelectric mechanism was first proposed by Einstein in 1905 by his extension to the photoelectric process of Planck's concept that interchanges of energy between radiation and matter take place in energy quanta of magnitude $h\nu$ (Sec. 5.7). Einstein assumed that a whole quantum $h\nu$ of radiant energy was absorbed by a single electron. This implies that not only is radiant energy emitted in quanta of energy $h\nu$ but also that this radiant energy is transmitted as a sufficiently localized energy packet to be absorbed as a unit. Such a quantum of electromagnetic radiation is called a *photon*. Einstein proposed further that if a minimum energy ϕ_0 is required to liberate an electron from the metal, the maximum kinetic energy K_{max} of photoelectrons freed by photons of energy $h\nu$ should be given by

$$K_{max} = h\nu - \phi_0 \tag{6.11}$$

This is the *Einstein photoelectric equation*. Only qualitative data were available at that time showing that his equation gave results of the right order of magnitude, but subsequently the equation received experimental verification (Fig. 6.6). Its validity has been extended to the x-ray region, where the frequencies are several thousand times the frequencies of visible light. Einstein's photoelectric equation played an enormous part in the development of the modern quantum theory.

By regarding light as a rain of corpuscles or photons of energy $h\nu$ and assuming that a whole photon can be absorbed by an electron, Einstein at once had an explanation of the principal properties of the photoelectric effect in metals. There could be no observable time lag in the photoelectric process; the current would be exactly proportional to the light intensity, i.e., to the number of photons striking per second;

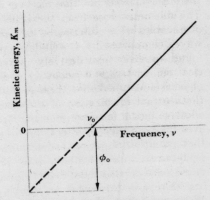

Fig. 6.6 Variation of maximum energy of photoelectrons with frequency of the exciting radiation.

and the velocities of the photoelectrons would be entirely independent of the light intensity, since the absorption of each photon would be a process quite independent of the absorption of other photons. The Einstein equation would follow naturally.

The difficulties with such a radical theory of light, however, are many. For one thing, if we regard light as a "shower" of photons, what can possibly be the meaning of *frequency?* There is nothing periodic about a falling raindrop, for example. As a matter of fact, in order to find the frequency of a beam of light, we *measure* (1) the velocity c of the light and (2) its wavelength λ *on the assumption that light consists of waves,* and then we *compute* the frequency as $\nu = c/\lambda$. *Thus we rely on the wave theory of light to give us the energy value $h\nu$ of a photon!* And there still remains the phenomenon of interference, which, since its discovery by Young in 1802, has defied explanation on any other basis than by assuming light to have wave properties. However, the experimental facts of photoelectricity are just as cogent as the phenomena of interference, and these *cannot be explained on the basis of the classical wave theory of light.*

Thus in 1920 the physicist could sum up the situation about like this: on one side of a seemingly impenetrable barrier is to be found a group of phenomena—such as interference, polarization, indeed, the whole electromagnetic theory—according to which he should say that light *must consist of waves.* On the other side of the fence is to be found another group—the photoelectric effect, and other phenomena which we shall consider in subsequent chapters—according to which he should say that light *must be corpuscular.* The situation thus created was perhaps the most puzzling one that has ever arisen in the whole history of physics. The dilemma arose from the belief of classical physicists that certain observables were either *waves* or *particles.* The abstract idealization of waves, supported by a well-developed mathematical theory, had been found to give quantitatively correct predictions for the behavior of electromagnetic (and acoustic) radiation in so many situations that physicists assumed that these radiations were indeed waves. Similarly the abstract idealization of the classical particle had proved to be an effective model for describing the motions of balls and electrons. However the photoelectric effect and other phenomena to be discussed show that radiation also has particle properties, and in Chap. 11 we present evidence that electrons have wave properties. A modern point of view might be expressed thus: there is in nature nothing which in every aspect can be regarded as either an ideal wave or as an ideal particle; rather, all the quantities of physics which were classically divided into wave or particle classes have both wave and particle properties.

The particle and wave aspects of electromagnetic radiation are not shown simultaneously, nor do they ordinarily appear in the same experiments, although there are experimental results which can be explained either in terms of a wave description or of a particle description. In 1928 Bohr suggested that the wave and particle properties of radiation are *complementary* and to explain the results of an experiment we must choose one interpretation or the other; it is not possible to apply both descriptions at the same time. This idea is known as the *Bohr complementarity principle*. It is applicable to electrons and other classical particles as well.

6.8   Properties of Photoelectric Emission

From Einstein's theory there arise several conclusions which have been verified by subsequent experiments.

1. The photoelectric current is directly proportional to the intensity of illumination on the emitting surface so long as there is no change in the spectral quality of the incident radiation. Experiments of Elster and Geitel in 1916 showed that this proportionality holds rigorously over an intensity range of 5×10^7 to 1. For a single frequency of incident light the number of photoelectrons emitted is proportional to the number of photons reaching the surface.

2. There is essentially no time lag between the arrival of the first photons and the emission of the photoelectrons. Measurements of Lawrence and Beams in 1927 showed that this lag is less than 3×10^{-9} s.

3. The maximum kinetic energy K_{max} of the photoelectrons ejected from a surface by a monochromatic beam increases with the frequency of the incident radiation as predicted by Eq. (6.11). A plot of K_{max} as a function of ν is a straight line with slope h (Fig. 6.6) and an intercept on the frequency axis of $\nu_0 = \phi_0/h$, as was shown by Millikan in 1916. Here ν_0 is the photoelectric threshhold frequency below which no photoelectrons are emitted. The critical frequency ν_0 has been found to vary considerably with the state of the surface, as does the photoelectric current in general. Usually ν_0 lies in the ultraviolet, but for the alkali metals and for barium and strontium it lies in the visible region; for potassium ν_0 lies in the red, and for cesium even in the infrared.

4. Most of the photoelectrons have kinetic energies less than K_{max}. Figure 6.7 shows data on the energy distribution of photoelectrons from aluminum published by Richardson and Compton in 1912. They used a small photoemitter located at the center of a spherical conductor, to which they applied the retarding potential V and measured the photocurrent as a function of retarding potential for several frequencies in the ultraviolet. For $\lambda = 2 \times 10^{-7}$ m ($\nu = 1.5 \times 10^{15}$ s^{-1}) no photoelectrons reach the spherical conductor when the retarding potential exceeds 2.3 V, and so K_{max} for this frequency is 2.3 eV.

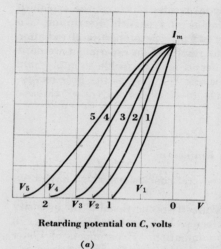

Curve	Wavelength, cm	Critical potential, V
1	0.0000313	$V_1 = 0.90$
2	0.0000275	$V_2 = 1.30$
3	0.0000254	$V_3 = 1.50$
4	0.000023	$V_4 = 1.90$
5	0.000020	$V_5 = 2.30$

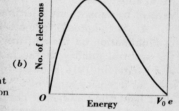

(b)

Fig. 6.7 (a) Variation of photoelectric current with retarding potential. (b) Energy distribution of photoelectrons.

5. *Variation of temperature* usually has little or no effect upon the photoelectric current, so long as the temperature does not exceed several hundred degrees Celsius and so long as no change occurs in the crystal structure or in the physical state of the metal. The alkalis, however, form an exception to this rule.

6. If *polarized light* is used, differences are commonly found as the plane of polarization is rotated, except, of course, at normal incidence. The effect is complicated, and its cause is in doubt. Especially interesting is the *selective effect* in the alkali metals. Over certain ranges of wavelength, the photoelectric current from these metals is much greater when the electric vector in the light has a component perpendicular to the surface than when it is parallel to the surface; in the case of certain sodium-potassium alloys, the ratio of the currents in the two cases may be 10:1 or 20:1 or even more.

6.9 Thermionic Emission

The thermionic current from an emitter increases very rapidly with increasing temperature. It depends also upon geometrical factors, which determine space-charge effects, and upon the potential of the emitter relative to its surroundings. From a given emitter operating at a given temperature, however, the thermionic current cannot be made to exceed a certain limiting value I, given by (Sec. 21.8)

$$I = AT^2 e^{-\phi/kT} \tag{6.12}$$

in which ϕ and A are constants depending on the emitting substance and on the state of its surface.

The quantity ϕ represents an energy; in the theoretical derivation it represents the *heat of emission* of the electrons (or thermions) at the absolute zero of temperature, i.e., the amount of energy required to extract an electron at 0°K. As calculated from thermionic data, ϕ ranges from 2 to 6 eV; ϕ is called the *work function.*

It was early assumed that thermions and photoelectrons come from the same source inside the metal, the two types of emission differing chiefly in the mechanism by which an electron acquires sufficient energy to enable it to escape from the emitter. On this view, the quantity ϕ in the thermionic equation ought to be at least approximately the same as $\phi = h\nu_0$ in the Einstein photoelectric equation.

It has not been easy to check this equation by experiment, owing to the many spurious disturbances inherent in both photoelectric and thermionic research. Reasonably comparable measurements have, however, been accomplished on a number of the more stable metals; a few values are as follows:

Metal	$h\nu_0$, eV	ϕ, eV
Platinum	6.30	6.27
Tungsten	4.58	4.52
Silver	4.73	4.08
Gold	4.82	4.42

The agreement between the values of the work function determined photoelectrically and those determined thermionically warrants the hypothesis that the photoelectrons and the thermionic electrons have a common origin.

Problems

1. An electron is accelerated through a potential difference V. It then passes between two long parallel plates with electric intensity **E** between

them.   Perpendicular to E is a magnetic induction **B** of strength such that the electron is not deflected as it passes between the plates.   Show that $e/m = E^2/2VB^2$.

2.   Electrons with a maximum kinetic energy of 2.3 eV are ejected from a surface by ultraviolet photons with $\lambda = 200$ mμ.   Calculate the energy of the incident photons in joules and electron volts.   What is the smallest energy which can free an electron from the surface?   What is the wavelength of a photon with this energy?

Ans: 9.9×10^{-19} J; 6.2 eV; 3.9 eV; 320 mμ

3.   Find the maximum kinetic energy of photoelectrons ejected from a potassium surface, for which $\phi_0 = 2.1$ eV, by photons of wavelengths 2000 and 5000 Å.   What is the threshold frequency ν_0 and the corresponding wavelength?

Ans: 4.1 eV, 0.4 eV; 5.08×10^{14} Hz, 5900 Å

4.   The photocurrent of a cell can be cut to zero by a minimum retarding potential of 2 V when monochromatic radiation of $\lambda = 250$ mμ is incident.   Find the work function, the stopping potential for $\lambda = 200$ mμ, and the threshold wavelength.

5.   An experiment on the photoelectric effect of cesium gives the results that the stopping potentials for $\lambda = 4358$ and 5461 Å are respectively 0.95 and 0.38 eV.   From these data, find h, the threshold frequency, and the work function of cesium.

Ans: 4.6×10^{14} Hz; 1.9 eV

6.   An experiment on the threshold wavelength for photoelectric emission from tungsten yields the value 270 mμ.   Find the photoelectric work function in joules and in electron volts.   Compute the maximum energy of photoelectrons ejected by photons of $\lambda = 200$ mμ.

Ans: 7.34×10^{-19} J; 4.59 eV; 1.61 eV

7.   An electron with kinetic energy K enters a uniform magnetic field of induction **B** which is perpendicular to the velocity of the electron.   If B is 0.005 Wb/m^2, find the radius of the electron's path when K is (a) 200 eV, (b) 200 keV, (c) 200 MeV.

Ans: (a) 9.54 mm; (b) 33 cm; (c) 134 m

8.   Compute the separation in angstrom units of the outer two lines of a normal Zeeman pattern for spectral lines of 3000 and 6000 Å wavelength in a magnetic induction of 1.8 Wb/m^2.

9.   A cathode-ray tube deflects 1500-eV electrons by use of a magnetic field which may be taken as uniform over a distance of 2 cm and zero elsewhere.   If the fluorescent screen is 25 cm beyond the deflecting coils, find the beam deflection on the screen for $B = 0.002$ Wb/m^2.

Ans: 8 cm

10. An electron is projected along the axis of a cathode-ray tube midway between two parallel plates with a speed of 1.5×10^7 m/s. The plates are 1.2 cm apart, 3 cm long, and have a potential difference of 120 V between them. Find (a) the angle with the axis at which the electron leaves the plates, (b) the distance of the electron from the axis as it leaves the plate, and (c) the distance from the axis at which it strikes a fluorescent screen 20 cm beyond the plates.

Ans: (a) 0.23 rad; (b) 0.35 cm; (c) 5.0 cm

11. A cathode-ray tube has electrostatic deflection plates 2 cm square, separated by 0.5 cm and 25 cm from the fluorescent screen. If the electrons have 1500 eV of kinetic energy when they pass between the plates, calculate the approximate beam deflection on the screen if 20 V is applied between deflector plates.

Ans: 7 cm

12. Show that the path of an electron after it has passed between the plates of Fig. 6.4 is a straight line which, if extended back, originates at the midpoint of the volume between the plates as though the incident electron went undeflected to that point and then changed direction discontinuously.

13. A photon with 13,600 eV of energy interacts with a hydrogen atom at rest and ejects the electron photoelectrically in the direction in which the photon was traveling. If 13.6 eV is required to eject the electron, find the speed of the photoelectron and the momentum and energy of the recoiling proton.

Ans: 6.7×10^7 m/s; 6.3×10^{-23} kg-m/s; 7.4 eV

14. In an apparatus designed by Busch in 1922 for measuring e/m, electrons fall through a potential difference V and pass through a conical aperture of half-angle ϕ. They enter a uniform axial magnetic field B supplied by a solenoid with their velocity vectors making an angle ϕ with the axis and spiral down the tube. If B is increased from zero,

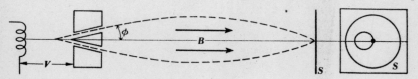

Fig. 6.8 Busch's arrangement for measuring e/m.

there come values of B at which the electrons refocus at the axis of the screen S. Show that the electrons are first focussed on S when $e/m = (8\pi^2 V \cos^2 \phi)/B^2 L^2$.   (See Fig. 6.8.)

15.   Prove that a free electron cannot absorb a photon photoelectrically by showing that the conservation of energy and the conservation of momentum cannot be satisfied simultaneously in such a process.

16.   (a) In Thomson's experimental arrangement for finding e/m narrow beam of electrons of velocity v passes between parallel plates of length l and separation d.   If a potential difference V is applied to the plates, show that the impact point of the beam on a fluorescent screen a distance D beyond the plates moves through a distance

$$s = \frac{e}{m} \frac{V}{d} \frac{l(D + l/2)}{v^2}$$

(b) Prove that the deflection of the beam is the same as though the electrons suffered a single abrupt deflection through an angle $\theta = s/(D + l/2)$ a distance $D + l/2$ from the screen (Fig. 6.4b).

chapter seven

X-rays

Probably no subject in all science illustrates better than x-rays the importance to the entire world of research in pure science. Within 3 months after Röntgen's fortuitous discovery, x-rays were being put to practical use in a hospital in Vienna in connection with surgical operations. The use of this new aid to surgery soon spread rapidly. Since Röntgen's time, x-rays have completely revolutionized certain phases of medical practice. However, had Röntgen deliberately set about to discover some means of assisting surgeons in reducing fractures, it is almost certain that he would never have been working with the evacuated tubes, induction coils, and the like, which led to his famous discovery.

In many other fields of applied science, both biological and physical, important uses have been found for x-rays. Transcending these uses in applied science are the applications of x-rays to such problems as the atomic and the molecular structure of matter and the mechanism of the interaction of

171

radiation with matter.   X-rays provide us with a kind of supermicroscope, by means of which we can "see" not only atoms and their arrangement in crystals but also even the interior of the atom itself.   Röntgen's discovery must be ranked among the most important.

7.1   The Discovery of X-rays

In the autumn of 1895, Wilhelm Konrad Röntgen, professor of physics at Würzburg, was studying the discharge of electricity through rarefied gases.   A large induction coil was connected to a rather highly evacuated tube (Fig. 7.1), the cathode C being at one end and the anode A at the side.   The tube was covered "with a somewhat closely fitting mantle of thin black cardboard."   With the apparatus in a completely darkened room, he made the accidental observation that "a paper screen washed with barium-platino-cyanide lights up brilliantly and fluoresces equally well whether the treated side or the other be turned toward the discharge tube."   The fluorescence was observable 2 m away from the apparatus. Röntgen soon convinced himself that the agency which caused the fluorescence originated at that point in the discharge tube where the glass walls were struck by the cathode rays.

Realizing the importance of his discovery, Röntgen at once proceeded to study the properties of these new rays—the unknown nature of which he indicated by calling them x-rays.   In his first communications he recorded, among others, the following observations:

1.   All substances are more or less transparent to x-rays.   For example, wood 2 to 3 cm thick is very transparent.   Aluminum 15 mm thick "weakens the

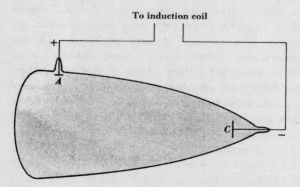

To induction coil

Fig. 7.1   Type of tube with which Röntgen discovered x-rays.

effect considerably, though it does not entirely destroy the fluorescence." Lead glass is quite opaque, but other glass of the same thickness is much more transparent. "If the hand is held between the discharge tube and the screen the dark shadow of the bones is visible within the slightly dark shadow of the hand."

2. Many other substances besides barium-platino-cyanide fluoresce—calcium compounds, uranium glass, rock salt, etc.

3. Photographic plates and films "show themselves susceptible to x-rays." Hence, photography provides a valuable method of studying the effects of x-rays.

4. X-rays are neither reflected nor refracted (so far as Röntgen could discover). Hence, "x-rays cannot be concentrated by lenses."

5. Unlike cathode rays, x-rays are not deflected by a magnetic field. They travel in straight lines, as Röntgen showed by means of pinhole photographs.

6. X-rays discharge electrified bodies, whether the electrification is positive or negative.

7. X-rays are generated when the cathode rays of the discharge tube strike any solid body. A heavier element, such as platinum, however, is much more efficient as a generator of x-rays than a lighter element, such as aluminum.

It is a tribute to Röntgen's masterly thoroughness that most of the basic properties of x-rays were described in the paper first announcing the discovery. Intense interest was aroused, and work on x-rays began at once in many laboratories both in America and in Europe. This early work is beautifully illustrative of the qualitative phase of development of a typical field of physics.

7.2 Production and Detection of X-rays

Until 1913, tubes for the production of x-rays were similar to a form suggested by Röntgen. A residual gas pressure of the order of 10^{-3} mm Hg provides, when voltage is applied, a few electrons and positive ions. These positive ions, bombarding the cathode, release electrons, which, hurled against the anode, give rise to x-rays. A curved cathode converges the electrons into a focal spot of desired shape and size. In this type of tube, known as the *gas tube*, the anode current, applied voltage, and gas pressure are more or less interdependent, and it is essential that the gas pressure be maintained at the desired value. Various ingenious devices were introduced for accomplishing this. In 1913, however, an important improvement was introduced by Coolidge. He evacuated the tube to the highest attainable vacuum and incorporated a hot spiral filament of tungsten to serve as a source of electrons. The filament was heated by an adjustable current from a battery. Thus the electron current could be controlled independently of the applied voltage.

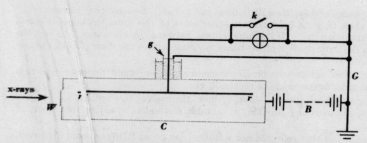

Fig. 7.2 An ionization chamber for measuring the intensity of a beam of x-rays or other ionizing radiation.

For quantitative measurements, the ionization method was early adopted. The discharging effect of x-rays upon charged bodies was traced to ionization of the molecules of the surrounding gas. The effect was found to increase rapidly with density and also to depend on the nature of the gas, the following being increasingly active in the order: H_2, CO, air, CO_2, ether vapor, CS_2. At first the rate of discharge of an electroscope was used in measuring the intensity of an x-ray beam, but later an *ionization chamber* was introduced. This is shown schematically in Fig. 7.2. C is a metal tube several centimeters in diameter, from about 20 to 100 cm long, and closed at both ends except for an opening or window W, over which may be placed a thin sheet of cellophane or aluminum for admitting the x-rays. A rod rr suitably supported by *good* insulating material, such as amber or quartz, is connected to an electrometer or a vacuum-tube amplifier. An electric field is maintained between the rod rr and the cylinder C by a battery B. The chamber may be filled with a heavy gas to make the arrangement more sensitive; argon or methyl bromide is often used. When x-rays enter the window W, the gas within the cylinder is made conducting, and because of the electric field between the cylinder and the rod, the latter acquires a charge at a rate which is a measure of the intensity of the x-ray beam. Nowadays a Geiger counter or a scintillation detector with appropriate electronic circuits may replace the ionization chamber.

7.3 *Wavelengths of X-rays*

Although the hypothesis that x-rays are a form of electromagnetic radiation was early suggested, it proved to be difficult to establish. Evidence for the *diffraction* of x-rays was reported in 1899 by Haga and Wind, who concluded that the wavelengths of x-rays must be of the order of

10^{-10} m, but their experiments were regarded by many as inconclusive. They used wedge-shaped slits only a few microns wide and observed a slight broadening of the image on a photographic plate. The first practical method for resolving x-ray beams according to wavelength developed out of a brilliant suggestion by Laue. The order of magnitude of x-ray wavelengths, as revealed by the diffraction experiment described above, is the same as the order of magnitude of the spacing of the atoms in crystals. Laue suggested, therefore, that a crystal, with its regular array of atoms, might behave toward a beam of x-rays in somewhat the same way as a ruled diffraction grating behaves toward a beam of ordinary light. Assume that plane electromagnetic waves traveling in a given direction fall upon a crystal. Each atom scatters some of the incident radiation. The wavelets scattered by different atoms combine, in general, in all sorts of phases and so destroy each other by interference. Laue argued, however, that for certain wavelengths and in certain directions the wavelets should combine in phase and so produce a strong diffracted beam. It would be expected, therefore, that such diffracted beams might be observed upon passing a heterogeneous x-ray beam through a crystal.

Such an experiment was performed by Friedrich and Knipping in 1912. A narrow pencil of x-rays was allowed to pass through a crystal beyond which was a photographic plate. After an exposure of many hours, it was found on developing the plate that, in addition to the interior central image where the direct beam struck the plate, there were

Fig. 7.3 Laue photograph of a tungsten crystal.

present on the plate many fainter but regularly arranged spots, indicating that the incident x-ray beam had been diffracted by the crystal in certain special directions, just as Laue had predicted (Fig. 7.3). In their original paper, Friedrich, Knipping, and Laue, from an analysis of a series of photographs of a crystal of zinc blende oriented at various angles with respect to the incident pencil, concluded that there were present in the x-ray beam wavelengths varying between 1.27 and 4.83 $\times$ 10^{-11} m. This result supported the *two* postulates underlying the experiment: (1) that x-rays are electromagnetic waves of definite wavelengths, and (2) that the atoms of a crystal are arranged in regular three-dimensional order, as suggested by the external symmetry of crystals.

This experiment marked the beginning of a new era in x-rays. Two new fields of investigation were at once opened up: (1) in x-rays, the study of spectra and the use of homogeneous beams in experiments on scattering, absorption, etc.; (2) the study of the arrangements of atoms or molecules in crystals. In the following sections, we shall confine our discussion to some of the more important aspects of the former field.

7.4 Bragg's Law

Immediately following the announcement by Friedrich, Knipping, and Laue of their successful experiment many investigators took up the study of the new phenomenon. A simple way of looking at the process of diffraction by a crystal grating was proposed by W. L. Bragg. He pointed out that through any crystal a set of equidistant parallel planes can be drawn which, among them, pass through *all* the atoms of the crystal. Indeed, a great many such families of planes can be drawn, the planes of each family being separated from each other by a characteristic distance. Such planes are called *Bragg planes*, and their separations, *Bragg spacings*. Traces of five families of Bragg planes are shown in Fig. 7.4.

If plane monochromatic waves fall upon the atoms in a Bragg plane, a wavelet of scattered radiation spreads out from each atom in all directions. There is just one direction in which, irrespective of the atomic distribution in the plane, the scattered wavelets meet in the same phase and constructively interfere with each other, viz., the direction of *specular reflection*[1] from the plane for which there is no path difference for radiation scattered by atoms in the plane. This follows from the ordinary Huygens construction as used for the reflection from a mirror. The

---

[1] For a regular array of atoms in one plane there are also other directions for which constructive interference could occur, but for these directions there is destructive interference of the wavelets from the set of parallel planes

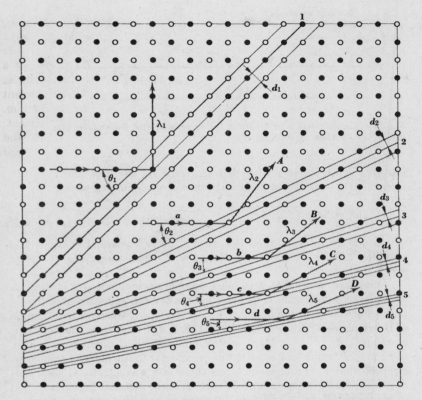

Fig. 7.4 Reflection of monochromatic beams of x-rays by a NaCl crystal when heterochromatic radiation is incident.

beam scattered in this direction may be thought of as "reflected" from the Bragg plane. But each Bragg plane is one of many regularly spaced, parallel planes; the beams reflected from these various parallel planes combine, in general, in different phases and so destroy each other by interference. Only if certain conditions of wavelength and angle of incidence of the beam on the planes are satisfied do the waves *from different planes* combine in the same phase and reinforce each other. In Fig. 7.5 the horizontal lines represent the traces of two successive Bragg planes spaced *d* apart. Consider a plane wave incident on these planes with θ the angle between the propagation direction of the incident beam and the planes; θ is known as the *glancing angle*, and the reflection condition requires that $\theta = \theta'$. For radiation scattered by the second plane to

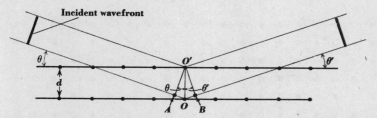

Fig. 7.5 Constructive interference of the waves scattered from atoms of a crystal occurs when $n\lambda = 2d \sin \theta$.

interfere constructively with that scattered by the first, the path difference AOB must be an integer number of wavelengths, whence

$$n\lambda = 2d \sin \theta \tag{7.1}$$

where n is an integer called the *order of the reflection*. Equation (7.1) together with the requirement that $\theta = \theta'$ constitutes *Bragg's law* for x-ray "reflection."

Suppose, now, that a parallel wave train containing a *continuous spectrum* of wavelengths is incident upon a crystal, as represented by the parallel arrows a, b, c, d in Fig. 7.4. In the figure, traces of five families of Bragg planes are shown, numbered 1, 2, 3, 4, 5, with their characteristic spacings d_1, d_2, ... Many other families of planes might be imagined, some perpendicular and some not perpendicular to the plane of the paper. Suppose that in the incident beam there is a wavelength λ_2 such that $n\lambda_2 = 2d_2 \sin \theta_2$, where n is an integer, d_2 is the distance between the set of planes numbered 2, and θ_2 is the glancing angle between the direction of the incident radiation and these planes. Then there will be reflected from this group of planes a beam A, of wavelength λ_2, which will proceed in the direction of the arrow A. Similarly, we may have reflected beams B, C, D, . . . in different directions in the plane of the paper, and also many other beams reflected from other families of planes in directions not in the plane of the paper. Each *Laue spot* in the experiment of Friedrich and Knipping may be interpreted as produced by such a reflected beam. In general, the most intense spots correspond to reflections from Bragg planes containing the greatest number of atoms on each plane.

7.5   The X-ray Spectrometer

W. H. Bragg and his son W. L. Bragg were responsible for the early development of the x-ray spectrometer, shown in Fig. 7.6a. X-rays from target T pass through two narrow slits S_1 and S_2 and fall at glancing angle θ on the cleavage face of a crystal K—rock salt, calcite, mica, gypsum, quartz, etc.—mounted on a table D, the angular position of which can be read. The reflected beam, which makes an angle 2θ with the incident beam, enters an ionization chamber C by means of which the intensity can be measured.

For photographic registration, the ionization chamber can be replaced by a photographic plate PP (Fig. 7.6b). With the crystal set at a glancing angle θ, the reflected beam will strike the plate at L (or at L', if the crystal is reversed). From the position O at which the direct beam strikes the plate, the distances OL and OA, and hence the angle 2θ, can be determined.

The distance d between the reflecting planes of a crystal such as NaCl is determined as follows. In the rock-salt crystal, Na and Cl ions occupy alternate positions in a cubic lattice similar to that in Fig. 7.4, which represents one plane of atoms. Taking the atomic weight of Cl as 35.45 and of Na as 22.99, we find the molecular weight of NaCl to be 58.44. Therefore, 58.4 kg of the NaCl contains $2N_A$ atoms, where N_A is Avogadro's number. Thus we find for the number of atoms n in

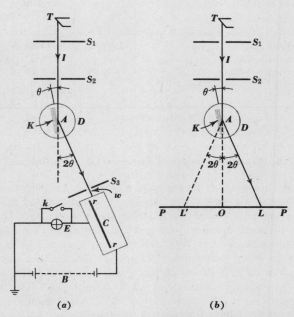

Fig. 7.6 An x-ray spectrometer using (a) ionization detection and (b) photographic detection.

1 m³ of rock salt $2N_A\rho/58.44$, where $\rho = 2164$ kg/m³, the density of crystalline NaCl.   If d is the distance between the center of one atom and the next along the edge of the cube, $1/d$ is the number of atoms in a row of atoms 1 m long, and $n = 1/d^3$.   From these two equations $d = 2.82 \times 10^{-10}$ m = 2.82 Å.

By using crystal gratings, x-ray wavelengths can be compared with a precision of a few parts in 10^5, which is more precise than Avogadro's number and the density and molecular weight of a typical crystal are known.   Therefore, x-ray wavelengths have often been reported in terms of X *units*, defined by taking the grating space of calcite at 18°C to be 3029.45 XU.   It is now known that the X unit is 0.0202 percent less than 10^{-13} m.   Table 7.1 lists some crystals commonly used in x-ray spectroscopy together with their grating spaces and the linear expansion per degree Celsius.

Table 7-1   *Grating spaces of some crystals used in x-ray spectroscopy*

Crystal	Grating Space d at 18°C		Change in d per Degree Celsius, XU or 10^{-13} m
	XU	Corrected, $\times 10^{-13}$ m	
Rock salt, NaCl	2814.00	2819.68	0.11
Calcite, $CaCO_3$	3029.45	3035.57	0.03
Quartz, SiO_2	4246.02	4254.60	0.04
Gypsum, $CaSO_2 \cdot SII_2O$	7584.70	7600.0	0.29
Mica	9942.72	9962.8	0.15

7.6   *Monochromatic Characteristic Radiation*

With a beam of x-rays from a platinum target incident on the cleavage face of a rock-salt crystal, W. H. Bragg rotated the crystal in steps of $\Delta\theta$ and the ionization chamber in steps of $2\Delta\theta$.   He plotted the curve of ionization current against glancing angle θ and found that the x-ray intensity did not vary uniformly with angle but rose at certain angles to a sharp maximum.   A curve similar to that shown in Fig. 7.7 was obtained.   A group of three maxima, a_1, b_1, and c_1, was observed at the respective angles θ of 9.9, 11.6, and 13.6°.   A second group of three maxima, a_2, b_2, and c_2, was observed at approximately double these angles.   Bragg interpreted the maxima a_2, b_2, and c_2 as second-order

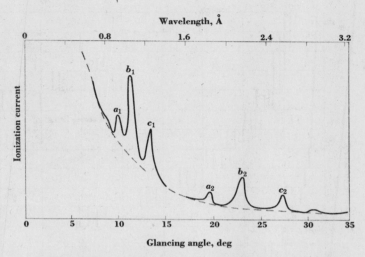

Fig. 7.7 Bragg's curve for the energy distribution in an x-ray spectrum, showing the characteristic lines *a, b, c.*

reflections of the lines a_1, b_1, and c_1. He computed their wavelengths, taking $n = 1$ for the lines a_1, b_1, and c_1 and $n = 2$ for the second-order lines a_2, b_2, and c_2, and obtained 0.97, 1.13, 1.32 Å, respectively.

Curves similar to Fig. 7.7 were obtained with other crystals, the only difference being that the maxima occurred at different glancing angles, indicating that each crystal had a characteristic grating space *d.* Bragg convinced himself, however, that these respective maxima for different crystals always represented the same monochromatic radiation, since, for example, the absorption in aluminum of peak b_1 was always the same, whatever the crystal used. In short, the peaks of the curve in Fig. 7.7 represent *spectral lines* the wavelengths of which are *characteristic of the target emitting the rays.* These monochromatic lines are superimposed on a *continuous* spectrum represented by the partially dotted line in the figure. Curves of the type shown in Fig. 7.7, therefore, represent the distribution of *energy* in the x-ray spectrum, continuous and characteristic combined, of an element.

The development of the Bragg spectrometer made it possible to make precise measurements of x-ray wavelengths. In 1913 Moseley undertook a systematic study of x-ray spectra emitted by targets of a number of elements of medium and high atomic number. His results were soon supplemented by data from other observers. X-ray spectra of a number of different elements are shown in Fig. 7.8. For each ele-

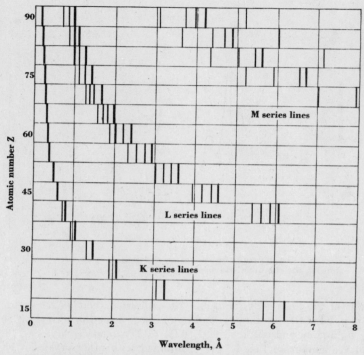

Fig. 7.8 Wavelengths of some of the more prominent characteristic lines; only a small fraction of *L* and *M* lines are shown.

ment the characteristic spectrum has a few lines at a relatively short wavelength, then an extended region in which there are no x-ray lines, followed in turn by a region in which there are a large number of lines, and so forth. The shortest wavelength and most penetrating lines emitted by an element are grouped in a narrow wavelength interval and are called the *K* lines; those in the next shorter wavelength region are called the *L* lines; for the heavier elements there are groups at still longer wavelengths called the *M* and *N* lines.

About 5 years before Moseley's work Barkla and his collaborators had clearly established the existence of characteristic radiation by absorption measurements.   A schematic diagram of their experimental arrangements is shown in Fig. 7.9.   A well-collimated beam of x-rays fell upon a block of material *E*, which absorbed the primary beam and

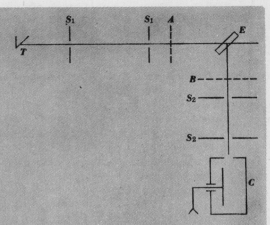

Fig. 7.9 Schematic arrangement for studying secondary radiation. Aluminum absorbing foils were inserted at *A* or at *B*.

emitted as secondary radiation its own characteristic x-rays. The fluorescent rays emitted at right angles to the primary beam were measured by an ionization chamber. The absorption coefficient of the radiation from the secondary emitter was determined by placing aluminum foils between *E* and the ionization chamber. A large number of materials were used as the secondary radiator; Barkla found that the greater the atomic weight, the more penetrating this secondary radiation. Even when the hardness of the primary beam was varied over a fairly broad range, the secondary radiation from a given material remained unchanged. However, when the voltage across the x-ray tube was reduced sufficiently, there came a point at which the characteristic radiation emitted by a radiator of high atomic number became much softer and more readily absorbed. With a further reduction in the hardness of the primary beam, the secondary radiation continued to have the same absorption coefficient. Thus, Barkla established the fact that secondary emitters of fairly high atomic number have characteristic radiation in at least two wavelength regions. He called the more penetrating rays the *K* radiation and the softer *L* radiation.

7.7 *Moseley's Law*

The *K* series of each element, as photographed by Moseley, appeared as two lines, the stronger of which was named $K\alpha$ and the weaker $K\beta$.

(Later measurements have shown that the $K\alpha$ line is actually a doublet and that the line originally known as $K\beta$ is also complex.) For any given element the $K\alpha$ line always has a longer wavelength than the $K\beta$ line. Moseley found, as Fig. 7.8 shows, that the wavelength of a given spectral line decreases as the atomic number of the target is increased. There is no indication of the periodicity so characteristic of many atomic properties.

Moseley discovered that the square root of the frequency of the $K\alpha$ line was nearly proportional to the atomic number of the emitter. Indeed, a plot of the square root of the frequency of almost any given x-ray line as a function of atomic number yields nearly a straight line. (Precise measurements have shown that the lines have a slight upward concavity.) Figure 7.10 shows the plot of the square root of the energies of the $K\alpha$ and $K\beta$ lines as a function of atomic number. A graph in which the square root of the frequency (usually multiplied by a constant) of an x-ray line is plotted as a function of atomic number is called a *Moseley plot.*

The frequency of the $K\alpha$ line for an element of atomic number Z can be represented very closely by the relation

$$\nu = 0.248 \times 10^{16}(Z - 1)^2 \qquad \text{Hz} \tag{7.2}$$

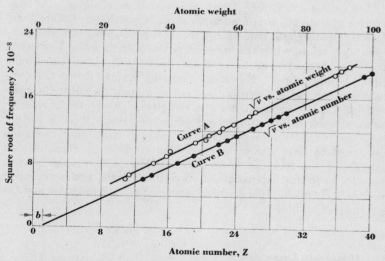

Fig. 7.10 Moseley's plot showing relation between the frequencies of x-ray lines and atomic number (Curve B), or atomic weight (Curve A).

For other lines it was found that an equation of the form

$$\nu = A(Z - \sigma)^2 \tag{7.2a}$$

gave reasonably good fits to experimental measurements. Here A and σ are constants for a given line. An explanation of Moseley's empirical relations was almost immediately available in terms of a theory developed by Bohr (Sec. 9.4).

One of the first important applications of Moseley's work was the resolution of several problems which were of great interest in connection with the periodic table. It was found that cobalt (at wt = 58.93) rather than nickel (at wt = 58.71) is element number 27. Similarly it was confirmed that argon (at wt = 39.948) should be placed before potassium (at wt = 39.102), as had been suggested by chemists. Still another contribution was the observation that element 43 was missing and that ruthenium has atomic number 44. The observation of the K-series characteristic spectrum of any element is generally regarded as conclusive proof of the presence of this element in the target material. Ever since Moseley's work, any claim for the discovery of a new element has available a definitive check in the form of the proper K x-ray spectrum, provided a large enough sample is available.

7.8   The Continuous X-ray Spectrum

Support for the hypothesis that electromagnetic radiation is emitted as photons of energy $h\nu$ came from the continuous x-ray spectrum. When high-energy electrons fall on a target, there is always a broad continuous spectrum of x-rays emitted as well as some characteristic lines. The general features of the continuous spectra from all sorts of target elements are essentially identical. Figure 7.11 shows the 1918 data of Ulrey for the continuous spectrum of wolfram (tungsten) for four potential differences applied to the x-ray tube. Here $I_\lambda \, d\lambda$ is the power emitted in the wavelength region between λ and $\lambda + d\lambda$.

Three features of the curves strike the eye immediately.

1. For each potential difference there exists a *short-wavelength limit;* i.e., the shortest wavelength emitted at a given tube potential difference is well defined, whereas on the long-wavelength side the intensity falls off gradually. In 1915 Duane and Hunt showed that the short-wavelength limit λ_{min} is inversely proportional to the potential difference V. Clearly the maximum kinetic energy K_{max} of an electron striking the target is Ve, and this is the largest energy that the electron can give up in producing electromagnetic radiation. If all this energy

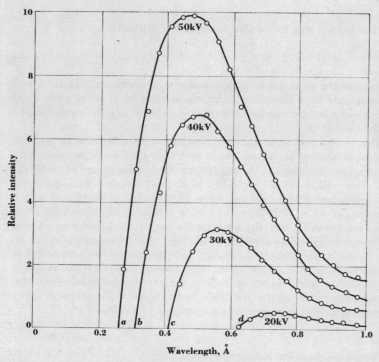

Fig. 7.11 Ulrey's curves for the distribution of energy in the continuous x-ray spectrum of tungsten.

goes to a single photon, the resulting frequency ν_{max} and the corresponding wavelength λ_{min} should, according to the Planck-Einstein theory, be given by

$$K_{max} = Ve = h\nu_{max} = \frac{hc}{\lambda_{min}} \tag{7.3}$$

This relation, known as the *Duane-Hunt law*, led to one of the earliest reliable determinations of Planck's constant by Duane and his collaborators from data on λ_{min} as a function of V.

Sometimes this phenomenon of the short-wavelength limit is called the *inverse photoelectric effect*. Here the kinetic energy of the incident electron is converted into radiant energy, while in the photoelectric effect radiant energy is converted, at least partially, into kinetic energy of an electron.   Whenever ϕ_0 is negligible compared with $h\nu$, Eqs. (6.11) and (7.3) take the same form.

Thus the quantum theory readily offers explanation of the short-wavelength limit of the continuous x-ray spectrum, a topic with which classical physics never dealt successfully.

2. The intensity of the continuous spectrum increases at all wavelengths as the potential difference across the tube is increased.   The total x-ray power emitted is roughly proportional to the square of the potential difference V across the tube and to the atomic number Z of the target material.   For ordinary x-ray potential differences the efficiency $\mathcal{E}$ of x-ray production, defined as the ratio of x-ray power emitted to the power dissipated by the electron beam, is given approximately by

$$\mathcal{E} \approx 1.4 \times 10^{-9} ZV$$

Most of the incident electrons in an x-ray tube do not lose all their kinetic energy in a single collision.   A typical electron undergoes several collisions in a thick target, emitting quanta which have wavelengths longer than λ_{min}.

3. As the potential difference V across the x-ray tube is increased, the wavelength at which the I_λ curve reaches its maximum moves toward shorter wavelengths.   Ulrey found for the data of Fig. 7.11 that $\lambda_m \sqrt{V} =$ const, where λ_m is the wavelength at which maximum energy is radiated.   However, data obtained with other tubes are not always consistent with this equation.   This is not surprising when one considers the complexity of the phenomenon.   Some of the incident electrons penetrate into the target and radiate photons which may be absorbed before they escape, and so forth.

Thus far we have been discussing the continuous x-rays from a thick target.   However, if the target is sufficiently thin—say of very thin gold foil—only a few of the electrons collide with atoms in it, most of them passing through the target undeviated.   Thus slowly moving electrons will not be present in a *thin* target to the same degree as in a *thick* one.   Accordingly, we expect that a greater proportion of the energy in the continuous spectrum from thin targets should lie near the λ_{min} limit than from thick targets.   This is in agreement with experiment.

In the continuous spectrum from a *very thin* target, experiment indicates, in agreement with the wave-mechanical computation of Sommerfeld, that the maximum of the energy-distribution curve occurs *at the limiting wavelength* λ_{min} *itself*.   On the short-wave side of λ_{min}, the curve drops abruptly to the axis of abscissas, whereas toward longer waves it falls nearly in proportion to $1/\lambda^2$, as illustrated in Fig. 7.12.   The curves for a thick target, as in Fig. 7.11, can be regarded as arising from

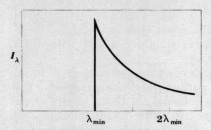

Fig. 7.12 Distribution of energy in the continuous x-ray spectrum from a very thin target.

the superposition of many elementary curves, such as that in Fig. 7.12, with various values of λ_{min}. In such a target it might be expected that there would be a *most probable* type of collision which would correspond to the peak or maximum of the energy-distribution curve.

7.9 X-ray Scattering, Classical

The phenomenon of x-ray scattering has played an important part in theories of modern physics and has been the object of many researches, both theoretical and experiment. The first theory, proposed by J. J. Thomson, considered a plane-polarized electromagnetic wave moving along the z axis with its electric vector in the x direction (Fig. 7.13).

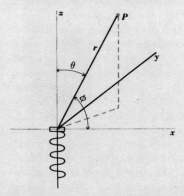

Fig. 7.13 An electron at the origin, set into harmonic oscillation by the electromagnetic wave from below, scatters a fraction of the incident intensity at angle θ.

If this wave interacts with a single free electron, the electron performs simple harmonic motion and in so doing reradiates. The acceleration of the electron is given by $a_x = -e\mathsf{E}/m$, where E is the electric intensity of the incident radiation. According to classical electromagnetic theory, an accelerated charge radiates electromagnetic waves with the electric vector at point P of amplitude

$$\mathsf{E}_\phi = \frac{ea \sin \phi}{4\pi\epsilon_0 c^2 r} \tag{7.4}$$

where r is the distance from the accelerated charge to P and ϕ is the angle between $\mathbf{r}$ and $\mathbf{a}$. When $a = -e\mathsf{E}/m$ is substituted, we have the magnitude of E_ϕ

$$\mathsf{E}_\phi = \frac{e^2 \mathsf{E} \sin \phi}{4\pi\epsilon_0 mc^2 r} \tag{7.5}$$

Since intensity is proportional to the square of the electric intensity, the ratio of the scattered intensity at P to the incident intensity is

$$\frac{I_\phi}{I} = \frac{\mathsf{E}_\phi{}^2}{\mathsf{E}^2} = \frac{e^4 \sin^2 \phi}{16\pi^2\epsilon_0{}^2 m^2 c^4 r^2} \tag{7.6}$$

If unpolarized radiation is incident on an electron, it is convenient to let $\mathbf{r}$ and the z axis fix the xz plane. The incident electric vector E can then be resolved into x and y components, which are equal on the average. For the x components $\sin \phi$ in Eq. (7.6) is equal to $\cos \theta$, while for the y components, $\sin \phi = 1$. The total intensity of the scattered wave at P is given by

$$I_s = \frac{e^4}{16\pi^2\epsilon_0{}^2 m^2 c^4 r^2}\left[\left(\frac{I}{2} \times 1\right) + \left(\frac{I}{2} \times \cos^2 \theta\right)\right]$$
$$= \frac{Ie^4}{32\pi^2\epsilon_0{}^2 m^2 c^4 r^2}(1 + \cos^2 \theta) \tag{7.7}$$

where θ is the angle between $\mathbf{r}$ and the z axis. Maxima in scattered intensity occur in the forward and backward directions; a minimum exists at $\theta = 90°$. By integrating Eq. (7.7) over a sphere of radius r surrounding the electron, the scattered power P_s is found to be

$$P_s = \int_0^\pi I_s\, 2\pi r^2 \sin \theta\, d\theta = \frac{8\pi}{3}\left(\frac{e^2}{4\pi\epsilon_0 mc^2}\right)^2 I \tag{7.8}$$

where $e^2/4\pi\epsilon_0 mc^2$ is the classical radius of the electron [Eq. (2.16)].

The ratio of the power scattered to the primary intensity is called the *scattering cross section* (or *scattering coefficient*) of the free electron, designated by σ_e.

$$\sigma_e = \frac{8\pi}{3}\left(\frac{e^2}{4\pi\epsilon_0 mc^2}\right)^2 \tag{7.9}$$

The scattering coefficient has the dimensions of area; of the radiation incident on a unit area, the electron scatters the amount which would fall on the area σ_e, which has the numerical value 0.666×10^{-28} m², or 0.666 barn (1 barn $= 10^{-24}$ cm² $= 10^{-28}$ m²). The Thomson (or classical) scattering coefficient is independent of wavelength. If each electron of an atom scatters independently, the atomic scattering cross section σ_a is the product of the atomic number of the atom and the electron cross section; $\sigma_a = Z\sigma_e$.

Thomson's theory predicts not only the intensity of the scattered beam but also its polarization. At $\theta = 0$ and $\theta = 180°$, the scattered beam from an unpolarized incident beam is unpolarized, but at $\theta = 90°$ the scattered beam is plane-polarized with electric vector in the y direction.

When a beam of x-rays of initial intensity I_0 passes through a thin layer of thickness Δx of any material, scattering by electrons in the layer changes the intensity by an amount

$$-\Delta I = \sigma I_0 \, \Delta x \tag{7.10}$$

where σ is called the *linear scattering coefficient* of the material. If each of the electrons of the material were to scatter independently of the others, the linear scattering coefficient would be

$$\sigma = n\sigma_e = \frac{8\pi}{3}\left(\frac{e^2}{4\pi\epsilon_0 mc^2}\right)^2 n \tag{7.11}$$

where n is the number of electrons per unit volume.

The ratio of σ to the density ρ of the scatterer is the *mass scattering coefficient* σ_m. For substances such as gases, the linear scattering coefficient depends, of course, on the density, but $\sigma_m = \sigma/\rho$ is constant. The first quantitative estimate of the number of electrons in an atom was made by Barkla and his coworkers in 1909 by applying Thomson's theory to data on the scattering of x-rays by carbon. They passed a beam of Mo $K\alpha$ x-rays through a graphite absorber and after correcting for photoelectric absorption found that the value of σ_m is approximately 0.02 m²/kg for carbon. If $\sigma = n\sigma_e$, then $\sigma_m = n\sigma_e/\rho$, and the number of electrons per kilogram is

$$\frac{n}{\rho} = \frac{\sigma_m}{\sigma_e} = \frac{0.02 \text{ m}^2/\text{kg}}{0.666 \times 10^{-28} \text{ m}^2/\text{electron}}$$
$$= 3 \times 10^{26} \text{ electrons/kg}$$

The number of carbon atoms per kilogram is the ratio of Avogadro's number N_A to the atomic weight A, or

$$\frac{N_A}{A} = \frac{6.02 \times 10^{26} \text{ atoms/kmole}}{12 \text{ kg/kmole}}$$
$$= 5 \times 10^{25} \text{ atoms/kg}$$

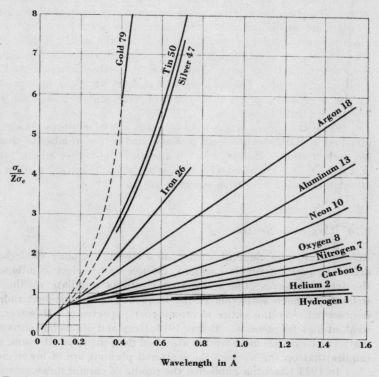

Fig. 7.14 Ratio of the scattering coefficients σ_a of various atoms to $Z\sigma_e$, the value which would pertain if each electron scattered independently with cross section σ_e. (*Reproduced by permission of D. Van Nostrand Company, Inc., from Fig. III-3 in A. H. Compton and S. K. Allison, "X rays in Theory and Experiment," 1935.*)

Hence the number of electrons per carbon atom must be 6. This determination was significant in the development of atomic theory, since it suggested the fact that the number of electrons per atom is less than the atomic weight and is perhaps about half of A.

Later measurements of scattering coefficients for $\lambda > 0.3$ Å showed that for heavier elements the scattering coefficient σ exceeds $n\sigma_e$ and hence the atomic scattering cross section σ_a is greater than $Z\sigma_e$ (Fig. 7.14). The excessive scattering is easily explained classically. When λ becomes comparable with the distances between electrons in an atom, they no longer scatter independently; the waves scattered by neighboring electrons are superposed more or less in the same phase, and so constructive interference occurs. As λ increases from a very small value, at first the inner (K) electrons scatter coherently; for longer wavelengths more electrons scatter nearly in phase. When λ is large compared to the atomic diameter, the relative phases of the waves from all Z electrons of an atom are small, and so the amplitude of the scattered wave from the atom is approximately proportional to Z and the scattered intensity to Z^2. The theory of coherent scattering was worked out by Rayleigh, the cooperative scattering of the atomic electrons often being referred to as *Rayleigh scattering*. The decrease in σ at short wavelengths is predicted by the relativistic, quantum-mechanical treatment of scattering due to Klein and Nishina (Sec. 24.2).

From the curves of Fig. 7.14 one sees what fortuitous choices were involved in Barkla's successful measurement of the number of electrons per carbon atom. Had he selected a scatterer of higher atomic number or used Cu $K\alpha$ radiation (1.5 Å) rather than Mo $K\alpha$ (0.707 Å), his calculation would have been substantially in error.

7.10   Compton Scattering

According to the classical theory of scattering, atomic electrons are driven to perform simple harmonic motion by the electric intensity of the incident waves; the accelerated electrons reradiate at the same frequency. Such scattering at the frequency of the incident radiation is observed over the entire electromagnetic spectrum. However, it is weak at high frequencies. Before 1914, Gray and others had shown that scattered x-rays are more readily absorbed than the incident beam, which implies that on the average the scattered photons are of lower energy.

In 1923 Compton published the results of careful measurements of the x-ray frequencies scattered by carbon atoms upon which monochromatic x-rays were incident. He found that the scattered beam contained *two frequencies*, one the same as that of the incident beam and the

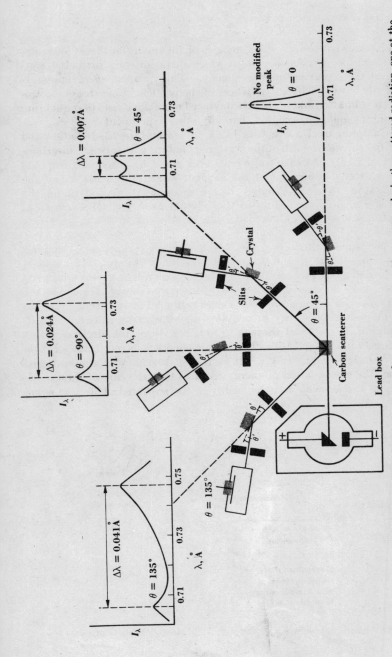

Fig. 7.15 The scattering of Mo $K\alpha$ x-rays ($\lambda = 0.707$ Å) at an angle θ gives rise to two peaks in the scattered radiation, one at the incident wavelength and the second at a wavelength greater by $\Delta\lambda = 0.024(1 - \cos\theta)$ Å. (*After A. W. Smith and J. N. Cooper,* "*Elements of Physics,*" *7th ed. Copyright 1964. McGraw-Hill Book Company. Used by permission.*)

193

second somewhat lower.   On the basis of the quantum theory of radiation, Compton derived a relation which quantitatively predicted scattering at the observed lower frequency.   This *Compton effect* represents one of the most conclusive evidences of the particle properties of electromagnetic radiation and probably convinced more physicists that quantum mechanics should be accepted than any other experiment.

Compton used a graphite block to scatter Mo $K\alpha$ radiation and measured the wavelength of the scattered photons with a Bragg spectrometer (Fig. 7.15).   At each finite scattering angle, he observed two peaks in the scattered beam, one at the incident wavelength 0.707 Å and the second at a wavelength longer by an amount $\Delta\lambda$ dependent on the scattering angle θ, according to the relation $\Delta\lambda = 0.024(1 - \cos\theta)$ Å. Compton subsequently showed that $\Delta\lambda$ is independent of the scattering material.

To explain the occurrence of the shifted component, Compton boldly applied the quantum picture of radiant energy.   He assumed that the scattering process could be treated as an elastic collision between a photon and an electron, governed by the two laws of mechanics, the conservation of energy and the conservation of momentum.   Let an incident photon of energy $h\nu$ collide with an electron initially at rest. The photon is scattered through an angle θ, while the electron recoils in a direction φ (Fig. 7.16).   The kinetic energy K given to the electron is $(m - m_0)c^2$ by Eq. (2.12).   If ν_θ is the frequency of the scattered photon, the conservation of energy requires that the sum of the kinetic energy of the electron and the energy of the scattered photon be equal

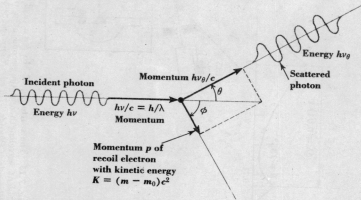

Incident photon

Energy $h\nu$

$h\nu/c = h/\lambda$
Momentum

Momentum $h\nu_\theta/c$

Energy $h\nu_\theta$

Scattered photon

θ

ϕ

Momentum p **of recoil electron with kinetic energy** $K = (m - m_0)c^2$

Fig. 7.16   Elastic collision of a photon with electron initially at rest.

to the energy of the incident photon, or

$$h\nu = h\nu_\theta + (m - m_0)c^2 \tag{7.12}$$

Each photon carries momentum equal to its energy $h\nu$ divided by the speed of light c, or h/λ. Since momentum is a vector quantity and is conserved, the x and y components (Fig. 7.16) must obey the equations

$$\frac{h\nu}{c} = \frac{h\nu_\theta}{c} \cos \theta + p \cos \varphi \tag{7.13a}$$

$$0 = \frac{h\nu_\theta}{c} \sin \theta + p \sin \varphi \tag{7.13b}$$

where p is the momentum of the recoil electron. (The plus sign appears with $\sin \varphi$ because it is assumed all angles are positive when measured counterclockwise from the x axis; thus in Fig. 7.16 φ is a negative angle.)

If we introduce the wavelength of the two photons through $\nu/c = 1/\lambda$ and recall that $p = m_0v/\sqrt{1 - (v/c)^2}$, we may rewrite Eqs. (7.13a, b) thus:

$$\frac{h}{\lambda} - \frac{h \cos \theta}{\lambda_\theta} = \frac{m_0v \cos \varphi}{\sqrt{1 - (v/c)^2}} \qquad \frac{h}{\lambda_\theta} \sin \theta = - \frac{m_0v \sin \varphi}{\sqrt{1 - v^2/c^2}}$$

Squaring these two equations and adding them eliminates φ and leaves

$$\frac{h^2}{\lambda^2} + \frac{h^2}{\lambda_\theta^2} - \frac{2h^2}{\lambda\lambda_\theta} \cos \theta = \frac{m_0^2v^2}{1 - (v/c)^2} = \frac{m_0^2c^2}{1 - (v/c)^2} - m_0^2c^2 \tag{7.14}$$

Similarly, Eq. (7.12) can be divided by c and written

$$\frac{h}{\lambda} - \frac{h}{\lambda_\theta} + m_0c = \frac{m_0c}{\sqrt{1 - (v/c)^2}} \tag{7.12a}$$

and squared to give

$$\frac{h^2}{\lambda^2} + \frac{h^2}{\lambda_\theta^2} - \frac{2h^2}{\lambda\lambda_\theta} + 2m_0ch \left(\frac{1}{\lambda} - \frac{1}{\lambda_\theta} \right) + m_0^2c^2 = \frac{m_0^2c^2}{1 - v^2/c^2} \tag{7.15}$$

Subtracting Eq. (7.14) from (7.15) yields

$$\frac{2h^2}{\lambda\lambda_\theta} (1 - \cos \theta) - 2m_0ch \left(\frac{1}{\lambda} - \frac{1}{\lambda_\theta} \right) = 0$$

or

$$\Delta\lambda = \lambda_\theta - \lambda = \frac{h}{m_0 c}(1 - \cos\theta)$$
$$= 0.02426(1 - \cos\theta) \quad \text{Å} \tag{7.16}$$

Thus, the wavelength shift predicted by Compton's theory is in excellent agreement with the experimental results. Compton scattering reveals the particle characteristics of photons.

The quantity $h/m_0 c$ is known as the *Compton wavelength* of the electron; it corresponds to the wavelength of a photon with energy equal to the rest energy $m_0 c^2$ of the electron. It is sometimes useful to extend this idea to other particles; thus the Compton wavelength of a proton of rest mass M_p is $h/M_p c$.

The Compton shift in wavelength, Eq. (7.16), depends on the scattering angle but is independent of the energy of the incident photon. On the other hand, the energy difference between incident and scattered photon increases rapidly as the energy of the incident photon is raised, as can be readily found by rewriting Eq. (7.16) in the form

$$\frac{1}{h\nu_\theta} - \frac{1}{h\nu} = \frac{1}{m_0 c^2}(1 - \cos\theta) \tag{7.16a}$$

In Compton's experimental work there appeared not only a modified line but also a line with the same wavelength as the primary. This unmodified line can be explained in terms of the photon concept as the result of a collision of a photon with a bound electron, held sufficiently tightly for an entire atom (or even an entire crystal) to recoil. In such a collision the predicted wavelength shift is given by replacing m_0 by the mass of the recoiling atom in Eq. (7.16) and is so small as to be unobservable for typical x-ray photons.

The substantially greater breadth of the modified line compared to the unmodified line can be explained satisfactorily in terms of the fact that most scattering electrons are not initially at rest but have a range of initial momenta and energies.

7.11   Compton Recoil Electrons

The recoil electrons predicted by Compton's theory were promptly looked for and found by C. T. R. Wilson and by Bothe. The kinetic energy $K = (m - m_0)c^2$ of the recoil electrons is given by Eq. (7.12)

$$K = (m - m_0)c^2 = h\nu - h\nu_\theta \tag{7.12b}$$

If ϵ represents $h\nu/m_0c^2$, it can be shown that

$$K = h\nu \, \frac{2\epsilon \cos^2 \varphi}{(1 + \epsilon)^2 - \epsilon^2 \cos^2 \varphi} \tag{7.17}$$

$$K = h\nu \, \frac{\epsilon(1 - \cos \theta)}{1 + \epsilon(1 - \cos \theta)} \tag{7.17a}$$

By 1927 Bless, using the magnetic spectrograph, showed that the observed values of K for the recoil electrons are in agreement with the theory. The relation between the scattering angles, φ for the electron and θ for the photon, is given by

$$\cot \varphi = (1 + \epsilon) \tan \frac{\theta}{2} \tag{7.18}$$

Compton's theory is based on the assumption that the recoil of the electron occurs simultaneously with the scattering of the photon and that both energy and momentum are conserved in the collision. The *simultaneity* and *conservation* requirements have been tested experimentally with great care by several observers by use of scintillation detectors and coincidence techniques. One detector fixes θ and measures $h\nu_\theta$ for the photon, while a second detector gives φ and K for the recoil electron. Within experimental error, for every Compton photon scattered an angle θ there is a recoil electron at angle φ given by Eq. (7.18) with kinetic energy given by Eq. (7.17b). The simultaneity time limit is less than 2×10^{-8} s in many experiments and in some cases less than 10^{-11} s.

Throughout this discussion the conventional assumption has been made that the recoil electron was initially unbound and at rest, an approximation which is tenable only when the kinetic energy acquired by the recoil electron is much greater than the energy with which it is bound to its atom. The effect of the binding is to reduce $\Delta\lambda$ slightly. In addition to the Compton process, in which the electron is freed from the atom, there are also *Smekal-Raman* processes, in which the frequency of the scattered radiation is changed by an amount which corresponds to transition between two bound states of the scattering atom.

The Compton or modified line is very broad at comparatively low frequencies. This breadth can be thought of as caused by the motion of the electrons in the atom. In our simple deduction of the Compton effect, the electron was assumed to be initially at rest; if it is assumed to have a component of velocity, positive or negative, in the direction of the incident radiation, the wavelength shift is different. Wave mechanics furnishes a probability distribution for the velocities of the atomic electrons, from which the broadening can be computed, in agreement with experiment. As ν is then increased further, the Compton line becomes narrower. Eventually, in any direction of scattering other than that of the incident beam, the unmodified line becomes weaker than the Compton line, sooner at large angles of scattering than at small angles, and sooner for heavy atoms than for light ones. Finally, only the modified line remains in appreciable intensity.

7.12   The Nature of Electromagnetic Radiation

The question as to the true nature of electromagnetic radiation arises in its sharpest form in contemplating the contrasting properties of x-rays. How can an entity exhibit the wave behavior that is evidenced in the diffraction of x-rays by crystals and, on the other hand, the particle properties that are revealed by the Compton effect?

According to quantum mechanics, as interpreted by Bohr and other theoretical physicists, the riddle can be solved only by extending to radiation a limitation upon the use of ordinary space-time conceptions. Bohr insisted that whenever an observation of any sort is made, its *immediate* result is always expressible in terms of familiar ideas of space and time, since these ideas have been developed out of human experience and any observation necessarily includes as its primary stage a certain experience by a human observer. But it does not follow that it will always be possible to construct a picture of the physical reality that causes these experiences in the same way in which we picture everyday objects. In classical theory, an electromagnetic field was assumed to have a certain character at every point in space and at every instant of time, as represented by certain values of the electric and magnetic vectors. According to the new view, such a conception of the electromagnetic field is valid at best as an approximation.

The essential significance of what we call a radiation field lies in its effects upon the motion of charged particles. It is possible for the field to exhibit contrasting characteristics under different circumstances. At one extreme, the action of the field takes the form with which we have become familiar in the Compton effect. Here a photon appears to bounce off an electron, thereby changing the momentum of the electron suddenly and discontinuously. The change of momentum δp will be definitely observable and measurable by the physicist only if it considerably exceeds the range of indeterminateness of the momentum already possessed by the particle. A clear-cut scattering process of the Compton type can occur only when *the wavelength of the radiation is shorter than the diameter of the region in which the particle may be supposed to be effectively located.* For an electron in an atom, this condition can be met only for wavelengths considerably shorter than the atomic diameter, such as the wavelengths of hard x-rays or γ rays.

In the Compton process there is no feature that can be regarded as a manifestation of an electric intensity in the wave. In order to obtain, at the opposite extreme, an action of radiation that can be interpreted in terms of the familiar electric and magnetic vectors, two conditions must be met. The experimental conditions must be such that it is possible to follow the test particle in effectively continuous motion along a path, so that its acceleration can be determined. It is also necessary that a segment of path which is sufficiently long to permit an adequate determination of the acceleration shall yet be short enough to ensure that along it the field vectors do not vary appreciably in value. Thus *the familiar action ascribable to electric and magnetic fields in the radiation is obtained only when the particle acted on is definitely located within a region much smaller than a wavelength.* This latter condition is satisfied, for example, by the electrons in a wire held parallel to the electric vector in long electromagnetic waves. In such a wire alternating electric current is observed to be produced, varying in phase with the electric vector. Ions in the upper atmosphere acted on by such waves furnish another example.

In such cases the *magnetic* field in the waves will also act upon the current or on the moving ions. It thus comes about that the *average* result is a force in the direction of propagation of the waves. This force constitutes an example of light pressure, and it can be regarded as analogous to the Compton recoil that occurs under other circumstances; but here there is a continuous rather than a discontinuous action.

Problems

1. Find the grating space for KBr and KCl, which have densities of 2750 and 1984 kg/m³ respectively.

Ans: 3.30 Å; 3.14 Å

2. Find the wavelength for which first-order Bragg reflection occurs at a 20° glancing angle from calcite. At which angle does second-order

reflection occur? What is the glancing angle for Mo $K\alpha$ radiation (0.707 Å) in first order?

> *Ans:* 2.076 Å; 43.2°; 6.7°

3. An x-ray tube with a copper target is operated at a potential difference of 25 kV. The glancing angle with a NaCl crystal for the Cu $K\alpha$ line is found to be 15.8°. Find the wavelength of this line and the glancing angle for photons at the short-wavelength limit.

> *Ans:* 1.54 Å; 5.1°

4. An x-ray tube with a tungsten target is operated with a dc power supply giving 15 mA at a potential difference of 60 kV. What is the short-wavelength limit of the x-rays produced? Estimate the approximate power radiated in x-rays.

> *Ans:* 0.207 Å; about 6 W

5. Find the short-wavelength limit for x-radiation from (*a*) a television tube operated at 10 kV, (*b*) an x-ray tube operated at 100 kV, (*c*) a 20-MeV betatron, and (*d*) a 1-GeV linear electron accelerator.

> *Ans:* 1.24 Å; 0.124 Å; 0.62 XU; 0.0124 XU

6. Find the short-wavelength limit for the radiation from (*a*) an x-ray tube operated at 30 kV, (*b*) a 50-MeV electron linac (linear accelerator), and (*c*) a 300-MeV synchrotron.

7. One of the best proofs for the discovery of a new element lies in obtaining the characteristic x-ray spectrum predicted for that element. Plot a Moseley graph for the $K\alpha$ line and predict to three significant figures the wavelength of this line for elements 85, 87, and 94.

> *Ans:* 0.150, 0.147, 0.135 Å

8. Assuming that all scattering is of the incoherent Thomson variety, find the atomic, mass, and linear scattering coefficients for copper. Judging by Fig. 7.14, at what wavelength would you expect these values to be approximately correct? At $\lambda = 0.707$ Å (Mo $K\alpha$) what would you estimate σ_a to be for Cu?

9. Unpolarized x-rays undergo Thomson scattering at a graphite block. Find the polarization of the radiation scattered in the xz plane at $\theta = 30°$. (The polarization is defined as

$$P = \frac{I_\perp - I_\parallel}{I_\perp + I_\parallel}$$

where $I_\perp$ and $I_\parallel$ refer to the intensities scattered with the electric vector respectively normal and parallel to the xz plane.)

10. X-rays of $\lambda = 0.085$ Å are scattered by carbon. At what angle will Compton scattered photons have a wavelength of 0.090 Å?

> *Ans:* 37.5°

11. A photon is Compton scattered by an electron through an angle of 90°. Find the energy of the scattered photon for incident-photon energies of 10 keV, 0.511 MeV, and 10 MeV.

Ans: 0.0098, 0.256, and 0.486 MeV

12. Show that the kinetic energy of a Compton scattered electron is given by Eq. (7.17) in terms of ϕ or by Eq. (7.17a) in terms of θ.

13. Find the smallest energy a photon may have and still transfer one-half of its energy to an electron initially at rest.

Ans: 256 keV

14. Show that the recoil angle for a Compton electron is given by

$$\cos \phi = \frac{(h\nu)^2 - (h\nu_\theta)^2 + p^2 c^2}{2h\nu pc}$$

where p is the momentum of the electron.

15. A cosmic-ray photon of energy $h\nu$ is scattered through 90° by an electron initially at rest. The scattered photon has a wavelength twice that of the incident photon. Find the frequency of the incident photon and the recoil angle of the electron.

Ans: 1.2×10^{20} Hz; 26.6°

16. Derive Eq. (7.18).

17. A photon of high energy $h\nu$ makes a head-on collision with a free electron at rest.

(a) Apply the conservation laws to this collision to show that $2h\nu = K + \sqrt{K^2 + 2m_0 c^2 K}$, where K is the kinetic energy of the recoil electron.

(b) If $h\nu = 10$ MeV, find K.

Ans: 9.75 MeV

18. For a photon which has undergone Compton scattering, show that the energy is

$$h\nu_\theta = \frac{h\nu}{1 + \epsilon(1 - \cos \theta)}$$

where $\epsilon = h\nu/m_0 c^2$.

19. A 0.511-MeV photon strikes an electron head on and is scattered straight backward. Find the energy of the scattered photon, the energy of the recoil electron, and the ratio v/c for the latter.

20. If in a Compton scattering the recoil electron moves off at angle ϕ, show that the photon recoils at an angle θ given by

$$\cos \theta = 1 - \frac{2}{(1 + \epsilon)^2 \tan^2 \phi + 1}$$

21. Verify that a freely moving electron cannot spontaneously emit or absorb a photon.

22. When a high-energy photon undergoes Compton scattering at an angle of 30°, the recoil electron is projected with $\phi = 30°$. Find the energy of the incident photon.

   Ans: 2.78 MeV

23. (a) The average intensity of a beam of plane-polarized electromagnetic radiation is $\frac{1}{2}c\epsilon_0 E_0^2$, where E_0 is the amplitude of the electric vector. Calculate the classical amplitude of vibration of a free electron under the influence of a beam of x-rays with wavelength 0.7 Å and an intensity of 20 W/m².

   (b) What power is reradiated as scattered x-radiation by the electron?

   (c) Find the intensity required to give the electron an amplitude of 0.05 Å.

24. A spectrometer to determine the energy of incident γ rays operates as follows. Collimated γ rays strike a thin target, and Compton recoil electrons proceeding in the direction of the original γ's are analyzed magnetically. (Note that the Compton scattering angle is then 180°.) If r is the radius of the electron's path and B is the magnetic induction, derive a relativistically correct equation for the γ-ray energy in terms of r, B, and fundamental constants.

25. In the inverse Compton effect a photon gains energy in an elastic collision with a high-energy electron. If in the laboratory reference frame, a 10-eV photon collided head on with a 500-MeV electron, find the energy of the Compton scattered photon in the laboratory system.

26. One means of producing monoergic photons with high energy (and nearly 100 percent polarization) is through the inverse Compton effect (see Prob. 25). If an intense laser beam of 6942-Å photons is made to collide head on with a beam of 20-GeV electrons, find (a) the energy of the incident photons in a reference frame moving with the electrons, and (b) the energy of the back-scattered photons in the laboratory reference frame. Some of the high-energy photons observed in cosmic rays may have gained energy through the inverse Compton process.

chapter eight

Radioactivity and the Nuclear Atom

The discovery of radioactivity by Henri Becquerel in 1896 marks the starting point of nuclear physics. As a result of his work and the investigations by the Curies, Rutherford and his collaborators, and many others, it became evident that this phenomenon involves the operation of forces of a different order of magnitude from any familiar to chemists or physicists of that time and that one was dealing with the actual transmutation of the elements, for which men had searched in vain since the time of the Egyptians. As a direct outgrowth of these studies came the identification of a host of materials possessing the remarkable property of spontaneously transforming from one into another, and with the unraveling of these successive transformations came the first clue to the intimate relation between the elements. Exploitation of the enormously energetic radiations which these substances emitted led to the discovery of the nucleus itself, to the nuclear model of the atom, and finally to the artifically induced transmutation of the stable elements.

Nuclear physics has progressed greatly since the time of these early experiments, and the startling developments in the field have often deserved the epithet "spectacular."

8.1   Earliest Developments

The discovery of radioactivity by Becquerel, although accidental, resulted directly from the discovery of x-rays. The apparent association between the production of x-rays and a fluorescence of the glass tube suggested that substances which were naturally fluorescent or phosphorescent might also produce such a penetrating radiation after exposure to light. As a part of a program to investigate this possibility, Becquerel prepared in February, 1896, a sample of the double sulfate of potassium and uranium, a well-known phosphorescent material, and placed it on a photographic plate enveloped in black paper, intending to expose it to sunlight. After waiting in vain for sunshine for some days, he decided to develop the plate, since the package had been exposed to some diffuse light and would therefore not be suitable for his experiment. To his surprise, the image of the crystals stood out clearly on the plate. Further experiments showed that the blackening of the plate was quite independent of exposure to light or of the previous treatment of the materials, and even compounds which exhibited no fluorescence could produce the effect, so long as they contained uranium. The magnitude of the effect was proportional to the atomic content of uranium in the compound, and extreme variations of temperature had no demonstrable influence. These observations led to the conclusion that *radioactivity*, as it was later named by Marie Curie, was not a molecular effect but had something to do with an atomic process, the like of which had not been seen in any chemical reaction.

The announcement of these results led to an intensive search for other substances possessing the same properties, and in 1898 Marie and Pierre Curie discovered polonium and radium. These materials, found in uranium minerals, were many times more active than uranium itself; radium is so active that its compounds are able to maintain themselves several degrees above the temperature of their surroundings. The radioactivity of thorium was found by Schmidt, and independently by Marie Curie, in 1898, and in 1900 the radioactive gases thoron and radon emanating from thorium and radium were reported by Rutherford and Dorn, respectively.

It was soon found that the rays from these various substances were mainly of two kinds—one a "soft" radiation, which could penetrate only thin paper or a few centimeters of air, and another which traveled some

tens or hundreds of centimeters in air and could penetrate thin foils of
metal.   These two radiations were called *α rays* and *β rays*, respectively.
The existence of still a third, much more penetrating component, called
γ radiation, was discovered by Villard in 1900.   The *α* and *β* rays were
shown by Becquerel to be capable of rendering air conducting and of
discharging a gold-leaf electroscope just as Röntgen had shown to be
the case for x-rays.

8.2   Alpha, Beta, and Gamma Rays

In 1899 Giesel and, independently, Becquerel found that the *β* rays could
be easily deflected in a magnetic field.   The arrangement used by
Becquerel is illustrated in Fig. 8.1.   A small lead container, open at
the top and containing a radioactive preparation, is placed on a photographic plate with a magnetic field parallel to the surface of the plate.
β rays emerging from the container in the plane of the figure are deflected
by the field through semicircles and register on the plate at various distances, depending on their momenta.   From the law of magnetic deflection, the distance at which particles of a given velocity *v* and charge *e*
strike the plate is

$$x = 2\rho = 2\,\frac{mv}{eB}$$

where *B* is the magnetic induction.   The quantity *Bρ*, called the *magnetic rigidity*, is a measure of the momentum divided by the charge.
Becquerel found a wide variation in *Bρ* even for a thin source, indicating
that the *β* rays are not homogeneous.   By placing absorbing foils on the

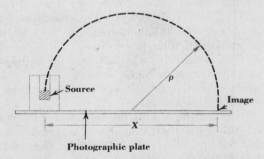

Fig. 8.1   In Becquerel's arrangement *β* rays from
a radioactive source were deflected in a semicircular
arc by a magnetic field into the plane of the paper.

plate, he found that the most easily deflected β particles were the most easily absorbed.   He soon found that the β particles could also be deflected electrostatically and set out to determine their specific charge e/m by a comparison of the magnetic and electric deflections.   A beam of particles traveling perpendicularly to an electrostatic field E suffers a displacement y perpendicular to the motion after traveling a length l in the field, where, by Eq. (6.6),

$$y = \frac{1}{2}\frac{Ee}{m}\left(\frac{l}{v}\right)^2$$

Inserting the value of $B\rho$ for the rays, Becquerel could obtain v, and hence also e/m.   He found a velocity of more than half the speed of light and obtained a value for e/m of about 10^{11} C/kg (the charge was negative).   This observation suggested that the β rays were to be identified with the electrons observed as the cathode rays by Thomson (only 3 years earlier) but showed that their velocities greatly exceed those of the cathode rays.   Later work by Kaufmann in 1902, who used simultaneous electric and magnetic deflections, gave a more accurate value of e/m and showed the variation of mass with velocity that was later to be explained by the special theory of relativity.   The enormous velocities observed, up to $0.96c$, emphasized again that radioactivity was no ordinary chemical phenomenon.

The magnetic deflection of the α rays was demonstrated by Rutherford in 1903.   How difficult they were to deflect may be judged from Rutherford's own figures, according to which, in a field of 1 Wb/m², the radius of curvature was some 39 cm, while that of the cathode rays from an ordinary tube would be only 0.01 cm!   The sense of the deflection indicated a positive charge.   Applying the methods of magnetic and electrostatic deflection, Rutherford obtained values of $v = 2.5 \times 10^7$ m/s and $e/m = 6 \times 10^7$ C/kg for the α particles of RaC' (Po²¹⁴), a derivative of radium.   Unlike the β particles, the α particles turned out to occur in groups with discrete velocities.   An improved evaluation showed that e/m for α particles was half that for hydrogen ions, which suggested identification with doubly ionized helium.   This suggestion was particularly attractive since it was well known that helium is always present in radioactive minerals.

In a series of experiments reported in 1908, Rutherford and Geiger measured directly the number of α particles emitted per gram of radium and the total charge carried by them, thus making possible a determination of the charge of the α particle.   The apparatus for the first measurement is illustrated in Fig. 8.2.   An evacuated vessel A contains a thin deposit of radium on a holder D.   When the stopcock F is opened,

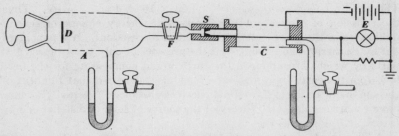

Fig. 8.2 Apparatus of Rutherford and Geiger for counting the number of α particles emitted by radium. Particles from the source *D* pass through aperture *S* and are registered in the counter *C*. (*Adapted from E. R. Rutherford, J. Chadwick, and C. D. Ellis, "Radiations from Radioactive Substances," Cambridge University Press, 1930.*)

the α particles pass through an aperture *S*, which is covered by a thin mica foil, into the counter *C*. The counter consists of a brass tube, provided with ebonite plugs through which a fine center wire passes. The tube is connected to the negative side of a high-voltage battery, the positive end of which is grounded, and the center wire is connected to an electrometer *E* provided with a high-resistance leak to ground. With the counter exhausted to a few centimeters pressure, the passage of a single α particle produces a momentary discharge which causes the electrometer to deflect. With this instrument, Rutherford and Geiger found that 3.4×10^{10} α particles were expelled per gram of radium per second (actually, from the amount of RaC' (Po^{214}) in equilibrium with 1 g of radium). In the second experiment, they determined by direct measurement the amount of charge collected on a plate suspended above the preparation in an evacuated vessel. β rays and secondary electrons were eliminated by putting the apparatus in a strong transverse magnetic field. The result was 1.05×10^{-5} C/kg-s. Dividing this number by the total number of α particles gave for the charge per α particle 3.1×10^{-19} C, which corresponds to two elementary charges. There could be little doubt that the α particles were actually helium ions. Even more conclusive evidence was obtained in 1909, when Rutherford and Royds detected spectroscopically the helium gas formed when α particles pass into a closed glass vessel.

The velocity obtained by Rutherford for the α particles from Po^{214} implies a very high energy indeed. The mass of the helium atom, obtained from the atomic weight and Avogadro's number, is

$$M = \frac{4.00}{6.02} \times 10^{-26} = 6.64 \times 10^{-27} \text{ kg}$$

The kinetic energy is then $\frac{1}{2}mv^2 = 1.23 \times 10^{-13}$ J $= 7.7$ MeV. Thus the emission of each α particle involves an energy change of several million times the energy release in a chemical reaction! It was pointed out by Rutherford and Soddy that the presence of only a few times 10^{-14} parts by weight of radium in the earth would compensate for the heat lost from the earth by radiation. They also suggested that the heat of the sun might be derived from radioactive substances. Actually we now have a more sophisticated theory of the source of solar and stellar energy (Sec. 25.22).

The third component of the radioactive radiations, the γ rays, proved less amenable to study than the α and β rays. The γ rays are much more penetrating, and they are not deflected by electric or magnetic fields. They do, however, produce ionization in air and fluorescence in many kinds of crystals; it is by these processes that they are ordinarily detected. It was conjectured early that they might be a form of electromagnetic radiation, like x-rays, but the proof of this hypothesis had to await development of a much more complete understanding of the interaction of electromagnetic radiation with matter.

8.3   Early Views on Atomic Structure

Speculations about the structure of the atom date from the early years of the nineteenth century. In 1815 Prout proposed the hypothesis that the atoms of all elements are composed of atoms of hydrogen; but this view failed of acceptance when it was established that the atomic weights of some elements differ markedly from multiples of the atomic weight of hydrogen.

The discovery of the electron in 1897 led to renewed interest in the internal structure of atoms by indicating that they must contain both negative electrons and positive charges. Two questions then arose: (1) How many electrons are there in an atom? (2) How are the electrons and the positive charges arranged? Between 1909 and 1911 Barkla obtained evidence from the scattering of x-rays, which indicated that the number of electrons in an atom of the lighter elements was equal to about half the atomic weight. As to the arrangement of the electrons, it seemed on the basis of classical ideas that two conditions must be met. (1) The ensemble of positive charges and electrons in an atom must be stable; the electrons, for example, must be held by forces of some sort in fixed positions of equilibrium about which they can vibrate, when disturbed, with the definite frequencies required to explain the characteristic line spectra of the elements. (2) Except when so disturbed, the electrons must be at rest, since otherwise they would emit radiation in accord with electromagnetic theory. The much larger mass of the atom as a whole was assumed to belong to the positive charges, which for this reason would vibrate very little.

A possibility considered by J. J. Thomson was that the positive electricity might be distributed continuously throughout a certain small region, perhaps with uniform density throughout a sphere. The electrons might then be embedded in the positive electricity, occupying normally certain positions of equilibrium and executing harmonic vibrations about these positions when slightly disturbed. Frequencies in the visible spectrum might thereby be emitted if the sphere of positive electricity were of the order of 10^{-10} m in radius. But Thomson was unable to show that these frequencies could be such as to form a series converging to an upper limit, and eventually his theory came into conflict with the experiments of Rutherford and his collaborators on the scattering of α particles, which are now to be described.

8.4 The Scattering of Alpha Particles by Atoms

The α rays from radioactive materials had been shown to be helium atoms which have lost two electrons. The α particles, whose initial velocity is of the order of 2×10^7 m/s, could be studied by means of the flashes of light, or scintillations, they produce on striking a zinc sulfide screen, the impact of a *single* particle producing a visible flash which is readily observed under a low-power microscope.

If a stream of α particles, limited by suitable diaphragms to a narrow pencil, be allowed to strike a zinc sulfide screen placed at right angles to the path of the particles, scintillations occur over a well-defined circular area equal to the cross section of the pencil. If, now, a thin film of matter, such as gold or silver foil, is interposed in the path of the rays, it is found that they pass quite readily through the foil but that the area over which the scintillations occur becomes larger and loses its definite boundary, indicating that some of the particles have been deflected from their original direction.

Qualitatively, it is easy to explain the origin of the forces which cause the deflection of the α particle. The particle itself has a twofold positive charge. The atoms of the scattering material contain charges, both positive and negative. In its passage through the scattering material, the particle experiences electrostatic forces the magnitude and direction of which depend on how near the particle happens to approach to the centers of the atoms past which or through which it moves.

If we assume the Thomson atomic model, the path of an α particle in passing through an atom might be as indicated in Fig. 8.3a, the net

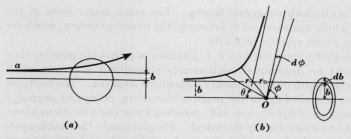

Fig. 8.3 Comparison of deflection of an α particle by (a) the Thomson atom and (b) the Rutherford nuclear atom.

result of the passage being to deflect the particle through a small angle θ. That θ should be small is made plausible by the arguments below.

The atom is electrically neutral, so that an α particle passing along a diameter of the atom should experience no net force from the positive charge and negligible deflections from those electrons which it misses widely. But if there were a direct hit on some electron, the recoil velocity of the electron could not exceed $2v$, where v is the initial speed of the particle, since even in an elastic collision the velocity of separation cannot exceed the velocity of approach. Hence the maximum momentum which can be transferred to an electron, and therefore lost by the α particle, is $2mv$, where m is the electron mass. Surely then the component of the momentum normal to its initial momentum must be less than $2mv$, and the maximum scattering angle θ_{max} must be less than $2mv/Mv$, where M is the mass of the α particle, or much less than 10^{-3} rad.

Finally the effect of the positive core cannot produce a large deflection. For any passage through the atom there are volume elements of positive charge symmetrically arranged on each side of the path, and the contributions from these cancel. There remains some positive charge elements for which there are no canceling elements, but even if one ignores the electrons embedded therein, the maximum deflection expected is of the order of 10^{-4} rad. The total deflection of a particle in passing through a number of atoms in a thin layer of scattering material varies according to the laws of probability. Such a process is called *multiple* or *compound scattering*. According to Rutherford, the number of α particles N_ϕ which, as a result of *multiple* scattering, should be scattered by such an atom through an angle ϕ *or greater* is given by $N_\phi = N_0 e^{-(\phi/\phi_m)^2}$, where N_0 is the number of particles for $\phi = 0$ and ϕ_m is the average deflection after passing through the scattering material.

Geiger had shown experimentally that the most probable angle of deflection of a pencil of α particles in passing through gold foil $\frac{1}{2000}$ mm thick is of the order of 1°.   It is evident, therefore, from the last equation, that the probability for scattering through large angles becomes vanishingly small; for 30°, it would be of the order of 10^{-13}.   Geiger showed that *the number of particles scattered through large angles was much greater than the theory of multiple scattering predicted.*   Indeed, Geiger and Marsden showed that 1 in 8000 α particles was turned through an angle of *more than* 90° by a thin film of platinum.   This so-called *reflection,* however, was shown to be not a surface phenomenon but a volume effect, since the number of particles turned through more than 90° increased, up to a certain point, with increasing thickness of the scattering foil.

8.5   The Nuclear Atom

a. Rutherford's Hypothesis   In a classic article published in 1911, Rutherford proposed a new type of atomic model capable of giving to an α particle a large deflection as a result of a *single* encounter.   He assumed that in an atom *the positive charge and most of the atomic mass are concentrated in a very small central region,* later called the nucleus, about which the electrons are grouped in some sort of configuration. Since normally atoms are electrically neutral, the positive charge on the nucleus must be Ze, where e is the numerical electronic charge and Z is an integer, characteristic of the kind of atom.

The deflection of an α particle by such a nucleus might be as illustrated in Fig. 8.3b.   Rutherford calculated the distribution of α particles to be expected as the result of *single-scattering* processes by atoms of this type.   According to the Rutherford scattering formula, derived in Appendix 8A, the number N_d of scattered α particles reaching a small detector of area A is given by

$$N_d = \frac{N_i n t A}{4R^2} \frac{4Z^2 e^4}{16\pi^2 \epsilon_0^2 m^2 v_0^4} \frac{1}{\sin^4 (\phi/2)} \tag{8.1}$$

where N_i = number of α particles incident on a foil of thickness t with n nuclei of atomic number Z per unit volume

R = distance of detector from foil

m = mass of incident α particles

v_0 = speed of incident α particles

ϕ = scattering angle

e = charge of a proton

ϵ_0 = permittivity of free space

This formula is subject to correction, however, for the motion induced in the scattering nucleus, which may be appreciable for light atoms.

According to Rutherford's formula, the number of particles striking a small detector with its sensitive face perpendicular to the direction of motion of the scattered particles should be:

1. Proportional to the square of the atomic number Z of the scattering nuclei
2. Proportional to the thickness t of the foil
3. Inversely proportional to the square of the kinetic energy of the particles, i.e., to v_0^4
4. Inversely proportional to the fourth power of the sine of half the scattering angle ϕ

b. Experimental Confirmation   These predictions were completely verified in 1913 by the experiments of Geiger and Marsden.   Their data are shown graphically in Fig. 8.4, in which the logarithm of the number of scintillations on the screen per minute is plotted as abscissa against the logarithm of $1/\sin^4 (\phi/2)$ as ordinate.   If these two quantities are proportional to each other, the points for each substance should lie on a straight line inclined at 45° with the axes.   The two lines in the figure are drawn at exactly 45°, and the observed points are seen to agree well with the predictions.   This is the more remarkable since the numbers of scintilla-

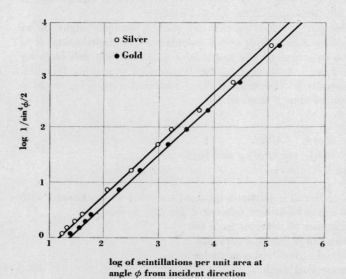

log of scintillations per unit area at
angle ϕ from incident direction

Fig. 8.4 Data of Geiger and Marsden on the scattering of α particles by thin foils.

tions varied in the experiment over a very wide range of values, the left-hand points on the plots representing 22 particles per minute for silver and 33 for gold, whereas the right-hand points represent 105,400 and 132,000, respectively.

Rutherford's prediction that the scattering should, for small thicknesses, be proportional to thickness was confirmed by the observations. Geiger and Marsden also showed that "the amount of scattering by a given foil is approximately proportional to the inverse fourth power of the velocity (inverse square of the energy) of the incident α particles," over a range of velocities such that the number of scattered particles varied as $1:10$.

c. Atomic Number   Geiger and Marsden further concluded, from a study of the variation of scattering with atomic weight and of the fraction of the total number of incident particles scattered through a given angle, (1) that the scattering is approximately proportional to the atomic weight of the scatterer over the range of elements from carbon to gold, and (2) that "*the number of elementary charges composing the center of the atoms is equal to half the atomic weight.*" This second conclusion was in agreement with Barkla's experiments on the scattering of x-rays. Thus, carbon, nitrogen, and oxygen should have, respectively, 6, 7, and 8 electrons, around a nucleus containing, in each case, an equal amount of positive charge. These elements are, respectively, the sixth, seventh, and eighth elements in the periodic table. The hypothesis was natural, therefore, that the number of electrons in the atom, *or the number of units of positive charge on its nucleus*, is numerically equal to the ordinal number which the atom occupies in the series of the elements, counting hydrogen as the first. Thus originated the concept of the *atomic number* of an element, which we may think of variously as (1) the ordinal number of the element in the series of the elements starting with $Z = 1$ for hydrogen, or (2) the positive charge carried by the nucleus of the atom, in terms of the numerical electronic charge e as a unit, or (3) the number of electrons surrounding the nucleus in the neutral atom.

These experiments of Geiger and Marsden so completely confirmed the conclusions which Rutherford had reached by postulating the nuclear type of atom that in spite of certain weighty objections his atomic model was at once adopted.

d. Some Difficulties   The objections to the nuclear type of atom were based on questions of stability. No stable arrangement of positive and negative point charges at rest can be invented; it can be proved, in fact, that such an arrangement is impossible (Earnshaw's theorem). An electron might be imagined to revolve around a nucleus as the earth revolves around the sun. But according to the laws of classical electromagnetic theory, it would *radiate energy*. The system will, therefore, "run down"; the electron will approach the nucleus along a spiral path and will *give out radiation of constantly increasing frequency*. This does not agree with the observed emission of spectral lines of fixed frequency.

It was at this point that Bohr introduced his epoch-making theory of the structure of the atom and of the origin of spectra, described in Chap. 9. His theory constituted an extension of Planck's theory of quanta to Rutherford's nuclear atom, in an attempt both to remove the difficulties of the nuclear model and to explain the origin of atomic spectra.

8.6   Radioactive Transformations

The hypothesis that there exists a genetic relation between some of the radioactive elements was advanced by Rutherford and Soddy in 1903, on the basis of their work on the decay and production of chemically separable activities. A typical example is afforded by a study of the γ rays of radium. Radium solutions continually evolve a gas, called radon. If the radon gas is completely pumped away from a radium source which has been standing in a sealed vessel, it is found that after

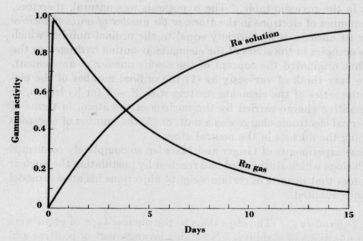

Fig. 8.5 Growth and decay of the γ-ray activities associated with (1) a radium solution from which all radon gas has been pumped and (2) the radon gas. (The initial transient due to the growth of Po^{218} and Pb^{214} is exaggerated in the figure; the γ rays come from Bi^{214}.)

a short time the gas exhibits all the γ activity (measured by absorbing out the α and β components), while the solution has none. If the vessels containing the radon and the radium are left standing separately for a few weeks, it is found that the latter now is γ active again and the former is not. More detailed measurements of the γ-ray activity of the two samples as a function of time lead to the curves shown in Fig. 8.5, where it may be seen that aside from certain transient effects at the very beginning, the activity of the radon falls exponentially, reaching half-value in about 3.8 days. The γ-ray activity of the radium solution rises exponentially, reaching half value in 3.8 days; the sum of the two curves is constant in time. That two activity curves relating to chemically different substances should bear such close similarity suggests the hypothesis that one member (Rn) is produced by spontaneous transmutation of the other. This is indeed the case. [Actually the γ rays in this case come not directly from the radon, but from RaC (Bi^{214}), which appears after three rapid subsequent disintegrations. The least rapid of these transmutations has a half-life of 26.8 min, so that there is a transient effect of this magnitude in the growth and decay curves of Fig. 8.5.]

a. Theory of Spontaneous Transformations   The essential feature of the transformation theory lies in the assumption that each radioactive nucleus has a definite probability of disintegrating in a given time, a probability which depends on the kind of nucleus but which is constant for all nuclei of a given kind. If we have N such nuclei, the number disintegrating in a time dt is $dN = -N\lambda \, dt$, where λ is the *disintegration constant* and the minus sign indicates that N is decreasing with time. The number of atoms decaying per unit time $-dN/dt$ is called the *activity* or *disintegration rate* of the sample. If the sample is isolated so that no new atoms are being added,

$$- \frac{dN}{dt} = \lambda N \tag{8.2a}$$

and

$$N(t) = N(0)e^{-\lambda t} \tag{8.2b}$$

where $N(0)$ is the number present at $t = 0$. The time at which half of the original atoms have disintegrated, called *half-life*, is given by

$$\frac{N(t_{\frac{1}{2}})}{N(0)} = \frac{1}{2} = e^{-\lambda t_{\frac{1}{2}}} \qquad t_{\frac{1}{2}} = \frac{\ln 2}{\lambda} = \frac{0.693}{\lambda}$$

The *mean life*, or life expectancy, of the average atom is given by

$$t_m = \frac{\int_0^\infty t e^{-\lambda t}\, dt}{\int_0^\infty e^{-\lambda t}\, dt} = \frac{1}{\lambda} \tag{8.3}$$

The behavior of an isolated radon sample with a given initial number of atoms is thus accounted for by a simple decay, with disintegration constant $\lambda(\text{Rn}) = 0.693/t_{\frac{1}{2}} = 0.21 \times 10^{-5}\ \text{s}^{-1}$.

If now we consider the production of radon in the vessel containing radium, we have a somewhat different situation. The number of radium atoms disintegrating in time dt is $dN_1 = -\lambda_1 N_1\, dt$, where λ_1 is the disintegration constant of radium. Each of these disintegrations produces a radon atom; in the same time, however, a number $\lambda_2 N_2\, dt$ of radon atoms disintegrate (the subscript 2 refers to radon), so that the total change in the amount of radon is

$$dN_2 = +\lambda_1 N_1\, dt - \lambda_2 N_2\, dt = [\lambda_1 N_1(0)e^{-\lambda_1 t} - \lambda_2 N_2]\, dt$$

If at time $t = 0$ there is no radon present, the integral of this equation is

$$N_2(t) = N_1(0)\, \frac{\lambda_1}{\lambda_2 - \lambda_1}\, (e^{-\lambda_1 t} - e^{-\lambda_2 t}) \tag{8.4}$$

and the ratio of the activities is

$$\frac{\lambda_2 N_2(t)}{\lambda_1 N_1(t)} = \frac{\lambda_2}{\lambda_2 - \lambda_1}\, (1 - e^{-(\lambda_2 - \lambda_1)t}) \tag{8.5}$$

In the present case the parent, radium, has a relatively long half-life (1622 years), and so we may ignore λ_1 compared to λ_2. We then obtain

$$N_2(t) = N_1(t)\, \frac{\lambda_1}{\lambda_2}\, (1 - e^{-\lambda_2 t}) \tag{8.5a}$$

From this relation we see that the growth of radon from radium will exhibit a time constant characteristic of the decay of radon, as observed in Fig. 8.5.

When $\lambda_2 > \lambda_1$ (as in this case), a state of *transient equilibrium* is reached when the exponential of Eq. (8.5) becomes negligible. After a time long compared to $1/(\lambda_2 - \lambda_1)$, the number of daughter nuclei in equilibrium with the parent approaches the constant value $N_2 = N_1\lambda_1/\lambda_2$;

thus

$$N_2\lambda_2 = N_1\lambda_1 \tag{8.6}$$

This relation, which is general for any radioactive chain provided only that the system has been undisturbed for a time long compared with the half-life of any of the products, expresses the fact that at equilibrium the rate of decay of any radioactive product is equal to its rate of production from the previous member of the chain. A system which has reached this condition is said to be in *secular equilibrium*.

The picture of the radioactive chains which emerged from Rutherford and Soddy's considerations is the following. A radioactive atom disintegrates by the emission of an α particle, thereby losing four units of mass and changing its chemical properties. The product then further disintegrates either by α or β emission, each disintegration leading to a different element, until finally a stable substance is formed, ending the chain. A little later, when the importance of the nuclear charge in determining the chemical behavior of an element was realized, it was established that the α transformation removes four units of mass and two units of charge, producing an element two steps down in the periodic table, while the β disintegration removes one negative charge and essentially no mass, producing an element one step higher in the table. This picture has found confirmation in studies of the chemical properties of the various products. Although the amounts of material available for such studies are ordinarily too minute for conventional analytical techniques, the fact that the substances are radioactive makes it possible to carry out the usual chemical manipulations and to trace the material under investigation through the various stages by means of the radiation emitted. Thus, Rutherford and his coworkers were able in 1903 to determine the chemical and physical properties of radon gas, including its condensation temperature, with samples amounting to small fractions of a microgram.

b. Radioactive Isotopes   Further investigations of the radioactive transformations led to the identification of a great number of such successive transformations and to the discovery of many new radioactive substances. A particularly important result of this work was the observation for the first time of *isotopes*, atoms having identical chemical properties but differing in atomic weights. The discovery of isotopes came incidentally to a search for the origin of radium. The fact that radium has a half-life which is short on a geological time scale suggested that it must be a product of some longer-lived material, probably uranium, since radium is always found in uranium minerals. In the course of a search for this connection, Boltwood discovered in 1906 a long-lived substance which appeared to be the intermediate between uranium and radium. The interesting feature of his report was that this substance, which he named *ionium*, appeared to be chemically similar to thorium. The matter was followed up by Marckwald and Keetman in 1909, and later by von Welsbach, with the astonishing result that ionium and thorium are chemically identical, although their radioactive properties are quite different. Within a year several other such cases were found, and it was established that the difference in radioactive properties was associated with a difference in atomic weight. The importance of these findings was emphasized by Soddy in 1910, when he pointed out that the existence among radioactive elements—and, as he presumed to be the case, among stable elements as well—of atoms having various masses but identical chemical properties provided a natural explanation of the nonintegral atomic weights of certain elements and reopened the question of whether all atoms might not be built out of some common constituents.

The hypothesis that the elements are constructed from hydrogen atoms had been advanced by the English chemist Prout in 1815, some 10 years after Dalton's formulation of the atomic hypothesis, from the observation that many atomic weights appeared to be integral multiples of that of hydrogen. Prout's suggestion fell into disrepute, however, when with more precise measurements it developed that there were deviations from the whole-number relation in many cases. Finally, when the atomic weight of chlorine (35.5) was shown to deviate by a full half-integer, further discussion of the point was generally abandoned. With the discovery of isotopes, such deviations could be accounted for by assuming an appropriate mixture, and one could again consider the possibility of some elemental constituents of atoms. The masses of isotopes are in fact very nearly (though not quite) integral multiples of a single number; there is now good evidence that the hydrogen nucleus (not the atom) is one of the building blocks of nuclei.

In modern terminology, each nucleus is characterized by a definite charge number Z, a multiple of the protonic charge, and a definite mass number A. The mass number is the integer nearest the actual isotopic mass, expressed in a scale where the mass of carbon 12 equals 12 exactly. The nuclear charge determines the number of electrons in the neutral atom and hence the gross atomic properties, but such specifically nuclear properties as radioactivity depend just as much on A as on Z. A nuclear species with a given Z and A is called a *nuclide;* nuclides with the same Z and different A are isotopes, and those with the same A but different Z are called *isobars.* It is general practice to represent each nuclide by a symbol such as $_{88}Ra^{224}$ in which the letters refer to the chemical name, the subscript gives the atomic number and the superscript the mass number. The subscript is redundant and is often omitted.

c. Radioactive Chains Among the heavy elements there are four independent radioactive chains (Table 8.1). The first to be investigated starts with U^{238} and ends with Pb^{206}. The mass number changes only by four units from each α decay; consequently all members of the *uranium*

Table 8.1 The four radioactive series

Thorium (4n) Series		Neptunium (4n + 1) Series		Uranium (4n + 2) Series		Actinium (4n + 3) Series	
${}_{92}U^{236}$ α 2.39×10^7 y		${}_{93}Np^{237}$ α 2.25×10^6 y		${}_{92}U^{238}$ α 4.51×10^9 y		${}_{94}Pu^{239}$ α 24,300 y	
${}_{90}Th^{232}$ α 1.39×10^{10} y		${}_{91}Pa^{233}$ β 27.4 d		${}_{90}Th^{234}$ β 24.5 d		${}_{92}U^{235}$ α 7.07×10^8 y	
${}_{88}Ra^{228}$ β 6.7 y		${}_{92}U^{233}$ α 1.63×10^5 y		${}_{91}Pa^{234}$ β 1.14 m		${}_{90}Th^{231}$ β 24.6 h	
${}_{89}Ac^{228}$ β 6.13 h		${}_{90}Th^{229}$ α 7000 y		${}_{92}U^{234}$ α 2.33×10^5 y		${}_{91}Pa^{231}$ α 3.2×10^4 y	
${}_{90}Th^{228}$ α 1.90 y		${}_{88}Ra^{225}$ β 14.8 d		${}_{90}Th^{230}$ α 8.3×10^4 y		Ac^{227} α 1.2%	21 y β 98.8%
${}_{88}Ra^{224}$ α 3.64 d		${}_{89}Ac^{225}$ α 10.0 d		${}_{88}Ra^{226}$ α 1622 y		Fr^{223} β 21 m	Th^{227} α 18.9 d
${}_{86}Rn^{220}$ α 54.5 s		${}_{87}Fr^{221}$ α 4.8 m		${}_{86}Rn^{222}$ α 3.825 d		${}_{88}Ra^{223}$ α 11.2 d	
Po^{216} α 99.99%	158 s β 0.013%	${}_{85}At^{217}$ α 0.020 s		Po^{218} α 99.97%	3.05 m β 0.03%	${}_{86}Rn^{219}$ α 3.92 s	
Pb^{212} β 10.6 h	At^{216} α 300 μs	Bi^{213} α 2%	47 m β 98%	Pb^{214} β 26.8 m	At^{218} α 1.3 s	Po^{215} α 100 − %	1830 μs β 5×10^{-4}%
Bi^{212} α 33.7%	60.5 m β 66.3%	Tl^{209} β 2.2 m	Po^{213} α 4.2 μs	Bi^{214} α 0.04%	19.7 m β 99.96%	Pb^{211} β 36.1 m	At^{215} α 10 μs
Tl^{208} β 3.1 m	Po^{212} α 0.3 μs	${}_{82}Pb^{209}$ β 3.3 h		Tl^{210} β 1.32 m	Po^{214} α 150 μs	Bi^{211} α 99.68%	2.16 m β 0.32%
${}_{82}Pb^{208}$ Stable		${}_{83}Bi^{209}$ Stable		${}_{82}Pb^{210}$ β 22 y		Tl^{207} β 4.76 m	Po^{211} α 5000 μs
				Bi^{210} α 5×10^{-5}%	5.0 d β 100 − %	${}_{82}Pb^{207}$ Stable	
				Tl^{206} β 4.23 m	Po^{210} α 140 d		
				${}_{82}Pb^{206}$ Stable			

series have mass number A equal to $4n + 2$, where n is integer. The last to be studied is the neptunium $(4n + 1)$ series, all members of which have a half-life so short that only small amounts are found in nature; consequently they are artificially produced. As found in minerals which have not been subjected to chemical separation, a given chain is in secular equilibrium, and the number of atoms of any radioactive product present is simply proportional to its half-life. For example, the amount of Ra^{226} per kilogram of U^{238} in any ore is

$$\frac{1622 \times 226}{4.5 \times 10^9 \times 238} = 3.4 \times 10^{-7} \text{ kg}$$

In the chemical reduction of the ore, radium is separated both from the uranium and from all other members of the series, and the secular equilibrium is destroyed. In the course of a few days, the next product, Rn, is built up again, and with it, all the short-lived products, Po^{218}, Pb^{214}, Bi^{214}, Po^{214}. The half-life of Pb^{210}, however, is 22 years, so that only quite old radium samples will contain equilibrium amounts of Pb^{210}, Bi^{210}, and Po^{210}.

d. Half-life of Radium   The half-life of such a long-lived substance as radium is rather difficult to determine directly; instead, the disintegration constant is measured by counting the number of disintegrations per second from a known number of Ra atoms. In the early experiments of Rutherford and Geiger, the number of α particles emitted per second from the Po^{214} in equilibrium with the radon drawn off a known quantity of radium was counted (α particles from the other members of the chain have lower energies and can be cut out by an absorbing foil). In a later measurement the number of α particles emitted per second by a thin weighed deposit of freshly separated radium was determined directly and the value corrected for the growth of the decay products during the experiment. The number of disintegrations obtained was $3.608 \pm 0.028 \times 10^{10}$ α particles per second per gram, giving

$$\lambda = \frac{3.608 \times 10^{10} \times 226.096}{6.0247 \times 10^{23}} = 1.354 \times 10^{-11} \text{ s}^{-1}$$

and $t_{\frac{1}{2}} = 0.693/\lambda = 1622 \pm 13$ years. The standard unit of radioactivity is the *curie*, formerly defined as the amount, i.e., the activity, of radon in equilibrium with 1 g of radium. With the general application of the term in recent years to activities of other substances, the definition of the curie has been changed to mean an activity of exactly 3.7×10^{10} disintegrations per second. Another unit, the *rutherford*, which was proposed to mean exactly 10^6 disintegrations per second, has failed of general adoption.

e. Age of Minerals   The half-life of U^{238}, determined by the procedure just described, is 4.51×10^9 years. As it is reasonably certain that U^{238} is not itself derived from a longer-lived material, it would appear that this substance—and presumably all other elements as well— was formed not very many times this number of years ago. It is possible to make fairly accurate determinations of the age of uranium-containing minerals by several independent methods. To illustrate the principle involved, we consider the production of helium from U^{238}. For each atom of U^{238} which has decayed since the mineral was formed, eight helium atoms have been produced, presumably remaining occluded in the mineral. If the number of atoms of U^{238} originally present was N_0 and the number present at time t is N, the number of helium atoms produced is

$$N(\text{He}) = 8(N_0 - N) = 8N(e^{\lambda t} - 1)$$

where λ is the disintegration constant of U^{238}. A measurement of the ratio of helium to U^{238} content then permits calculation of t. In practice it is of course necessary to take into account the production of He by other radioactive materials present, such as U^{235} and Th^{232}. Also the loss of helium is a serious problem in rocks containing high concentrations of uranium. Other methods, involving measurement of the isotopic composition of the lead in uranium-containing minerals, are less subject to this difficulty. The age of the earth, i.e., the time which has elapsed since the oldest known minerals were formed, as estimated from such determinations, is 4.5×10^9 years.

Appendix 8A - The Rutherford Scattering Law

While the earliest application of Rutherford scattering was for α particles, the same theory has found wide use in treating the scattering of charged particles, such as protons and electrons, by the electrostatic fields of other charged centers. Although we treat below the scattering of α particles by nuclei, the equations developed are applicable to coulomb scattering for other particles. Let the charge on the incident particle be ze and its mass m, while the nucleus bears charge Ze and has a mass so much greater than the incident particle that, for a first approximation, it may be assumed to remain at rest. Let the α particle approach at speed v_0 along a line passing at a distance b from the nucleus; its energy is thus $mv_0{}^2/2$, and its angular momentum about the nucleus at O is mvb. Let r and θ be polar coordinates for the α particle with origin at O, θ being measured from a line drawn from O in a direction toward the distantly approaching particle (Fig. 8.3b), and let $\dot{r}$ and $\dot{\theta}$ be the time derivatives of r and θ respectively.

Then the particle has perpendicular components of velocity $\dot{r}$ and $r\dot{\theta}$, and its angular momentum about O at any time is $mr^2\dot{\theta}$. It also has potential energy due to repulsion by the nucleus of magnitude $zZe^2/4\pi\epsilon_0 r$ and initially zero. Thus, adding the kinetic energy, we have from the laws of conservation of energy and of angular momentum

$$\tfrac{1}{2}m(\dot{r}^2 + r^2\dot{\theta}^2) + \frac{zZe^2}{4\pi\epsilon_0 r} = \frac{1}{2}\,mv_0{}^2 \tag{8A.1}$$

$$mr^2\dot{\theta} = mv_0 b \tag{8A.2}$$

By eliminating $d\theta/dt$ it follows that

$$\dot{r} = -v_0\left(1 - \frac{2q}{r} - \frac{b^2}{r^2}\right)^{\tfrac{1}{2}} \tag{8A.3a}$$

$$q = \frac{zZe^2}{4\pi\epsilon_0 mv_0{}^2} \tag{8A.3b}$$

in which the negative sign is chosen because, during the approach, $dr/dt < 0$. Dividing this into Eq. (8A.2) and noting that $\dot{\theta}/\dot{r} = d\theta/dr$, we have

$$\frac{d\theta}{dr} = -\frac{b}{r^2}\left(1 - \frac{2q}{r} - \frac{b^2}{r^2}\right)^{\tfrac{1}{2}} \tag{8A.4}$$

The integral of this equation that vanishes at $r = \infty$, as is easily verified by substitution in Eq. (8A.4), is

$$\theta = \cos^{-1}\left[\frac{b}{\sqrt{b^2 + q^2}}\left(1 - \frac{2q}{r} - \frac{b^2}{r^2}\right)^{\tfrac{1}{2}}\right] - \cos^{-1}\frac{b}{\sqrt{b^2 + q^2}}$$

At the point of closest approach to the nucleus, $dr/dt = 0$. Hence, by (8A.3), the radical must vanish; then $\theta = \theta_0 = \pi/2 - \cos^{-1}(b/\sqrt{b^2 + q^2})$. The second half of the path, as the particle recedes, is symmetrical with the first half. Hence, the total increase in θ is $2\theta_0$; and the final deflection of the total change in the direction of motion of the particle is

$$\phi = \pi - 2\theta_0 = 2\cos^{-1}\frac{b}{\sqrt{b^2 + q^2}} \tag{8A.5a}$$

therefore

$$b = q \cot \frac{\phi}{2} \tag{8A.5b}$$

From Eq. (8A.5b) we see that the scattering angle depends on the *impact parameter b*, with large-angle scattering associated with small b. If the impact parameter is increased to $b + db$, the scattering angle ϕ changes by $d\phi$ where, by Eq. (8A.5b),

$$db = -\frac{q}{2} \frac{d\phi}{\sin^2 (\phi/2)} \tag{8A.6}$$

To find the statistical distribution of the α particles scattered by a foil of thickness t which contains n nuclei per unit volume, let N_0 α particles per unit area approach the foil along parallel paths normal to the foil. Then the number of incident particles per unit area which are incident along lines passing at a distance between b and $b + db$ of some nucleus is $N_0(2\pi b\, db)nt$, provided the foil is thin enough. These particles are scattered through angles between ϕ and $\phi + d\phi$. Hence the number of α particles per unit area scattered into the angular cone between ϕ and $\phi + d\phi$ is

$$dN = N(\phi)\, d\phi = N_0 nt\pi \left(\frac{zZe^2}{4\pi\epsilon_0 mv_0^2} \right)^2 \frac{\cos (\phi/2)}{\sin^3 (\phi/2)}\, d\phi \tag{8A.7}$$

In an experiment one typically measures the number of particles scattered at angle ϕ which strike a detector of area A a distance R from the small area element of a scattering foil. It is assumed that the useful foil is small enough for all points to be essentially equidistant from the detector. Let the number N_i represent the number of α particles incident on the element of foil; clearly, N_i is N_0 times the effective area of the foil. Of the particles scattered between ϕ and $\phi + d\phi$ the detector intercepts the fraction $A/(2\pi R \sin \phi)(R\, d\phi)$ (Fig. 8A.1). From Eq.

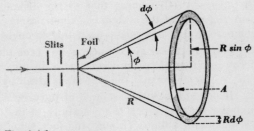

Fig. 8A.1

(8A.7) and the fact that $\sin \phi = 2 \sin (\phi/2) \cos (\phi/2)$, we find the number N_d of α particles reaching the detector to be

$$N_d = \frac{N_i n t A}{4R^2} \frac{z^2 Z^2 e^4}{16\pi^2 \epsilon_0^2 m^2 v_0^4} \frac{1}{\sin^4 (\phi/2)} \tag{8A.8}$$

Problems

1. (a) How many α and β particles are emitted in the radioactive chain starting with $_{94}Pu^{242}$ and ending with $_{82}Pb^{206}$?

Ans: 9, 6

(b) If one starts with $_{93}Np^{237}$, what nuclide remains after six α and three β decays?

Ans: $_{84}Po^{213}$

2. The half-life of $_{83}Bi^{210}$ is 5.0 days. What is the disintegration constant? What fraction of an isolated sample remains after 2.0 days? 10 days? 30 days? What is the mean life of a Bi^{210} atom?

Ans: 0.139 day^{-1}; 0.76; 0.25; 0.016; 7.2 days

3. Calculate the distance of closest approach of a 5-MeV α particle in a head-on collision with a gold nucleus. If the Au nucleus has a radius of about 7×10^{-15} m, find the minimum energy which an α particle must have to "touch" the Au nucleus if the α particle has a radius of about 2×10^{-15} m. How might one tell experimentally that α particles have come close enough to the Au nucleus to bring nuclear forces into play?

Ans: 4.6×10^{-14} m; 25 MeV

4. If an α particle of kinetic energy K makes a head-on elastic collision with a lead nucleus initially at rest, calculate the recoil energy of the Pb nucleus as a fraction of K. What is the distance of closest approach, assuming that K is not large enough to permit the α particle to touch the Pb nucleus. (Remember that at the distance of closest approach the center of mass of the system has kinetic energy.) How does this distance compare to that one calculates if the Pb nucleus remains at rest?

5. Show that if a given nucleus, for example, Po^{218} in Table 8.1, can decay in either of two processes with disintegration constants λ_1 and λ_2, $N(t) = N_0 e^{-(\lambda_1+\lambda_2)t}$ and the half-life is given by $(\ln 2)/(\lambda_1 + \lambda_2)$.

6. If one starts with a sample of pure Bi^{211} (see Table 8.1) at $t = 0$, find the time at which the β activity of the daughter Tl^{207} will attain its maximum value. (Ignore the fact that 0.32 percent of Bi^{211} does not decay to Tl^{207}.)

7. If a sample contains a mixture of two radioactive isotopes, for example, F^{20} and F^{21}, of the same element, show that the ratio r of the activity of the shorter-lived isotope to that of the longer-lived is $r = r_0 e^{-(\lambda_1 - \lambda_2)t}$, where λ_1 is the disintegration constant of the shorter-lived isotope.

8. Radon for medical purposes is pumped off from 5 g of radium every 48 hours. Calculate how many curies of Rn are obtained from each pumping.

Ans: 1.5 curies

9. Show that the fraction of the incident α particles scattered through an angle greater than ϕ by a thin foil is given by

$$\pi n t \left(\frac{2Ze^2}{4\pi\epsilon_0 m v_0{}^2} \right)^2 \cot^2 \frac{\phi}{2}$$

10. α particles from a Ra^{226} source with energy 4.79 MeV are scattered from gold film 0.6μ thick. If a collimated beam of 4×10^5 α particles per second is incident, the detector screen has an area of 1.2 cm², and the detector is 0.6 m from the foil, calculate the expected counting rate for scattering angles of 30, 60, 90, and 135° in a vacuum.

Ans: 0.15, 0.011, 0.0027, 9.2×10^{-4} s⁻¹

11. (a) In the Thomson atom the positive charge was distributed uniformly over a sphere of radius R, the atomic radius. Show that the electric intensity inside such a sphere is $Zer/4\pi\epsilon_0 R^3$ when $r \leq R$.

(b) Show that an electron at the center of this sphere is in stable equilibrium and, if displaced, describes simple harmonic motion of frequency

$$\nu = \frac{1}{2\pi} \sqrt{\frac{Ze^2}{4\pi\epsilon_0 m R^3}}$$

(c) Calculate this frequency for hydrogen and compare it to the highest frequency of the hydrogen spectrum.

Ans: 6.6×10^{15} Hz, about double the frequency of the Lyman series limit

12. Prove that the maximum possible deflection of an α particle by a Thomson atom is on the order of 10^{-4} rad by considering (a) the interaction with one of the electrons in the atom (*Hint:* Show that the maximum speed the electron could get in an elastic collision is twice the speed of the α particle and calculate the deflection of the α if the electron could have this speed normal to the path of the α) and (b) the interaction with the positive charge spread uniformly over a sphere of 1 Å radius. *Hint:* If R is the radius, the maximum force occurs when the α particle is a distance R from the center of the atom; calculate the transverse momentum given the α particle if this maximum force acted during the entire time the α particle needed to go a distance $2R$.

13. An example of transient equilibrium occurs in the decay series $_{88}\text{Ra}^{228} \xrightarrow{\beta} {}_{89}\text{Ac}^{228} \xrightarrow{\beta} {}_{90}\text{Th}^{228}$, in which the half-lives are respectively 6.7 years, 6 hours, and 1.9 years. Show that, when viewed on a time scale measured in years, the half-life of the Ac^{228} is so short that its activity is always equal to that of the Ra^{228}. If one starts with a sample of pure Ra^{228}, show that the activity of the Th becomes equal to that of the Ra in 4.8 years. Verify that the activity of the Th is maximum at 4.8 years. Show that the activity ratio of the Th to the Ra is 1.39 when transient equilibrium is achieved.

14. If there exists a series of radioactive decays arising from an initially pure parent isotope, extend the discussion of Sec. 8.6 to show that

$$N_3(t) = N_1(0)\lambda_1\lambda_2 \left[\frac{e^{-\lambda_1 t}}{(\lambda_2 - \lambda_1)(\lambda_3 - \lambda_1)} + \frac{e^{-\lambda_2 t}}{(\lambda_1 - \lambda_2)(\lambda_3 - \lambda_2)} + \frac{e^{-\lambda_3 t}}{(\lambda_1 - \lambda_3)(\lambda_2 - \lambda_3)} \right]$$

chapter nine

Spectral Lines and the Bohr Model

The Rutherford nuclear atom, proposed in 1911, was a great step forward. With the existence of electrons well established and with convincing evidence from the Zeeman effect that electrons were intimately involved in the emission of spectral lines from excited atoms, physicists inevitably sought a model of the atom in terms of which one might understand atomic spectra and the electron distribution. In 1913 Bohr suggested a specific model of the hydrogen atom which, in spite of certain failings, proved to be a major step toward an understanding of atomic structure. But before Bohr's theory is discussed, it is desirable to know some pertinent facts regarding atomic spectra.

9.1 The Balmer Series

The foundations of spectral analysis were laid by Kirchhoff and Bunsen about 1859. In the next quarter century a wealth of unexplained data on the spectral lines emitted by excited elements was accumulated. As soon as dependable wavelength measurements became available, numerous investigators, reasoning from the analogy of overtones in acoustics, sought for harmonic relations in the lines found in the spectrum of a given element. This search proved fruitless, but certain relations of a different type were discovered. Liveing and Dewar, about 1880, emphasized the physical similarities occurring in the spectra of such elements as the alkali metals. They called attention to the successive *pairs* of lines in the arc spectrum of·sodium and pointed out that these pairs were alternately *sharp* and *diffuse* and that they were more closely crowded together toward the short-wavelength end of the spectrum, suggesting some kind of series relation, which, however, they were unable to discover. In 1883, Hartley discovered an important numerical relationship between the components of doublets or triplets in the spectrum of a given element. If frequencies, instead of wavelengths, are used, Hartley found that *the difference in frequency between the components of a multiplet*, i.e., doublet or triplet, *in a particular spectrum is the same for all similar multiplets of lines in that spectrum.*

Hartley's law made it possible to isolate from the large number of lines in any given spectrum those groups of lines which were undoubtedly related. Figure 9.1 shows the lines in the zinc spectrum from about 2500 to 2800 Å. Two triplets having constant frequency differences are marked by crosses. This same spectral region, however, contains another overlapping series of triplets, which are designated by circles. Sorting out these related lines required a great deal of diligent and patient study.

The beginning of our knowledge of spectral-series formulas dates from the discovery by Balmer, in 1885, that the wavelengths of the nine then known lines in the spectrum of hydrogen could be expressed very closely by the simple formula

$$\lambda = b \frac{n^2}{n^2 - 4} = 3645.6 \frac{n^2}{n^2 - 4} \quad \text{Å} \tag{9.1}$$

where n is a variable integer which takes on the successive values 3, 4,

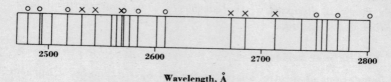

Wavelength, Å

Fig. 9.1 Some lines in the ultraviolet spectrum of zinc. The lines marked X belong to one series of triplets, lines marked O belong to another series.

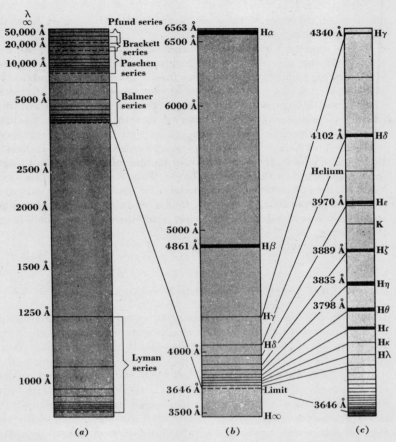

Fig. 9.2 (a) The spectrum of atomic hydrogen consists of the Lyman series in the ultraviolet, the Balmer series in the visible, and several series in the infrared region. (b) The Balmer series in greater detail. (c) A portion of the spectrum of the star ζ Tauri showing more than 20 lines of the Balmer series. (*From A. W. Smith and J. N. Cooper, "Elements of Physics," 7th ed. Copyright 1964. McGraw-Hill Book Company. Used by permission.*)

5, . . . for, respectively, the first (beginning at the red), second, third, . . . line in the spectrum. Figure 9.2b and c shows the Balmer lines of hydrogen.

In 1906 Lyman studied the ultraviolet spectrum of hydrogen and found a series which now bears his name, and 2 years later Paschen found a hydrogen series in the near infrared. These series are indicated in Fig. 9.2a. Formulas similar to that for the Balmer series predict the positions of the lines of these series and will be developed from the Bohr theory in Sec. 9.4.

The impetus which Balmer's discovery gave to work in spectral series is another illustration of the highly convincing nature of relations which are expressible in quantitative form. Soon after the publication of his work, intensive investigations in spectral series were initiated by Kayser and Runge and by Rydberg. It soon became evident that frequency ν is more fundamental than wavelength. We do not, however, measure frequencies directly; laboratory measurements yield wavelengths. Therefore it is customary in spectroscopy to use, instead of the frequency itself, the *wavenumber*, or number of waves per unit length in vacuum, which we shall denote by $\bar{\nu}$. Thus $\bar{\nu} = 1/\lambda_{vac}$. (The wavelength in air is less than that in vacuum for visible light by somewhat less than 1 part in 3000.) For more than a century it has been customary to express wavenumbers in *waves per centimeter*, and we shall follow this practice.

9.2   Spectral Series and Their Interrelations

In announcing his discovery, Balmer had raised the question whether his formula might be a special case of a more general formula applicable to other series of lines in other elements. Rydberg set out to find such a formula. Using the comparatively large mass of wavelength data then available for alkalis, Rydberg isolated other series of doublets of constant frequency difference, according to Hartley's law of constant wavenumber separation. In all cases, these series showed a tendency to converge to some limit in the ultraviolet. He could distinguish two types of such series: a type in which the lines were comparatively sharp, which he called, therefore, *sharp* series; and a type which, because the lines were comparatively broad, he called *diffuse* series. Both types occurred in the arc spectrum of the same element. In many arc spectra, he found also a third type of series in which the doublet spacing *decreased* as the frequency or ordinal number of the line increased, as if tending to vanish at the convergence limit of the series; these he called *principal* series, because they commonly contained the brightest and most persistent lines in the spectrum. By "line" in this connection is meant a complex of actual lines forming a doublet, which is seen as a single line in a spectroscope of moderate resolving power. The chief lines of the principal, sharp, and diffuse series of sodium are plotted in Fig. 9.3, marked *p*, *s*, *d*, respectively.

Eventually Rydberg found that many observed series could be

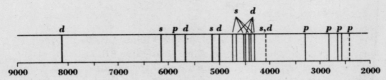

Fig. 9.3 Lines in the spectrum of neutral sodium: p = principal, s = sharp, d = diffuse series. Dotted lines indicate series limits. The first line of the sharp series at 11,393 Å is not shown.

represented closely by an equation of the form

$$\bar{\nu}_m = \bar{\nu}_\infty - \frac{R}{(m + \mu^2)} \tag{9.2}$$

μ and $\bar{\nu}_\infty$ being constants which vary from one series to another. Obviously, by properly choosing the ordinal number m of the lines, μ can always be made less than 1 in such a formula. The constant $\bar{\nu}_\infty$ represents the high-frequency limit to which the lines in the series ultimately converge. The Balmer formula is a special case of this more general *Rydberg formula;* for Eq. (9.2) can be written

$$\bar{\nu} = \frac{1}{\lambda} = \frac{1}{b} - \frac{4}{bm^2} = \bar{\nu}_\infty - \frac{R}{m^2} \tag{9.2a}$$

with $\bar{\nu}_\infty = 1/b$, $R = 4/b$, $m = 3, 4, 5, \ldots$. This is of the type of Eq. (9.2) with $\mu = 0$.

The constant R in Eq. (9.2), now called the Rydberg constant, was found to have the same value for a large group of series for each substance, and very nearly the same for all substances. Its slight variation from one atom to another is now known to be due to differences in the atomic weight, the effect of which can be calculated theoretically (Sec. 9.4). It is thus possible to calculate what R would be for an atom of infinite mass; this value is denoted by R_∞, the value for a particular kind of atom being denoted by another appropriate suffix. The values as found from spectroscopic observation are:

For hydrogen atoms, H¹

$$R_{\rm H} = 109{,}677.58 \text{ cm}^{-1} = 1.0967758 \times 10^7 \text{ m}^{-1}$$

For deuterium atoms, H²

$$R_{\rm D} = 109{,}707.42 \text{ cm}^{-1}$$

For helium, He⁴

$$R_{\rm He} = 109{,}722.27 \text{ cm}^{-1}$$

For atoms of infinite mass

$$R_\infty = 109{,}737.31 \text{ cm}^{-1}$$

9.3 Spectral Terms

Rydberg noticed also several remarkable relationships between different spectral series belonging to the same element. He noted that the *sharp and the diffuse series have a common convergence limit*. Next, he observed that this common limit was equal to a *term* in the formula for the *principal* series, viz., the term $R/(m + \mu)^2$ with m set equal to 1, and that a similar equality held between the *limit* of the *principal* series itself and the variable *term*, for $m = 1$, in the formula for the *sharp* series.

Because of these remarkable relations, the formulas for the three series in question could be rewritten in the following form:

$$\bar{\nu}_p = \frac{R}{(1 + S)^2} - \frac{R}{(m + P)^2} \qquad m = 1, 2, \ldots \qquad (9.3a)$$

$$\bar{\nu}_s = \frac{R}{(1 + P)^2} - \frac{R}{(m + S)^2} \qquad m = 2, 3, \ldots \qquad (9.3b)$$

$$\bar{\nu}_d = \frac{R}{(1 + P)^2} - \frac{R}{(m + D)^2} \qquad m = 2, 3, \ldots \qquad (9.3c)$$

Here we have indicated that, at least for the alkali metals, m starts from 1 in the principal series but from 2 in the others.

Two additional relations are apparent from the formulas. If we set $m = 1$ in the formula for $\bar{\nu}_s$, we obtain the same number as the value of $\bar{\nu}_p$ for $m = 1$, but with reversed sign! Furthermore, the difference between the limit of the principal series and the common limit of the other two series is just the wavenumber of the first line of the principal series (the *Rydberg-Schuster law*, enunciated in 1896 by Rydberg and independently in the same year by Schuster).

These formulas suggested to Rydberg the possibility that the first term on the right might also vary, in the same manner as the second, thus giving rise to additional series of lines; e.g., we might expect to find a series represented by the formula

$$\bar{\nu} = \frac{R}{(2 + S)^2} - \frac{R}{(m + P)^2} \qquad m = 3, 4, \ldots$$

Lines or series of this sort were discovered later by Ritz. Such lines are called *intercombination lines* or *series*, and the possibility of their occurrence is known as the *Ritz combination principle*. Many examples of them are now known.

The most significant features about atomic spectra thus seem to be the following:

1. The wavenumber of each line is conveniently represented as the difference between two numbers. These numbers have come to be called *terms*.
2. The terms group themselves naturally into ordered sequences, the terms of each sequence converging toward zero.
3. The terms can be combined in various ways to give the wavenumbers of spectral lines.
4. A series of lines, all having similar character, results from the combination of all terms of one sequence in succession with a fixed term of another sequence. Series formed in this manner have wavenumbers which, when arranged in order of increasing magnitude, converge to an upper limit.

The simple picture presented here requires considerable extension. In writing Rydberg formulas for spectral "lines," we have ignored the fine structure of the lines, by means of which series were first picked out; *singlet* series, in which each line is single, are known in many elements, but more commonly the lines form doublets, triplets, or groups of even more components. In such cases, a separate Rydberg formula must be written for each component line.

9.4 The Bohr Theory

Niels Bohr, working in Rutherford's laboratory at Manchester in 1913, succeeded in combining Rutherford's nuclear atom with Planck's quanta and Einstein's photons in a way that predicted the observed spectral lines of hydrogen. It had long been recognized that coulomb electrostatic attraction, identical with gravitational attraction in spatial properties, would lead to stable circular or elliptical orbits for an electron bound to a positive nucleus if this were the only force involved; but classical electromagnetic theory predicts that an accelerated charge radiates energy and, as a consequence, would quickly spiral into the nucleus. However, Planck's original quantum theory (Sec. 5.6) involved two essential features:

1. An oscillator can exist only in one of a number of discrete quantum states, and to each of these states there corresponds a definite allowed value of its energy.
2. No radiation is emitted while the oscillator remains in one of its quantum states, but it is capable of jumping from one quantum state into another one of lower energy, the energy lost in doing so being emitted in the form of a pulse or quantum of radiation.

Successful applications of the first of these assumptions had already been made in other fields, notably by Einstein and especially by Debye in the specific heat of solids, which will be discussed in Chap. 20. Nicholson had also attempted to apply the theory to spectra, and with some success, but he was unable to make it yield a series of lines converging to a limit. Bohr discovered how to apply similar ideas to a hydrogen atom of the Rutherford type and succeeded in arriving at a theoretical formula for its spectrum that agrees with observation.

Bohr assumed that an electron in the field of a nucleus was not capable of moving along every one of the paths that were possible according to classical theory but was restricted to one of a certain set of allowed paths. While it was so moving, he assumed that it did not radiate, contrary to the conclusion from classical theory, so that its energy remained constant; but he assumed that the electron could jump from one allowed path to another one of lower energy and that when it did so, radiation was emitted containing energy equal to the difference in the energies corresponding to the two paths.

As to the *frequency* of the emitted radiation, he considered several alternative hypotheses but finally adopted the same assumption that Planck had made for his oscillators. That is, if E_i and E_f are the energies of the atom when moving in its initial and final paths, respectively, the *frequency* ν of the radiation emitted is determined by the condition that

$$h\nu = E_i - E_f \tag{9.4}$$

where h is Planck's constant. This assumption had the additional advantage of agreeing with that made by Einstein in arriving at his highly successful photoelectric equation (Sec. 6.7). For the latter reason, Eq. (9.4) came later to be known as the *Einstein frequency condition*.

Concerning the formulation of the condition that determines the allowed paths, Bohr was also in doubt. While the electron remains in one of its *stationary states*, as he called them, he supposed it to revolve in a circular or elliptical orbit about the nucleus in accordance with the classical laws of mechanics. But what fixes the size and shape of this orbit? In the end, he postulated that the orbit is a circle, with the nucleus at its center, and of such size that the *angular momentum of the electron about the nucleus is an integral multiple of the natural unit $h/2\pi$*, which we write $\hbar$. (This is not true, as we shall see subsequently.)

The allowed values of the energies of the stationary states and of the spectral lines expected on the basis of Bohr's theory are easily deduced for a system consisting of a single electron and a nucleus of charge Ze from his three basic assumptions:

1. *The electron revolves in a circular orbit with the centripetal force supplied by the coulomb interaction between the electron and the nucleus,*

$$\frac{Ze^2}{4\pi\epsilon_0 r^2} = m\,\frac{v^2}{r} = m\omega^2 r \tag{9.5}$$

where r = radius of allowed orbit

v = electron velocity

ω = corresponding angular velocity

2. *The angular momentum A of the electron takes on only values which are integer multiples of $\hbar$.*

$$A = mvr = m\omega r^2 = n\hbar \tag{9.6}$$

3. *When an electron makes a transition from one allowed stationary state to another, the Einstein frequency condition $h\nu = E_i - E_f$ is satisfied.*

Eliminating v (or ω) from Eqs. (9.5) and (9.6), we find the allowed radii for the orbits to be (Fig. 9.4)

$$r_n = 4\pi\epsilon_0 \frac{\hbar^2 n^2}{me^2 Z} \tag{9.7}$$

Thus the radii of successive allowed orbits are proportional to n^2, or to 1, 4, 9, 16 . . . . For hydrogen $Z = 1$, and the radius of the smallest allowed circle, known as the *Bohr radius*, is given by

$$a_B = r_1(Z = 1, n = 1) = 4\pi\epsilon_0 \frac{\hbar^2}{me^2} = 0.5292 \times 10^{-10} \text{ m} \tag{9.8}$$

The diameter of the orbit is thus close to 1 Å, which agrees well with estimates from kinetic theory. This is a first indication that the new theory may be able to explain the apparent sizes of molecules.

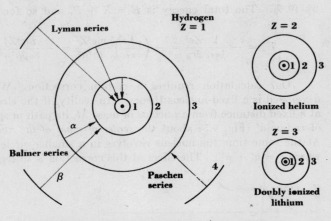

Fig. 9.4 The smallest Bohr orbits for three atoms, drawn to a common scale.

By Eq. (9.6), we have $v = n\hbar/mr$, and so the velocity of the electron in the nth orbit is

$$v_n = \frac{1}{4\pi\epsilon_0} \frac{e^2 Z}{\hbar n} = \frac{e^2 Z}{2\epsilon_0 h n} \tag{9.9}$$

The ratio α of the speed of the electron in the first Bohr orbit of hydrogen to the speed of light c is known as the *fine-structure constant;* its value is given by

$$\alpha = \frac{v_1}{c} = \frac{1}{4\pi\epsilon_0} \frac{e^2}{\hbar c} = \frac{1}{137.0377} \tag{9.10}$$

The energy of the electron is partly kinetic and partly potential. If we call the energy zero when the electron is at rest at infinity, its potential energy in the presence of the nucleus is

$$P = -\frac{1}{4\pi\epsilon_0} \frac{Ze^2}{r}$$

Its kinetic energy is

$$K = \tfrac{1}{2}mv^2 = \tfrac{1}{2}mr^2\omega^2 = \frac{1}{4\pi\epsilon_0} \frac{Ze^2}{2r}$$

by (9.5). The total energy is $E = K + P$, and so for the nth state

$$E_n = -\frac{1}{4\pi\epsilon_0} \frac{Ze^2}{2r_n} = -\left(\frac{1}{4\pi\epsilon_0}\right)^2 \frac{me^4 Z^2}{2\hbar^2 n^2} = -\frac{me^4 Z^2}{8\epsilon_0^2 h^2 n^2} \tag{9.11}$$

Our calculation requires a certain correction. What we have developed is a fixed-nucleus theory. In reality, if the electron revolves at a fixed distance from a nucleus of mass M, its path in space is a circle of radius a' (Fig. 9.5) about the *center of mass of the combined system*. At the same time the nucleus revolves in a smaller circle of radius a'', where $r = a' + a''$. The effect of this correction is to replace the mass

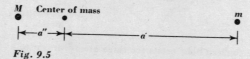

Fig. 9.5

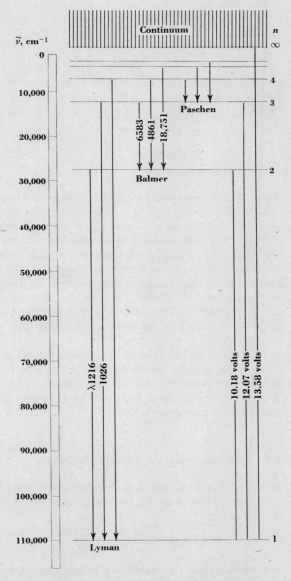

Fig. 9.6 The lower energy levels of hydrogen and a few of the lines arising from transitions between the levels. Potential differences between levels are shown at the right.

m by the *reduced mass*

$$m_r = \frac{m}{1 + m/M} \tag{9.12}$$

If we insert appropriate values for the constants of Eq. (9.11), we have

$$E_n = -\frac{2.176 \times 10^{-18} Z^2}{n^2} \quad \text{J}$$

$$= -\frac{13.58 Z^2}{n^2} \quad \text{eV} \tag{9.11a}$$

Thus an electron bound to a proton can have any one of a *discrete set of allowed energies* (Fig. 9.6). In addition, Bohr assumed that an unbound electron near the proton could move with *any value whatever of positive energy*. In such a case the orbit would be a hyperbola. Consequently, we have, in all, a *discrete set* of allowed *negative* energies, or energy levels, converging to the value zero and, from zero up, a *continuum* of allowed positive energies. Since the zero value of the energy corresponds to the electron at rest at infinity, *the numerical value of the (negative) energy of the normal state also represents the least energy required to ionize the atom by complete removal of the electron.* For hydrogen this *ionization energy* is 13.58 eV.

A final remark should be added to forestall misunderstanding. The orbital angular momenta assigned to the various quantum states in the Bohr theory are not the correct ones, as we shall see in Chap. 14. The Bohr theory does, however, lead to the observed energy levels (except for a minor correction).

9.5 The Spectrum of Atomic Hydrogen

When an electron makes a radiative transition from an initial state n_i to a lower final state n_f, the resulting photon has, by Eq. (9.4), energy

$$h\nu = \left(\frac{1}{4\pi\epsilon_0}\right)^2 \frac{m_r e^4}{2\hbar^2} \left(\frac{1}{n_f^2} - \frac{1}{n_i^2}\right) Z^2 \tag{9.13}$$

and *wavenumber* $\bar{\nu}$ (number of waves per unit length)

$$\bar{\nu} = \frac{1}{\lambda} = \frac{\nu}{c} = \frac{m_r e^4}{8\epsilon_0^2 h^3 c} \left(\frac{1}{n_f^2} - \frac{1}{n_i^2}\right) Z^2 = R\left(\frac{1}{n_f^2} - \frac{1}{n_i^2}\right) Z^2 \tag{9.14}$$

where $R = m_r e^4 / 8\epsilon_0^2 h^3 c$ is the Rydberg constant. Bohr's calculation of R in terms of fundamental constants was a dramatic triumph of his theory.

For hydrogen with $Z = 1$ the energies and wavenumbers of the Lyman series photons are given by Eqs. (9.13) and (9.14), respectively, by setting $n_f = 1$ and letting n_i take values 2, 3, 4, ... Since $n = 1$ corresponds to the lowest allowed energy state, or *ground state*, of hydrogen, we see that the lines of the Lyman series arise from electron transitions from higher, or *excited*, states to the ground state (Fig. 9.5).

The Balmer series arises from electron transitions to the state with $n_f = 2$ from initial states characterized by $n_i = 3, 4, 5, ...$ Similarly, the Paschen series results from transitions from states with $n_i = 4, 5, 6, . . .$ to a final state $n_f = 3$. The Bohr theory successfully predicted the wavelengths of all atomic-hydrogen lines which had been discovered and of lines yet to be observed. The Brackett series, found in 1922, lies in the infrared and is associated with transitions with $n_f = 4$. In 1924 the Pfund series with $n_f = 5$ was discovered in the far infrared.

9.6 Ionized Helium

A helium atom which has lost a single electron resembles a hydrogen atom, except that $Z = 2$ and the nucleus is nearly four times as heavy. The spectrum emitted by such atoms is known as the *spark spectrum* of helium (or the helium II spectrum), because it is emitted much more strongly when the helium is excited by a spark than when it is excited by an arc. The *arc spectrum* of helium, emitted by the neutral atom, will be considered later (Sec. 17.4).

Putting $Z = 2$ in Eq. (9.14), we see that *ionized helium should emit the same spectrum as hydrogen except that all frequencies are four times as great*, or all wavelengths a quarter as great.

This conclusion from the theory agrees with observation except for a slight numerical discrepancy due to the fact that the reduced mass m_r is slightly greater for He than for H. So precisely can the wavelengths of spectral lines be measured that the experimental value of the Rydberg constant for hydrogen and for helium can be used to make one of the better determinations of the ratio of the mass of the proton to the mass of the electron.

In a similar way, *doubly ionized lithium* is found to emit the hydrogen spectrum with all frequencies multiplied (almost exactly) by 9; *trebly ionized beryllium* emits them increased in the ratio 16; and so on. The first line of the Lyman series for quadruply ionized boron ($Z = 5$) has been found by Edlen at 48.585 Å, with a frequency 25.04 times that of the first Lyman line of hydrogen. The similarity of spectra of atoms and ions which have identical numbers of electrons extends throughout the periodic table. The term *isoelectronic sequence* is used to refer to a sequence of atoms and ions having the same number of electrons; thus H, He$^+$, Li^{++}, Be^{3+}, B^{4+}, and so forth constitute the one-electron isoelectronic sequence.

9.7 Moseley's Law

In 1913, the same year in which Bohr published his theory and Bragg reported the development of the x-ray spectrometer, Moseley undertook the systematic study of characteristic x-rays discussed in Sec. 7.7. In searching for a relation between the frequency of the $K\alpha$ line and some property of the atom in which the line originated, Moseley first observed that the frequency did not vary uniformly with the atomic weight (curve A, Fig. 7.10). In Bohr's theory, however, which had just been proposed, the charge on the nucleus played a fundamental role. According to Eq. (9.13) the frequency of a spectral line for a one-electron atom is given by

$$\nu = \frac{me^4 Z^2}{8\epsilon_0^2 h^3}\left(\frac{1}{n_f^2} - \frac{1}{n_i^2}\right) \tag{9.13a}$$

If we set $n_f = 1$ and $n_i = 2$ in Eq. (9.13a), we obtain

$$\nu = 0.246 \times 10^{16} Z^2 \text{ Hz}$$

Except for the slight correction to Z, this equation is almost identical to Eq. (7.2). This agreement suggests that the $K\alpha$ line arises from an electron transition from an initial state $n_i = 2$ to a final state $n_f = 1$.

It seems remarkable that a formula derived for a one-electron atom should be applicable to an atom such as copper with its 29 electrons. However, as we shall see in Chap. 15, we may think of an atom as having at most two electrons with $n = 1$. To excite K x-ray lines one of these electrons must be somehow ejected. If an electron from a state with $n = 2$ makes a radiative transition to fill this vacancy, a $K\alpha$ photon is emitted. In terms of the Bohr picture the other electrons are in orbits of greater radius and therefore affect the radiated energy little.

A Moseley plot of $\sqrt{\nu}$ as a function of Z for the $K\beta$ line had a somewhat larger slope than that for the $K\alpha$ line, but roughly the same intercept on the atomic-number abscissa. Moseley found that the slope agreed well with that predicted by Eq. (9.13a) with $n_f = 1$ and $n_i = 3$.

Thus within a year of its publication the Bohr theory achieved another great success.

9.8   The Bohr Magneton

An electron moving in a Bohr orbit should possess a magnetic moment. A loop bearing a current I has a magnetic moment μ equal to the product of I and the area of the loop.   For the nth Bohr orbit the area is $\pi r_n{}^2$, while the current (charge passing per unit time) is $-e$ multiplied by the number of revolutions made by the electrons per unit time.   Thus

$$\mu = I(\pi r_n{}^2) = \frac{-ev}{2\pi r_n}\,\pi r_n{}^2 = \frac{-e}{2}\,v r_n \qquad (9.15)$$

where v is the velocity of the electron.   If we multiply both numerator and denominator by m and recall that $m v r_n$ is the orbital angular momentum, we have

$$\mathbf{\mu}_l = \frac{-e}{2m}\,\mathbf{A}_l \qquad (9.16)$$

where $\mathbf{\mu}_l$ is the vector magnetic moment and $\mathbf{A}_l$ the vector angular momentum and the negative sign shows these two vectors to be in opposite directions for a negative revolving charge (Fig. 9.7).   The subscript l is used here and later to designate "orbital" to distinguish it from the corresponding quantities we shall associate with electron spin (Sec. 9.9).   The ratio $\mathbf{\mu}_l/\mathbf{A}_l\ (=\,-e/2m)$ is called the *gyromagnetic ratio* for orbital motion.

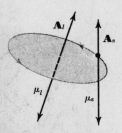

Fig. 9.7 The orbital and spin angular momenta of an electron with the associated magnetic moments.

For a Bohr hydrogen atom in its ground state $n = 1$, and by Eq. (9.6) $(A_l)_{n=1} = \hbar$. The magnetic moment predicted by Eq. (9.16) for this orbit is called the Bohr magneton, given by

$$\mu_B = \frac{e\hbar}{2m} \tag{9.17}$$

It has the numerical value 9.27×10^{-24} A-m².

Although quantum mechanics shows that the orbital motion of a hydrogen electron in its ground state contributes no magnetic moment, it turns out that electronic magnetic moments are conveniently expressed in units of Bohr magnetons, and we shall use them repeatedly in later chapters.

The following feature of classical electromagnetics is of interest. Suppose that the magnetic moment $\mathbf{\mu}$ of a body is proportional to its vector angular momentum $\mathbf{A}$ about its center of mass. Draw OP (Fig. 9.8) to represent $\mathbf{A}$, making an angle θ with an axis OB drawn in the direction of a magnetic field $\mathbf{B}$. Then, during time dt, the torque on the magnet will add to $\mathbf{A}$ a vector increment $d\mathbf{A}$ of magnitude $\mu B \sin \theta \, dt$ in a direction perpendicular to the plane POB. This increment will change the direction of $\mathbf{A}$ but not its magnitude, so that $\mathbf{A}$ precesses about OB in such a way that the end of the vector OP revolves in a circle of radius $A \sin \theta$; since during dt it revolves through an angle $dA/(A \sin \theta) = \mu B \sin \theta \, dt/(A \sin \theta)$ or $\mu B \, dt/A$, it revolves at an angular

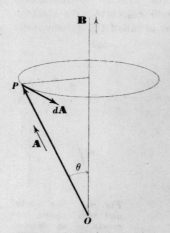

Fig. 9.8 Precession of a magnetic gyro in a magnetic field.

rate

$$\omega = \frac{\mu B}{A} \qquad \text{rad/s} \tag{9.18}$$

For an electron in orbital motion, its orbital angular momentum A_l of magnitude $-2m\mu/e$ by Eq. (9.16) will thus precess at the numerical rate

$$\omega_l = \frac{eB}{2m} \tag{9.19}$$

The angular frequency ω_l is known as the *Larmor precessional frequency*. According to a theorem due to Larmor, the introduction of a magnetic field modifies the classical motion of a set of electrons in the field of a nucleus chiefly as if a uniform rotation at angular frequency ω_l were superposed upon their original motion. The angular momentum of the electrons about an axis parallel to **B** is thereby slightly altered, as a result of inductive action during the production of the field, but this change is so small compared to the values of A_l due to orbital motion that it may be ignored in the present connection.

9.9 Electron Spin

In 1925 Goudsmit and Uhlenbeck pointed out that certain features in atomic spectra, such as the doublet fine structure of the alkali elements, could be explained if it was assumed that the electron "spins" about an axis through its center of mass and that it has both *angular momentum* and a *magnetic moment* associated with this rotation. In the Bohr model with spin the electron is closely analogous to the earth, which rotates about its axis each sidereal day and revolves about the sun each year. The spin angular momentum A_s of the electron has a constant magnitude

$$A_s = \sqrt{\tfrac{1}{2}(\tfrac{3}{2})}\,\hbar = \sqrt{s(s+1)}\,\hbar \tag{9.20}$$

where s has the value $\frac{1}{2}$.

Any measurement of the component of the spin angular momentum parallel to some prescribed z axis yields the result that

$$(A_s)_z = m_s\hbar = \pm\tfrac{1}{2}\hbar \tag{9.20a}$$

where m_s is called the *spin quantum number*, for which the allowed values are $\frac{1}{2}$ and $-\frac{1}{2}$. Thus, by virtue of its spin, the electron possesses two *internal* quantum states, characterized by the two values of m_s.

Associated with the spin angular momentum $\mathbf{A}_s$ there is a magnetic moment

$$\mathbf{\mu}_s = g_s \frac{e}{2m} \mathbf{A}_s = -2 \frac{e}{2m} \sqrt{\left(\frac{1}{2}\right)\left(\frac{3}{2}\right)} \hbar \mathbf{1}_s \qquad (9.21)$$

where g_s is a number[1] known as the *spin g factor* and $\mathbf{1}_s$ is a unit vector in the direction of $\mathbf{A}_s$. If a magnetic field defines the z direction, the quantization rule, Eq. (9.20a), requires that the z component of $\mathbf{\mu}_s$ have one of the two values

$$(\mathbf{\mu}_s)_z \approx \mp 2 \frac{e}{2m} \frac{\hbar}{2} \approx \mp \frac{e\hbar}{2m} \qquad (9.21a)$$

Comparison with Eq. (9.17) shows that this is essentially 1 Bohr magneton. The interaction of the spin magnetic moment of an electron with the magnetic field arising from its orbital motion plays an important role in the fine structure of atomic energy levels.

9.10   The Bohr Correspondence Principle

Bohr's theory represents a remarkable combination of principles taken over from classical theory with postulates radically at variance with that theory. He solved the old problem of stability merely by *postulating* that the cause of instability, the emission of radiation and the accompanying radiation reaction, did not exist so long as the electron remained in one of its allowed stationary orbits. The electron could thus remain in its stationary state of lowest energy indefinitely, without spiraling down into the nucleus, as classical theory would require. But the problem of stability was solved only at the expense of throwing away the only picture we had of the mechanism by which the atom could emit spectral lines. For Bohr's postulates provide no picture of the sequence of events *during transitions between orbits*.

A hybrid theory of this sort was widely felt to be unsatisfactory, but it was astonishingly successful. Later we describe the wave-mechanical theory that has replaced the Bohr theory. This theory leads very nearly to the same set of energy values for the hydrogen

[1] Actually, $g_s = 2.00230$; see Sec. 18.3.

atom, but it suggests a quite different picture of the behavior of the electron while in a quantum state. In this theory, it is only about half true that the electron is in motion in the atom, even when it is in one of its higher quantum states; at least, it cannot be said to follow a definite orbit. Because of the abstractness of wave mechanics the original simple Bohr picture retains something more than mere historical interest, and it offers a bridge between classical physics and modern quantum mechanics.

Classically an electron in a circular orbit radiates at the frequency of its motion. In Bohr's theory the emitted spectral frequencies are distinct from the *frequency of orbital revolution* of the electron. This latter frequency can be found by solving Eqs. (9.5) and (9.6) for ω, from which

$$\nu_{orb} = \frac{\omega}{2\pi} = \frac{me^4 Z^2}{4\epsilon_0^2 h^3} \frac{1}{n^3} \tag{9.22}$$

The emitted frequency for a transition between states n_i and n_f is given by Eq. (9.13a), which may be written

$$\nu = \frac{me^4 Z^2}{4\epsilon_0^2 h^3} \frac{n_i + n_f}{2n_i^2 n_f^2} (n_i - n_f)$$

It can be shown that, since $n_i > n_f$,

$$\frac{1}{n_i^3} < \frac{n_i + n_f}{2n_i^2 n_f^2} < \frac{1}{n_f^2} \tag{9.23}$$

Hence, when $n_i - n_f = 1$, the frequency ν of the emitted radiation is intermediate between the frequencies of orbital revolution in the initial and final states; for large n the emitted and orbital frequencies are almost the same. Furthermore, putting in succession $n_i - n_f = 2, 3, \ldots$, we have an approximation to various harmonic overtones of a fundamental frequency. Thus, quantum jumps in which $\Delta n > 1$ correspond to the overtones in the case of classical vibrations. The fact that *in the limit characterized by large quantum numbers classical physics and quantum physics lead to the same predictions* was pointed out by Bohr in 1923 and is known as the *Bohr correspondence principle*. The requirement that the prediction of any quantum-mechanics calculation reduce to that of classical physics in the appropriate limit was a great heuristic aid in the early development of quantum mechanics before the logical foundations were well established.   It continues to be a useful test for some theories.

9.11   *Extension of Bohr's Theory*

In his original paper, Bohr remarked that the orbit of the electron in a hydrogen atom might be an ellipse instead of a circle.   A detailed theory of elliptical orbits was developed by Sommerfeld several years later.   The geometrical ideas involved still possess a certain interest.

According to the laws governing motion under an inverse-square force, an elliptical orbit has one of its foci at the center of mass and the energy of the system depends only on the major axis of the ellipse. The orbit lies in a fixed plane, so that the motion can be described by means of polar coordinates r, θ, with the origin at the nucleus.   Then, as θ increases through 2π, r increases from its minimum to a maximum, after which it decreases again to the minimum (Fig. 9.9).

To select from the infinity of classically allowed ellipses Sommerfeld generalized the quantum condition which had been used by Planck and by Bohr.   We saw in Sec. 5.6 that Planck assumed that the ellipse arising from a plot of p as a function of x for a harmonic oscillator (Fig. 5.5) had an area equal to an integer times h.   This area may be represented as $\oint p\, dx$, where $\oint$ means integration over a complete cycle.   Therefore, Planck's condition for a quantum state of an oscillator can be expressed as

$$\oint p\, dx = nh \tag{9.24}$$

This rule is easily modified to apply to a Bohr orbit.   If we choose as coordinate of the electron its angular displacement θ (Fig. 9.9), the generalized momentum p_θ corresponding to θ is the angular momen-

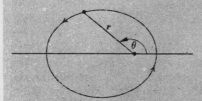

Fig. 9.9 An electron revolving in an elliptical path about a proton at one focus.

tum mvr, and we see that

$$\oint p_\theta \, d\theta = \int_0^{2\pi} mvr \, d\theta = nh \tag{9.25}$$

gives just the Bohr condition, Eq. (9.6). Wilson (1915) and Sommerfeld (1916) independently concluded that for any generalized coordinate q and its conjugate momentum p_q, the quantum condition

$$\oint p_q \, dq = n_q h \tag{9.26}$$

was applicable. This condition, known as the *Wilson-Sommerfeld quantization* rule, was widely used in the expansion of the older quantum theory. (It is now known that it is not a generally valid condition.) For his elliptic orbits Sommerfeld therefore took

$$\oint p_r \, dr = n'h \tag{9.27a}$$
$$\oint p_\theta \, d\theta = kh \tag{9.27b}$$

where the integers n' and k are known respectively as the *radial* and *azimuthal quantum numbers*.

Sommerfeld assumed that n' could be 0, 1, 2, 3, . . . while the azimuthal quantum number k took on values 1, 2, 3, . . . , with 0 excluded since that would correspond to harmonic oscillation of the electron directly through the nucleus. He showed that the orbits allowed by Eqs. (9.27a, b) had semimajor axes $(n' + k)^2 a_B/Z$ and semiminor axes $(n' + k)k a_B/Z$, where a_B is the Bohr radius. The allowed energies in the nonrelativistic case depended only on the total quantum number $n(= n' + k)$ and are the same as those of the Bohr theory. Thus up to this point nothing is gained except a greater variety of orbital shapes. As an example, for $n = 3$, we can have the Bohr circle with $k = 3$ and $n' = 0$ or either one of two ellipses with major axis equal to the diameter of the circle, with $k = 2$, $n' = 1$ or $k = 1$, $n' = 2$, respectively. These orbits are shown to scale in Fig. 9.10 labeled also with wave-mechanical and spectroscopic symbols (Sec. 15.1).

Sommerfeld then showed, however, that, if allowance is made for the known *variation of electronic mass with speed*, the energy of the elliptical motion is slightly different from that of the circular motion, and in this way he arrived at a splitting, or fine structure, of the levels of one-electron atoms, which appeared to be in quantitative agreement with observation. However, the idea of sharp orbits is not in accord

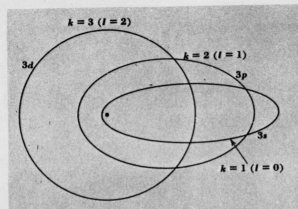

Fig. 9.10   The Bohr-Sommerfeld orbits for $n = 3$.

with modern quantum mechanics, which we believe to be the correct approach and which is capable of predicting the hydrogen fine structure in much greater detail than Sommerfeld's historic elliptic orbits could. In the meantime, Bohr and others were endeavoring to extend the theory to atoms containing more than one electron. Little quantitative success was achieved in this direction. In particular, no plausible arrangement of orbits could be devised for the two electrons of helium which would give the correct value for the first ionization potential (24.6 V). In a broader way, Bohr endeavored to understand the known variation of chemical and physical properties from atom to atom through the periodic table, but without great success. The truth is that two principles essential for the understanding of complex atoms were at that time unknown—the Pauli exclusion principle and the phenomenon of electron spin. Accordingly, we shall follow the old quantum theory no further.

Problems

1.   Calculate the wavelengths of the first two lines and of the series limit for the Balmer series of ionized helium.

   Ans: 164, 122, 91 mμ

2.   Find the energies of the photons and the wavelengths of the first three lines of the Paschen series for hydrogen.

   Ans: 0.66 eV, 1.88 μ; 0.97 eV, 1.28 μ; 1.13 eV, 1.09 μ

3. How many revolutions does an electron in the first Bohr orbit of hydrogen make each second? To what current does this correspond? Find the magnetic induction at the proton due to the motion of the electron.

Ans: 6.6×10^{15} rps; 1.05 mA; 12.5 Wb/m^2

4. Hydrogen atoms are in an excited state with $n = 5$. In terms of the Bohr model how many spectral lines can be radiated as these atoms return to the ground state? Find the energies of the lines with the longest and shortest wavelengths.

5. What is the binding energy of the single electron of a B^{4+} ion? Find the wavelength of the photon emitted by such an ion when the electron goes from the first excited state to the ground state.

6. Find the energy in electron volts required to strip a calcium atom ($Z = 20$) of its last electron, assuming the other 19 have been removed. How does this compare with the energy required to excite the K x-ray lines of Ca, which is about 3.7 keV? Why the difference?

Ans: 5.4 keV; other electrons change potential

7. Find the frequency of a photon emitted when a hydrogen atom makes a transition from a state with $n = 30$ to one with $n = 29$. Find the fractional difference between this frequency and the orbital frequency of the electron in its initial state.

8. Hydrogen atoms are initially in the $n = 4$ energy level. How many different photon energies will be emitted if the atoms go to the ground state, with all possible transitions represented? Assuming that from any given excited state all possible downward transitions are equally probable (not true!), what is the total number of photons emitted if there are 600 atoms initially in the $n = 4$ state?

9. Calculate the energy and wavelength of a photon which is emitted when a hydrogen atom initially at rest makes a transition between states characterized by $n = 6$ and $n = 1$. Find the speed with which the H atom recoils.

10. What transition in singly ionized helium leads to a line with almost the same wavelength as that of the first line of the hydrogen Lyman series? Using the appropriate Rydberg constants, find the wavelength difference between these two lines. Which of the lines has the shorter wavelength?

11. Show that, on the basis of the approximations which led to Eq. (9.13a), the slope of the Moseley plot for $K\beta$ is expected to be $\sqrt{\frac{32}{27}}$ times that for $K\alpha$.

12. Show that if the electron and nucleus of a Bohr atom revolve about the center of mass, the angular momentum of the system and the allowed radii and energies are all given by expressions identical to those of Sec. 9.4 except for the replacement of m by the reduced mass m_r.

13. A hydrogen atom in a state with $n = 6$ is moving with a speed of 5×10^7 m/s when it enters a uniform magnetic field with induction B. What is the minimum value of B which can ionize the atom if this can be done by exerting a disruptive force greater than or equal to the coulomb attraction?

14. A hydrogen atom initially at rest and in its ground state absorbs a 100-eV photon. If the ejected photoelectron moves in the same direction as that in which the incident photon was moving, find (a) the kinetic energy and speed of the photoelectron, and (b) the momentum and energy of the recoiling proton.

15. Consider an electron rotating about a nucleus of charge Ze in a circular path of radius 1 Å. Classically, such an electron should radiate energy at the rate

$$\frac{2}{3} \frac{e^2 a^2}{4\pi\epsilon_0 c^3}$$

where a is the acceleration of the charge. If this were true, find the initial frequency and wavelength emitted. Show that this electron would spiral into a nucleus of radius 10^{-14} m in a time of about $10^{-10}Z^{-1}$ s.

Ans: $\lambda = 1.18 \times 10^{-7}Z^{-\frac{1}{2}}$ m

16. The Wilson-Sommerfeld quantization rule (not always valid!) states that $\oint p \, dq = nh$, where n is an integer. If an elastic bouncing ball follows this law, show that the allowed energies are quantized with the eigenvalues $E = (9mg^2h^2n^2/32)^{\frac{1}{3}}$.

17. A hydrogen atom at rest is struck by a second H atom with kinetic energy K. Show that the minimum K which is required to raise one of the atoms to its first excited state is 1.5 times the ionization energy.

18. Calculate the smallest kinetic energy which an electron may have and still excite a hydrogen atom initially at rest. What minimum kinetic energy is necessary to ionize the H atom?

19. From the values of the Rydberg constant for helium and for hydrogen (and from the ratio of the atomic weights of H and He), find the ratio of the mass of the proton to that of the electron.

chapter ten

Particles and Interactions

The hydrogen atom is composed of a proton serving as nucleus and a single electron. It was once hoped that perhaps all matter could be built up from protons and electrons alone. However, this is not the case. In this chapter we introduce some of the other particles of modern physics and discuss qualitatively the interactions between them.

10.1   Alpha and Gamma Spectra of Radioelements

The study of the energies of the photons emitted by excited atoms is a primary source of information regarding the allowed energy states of the electrons. Similarly, measurement of the energies of the α, β, and γ rays emitted by radioactive elements yields important clues for determining the energy-level scheme of nuclei.

a. Alpha-particle Spectra   The energies of α particles can be determined from their deflection in a magnetic field, once their specific charge e/m is known.   If the intensity of the source suffices, the measurement can be made in a straightforward manner, defining a narrow beam with a system of slits and observing the deflection in a uniform field by means of a photographic plate.   For a precision determination, a very thin source must be used, in order to minimize the energy loss of the particles in emerging from the source.   Magnetic spectrographs employing special geometry or specially shaped fields which focus rays emerging from the source over a range of angles are often used when only weak sources are available.   A simple example of this type, the semicircular spectrograph, is illustrated in Fig. 10.1.   With a uniform field perpendicular to the plane of the figure, trajectories starting at angles differing even by several degrees come together again after 180° deflection and a reasonably sharp line image of the point source is formed.   If the magnetic field is made to vary as the inverse square root of the radius, rays emerging in a cone are brought to a point focus after a 255° deflection.

It was discovered rather early that α particles from radioactive substances occur in groups, each quite homogeneous in energy.   In Fig. 10.2 is shown, for example, the magnetic spectrum of Rn^{222}, $RaA(Po^{218})$, and $RaC'(Po^{214})$ obtained with a radon source.   For a time it was thought that all α particles from a given radioactive species had the same energy, and the observed complexity was attributed to the presence of several members of a radioactive chain in the source.   Later work showed, however, that this was not the case and that many nuclides emit more than one group.   In the simpler case, where only one group is emitted—radon is an example—the α particle is believed to result from a transition from the normal, or ground, state of the parent nucleus to that of the daughter, with a release of energy equal to the kinetic

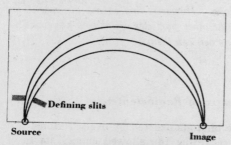

Fig. 10.1 Principle of the magnetic spectrograph with semicircular focusing. There is a uniform magnetic field perpendicular to the plane of the figure.

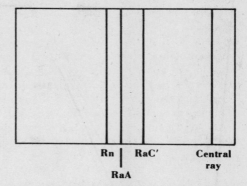

Rn | RaC′   Central ray

RaA

Fig. 10.2 Magnetic spectrum of α particles from a source of radon and its daughters. The line at the right is from the undeflected beam, obtained with no magnetic field. [*S. Rosenblum, J. Phys. Radium,* 1:438 (1930).]

energy of the α particle plus the energy of recoil. In a more complicated case, like that of Bi^{212} (see Fig. 10.3), the appearance of groups having less than the maximum energy is attributed to transitions to excited quantum states of the residual nucleus (Tl^{208}), which then release their excitation energy in the form of γ radiation. The decay of these levels to the ground state is thus quite analogous to the decay of excited atomic states, and the γ spectrum corresponds to the line spectrum emitted by an excited atom. The correctness of this conclusion can be established by comparing the γ-ray energies and α energies: a short-range α particle should be accompanied by one or more γ rays, such that the total energy equals that of the longest-range α particles. Such a comparison is exhibited in Table 10.1.

Table 10.1 Alpha particles and gamma rays of ThC†

α-particle Energy, MeV	Disintegration Energy, MeV	γ-ray Energy, MeV	Sum, MeV
6.0837	6.2007		6.2007
6.0445	6.1607	0.03995	6.2007
5.7621	5.8729	0.3267	6.1996
5.6202	5.7283	0.4709	6.1992
5.6012	5.7089	0.4908	6.1997

† In the second column a correction has been made for the kinetic energy of the recoil nucleus.

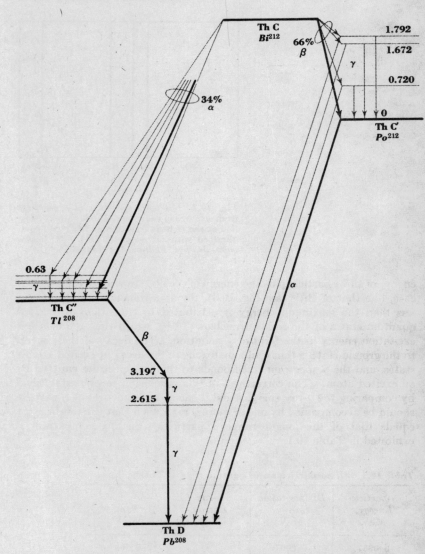

Fig. 10.3 Radiations observed in the two branches of the Bi²¹²-Pb²⁰⁸ chain. The vertical scale is proportional to the energy released. Strong transitions are indicated by heavy arrows; light horizontal lines represent excited nuclear levels.

In ThC' (Po^{212}), formed by β decay of ThC, there occur, in addition to the main group of 8.776-MeV α particles which are believed to represent transitions between the ground states of ThC' and ThD, a very few α particles of considerably higher energy. These long-range α particles come from excited states of the *parent* nucleus, ThC'. This behavior results from the fact that γ radiation is a relatively slow process on a nuclear time scale, and, given sufficient energy, the α particles can escape before the parent nucleus can become deexcited. The lifetime of ThC' is only a fraction of a microsecond.

b. Gamma Rays   Nuclear γ-ray spectra consist typically of sharp lines; in the naturally radioactive elements they are invariably associated either with α or β emission. When a nucleus emits an α or β particle, it is likely that the residual nucleus is left in an excited state. Ordinarily, when it goes to a more stable configuration, the energy released is radiated as a γ ray. The sharpness of both α- and γ-ray lines is consistent with the idea that excited states of nuclei ordinarily have sharp, well-defined energies.

If the excited state from which a given nucleus emits a γ ray is long-lived, all the γ rays arising from a transition to the ground state have almost the same energy. For example, the 93-keV photon from Zn^{67} has a resonance width of 4.8×10^{-11} eV. Thus, when emitting or absorbing photons, nuclei are among the most "sharply tuned" oscillators in nature. A free Zn^{67} nucleus emitting a γ ray recoils (momentum is conserved!) with an energy which, while small compared with photon energy, is large compared with the width of the γ-ray line. Consequently, the photon does not have enough energy to raise another Zn^{67} atom from the ground state to the excited state. In 1958 Mössbauer discovered that if both emitting and absorbing atoms were tightly bound in crystal lattices, the full excitation energy could be emitted and absorbed. (More exactly, the crystals as a whole take up the recoil momenta, and the recoil energies are negligible.) The discovery of the *Mössbauer effect* made it possible for physicists to exploit the extraordinary sharpness of γ-ray lines. The 14-keV line from excited Fe^{57}, a widely used Mössbauer source, is so sharp that the doppler shift is observable for speeds of the order of 1 cm/min.

10.2   Beta Rays and the Antineutrino

β particles are electrons which may be ejected from nuclei with energies up to several MeV. If the intensity of the source suffices, the energies of the β rays can be measured by a β-ray spectrometer. The spectrum

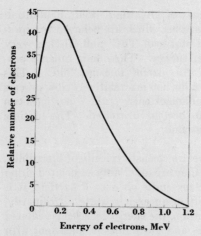

Fig. 10.4 β-ray spectrum of Bi²¹⁰ showing the relative number of electrons emitted in a constant-energy interval ΔE as a function of the energy of the electrons.

of the β rays from a $_{82}$Bi^{210} source, analyzed by use of a magnetic field, is shown in Fig. 10.4. The curve shows a broad and continuous range of β energies. One might argue that this distribution represents β transitions to a large number of excited quantum states of the daughter nucleus, but then the corresponding γ spectrum would exhibit a continuous distribution whereas the γ rays are in fact discrete. Furthermore, β spectra entirely unaccompanied by γ rays are not rare. It was early suggested that the observed β spectrum was distorted in some way; e.g., it could be conceived that the β particles were actually homogeneous as they left the nucleus but that they lost varying amounts of energy as they went through the atomic electron cloud. That this is not true was established by calorimeter measurements by Ellis and Wooster in 1927. Using a calorimeter with walls sufficiently thick to ensure that no radiation could be detected outside, they established that the energy absorbed in the calorimeter per disintegration corresponds to the average β-ray energy. Thus they were forced to conclude either that the β transitions leading to values other than the average violate the principle of conservation of energy or that some particle is emitted which is neither absorbed in the calorimeter nor detectable outside. The first suggestion was seriously discussed by Bohr on the assumption that the conservation of energy might apply only in a statistical sense to nuclear processes. This distasteful hypothesis met its fate in an analysis by

Ellis of the ThC → ThD disintegration energies. This transition, shown in Fig. 10.3, can occur in two ways: ThC goes to ThC′ by β emission, with a maximum energy of 2.253 MeV; the decay of ThC′ then proceeds mainly by an α transition of 8.946 MeV (including the recoil energy), giving a total (maximum) energy release along this path of 11.199 MeV. On the other branch we have first an α transition, of energy 6.203 MeV, followed by a β particle of maximum energy 1.798 MeV, followed by γ transitions (in ThD) aggregating 3.197 MeV, giving a total energy change of 11.198 MeV. Thus the energy balance is accurately accounted for by the assumption that the energy release in a β decay process corresponds always to the *maximum* β-ray energy. This result has been amply confirmed in later experiments on artificially produced radioactive nuclei. In the β decay of $B^{12} \rightarrow C^{12}$, for example, the known mass difference implies an energy release of 13.370 MeV, in reasonable agreement with the observed maximum β-ray energy of 13.43 ± 0.06 MeV. The average β-ray energy in this case is less than 7 MeV. It has also been established that not only energy but also linear momentum and angular momentum are inadequately accounted for by the observed β particles.

If some other particle were emitted simultaneously with the electron, this particle would share the total energy available and the β spectrum would be continuous. In 1930 Pauli suggested that there was indeed an uncharged particle of zero or small rest mass which accompanied each electron from a nucleus. This particle was originally called the *neutrino*, meaning "little neutral one," but for reasons to be developed later, it seems preferable to call it the *antineutrino*. Then the β spectrum of Fig. 10.4 arises from the reaction

$$_{82}Bi^{210} \rightarrow {}_{83}Po^{210} + {}_{-1}\beta^0 + \bar{\nu}$$

where the antineutrino is represented by $\bar{\nu}$ (not to be confused with wavenumber). For many years all evidence for the existence of such a particle was indirect, in that energy, momentum, and angular momentum could be conserved only if there were an elusive and undetected third particle on the right of the reaction above. However, in 1953 and subsequently, certain experiments involving extremely sensitive detecting apparatus have strongly supported the antineutrino hypothesis.

10.3 Masses of Atoms

In the radioactive chain of Sec. 8.6 we saw that isotopes of two chemically identical species may have entirely different radioactive properties. Isotopes have the same value of nuclear charge Ze but different atomic

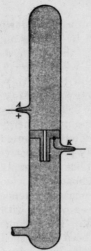

Fig. 10.5   Schematic diagram of Goldstein's discharge tube.

masses; thus the mass is of major importance in determining the properties of the nucleus.

The first step in the development of direct, precision mass-measuring techniques for atomic masses came with the discovery of the *canal rays* in the Crookes discharge tube, by Goldstein in 1886.   Goldstein's tube is schematically illustrated in Fig. 10.5.   In a glass envelope are provided two electrodes: the anode *A*, a simple side connection, and the cathode *K*, which is perforated by a long canal.   When the tube is evacuated to the point where the Crookes dark space is well developed, one sees the cathode rays streaming upward and, in addition, a stream of rays issuing downward from the canal in the cathode.   If the residual gas in the tube is neon, the ionization due to the cathode rays produces a pale blue light, while the canal rays are a brilliant red.   Unlike the cathode rays (electrons), the canal rays are quite difficult to deflect with a magnetic field.   It was not, in fact, until 1898 that Wien exhibited the effect with a powerful electromagnet, and showed, by simultaneous application of electrostatic and magnetic deflections, that the values of e/m were of the order of those observed for ions in electrolytic cells. Because he operated at too high pressures, Wien was unable to make a precise determination, and found a continuous spread of e/m values.

In a series of experiments initiated in 1906, J. J. Thomson made a careful study of the formation and properties of the canal rays.   His apparatus is shown diagrammatically in Fig. 10.6.   The large discharge tube *B* is provided with an anode (not shown) and a cathode *K*, perforated with an extremely fine hole.   (In some of his later experiments, Thomson made this canal by ruling a scratch in a piece of flat steel and laying another piece on top of it.) The canal rays, or, as Thomson came to call them, the positive rays, stream out of the canal into the space C, where they strike a screen covered with a scintillating material or a photographic plate P. A magnet M_1M_2 with electrically insulated pole pieces A_1A_2 provides a strong magnetic field, and by electrical connection to A_1 and A_2, an electrostatic field parallel to the magnetic field can be applied. Thomson used pressures of the order of 10^{-3} mm Hg and discharge voltages from 1000 to 20,000 V.

In passing through the magnetic and electrostatic fields at A_1A_2, the particle suffers simultaneous deflections along the z axis (perpendicular to the plane of the drawing) and along the y axis. The displacement at the end of the gap due to the magnetic deflection is

$$z' = \frac{1}{2} \frac{Bev}{m} \left(\frac{l}{v}\right)^2 \tag{10.1}$$

and the displacement due to the electrostatic deflection is

$$y' = \frac{1}{2} \frac{Ee}{m} \left(\frac{l}{v}\right)^2 \tag{10.2}$$

where l is the length of the gap. Since the rays travel in straight lines after leaving the gap, the displacement on the plate PP is proportional to these quantities. To obtain the shape of the trace on the plate, we eliminate the velocity, obtaining

$$z^2 = C \frac{B^2}{E} \frac{e}{m} y \tag{10.3}$$

where C is a constant depending on l and other geometrical features. This equation represents a parabola, with vertex at the origin; ions of

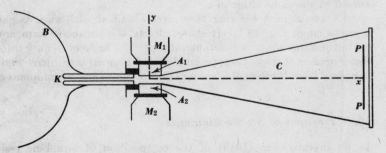

Fig. 10.6 Thomson's positive-ray spectrograph.

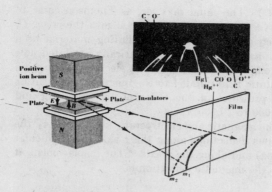

Fig. 10.7 Parabolic traces on a film are formed by positive ions of the same charge-to-mass ratio but different velocities when deflected by parallel electric and magnetic fields. Reversal of the magnetic field produced the other half of the parabolas in the insert.

various velocities but with the same value of m/e lie at various points on the parabola, the highest velocities nearest the vertex. If we consider ions of only a single charge, the highest velocity will be given by

$$\tfrac{1}{2}mv_{\max}^2 = Ve$$

where V is the potential across the discharge tube; inserting this in the expression for y' above, we see that

$$y'_{\min} = \frac{1}{4}\frac{E}{V}l^2$$

Thus all parabolas, regardless of the value of m/e which they represent, cut off at the same value of y.

A photograph showing the traces obtained with various gases is reproduced in Fig. 10.7. It shows clearly the parabolic form and the abrupt termination at a constant value of y. *The fact that these traces are sharp constitutes the first experimental proof that atoms have discrete masses— or at least discrete values of e/m—and are not spread over a continuous range.*

10.4   Isotopes of Stable Elements

In an investigation (1913) of the composition of liquid-air residues, Thomson discovered, in addition to the neon line at the position expected for mass 20, a faint, just resolved line at mass 22. After a number of tests to eliminate other possible sources of such a line, he concluded that this line was due to another component of neon, with mass 22. This was the first observation of the isotopic constitution of a stable element and verified the prediction made by Soddy that elements with atomic weights differing markedly from integral values would be found to have a complex isotopic constitution.

The subsequent development of mass-measuring techniques in the hands of Aston, and later Dempster, Bainbridge, Mattauch, Nier, and others, has enormously improved the sensitivity and precision of the method. Fundamental to all these improvements has been the use of focusing principles which greatly increase the intensity available at the detector. The development of more efficient ion sources has also been a matter of great importance in increasing the usable ion currents and extending the techniques to materials which are not available in gaseous form. The applications of the mass spectrometer can be generally divided into three categories: the identification of isotopes, the determination of the relative abundances, and the precise measurement of their masses.

Coupled with the increase in sensitivity gained in the course of 50 years' development of the mass spectrograph has also come an enormous increase in resolution and accuracy. Like chemical atomic weights, the masses of isotopes are seldom expressed in absolute units but are given in terms of an adopted standard. The standard for the *unified atomic mass unit*[1] is based on the most abundant isotope of carbon with the mass of a C^{12} atom taken as exactly 12 of these units (abbreviated u by international agreement). The absolute value of this mass unit in kilograms can be obtained from the Avogadro number (which, is, in turn, derived from several other basic physical constants). Thus

$$1 \text{ u} = \frac{1}{N_A} = 1.6604 \times 10^{-27} \text{ kg}$$

and the rest energy of 1 u is 931.48 MeV.

The most precise direct mass measurements are now made by the so-called doublet method, in which two ions of the same nominal mass are compared. Figure 10.8 shows the $(He^4)^+$-$(H_2^2)^+$ doublet; the left-hand line is due to singly ionized helium atoms—4 u and one charge—and the right-hand line is from singly ionized deuterium molecules—also 4

[1] The *atomic mass unit* was originally defined as one-sixteenth the mass of an O^{16} atom and formed the basis for the *physical scale* of atomic weights. It differed slightly from the *chemical scale*, which was based on natural oxygen with its small amounts of O^{17} and O^{18}. The unified atomic mass unit is 0.0318 percent larger than the atomic mass unit based on O^{16}.

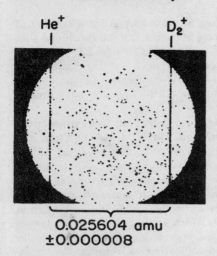

He⁺   D₂⁺

0.025604 amu
±0.000008

Fig.. 10.8 Mass spectrograph record of $(He^4)^+$—$(H_2{}^2)^+$ doublet; in unified atomic mass units the doublet separation is 0.025596 u. [*H. Ewald, Z. Naturforsch.*, 5a, p. 1 (1950).]

u and one charge.   The center-to-center spacing can be determined to an accuracy of about 1 part in 3000.   If the dispersion of the spectrometer is known to the same accuracy, the value of the He⁴ mass relative to that of the deuteron can be deduced with an accuracy of 0.000008 u. By use of such other doublets as $(H_2^1)^+$-$(H^2)^+$, $(C^{12}H_4^1)^+$-$(O^{16})^+$, and $(H_3^2)^+$-$(C^{12})^{++}$ the mass can be linked to the standard C^{12}, using only mass-difference measurements.   A few masses of common nuclides obtained in this way are given in Table 10.2.

Between 1919, when Aston's first spectrograph was completed, and 1922, the isotopic constitution of 27 elements was investigated, and in the following 2 years 26 more were reported.   A table published by Bainbridge in 1953 lists over 280 naturally occurring isotopes, most of which have been discovered by the mass-spectrographic techniques. In every case where the chemically determined atomic weight differs greatly from a whole number, the element is found to consist of more than one isotope.   We may anticipate our later discussion of the exact masses by saying that the isotopic masses observed are not quite integral multiples of the atomic mass unit, but all stable isotopes have masses which lie within less than 0.1 u of an integer; it is this integer that we call the mass number A.

Table 10.2   *Masses in unified atomic mass units;*
$C^{12} = 12.000000$

Nuclide	Mass	Nuclide	Mass	Nuclide	Mass
Neutron	1.008665	$_3Li^6$	6.015126	$_9F^{19}$	18.998405
Proton	1.007276	$_3Li^7$	7.016005	$_{11}Na^{23}$	22.989771
$_1H^1$	1.007825	$_4Be^9$	9.012186	$_{13}Al^{27}$	26.981539
$_1H^2$	2.014102	$_5B^{10}$	10.012939	$_{29}Cu^{63}$	62.929592
$_1H^3$	3.016050	$_5B^{11}$	11.009305	$_{47}Ag^{107}$	106.905094
$_2H^3$	3.016030	$_7N^{14}$	14.003074	$_{79}Au^{197}$	196.966541
$_2He^4$	4.002603	$_8O^{16}$	15.994915	$_{92}U^{238}$	238.050770

10.5   The Nucleons

Since all nuclei have charges which are integral multiples of the protonic charge, it is reasonable and fruitful to assume that a nucleus of atomic number Z contains Z protons. As we saw above, all atomic masses are very nearly an integral number A of atomic mass units, but A is roughly twice Z. In 1920 Rutherford suggested that there might be a particle of roughly the proton mass but no charge. This particle, known as the *neutron*, was discovered in 1932, and we now think of any nucleus as being a collection of Z protons and $A - Z$ neutrons.

Protons and neutrons are referred to as *nucleons*. In theoretical treatment of nuclear structure it is common to regard the proton and neutron as different charge states of "the nucleon." The proton and the neutron have comparable sizes with radii of the order of 2×10^{-15} m; both have intrinsic spin angular momentum such that the component in the direction of an applied external field is $+\hbar/2$ or $-\hbar/2$.

Associated with the spin of the proton there is a small magnetic moment (Sec. 9.8)

$$\mu_p = 1.521 \times 10^{-3} \text{ Bohr magneton} = 2.79275 \frac{e\hbar}{2M_p} \qquad (10.4)$$

where M_p is the proton mass. It is customary to measure nuclear magnetic moments in terms of the unit $e\hbar/2M_p$, which is called the *nuclear magneton*. Similarly the neutron, with no net charge, possesses a magnetic moment comparable in magnitude but opposite in sign to that of the proton;

$$\mu_n = -1.91315 \text{ nuclear magnetons} \qquad (10.4a)$$

In addition to mass, each nucleon contributes spin angular momentum to a nucleus, but this is a vector quantity, and the resultant nuclear spin is usually not large—indeed, it is typically zero for nuclei with even numbers of both protons and neutrons.   Similarly, each nucleon contributes to the magnetic moment of the nucleus; however, the resultant magnetic moment cannot be predicted with precision by adding the individual magnetic moments.

The neutron is slightly more massive than the proton

$$M_p = 1.6725 \times 10^{-27} \text{ kg} = 1.00728 \text{ u};$$
$$M_n = 1.6748 \times 10^{-27} \text{ kg} = 1.00866 \text{ u};$$

$M_p c^2 = 938$ MeV; $M_n c^2 = 939.5$ MeV. A free neutron is unstable, decaying by β emission with a half-life of 12 min in the reaction

$$_0n^1 \rightarrow {}_1H^1 + {}_{-1}e^0 + \bar{\nu} \tag{10.5}$$

Although nuclei are exceedingly small, several types of measurements have been devised to determine nuclear radii.   While there are minor differences in the empirical values, all methods agree on the order of magnitude.   The volumes of all nuclei are roughly proportional to the number of nucleons A.   If we assume a roughly spherical shape, we find that the radius of a nucleus of mass number A is given by

$$r = 1.2 \times 10^{-15} \sqrt[3]{A} \quad \text{m} \quad = 1.2 \sqrt[3]{A} \quad \text{fermi} \tag{10.6}$$

The heaviest of all common nuclei is that of uranium 238, for which the nuclear radius is somewhat less than 10^{-14} m, as compared with the typical atomic radius of about 10^{-10} m.   The densities of nuclei of all kinds are roughly equal, about 2×10^{17} kg/m^3.

10.6   The Positron

In 1932 Anderson found tracks of a particle with the same mass as the electron but positive charge in cloud-chamber studies of cosmic rays. These particles, called *positrons*, appear when high-energy photons ($h\nu > 1.02$ MeV) interact with nuclei; each positron is accompanied by an electron.   The reaction may be written

$$h\nu \rightarrow {}_1m_0c^2 + {}_{-1}m_0c^2 + K^+ + K^- \tag{10.6a}$$

where $_1m$ and $_{-1}m$ are respectively the masses of the positron and electron and K^+ and K^- are their kinetic energies. The nucleus in the field of which the interaction takes place does not appear in (10.6a), but the involvement of another particle is required to satisfy the conservation of momentum.

In this reaction we have the conversion of the electromagnetic energy of the photon into the rest masses of the positron and electron and their kinetic energies. The process is known as *pair production*. The rest energy of the positron is 0.511 MeV, equal to that of the electron, so that the threshhold energy for pair production is 1.02 MeV. Only photons with energy greater than 1.02 MeV can create pairs; any additional energy of the photon is shared between the particles as kinetic energy.

Positrons exist for a short time. As they move through matter, they lose energy rapidly. Eventually they combine with an electron in a process called *pair annihilation*, with the conversion of their energies into electromagnetic energy. As the positron slows down, it may pick up an electron and form *positronium*, a hydrogen-like "atom" in which the two particles "revolve" about their center of mass. Usually formed in a state of high angular momentum, the positronium makes successive radiative transitions until the ground state with energy of 6.7 eV is reached. If pair annihilation occurs when the linear and angular momenta of the positronium are zero, two 0.511-MeV photons are emitted in precisely opposite directions; only in that way can the conservation laws be satisfied for two-photon decay (see Sec. 10.10).

The existence of the positron had been predicted by Dirac in 1928 in his development of a theory of the electron which was invariant under Lorentz transformation. According to this theory an electron has energy $\pm \sqrt{(m_0c^2)^2 + p^2c^2}$, where p is the momentum; thus negative as well as positive energies are allowed. These ideas can perhaps be made a little clearer by reference to the diagrams of Fig. 10.9. Here we represent the possible energy states of a free electron by a continuum, extending from $E = +m_0c^2$ upward and from $E = -m_0c^2$ downward. The lower levels are all filled, and their effect merely adds a constant (infinite) charge which is undetectable. In Fig. 10.9b a γ ray of energy $h\nu$ ($>2m_0c^2$) raises an electron in a state of energy $-E_1$ to a state of positive energy $+E_2$: the electron now has positive kinetic energy $K_2 = h\nu - E_1 - m_0c^2$, and the "hole" behaves like a positron of (positive) kinetic energy $E_1 - m_0c^2$. Since momentum cannot be conserved in the interaction of a γ quantum and two electrons. the event must take place near another particle (usually a nucleus) which can take up the excess momentum. The interaction with a nucleus takes place through the coulomb field, and calculation gives a roughly Z^2 dependence for the cross section for

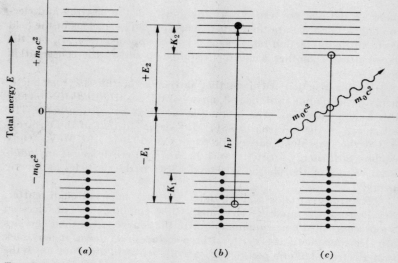

Fig. 10.9 Transitions between positive and negative energy states of free electrons.

pair formation. The positron so created can interact with atomic electrons through its electrostatic field, gradually losing its kinetic energy until it finally undergoes an annihilation process with a (negative) electron at rest (Fig. 10.9c). Momentum can be conserved in this process if two quanta, each of energy m_0c^2, are emitted. One-quantum annihilation, resulting in a single γ ray of energy $2m_0c^2$, is also possible if the electron is strongly bound to a nucleus. Positrons can be annihilated in flight through collisions with electrons, but the probability falls off rapidly as the velocity increases. The lifetime of a very slow positron in lead is about 10^{-10} s. The positron is called the *antiparticle* of the electron or the *antielectron*, since a positron and an electron can annihilate one another.

10.7 Muons and Pions

In 1936 Anderson and Neddermeyer, in cloud-chamber photographs of cosmic-ray events, found evidence of particles of mass 207 times that of the electron. These particles, called *muons*, may bear a charge of $+e$ or $-e$. Both the μ^+ and μ^- particles are short-lived, decaying by the reactions

$$\mu^- \longrightarrow \beta^- + \bar{\nu} + \nu_\mu \qquad\qquad (10.7a)$$

$$\mu^+ \longrightarrow \beta^+ + \nu + \bar{\nu}_\mu \qquad\qquad (10.7b)$$

where the overbar implies an antiparticle and ν_μ is the *mu neutrino*, a particle similar to the neutrino but distinguishable from it. A decade later another particle of intermediate mass, called the *pion* or π *meson*, was discovered. Pions may be positive, negative, or neutral. The mass of a charged pion is 273 electron masses, while that of the neutral pion is 264 electron masses. Negative pions are strongly attracted to nuclei and are quickly captured in dense materials; positive pions are repelled by the nuclear charge and are less likely to be captured. Isolated pions of all varieties are unstable and typically decay by the reactions

$$\pi^- \longrightarrow \mu^- + \bar{\nu}_\mu \qquad\qquad (10.8a)$$

$$\pi^+ \longrightarrow \mu^+ + \nu_\mu \qquad\qquad (10.8b)$$

$$\pi^0 \longrightarrow 2 \ 67\text{-MeV} \ \gamma \ \text{rays} \qquad\qquad (10.8c)$$

Pions are intimately linked to the attractive force between nucleons.

Other mesons, in addition to pions, have been discovered (see table inside rear cover).

10.8 Particles and Antiparticles

Dirac's theory anticipating the positron (Sec. 10.6) could be interpreted as implying a particle identical to the proton except for a negative charge. The existence of this particle, the *antiproton*, was established in 1955 by Segrè, Chamberlain, and their collaborators. Antiprotons were produced by bombarding protons in a target with 6-GeV protons, thereby inducing the reaction

$$p + p \ (+ \text{ energy}) \longrightarrow p + p + p + \bar{p} \qquad\qquad (10.9)$$

The kinetic energy of the bombarding proton is converted to a proton-antiproton pair plus the kinetic energy of the four residual particles. As soon as the antiproton is slowed down, it is annihilated by a proton; in a typical annihilation reaction the rest mass of the annihilating pair appears as five pions and their kinetic energy (Fig. 10.10).

$$p + \bar{p} \longrightarrow \pi^+ + \pi^- + \pi^+ + \pi^- + \pi^0 \qquad\qquad (10.9a)$$

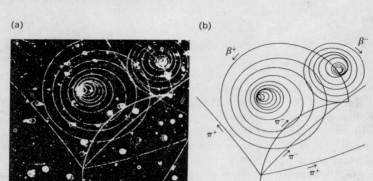

(a) **(b)**

Fig. 10.10 (a) Bubble-chamber photograph showing the path of an anti-proton entering from below. Near the bottom of the picture the antiproton annihilated a proton, forming two positive pions, two negative pions, and a neutral pion, which leaves no track. In the upper quadrant of the picture a negative pion interacted with a proton. Among the resulting products were a positron, which formed the counterclockwise spiral at top center, and an electron, which produced the smaller clockwise spiral. (b) Explanatory sketch. (*Courtesy of Lawrence Radiation Laboratory, University of California.*)

One year later the antineutron $\bar{n}$ was discovered at the same laboratory. Since the neutron bears no charge, the antineutron is also neutral. It is quickly annihilated, either by a proton or a neutron, usually with the production of several pions. If an antineutron is not annihilated by a nucleon, it decays by the reaction $\bar{n} \rightarrow \bar{p} + \beta^+ + \nu$.

When a beam of high-energy protons bombards a target, many kinds of particles may appear in addition to those already introduced. The more common "elementary" particles are listed in the table inside the rear cover. They fall into four natural divisions: (1) photons, with spin (quantum number) 1 and zero rest mass; (2) leptons (light particles), with spin $\frac{1}{2}$ and rest mass between 0 and 210 electron masses; (3) mesons, with 0 or integer spin and rest mass greater than 260 electron masses; (4) baryons (heavy particles), with masses greater than 1800 electron masses and with half-integral spin. It appears that all baryons, leptons, and charged mesons have distinct antiparticles. (Of course, any charged antiparticle bears a charge opposite to that of the corresponding particle.) Photons and the neutral π^0 and η mesons do not possess distinct antiparticles but are said to be their own antiparticles. However, there are two distinct neutral K mesons which form a particle-antiparticle pair, just as the neutron and antineutron do.

In atomic physics it has long been useful to consider an atom as composed of extranuclear electrons and a nucleus formed of protons and neutrons. There is reason to believe that a positron and an antiproton could form an atom of antihydrogen, which would have a spectrum similar to that of ordinary hydrogen. Indeed, from a collection of antiprotons, antineutrons, and positrons a world of antimatter might be constructed which could be indistinguishable from our world so long as everything were made of antiparticles. However, if some of this *contraterrene* matter were to come in contact with ordinary matter, particle-antiparticle annihilation would occur with a tremendous release of energy. At one time it was thought that if we knew a decay mode of any particle, we could immediately write a corresponding possible decay mode for its antiparticle by changing each particle in the reaction to its antiparticle [just as was done for the muon decays in Eq. (10.7a, b)]. The replacement of all particles in a reaction by its antiparticle (including neutrals) is known as *charge conjugation*. It is now known that the principle of charge-conjugation invariance is violated in so-called *weak interactions*.

All leptons and baryons have half-integral ($\frac{1}{2}$, $\frac{3}{2}$, . . .) spin quantum numbers, and all obey Fermi-Dirac statistics (Chap. 20). If we assign to each baryon the *baryon number* $A = 1$, to each antibaryon $A = -1$, and to other particles $A = 0$, we find that in every reaction the total baryon number is conserved. Thus in Eq. (10.9) the total baryon number remains 2, while in Eq. (10.5) it is 1. Note that for any nuclide, such as C^{12}, the superscript represents the baryon number. As far as is known, the *conservation of baryons* is always obeyed in every sort of particle decay and nuclear reaction. Baryons more massive than the proton are known as *hyperons;* all except neutrons in nuclei are unstable.

Similarly, if we assign a lepton number of 1 to the μ^-, e^-, ν, and ν_μ particles and a lepton number of -1 to μ^+, e^+, $\bar{\nu}$, and $\bar{\nu}_\mu$, we find that there is a similar conservation of lepton number in all reactions. In Eq. (10.5) we have total lepton number zero, while in Eqs. (10.7a, b) the lepton numbers are 1 and -1 respectively. We now see why the particle accompanying the electron in a β decay is called an antineutrino. Before the decay we have no leptons; after the decay there is one lepton and one antilepton, or a net zero lepton.

On the other hand, photons (spin $\hbar$) and mesons (spin quantum number 0 or integer) are *not conserved* in reactions. They obey different statistics (Bose-Einstein) than baryons and leptons, as we shall see in Chap. 20.

In addition to listing many of the particles of modern physics, the table inside the rear cover shows the dominant decay modes for the unstable particles. Since neutral particles do not leave a trail of ions in cloud chambers or bubble chambers, their paths are not visible in photographs of the chamber. The presence of such particles is inferred from indirect evidence such as the subsequent appearance of charged particles (arising either from the decay of the neutral one or from a reaction induced by the latter) or from the failure of one or more of the conservation laws. In view of the important role of conservation principles in physics, we set out below those laws which we shall hereafter assume are applicable to all processes.

10.9   The Conservation Laws

From classical physics come the following laws:

1.   Conservation of momentum
2.   Conservation of energy
3.   Conservation of angular momentum
4.   Conservation of charge

For any isolated system each of these four quantities remains constant. (As we saw in Chap. 3, the conservation of momentum and of energy may be regarded as a single law if energy/c is regarded as the time component of the momentum four-vector for the system.)

Further, we introduce the particle-conservation laws:

5.   Conservation of baryons.   *In any reaction the total baryon number is constant.*
6.   Conservation of leptons.   *In any reaction the total lepton number remains constant.*

There are, in addition, more sophisticated conservation laws for which additional background must be supplied before they can be properly introduced.

10.10   Interactions between Particles

The particles of modern physics interact with one another to form composites, some stable (like an oxygen atom) and others transient (like positronium). It appears that the forces involved in the operations of the physical universe can be classified into four types; in the order of increasing relative strength they are (1) gravitational interactions, (2) the weak interaction, which governs the decay of many unstable nuclei and other particles, such as muons, (3) the electromagnetic interactions, which act on all charged particles and which involve the photon, the fundamental unit of electromagnetic radiation, and (4) the nuclear (or strong) interaction.

It is interesting that it is the weakest force which was first understood quantitatively. The gravitational force is always attractive, its range is in principle unlimited, and the gravitational field at the earth's surface is a dominant factor in the motions of macroscopic bodies. The gravitational force F_G between two particles of masses m_1 and m_2 is given by Newton's law of universal gravitation $F_G = Gm_1m_2/r^2$, where $G = 6.670 \times 10^{-11}$ N-m²/kg² and r is the distance between the particles. The fabulous success of this law and of the generalized newtonian mechanics in dealing with both astronomical and terrestrial bodies is one of the great heritages from classical physics.

The next force to be understood quantitatively was the electrostatic contribution to the electromagnetic interaction, given for two charges q_1 and q_2 in vacuum by Coulomb's law $F_E = q_1q_2/4\pi\epsilon_0r^2$, where ϵ_0 is the permittivity of free space. The electrostatic interaction has the same variation with distance as the gravitational, but for two protons it is more than 10^{36} times as great. Clearly, gravitational forces between charged particles on the atomic scale are quite negligible. Two charges in motion relative to a newtonian frame interact not only through the coulomb field but also through a magnetic interaction. Thus, between such a pair of charges there exists a velocity-dependent interaction in addition to the electrostatic one. Furthermore, in classical electromagnetism an accelerated charge must radiate and is therefore subjected to a radiation reaction force which is acceleration-dependent. The forces between electrons and nuclei in atoms and the forces between atoms (which lead to molecules, liquids, and solids) are electromagnetic in nature, but classical mechanics failed in treating such systems. This failure led to the development of quantum mechanics.

The nuclear (or strong) interaction, which binds protons and neutrons into nuclei, has a very short range. At a distance of 10^{-15} m it is about 100 times the magnitude of the electrostatic reaction between fundamental charge units; however, it decreases rapidly with separation. All elementary particles except the photon and leptons participate in the *strong reaction;* the strongly interacting particles are known as *hadrons.* The precise quantitative force law for hadrons is not known. In this respect there is a sharp contrast between the situation in nuclear physics today and that of atomic physics 50 years ago. Then the force laws were known, but an adequate mechanics had not been developed; now quantum mechanics is believed adequate, but the force laws are not known.

Finally comes the *weak interaction* (roughly 10^{-14} times the strength of the strong reaction) which governs the decay of unstable hadrons, such as the neutron or the Λ particle ($\Lambda \rightarrow p + \pi^-$). The latter has a half-life of about 10^{-10} s, very long compared with times of about 10^{-24} s associated with the strong reactions. The weak reaction, like the strong one, has a very short range.

Some particles are coupled to all four interactions. For example, the proton is a strongly interacting nucleon which, by virtue of its charge, also participates in electromagnetic interaction and, by virtue of its mass, in the gravitational interaction. Further, it is involved in positron emission from a radioactive nuclei, whereby it is transformed into a neutron, neutrino, and a positron through the weak interaction. The electron and the charged muon participate in all but the strong interaction.

10.11 *Preternatural Atoms*

The hydrogen atom consists of an electron bound to a proton through the electromagnetic interaction; other atoms are composed of a positive nucleus binding enough electrons to balance the nuclear charge. However, all charged particles participate in the electromagnetic reaction, and there exist a number of *preternatural atoms* formed by various combinations of charged particles.

Positronium (Sec. 10.6) is an example of such an "atom"; it consists of a positron and an electron bound in a hydrogen-like system. There are two forms of positronium: *parapositronium*, in which the z components of spin angular momenta of the particles are opposite, and *orthopositronium*, in which they are in the same direction. In the ground state neither form has any orbital angular momentum. Consequently, in the ground state parapositronium has no angular momentum, while ortho-positronium has one unit. The former, with a half-life of 8×10^{-9} s, usually decays by two-photon emission. In a frame in which the center of mass is at rest these photons must be emitted in opposite directions with equal momenta and energies of 0.511 MeV. Each photon bears one unit of angular momentum which add vectorially to give zero. For orthopositronium it is impossible to satisfy the conservation laws by two-photon decay, and annihilation ordinarily occurs with three photons produced.

In 1949 Chang reported that for every negative muon stopped in lead, an average of roughly three photons were emitted with energies between 1 and 5 MeV. These photons arise from muon transitions between states directly analogous to the Bohr-Sommerfeld levels. As a muon passes through matter, it loses kinetic energy and may be captured in a Bohr orbit about the nucleus, thus forming a *muonic atom* with a nucleus of charge Z, a muon, and $Z - 1$ electrons. The captured muon makes a series of transitions, reaching the ground state ($n = 1$) in about 10^{-13} s. Thereafter the muon either decays, Eq. (10.7a), or is captured by a nucleon in the reaction $\mu^- + p \rightarrow n + \nu_\mu$.

If for a moment we consider the nucleus as a point charge, the radius r of a muonic Bohr orbit would be $\frac{1}{207}$ the radius of the corresponding electron orbit. For lead one finds $r = 3 \times 10^{-15}$ m for $n = 1$, while Eq. (10.6) gives a radius for the lead nucleus just under 10^{-14} m. If this value is accepted, the muon $n = 1$ orbit lies well within the nucleus, while the $n = 2$ orbit lies just outside. Experimentally one finds that the $n = 2$ to $n = 1$ muon transition in lead leads to a photon of about 6 MeV energy, whereas for a point nucleus the predicted energy is 16 MeV. Thus the finite nuclear size results in a large energy reduction for what might be called the muonic $K\alpha$ line for lead. Study of muonic atoms can yield important information regarding (1) nuclear radii, (2) nuclear quadrupole moments, (3) the magnetic moment of the muon, (4) possible nonuniformity of nuclear charge distribution, and (5) nuclear polarizability and compressibility. For probing the nuclear electric field the μ^- is an ideal test particle. Its interaction with nucleons is exceedingly weak, and in its brief half-life the μ^- may traverse several meters of nuclear matter with average density of 10^{17} kg/m³.

X-rays have been observed not only from muonic atoms but also from the capture of π^- and K^- particles. Since pions and kaons participate in the strong interaction, they are quickly absorbed by nuclei, so that a π^- meson, for example, is ordinarily absorbed before it reaches the $n = 1$ level except in the case of very light nuclei.

A positive muon may capture an electron to form an "atom" of *muonium*, a light isotope of hydrogen with the muon replacing the proton. The energy levels are very close to those of hydrogen, differing slightly because of the smaller reduced mass (and, in detail, because the muon magnetic moment is 3.183 times that of the proton). Still other "atoms" can be formed with an electron and one of the other positive particles, for example, π^+, K^+, Σ^+; in each case the life time of the atom is essentially that of the positive particle.

Problems

1. A collimated beam of singly ionized carbon atoms with 4 keV kinetic energy enters a region where there is a uniform magnetic induction of 0.25 Wb/m² perpendicular to the velocity. If the ions travel a semicircular path to a photographic plate, find the radii of the paths of C^{12} and C^{13} ions and their separation at the plate.

Ans: 12.62 cm; 13.14 cm; 1.04 cm

2. Imagine a model of an electron as a solid sphere of radius 5×10^{-14} m. Find the angular velocity and tangential velocity of such a sphere if its angular momentum is $\sqrt{3}\hbar/2$.

Ans: 1.0×10^{23} rad/s; 5×10^9 m/s ($\sim 17c$)

3. Show that pair production and pair annihilation with only a single photon emitted cannot occur unless some other particle is involved in addition to the pair and the photon.

4. Take mass-spectrographic doublets

$$C_3H_8\text{-}CO_2 = a$$
$$CO_2\text{-}CS = b$$
$$C_6H_4\text{-}CS_2 = d$$

Show that if the mass of C^{12} is 12 exactly, the mass of hydrogen is $H = 1 + (2a + 2b - d)/12$.

5. The triton ($_1H^3$) is unstable, emitting a β ray with a maximum energy of 18.5 keV. Compute the expected maximum energy from the masses of H^3 and He^3 in Table 10.2.

6. Assuming that a deuteron can be crudely approximated as a proton and a neutron with centers 2.5×10^{-15} m apart and a binding energy of 2.2 MeV, compare the energy associated with the strong (nuclear) interaction (a) to the electrostatic potential energy of two protons the same distance apart and (b) to the gravitational potential energy of two nucleons. What is the ratio of the coulomb repulsive force between two protons to the gravitational attractive force?

7. A cosmic-ray photon with $\lambda = 0.00400$ Å interacts with a lead nucleus to form a pair in a cloud chamber. There is a uniform magnetic field of 0.050 Wb/m², which happens to be normal to the velocity vector of the electron.

(a) If the electron ends up with 40 percent of the total kinetic energy of the pair, find the radius of its path in the field.

(b) If, as the positron loses its kinetic energy, it captures an electron in the fourth Bohr orbit, calculate the energy of the photons emitted if the ground state is reached in three transitions.

Ans: (a) 8.3 cm; (b) 0.33, 0.94, 5.1 eV

8. An excited Fe^{57} nucleus emits a γ photon with an energy of 14.4 keV. If it was initially at rest and free, calculate the recoil energy of the nucleus. How many times as great is this recoil energy than the width of the γ-ray line, which is found to be 4.6×10^{-9} eV? At what speed would a second Fe^{57} nucleus in its ground state have to approach the emitting nucleus if the second nucleus is to absorb the photon and end up at rest in its excited state?

9. A nucleus is unstable against α decay if its mass exceeds the sum of the mass of an α particle and the mass of the residual nucleus.

(a) If the mass of Th^{234} is 234.043583 u, calculate the maximum kinetic energy of the α particles from U^{238}.

(b) Show that the maximum kinetic energy in MeV of the α particles from any α emitter initially at rest is given by

$$K = 931.48(_zM^A - _{z-2}M^{A-4} - _2He^4)\left(1 - \frac{4}{A}\right)$$

where the masses of the atoms are in unified atomic mass units.

10. A positron and an electron form a positronium "atom" in a state of high excitation. The positronium emits photons as it undergoes transitions to lower energy states. Calculate the energies of photons from transitions in positronium from $n = 5$ to $n = 3$ and from $n = 3$ to $n = 1$.

11. Show that the conservation laws forbid the annihilation of isolated orthopositronium in a two-photon process but that two-photon annihilation of parapositronium satisfies the conservation laws.

12. In a three-photon annihilation of positronium it is observed that a 0.42-MeV photon goes in the $-z$ direction and a second photon goes off at 37° with the z axis. If the positronium was initially at rest, find the energies of the second and third photons and the angle the latter makes with the z axis.

Ans: 0.35 and 0.25 MeV; 56°

13. In a cloud chamber a positron with kinetic energy K^+ and momentum p^+ and an electron with kinetic energy K^- and momentum p^- are produced when a high-energy γ photon interacts with a nucleus. Find the velocity of the reference frame S' in which the pair has zero momentum. Compute the kinetic energy of the positron in the S' frame.

14. A negative muon is captured by the electric field of a lead nucleus to form a muonic atom. If the muon is in a state with $n = 5$ and the Bohr model is assumed, calculate the energy of the state and the orbit radius. If the muon makes a transition to the state with $n = 2$, find the energy of the photon emitted and the new orbit radius.

15. For what nuclide would the Bohr radius for a negative pion be most nearly equal to the radius of the nuclide if the latter is given by Eq. (10.6)? (Assume that $A = 2Z$, a fair approximation for the case in point.) On the basis of the Bohr model what is the energy of the pion for this situation?

16. Compute the maximum energy of a positron arising from the decay of a positive muon at rest. Assume the rest mass of both neutrinos to be zero.

17. Find the momentum carried off by the mu neutrino in the decay of a pion at rest.

Ans: 1.57 × 10⁻²⁰ kg-m/s

18. Pions created in a high-energy accelerator have a speed $v = 0.800c$ in the laboratory frame. In this frame (a) what is the apparent half-life of the pions and (b) what is the distance traversed by a pion during this time?

19. If the decaying particle is initially at rest, calculate the energies of the γ photons emitted in each of these reactions: (a) $\pi^0 \to \gamma + \gamma$; (b) $\eta \to \gamma + \gamma$; (c) $\Sigma^0 \to \Lambda + \gamma$.

20. Stars resulting from a nuclear disintegration in a photographic emulsion often involve the emission of positive and negative pions. Assuming that the pions originate at the surface of an Ag nucleus (radius $\sim 5 \times 10^{-15}$ m), estimate the minimum energy a π^- must have to escape from the nucleus. Explain why positive pions observed in the emulsion have energies above 4 MeV, while many negative pions have much lower energies.

21. Among the many possible reactions when an antiproton is captured by a deuteron is $\bar{p} + d \to n + \pi^0$. If both the $\bar{p}$ and the deuteron were initially at rest, find the total energies of the π^0 and the neutron after the interaction.

Ans: 1.25 GeV, 1.56 GeV

chapter eleven

Wave Properties of Particles ————————————

One must approach the subject of this chapter philosophically, prepared
to accept conclusions which at first thought seem at variance with our senses
and with a belief that has persisted almost unquestioned from the time of
the Greeks, viz., that matter is made up of particles.   We have seen in Chap. 7
that light possesses both undulatory and corpuscular characteristics.   But
even so, we might say, light differs from matter; whereas we can determine
the nature of light only by indirect observation, matter we can see.   We
observe directly that a handful of sand is made up of real particles.   In
ordinary experience they certainly do not exhibit wavelike characteristics.
The particles of sand which we see so clearly are, however, made up of
molecules or atoms, and these of electrons and protons and neutrons, none
of which we can see directly any more than we can light waves or photons.
It is with these so-called particles, evidence concerning which is almost as
indirect as with photons, that the wave theory of matter is primarily con-

cerned.   We attempt in this chapter to give a brief introduction to this most important subject.   Our purpose will be to show how the concept of matter waves can be developed from previous concepts of classical and quantum physics and to summarize the pertinent experimental evidence.

11.1   Matter Waves

With the discovery of the law of the conservation of energy toward the middle of the nineteenth century, it became accepted by physicists that the physical universe is made up of two great entities, viz., matter and energy, each of which is conserved.   These two great conservation laws provided much of the foundation upon which classical physics was built.   By 1900, the corpuscular nature of matter had become firmly established; likewise the undulatory nature of light.   By 1910, furthermore, Planck's quantum theory and the Einstein photoelectric equation together with various lines of experimental evidence had made it clear that light itself possesses corpuscular characteristics.   By 1920 the dual nature of radiant energy was generally recognized, although not understood.

During all this time, there was no suggestion that matter was anything but corpuscular.   But, in 1924, Louis de Broglie made the bold suggestion that particles of matter, and in particular electrons, might possess certain undulatory characteristics, so that they, too, might exhibit a dual nature.   He suggested a way in which the undulatory characteristics of electrons might furnish a new basis for the quantum theory.   The reasoning used might be paraphrased as follows.   (1) Nature loves symmetry.   (2) Therefore the two great entities, matter and energy, must be mutually symmetrical.   (3) If (radiant) energy is undulatory and/or corpuscular, matter must be corpuscular and/or undulatory.

In his first paper, de Broglie was concerned primarily with developing a theory of light in terms of *light quanta*, or photons.   If the energy of the light is concentrated in photons, how are the phenomena of interference to be understood?   There must be waves of *some sort* associated with the photons, in order to account for the observed interference effects. We can no longer suppose that the energy is spread out over these waves, as in classical theory; nevertheless, the waves must somehow determine where, in an interference pattern, the photons can produce effects by being absorbed.   The details of interference patterns depend largely upon the phase relations of the waves; hence, de Broglie called the latter *phase waves*.   He assumed their frequency ν to be such that the energy in a photon equals $h\nu$.

A material particle carries energy; according to de Broglie, the basic idea of the quantum theory is the impossibility of imagining an isolated quantity of energy without associating with it a certain frequency.   Material particles ought, therefore, like photons, to be accompanied by phase waves of some sort; and these waves ought, under suitable circumstances, to give rise to interference effects.   Furthermore, the waves associated with a particle should have a frequency equal to the energy of the particle divided by h.   De Broglie proceeded to extend wave-particle dualism to material particles in the only relativistically invariant way possible.   He included the rest energy $m_0 c^2$ in the energy, but we, following Schrödinger, constructed first a nonrelativistic theory. We assume that any material particle of mass m moving at speed v has associated with it waves of frequency ν given by

$$h\nu = \tfrac{1}{2}mv^2 + P = E \tag{11.1}$$

where h is Planck's constant and P is the potential energy of the particle.

The physical nature of these waves was left indefinite by de Broglie. We cannot suppose that a material particle is just a group of waves; for then its mass and energy (and also its charge if it has one) would be spread out over these waves and would, in consequence, soon become scattered widely in space.   It is of the nature of waves to diverge toward all sides.   We should not be disturbed, however, by the impossibility of visualizing the waves.   We should remember our experience in optics. Using classical theory, it was easy to picture light as wave motion; but if we retain this concept, it is very difficult to picture a beam of light as a moving stream of photons.   Similarly, as long as we retain the particle concept of an electron or proton, we cannot form a concrete picture of the accompanying waves.   Perhaps, even, they are only mathematical waves, so to speak, a device that we employ for the purpose of making calculations and predicting the results of observation.   In adopting this standpoint, we are doing only what has been done, in various ways, many times previously in physical science.   Strictly speaking, we have no very exact knowledge of the fundamental nature of a magnetic field, and yet we do not hesitate to use the symbol **B** with all due familiarity.

11.2   Mechanics as Geometrical Optics of the Waves

If waves are associated with all material particles and play a part in determining their motion, there should be a parallelism between the laws of mechanics and the laws of wave motion.   For the particles certainly obey the laws of mechanics, at least in some cases. It was pointed out long ago by Hamilton that there does exist a close parallelism between the laws of mechanics and the laws of ordinary *geometrical optics*. De Broglie suggests, accordingly, that the familiar laws of particle mechanics may represent an approximation which is valid under such circumstances that the laws of geometrical optics hold for the matter waves.

One of the fundamental laws of ordinary geometrical optics is that, in a homogeneous medium, light travels in straight lines, or rays. This assumption is very nearly correct as long as the lateral dimensions of the beam are large compared with the wavelength of the light. Under these conditions one might say that the particle characteristics of light appear to predominate, although the rectilinear propagation of light is also consistent with the wave theory. When, however, the cross section of the beam is of the same order of magnitude as the wavelength of the light, rectilinear propagation no longer holds, diffraction phenomena are observed, and undulatory characteristics of light predominate. If we carry the similarity over to mechanics, might we not expect that, for very small particles of matter, the ordinary laws of mechanics—found by Newton to be applicable to the phenomena of macromechanics—would fail, and, by analogy with light, we should find that matter shows undulatory properties in the realm of micromechanics?

If we assume that the phenomena of ordinary mechanics constitute those of geometrical optics for matter waves, we can find the properties these waves must have in order that, under suitable conditions, the laws of mechanics may hold. Geometrical optics is based upon two laws, those of reflection and of refraction. The law of reflection as applied to matter waves presents no difficulty, for it is essentially the same as the law governing the rebound of a elastic body from a hard wall. Refraction corresponds to the deflection of a particle by the action of forces.

To take a mechanical phenomenon which imitates a simple case of refraction in geometrical optics, let a pencil of electrons from a suitable gun G (Fig. 11.1) enter through orifice a an enclosed metal box A, the potential of which relative to the filament of the gun is V. Let these electrons emerge from A through orifice b and enter through c another box B, which is maintained at a potential $V + \Delta V$. The electric field between the two boxes changes the component of velocity of the electrons perpendicular to the adjacent surfaces, and the electrons enter B with a change in their direction of motion. Let v_A and v_B (Fig. 11.1) be the velocities of the electrons in A and B, respectively, and θ_A, θ_B the angles between these directions and the normal to the box faces at b and c. Since the electric field does not change the horizontal component

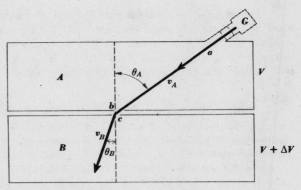

Fig. 11.1 The "refraction" of a pencil of electrons.

of velocity,

$$v_A \sin \theta_A = v_B \sin \theta_B \qquad \frac{\sin \theta_A}{\sin \theta_B} = \frac{v_B}{v_A}$$

Now if we were dealing with light waves undergoing refraction, or any other kind of wave, the relation would be

$$\frac{\sin \theta_A}{\sin \theta_B} = \mu = \frac{u_A}{u_B}$$

where μ is the relative refractive index of the two media and u_A, u_B are the corresponding phase velocities. Comparison of the last two equations gives the result $u_A/u_B = v_B/v_A$. We may conclude that if matter waves follow the electron along its path, *the wave speed u is inversely proportional to the speed v of the electron*, or

$$u = \frac{b}{v} \tag{11.2}$$

where b is a constant during the motion that remains to be discovered. Since if E is the total energy of the electron and P its potential energy (here $-eV$), we have $\frac{1}{2}mv^2 = E - P$, and since for matter waves we assume that $E = h\nu$, we can also write (11.2) in the form

$$u = b \left[\frac{m}{2(E - P)} \right]^{\frac{1}{2}} = b \left[\frac{m}{2(h\nu - P)} \right]^{\frac{1}{2}} \tag{11.2a}$$

Here b may be different for different values of E or ν.

The same equation must hold for matter waves accompanying any kind of particle, although perhaps with a different value of b. It is evident that the wave speed u will vary along the path whenever P varies. Furthermore, u will vary with the frequency ν, so that the waves will exhibit the phenomenon of dispersion, even in free space, where $P = 0$.

Parenthetically, it should be remarked that the well-known electron microscope does not depend for its operation upon such wave properties of the electrons. In this microscope the electrons move essentially according to classical mechanics, although their motion is sometimes treated in terms of the equivalent refractive index described above. The only connection that matter waves have with the electron microscope is that their wavelength sets a limit to the possible resolving power that can be attained, just as the resolving power of ordinary microscopes is limited by the finite wavelength of light. The wavelength of the electron waves, however, we shall presently find to be so small that the resolving power of existing electron microscopes is limited by other factors.

11.3   Phase and Group Velocity

The *wavelength* to be expected for matter waves can be discovered by developing another suggestion—likewise due to de Broglie. It follows from Eq. (11.2) that an individual wave cannot stay with the particle permanently, since u and v usually differ. Whenever the wave velocity (or phase velocity) varies with frequency, a *finite group* of waves moves with a velocity different from the phase velocity. This phenomenon is easily observed on water. Close inspection of a group of waves advancing over a water surface will show that the individual waves advance twice as fast as the group as a whole; new waves continually arise at the rear of the group, pass through it, and die out at the front. Let us assume, therefore, with de Broglie, that the *group velocity* of matter waves is equal to the *particle velocity*, so that a group of them can accompany the particle in its motion.

An ideal plane wave traveling in the x direction can be represented by the equation

$$\Psi = A_0 \cos 2\pi \left(\nu t - \frac{x}{\lambda} \right) = A_0 \cos (\omega t - kx) \tag{11.3}$$

where A_0 is the amplitude, $\omega = 2\pi\nu$, and k, the propagation number, is $2\pi/\lambda$. The phase velocity is $u = \nu\lambda = \omega/k$ for this wave.

When the phase velocity varies with wavelength, a finite group of waves and energy is propagated at the group velocity v. To find the group velocity for waves of angular frequency near ω, it is sufficient to consider two angular frequency components which we select to be $\omega - \Delta\omega$ and $\omega + \Delta\omega$. Let the two components have the same amplitude A and be given by

$$\Psi_1 = A \cos [(\omega - \Delta\omega)t - (k - \Delta k)x]$$
$$\Psi_2 = A \cos [(\omega + \Delta\omega)t - (k + \Delta k)x]$$

Adding gives the resulting displacement

$$\Psi = A \{\cos [(\omega - \Delta\omega)t - (k - \Delta k)x]$$
$$+ \cos [(\omega + \Delta\omega)t - (k + \Delta k)x]\}$$
$$= 2A \cos (\omega t - kx) \cos (\Delta\omega t - \Delta kx) \qquad (11.4)$$

where we have made use of the fact that

$$\cos (\theta \pm \varphi) = \cos \theta \cos \varphi \mp \sin \theta \sin \varphi$$

letting $\theta = \omega t - kx$ and $\varphi = \Delta\omega t - \Delta kx$.

The two cosine terms show the presence of beats in the resulting wave (Fig. 11.2). For a constant phase point, $\omega t - kx = \text{const}$ and $dx/dt = \omega/k = \nu\lambda$ is the phase velocity u.

Setting the argument of the second cosine equal to a constant gives $\Delta\omega t + \Delta kx = \text{const}$, and $dx/dt = \Delta\omega/\Delta k = \Delta f \Delta\lambda$ is the velocity of the wave envelope, or the *group velocity* v. For a particular frequency

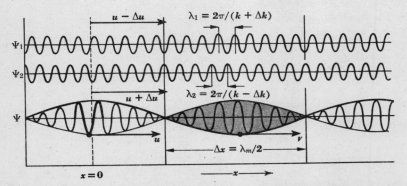

Fig. 11.2 The phase and group velocities of two waves differing slightly in wavelength and frequency.

ν, the group velocity v is

$$v = \lim_{\Delta k \to 0} \frac{\Delta \omega}{\Delta k} = \frac{\partial \omega}{\partial k} = \frac{\partial \nu}{\partial (1/\lambda)} = -\frac{1}{\lambda^2} \frac{\partial \nu}{\partial \lambda} \qquad (11.5)$$

Since $\omega = ku$, we may also write

$$v = \frac{\partial \omega}{\partial k} = \frac{\partial (ku)}{\partial k} = u + k \frac{\partial u}{\partial k} = u - \lambda \frac{\partial u}{\partial \lambda} \qquad (11.6)$$

Here partial derivatives are written to indicate that u and v may vary with other factors than λ; for example, the refractive index may not be uniform.

In Fig. 11.2 the wavelets travel with phase velocity u and have wavelength essentially λ and frequency essentially ν. The modulation envelope, on the other hand, has the group velocity v and the number of beats per unit time is the frequency difference $2\Delta\nu$, so that the modulation-wave frequency is $\Delta\nu$. Consequently, the wavelength λ_m of the modulation wave is $v/\Delta\nu$, and the modulation wavenumber $k_m = 2\pi/\lambda_m$ is just Δk by Eq. (11.4). Therefore, $\lambda_m = 2\pi/\Delta k$.

11.4   The de Broglie Wavelength

Following de Broglie, we seek a relationship between the velocity v of a particle of mass m and its wavelength λ, assuming that v corresponds to the group velocity. Differentiating Eq. (11.1) with P held constant yields

$$h \frac{\partial \nu}{\partial \lambda} = mv \frac{\partial v}{\partial \lambda}$$

Eliminating $\partial \nu/\partial \lambda$ with the aid of Eq. (11.5) yields

$$\frac{\partial v}{\partial \lambda} = -\frac{h}{m} \frac{1}{\lambda^2}$$

and integrating,

$$v = \frac{h}{m} \frac{1}{\lambda} + C$$

The value of C cannot be uniquely determined. Making the simplest assumption, that $C = 0$, we find

$$\lambda = \frac{h}{mv} = \frac{h}{p} \tag{11.7}$$

where $p = mv$, the momentum of the particle.   Then

$$b = uv = \nu\lambda v = \frac{h\nu}{m}$$

Wavelengths given by Eq. (11.7) are known as *de Broglie wavelengths*.   The equation holds for photons as well as for material particles; for the momentum of a photon is $h\nu/c = h/\lambda$.   Equation (11.7) combines corpuscular and undulatory concepts in a very intimate way; for λ has a clean-cut meaning only in connection with a *wave* theory, and p, the momentum, is most naturally associated with a moving *particle*.

We can now compute the wavelengths to be expected for electron waves, atom waves, or molecule waves.   For an electron starting from rest and accelerated through a modest potential difference V, the kinetic energy $K = Ve = \frac{1}{2}mv^2$ and the momentum $p = mv = \sqrt{2meV}$, from which

$$\lambda = \frac{h}{p} = \frac{h}{\sqrt{2meV}} \tag{11.8}$$

For 100-V electrons, $\lambda_e = 1.23$ Å; for 10,000-V electrons, $\lambda_e = 0.123$ Å. When $K = Ve$ is not negligible compared to m_0c^2, Eq. (11.8) is no longer valid and must be replaced by

$$\lambda = \frac{h}{p} = \frac{h}{\sqrt{2m_0K}}\left(1 + \frac{K}{2m_0c^2}\right)^{-\frac{1}{2}} \tag{11.8a}$$

The de Broglie wavelength can be calculated from Eq. (11.8a) [or (11.8) when $K \ll m_0c^2$] for any particle, such as a proton, a helium atom, or even a baseball.   The larger the mass, the shorter the wavelength at given speed.   De Broglie's speculations aroused great interest.   Soon experiments designed to test the hypothesis that electrons exhibit wave properties were launched by several physicists.   On the theoretical front Heisenberg and Schrödinger discovered how to develop mathematical theories which, with additions by Born, Dirac, and others, have become the highly successful quantum mechanics of the present day.

11.5   *Experiments on Electron Waves*

a. Reflection from a Crystal   Experiments on electron diffraction were reported by Davisson and Germer 3 years after de Broglie's first paper appeared.   They were studying the reflection of electrons from a nickel target and accidentally subjected a target to heat treatment that transformed it into a group of large crystals.   Anomalies then appeared in the reflection from it.   Following up this lead, they prepared a target consisting of a single crystal of nickel and bombarded its surface at normal incidence by a narrow pencil of low-voltage electrons; by means of a suitable collector of small aperture, they studied the distribution in angle of the electrons reflected from the crystal.   In this reflected beam, they found striking maxima and minima, which they were able to explain in terms of diffraction of the electron waves.

The diffraction of such waves by a crystal is very similar to the diffraction of x-rays.   Crystallographic studies by means of x-rays show that the nickel crystal is of the face-centered cubic type, as shown in Fig. 11.3; i.e., the crystal can be imagined to be constructed of cubical unit cells having an atom at each corner and one in the center of each face, with none inside the cube.   The atoms are indicated by circles in the figure, certain ones being joined by lines in order to outline the unit cells.   The length a_0 of the edge of the unit cube is 3.51 Å.   Figure 11.3*b* shows a face cut on the crystal at right angles to one of the diagonals of the cube.

From the surface layer of atoms in the triangular face of Fig. 11.3*c*, it is readily seen that these atoms are arranged in rows parallel to one

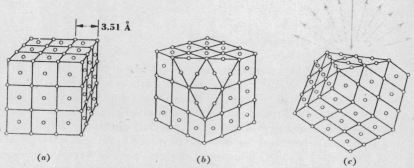

Fig. 11.3   (*a*) A nickel crystal showing face-centered cubic structure.   (*b*) Same, with a face cut at right angles to a diagonal.   (*c*) An incident beam of electrons is scattered backward.

side of the triangle. The distance D between these rows is 2.15 Å. We may regard these rows of atoms as equivalent to the lines of a plane grating of grating space D. Radiation of wavelength λ incident normally on such a grating, the plane of incidence being taken normal to one side of the triangle, should be diffracted, as is light from a reflection grating, according to the well-known grating law

$$n\lambda = D \sin \phi \tag{11.9}$$

where ϕ is the angle between the normally incident beam and the diffracted beam and n is the order of diffraction. The crystal is equivalent to a series of such plane gratings piled one above the other. Radiation penetrating to, and diffracted backward from, any one of these underlying layers is combined with that diffracted from other layers, with the result that a diffracted beam is observed at the angle ϕ unless the beams from the several layers happen to destroy each other completely.

In the apparatus used by Davisson and Germer, the electron gun consists of a heated filament emitting electrons, an accelerating field to give the electrons any desired velocity, and a series of collimating apertures to produce a (nearly) unidirectional beam. This monovelocity beam of electrons strikes the surface of the crystal at normal incidence, and the electrons are reflected, or scattered, in all directions. A collector for measuring the reflected electrons is so arranged that it can be adjusted to any angular position with respect to the crystal. The collector has two walls insulated from each other, between which a retarding potential is applied so that only the fastest electrons—those possessing nearly the incident velocity—can enter the inner chamber and be measured by the galvanometer. The crystal can be turned about an axis parallel to the axis of the incident beam, and thus any azimuth of the crystal may be presented to the plane defined by the incident beam and the beam entering the collector.

If a beam of low-voltage electrons is incident on the crystal, turned at any arbitrary azimuth, and the distribution of the scattered beam is measured as a function of the colatitude—the angle between the incident beam and the beam entering the collector—a curve similar to that in Fig. 11.4a is obtained, which refers to incident 36-V electrons. If now the crystal is turned to the A azimuth, the distribution curve for 40-V electrons (Fig. 11.4b) shows a slight hump at about colatitude 60°. With increasing voltage this hump moves upward and develops into a spur which becomes most prominent at 54 V (Fig. 11.4e) at which voltage the colatitude of the spur is 50°. At higher voltages the spur gradually disappears.

The spur in its most prominent state of development offers con-

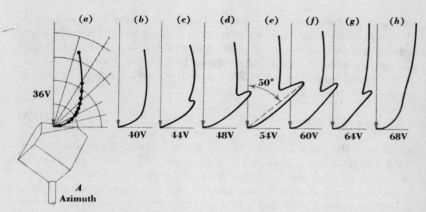

Fig. 11.4 The development of the diffraction beam in the *A* azimuth, showing maximum length of spur for 54-eV electrons at colatitude 50°.

vincing evidence for the existence of electron waves.   From Eq. (11.5) the de Broglie wavelength of 54-V electrons should be

$$\lambda_e = \frac{12.27}{\sqrt{54}} = 1.67 \text{ Å}$$

From Eq. (11.9) we find as the observed wavelength

$$\lambda_e = D \sin \phi = 2.15 \sin 50° = 1.65 \text{ Å}$$

The two values of λ_e are in excellent agreement.

A spur was also found at 55° colatitude with 181-V electrons. This was interpreted as a second-order beam.   For 181-V electrons, $\lambda_e = 0.91$ Å, whereas second-order diffraction gives 0.88 Å.   Over 20 such beams were reported, in three different azimuths.   With improved apparatus, the agreement between observed and de Broglie values of wavelength was better than 1 percent.

b. Transmission through a Crystal   Given sufficient energy, electrons pass readily through thin films of matter such as metal foil or thin mica. If an electron beam has wave properties, then in its passage through matter we should observe some or all of the phenomena characteristic of the similar passage of x-rays.

Soon after the discovery of electron waves, Kikuchi succeeded in obtaining electron diffraction patterns by passing a pencil of 68-kV electrons through a thin mica crystal.   These observations are the exact analog of the first experiments on the diffraction of x-rays by Friedrich,

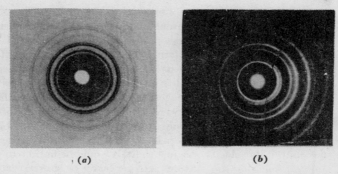

(a)   *(b)*

Fig. 11.5 (a) Diffraction pattern produced by passing x-rays through a thin film of aluminum. (*Courtesy of Mrs. M. H. Read, Bell Telephone Laboratories.*) (b) Diffraction pattern produced by passing 48-keV electrons through a silver foil 500 Å thick. (*Courtesy of Dr. L. H. Germer, Bell Telephone Laboratories.*)

Knipping, and Laue. About the same time G. P. Thomson (son of J. J. Thomson) passed a comparable pencil of electrons through thin foils of polycrystalline materials. Such foils are composed of a large number of randomly oriented microcrystals, and the observed diffraction patterns consist of a series of concentric rings instead of spots (Fig. 11.5). The similarity between the patterns for electrons and those for x-rays is striking.

Measurements of electron wavelengths have now attained such a degree of precision that they are regarded as a valuable source of information on the values of the fundamental constants. If the speed v of the electrons is measured by a kinematical device, observations of λ or of h/mv furnish values of h/m; whereas if the speed is calculated from an accelerating voltage V, so that $mv^2/2 = eV$, observations of λ are best regarded as furnishing values of $h/\sqrt{em}$.

11.6   Diffraction of Neutrons and Molecules

Even molecules exhibit wave properties. A molecular beam may be formed by allowing molecules of a gas to stream out of an enclosure through a small hole or slit into an evacuated chamber. Two difficulties have to be overcome, however, which do not arise in working with electrons. The molecules issue with a maxwellian distribution of velocities, whereas for diffraction experiments a beam of uniform velocity is desirable, corresponding to monochromatic waves. Then, too, neutral molecules are very much harder to detect than charged particles.

Especially interesting is an experiment performed in 1931 by Estermann, Frisch, and Stern. They managed to select a beam of helium molecules or atoms having fairly uniform velocities by passing the beam through narrow slits in two parallel circular disks placed 3 cm apart, rigidly connected together and rotated about their common axis. The slits in the disks were adjusted so as to be opposite each other. An atom, after passing through a slit in the first disk, would arrive at the second disk too late to pass through the corresponding slit in that disk; but, if its velocity was just right, it would be in time to pass through the *next following slit*. Atoms moving faster or slower would arrive too soon or too late and would be stopped by the disk. After leaving the second disk, the beam fell upon the surface of a lithium fluoride crystal, by which it was reflected or diffracted. To measure the intensity of the beam diffracted in a given direction, the helium atoms were allowed to pass through a small hole into a chamber where they accumulated until a certain pressure of helium was reached (of the order of 10^{-6} cm Hg). This pressure was measured by the cooling effect of the helium on an electrically heated metal strip, owing to conduction of heat through the helium; the electric resistance of the strip was measured with a Wheatstone bridge as an indication of its temperature.

A strong diffracted beam of helium atoms was observed in addition to the regularly reflected beam. In the most precise measurement, the maximum in the diffracted beam was found at 19.45°, corresponding to a wavelength of $\lambda = 0.600$ Å as calculated from the usual formula for diffraction by a crystal. From the dimensions and rate of rotation of the disk, the velocity of the helium atoms was calculated to be 1.635 $\times$ 10^3 m/s; the corresponding de Broglie wavelength is, by Eq. (11.7), 0.609 Å.

Neutrons in nuclear reactors which have experienced a large number of collisions have a maxwellian velocity distribution characteristic of the reactor temperature. A typical neutron at room temperature has an energy of about 0.025 eV and a wavelength of about 1.8 Å. Such neutrons undergo *Bragg reflection* at the crystal planes of typical crystals at appropriate angles. The de Broglie wavelength of the neutron makes possible the use of neutron spectrometers and permits the physicist to obtain beams of almost monoergic neutrons.

11.7   Electrons and the Wave Function

As we saw in Sec. 11.5, beams of electrons and of photons behave in essentially identical ways when they pass through thin foils of various materials. To further emphasize the common behavior of electron waves and photons of the same wavelength, let us consider the following experiment. Imagine a pair of slits (Fig. 11.6) through which we can alternatively send beams of photons and of electrons of comparable wavelength from either an electron gun or a photon source. Assuming that the wavelengths and dimensions of the apparatus have been suitably chosen to emphasize the wave aspects of the electrons and photons, we observe on a photographic plate (or other suitable detector) the pattern shown in Fig. 11.6b for both electrons and photons. In terms of the classical particle picture of an electron the expected pattern would consist simply of spots at P_1 and P_2, corresponding to electrons which pass directly through the slits and impinge on the detecting plate. Instead, we find photons and electrons exhibiting the same wave properties.

We should obtain identical distribution of intensities if we used a little scintillator which moved along the plate in Fig. 11.6a and flashed each time an electron or photon struck it. The individual scintillations would be clear evidence that the photons or electrons had particle characteristics, since classical waves would produce a uniform glow proportional to the intensity, not individual flashes of light.

The similarity of behavior of electrons and photons is further illustrated when we consider what happens if we close one of the slits. If only slit 1 is open, for either electrons or photons, the observed intensity pattern corresponds to curve I_1 of Fig. 11.6c. Similarly if slit 1 is closed and slit 2 open, the observed pattern resembles I_2. The pattern with

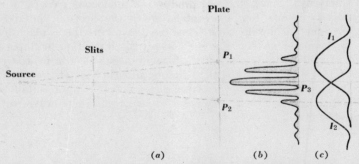

Fig. 11.6 (a) If electrons behaved like classical particles, the pattern expected for electrons from the source passing through two slits would be spots at P_1 and P_2. (b) The actual pattern observed on the plate is that expected from classical waves. (c) Patterns expected from waves through each slit if the other slit is closed.

both slits open is not the sum of the patterns of the two individual slits. At this point there arises a grievous dilemma for the person accustomed to thinking in terms of classical physics. He naturally assumes that if an electron arrives at a detector when both slits are open, the electron must have gone either through the first slit or through the second. Even if one counts the arrival of individual particles by a scintillation detector, one obtains the curve of Fig. 11.6b, not the sum of the curves of Fig. 11.6c. If the electron or photon goes through only one slit, how can opening or closing the second slit possibly affect the trajectory? How could opening the second slit reduce the probability of any particle's arriving at a specific point (such as P_3 on Fig. 11.6b) to zero, when each slit alone makes a positive contribution to the intensity at this point?

This result is paradoxical only when we insist on a rigid classical view that electrons *are* particles. What is important is not what name they bear but how they react in experiments. "Particle" is a convenient name for an object which we can localize, but we use another name to describe its behavior in a diffraction experiment. The fact confronting us here is that an entity which can be located in space in some experiments can exhibit wave properties in others. Indeed, there is a continuum between the two where neither property is well defined. The entire manifold has its analogy in classical optics, where one speaks of long wave trains with precisely defined wavelengths, or short light pulses, which can be located in space but for which wavelength cannot be defined. That this analogy is relevant to matter is the result of experiments; one makes progress in physics by explaining what actually happens, not by explaining what one thinks ought to happen but does not.

Classical wave theory explains the interference pattern of Fig. 11.6b satisfactorily. Since electrons produce an analogous pattern, we shall tentatively assume that the explanation of the electron "interference" pattern can be developed in close analogy with the explanation for electromagnetic waves. For such waves the instantaneous intensity is proportional to the square of the instantaneous electric field strength. When two waves interfere, the resulting intensity is not the sum of the two intensities; rather the resulting displacement is the resultant of the individual displacements. Consider two electromagnetic waves with the electric vector **E** given by

$$E_1 = A_1 \cos 2\pi \left(\frac{x}{\lambda} - \nu t \right)$$

$$E_2 = A_2 \cos \left[2\pi \left(\frac{x}{\lambda} - \nu t \right) - \phi \right]$$

(11.10)

which interfere to give a resultant displacement

$$E = E_1 + E_2 = A_1 \cos 2\pi \left(\frac{x}{\lambda} - \nu t\right)$$
$$+ A_2 \cos 2\pi \left(\frac{x}{\lambda} - \nu t\right) \cos \phi + A_2 \sin 2\pi \left(\frac{x}{\lambda} - \nu t\right) \sin \phi$$
$$= \sqrt{A_1^2 + A_2^2 + 2A_1A_2 \cos \phi} \cos \left[2\pi \left(\frac{x}{\lambda} - \nu t\right) + \delta\right] \quad (11.11)$$

where $\delta = A_2 \sin \phi/(A_1 + A_2 \cos \phi)$. Thus the time average of the intensity over a full cycle is proportional to $A_1^2 + A_2^2 + 2A_1A_2 \cos \phi$. The corresponding intensity of wave 1 alone is proportional to A_1^2, and that due to wave 2 is proportional to A_2^2. We see that the resulting intensity differs from the sum of the individual intensities by $2A_1A_2 \cos \phi$, which is known as the *interference term*.

In the wave description of electromagnetic waves the intensity is proportional to the square of the electric intensity E, while in the particle description the intensity is proportional to the number of photons N passing through a unit area in unit time. In free space the intensity I of a beam with a single frequency ν is

$$I = c\epsilon_0 E^2 = Nh\nu \quad (11.12)$$

where c = speed of light
ϵ_0 = permittivity of free space
h = Planck's constant
Clearly the photon flux N is proportional to E^2, since all other quantities are constant. When N is very large, we can measure E and expect to find E^2 a valid measure of N. Under this high-flux condition a scintillation detector would show continuous and essentially constant response with the discrete nature of the photons washed out. However, as the flux is reduced until only one photon arrives each second, the particle nature becomes prominent and the wave treatment ceases to give a useful description of the beam. However, the average value of E^2 over a reasonable time interval is a meaningful measure of the photon flux N; or alternatively, the value of $E^2 \, dV$ is proportional to the probability of finding a photon in an infinitesimal volume element dV.

We develop the analogy between photon and electron beams by assuming that the electrons in a beam traveling in the x direction with constant momentum p_x are controlled by "pilot" waves with a wave function Ψ which plays the same role for electrons that the electric intensity plays for photons. Following de Broglie, we take

$$\Psi = Ae^{2\pi i(x/\lambda - \nu t)} = Ae^{i(p_x x - Et)/\hbar} \tag{11.13}$$

where we have made use of $\lambda = h/p$ and $E = h\nu$. This relation is analogous to Eq. (11.10), which is just the real part of Eq. (11.13). For the electron beam we are discussing Ψ is a complex number, while the electric intensity E is an observable quantity and hence real. That there should be differences in the wave descriptions of electrons and photons is not at all surprising, since the electron has rest mass, charge, and a different spin than the photon. Electrons and photons have analogous behavior in some ways but are very different in others.

The question at once arises: What is the physical significance of the mysterious Ψ? An answer sometimes given to this question is that Ψ is merely an auxiliary mathematical quantity that is introduced to facilitate computations relative to the results of experiment. For example, in the experiments described above on the diffraction of electrons by a crystal, the experimenter sets up an electron gun that fires electrons of a certain energy at a crystal. A detector placed at a certain angle gives indications which are taken to mean that electrons are being received. In order to develop a quantitative theory for such observations, we *assume* that a beam of electron waves of frequency $\nu = E/h$ falls on the crystal, and we *calculate* the intensity of the waves scattered in the direction of the detector. From this we infer, following certain rules, what the indication of the detector should be. In order to make such calculations, it is not really necessary to attach any physical significance at all to the mathematical symbol Ψ.

There is much to be said for such a view. After all, observational results are the primary material of physics; the purpose of theories is to correlate these results, to group them into those regularities of experience which we call laws, and to predict the results of new experiments. Even the motion of the electron as a particle is only an auxiliary concept, introduced for convenience in describing and interpreting observational results. Such auxiliary concepts as particles constitute convenient aids to thinking, and most physicists find it advantageous to make use of them. Whereas some experimental results are conveniently understood in terms of the electrons as particles, however, others can be understood only in terms of the interference of waves. It is worthwhile, therefore, to go as far as we can toward assigning some physical significance to these waves or to the quantity Ψ in terms of which the mathematical theory of the waves is expressed. It turns out that this can be done, according to circumstances, in two rather different ways.

1. In dealing with an experiment on scattering such as that just

described, a beam of waves of indefinite total length is commonly assumed, represented by a Ψ like that in Eq. (11.3). The *square of the absolute value of* Ψ, $|\Psi|^2$, *is then taken to be proportional* (for given particle energy) *to the number of particles in the beam that cross unit area per second*, the unit area being taken perpendicular to the direction of motion.[1] Thus, in an experiment on the diffraction of electrons, let n_0 electrons cross unit area per second in the incident beam and let Ψ_0 denote the mathematical expression assumed for Ψ in the waves representing this beam. Then, if calculation by means of wave mechanics gives Ψ_d as the value of Ψ in a certain direction of diffraction, the theoretical value for the intensity of the diffracted beam in that direction, defined as the number of diffracted electrons crossing unit area per second, is

$$n_d = \frac{|\Psi_d|^2}{|\Psi_0|^2}\, n_0 \tag{11.14}$$

[provided the incident and the diffracted beams move in regions where P in (11.1) has the same value]. This procedure corresponds exactly to the method of handling similar problems in optics. The intensity of a beam of light, which might be defined as the number of photons crossing unit area per second, is proportional (in a given medium) to E^2, the square of the electric vector. The principal difference here is that, Ψ being a complex number, we must use $|\Psi|^2$ instead of Ψ^2. Usually $\Psi^*\Psi$ is written instead of $|\Psi|^2$, Ψ^* denoting the complex conjugate of Ψ.

Returning now to the problem of the two-slit interference pattern, we see that if Eq. (11.13) is indeed applicable to electrons, we should expect the curve of Fig. 11.6b. At any point on the screen there is a contribution Ψ_1 from the first slit and a contribution Ψ_2 from the second slit, so that the resulting wave-function probability amplitude is $\Psi = \Psi_1 + \Psi_2$. The probability of an electron's arriving at this point on the screen is proportional to $\Psi^*\Psi$. If there are places where $\Psi_1 = -\Psi_2$, $\Psi^*\Psi$ goes to zero and an electron interference pattern is to be expected.

2. There are other cases, as in the photoelectric effect, in which one wishes to follow the flight of a single electron or other particle. Then Ψ is taken to refer to a single particle. Usually, in a given case, values of Ψ appreciably different from zero occur only within some finite region. A solution of the wave equation of this latter type is sometimes called a *wave packet*. In such a case it is natural to ask: Where is the particle in relation to the wave packet? The accepted answer may be stated in terms of probabilities. The *position of the particle* is usually considered

[1] This flux relation is special for the case in hand. See Sec. 12.7 for a general relation.

not to be defined any more closely than is indicated by the values of Ψ. At any given instant, if a suitable observation were made, the particle might be found at any point where Ψ is different from zero; the *probability* of finding it in the neighborhood of a given point is proportional to the value of $|\Psi|^2$ at that point. More exactly, the *probability density* at any point is represented by $|\Psi|^2$; the *probability of finding the particle within any element of volume dx dy dz* is

$$\Psi^*\Psi \, dx \, dy \, dz \tag{11.15}$$

The wave scalar Ψ itself is sometimes called a *probability amplitude* for position of the particle.

The interpretation just stated imposes upon Ψ a certain mathematical requirement. For the *total probability* of finding the particle *somewhere* is, of course, unity. Hence Ψ must satisfy the condition that

$$\iiint\Psi^*\Psi \, dx \, dy \, dz = 1 \tag{11.15a}$$

the triple integral extending over all possible values of x, y, and z. A Ψ satisfying this requirement is said to be *normalized* (to unity).

There are also interesting relations between a wave packet and the velocity or momentum of the particle. A detailed study brings out the following features. Suppose the particle is in free space with $P = 0$, so that $E = p^2/2m$. Such a particle cannot be represented by Eq. (11.13), which denotes an infinite train of waves, since it makes

$$\iiint\Psi^*\Psi \, dx \, dy \, dz = \infty$$

unless $A \equiv 0$. Thus we conclude that *it is not possible for a particle to have a perfectly definite momentum or velocity.* We can, however, have as close an approach as we please to a definite momentum, for Ψ may be assumed to be indistinguishable from (11.13) over as large a region as desired, sinking to zero outside the boundary of this region; the constant A can then be chosen small enough so that $\iiint\Psi^*\Psi \, dx \, dy \, dz = 1$. But then the position of the particle is not well defined, since $\Psi^*\Psi$ is finite over a considerable volume.

11.8   The Heisenberg Uncertainty Principle

De Broglie's proposal that the motion of a particle with velocity v is controlled by pilot waves with group velocity v suggests that there is a limit beyond which we cannot determine simultaneously both the momentum and the position of the particle. Suppose that we use the two waves of Fig. 11.2 to represent the pilot wave for a particle, assuming that the particle is somewhere in the shaded half wave of the envelope. Then the uncertainty in position Δx corresponds to $\lambda_m/2$, from which $\Delta x = \pi/\Delta k$ (Fig. 11.2). From $k = 2\pi/\lambda = 2\pi p/h$, we have $\Delta k = (2\pi\,\Delta p)/h$. The propagation numbers of the two waves differ by $2\Delta k$, so that the uncertainty in the x component of the momentum $\Delta p_x = (2h\,\Delta k)/2\pi$, whence

$$\Delta x\,\Delta p_x \approx h \tag{11.16}$$

It is not surprising that our choice of two beating waves of slightly different frequency is far from an optimum way in which to choose the pilot waves for the motion of a particle. It gave us (Fig. 11.2) a beat pattern which resembles a row of beads. It is possible (Sec. 12.5) to eliminate all but one "bead" by adding other frequencies (or k's) and thus devise a wave packet which can be localized to any Δx we choose. But even when a wave packet of optimum shape is selected, it is impossible to reduce both Δx and Δp_x to zero simultaneously.[1] Indeed, regardless of the wave packet chosen,

$$\Delta x\,\Delta p_x \geq \frac{\hbar}{2} \tag{11.17}$$

This fact was first enunciated in 1927 by Heisenberg, who called it the principle of *Unbestimmtheit*. This term has been variously translated as indeterminacy, indefiniteness, or uncertainty.

The indefiniteness we have just found to exist in the values of certain mechanical magnitudes associated with a particle, such as its position or momentum, is a fundamental feature of wave mechanics. The indefiniteness in position can be minimized by making the wave packet very small (Ψ practically zero except within a very small region); but in that case it can be shown that the packet will spread rapidly because a broad range of propagation numbers is required. Thus, a small packet means a large indefiniteness in momentum and velocity. On the other hand, if we give to Ψ a form like that in Eq. (11.13) over a large region, in order to fix the velocity and momentum of the particle within narrow limits, there is a large indefiniteness in the position.

The uncertainty principle has an analog in the field of optics. A single sinusoidal wave of light of wavelength λ represents a certain

[1] Unfortunately we need certain developments in Chap. 12 to define Δx and Δp_x quantitatively. Until this is done in Sec. 12.6, the quantitative aspect of Eq. (11.16) must remain vague.

amount of energy that is closely localized in space, but it does not constitute monochromatic light; for, upon passing through a spectroscope, it will spread widely in the spectrum (see Fig. 11.7). To have an approach to monochromatic light, we must have a train of many waves; and this means a corresponding dispersal of the energy in space. In general, any wave packet can be expanded in terms of wave trains like (11.13), just as any patch of light waves can be resolved into monochromatic trains. This amounts to representing Ψ by a Fourier integral. When this has been done in a suitable way, the coefficients of the various wave trains in the expansion constitute a probability amplitude for momentum; i.e., the square of the absolute value of any coefficient gives the probability that a suitable observation would reveal the particle as moving in the direction and with the momentum or velocity that is

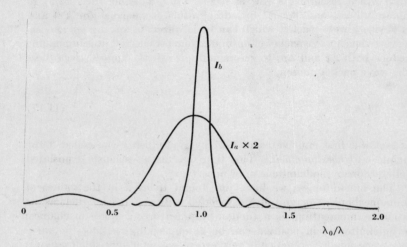

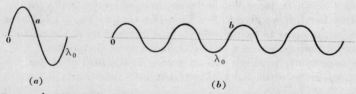

Fig. 11.7 A single sinusoid (a) of wavelength λ_0 analyzed by a spectrograph has wavelength components over a broad region (curve I_a). The analysis of a four-wave train (b) leads to a much higher peak at λ_0 (curve I_b).

associated with the corresponding wave train. Thus, a particle does not have either a sharply defined position or a sharply defined momentum.

The conclusion that a wave packet cannot represent a particle as having at the same time a definite position and a definite momentum might seem to be in conflict with the fact that, in practice, both position and momentum are capable of measurement; e.g., from two snapshots of a rifle bullet, its position and velocity at a given instant can both be calculated. Heisenberg pointed out, however, that this is possible only because, on the scale of observation used in ordinary physical measurement, the indeterminacy required by Eq. (11.17) is so minute as to be lost in the experimental errors.

It is quite otherwise for an electron or a molecule. Consider, for example, how an electron might be located with *atomic* precision. We might use a microscope; but then we should have to use light of extremely short wavelength in order to secure sufficient resolving power. To distinguish positions 10^{-11} m apart, for example, we should have to use γ rays. Under these circumstances, however, the effect of the light on the electron cannot be neglected. If we are to "see" the electron, at least one photon must bounce off it and enter the microscope. In rebounding from the electron this photon will give it a strong Compton kick. Thus at the instant when we locate the electron, its momentum undergoes a discontinuous change. Furthermore, there is an indefiniteness about the magnitude of this change, for it will vary according to the direction in which the scattered photon leaves the scene of action. We cannot limit closely the range of possible directions for the scattered photons that enter the microscope, by stopping down the aperture, without a serious loss of resolving power. A quantitative analysis of such observations leads again to Eq. (11.17).

It appears, then, that we cannot *at the same time* assign to an electron or other small particle, in terms of actual or possible observations, a definite position and a definite momentum (or energy). Thus, an assertion that both the position and the momentum of a particle have simultaneously certain precise values is a statement devoid of physical meaning; for, since 1900, it has become increasingly accepted as a principle of physics that only those magnitudes which can be observed, directly or indirectly, have physical significance. Our classical notion of a particle as, something that can move along a sharply defined path, having at each instant a definite position and velocity, is therefore not fully applicable to electrons or protons or atoms or molecules. These small bits of matter may be said to have some particle properties, but they also possess certain wave properties, so that, in the classical sense of the words, they are neither true particles nor true waves. C. G. Darwin proposed calling them *wavicles*.

The Heisenberg uncertainty relationship applies to any pair of variables which are canonically conjugate in the hamiltonian formulation of mechanics. In addition to Δx and Δp_x, we have $\Delta y\,\Delta p_y \geq \hbar/2$, $\Delta z\,\Delta p_z \geq \hbar/2$ for cartesian coordinates; $\Delta E\,\Delta t \geq \hbar/2$, where ΔE is the uncertainty in energy and Δt that in time; $\Delta\theta\,\Delta A_\theta \geq \hbar/2$, where ΔA_θ is the uncertainty associated with the angular momentum A_θ and $\Delta\theta$ the indefiniteness in the corresponding angle; and so forth.

11.9   The Schrödinger Wave Equation

If a particle has wave properties, as de Broglie proposed, it is expected that there should be some sort of wave equation which describes the behavior of the wave function. The usual basis of the mathematical theory of wave motion is furnished by a *differential equation*. Thus an electromagnetic wave moving in the x direction can be described by

$$\frac{\partial^2 \mathsf{E}_y(x,t)}{\partial x^2} = \frac{1}{c^2}\frac{\partial^2 \mathsf{E}_y(x,t)}{\partial t^2}$$

in which E_y is the y component of the electric intensity.

We cannot follow a logical development from first principles to find the differential equations for matter waves, but we can guess a wave equation in harmony with the properties which we find the waves to have and then test our guess by comparison of experimental data with predictions deduced from this wave equation. Let us assume that Eq. (11.13) is an appropriate representation of a plane wave moving in the x direction, so that

$$\Psi = Ae^{i(p_x x - Et)/\hbar} \tag{11.13}$$

in which E and p for the particle "piloted" by Ψ are related by

$$\frac{p_x{}^2}{2m} + P = E \tag{11.18}$$

where P is the potential energy of the particle and $p_x{}^2/2m$ is its kinetic energy. Whatever the correct differential equation for Ψ may be, it should be such that any allowable function Ψ expressed in the form of Eq. (11.13) is a solution of it, regardless of the value of E or p. Let us endeavor, therefore, to find by trial a differential equation such that, when the expression just given for Ψ is substituted in it, E and p cancel out in consequence of the known relationship expressed in Eq. (11.18).

A few derivatives of Ψ as obtained from Eq. (11.13) are

$$\frac{\partial \Psi}{\partial x} = \frac{i}{\hbar} p_x \Psi \tag{11.19a}$$

$$\frac{\partial^2 \Psi}{\partial x^2} = -\frac{1}{\hbar^2} p_x{}^2 \Psi \tag{11.19b}$$

$$\frac{\partial \Psi}{\partial t} = -\frac{i}{\hbar} E \Psi \tag{11.19c}$$

If Eq. (11.18) is multiplied by Ψ to give $p^2 \Psi / 2m + P\Psi = E\Psi$ and use is made of Eqs. (11.18b, c), we have the equation, first obtained by Schrödinger in 1926,

$$-\frac{\hbar^2}{2m} \frac{\partial^2 \Psi}{\partial x^2} + P\Psi = -\frac{\hbar}{i} \frac{\partial \Psi}{\partial t} \tag{11.20}$$

Equation (11.20) is Schrödinger's famous *wave equation containing the time.* The agreement of results deduced from this equation justifies the belief that the equation is valid for the matter waves associated with a particle of mass m as long as relativistic effects can be neglected. The equation is remarkable among the differential equations of mathematical physics in that it contains $i = \sqrt{-1}$. But for this factor it would resemble the equation for the flow of heat in a solid body, which likewise contains the first derivative with respect to time. The motion of matter waves combines, in fact, some of the features of heat flow in a solid with other features resembling the propagation of mechanical disturbances, such as sound waves.

Let us consider Eq. (11.20) for the case in which E, p_x, and P are all constants satisfying Eq. (11.18) and A is a constant representing the wave amplitude. If $E > 0$ and $p > 0$, this solution of the wave equation represents plane waves of Ψ traveling toward $+x$; for at a point moving with velocity E/p_x, the value of Ψ remains constant. E and p_x represent energy and momentum of the associated particle, respectively; the frequency ν and the wavelength λ of the waves are easily seen to be $\nu = E/h$, $\lambda = h/p_x$ in accordance with conclusions reached previously. If $p_x < 0$ but $E > 0$, both the particle and the waves are moving toward $-x$.

If E is negative, as may happen if P is sufficiently negative, the phase waves and the particle travel in opposite directions. We can always make this so, if we wish, by adjusting the arbitrary additive constant that occurs in P, the potential energy. The possibility of reversing the direction of motion of matter waves by a mere mathematical change of this sort is a clear indication that they cannot be ordinary real waves!

When the potential energy P is a function of x only and the total energy E is constant, we may write

$$\Psi(x,t) = \psi(x)f(t) \tag{11.21}$$

and substitute this into Eq. (11.20). Upon dividing through by $f(t)$ and making use of Eq. (11.19c) we have

$$-\frac{\hbar^2}{2m}\frac{d^2\psi}{dx^2} + P\psi = E\psi \tag{11.22}$$

which is the one-dimensional *time-independent* Schrödinger equation, also known as the *Schrödinger wave equation*.

For motion in three dimensions Ψ is a function of x, y, z, and t; the Schrödinger equation becomes

$$-\frac{\hbar^2}{2m}\left(\frac{\partial^2\Psi}{\partial x^2} + \frac{\partial^2\Psi}{\partial y^2} + \frac{\partial^2\Psi}{\partial z^2}\right) + P\Psi = -\frac{\hbar}{i}\frac{\partial\Psi}{\partial t} \tag{11.23}$$

or

$$-\frac{\hbar^2}{2m}\nabla^2\Psi + P\Psi = -\frac{\hbar}{i}\frac{\partial\Psi}{\partial t} \tag{11.23a}$$

It was early noticed that Schrödinger's equation can be obtained from the classical hamiltonian expression for the energy of the moving particle by substituting suitable *differential operators* for energy and momentum. For example, Eq. (11.19a) can be written

$$p_x\Psi = \frac{\hbar}{i}\frac{\partial\Psi}{\partial x}$$

from which we are led to the successful approach of replacing p_x by the operator

$$\hat{p}_x = \frac{\hbar}{i}\frac{\partial}{\partial x} \tag{11.24a}$$

where the "hat" is used to distinguish an operator from the physical quantity. Similarly, analogy and Eq. (11.19c) to make plausible the differential operators

$$\hat{p}_v = \frac{\hbar}{i} \frac{\partial}{\partial y} \tag{11.24b}$$

$$\hat{p}_z = \frac{\hbar}{i} \frac{\partial}{\partial z} \tag{11.24c}$$

$$\hat{E} = -\frac{\hbar}{i} \frac{\partial}{\partial t} \tag{11.24d}$$

The total energy of a particle in the hamiltonian form is

$$H = \frac{1}{2m} (p_x{}^2 + p_y{}^2 + p_z{}^2) + P = E \tag{11.25}$$

where H denotes the hamiltonian and P the potential energy. If we let

$$\hat{H} = -\frac{\hbar^2}{2m} \left(\frac{\partial^2}{\partial x^2} + \frac{\partial^2}{\partial y^2} + \frac{\partial^2}{\partial z^2} \right) + P \tag{11.26}$$

and we multiply Eq. (11.25) by Ψ, we have, upon replacing H, p_x, E, etc., by their operators

$$\hat{H}\Psi = E\Psi \qquad \text{or} \qquad -\frac{\hbar^2}{2m} \nabla^2\Psi + P\Psi = -\frac{\hbar}{i} \frac{\partial \Psi}{\partial t}$$

Thus we have recovered the Schrödinger equation by a procedure which lies close to Schrödinger's original line of thought.

Even though we have not been able to derive Schrödinger's equation, its fabulous success in dealing with problems of modern physics justifies our use of it in nonrelativistic applications, some of which we consider in subsequent chapters.

Problems

1. The ionization potential of helium is 24.6 V. Find the de Broglie wavelength of an electron of 24.6 eV kinetic energy. What is the ratio of this wavelength to the radius of the first Bohr orbit calculated on the assumption that the effective nuclear charge is given by 13.6 $Z_{eff}^2 = 24.6$?

Ans: 2.48 Å; 6.3

2. Assume that a particle cannot be confined to a spherical volume of diameter less than the de Broglie wavelength of the particle. Estimate the minimum kinetic energy a proton confined to a nucleus of diameter 10^{-14} m may have. What kinetic energy would an electron have to possess if it were confined to this nucleus? (This is one of the arguments against the idea that free electrons exist in nuclei.)

Ans: 8.2 MeV; 124 MeV

3. The half-life of a certain excited state is about 8 ns. If this is essentially the uncertainty Δt for photon emission, calculate the uncertainty in frequency $\Delta \nu$, assuming that $\Delta E \, \Delta t \approx h$. Find $\Delta \nu / \nu$ if the photons have $\lambda = 500$ mμ.

　Ans: 1.25×10^8 Hz; 2×10^{-7}

4. The Heisenberg uncertainty principle suggests that a particle of momentum p cannot be confined by a central force to a circle of radius r less than $\hbar/p$. Assuming that $r = \hbar/p$, write the total energy $E = K + P$ as a function of r for a hydrogen electron. Find the value of r for which E is minimum and show that it is equal to the Bohr radius. Sketch E, K, and P as functions of r.

5. Imagine that the angular position of an electron initially in the ground state of hydrogen is determined by an experiment to an accuracy $\Delta \varphi = 0.2$ rad. If $\Delta A_\varphi \, \Delta \varphi \approx \hbar$, calculate the uncertainty in the angular momentum A_φ and compare it with the angular momentum postulated for the first Bohr orbit. Assuming for the moment that the electron continues in an orbit with the Bohr radius, find the uncertainty in the kinetic energy K and compare it to the ionization energy. Is the electron still in the ground state? (Remember the magnitude of a quantity is at least as great as the uncertainty in the quantity.)

　Ans: $5\hbar$; $\Delta K > 35 E_{\text{ion}}$

6. An electron has a de Broglie wavelength of 0.15 Å. Compute the phase and group velocities of the de Broglie waves. Find the kinetic energy of the electron.

7. Compute the kinetic energy of an electron which has a de Broglie wavelength equal to the wavelength of a Mo $K\alpha$ photon (0.707 Å).

8. Find the kinetic energy of a neutron which has a wavelength of 3 Å. At what angle will such a neutron undergo first-order Bragg reflection from a calcite crystal for which the grating space is 3.036 Å?

9. Calculate the de Broglie wavelength of (*a*) an electron, (*b*) a muon, (*c*) a proton, and (*d*) a Hg^{200} atom if each has a kinetic energy of 25 eV.

10. Calculate the de Broglie wavelength for an average helium atom in a furnace at 400°K for which the kinetic energy is $3kT/2$.

　Ans: 0.66 Å

11. In an experiment the wavelength of a photon is measured to an accuracy of one part per million; that is, $\Delta \lambda / \lambda = 10^{-6}$. What is the approximate minimum uncertainty Δx in a simultaneous measurement of the position of the photon in the case of (*a*) a photon with $\lambda = 6000$ Å and (*b*) an x-ray photon with $\lambda = 5$ Å?

12. The radius of a typical atom is approximately 10^{-10} m = 1 Å. To locate the nucleus or an electron within a distance of 5×10^{-12} m by use of electromagnetic waves, the wavelength must be at least this small. Calculate the energy and momentum of a photon with $\lambda = 5 \times 10^{-12}$ m. If Δx is 5×10^{-12} m for an electron, what is the corresponding uncertainty in its momentum? This represents one of Heisenberg's gedanken experiments illustrating the uncertainty principle.

13. A nucleon, either proton or neutron, is confined to a nucleus of radius 4×10^{-15} m. If this implies an uncertainty $\Delta x \geq 8 \times 10^{-15}$ m, what is the corresponding uncertainty in p_x? Since the momentum must be at least as big as the uncertainty Δp_x, estimate the kinetic energy and the speed of a nucleon by taking $p = \Delta p_x$. (Note that these are low values since we have overestimated Δx and underestimated p.)

14. According to the theory of Abbe, a microscope can resolve the images of two objects if the objects are separated by a distance $s = \lambda/(2n \sin \alpha)$, where n is the index of refraction of the medium between the objects and the microscope objective and α is the half-angle subtended at the object by the objective. (In practice, $n \sin \alpha$ has an upper limit of about 1.6.) If $n \sin \alpha$ is 1.5 for the microscope in question and we wish to resolve objects separated by 0.5 Å, find the minimum energy required for (a) photons and (b) electrons. What are the corresponding momenta?

15. A beam of 36-eV electrons is incident normally on a thin polycrystalline metal film for which the grating space a is 2.5 Å. If the depth ϕ of the potential well representing the metal is 4 eV, find the angular diameter of the smallest ring of the diffraction pattern formed by electrons passing through the foil.

16. Derive Eq. (11.8a) and show that (a) for $K/m_0c^2 \ll 2$,

$$\lambda = \frac{h}{\sqrt{2m_0K}} \left(1 - \frac{K}{4m_0c^2} + \frac{3}{32} \frac{K^2}{m_0^2c^4} + \cdots \right)$$

and (b) for $K/m_0c^2 \gg 2$,

$$\lambda = \frac{ch}{K} \left(1 - \frac{m_0c^2}{K} + \frac{3m_0^2c^4}{2K} + \cdots \right)$$

17. Show that Eq. (11.9) is equivalent to Bragg's law $n\lambda = 2d \sin \theta$. *Hint:* Show that $d = D \sin (\phi/2)$, while $\sin \theta = \cos (\phi/2)$.

18. How long does the uncertainty principle allow a π° meson to exist as a $p - \bar{p}$ pair?

chapter twelve

Wave Mechanics I
Free States

Modern quantum mechanics has led to the solution of many problems which were classically intractable. The Schrödinger formulation is particularly useful for our purposes, and we begin by considering the wave-mechanical solutions for the simplest one-dimensional examples of electron motion.

12.1 Electron Beam in a Field-free Space

Consider a region in which the potential energy P of an electron is constant, so that there are no forces acting. Let a uniform beam of electrons be traveling in the x direction, each electron having momentum p_x and total energy $E = p_x^2/2m + P$. For this beam the Schrödinger wave equation is

$$-\frac{\hbar^2}{2m}\frac{d^2\psi}{dx^2} + P\psi = E\psi \tag{11.22}$$

where ψ is the wave function such that the probability of finding an electron between x and $x + dx$ is $\psi^*\psi\, dx$. From Eq. (11.22) we have

$$\frac{d^2\psi}{dx^2} + \frac{2m(E - P)}{\hbar^2}\,\psi = 0 \tag{12.1}$$

which is the simple-harmonic-motion equation, since the coefficient of ψ is a positive constant. The general solution of Eq. (12.1) is

$$\psi = C_1 \exp \frac{i\,\sqrt{2m(E - P)}\,x}{\hbar}$$
$$+ C_2 \exp\left[-\frac{i\,\sqrt{2m(E - P)}\,x}{\hbar}\right] \tag{12.2}$$

The complete wave function $\Psi(x,t) = \psi(x)f(t)$ satisfies Eq. (11.11c), from which

$$\frac{df(t)}{dt} = -\frac{i}{\hbar}\,Ef \tag{12.3a}$$

and

$$f(t) = Ce^{-iEt/\hbar} \tag{12.3b}$$

Thus we find that

$$\Psi(x,t) = A \exp \frac{i[\sqrt{2m(E - P)}\,x - Et]}{\hbar}$$
$$+ B \exp\left\{-\frac{i[\sqrt{2m(E - P)}\,x + Et]}{\hbar}\right\} \tag{12.4}$$

is the general solution of Eq. (11.20). However, the second term of Eq. (12.4) represents a wave moving to the left, while we have postulated that the beam associated with Ψ is moving in the $+x$ direction. Therefore B is zero and

$$\Psi(x,t) = A \exp \frac{i[\sqrt{2m(E - P)}\,x - Et]}{\hbar} = Ae^{i(p_x x - Et)/\hbar}$$
$$= A \cos \frac{p_x x - Et}{\hbar} + iA \sin \frac{p_x x - Et}{\hbar} \tag{12.5}$$

[Note that we have retrieved Eq. (11.13).]

The probability of finding an electron between x and $x + dx$ is given by the one-dimensional form of Eq. (11.15)

$$\mathcal{P} \, dx = \Psi^*\Psi \, dx = A^2 \, dx \qquad (12.6)$$

As expected for a uniform electron beam, the probability $\mathcal{P}$ is not a function of x. The constant A is related to the intensity of the beam. If there are on the average n electrons in a length L, we must have $A^2 = n/L$ since

$$\int_{x}^{x+L} \Psi^*\Psi \, dx = \int_{x}^{x+L} A^2 \, dx = A^2 L = n$$

12.2   The Step Barrier

Consider next a uniform beam of electrons of energy E coming from the left of Fig. 12.1. Let the potential energy P be 0 to the left of the origin and P_0 to the right. Such a rectangular barrier does not exist in nature, but it is reasonably approximate to the instantaneous potential difference between the dees of a cyclotron or to the potential barrier at the surface of a metal. For a monoergic electron beam the time dependence of $\Psi(x,t)$ is given by Eq. (12.3b) for all x and t; therefore, our first problem is to find the spatial wave function $\psi(x)$, which must satisfy Eq. (11.22).

Region 1

$$-\frac{\hbar^2}{2m}\frac{d^2\psi}{dx^2} + 0 = E\psi \qquad \text{for } -\infty \leq x < 0 \qquad (12.7a)$$

Region 2

$$-\frac{\hbar^2}{2m}\frac{d^2\psi}{dx^2} + P_0\psi = E\psi \qquad \text{for } 0 \leq x \leq \infty \qquad (12.7b)$$

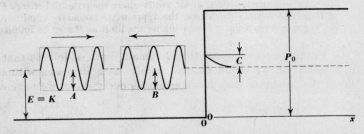

Fig. 12.1 When a beam of free electrons with kinetic energy $K = E$ is incident on an ideal potential barrier of height $P_0 > E$, the beam is reflected. Shown in the rectangles are small sections of the real part of the incident and reflected waves.

The general solution for region 1 is

$$\psi_1 = A \exp \frac{i \sqrt{2mE}\, x}{\hbar} + B \exp\left(- \frac{i \sqrt{2mE}\, x}{\hbar}\right)$$
$$= Ae^{ik_1 x} + Be^{-ik_1 x} \tag{12.8}$$

where A and B are constants to be determined and $k_1 = \sqrt{2mE}/\hbar$. The term at the right represents particles moving to the left, which arise from reflection at the barrier. For region 2 $(0 \leq x \leq \infty)$ we must distinguish between two situations, according as $K = E - P_0$ is negative or positive. If it is negative, classical physics would forbid the electrons (or other particles) to enter region 2, while if $E - P_0 = K$ is positive, classical physics predicts that all the electrons would enter and pass freely through. Wave mechanics makes different predictions.

CASE 1. $E - P_0 < 0$ If P_0 exceeds E, the kinetic energy K is negative. Nevertheless, ψ is not zero in region 2 but is given by the solution to Eq. (12.7b), which is

$$\psi_2 = C \exp\left[- \frac{\sqrt{2m(P_0 - E)}\, x}{\hbar} \right]$$
$$+ D \exp \frac{\sqrt{2m(P_0 - E)}\, x}{\hbar} = Ce^{-\alpha x} + De^{\alpha x} \tag{12.9}$$

where C and D are constants to be determined and $\alpha = \sqrt{2m(P_0 - E)}/\hbar$.

Next we place two requirements on ψ and its derivatives in order that the probability $\psi^*\psi\, dx$ be finite and single-valued so that the wave function represents a specific physical situation:

1. The wave function ψ is finite, continuous, and single-valued.
2. The first derivative (or gradient in several dimensions) of ψ is finite, continuous, and single-valued at all points where the potential energy P is finite; if P goes to infinity anywhere, the appropriate boundary condition is found by starting with a finite P and taking the limit as P goes to infinity.

From the requirement that ψ remain finite, it follows that $D = 0$; if it were not, ψ_2 would become infinite as x increases without limit. From the condition that ψ is continuous everywhere, we have at $x = 0$

$$A + B = C \tag{12.10}$$

while the requirement that $d\psi/dx$ be continuous at $x = 0$ yields

$$ik_1(A - B) = -\alpha C \tag{12.11}$$

If we solve Eqs. (12.10) and (12.11) for B and C in terms of A, we find that the amplitude of the reflected wave B is related to that of the incident wave A by

$$B = \frac{k_1 - i\alpha}{k_1 + i\alpha} A \qquad (12.12)$$

while

$$C = \frac{2k_1}{k_1 + i\alpha} A \qquad (12.13)$$

The wave entering region 2 is exponentially damped (Fig. 12.1); although there is a finite probability of finding electrons in the classically forbidden region, there is no steady transmission of particles there. In the steady state all incident particles are ultimately reflected. To see that this is true, it is convenient to write complex quantities in polar form, making use of the fact that $k_1 \pm i\alpha = \rho e^{\pm i\delta}$, where $\rho^2 = k_1{}^2 + \alpha^2$ and $\tan \delta = \alpha/k_1$; Eq. (12.12) then becomes

$$B = \frac{\rho e^{-i\delta}}{\rho e^{i\delta}} A = e^{-2i\delta} A \qquad (12.12a)$$

where $\delta = \tan^{-1}[(P_0 - E)/E]^{\frac{1}{2}}$.

Substitution of Eq. (12.12a) into (12.8) yields

$$\psi_1 = A(e^{ik_1x} + e^{-2i\delta}e^{-ik_1x}) = Ae^{-i\delta} \cos\left(\frac{\sqrt{2mE}\,x}{\hbar} + \delta\right) \qquad (12.13a)$$

which represents a standing wave in which equal numbers of particles are going left and right. This is just the classical behavior of electromagnetic waves at a metal surface.

CASE 2. $E - P_0 > 0$ When the total energy exceeds the potential energy (Fig. 12.2), the kinetic energy K is positive and region 2 is classically accessible to the electrons. In this case the general solution of Eq. (12.7b) is

$$\psi_2 = Ge^{ik_2x} + He^{-ik_2x} \qquad (12.14)$$

where $k_2 = \sqrt{2m(E - P_0)}/\hbar$. Here the first term represents a wave traveling to the right and the second a wave traveling to the left. Since we have assumed that we have electrons initially incident only from the left, we set $H = 0$.

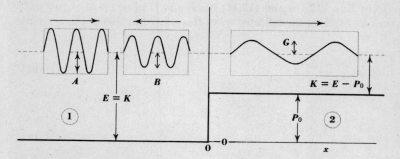

Fig. 12.2   When a beam of free electrons with kinetic energy $K = E$ is incident on an ideal potential barrier of height $P_0 < E$, part of the beam is reflected and part transmitted.   Small sections of the real part of the incident, reflected, and transmitted waves are shown in the rectangles.

Applying the requirements that at $x = 0$ both ψ and its derivative are continuous gives

$$A + B = G \tag{12.15a}$$
$$k_1 A - k_1 B = k_2 G \tag{12.15b}$$

from which the amplitudes of the reflected and transmitted waves in terms of that of the incident wave are

$$B = \frac{k_1 - k_2}{k_1 + k_2} A \tag{12.16}$$

and

$$G = \frac{2k_1}{k_1 + k_2} A \tag{12.17}$$

The number of reflected electrons per unit length is proportional to B^2 (Sec. 11.7), while the corresponding number of incident electrons is proportional to A^2.   Therefore the fraction of the incident electrons reflected, or the *reflection coefficient R*, is given by

$$R = \frac{|B^2|}{|A^2|} = \frac{(k_1 - k_2)^2}{(k_1 + k_2)^2} \tag{12.18}$$

Even if P_0 is negative, so that the electron is accelerated into region 2, there is a reflection coefficient which increases as the size of the step increases.   The situation is directly analogous to that in optics, where there is a reflected beam for normal incidence at any interface at which the optical properties change, regardless of whether the light is speeded up or slowed down as it enters the second medium.   [Indeed, if $k_1 = 2\pi/\lambda_1$ for the light in the first medium and $k_2 = 2\pi/\lambda_2$, Eq. (12.18) gives the reflection coefficient for light incident normally on the interface between two transparent media.]

The transmission coefficient T is the fraction of the incident electrons transmitted across the barrier.   To compare beam intensities in two regions where the particles have different velocities, we must note that the beam intensity is the number of particles crossing unit area per unit time, and this involves the product of the particle velocity and the number of particles per unit volume.   In the nonrelativistic case the particle velocity is given by $\sqrt{2K/m}$.   Therefore in this case

$$T = \sqrt{\frac{E - P_0}{E}} \frac{|G^2|}{|A^2|} = \frac{k_2}{k_1} \frac{4k_1^2}{(k_1 + k_2)^2} = \frac{4k_1 k_2}{(k_1 + k_2)^2} \qquad (12.19)$$

Of course, conservation of particles requires that $T + R = 1$.

12.3 Barrier Penetration

The possibility in quantum physics that a particle may penetrate into a region where its total energy E is less than its potential energy P leads to the tunneling of particles through thin barriers which are classically impenetrable.   Suppose that a beam of electrons is incident from the left on the rectangular barrier of high P_0 and width w of Fig. 12.3.

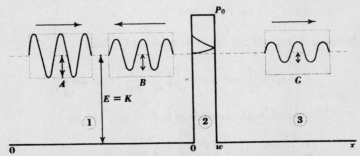

Fig. 12.3   When a beam of free electrons with kinetic energy $K = E$ is incident on a thin ideal barrier of height $P_0 > E$, a beam is transmitted through the barrier.   Small sections of the real part of the incident, reflected, and transmitted beams are shown in the rectangles.

Since $P = 0$ in regions 1 and 3 the Schrödinger wave equation for both regions is Eq. (12.7a), while for region 2 it is (12.7b). As we found in the preceding section, the solutions for a region of constant potential energy are sinusoidal when $E > P$ (kinetic energy positive) and exponentials for $E < P$ (K negative). The general solutions are:

Region 1: $-\infty < x < 0$

$$\psi_1 = A \exp \frac{i \sqrt{2mE}\, x}{\hbar} + B \exp \left(\frac{-i \sqrt{2mE}\, x}{\hbar} \right) \tag{12.20a}$$

Region 2: $0 \le x \le w$

$$\psi_2 = C \exp \frac{\sqrt{2m(P_0 - E)}\, x}{\hbar}$$
$$+ D \exp \left[\frac{-\sqrt{2m(P_0 - E)}\, x}{\hbar} \right] \tag{12.20b}$$

Region 3: $w \le x < \infty$

$$\psi_3 = G \exp \frac{i \sqrt{2mE}\, x}{\hbar} + H \exp \left(\frac{-i \sqrt{2mE}\, x}{\hbar} \right) \tag{12.20c}$$

We immediately set $H = 0$ since no particles are coming from the right. However, B and C are not zero, since we have reflections at both surfaces. If we let $k_1 = \sqrt{2mE}/\hbar$ and $\alpha = \sqrt{2m(P_0 - E)}/\hbar$, applying the conditions that ψ and its derivatives are continuous at $x = 0$ and $x = w$ leads to the four equations

$$A + B = C + D \tag{12.21a}$$
$$ik_1 A - ik_1 B = \alpha C - \alpha D \tag{12.21b}$$
$$C e^{\alpha w} + D e^{-\alpha w} = G e^{ik_1 w} \tag{12.21c}$$
$$\alpha C e^{\alpha w} - \alpha D e^{-\alpha w} = ik_1 G e^{ik_1 w} \tag{12.21d}$$

Solution of Eq. (12.21) for A in terms of G yields

$$A = G e^{ik_1 w} \left[\cosh \alpha w + \frac{i}{2} \left(\frac{\alpha}{k_1} - \frac{k_1}{\alpha} \right) \sinh \alpha w \right] \tag{12.22}$$

The transmission coefficient is $T = |G^2|/|A^2|$ or

$$\frac{1}{T} = \frac{|A^2|}{|G^2|} = 1 + \frac{1}{4} \left(\frac{\alpha}{k_1} + \frac{k_1}{\alpha} \right)^2 \sinh^2 \alpha w$$
$$= 1 + \frac{P_0^2}{4E(P_0 - E)} \sinh^2 \frac{\sqrt{2m(P_0 - E)}\, w}{\hbar} \tag{12.22a}$$

Since the hyperbolic sine increases rapidly with its argument, the transmission coefficient T decreases correspondingly with w. The sensitive dependence of the transmission coefficient on the height of the barrier is also of importance.

12.4 The "Square" Well: Free States

Let us replace the barrier of Fig. 12.3a by a well (Fig. 12.4). It proves convenient to place the zero of potential energy at the bottom of the well; $P = P_0$ for $-\infty < x < 0$ (region 1) and for $w < x < \infty$ (region 3), and $P = 0$ for $0 \leq x \leq w$ (region 2). Again we begin with the Schrödinger wave equation

$$-\frac{\hbar^2}{2m}\frac{d^2\psi}{dx^2} + P\psi = E\psi \tag{11.22}$$

for which the general solutions in the three regions are:

Region 1

$$\psi_1 = A\exp\frac{i\sqrt{2m(E-P_0)}\,x}{\hbar} + B\exp\left[\frac{-i\sqrt{2m(E-P_0)}\,x}{\hbar}\right]$$
$$= Ae^{ik_1x} + Be^{-ik_1x}$$

Region 2

$$\psi_2 = C\exp\frac{i\sqrt{2mE}\,x}{\hbar} + D\exp\left(\frac{-i\sqrt{2mE}\,x}{\hbar}\right)$$
$$= Ce^{ik_2x} + De^{-ik_2x}$$

Region 3

$$\psi_3 = G\exp\frac{i\sqrt{2m(E-P_0)}\,x}{\hbar} + H\exp\left[\frac{-i\sqrt{2m(E-P_0)}\,x}{\hbar}\right]$$
$$= Ge^{ik_1x} + He^{-ik_1x}$$

If a beam of particles is incident from the left and none from the right, $H = 0$. The requirement that ψ and its derivatives be con-

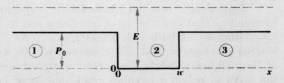

Fig. 12.4 A square well of depth P_0 and width w.

tinuous gives for the boundaries $x = 0$ and $x = w$

$$A + B = C + D \qquad (12.23a)$$

$$ik_1 A - ik_1 B = ik_2 C - ik_2 D \qquad (12.23b)$$

$$Ce^{ik_2 w} + De^{-ik_2 w} = Ge^{ik_1 w} \qquad (12.23c)$$

$$ik_2 Ce^{ik_2 w} - ik_2 De^{-ik_2 w} = ik_1 Ge^{ik_1 w} \qquad (12.23d)$$

From these four equations one finds that

$$A = Ge^{ik_1 w}\left[\cos k_2 w - \frac{i}{2}\left(\frac{k_1}{k_2} + \frac{k_2}{k_1}\right) \sin k_2 w \right] \qquad (12.24)$$

The transmission coefficient T can be found from

$$\frac{1}{T} = \frac{|A^2|}{|G^2|} = 1 + \frac{1}{4}\left(\frac{k_1}{k_2} - \frac{k_2}{k_1}\right)^2 \sin^2 k_2 w$$

$$= 1 + \frac{1}{4}\frac{P_0^2}{E(E - P_0)} \sin^2 \frac{\sqrt{2mE}\, w}{\hbar} \qquad (12.25)$$

Equation (12.25) shows the interesting property that the transmission coefficient rises and falls as the energy E is increased, reaching unity when the argument of the sine is an integer times π.

This result permits us to make a qualitative explanation of the *Ramsauer effect*, discovered in 1920 and inexplicable classically. Ramsauer found that when he passed a beam of monoergic electrons through certain noble gases such as argon, krypton, and xenon, there was almost no scattering for electron energies of 1 to 2 eV, while it was great for both lower and higher energies. Of course, an argon atom is only roughly approximated by a square potential well, but there is an attractive potential as the incident electron enters the atom because the outer electrons no longer screen the nucleus. Argon atoms are almost transparent to 2-V electrons, because for that E the transmission coefficient given by the analog of Eq. (12.25) approaches unity.

Some insight into the reason for the variation of T may be gained by recalling that both the wave function ψ and its derivative $d\psi/dx$ must be continuous at both $x = 0$ and $x = w$. As E is varied, the wavelengths are changed, both inside and outside the well. Only if the width of the well has precisely the correct relationship to the wavelength can the wave in region 3 have an amplitude equal to that of the incident wave in region 1.

12.5   Wave Packets and the Momentum Representation

Thus far we have been considering uniform beams of particles such as electrons; we now turn our attention to a single electron, which may be regarded as a more or less localized entity. Let us assume that at time $t = 0$ a single noninteracting electron is located near $x = x_0$ with the probability $\mathcal{P} \, dx$ of finding it between x and $x + dx$ being given by a known function such as that of Fig. 12.5. The curve shown is associated with the particular wave function

$$\Psi(x,0) = Ae^{-(x-x_0)^2/2a^2}e^{ip_0x/\hbar} \tag{12.26}$$

where $A = a^{-\frac{1}{2}}\pi^{-\frac{1}{4}}$ if Ψ is normalized so that $\int_{-\infty}^{\infty} \Psi^*\Psi \, dx = 1$. For this case a is the half-width of the curve at the points where $\mathcal{P}(x,0)$ has $1/e$ of its maximum value. In the discussion which follows, we shall develop the theory in terms of a general Ψ and then use Eq. (12.26) as a particular example for carrying out calculations.

In general, if a particle is associated with a normalized wave function $\Psi(x,t)$, we can specify the average value of x for the particle by giving each possible value of x a weight associated with the probability that the particle is between x and $x + dx$ and integrating over all possible values of x. The value so obtained is called the *expectation value* for x, which we shall represent by $\langle x \rangle$. At any instant t

$$\langle x \rangle = \int_{-\infty}^{\infty} \Psi^*(x,t)x\Psi(x,t) \, dx \tag{12.27}$$

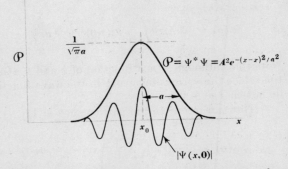

Fig. 12.5 The function $\mathcal{P}(x)$ for a gaussian wave packet where $\mathcal{P}(x)$ represents the probability of finding the particle between x and $x + dx$ at time $t = 0$. The finer curve shows the real part of $\Psi(x,0)$ for the wave function of Eq. (12.26).

For the particular Ψ of Eq. (12.26), at $t = 0$,

$$\langle x \rangle = \int_{-\infty}^{\infty} \frac{1}{a\pi^{\frac{1}{2}}} \, x e^{-(x-x_0)^2/a^2} \, dx = x_0 \tag{12.27a}$$

Any function, such as (12.26), appropriate as a wave function for a single noninteracting (i.e., potential energy $P = 0$) particle can be represented as an infinite sum of sinusoidal wave functions through a Fourier integral. Clearly for any value of p the simple wave

$$\Psi_s(x,t) = (2\pi\hbar)^{-\frac{1}{2}}\phi(p)e^{i(px-Et)/\hbar} \tag{12.28}$$

satisfies the Schrödinger equation (11.20) in any region where $P = 0$. Here $\phi(p)$ is any function of p not containing x or t, and E is a function of p, since $E = p^2/2m$. Since the sum of any number of such solutions is also a solution, we can multiply (12.28) by dp and integrate to obtain a general solution

$$\Psi_g(x,t) = \int_{-\infty}^{\infty} (2\pi\hbar)^{-\frac{1}{2}}\phi(p)e^{i(px-Et)/\hbar} \, dp \tag{12.29}$$

where $(2\pi\hbar)^{-\frac{1}{2}}$ is a normalizing constant. Then

$$\Psi_g(x,0) = \int_{-\infty}^{\infty} (2\pi\hbar)^{-\frac{1}{2}}\phi(p)e^{ipx/\hbar} \, dp \tag{12.29a}$$

and by the Fourier integral theorem

$$\phi(p) = (2\pi\hbar)^{-\frac{1}{2}} \int_{-\infty}^{\infty} \Psi_g(x,0)e^{-ipx/\hbar} \, dx \tag{12.30}$$

Note that although Ψ is a function of x and t, $\phi(p)$ is not a function of time since the various momenta in the wave function for a free particle do not change in time.

Now we return to the particular Ψ of Eq. (12.26) and write

$$\phi(p) = (2\pi\hbar)^{-\frac{1}{2}} \int_{-\infty}^{\infty} \frac{1}{a^{\frac{1}{2}}\pi^{\frac{1}{4}}} \, e^{-(x-x_0)^2/2a^2}e^{ip_0x/\hbar}e^{-ipx/\hbar} \, dx$$

$$= \frac{1}{\pi^{\frac{1}{4}}}\sqrt{\frac{a}{\hbar}} \, e^{-(p-p_0)^2a^2/2\hbar^2}e^{-i(p-p_0)x_0/\hbar} \tag{12.30a}$$

A knowledge of either $\Psi(x,0)$ or of $\phi(p)$ permits us to determine the other, so that either Ψ or $\phi(p)$ specifies the state of the particle. Since $\Psi(x,0)$ can be any normalized function, every possible wave function for a noninteracting particle can be written in the form of Eq. (12.29) by application of Eq. (12.30). The factor $(2\pi\hbar)^{-\frac{1}{2}}$ in Eq. (12.30) was introduced so that the momentum wave function $\phi(p)$ is normalized:

$$\int_{-\infty}^{\infty} \phi^*(p)\phi(p)\,dp = 1 \tag{12.31}$$

The interpretation of $\phi^*(p)\phi(p)\,dp$ is analogous to our interpretation of $\psi^*\psi\,dx$; viz., $\phi^*(p)\phi(p)\,dp$ is *the probability that the momentum of the particle lies between p and $p + dp$.*

From Eqs. (12.30a) and (12.31) we see that the probability distribution in momentum is of the same form as that in space for $t = 0$, except that the half-width at $1/e$ of the maximum is $\hbar/a$. The expectation value for p can be found in either of two ways: (1) in the momentum representation

$$\langle p \rangle = \int_{-\infty}^{\infty} \phi^*(p)p\phi(p)\,dp = \frac{1}{\sqrt{\pi}}\frac{a}{\hbar}\int_{-\infty}^{\infty} pe^{-(p-p_0)^2a^2/\hbar^2}\,dp = p_0 \tag{12.32}$$

and (2) in the coordinate representation

$$\langle p \rangle = \int_{-\infty}^{\infty} \Psi^*(x,t)\hat{p}\Psi(x,t)\,dx$$

To express p as a function of x, use is made of Eq. (11.24a), from which

$$\hat{p}\Psi = \frac{\hbar}{i}\frac{d\Psi}{dx}$$

This yields

$$\langle p \rangle = \int_{-\infty}^{\infty} \Psi^*(x,t)\frac{\hbar}{i}\frac{\partial\Psi(x,t)}{\partial x}\,dx = p_0 \tag{12.32a}$$

We now turn to the question of how the motion of the electron, described at $t = 0$ by Eq. (12.26), proceeds in time. By substituting Eq. (12.30a) in (12.29) and making use of the fact that $p^2/2m = E$, we have

$$\Psi(x,t) = \sqrt{\frac{a}{2\pi}} \frac{1}{\hbar\pi^{\frac{1}{4}}} \int_{-\infty}^{\infty} e^{-(p-p_0)^2 a^2/2\hbar^2} e^{-i(p-p_0)x_0/\hbar} e^{i(px-p^2t/2m)/\hbar} \, dp$$

$$= \sqrt{\frac{a}{2\pi}} \frac{1}{\hbar\pi^{\frac{1}{4}}} e^{ip_0 x_0/\hbar} \int_{-\infty}^{\infty} e^{-(p-p_0)^2 a^2/2\hbar^2} e^{ip(x-x_0)/\hbar} e^{-ip^2 t/2m\hbar} \, dp$$

$$= \frac{1}{\pi^{\frac{1}{4}}} \left(a + \frac{i\hbar t}{ma} \right)^{-\frac{1}{2}} e^{i(p_0 x - p_0^2 t/2m)/\hbar}$$

$$\times \exp\left[-\frac{(x - x_0 - p_0 t/m)^2 (1 - i\hbar t/ma^2)}{2(a^2 + \hbar^2 t^2/m^2 a^2)} \right] \quad (12.33)$$

from which the probability distribution is

$$\mathcal{P}(x,t) = \Psi^* \Psi$$

$$= \frac{1}{\sqrt{\pi}} \left(a^2 + \frac{\hbar^2 t^2}{m^2 a^2} \right)^{-\frac{1}{2}} \exp\left[-\frac{(x - x_0 - p_0 t/m)^2}{a^2 + \hbar^2 t^2/m^2 a^2} \right] \quad (12.34)$$

The probability distribution remains gaussian as it moves with a velocity p_0/m, but it broadens as time goes on, corresponding to an increasing uncertainty of position.

In 1927 Ehrenfest proved that in quantum mechanics *a particle moving in a potential field P(x) obeys the classical laws of motion*

$$\frac{d\langle p_x \rangle}{dt} = -\frac{\partial P(x)}{\partial x} \quad \text{and} \quad \langle p_x \rangle = m \frac{d\langle x \rangle}{dt}$$

in the limit that the uncertainties in momentum and position may be ignored. Thus, in the macroscopic limit wave mechanics yields the same predictions as newtonian physics. Proof of *Ehrenfest's theorem* is left to the problems.

A wave function such as that of Eq. (12.33), used to describe the motion of a particle for which the position and momentum cannot be known exactly, is called a *wave packet*. It is formed by superimposing an infinite number of sinusoidal waves to yield a resultant which is zero everywhere except for a restricted region in which the particle is to be found.

12.6   The Heisenberg Uncertainty Principle

When the Heisenberg principle of indeterminacy was introduced in Sec. 11.8, no clear definition was given of Δx and Δp_x, the uncertainties in coordinate and conjugate momentum. Heisenberg originally discovered the principle from a mathematical formulation, and we now treat it more quantitatively. It is possible to define Δx and Δp_x in various ways; perhaps the most useful definition arises from the following considerations. Imagine a series of measurements of x to be made on many identical systems [such as one described by the wave packet of Eq. (12.26)] with the result that a number of values of x grouped around x_0 are obtained. We define Δx by the relation $\Delta x = [\Sigma(x - x_0)^2]^{\frac{1}{2}}$, which is known as the *standard deviation*. More generally, in terms of expectation values,

$$(\Delta x)^2 = \langle (x - \langle x \rangle)^2 \rangle = \langle x^2 \rangle - \langle 2x\langle x \rangle \rangle + \langle x \rangle^2$$
$$= \langle x^2 \rangle - 2\langle x \rangle^2 + \langle x \rangle^2 = \langle x^2 \rangle - \langle x \rangle^2 \qquad (12.35a)$$
$$(\Delta p)^2 = \langle (p - \langle p \rangle)^2 \rangle = \langle p^2 \rangle - \langle p \rangle^2 \qquad (12.35b)$$

Thus Δx and Δp are the root-mean-square deviations from the mean. For $\Psi(x,0)$ of Eq. (12.26), we have after using Eq. (12.27a),

$$(\Delta x)^2 = \langle (x - x_0)^2 \rangle$$
$$= \frac{1}{a\sqrt{\pi}} \int_{-\infty}^{\infty} (x - x_0)^2 e^{-(x-x_0)^2/a^2}\, dx = \frac{a^2}{2} \quad (12.36a)$$

where we have made use of integral I_2 of Appendix 4A. Similarly

$$(\Delta p)^2 = \langle (p - p_0)^2 \rangle$$
$$= \frac{1}{\sqrt{\pi}} \frac{a}{\hbar} \int_{-\infty}^{\infty} (p - p_0)^2 e^{-(p-p_0)^2 a^2/\hbar^2}\, dp = \frac{1}{2}\left(\frac{\hbar}{a}\right)^2 \quad (12.36b)$$

Therefore $\Delta p\, \Delta x = \hbar/2$.

It can be shown that no wave packet can be constructed which has a smaller value of $\Delta p\, \Delta x$ [as defined in Eqs. (12.35)] than the wave packet of Eq. (12.26). If we put $x - x_0 = \Delta x = a/\sqrt{2}$ into Eq. (12.26), we find that $\mathcal{P}(x_0 \pm \Delta x, 0) = 0.6065\mathcal{P}(x_0,0)$ and that the probability of the particle's being found between $x_0 - \Delta x$ and $x_0 + \Delta x$ at $t = 0$ is 0.68.

12.7 Probability Stream Density

A stream of protons or electrons constitutes a flow of charged particles, and hence an electric current. How can such a flow be quantitatively described in terms of wave functions?

   To find the quantum-mechanical formula for expressing **particle flow** we begin with classical relations. Let $\rho(x,y,z,t)$ be the charge density and consider the charge in a very small volume element $\Delta x\, \Delta y\, \Delta z$ (Fig. 12.6), first for the case that all velocities are along the x axis. The net charge per unit time entering $\Delta x\, \Delta y\, \Delta z$ at the left is ρv_x evaluated at $x = a$ times $\Delta y\, \Delta z$, while that leaving at the right is

$$(\rho v_x)_{a+\Delta x}\, \Delta y\, \Delta z = [\rho v_x + \partial(\rho v_x)/\partial x\, \Delta x]\, \Delta y\, \Delta z$$

The charge in $\Delta x\, \Delta y\, \Delta z$ is, of course, $\rho\, \Delta x\, \Delta y\, \Delta z$. If more charge leaves per second than arrives, ρ decreases in time, and conservation of charge requires that

$$(\rho v_x)_{a+\Delta x}\, \Delta y\, \Delta z - (\rho v_x)_a\, \Delta y\, \Delta z = \frac{\partial(\rho v_x)}{\partial x}\, \Delta x\, \Delta y\, \Delta z$$

$$= -\frac{\partial \rho}{\partial t}\, \Delta x\, \Delta y\, \Delta z \quad (12.37)$$

from which

$$\frac{\partial(\rho v_x)}{\partial x} = -\frac{\partial \rho}{\partial t} \tag{12.38a}$$

in one dimension. If we extend the reasoning to three dimensions, we have

$$\nabla \cdot \rho \mathbf{v} = -\frac{\partial \rho}{\partial t} \tag{12.38b}$$

which is the *equation of continuity*, applicable to any flow process in which ρ represents the density of something that is conserved.

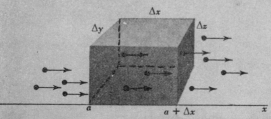

Fig. 12.6 Charge entering and leaving a small cubical volume element $\Delta x\, \Delta y\, \Delta z$ for a one-dimensional flow.

To find the quantum-mechanical analogs of Eqs. (12.37) and (12.38) we identify ρ with $q\Psi^*\Psi$ where q is the charge borne by a particle and $\Psi^*\Psi\, dx\, dy\, dz$ is the probability at time t of finding a particle in the volume element $dx\, dy\, dz$. (Note that $\int\int\int\Psi^*\Psi\, dx\, dy\, dz$ over all space is normalized to give the total number of particles enclosed.) We assume that there is a quantity S which represents the probability stream density of the particles in such a way that qS plays the role of $\rho\mathbf{v}$ in the equation of continuity. We seek S in terms of Ψ by writing the equation of continuity as

$$\nabla \cdot q\mathbf{S} = -\partial(q\Psi^*\Psi)/\partial t$$

or

$$\nabla \cdot \mathbf{S} = -\frac{\partial}{\partial t}(\Psi^*\Psi) \tag{12.39}$$

With the aid of Eq. (11.23a), we obtain

$$\nabla \cdot \mathbf{S} = -\frac{\partial \Psi^*}{\partial t}\Psi - \Psi^*\frac{\partial \Psi}{\partial t}$$

$$= -\frac{\hbar}{2mi}(\Psi \nabla^2\Psi^* - \Psi^* \nabla^2\Psi) \tag{12.40}$$

or

$$\mathbf{S} = \frac{\hbar}{2mi}(\Psi^* \nabla\Psi - \Psi \nabla\Psi^*) \tag{12.41}$$

The quantity S is a vector *probability current density* for the flow of particles; if each particle bears charge q, the corresponding probability current density is qS. If A denotes any surface area enclosing a volume $\mathcal{U}$, the integral of the normal component of S over the surface is equal to the time rate of decrease of probability of finding a particle within the volume; thus the left-hand integral represents the rate at which probability is streaming outward across the closed surface A, while the right-hand integral represents the rate at which the probability is decreasing within the volume.

When Ψ has the form of Eq. (11.13), then $\Psi^* = A^*e^{i(Et - p_x x)/\hbar}$ and

$$S = \frac{p}{m}|A|^2$$

where S is the magnitude of **S**.   Thus the significance of the beam represented by Eq. (11.13) can be based upon the vector **S** by assuming the number of particles crossing unit area per second is $n = p|A|^2/m$, where p/m is, of course, the speed of the particles.

_____***Problems***

1.   Show that the sum of the reflection and transmission coefficients is equal to unity for the step barrier of Sec. 12.2.

2.   Show that the reflection coefficient R for the step barrier of Sec. 12.2 is given by

$$R = \left(\frac{\sqrt{1 - P_0/E} - 1}{\sqrt{1 - P_0/E} + 1}\right)^2$$

and thus that $\sqrt{1 - P_0/E}$ plays the role of relative index of refraction in optical reflection at normal incidence.   Prove that so long as E is a constant greater than P_0, the greater $|P_0|$ is for the step barrier, the greater the reflection coefficient, whether P_0 is positive or negative.

3.   Solve Eqs. (12.21) for A in terms of G and then derive Eq. (12.23).

4.   Find the reflection coefficient $|B/A|^2$ for the barrier-penetration situation of Sec. 12.3 (Fig. 12.3) and show that the sum of the reflection and transmission coefficients is equal to 1.

5.   Find the reflection and transmission coefficients for the barrier of Fig. 12.3 for the case where $E > P_0$.   Show that the transmission coefficient goes to unity for certain values of the energy.   Explain how this can happen when there is reflection from both steps of the barrier.

6.   Find the reflection coefficient R for the square well of Fig. 12.4. Show that $R + T = 1$.   Find the values of E for which reflection is maximum if P_0 and w are fixed and calculate the largest value of R.

7.   A stream of electrons with energy $E > P_L$ (Fig. 12.7) is incident from $x = -\infty$ on the potential barrier shown.

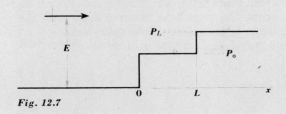

Fig. 12.7

(a) Write the general solutions of the Schrödinger wave equation in regions 1, 2, and 3.

(b) If the amplitude of the incident wave in region 1 is A, find sufficient equations to evaluate all the other constants introduced in part (a) in terms of A and other given quantities. (It is not necessary to solve these equations.)

8. Repeat Prob. 7 for the case where $P_0 < E < P_L$.

9. Show that for any dynamical function $G(x, p_x)$,

$$\frac{d\langle G \rangle}{dt} = \int \Psi^* \frac{i}{\hbar} (\hat{H}\hat{G} - \hat{G}\hat{H}) \Psi \, dx$$

Hint: Write $\langle G \rangle$, differentiate, and use

$$\hat{H}\Psi = \hat{E}\Psi = -\frac{i}{\hbar} \frac{\partial \Psi}{\partial t}$$

10. Ehrenfest's theorem states that for a particle for which the uncertainty in position and momentum can be neglected, the quantum-mechanical prediction of the motion agrees with the classical prediction. Show that this theorem is valid for a one-dimensional motion for which $P = P(x)$ by showing (a) that

$$\frac{d\langle x \rangle}{dt} = \int \Psi^* \frac{i}{\hbar} (\hat{H}x - x\hat{H}) \Psi \, dx = \frac{\langle p_x \rangle}{m}$$

and (b) that

$$\frac{d\langle p_x \rangle}{dt} = \left\langle -\frac{dP}{dx} \right\rangle$$

chapter thirteen

Wave Mechanics II
Bound States

When a particle is restricted to moving in a limited region, quantum mechanics yields the prediction that only certain discrete values of the energy are possible. In this chapter the allowed quantum states for the harmonic oscillator and for a particle in a box are studied, while in Chap. 14 the problem of an electron bound to a nucleus is treated.

13.1 Stationary or Quantum States

Consider an electron or other particle bound to some region by a force field such that the particle cannot escape classically; i.e., its total energy E is too small for the particle to reach infinity. In the classically allowed region E is greater than the potential energy P, assumed to be a function of position only; where $P > E$, the particle has negative kinetic energy K.   To treat such cases, Schrödinger suggested seeking solutions of the wave equation which represent standing waves.   In such waves the phase of the vibration is everywhere the same, whereas in running waves there exists, at any moment, a progression of phase along the wave train.   In a mathematical expression for representing standing waves, therefore, the time must occur in a separate factor.

As we saw in Sec. 11.9, when P is not a function of the time, we may write

$$\Psi(x,t) = \psi(x)f(t) = \psi(x)e^{-iEt/\hbar} \tag{13.1}$$

where ψ satisfies the Schrödinger wave equation

$$-\frac{\hbar^2}{2m}\frac{d^2\psi}{dx^2} + P(x)\psi(x) = E\psi(x) \tag{11.22}$$

Not every mathematically correct solution of the Schrödinger wave equation is physically acceptable.   Intimately tied to various physical conditions required to give the solutions reasonable meaning are certain mathematical conditions which we shall introduce as the need arises.

13.2   General Solutions for the Square Well

Let a single particle of mass m be bound in the square potential well of Fig. 13.1 for which $P = 0$ for $0 \leq x \leq L$ and $P = P_0$ for $x < 0$ and $x > L$.   The energy E of the particle is less than P_0 if classically the

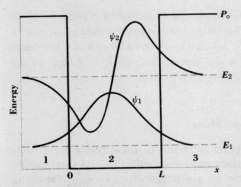

Fig. 13.1   A square well of depth P_0.   Also shown are the two lowest energy levels and the associated wave functions ψ_1 and ψ_2.

particle is confined to the well. This is mathematically a particularly simple case of a bound particle for which a complete solution can readily be found. The solution proves useful for treating such topics as gases and electrons in metals, as we shall see. Qualitative features of bound-particle problems in general are well illustrated by this example.

For regions 1 and 3 the Schrödinger wave equation is

$$-\frac{\hbar^2}{2m}\frac{d^2\psi}{dx^2} + P_0\psi = E\psi \qquad \text{or} \qquad \frac{d^2\psi}{dx^2} = \frac{2m}{\hbar^2}(P_0 - E)\psi \qquad (13.2)$$

Since $P_0 > E$, the general solution is

Region 1

$$\psi_1 = A\exp\frac{\sqrt{2m(P_0 - E)}\,x}{\hbar}$$
$$+ B\exp\left[-\frac{\sqrt{2m(P_0 - E)}\,x}{\hbar}\right] = Ae^{\alpha x} + Be^{-\alpha x} \qquad (13.3)$$

Region 3

$$\psi_3 = G\exp\frac{\sqrt{2m(P_0 - E)}\,x}{\hbar}$$
$$+ H\exp\left[-\frac{\sqrt{2m(P_0 - E)}\,x}{\hbar}\right] = Ge^{\alpha x} + He^{-\alpha x} \qquad (13.4)$$

where A, B, G, and H are constants to be evaluated and we have put $\sqrt{2m(P_0 - E)}/\hbar = \alpha$. Note that in regions where P is a constant greater than the total energy, the solutions of the Schrödinger wave equation are exponential functions.

For region 2 the wave equation is

$$-\frac{\hbar^2}{2m}\frac{d^2\psi}{dx^2} + 0 = E\psi \qquad \text{or} \qquad \frac{d^2\psi}{dx^2} = -\frac{2mE}{\hbar^2}\psi \qquad (13.5)$$

This is the simple-harmonic-motion differential equation, for which the general solution is

$$\psi_2 = C'\exp\frac{i\sqrt{2mE}\,x}{\hbar} + D'\exp\left(-\frac{i\sqrt{2mE}\,x}{\hbar}\right)$$
$$= C\sin\frac{\sqrt{2mE}\,x}{\hbar} + D\cos\frac{\sqrt{2mE}\,x}{\hbar}$$
$$= C'e^{ikx} + D'e^{-ikx} = C\sin kx + D\cos kx \qquad (13.6)$$

where C', D', C and D are constants and $k = \sqrt{2mE}/\hbar$. Observe that for regions of constant P less than E, the general solutions of the wave equation are oscillating functions which may be expressed equally well in terms of imaginary exponentials or by trigonometric functions.

We have postulated (Sec. 11.7) that the probability of finding our particle between x and $x + dx$ at any time t is given by $\Psi^*(x,t)\Psi(x,t)\ dx$. If we sum this probability over all possible values of x, we have for this problem the normalization condition $\int_{-\infty}^{\infty} \Psi^*\Psi\ dx = 1$, since there is only one particle and it is somewhere. This fact enables us to show that B and G of Eqs. (13.3) and (13.4) must be zero. Since $Be^{-\alpha x}$ in region 1 is an exponentially increasing function as x goes from 0 to $-\infty$, the normalization condition can hold only if $B = 0$; a similar argument shows that $G = 0$. Thus in both regions 1 and 3 the only allowed wave functions are exponentially decreasing as we leave the potential well; $\psi_1 = Ae^{\alpha x}$ and $\psi_3 = He^{-\alpha x}$.

If $\Psi^*\Psi\ dx = \psi^*\psi\ dx$ is to represent the probability of finding the particle between x and $x + dx$, the reasonable expectation that this probability be finite, single-valued, and continuous can be met by imposing on ψ the following condition:

1. *ψ is finite, single-valued, and continuous.*

Further, we shall find that if the momentum of the particle is to be finite and continuous, the first derivatives of ψ must also be finite and continuous. It turns out that the momentum does change discontinuously at an infinite potential barrier—a not uncommon situation in which an unrealizable physical condition forces an unnatural mathematical requirement. We therefore add the following condition:

2. *$d\psi/dx$ is continuous* (except when P goes to infinity).

When we apply conditions 1 and 2 at the boundaries $x = 0$ and $x = L$ for the particle in the square well, we obtain the equations

$$A = D \tag{13.7a}$$

$$\alpha A = kC \tag{13.7b}$$

$$C \sin kL + D \cos kL = He^{-\alpha L} \tag{13.7c}$$

$$kC \cos kL - kD \sin kL = -\alpha He^{-\alpha L} \tag{13.7d}$$

These equations, together with the normalization condition, permit the determination of all constants. It is simplest to consider first the artificial but highly important case of an infinite well, for which P_0 is equal to infinity.

13.3   The Infinite Square Well

When P_0 goes to infinity, α becomes infinite. If Eqs. (13.7) are to be satisfied, kC must be finite, and so A must be zero. Since $A = D$, $D = 0$. Since $He^{-\alpha L}$ is zero, $C \sin kL = 0$. [Equation (13.7d) is not applicable.]

If C were zero, ψ would be zero everywhere, which is incompatible with the normalization condition, Eq. (11.15a). Therefore, we conclude that $\sin kL = 0$, or $kL = n\pi$, where n is an integer. Since $k = \sqrt{2mE}/\hbar$, *acceptable solutions of the Schrödinger wave equation exist only for certain discrete values of the energy* satisfying the relation

$$E_n = \frac{n^2 h^2}{8mL^2} \tag{13.8}$$

An important prediction of quantum mechanics, quite at variance with classical physics, is that the particle may have for its energy only certain specific values. These are known as *eigenvalues* (*eigen* means *proper* or *characteristic*); only the eigenvalues permit acceptable solutions of the Schrödinger equation. For the infinite square well there is an eigenvalue given by Eq. (13.8) for every integer n, known as the *quantum number*.

Of special interest is the fact that the energy of the particle cannot be zero, since $n = 0$ gives $\psi = 0$ everywhere—hence no particle. The lowest possible energy, or *zero-point energy*, in this case corresponds to $n = 1$. This is directly connected to the Heisenberg uncertainty principle, since if the position of the particle is certain within $\Delta x = L/2$, there must be an uncertainty in the momentum not less than that given by Eq. (11.17). In the first eigenstate, $E_1 = h^2/8mL^2 = p_1{}^2/2m$, from which p_1, the momentum in the first eigenstate, is $\pm h/2L$. Since the momentum can be in either direction in the well, the uncertainty Δp_x is essentially $h/2L$. Therefore, $\Delta x\, \Delta p_x \approx h/4$, a result entirely compatible with the uncertainty principle.

Corresponding to each eigenvalue (and quantum number n) there is an eigenfunction given for $0 \leq x \leq L$ by

$$\psi_n = C_n \sin k_n x = C_n \sin \frac{n\pi}{L} x \tag{13.9}$$

and for $-\infty < x \leq 0$ and $x \geq L$ by $\psi_n = 0$.

From the normalization condition (11.15a) we can obtain C, since $\int_{-\infty}^{\infty} \psi^* \psi\, dx = \int_0^L C^2 \sin^2 (n\pi x/L)\, dx = C^2 L/2 = 1$, whence $C = \sqrt{2/L}$.

Therefore the normalized eigenfunctions are

$$\psi_n = \sqrt{\frac{2}{L}} \sin\left(\frac{n\pi}{L}x\right) \qquad \text{for } 0 \leq x \leq L \qquad (13.9a)$$

The eigenfunctions correspond to those of a vibrating string of length L and fixed ends. A few of them are shown in Fig. 13.2, along with the corresponding eigenvalues. A glance at the first eigenfunction shows that the probability $|\psi|^2\, dx$ of finding the particle in a small element of length dx is much greater at the center of the well than at the edges. This again is in sharp contrast to the classical case, for which the particle would move between barriers with constant speed and would be equally likely to be found near the walls or near the center. It is a surprising prediction of quantum mechanics that the particle should "avoid" the walls.

As one goes to higher quantum number n, ψ oscillates more rapidly, and thus the probability of the particle's being found in any length dx becomes more uniform across the well in accord with the correspondence

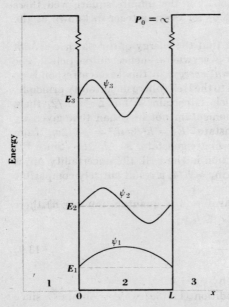

Fig. 13.2 The first three energy levels and the corresponding wave functions for an infinite square well.

principle. Even so, ψ remains zero at the walls. It should be noted that the number of zeros of ψ other than those at the boundary is given by $n - 1$.

13.4   The Finite Square Well

We have argued in Sec. 13.2 that in regions 1 and 3 the wave function is given by $\psi_1 = Ae^{\alpha x}$ and $\psi_3 = He^{-\alpha x}$, declining exponentials as we go away from the well. Inside the well ψ_2 is represented by oscillating trigonometric functions. If ψ and its derivative are continuous at the boundaries, the trigonometric function must be decreasing in magnitude as the boundary is approached (Fig. 13.1). When P_0 is large compared to E, Eqs. (13.7) require that A, D, and H be small but not zero. The resulting eigenfunctions bear close resemblance to those for an infinite well, except that there is a small exponential tail outside the well, showing some penetration of the particle into the classically forbidden regions of negative kinetic energy.

The wavelength of the eigenfunctions for the finite well are somewhat greater than those for the infinite well. Longer wavelength corresponds to lower momentum and to lower energy. Thus Eq. (13.8) predicts somewhat too great an energy for each eigenvalue of the finite well.

As the difference $P_0 - E$ decreases, penetration into the classically forbidden regions increases; that is, A and H of Eqs. (13.7) increase, and α decreases. Further, the number of bound states decreases, since as n increases, the energy of the bound state rises, and for a square well with $P_0 - E$ finite, the number of bound states is finite. Once E exceeds P_0, the particle is free, and any energy is an eigenvalue.

13.5   Expectation Values

In problems such as ones involving an electron in a square well (or bound to a proton to form a hydrogen atom), quantum mechanics yields definite values for some quantities, e.g., the energy of a given state, but not for others, e.g., the position of the particle or its momentum. As we have seen in Sec. 12.5, the expectation value for the x coordinate of a particle, which is the average value of x obtained by a large number of independent measurements on identical systems, is

$$\langle x \rangle = \int_{-\infty}^{\infty} \Psi^*(x,t)\, x \Psi(x,t)\, dx \qquad (12.27)$$

where Ψ is a normalized wave function. In a similar way the expectation value of any function $f(x)$ is given by

$$\langle f(x) \rangle = \int_{-\infty}^{\infty} \Psi^*(x,l) f(x) \Psi(x,l) \, dx \tag{13.10}$$

Thus, if the potential energy is a function $P(x)$, the expectation value for P is

$$\langle P \rangle = \int_{-\infty}^{\infty} \Psi^*(x,l) P(x) \Psi(x,l) \, dx \tag{13.10a}$$

As we saw in Sec. 12.5, to obtain the expectation value of the momentum we must express p_x as a function of x, but the uncertainty principle specifically denies that we can know p_x and x exactly and simultaneously. To determine $\langle p_x \rangle$, we replace $f(x)$ by the appropriate operator $\hat{p}_x$ [Eq. (11.24a)] to obtain

$$\langle p_x \rangle = \int_{-\infty}^{\infty} \Psi^*(x,l) \hat{p}_x(x) \Psi(x,l) \, dx$$
$$= \int_{-\infty}^{\infty} \Psi^*(x,l) \frac{\hbar}{i} \frac{\partial \Psi(x,l)}{\partial x} \, dx \tag{13.11}$$

13.6 Differential Operators

In three dimensions the operator for the momentum $\mathbf{p}$ is the vector differential operator

$$\hat{\mathbf{p}} = \frac{\hbar}{i} \nabla \tag{13.12a}$$

and

$$\hat{p}^2 = -\hbar^2 \nabla^2 \tag{13.12b}$$

In the nonrelativistic case the kinetic energy K of a particle is $p^2/2m$, and the operator for K is

$$\hat{K} = -\frac{\hbar}{2m} \nabla^2 \tag{13.13}$$

For a one-dimensional problem

$$\langle K \rangle = \int_{-\infty}^{\infty} \Psi^*(x,l) \left(-\frac{\hbar^2}{2m} \frac{\partial^2}{\partial x} \right) \Psi(x,l) \, dx \tag{13.14}$$

which can be readily integrated by parts to obtain

$$\langle K \rangle = -\frac{\hbar^2}{2m} \left(\left[\Psi^* \frac{\partial \Psi}{\partial x} \right]_{-\infty}^{\infty} - \int_{-\infty}^{\infty} \frac{\partial \Psi}{\partial x} \frac{\partial \Psi^*}{\partial x} \, dx \right)$$

The normalization condition $\int_{-\infty}^{\infty} \Psi^* \Psi \, dx = 1$ requires that Ψ^* be zero at $x = \pm \infty$, and therefore

$$\langle K \rangle = \frac{\hbar^2}{2m} \int_{-\infty}^{\infty} \frac{\partial \Psi}{\partial x} \frac{\partial \Psi^*}{\partial x} \, dx \tag{13.15}$$

Thus the expectation value of the kinetic energy is proportional to the integral of $|\partial \Psi / \partial x|^2$.

The operator for *angular momentum* is of great interest in atomic theory. The classical expression for the angular momentum A_z of a particle about the z axis is

$$A_z = (\mathbf{r} \times \mathbf{p})_z = xp_y - yp_x$$

Replacement of p_y and p_x gives as the corresponding operator in wave mechanics

$$\hat{A}_z = \frac{\hbar}{i} \left(x \frac{\partial}{\partial y} - y \frac{\partial}{\partial x} \right) \tag{13.16}$$

If polar coordinates are introduced with the z axis as axis, then

$$x = r \sin \theta \cos \phi \qquad y = r \sin \theta \sin \phi \qquad z = r \cos \theta$$

$$\frac{\partial}{\partial \phi} = \frac{\partial x}{\partial \phi} \frac{\partial}{\partial x} + \frac{\partial y}{\partial \phi} \frac{\partial}{\partial y} = x \frac{\partial}{\partial y} - y \frac{\partial}{\partial x}$$

Thus the operator for the z component of angular momentum can also be written

$$\hat{A}_z = \frac{\hbar}{i} \frac{\partial}{\partial \phi} \tag{13.16a}$$

Analogous expressions are found for the x and y components. The operator for the square of the resultant angular momentum is, then,

$$-\hbar^2 \left[\left(y \frac{\partial}{\partial z} - z \frac{\partial}{\partial y} \right)^2 + \left(z \frac{\partial}{\partial x} - x \frac{\partial}{\partial z} \right)^2 + \left(x \frac{\partial}{\partial y} - y \frac{\partial}{\partial x} \right)^2 \right] \tag{13.17}$$

In wave mechanics physical magnitudes in general are represented by appropriate operators. The ultimate test of the validity of the operator assumed to represent a given magnitude will lie in a demonstration that the values of this magnitude as predicted by wave mechanics are in agreement with observation. The accepted condition that any physical magnitude Q has a definite value or is quantized is that, if $\hat{Q}$ denotes the corresponding wave-mechanical operator and Ψ the wave function for the system to which Q belongs, then

$$\hat{Q}\Psi = Q_n\Psi \tag{13.18}$$

where Q_n is a real number, the value of Q. Solutions of this equation satisfying appropriate boundary conditions represent stationary or quantum states for the magnitude Q; and the set of allowed values of Q_n includes the only correct values that can be obtained in an experimental measurement of Q.

The requirement that the result of a measurement be a real number admits only real solutions for expectation values and eigenvalues. This limits the type of function which can serve as operators for physical quantities to those which satisfy the equation $\langle Q \rangle = \langle Q \rangle^*$; that in turn imposes the condition that $\int \Psi^* \hat{Q}\Psi \, dx \, dy \, dz = \int \Psi(\hat{Q}\Psi)^* \, dx \, dy \, dz$. An operator which satisfies the latter equation is said to be *hermitian;* operators for physical observables in quantum mechanics must be hermitian to guarantee that the expectation value of the observable will be real.

Two operators $\hat{Q}$ and $\hat{F}$ are said to commute when $\hat{Q}\hat{F} - \hat{F}\hat{Q} = 0$. Hermitian operators do not in general commute; for example,

$$x \frac{\hbar}{i} \frac{\partial \psi}{\partial x} - \frac{\hbar}{i} \frac{\partial}{\partial x} x\psi = -\frac{\hbar}{i} \psi$$

The operator $\hat{C} = \hat{Q}\hat{F} - \hat{F}\hat{Q}$ is the *commutator operator* of $\hat{Q}$ and $\hat{F}$. The equation above tells us that $\hat{C}(x,p_x) = -\hbar/i$. On the other hand $\hat{C}(y,p_x) = 0$, so that y and p_x do indeed commute. The commutator operator $\hat{C}(Q,F)$ is often written $[Q,F]$.

13.7   The Rectangular Box

As a first example of a three-dimensional problem in wave mechanics, consider a rectangular box (Fig. 13.3) with sides of lengths L_x, L_y, L_z. Let the potential energy P inside the box be zero and that outside be

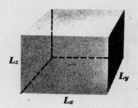

Fig. 13.3 A rectangular box with sides L_x, L_y, L_z.

infinite. Then outside the box $\psi = 0$, while inside

$$-\frac{\hbar^2}{2m} \nabla^2 \psi + 0 = E\psi \tag{13.19a}$$

or

$$\frac{\partial^2 \psi}{\partial x^2} + \frac{\partial^2 \psi}{\partial y^2} + \frac{\partial^2 \psi}{\partial z^2} = -\frac{2mE}{\hbar^2} \psi \tag{13.19b}$$

If we let $\sqrt{2mE}/\hbar = k$, assume that $\psi(x,y,z)$ is separable so that $\psi(x,y,z) = X(x)Y(y)Z(z)$, and divide Eq. (13.19b) by ψ, we obtain

$$\frac{1}{X}\frac{d^2X}{dx^2} + \frac{1}{Y}\frac{d^2Y}{dy^2} + \frac{1}{Z}\frac{d^2Z}{dz^2} = -k^2 \tag{13.20}$$

Here we have the sum of a function of x alone, a function of y alone, and a function of z alone equal to a constant. This can be true only if each term on the left is separately constant, and so we write

$$\frac{1}{X}\frac{d^2X}{dx^2} = -k_x{}^2 \tag{13.21a}$$

$$\frac{1}{Y}\frac{d^2Y}{dy^2} = -k_y{}^2 \tag{13.21b}$$

$$\frac{1}{Z}\frac{d^2Z}{dz^2} = -k_z{}^2 \tag{13.21c}$$

where the negative sign is chosen for the constants for a reason explained below. Of course, $k_x{}^2 + k_y{}^2 + k_z{}^2 = k^2$ to satisfy Eq. (13.20).

A general solution for (13.21a) is

$$X = A_x \sin k_x x + B_x \cos k_x x \tag{13.22}$$

where A_x and B_x are constants. At $x = 0$ continuity of ψ requires that $X(0)$ be 0, and so B_x must be zero. Also $X(L_x) = 0$, so that

$$A_x \sin k_x L_x = 0 \tag{13.23}$$

from which it follows that $k_x L_x = n_x \pi$, where n_x is an integer. Thus we have

$$X = A_x \sin \frac{n_x \pi x}{L_x} \tag{13.24a}$$

We can now see why the separation constant was written $-k_x{}^2$. If the constant had been chosen positive, the solution for (13.21a) would have been exponential functions, and it would have been impossible to satisfy the conditions that $X(0) = 0$ and $X(L_x) = 0$. Similar developments for Y and Z give

$$Y = A_y \sin \frac{n_y \pi y}{L_y} \tag{13.24b}$$

$$Z = A_z \sin \frac{n_z \pi z}{L_z} \tag{13.24c}$$

and

$$\psi = A \sin \frac{n_x \pi}{L_x} x \sin \frac{n_y \pi}{L_y} y \sin \frac{n_z \pi}{L_z} z \tag{13.25}$$

where $A = \sqrt{8/L_x L_y L_z}$ if the normalization is for a single particle.

The eigenvalues for the energy may be found from

$$k^2 = k_x{}^2 + k_y{}^2 + k_z{}^2 = \pi^2 \left(\frac{n_x{}^2}{L_x{}^2} + \frac{n_y{}^2}{L_y{}^2} + \frac{n_z{}^2}{L_z{}^2} \right) = \frac{2mE}{\hbar^2}$$

or

$$E = \frac{h^2}{8m} \left(\frac{n_x{}^2}{L_x{}^2} + \frac{n_y{}^2}{L_y{}^2} + \frac{n_z{}^2}{L_z{}^2} \right) \tag{13.26}$$

The lowest possible energy, or ground-state energy, corresponds to $n_x = 1$, $n_y = 1$, $n_z = 1$, and it depends on the dimensions of the box. For an electron in a box with sides of the order of 1 Å, the smallest energy is of the order of electron volts.

For a cubical box $L_x = L_y = L_z = L$, and the energy eigenvalues are

$$E = \frac{h^2}{8mL^2}(n_x{}^2 + n_y{}^2 + n_z{}^2) \tag{13.26a}$$

The ground-state energy E_{111} is $3h^2/8mL^2$. The next lowest energy corresponds to $n_x = 2$, $n_y = 1$, $n_z = 1$ or $n_x = 1$, $n_y = 2$, $n_z = 1$, or $n_x = 1, n_y = 1, n_z = 2$. Thus there are three eigenfunctions, Eq. (13.25), which give rise to the same energy, and we say the first excited state is *triply degenerate*. In general, two or more eigenfunctions are *degenerate* when they lead to the same eigenvalue for the energy. A ninefold degeneracy occurs for the energy $38h^2/8mL^2$, which can be associated with $n_x n_y n_z$ given by (611), (161), (116), (532), (523), (352), (325), (253), (235). The degeneracy can be removed for these low-lying states by destroying the cubical symmetry of the box; if L_x, L_y, and L_z are slightly different, the energies for these states no longer coincide.

13.8   The Harmonic Oscillator

A problem of great importance is that of the simple harmonic oscillator, not only because it is one of the relatively few problems which can be solved exactly, but also because the methods used have broad application. There are many oscillating systems in nature which can be treated (at least approximately) in terms of simple harmonic oscillations.

Consider a particle of mass m bound to the origin by a force $F = -\beta x$, where β is a constant. For the one-dimensional problem, we choose the potential energy $P = 0$ at the origin and then the potential-energy function $P(x) = \frac{1}{2}\beta x^2$ (Fig. 13.4). The Schrödinger wave equation $\hat{H}\psi = (p^2/2m + P)\psi = E\psi$ for this problem becomes

$$-\frac{\hbar^2}{2m}\frac{d^2\psi}{dx^2} + \frac{1}{2}\beta x^2\psi = E\psi \tag{13.27}$$

The solution of Eq. (13.27) is carried out in Appendix 13A, where it is shown that Eq. (13.27) can be satisfied only for discrete energies

$$E_n = (n + \tfrac{1}{2})h\nu \tag{13.28}$$

where $n = 0, 1, 2, \ldots$ and ν is the classical frequency of the oscillator given by $\sqrt{\beta/m}/2\pi$. As is usual in quantum mechanics, bound states for the particle exist only for discrete values of the energy. The ground (or

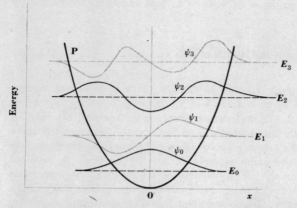

Fig. 13.4 The potential energy of a simple harmonic oscillator as a function of displacement (heavy curve) and the first four energy levels together with the associated wave functions.

lowest) energy state of the harmonic oscillator has $n = 0$ and $E_0 = \frac{1}{2}h\nu$. This energy, the zero-point energy, has no direct classical analog. The requirement of finite zero-point energy for a harmonic oscillator is of importance in thermodynamics and is an inevitable consequence of the Heisenberg uncertainty principle.

Corresponding to each eigenvalue E_n is the normalized eigenfunction

$$\psi_n(x) = \left(\frac{\alpha}{\sqrt{\pi}\, 2^n n!}\right)^{\frac{1}{2}} e^{-\alpha^2 x^2/2} H_n(\alpha x) \tag{13.29}$$

where $\alpha = \sqrt[4]{m\beta/\hbar^2}$ and $H_n(\alpha x)$ represents the nth Hermite polynomial (Appendix 13A). The first five Hermite polynomials are

$$H_0(\alpha x) = 1 \qquad H_1(\alpha x) = 2\alpha x \qquad H_2(\alpha x) = 4\alpha^2 x^2 - 2$$
$$H_3(\alpha x) = 8\alpha^3 x^3 - 12\alpha x \qquad H_4(\alpha x) = 16\alpha^4 x^4 - 48\alpha^2 x^2 \tag{13.30}$$
$$+ 12$$

and the first two eigenfunctions become

$$\psi_0(x) = \left(\frac{\alpha}{\sqrt{\pi}}\right)^{\frac{1}{2}} e^{-\alpha^2 x^2/2} \tag{13.29a}$$

$$\psi_1(x) = \left(\frac{\alpha}{2\sqrt{\pi}}\right)^{\frac{1}{2}} 2\alpha x e^{-\alpha^2 x^2/2} \tag{13.29b}$$

The first four eigenfunctions are sketched in Fig. 13.4. Two features deserve special mention. First is the penetration of the particle into the classically forbidden region. Second is the fact that, particularly for the lower states, the probability of finding the particle in any element dx of the allowed range is radically different from the classical predictions. Classically it would be most likely to find the particle near the end of its motion, where its speed is minimum, and least likely at the equilibrium position. Quantum mechanics however predicts that for the ground state the particle is most likely to be found near the equilibrium position. As one goes to higher values of n the probability distribution gradually becomes closer to the classical one, as is shown in Fig. 13.5.

From Fig. 13.4 as well as from Eqs. (13.29) and (13.30) it is clear that ψ is an even function $[\psi(x) = \psi(-x)]$ when n is even and an odd function $[\psi(x) = -\psi(-x)]$ when n is odd. Such symmetry is typical of standing wave eigenfunctions arising from a symmetrical potential energy function; the ψ functions have alternate even and odd symmetry relative to a reflection of the coordinate at the "force" center. The functions are said to exhibit even or odd *parity* according as they are even or odd. If in Secs. 13.3 and 13.4 the origin had been located at the center of the

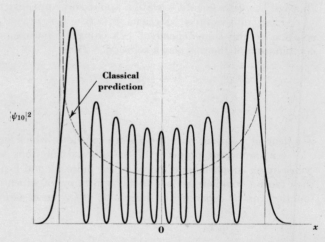

Fig. 13.5 The probability density function $\psi_{10}^{*}\psi_{10}$ for the $n = 10$ state of a harmonic oscillator. The vertical lines show the classical amplitude for the energy $(10 + \frac{1}{2})h\nu$, and the dashed curve shows the classical probability distribution function for finding the particle between x and $x + dx$. *(Modified from L. Pauling and E. B. Wilson, Jr., "Introduction to Quantum Mechanics." Copyright 1935. McGraw-Hill Book Company. Used by permission.)*

potential well, the resulting eigenfunctions would also have been alternatively of even and odd parity.

13.9 Properties of Eigenfunctions

Let $\psi_n(x)$ and $\psi_j(x)$ be any two normalized nondegenerate eigenfunctions of the Schrödinger equation for a bound particle with corresponding eigenvalues E_n and E_j. The eigenfunctions have the important property that

$$\int_{-\infty}^{\infty} \psi_j^* \psi_n \, dx = 0 \qquad \text{if } j \neq n \tag{13.31}$$

and are therefore said to be *orthogonal*. The orthogonality relation and the normalization condition are often combined into the relation

$$\int_{-\infty}^{\infty} \psi_j^* \psi_n \, dx = \delta_{ij}$$

where δ_{ij}, known as the *Kronecker delta*, is 0 for $j \neq n$ and 1 for $j = n$. Proof of the orthogonality relation appears in Appendix 13B.

Each of the wave functions $\Psi_n(x,t) = \psi_n(x)e^{-iE_n t/\hbar}$ is a particular solution of the time-dependent Schrödinger equation, and any linear combination of them is also a solution. Thus

$$\Psi(x,t) = a_0 \psi_0 e^{-iE_0 t/\hbar} + a_1 \psi_1 e^{-iE_1 t/\hbar} + \cdots$$
$$= \sum_{n=0}^{\infty} a_n \psi_n e^{-iE_n t/\hbar} \tag{13.32}$$

is a general solution; here the a's are constants which may be complex.

An important property of the allowed solutions $\Psi(x,y,z,t)$ of the Schrödinger equation [and indeed of the more general Eq. (13.18)] is that they form a complete set of functions in terms of which any given wave function can be expanded in the analog of a Fourier series; thus

$$\Psi = \sum_n a_n \Psi_n \tag{13.32a}$$

In some cases a continuum of possible values of E occurs, and then the sum must be supplemented or even replaced entirely by an integral analogous to a Fourier integral. If all functions are properly normalized, $|a_n|^2$ equals the probability of obtaining the value E_n as the result of a measurement of E performed upon a system initially in the condition represented by Ψ. It may also be remarked that if the entire setup and measurement are repeated many times, the average of the observed values of E will be very close to the *expectation* of E, defined as

$$\langle E \rangle = \int \Psi^* \hat{E} \Psi \, dq = \int \Psi^* i\hbar \frac{\partial \Psi}{\partial t} \, dq$$

where dq includes all the coordinates involved in Ψ. This is easily verified by substituting the series for Ψ and recalling that $\int \Psi_j^* \Psi_n \, dq = \delta_{jn}$.

In one dimension the probability of finding a particle in a given element of length dx is $\Psi^* \Psi \, dx$. When the particle is in a definite quantum state $j, \Psi_j^* \Psi_j = \psi_j^* \psi_j = $ const, so that the probability distribution does not change with time. However, when more than one a_n of Eq. (13.32) is nonzero, the particle is in a mixed state represented by the wave packet of (13.32). In this case the probability density $|\Psi|^2$ varies with time, and the particle can be regarded as moving around. The a's have the significance that $|a_n|^2$ is the probability of finding the particle in state n, provided some experimental means exists of selecting particles in that quantum state. [In certain problems, however, continuous ranges of allowed values of E also occur; in such cases, in Eq. (13.32), the sum over the states with discrete E's is supplemented by an integral analogous to a Fourier integral.] That $|\Psi|^2$ does indeed vary in time can be seen by carrying out the indicated evaluation

$$\Psi^*(x,t)\Psi(x,t) = \sum_n a_n^* a_n \psi_n^*(x) \psi_n(x)$$
$$+ \sum_{j \neq n} \sum_n a_j^* a_n \psi_j^* \psi_n e^{i(E_j - E_n)t/\hbar} \quad (13.33)$$

which is a function of the time.

13.10 Transitions between States

In classical electromagnetic theory an accelerated charge radiates energy at a rate which is proportional to the square of its acceleration. In quantum mechanics the uncertainty principle tells us that the position and velocity of the charge cannot both be known with unlimited precision. In going from classical to quantum theory one obtains valid predictions if one replaces the classically well-defined position coordinate x by its expectation value $\langle x \rangle$, and we shall use this semiclassical approach because of its relative simplicity.

Consider an electron in a stationary eigenstate such as we have discussed for a harmonic oscillator or for a square well. In either case, for level n, $\Psi_n(x,t) = \psi_n(x)e^{-iE_nt/\hbar}$, and the expectation value for x is

$$\langle x \rangle = \int_{-\infty}^{\infty} \Psi^* x \Psi \, dx = \int_{-\infty}^{\infty} \psi_n^* x \psi_n \, dx \tag{13.34}$$

The expectation value of x is not a function of time. Therefore $d^2\langle x \rangle/dt^2$ is zero, and no power is radiated. *An electron in a stationary state does not radiate.* In Bohr's half-classical, half-quantum theory he postulated that there was no radiation from an electron moving in a Bohr orbit, a clear violation of electromagnetic theory. In a full quantum-mechanical treatment it is the concept of sharp orbits which is set aside rather than electrodynamics.

But if a particle in a stationary state does not radiate, how do radiative transitions arise? To answer this we recall that the particle is not alone in the universe; sooner or later some photon (or atom or other particle) will approach and interact, thereby perturbing the energy so that the particle is no longer in a stationary state. Assume that the particle was initially in state i and that at $t = 0$ the perturbing interaction was introduced. For $t > 0$ the wave function for the particle is given by Eq. (13.32) in terms of the possible wave functions for the unperturbed system. The values of a_n are functions of time; for $t < 0$, $a_i = 1$ and all others were zero. The uncertainty principle does not permit us to know the energy exactly and hence the values of the a_n's as functions of time, but while the particle is in the mixed state, the expectation value of x, as given by Eq. (13.34), becomes

$$\langle x \rangle = \int_{-\infty}^{\infty} \left(\sum_n a_n^* a_n \psi_n^* x \psi_n + \sum_j \sum_{j \neq n} a_j^* a_n \psi_j^* x \psi_n e^{i(E_j - E_n)t/\hbar} \right) dx \tag{13.35}$$

Clearly, the first sum in the integral is not a function of time, but the second one is.

If we assume that the a's are all very near zero except $a_i^* \approx 1$ for the initial state and $a_f \approx 1$ for the final state, the only term that gives rise to a time-dependent variation of $\langle x \rangle$ is $\left(\int_{-\infty}^{\infty} \psi_i^* x \psi_f \, dx \right) e^{i(E_i - E_f)t/\hbar}$, which we shall designate $\langle x \rangle_{if}$. The product of the charge q on the particle and $\langle x \rangle_{if}$ is the x component of the oscillating electric dipole moment for which the frequency is given by the Einstein condition $\nu = (E_i - E_f)/h$. From

$$q\langle x \rangle_{if} = q \left(\int_{-\infty}^{\infty} \psi_i^* x \psi_f \, dx \right) e^{i(E_i - E_f)t/\hbar} \tag{13.36}$$

we observe that only if $\int_{-\infty}^{\infty} \psi_i^* x \psi_f \, dx$ is nonzero can there be a radiative transition between states i and f associated with the x component of the dipole moment. (Of course, the electric dipole moment is a vector, the y and z components of which can be calculated in a manner quite analogous to that shown above.) If all three components are zero, electric-dipole transitions between states i and f do not occur.

For example, in the case of the one-dimensional harmonic oscillator, dipole radiation is emitted only when n decreases by 1 and absorbed only when n increases by 1. Thus the harmonic oscillator obeys the selection rule $\Delta n \doteq \pm 1$, and the only frequency emitted or absorbed is the classical oscillator frequency $\nu = \sqrt{\beta/m}/2\pi$.

The oscillating electric dipole is the simplest classical system which radiates electromagnetic waves. We shall find that the most prominent lines of atomic spectra are electric-dipole lines, although weaker lines arising from magnetic dipoles, electric quadrupoles, and higher multipoles are also observed.

13.11 Perturbation Theory

In most cases the wave equation cannot be solved exactly in terms of familiar mathematical functions. Resort must then be had either to numerical integration or to approximate methods. The best-known method of approximation, and one of wide usefulness, is that known as *perturbation theory*. In this method certain minor terms are at first omitted from the wave equation, the simplified equation thus obtained is solved, and corrections to the energy and to the wave functions are then calculated to represent the effect of the terms that were omitted. The name arose from the analogous procedure of the astronomers, who first imagine each planet to move in an elliptical orbit about the sun and then calculate the perturbations of these motions caused by the attractions of the other planets.

Let us suppose that the exact nondegenerate eigenfunctions ψ_n and eigenvalues E_n are known for some problem, e.g., the simple harmonic oscillator, so that

$$-\frac{\hbar^2}{2m}\frac{d^2\psi_n}{dx^2} + P(x)\psi_n(x) = E_n\psi_n(x) \tag{13.37}$$

Suppose that now the potential energy is changed very slightly to

$$P'(x) = P(x) + f(x)$$

Then the wave equation takes the form

$$-\frac{\hbar^2}{2m}\frac{d^2\psi'_n}{dx^2} + [P(x) + f(x)]\psi'_n = E'_n\psi'_n \tag{13.38}$$

Since the potential energy has been changed very little, we expect that the new eigenfunctions ψ'_n and the new eigenvalues E'_n will differ only slightly from ψ_n and E_n. Since the ψ_n's represent a complete set in terms of which any reasonably smooth function can be expanded, we express the amount by which ψ'_n differs from ψ_n by the series $\sum\limits_{j=1}^{\infty} b_j\psi_j$, so that

$$\psi'_n = \psi_n + \sum_j b_j\psi_j \tag{13.39}$$

where the b_j's are all small if ψ'_n is almost equal to ψ_n. Substituting Eq. (13.39) in Eq. (13.38) and using Eq. (13.37) repeatedly, we obtain

$$E_n\psi_n + \Sigma b_j E_j\psi_j + f(x)\psi_n + f(x)\Sigma b_j\psi_j = E'\psi_n + E'_n\Sigma b_j\psi_j \tag{13.40}$$

Here, however, the term $f(x)\Sigma b_j\psi_j$ is of the second order of small quantities, since $f(x)$ and b_j are both of the first order. For a first correction we drop this term.

Next we multiply the resulting equation by $\psi_n^* \, dx$ and integrate over the full range of x, remembering that $\int\psi_n^*\psi_j \, dx = \delta_{nj}$. This gives

$$(1 + b_n)E_n + \int_{-\infty}^{\infty} \psi_n^* f(x)\psi_n \, dx = (1 + b_n)E'_n$$

$$E'_n - E_n = (1 + b_n)^{-1}\int_{-\infty}^{\infty} \psi_n^* f(x)\psi_n \, dx$$

Here the integral is a small quantity of the first order; hence b_n produces an effect of second order and for consistency must be dropped. We thus find for the first-order correction to the energy

$$E'_n - E_n = \int_{-\infty}^{\infty} \psi_n^* f(x)\psi_n \, dx \tag{13.41}$$

The first-order perturbation energy $E'_n - E_n$ is just the perturbation unction averaged over the corresponding unperturbed state.

To find the b_j's, we return to Eq. (13.40), with the second-order term omitted as before, multiply it through by $\psi_k^* \, dk$ with $k \neq n$, and integrate as before, obtaining

$$(E_k - E_n')b_k + \int_{-\infty}^{\infty} \psi_k^* f(x)\psi_n \, dx = 0 \tag{13.42}$$

Here we may replace E_n' by E_n, since the error thereby introduced, $(E_n' - E_n)b_k$, is of second order. Thus for a first-order correction

$$b_k = \frac{1}{E_n - E_k} \int_{-\infty}^{\infty} \psi_k^* f(x)\psi_n \, dx \tag{13.42a}$$

The coefficient b_n remains arbitrary and may be adjusted so as to normalize ψ_n'.

The integral $\int \psi_k^* f(x)\psi_n \, dx$ is called a matrix element of $f(x)$ with respect to the functions ψ_k and ψ_n. Its possible values could be written in the form of a matrix with its rows numbered by k and its columns by n. Components for $n = k$ are called diagonal components, i.e., located on the leading diagonal of the matrix. Thus the first-order correction to E_n is given by the diagonal matrix component of $f(x)$ with respect to ψ_n. This is an important result obtained from perturbation theory.

13.12 Emission and Absorption of Radiation

One of the most important properties of atoms and molecules is their ability to emit or absorb radiant energy. In classical mechanics, emission of radiation results from the electromagnetic field emitted by accelerated electric charges, whereas absorption results from work done on the charges by forces exerted on them by the electric vector of an incident field. In wave mechanics, on the other hand, both emission and absorption are associated with transitions of atoms or molecules between quantum states of different energies.

a. Transition Probabilities and Mean Life   Let the quantum states for an atom all be numbered off in a single series. Then, when an atom is in state n, there is a certain probability that during an interval of time dt it will jump spontaneously into another state j, with the emission of a photon of frequency $(E_n - E_j)/h$. This probability is denoted by $A_{nj} \, dt$. Out of a large number N of atoms in state n, NA_{nj} jump per second into state j.

Consider the history of N_0 atoms that start in state n. If N of them are still in that state after a time t, then during time dt a number $-dN$

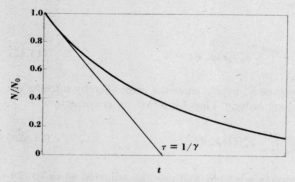

Fig. 13.6 The exponential decrease by radiative transitions.

will leave by spontaneous radiative transitions, where

$$dN = -N\gamma \, dt \qquad \gamma = \Sigma(j) A_{nj}$$

Here $\Sigma(j)$ denotes a sum over all states into which a spontaneous jump can occur out of state n, i.e., over all states having lower energy. In a particular case, some A_{nj}'s may, of course, be zero. Integrating (see Fig. 13.6), $N = N_0 e^{-\gamma t}$.

There is no absolute limit to the length of time that an individual atom may remain in a given quantum state. The *average* time τ_n spent by the atoms in state n is called the *mean life* of an atom in that state. To find it, consider N_0 atoms that have just entered this state. Then, during each dt, $N\gamma \, dt$ of them leave the state after spending a time t in it; and, inserting the value of N from the last equation, we can write

$$\tau_n = \frac{1}{N_0} \int_0^\infty t(N_0 e^{-\gamma t}\gamma \, dt) = \gamma \int_0^\infty t e^{-\gamma t} \, dt = \frac{1}{\gamma}$$

Hence

$$\tau_n = \frac{1}{\gamma} = [\Sigma(j) A_{nj}]^{-1} \tag{13.43}$$

For atomic states involved in visible or ultraviolet emission, τ_n is commonly of the order of 10^{-8} s; in the x-ray region it is much smaller.

b. The Einstein coefficients   In 1916 Einstein considered the emission and absorption of radiation by quantized systems such as atoms and molecules. Assume that there are many identical systems with energy states E_n and E_j which emit and absorb radiation of frequency

$$\nu_{nj} = \frac{E_n - E_j}{h}$$

(note that here $E_n > E_j$). They are immersed in a bath of black-body radiation such that $U_\nu\, d\nu$ represents the energy density of radiation with frequencies between ν and $\nu + d\nu$. Einstein assumed that the probability a system in state j will absorb a photon is proportional to $U_\nu(\nu_{nj})$; thus the probability is $B_{jn}U_\nu(\nu_{nj})$, where B_{jn} is *Einstein's coefficient of absorption*. Einstein's great insight led him to write the probability that a system in state n will emit a photon $h\nu_{nj}$ as the sum of the transition probability A_{nj} discussed above and an induced transition probability proportional to $U_\nu(\nu_{nj})$. Thus the probability of emission is

$$A_{nj} + B_{nj}U_\nu(\nu_{nj})$$

where A_{nj} and B_{nj} are respectively the *Einstein coefficients of spontaneous and induced emission*.

In equilibrium the number of systems N_n in state n and the number N_j in state j remain constant. The number of transitions per unit time from n to j is equal to the number from j to n, whence

$$N_n[A_{nj} + B_{nj}U_\nu(\nu_{nj})] = N_j B_{jn}U_\nu(\nu_{nj})$$

However, by Eq. (4.20), we have

$$N_j = N_n e^{(E_n - E_j)/kT} = N_n e^{h\nu_{nj}/kT}$$

and therefore

$$U_\nu(\nu_{nj}) = \frac{A_{nj}}{B_{jn}e^{h\nu_{nj}/kT} - B_{nj}} \tag{13.44}$$

Comparison of Eq. (13.44) with Eq. (5.21) for blackbody radiation density reveals that the three Einstein coefficients must be related by the equations

$$B_{jn} = B_{nj} \tag{13.45a}$$

$$A_{nj} = \frac{8\pi h\nu^3}{c^3} B_{nj} \tag{13.45b}$$

The fact that photons of frequency ν_{nj} not only raise systems in state

j to state n, but also induce systems in state n to make radiative transitions to state j, is widely exploited in modern technology.   The great intensities and coherence available in laser and maser beams are achieved by increasing the population of a suitable energy state greatly over its thermal-equilibrium value and stimulating transitions to a lower level.

Appendix 13A - The Harmonic Oscillator

The Schrödinger equation for the harmonic oscillator, Eq. (13.27), may be written

$$\frac{d^2\psi}{dx^2} = \frac{2m}{\hbar^2}(\tfrac{1}{2}\beta x^2 - E)\psi = \left(\frac{m\beta}{\hbar^2} - \frac{2mE}{\hbar^2}\right)\psi \qquad (13A.1)$$

We now make the substitutions

$$y = \sqrt[4]{\frac{m\beta}{\hbar^2}}\,x = \alpha x \qquad (13A.2a)$$

$$\epsilon = \frac{2E}{\hbar}\sqrt{\frac{m}{\beta}} \qquad (13A.2b)$$

to obtain the dimensionless equation

$$\frac{d^2\psi}{d^2y^2} = (y^2 - \epsilon)\psi \qquad (13A.3)$$

For a sufficiently large value of y, $y^2 \gg \epsilon$, and we may write Eq. (13A.3) in the asymptotic form $d^2\psi/dy^2 = y^2\psi$, which is satisfied by $\psi = e^{\pm y^2/2}$. To satisfy the normalization condition ψ must vanish as y goes to infinity; therefore we reject the positive sign and seek an exact solution of Eq. (13A.3) of the form $\psi(y) = Ce^{-y^2/2}H(y)$, where $H(y)$ is a function to be determined and C is a normalization constant.   Inserting this into Eq. (13A.3) reveals that $H(y)$ must satisfy

$$\frac{d^2H}{dy^2} - 2y\frac{dH}{dy} + (\epsilon - 1)H = 0 \qquad (13A.4)$$

One way of determining $H(y)$ is to assume that it can be written as a power series in y:

$$H(y) = a_0 + a_1y + a_2y^2 + \cdots = \sum_{n=0}^{\infty} a_n y^n \qquad (13A.5)$$

where the a's are constant coefficients yet to be found. Substitution of Eq. (13A.5) into Eq. (13A.4) yields

$$\sum_{n=0}^{\infty} y^n[(n+1)(n+2)a_{n+2} - 2na_n + (\epsilon - 1)a_n] = 0 \qquad (13A.6)$$

which can be true for every y only if each coefficient in the expansion is identically zero, or

$$a_{n+2} = \frac{2n+1-\epsilon}{(n+1)(n+2)} a_n \qquad (13A.7)$$

This *recursion formula* permits one to write $H(y)$ in terms of two coefficients a_0 and a_1. However, the normalization condition $\int_{-\infty}^{\infty} \psi^*\psi \, dy = 1$ can be satisfied only if $H(y)$ terminates at some value of n (Prob. 12). This happens if two conditions are satisfied:

1. For some specific n, $a_{n+2} = 0$, which occurs if

$$2n + 1 = \epsilon = \frac{2E}{\hbar} \sqrt{\frac{m}{\beta}}$$

2. If the n of condition 1 is odd, a_0 must be zero; if this n is even, a_1 must be zero.

Condition 1 restricts the allowed values of the energy of the oscillator to those given by $E_n = (n + \frac{1}{2})\hbar \sqrt{\beta/m}$. The classical frequency of the harmonic oscillator is $\nu = \sqrt{\beta/m}/2\pi$, and so the allowed energies are

$$E_n = (n + \tfrac{1}{2})h\nu \qquad (13A.8)$$

Condition 2 limits the eigenfunctions ψ_n to be either even or odd functions of y. The functions $H_n(y)$, known as the *Hermite polynomials*, are even or odd as n is even or odd. They can be obtained from the *generating function*

$$S(y,s) = e^{-s^2+2sy} = \sum_{n=0}^{\infty} \frac{H_n(y)}{n!} s^n \qquad (13A.9)$$

or alternatively from

$$H(y) = (-1)^n e^{y^2} \frac{d^n(e^{-y^2})}{dy^n} \qquad (13A.10)$$

Appendix 13B - Proof of the Orthogonality Relation

Let $\psi_n(x)$ and $\psi_j(x)$ be normalized eigenfunctions of the Schrödinger wave equation so that

$$-\frac{\hbar^2}{2m}\frac{d^2\psi_n}{dx^2} + P\psi_n = E_n\psi_n \tag{13B.1a}$$

$$-\frac{\hbar}{2m}\frac{d^2\psi_j}{dx^2} + P\psi_j = E_j\psi_i \tag{13B.1b}$$

Since everything in both equations is real except possibly the eigenfunctions, the complex conjugate of ψ_j obeys the equation

$$-\frac{\hbar^2}{2m}\frac{d^2\psi_j^*}{dx^2} + P\psi_j^* = E_j\psi_j^* \tag{13B.2}$$

If we multiply (13B.1a) by ψ_j^* and (13B.2) by ψ_n and subtract, we have

$$-\frac{\hbar^2}{2m}\left(\psi_j^*\frac{d^2\psi_n}{dx^2} - \psi_n\frac{d^2\psi_j^*}{dx^2}\right)$$
$$+ P(\psi_j^*\psi_n - \psi_n\psi_j^*) = (E_n - E_j)\psi_j^*\psi_n \tag{13B.3}$$

If we integrate both sides from $x = -\infty$ to $x = \infty$, we find

$$\int_{-\infty}^{\infty}\left(\psi_j^*\frac{d^2\psi_n}{dx^2} - \psi_n\frac{d^2\psi_j^*}{dx^2}\right)dx = \int_{-\infty}^{\infty}\frac{d}{dx}\left(\psi_j^*\frac{d\psi_n}{dx} - \psi_n\frac{d\psi_j^*}{dx}\right)dx$$
$$= -\frac{2m}{\hbar^2}(E_n - E_j)\int_{-\infty}^{\infty}\psi_j^*\psi_n\,dx$$

or

$$\left[\psi_j^*\frac{d\psi_n}{dx} - \psi_n\frac{d\psi_j^*}{dx}\right]_{-\infty}^{\infty} = \frac{2m}{\hbar^2}(E_j - E_n)\int_{-\infty}^{\infty}\psi_j^*\psi_n\,dx$$

The normalization condition requires that ψ_n and ψ_j^* go to zero at $x = \pm\infty$, and therefore $\int_{-\infty}^{\infty}\psi_j^*\psi_n\,dx = 0$ if $E_j \neq E_n$. If ψ_j^* and ψ_n are degenerate so that $E_j = E_n$, the situation becomes more complicated; nevertheless it is always possible to make degenerate eigenfunctions orthogonal.

Problems

1. Solve the problem of the infinitely deep square well with the origin at the center of the well and $P = 0$ for $-L/2 \leq x \leq L/2$. Show that

the solutions are alternately of even and odd parity. Prove that the eigenvalues of the energy are the same as those given by Eq. (13.8).

2. Find the commutator of the operators (a) $x \, d/dx$ and $x^2 \, d/dx$, (b) energy $\hat{E}$ and time t.

3. (a) Find the five lowest kinetic energies which are allowed for an electron in a cubical box 1 cm on a side.

   (b) Repeat for a cubical box 3 Å on a side. (Note that this is roughly the average volume occupied by an atom in a solid.)

   Partial Ans: (a) $E_{111} = 1.13 \times 10^{-14}$ eV; $E_{222} = 4.53 \times 10^{-14}$ eV; (b) $E_{111} = 12.6$ eV

4. Find the eigenvalues and eigenfunctions for the operator d^2/dx^2 for the interval $0 \le x \le L$ if the functions $u_i(x)$ on which it operates satisfy the boundary conditions $du/dx = 0$ at $x = 0$ and $u = 0$ at $x = L$.

5. Show that $u(x) = Ae^{-\alpha x}$ is an eigenfunction of the operator

$$\hat{C} = \frac{d^2}{dx^2} + \frac{2}{x}\frac{d}{dx} + \frac{2b}{x}$$

and determine the corresponding eigenvalue.

   Ans: b^2

6. (a) Show that if the hamiltonian operator can be written as the sum of terms, each of which depends on a different set of coordinates so that $\hat{H} = \hat{H}_a(x_1,y_1,z_1) + \hat{H}_b(x_2,y_2,z_2) + \hat{H}_c(r,\theta,\varphi) + \cdots$, the eigenfunctions can be written as a product of terms

$$\psi = \psi_a(x_1,y_1,z_1) \; \psi_b(x_2,y_2,z_2) \; \psi_c(r,\theta,\varphi) \cdots$$

   (b) Show further that the energy corresponding to ψ is $E_a + E_b + E_c + \cdots$, where E_a is the energy eigenvalue associated with ψ_a, etc.

7. A particle of mass m is bound in a square well of width W and height P_0. The eigenfunction for the ground state is given by

$$\psi(x) = B \cos\left(\frac{4\pi x}{5W}\right)$$

for $-W/2 \le x \le W/2$ and by $\psi = Ae^{-|\alpha x|}$ outside the well.

   (a) Taking m and W as given, find α, E and P_0. Compare the value of E_1 with that of the infinite square well.

   (b) Find A in terms of B and the given quantities.

   (c) Set up the integral from which B can be calculated.

   Ans: (a) $7.72/W$, $2h^2/25mW^2$, $21h^2/25mW^2$; (b) $14.7B$

8. A particle of mass m moves with potential energy $P(x) = \infty$ for $x < 0$ and $\frac{1}{2}\beta x^2$ for $x \geq 0$. Find the eigenfunctions and the allowed energies.

9. Find the expectation values of the kinetic and potential energies of a harmonic oscillator in its ground state.

Ans: $h\nu/4$, $h\nu/4$

10. A three-dimensional harmonic oscillator moves in the potential $P(x,y,z) = \frac{1}{2}m\omega^2(x^2 + y^2 + z^2)$. Separate the Schrödinger equation and show that the allowed energies are given by $E = (n + \frac{3}{2})\hbar\omega$.

11. Is $u(x) = xe^{-x^2/2}$ an eigenfunction of $d^2/dx^2 - x^2$? If it is, find the corresponding eigenvalue.

12. Show that unless $H(y)$ terminates for some value of n,

$$\psi = e^{-y^2/2} \sum_{n=0}^{\infty} a_n y^n$$

goes to infinity for $y \to \infty$. *Hint:* Show that $H(y)$ behaves like e^{y^2} for large y.

13. Show that when $\Psi(x,t) = \Sigma a_n \Psi_n(x,t)$, $\langle E \rangle = \Sigma a_n^* a_n E_n$ and thus that $a_n^* a_n$ is the probability of obtaining the value E_n as the result of a single measurement of the total energy E and hence is the probability of finding the particle in the eigenstate n.

14. A particle is confined to move in the x and y directions under the influence of the potentials $P(x,y) = P_0(x) + P_1(y)$, where $P_0(x) = \beta x^2/2$ and $P_1(y) = 0$ for $-a \leq y \leq a$ and $P_1(y) = \infty$ for $y < -a$ and $y > a$.

(*a*) Find the eigenfunctions by solving the Schrödinger wave equation.

(*b*) Find the allowed energies for the particle.

15. It is desired to calculate the expectation value of the product of position x and momentum p_x. Show that neither

$$\langle xp_x \rangle = \int_{-\infty}^{\infty} \Psi^* x \left(\frac{\hbar}{i}\frac{\partial}{\partial x}\right) \Psi \, dx$$

nor

$$\langle xp_x \rangle = \int_{-\infty}^{\infty} \Psi^* \left(\frac{\hbar}{i}\frac{\partial}{\partial x}\right) x\Psi \, dx$$

is acceptable since both lead to imaginary values. Show that

$$\langle xp \rangle = \frac{1}{2} \int_{-\infty}^{\infty} \Psi^* \left[x\left(\frac{\hbar}{i}\frac{\partial}{\partial x}\right) + \left(\frac{\hbar}{i}\frac{\partial}{\partial x}\right)x \right] \Psi \, dx$$

leads to a real value. Does $\langle xp \rangle = \langle x \rangle \langle p \rangle$? *Hint:* Integrate by parts.

16. A electron is bound in an infinite square well such that $P = 0$ for $-L/2 \leq x \leq L/2$ and infinity elsewhere. If at some instant the particle is in the state represented by $\psi(x) = \sqrt{2/L}$ for $|x| < L/2$ and 0 for $|x| > L/2$, calculate the probability that a measurement of the energy will give the result (a) $h^2/8mL^2$ and (b) $h^2/2mL^2$.

17. A particle is in the potential field $P(x) = \frac{1}{2}\beta x^2$ of a classical simple harmonic oscillator. Calculate the probability of finding the particle outside the classical limits of its motion if it is in the ground state.

    Ans: 0.16

18. A series of measurements on a one-dimensional system leaves the system with the wave function $\psi(x) = A \sin \pi x/L$ for $-L \leq x \leq L$ and zero elsewhere.

    (a) Find the normalization constant A.

    (b) Compute the probability that a measurement of the momentum p_x for the system yields the value p_n, where p_n is an allowed momentum of the system. Plot the probability against $p_n L/\hbar$ and show that $\Delta p_x \Delta x \approx \hbar$.

    (c) Calculate the expectation value $\langle p \rangle$ and compare with the graph of part (b).

19. Show that (a) $yH_n(y) = nH_{n-1}(y) + \frac{1}{2}H_{n+1}(y)$, (b) $dH_n(y)/dy = 2nH_{n-1}(y)$, and (c) $dH_n(y)/dy = 2yH_n(y) - H_{n+1}(y)$.

20. An electron moves in the potential well $P(x) = -\delta$ for $-a \leq x < 0$ and $P(x) = \delta$ for $0 < x \leq a$ (Fig. 13.7). Use first-order perturbation theory to compute the first four energy levels. Set up the expression for the first-order expansion coefficients for the lowest energy state.

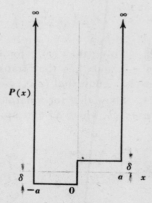

Fig. 13.7

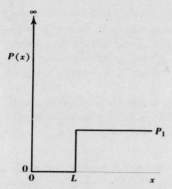

Fig. 13.8

21.   A particle is in the potential field shown in Fig. 13.8.

(*a*) Solve the Schrödinger wave equation and find the eigenfunctions for the bound states.

(*b*) Prove that the eigenenergies of the bound states are solutions of the   equation $-R = k \cot kL$,   where $k = \sqrt{2mE}/\hbar$ and $R = \sqrt{2m(P_1 - E)}/\hbar$.

(*c*) What condition is placed on P_1 by the requirement that there be only one bound state for the system?

22.   Prove that the Hermite functions $\psi_n = N_n e^{-v^2/2} H_n(y)$ are orthogonal and show that the normalization constant which makes $\int_{-\infty}^{\infty} \psi_n{}^2(x)\, dx = 1$ is

$$N_n = \left(\frac{\alpha}{\sqrt{\pi}\, 2^n n!} \right)^{\frac{1}{2}}$$

23.   Prove that $\langle x \rangle_{nm}$ for the harmonic oscillator is zero unless $n = m \pm 1$ and thus that transitions occur only between adjacent energy levels.   Show that $\langle x \rangle_{n,n+1} = \sqrt{(n + 1)/2\alpha^2}$ while $\langle x \rangle_{n,n-1} = \sqrt{n/2\alpha^2}$.

24.   Show that for a harmonic oscillator in its *n*th eigenstate $\Delta x\, \Delta p_x = (n + \frac{1}{2})\hbar$, where Δx and Δp_x are defined in Sec. 12.6.

chapter fourteen

Wave Mechanics III
The Hydrogen Atom

An understanding of the structure of the hydrogen atom is of primary importance, not so much because hydrogen is the simplest atom and was the first to be treated, but because the treatment of more complex atoms is typically based on hydrogenic wave functions. The Balmer formula yields the wavelengths of the visible lines of hydrogen, and the Bohr model provides a more complete prediction of energy states and spectral lines, but neither forms a usable foundation for understanding the spectra, structure, and interactions of other atoms.

14.1 The Schrödinger Equation for a One-electron Atom

The hydrogen atom is a system consisting of an electron and a proton bound by electrostatic attraction. Similarly an ionized helium atom and a doubly ionized lithium atom are composed of a positive nucleus and a single electron. If r is the distance between particles, the electrostatic potential energy P for the system is

$$P = -\frac{Ze^2}{4\pi\epsilon_0 r} \tag{14.1}$$

where Ze = nuclear charge
  $-e$ = electronic charge
  ϵ_0 = permittivity of free space

Let M and m be respectively the masses of the nucleus located at (x_1,y_1,z_1) and the electron at (x_2,y_2,z_2). The Schrödinger equation for this two-particle system is

$$-\frac{\hbar^2}{2M}\left(\frac{\partial^2\psi_T}{\partial x_1^2} + \frac{\partial^2\psi_T}{\partial y_1^2} + \frac{\partial^2\psi_T}{\partial z_1^2}\right)$$
$$-\frac{\hbar^2}{2m}\left(\frac{\partial^2\psi_T}{\partial x_2^2} + \frac{\partial^2\psi_T}{\partial y_2^2} + \frac{\partial^2\psi_T}{\partial z_2^2}\right) + P\psi_T = E\psi_T$$

where ψ_T is a function of x_1, y_1, z_1, x_2, y_2, z_2; the subscript T of ψ_T and E_T means total, indicating that they refer to the complete system of nucleus and electron. We now replace the six variables by the three coordinates (x_c,y_c,z_c) of the center of mass of the system and the three coordinates of the electron relative to the proton by making the substitutions

$$x_c = \frac{Mx_1 + mx_2}{M + m}$$
$$y_c = \frac{My_1 + my_2}{M + m}$$
$$z_c = \frac{Mz_1 + mz_2}{M + m}$$

$r\sin\theta\cos\varphi = x_2 - x_1$, $r\sin\theta\sin\varphi = y_2 - y_1$, and $r\cos\theta = z_2 - z_1$. The Schrödinger equation then becomes

$$-\frac{\hbar^2}{2(M+m)}\left(\frac{\partial^2\psi_T}{\partial x_c^2} + \frac{\partial^2\psi_T}{\partial y_c^2} + \frac{\partial^2\psi_T}{\partial z_c^2}\right) - \frac{\hbar^2(M+m)}{2Mm}\left[\frac{1}{r^2}\frac{\partial}{\partial r}\left(r^2\frac{\partial\psi_T}{\partial r}\right)\right.$$
$$\left. + \frac{1}{r^2\sin\theta}\frac{\partial}{\partial\theta}\left(\sin\theta\frac{\partial\psi_T}{\partial\theta}\right) + \frac{1}{r^2\sin^2\theta}\frac{\partial^2\psi_T}{\partial\varphi^2}\right] - \frac{Ze^2}{4\pi\epsilon_0 r}\psi_T = E_T\psi_T$$

This equation can be separated by expressing $\psi_T(x_c,y_c,z_c,r,\theta,\varphi)$ as the product of a function ψ_c of x_c, y_c, z_c alone and a function ψ of r, θ, φ only. Upon substituting $\psi_T(x_c,y_c,z_c,r,\theta,\varphi) = \psi_c(x_c,y_c,z_c)\psi(r,\theta,\varphi)$ in the Schrödinger equation and dividing through by $\psi_c\psi$, it is found that the resulting equation is the sum of two parts, one of which depends only on x_c, y_c, z_c and the other only on r, θ, φ. Therefore each part must be equal to a constant, and we obtain the two equations

$$-\frac{\hbar^2}{2(M+m)}\left(\frac{\partial^2\psi_c}{\partial x_c{}^2} + \frac{\partial^2\psi_c}{\partial y_c{}^2} + \frac{\partial^2\psi_c}{\partial z_c{}^2}\right) = E_{tr}\psi_c \tag{14.2}$$

$$-\frac{\hbar^2(M+m)}{2Mm}\left[\frac{1}{r^2}\frac{\partial}{\partial r}\left(r^2\frac{\partial\psi}{\partial r}\right) + \frac{1}{r^2\sin\theta}\frac{\partial}{\partial\theta}\left(\sin\theta\frac{\partial\psi}{\partial\theta}\right)\right.$$
$$\left. + \frac{1}{r^2\sin^2\theta}\frac{\partial^2\psi}{\partial\varphi^2}\right] - \frac{Ze^2}{4\pi\epsilon_0 r}\psi = E\psi \tag{14.3}$$

where $E = E_T - E_{tr}$.

Equation (14.2) represents the motion of the center of mass, which behaves like a free particle of mass $M + m$ with translational kinetic energy E_{tr}. For our purposes the translational energy of the atom is not of interest; E_{tr} can take on any positive value (or zero).

Equation (14.3) is the wave equation of a particle of mass, $Mm/(M + m)$ under the influence of the potential function of Eq. (14.1). The quantity

$$m_r = \frac{Mm}{M+m} = \frac{m}{1+m/M} \tag{14.4}$$

is just the *reduced mass* of the electron (see Sec. 9.4). Indeed, in the limit as M goes to infinity, m_r approaches m and Eq. (14.3) becomes the wave equation of an electron bound to a fixed origin with the potential energy of Eq. (14.1). It should be observed further that Eq. (14.3) may be written

$$-\frac{\hbar^2}{2m_r}\nabla^2\psi - \frac{Ze^2}{4\pi\epsilon_0 r}\psi = E\psi \tag{14.3a}$$

since the quantity in brackets in Eq. (14.3) is ∇^2 in spherical polar coordinates.

14.2 Separation of Variables

The advantage of writing the Schrödinger equation for the one-electron atom in spherical coordinates is that in this form it can be readily separated into three equations, each involving a single coordinate. To do this we assume that

$$\psi(r,\theta,\varphi) = R(r)\Theta(\theta)\Phi(\varphi)$$

If we substitute this equation into Eq. (14.3) and multiply by

$$2m_r r^2 \sin^2 \theta / (\hbar^2 R \Theta \Phi)$$

we obtain

$$\frac{\sin^2 \theta}{R} \frac{d}{dr}\left(r^2 \frac{dR}{dr}\right) + \frac{\sin \theta}{\Theta} \frac{d}{d\theta}\left(\sin \theta \frac{d\Theta}{d\theta}\right) + \frac{1}{\Phi} \frac{d^2\Phi}{d\varphi^2}$$
$$+ \frac{2m_r r^2 \sin^2 \theta}{\hbar^2}\left(\frac{Ze^2}{4\pi\epsilon_0 r} + E\right) = 0 \quad (14.5)$$

The third term of Eq. (14.5) is a function of φ alone, while none of the other terms involves φ at all. If Eq. (14.5) is to hold for all values of φ, this third term must be a constant, which we shall designate as $-m_l^2$, and Φ must obey the equation

$$\frac{d^2\Phi}{d\varphi^2} + m_l^2 \Phi = 0 \quad\quad\quad (14.6)$$

If we substitute $-m_l^2$ for the third term in Eq. (14.5), divide by $\sin^2 \theta$, and rearrange, we arrive at the result that

$$\frac{1}{R} \frac{d}{dr}\left(r^2 \frac{dR}{dr}\right) + \frac{2m_r r^2}{\hbar^2}\left(\frac{Ze^2}{4\pi\epsilon_0 r} + E\right) = \frac{m_l^2}{\sin^2 \theta}$$
$$- \frac{1}{\Theta \sin \theta} \frac{d}{d\theta}\left(\sin \theta \frac{d\Theta}{d\theta}\right) \quad (14.6a)$$

Here the left side is a function of r alone, while the right depends solely on θ. Therefore, both sides must be equal to some constant which is written as $l(l + 1)$ for reasons which will appear later. If we equate both sides to $l(l + 1)$ and rearrange, we find that

$$\frac{1}{\sin \theta} \frac{d}{d\theta}\left(\sin \theta \frac{d\Theta}{d\theta}\right) + \left[l(l + 1) - \frac{m_l^2}{\sin^2 \theta}\right] \Theta = 0 \quad (14.7)$$

$$\frac{1}{r^2} \frac{d}{dr}\left(r^2 \frac{dR}{dr}\right) + \left[\frac{2m_n}{\hbar^2}\left(\frac{Ze^2}{4\pi\epsilon_0 r} + E\right) - \frac{l(l + 1)}{r^2}\right] R = 0 \quad (14.8)$$

In Eqs. (14.6) to (14.8) we have three differential equations, each of a single variable, from which R, Θ, and Φ can be found. Once they are known, we can immediately write ψ, which is $R(r)\Theta(\theta)\Phi(\varphi)$.

14.3   The Wave Functions and Energy Levels

A solution of Eq. (14.6) may be written immediately

$$\Phi(\varphi) = Ce^{im_l\varphi}$$

where C is a constant of integration. The condition that the wave function be single-valued (Sec. 12.2) requires that m_l be 0 or a positive or negative integer, since increasing φ by 2π brings one back to the same point in space, so that $\Phi(\varphi_1) = \Phi(\varphi_1 + 2\pi)$. If we let m_l take on the values 0 and all positive and negative integers, and if we normalize $\Phi(\varphi)$ so that $\int_0^{2\pi} \Phi^*\Phi \, d\varphi = 1$, we have finally

$$\Phi_{m_l} = \frac{1}{\sqrt{2\pi}} e^{im_l\varphi} \qquad m_l = 0, \pm 1, \pm 2, \ldots \qquad (14.9)$$

where m_l is known as the *magnetic quantum number*.

To solve Eq. (14.7) it is customary to let $w = \cos\theta$. Upon making this substitution, Eq. (14.7) becomes

$$\frac{d}{dw}\left[(1 - w^2) \frac{d\Theta}{dw} \right] + \left[l(l + 1) - \frac{m_l^2}{1 - w^2} \right] \Theta = 0 \qquad (14.7a)$$

for which finite, well-behaved solutions are found only if l is an integer equal to or greater than m_l. The integer l is called the *orbital* (or *azimuthal*) quantum number. The solutions of Eq. (14.7a) are the *associated Legendre polynomials;* we write $\Theta_{lm}(\theta) = N_{lm}P_l^m(\cos\theta)$, where N_{lm} is the appropriate normalization constant and P_l^m is the associated Legendre polynomial for the chosen l and m_l. Values of $P_l^m(\cos\theta)$ for a few values of l and m_l are listed below.

$$P_0^0 = 1 \qquad\qquad\qquad P_1^1 = \sin\theta$$
$$P_1^0 = \cos\theta \qquad\qquad\quad P_2^1 = 3\sin\theta\cos\theta$$
$$P_2^0 = \tfrac{1}{2}(3\cos^2\theta - 1) \qquad P_2^2 = 3\sin^2\theta$$
$$P_3^0 = \tfrac{1}{2}(5\cos^3\theta - 3\cos\theta) \qquad P_3^1 = \tfrac{3}{2}\sin\theta(5\cos^2\theta - 1)$$
$$P_3^2 = 15\sin^2\theta\cos\theta \qquad P_3^3 = 15\sin^3\theta$$

Next we consider the solution of Eq. (14.8) with attention focused on bound states (E negative). In Appendix 14A it is shown that bound-state wave functions are possible only for energies given by

$$E_n = -\frac{m_r e^4 Z^2}{32\pi^2\epsilon_0^2 h^2 n^2} = -\frac{m_r e^4 Z^2}{8\epsilon_0^2 h^2 n^2}$$
$$= -\frac{13.6Z^2}{n^2} \qquad \text{eV} \qquad n = 1, 2, 3, \ldots \qquad (14.10)$$

which are just the values previously found by Bohr. In addition, any positive value of E is allowed.

The radial wave function $R(r)$ for a given value of the radial (or total) quantum number n and the orbital quantum number l is the product of (1) a normalization constant, (2) $e^{-\alpha r/2}$, where $\alpha = 2Z/na_B$ (a_B is the Bohr radius), and (3) a polynomial in αr. The complete wave function $\psi(r,\theta,\varphi)$ is accordingly given by $\psi_{nlm}(r,\theta,\varphi) = N_{nl}e^{-\alpha r/2}$ (polynomial in αr) (polynomial in $\cos\theta$) $e^{im_l\varphi}$, where N_{nl} is the normalization constant such that $\iiint \psi^*\psi \, d\upsilon = 1$ ($d\upsilon$ is a volume element).

As examples, the following relations give the wave functions for the states of lowest energy in terms of the Bohr radius a_B ($= 4\pi\epsilon_0\hbar^2/m_re^2$):

$$\psi_{100} = \frac{1}{\sqrt{\pi}}\left(\frac{Z}{a_B}\right)^{\frac{3}{2}} e^{-Zr/a_B} \qquad \begin{aligned} n &= 1 \\ l &= 0 \\ m_l &= 0 \end{aligned}$$

$$\psi_{200} = \frac{1}{2\sqrt{2\pi}}\left(\frac{Z}{a_B}\right)^{\frac{3}{2}}\left(1 - \frac{Zr}{2a_B}\right)e^{-Zr/2a_B} \qquad \begin{aligned} n &= 2 \\ l &= 0 \\ m_l &= 0 \end{aligned}$$

$$\psi_{210} = \frac{1}{2\sqrt{2\pi}}\left(\frac{Z}{a_B}\right)^{\frac{3}{2}}\frac{Zr}{2a_B}e^{-Zr/2a_B}\cos\theta \qquad \begin{aligned} l &= 1 \\ m_l &= 0 \end{aligned}$$

$$\psi_{211} = \frac{1}{4\sqrt{\pi}}\left(\frac{Z}{a_B}\right)^{\frac{3}{2}}\frac{Zr}{2a_B}e^{-Zr/2a_B}\sin\theta\, e^{\pm i\varphi} \qquad m_l = \pm 1$$

$$\psi_{300} = \frac{1}{9\sqrt{3\pi}}\left(\frac{Z}{a_B}\right)^{\frac{3}{2}}\left[3 - 6\frac{Zr}{3a_B} + 2\left(\frac{Zr}{3a_B}\right)^2\right]e^{-Zr/3a_B}$$

$$\begin{aligned} n &= 3 \\ l &= 0 \\ m_l &= 0 \end{aligned} \qquad (14.11)$$

$$\psi_{310} = \frac{\sqrt{2}}{9\sqrt{\pi}}\left(\frac{Z}{a_B}\right)^{\frac{3}{2}}\frac{Zr}{3a_B}\left(2 - \frac{Zr}{3a_B}\right)e^{-Zr/3a_B}\cos\theta$$

$$\begin{aligned} l &= 1 \\ m_l &= 0 \end{aligned}$$

$$\psi_{311} = \frac{1}{9\sqrt{\pi}}\left(\frac{Z}{a_B}\right)^{\frac{3}{2}}\frac{Zr}{3a_B}\left(2 - \frac{Zr}{3a_B}\right)e^{-Zr/3a_B}\sin\theta\, e^{\pm i\varphi}$$

$$m_l = \pm 1$$

$$\psi_{320} = \frac{1}{9\sqrt{6\pi}}\left(\frac{Z}{a_B}\right)^{\frac{3}{2}}\left(\frac{Zr}{3a_B}\right)^2 e^{-Zr/3a_B}(3\cos^2\theta - 1)$$

$$\begin{aligned} l &= 2 \\ m_l &= 0 \end{aligned}$$

$$\psi_{321} = \frac{1}{9\sqrt{\pi}} \left(\frac{Z}{a_B}\right)^{\frac{3}{2}} \left(\frac{Zr}{3a_B}\right)^2 e^{-Zr/3a_B} \sin\theta \cos\theta\, e^{\pm i\varphi}$$

$$m_l = \pm 1$$

$$\psi_{322} = \frac{1}{18\sqrt{\pi}} \left(\frac{Z}{a_B}\right)^{\frac{3}{2}} \left(\frac{Zr}{3a_B}\right)^2 e^{-Zr/3a_B} \sin^2\theta\, e^{\pm 2i\varphi}$$

$$m_l = \pm 2$$

14.4 Probability Density and Charge-cloud Density

The probability density corresponding to one of the wave functions $\psi_{nlm_l}(r,\theta,\varphi)$ is given by $\psi^*\psi$. The quantity $|\psi|^2\, d\upsilon$ represents the probability that if the electron could be located experimentally at any instant, it would be found in the volume element $d\upsilon$. It is not possible, however, to follow the electron in an orbital motion around the nucleus by means of a succession of such observations as the astronomers follow the planets around the sun; for one observation with a γ-ray microscope of sufficient revolving power would suffice, because of the Compton effect, to eject the electron from the atom.

There is no suggestion here of orbital motion, but often it is convenient to imagine the electronic charge to be distributed in space as sort of a charge cloud with a density

$$\eta = e\psi^*\psi$$

Many effects of the atom on its surroundings are approximately the same as if the atom actually contained a distribution of charge of density η. Rather than following a fixed orbit, an electron is said to occupy an atomic *orbital* determined by ψ_{nlm_l}.

Since φ appears in ψ as $e^{im_l\varphi}$, it is clear that $\psi^*\psi$ is not a function of φ. Consequently, when the electron is in one of its quantum states, the probability density (or charge-cloud density) has cylindrical symmetry about the chosen polar axis and $|\psi|^2 = R^2\Theta^2$. Polar graphs of the function $\Theta^2_{lm_l}$ (Fig. 14.1) reveal that the variation with θ, although symmetrical relative to the plane $\theta = \pi/2$, is large unless $l = 0$. The latter leads to a spherical distribution. The angular variation of charge density is of great interest when one approaches the problem of binding between atoms in terms of atomic orbitals.

For discussing the average radial variation in charge density let $\mathcal{P}_r\, dr$ denote the total probability (and $\eta\, dr$ the numerical amount of charge) contained between two spheres of radii r and $r + dr$ drawn about the nucleus as a center. The volume between the two spheres is propor-

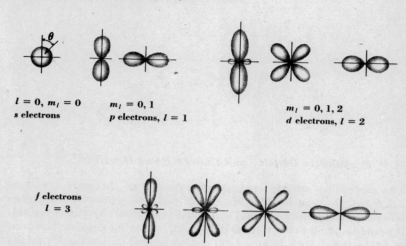

$l = 0, m_l = 0$
s electrons

$m_l = 0, 1$
p electrons, $l = 1$

$m_l = 0, 1, 2$
d electrons, $l = 2$

f electrons
$l = 3$

$m_l = 0$ 1   2   3

g electrons
$l = 4$

$m_l = 0$ 1   2   3   4

Fig. 14.1   The probability-density function $|\Theta_{lm_l}|^2$ as a function of θ for $l = 0, 1, 2, 3, 4$.   The length of a straight line from the origin to any point on a given curve is proportional to the probability that the electron is in the direction of that line. (The figures are not drawn to the same scale.)

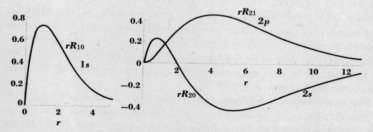

Fig. 14.2   The radial function rR_n for three quantum states of hydrogen as a function of r in Bohr radii.

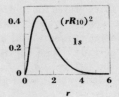

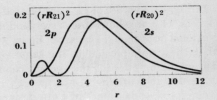

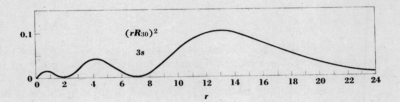

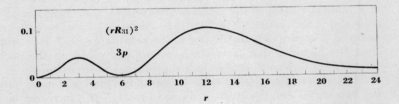

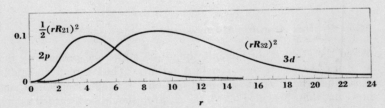

Fig. 14.3 The radial probability density $\mathcal{P}(r) = r^2 R_n{}^2$ for six quantum states of hydrogen as a function of r in Bohr radii. (The curve for the $2p$ orbital is drawn twice for comparison.)

tional to r^2. Hence, if R_{nl} is separately normalized so that

$$\int_0^\infty \mathcal{P}_r \, dr = \int_0^\infty r^2 R_{nl}{}^2 \, dr = 1$$

then

$$\mathcal{P}_r = r^2 R_{nl}{}^2 \qquad \text{and} \qquad \eta_r = e\mathcal{P}_r = er^2 R_{nl}{}^2$$

In Fig. 14.2 values of rR_{nl} are plotted for three states for hydrogen, and in Fig. 14.3 are shown the corresponding curves for $\mathcal{P}_r$ or η_r, which is proportional to $r^2R_{nl}{}^2$. The areas under the latter curves would all be equal if they were plotted to the same vertical scale. The curves are labeled in the notation of the spectroscopists, the letters s, p, d, f, . . . being used to indicate $l = 0, 1, 2, 3, . . .$, preceded by a number giving the value of n; thus values of l are designated by

0	1	2	3	4	5	6	7 . . .
s	p	d	f	g	h	i	k . . .

The reason for the choice of letters is stated in Sec. 17.2. All the radial factors R_{nl} show $n - l$ numerical maxima, with intervening points of zero value if $n - l > 1$.

Although nothing in the wave functions suggests the former Bohr orbits, a certain correspondence between them can be traced. In states with $l = n - 1$, the maximum value of the radial charge density η occurs at a value of r equal to the radius of the corresponding Bohr circle. In these states, as is evident from Fig. 14.3, the density forms a single broad hump; it is most widely distributed in the s states ($l = 0$). It is evident from the formulas that for s states ψ does not vanish at the nucleus ($r = 0$); this fact tends to give special properties to these states.

As n increases, the curves for $\mathcal{P}_r$ spread out more and more, just as the Bohr orbits did; the atom swells, so to speak. If Z increases, on the other hand, the atom shrinks in inverse ratio to Z.

14.5 Orbital Angular Momentum

When the atom is in a definite quantum state, the probability density $\Psi^*\Psi$ or $\psi^*\psi$ is everywhere constant in time. Thus it might be said in one sense that the electron is not in motion. Certain phenomena suggest, however, that, provided $l > 0$, the electron should be regarded as possessing *angular momentum* about the nucleus; this momentum is associated with magnetic effects as is that of the classical motion in an orbit. Accordingly, for convenience, we shall speak of the electron as "moving" in the field of the nucleus, and we sometimes refer to this motion as *orbital*. The value of the angular momentum furnishes a useful means of distinguishing between states that have the same energy.

When a rigid body is rotating about a fixed point, like a top about its point, there will be a certain axis through the fixed point about which the angular momentum of the body is a maximum. About any other axis, inclined at an angle θ to the line of maximum angular momentum, the angular momentum is equal to the maximum value multiplied by $\cos \theta$. The angular momentum can be represented by a vector drawn in the direction of the axis about which it is a maximum; the length of this vector can be made to represent the magnitude of the maximum angular momentum, and its direction is taken in the direction in which a right-handed screw would advance along the axis while turning about it in the same direction as the rotating body. The angular momentum about any other axis drawn through the fixed point is then represented by the component of this vector in the direction of that axis.

A similar treatment can be given to the angular momentum of a moving particle. In classical theory, the electron can be regarded as moving at every instant in a certain plane drawn through the nucleus; the vector representing its angular momentum about the nucleus is commonly drawn perpendicular to this plane (see Fig. 14.4). Let the orbital angular momentum be represented by $\mathbf{A}_l$.

Equation (13.17) gives the operator for the square of the angular momentum in cartesian coordinates. If one makes the substitutions $x = r \sin \theta \cos \varphi$, $y = r \sin \theta \sin \varphi$, and $z = r \cos \theta$, one obtains for this operator

$$\widehat{\mathbf{A}_l \cdot \mathbf{A}_l} = -\hbar^2 \left[\frac{1}{\sin \theta} \frac{\partial}{\partial \theta} \left(\sin \theta \frac{\partial}{\partial \theta} \right) + \frac{1}{\sin^2 \theta} \frac{\partial^2}{\partial \varphi^2} \right] \qquad (14.12)$$

If this operator is applied to the hydrogen wave function

$$\psi_{nlm_l} = R_{nl} \Theta_{lm_l} e^{im_l \varphi}$$

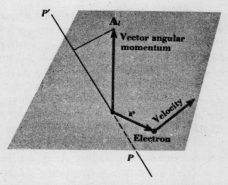

Fig. 14.4 Orbital vector angular momentum $\mathbf{A}_l = \mathbf{r} \times m\mathbf{v}$.

we obtain

$$\widehat{\mathbf{A}_l \cdot \mathbf{A}_l} R\Theta e^{im_l\varphi} = \hbar^2 \left[-\frac{1}{\Theta \sin\theta} \frac{\partial}{\partial\theta} \left(\sin\theta \frac{\partial\Theta}{\partial\theta} \right) + \frac{m_l^2}{\sin^2\theta} \right] R\Theta e^{im_l\varphi}$$

The quantity in the brackets is, however, just the right side of Eq. (14.6a), which has the value $l(l + 1)$. Thus $\widehat{\mathbf{A}_l \cdot \mathbf{A}_l}\psi = \hbar^2 l(l + 1)\psi$, so that the eigenvalues of $A_l^2 = \mathbf{A}_l \cdot \mathbf{A}_l$ are given by Eq. (13.18); $A_l^2 = l(l + 1)\hbar^2$, and the magnitude of the orbital angular momentum $\mathbf{A}_l$ has the possible values

$$A_l = \sqrt{l(l + 1)}\,\hbar \qquad l = 0, 1, 2, \cdots \tag{14.13}$$

According to quantum mechanics (and experiment!) the hydrogen electron in its ground state or any other state with $l = 0$ has no orbital angular momentum!

If the operator for angular momentum $(A_l)_z$ about the z axis as written in Eq. (13.16a) is applied to ψ_{nlm_l}, the result is

$$\widehat{(\mathbf{A}_l)}_z\psi_{nlm_l} = \frac{\hbar}{i} \frac{\partial}{\partial\varphi} \psi_{nlm_l} = m_l\hbar\psi_{nlm_l}$$

Thus the wave function ψ_{nlm_l} represents a state of the electron in the nuclear field in which its angular momentum about the axis of polar coordinates has the definite value

$$(A_l)_z = m_l\hbar \tag{14.14}$$

The angular momentum $(A_l)_z$ about the polar axis ranges in integral steps from a maximum of l units, each of magnitude $\hbar$ in one direction through zero to l units in the opposite direction. This is illustrated in Fig. 14.5 for the case in which $l = 2$, and the orbital angular momentum is $\sqrt{6}\,\hbar$. The vector representing $\mathbf{A}_l$ is shown in five alternate positions corresponding to $m_l = -2, -1, 0, 1, 2$. The orbital angular momentum $\mathbf{A}_l$ is allowed to take on only such orientations relative to a field defining the z axis such that the z component of $\mathbf{A}_l$ is an integer m_l times $\hbar$. The relationships involved here are sometimes referred to as *space quantization*.

If the operator representing a perpendicular component of the vector angular momentum $\mathbf{A}_l$ is applied to ψ_{nlm_l}, the result is not just ψ_{nlm_l} multiplied by a constant. It is a general consequence of the indeterminacy principle that, *whenever one component of a vector angular momentum has a definite value, other components do not*. Measurement

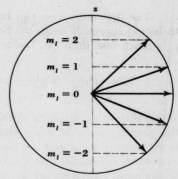

Fig. 14.5 For $l = 2$ the orbital-angular-momentum vector of magnitude $\sqrt{6}\,\hbar$ must be oriented such that its z component is $m_l\hbar$, where $m_l = 2, 1, 0, -1,$ or -2.

of a perpendicular component for an atom in the state represented by ψ_{nlm_l} would yield one of several alternative multiples of $\hbar$ (provided $l > 0$), with an average expectation of zero. This is an important novel feature of wave mechanics; for in classical mechanics $\mathbf{A}_l$ would have a fixed direction and all its components would have definite values (so long as external forces did not act on the system). The classical vector $\mathbf{A}_l$ might, however, have any orientation about the axis while its axial component had a given value $m_l\hbar$. Thus a quantum state is the analog not of one classical motion but of a class of motions. An absence of fixed orientation and of fixed values for perpendicular components would occur if the vector $\mathbf{A}_l$ were revolving about the axis at a fixed angle. It sometimes assists the memory to imagine such a "precession" to be occurring. It should not be forgotten, however, that in wave mechanics the indefinitenesses in question hold even in the absence of all external forces.

14.6 Relativistic Effects and Electron Spin

The theory described so far has been a nonrelativistic one suited to cases of small energy only. Like newtonian mechanics, it requires modification to bring it into harmony with the principles of relativity. There are two principal correction terms, one associated with introducing the relativistic expression for kinetic energy and another arising from electron spin.

The first correction can be calculated approximately from a perturbation approach, using the Schrödinger equation.   If the momentum p is much less than mc, the relativistic kinetic energy K is given by

$$K = (p^2c^2 + m^2c^4)^{\frac{1}{2}} - mc^2 = \frac{p^2}{2m} - \frac{p^4}{8m^3c^2} + \cdots \qquad (14.15)$$

When this value for K is substituted in the hamiltonian of Sec. 13.6, the quantity $p^4/8m^3c^2$ is a perturbation term.   When the operator $\hbar\nabla/i$ is substituted for p and the three-dimensional analog of Eq. (13.41) is evaluated, it is found that the relativistic correction ΔE_r must be added to the unperturbed energy E_n, where

$$\Delta E_r = E_n' - E_n = \frac{Rhc\alpha^2Z^4}{4n^4}\left(3 - \frac{4n}{l + \frac{1}{2}}\right) \qquad (14.16)$$

where R is the Rydberg constant and α the fine-structure constant.

In addition to this relativistic correction the spin of the electron, introduced by Goudsmit and Uhlenbeck (Sec. 9.9) before the advent of wave mechanics, appeared as an astonishing feature of a relativistic quantum theory developed in 1928 by Dirac.   He showed that the most natural way to bring wave-mechanical theory into harmony with the theory of relativity is to adopt quite a different wave equation and that when this is done, the new equation leads automatically to effects equivalent to those deduced from electron spin.   It need not be assumed that the electron is spinning or turning on its axis.   According to this theory of Dirac's, the electron does behave as if it had an internal angular momentum of the sort just described and an associated magnetic moment.   Furthermore, if a wave packet is formed representing an electron, there is in it, in general, something like a closed current, or eddy, of probability that can be regarded as analogous to an actual spinning motion.

In the Dirac theory there are four wave functions for an electron instead of one.   The complete Dirac theory is seldom used, however, in spectroscopic calculations.   A simplified approximate form usually suffices, in which only two functions, Ψ_1 and Ψ_2, are used, one for each direction of the spin moment.   For a first approximation, the effect of the spin magnetic moment upon the energy may be ignored.   When this is done, the separate wave functions for a one-electron atom may be simply the functions described in the last section, except that a fourth quantum number m_s is now to be added; a time-free wave function including spin may be denoted by the symbol $\psi_{nlm_lm_s}$.   Since m_s may have either of two values, the effect of spin in this approximation is to double the number of quantum states.

14.7 The Spin-Orbit Interaction

By virtue of its spin, the electron has a magnetic moment (Sec. 9.9) of magnitude $\mu_s = -2(e/2m) \sqrt{(\frac{1}{2})(\frac{3}{2})}\,\hbar$. We may then think of the electron as a small magnetic dipole of moment $\mathbf{\mu}_s$ directed opposite the spin. For simplicity consider a one-electron atom with nuclear charge Z. If the electron is moving relative to the nucleus, from the point of view of a reference frame in which the electron is momentarily at rest, the nucleus is moving with velocity $\mathbf{v}$ and produces therefore a magnetic field $\mathbf{B}_l$ at the location of the electron. By Eq. (3.34) the magnetic field seen by an electron by virtue of its own orbital motion through the electric field $\mathbf{E}_N$ of the nucleus is

$$\left| \mathbf{B}_l \right| = \left| -\frac{1}{c^2} (\mathbf{v} \times \mathbf{E}_N) \right| = \frac{A_l}{mec^2 r} \frac{dP}{dr} \tag{14.17}$$

where $P = -Ze^2/4\pi\epsilon_0 r$ and we have made use of the fact that the orbital angular momentum $\mathbf{A}_l = \mathbf{r} \times m\mathbf{v}$.

The energy of a magnetic dipole in a magnetic field varies with orientation, being least for the dipole moment parallel to the field. The orientation energy, referred to as energy zero when the magnetic moment and field are mutually perpendicular, is given by

$$E_{\mu B} = -\mu B \cos(\mu, B) = -\mathbf{\mu}_s \cdot \mathbf{B}_l$$

For the spinning electron in the field of the nucleus, we have by Eqs. (9.21) and (14.17),

$$E_{\mu B} = \frac{Ze^2}{4\pi\epsilon_0 m^2 c^2 r^3} \mathbf{A}_s \cdot \mathbf{A}_l \tag{14.18}$$

The calculation above is still incomplete because the electron is being accelerated, and hence the frame of reference in which it is momentarily at rest changes continually. It was shown by Thomas that this circumstance reduces $E_{\mu B}$ to one-half of the value given by Eq. (14.18). When this reduced value is introduced in the hamiltonian and the perturbation calculation is made, it is found that there is a shift ΔE_{ls} in energy due to the spin-orbit interaction of

$$\Delta E_{ls} = \frac{1}{2} \frac{Z^4 e^2 \mathbf{A}_s \cdot \mathbf{A}_l}{4\pi\epsilon_0 m^2 c^2 a_B^3 n^3 l(l+\frac{1}{2})(l+1)} \tag{14.19}$$

where a_B is the Bohr radius ($l \neq 0$).

Fig. 14.6 As a result of the spin-orbit interaction, A_l and A_s precess around A_j, which remains fixed in space in the absence of torques acting on it.

The torque $\mathbf{\mu}_s \times \mathbf{B}_l$ does not change the magnitude of the spin momentum $\mathbf{A}_s$ but alters its direction. The resultant angular momentum, $\mathbf{A}_l + \mathbf{A}_s = \mathbf{A}_j$, remains fixed both in magnitude and in direction in the absence of an imposed magnetic field; meanwhile the vectors $\mathbf{A}_l$ and $\mathbf{A}_s$ precess around it (Fig. 14.6).

14.8 The Quantum Number j

The resultant angular momentum $\mathbf{A}_j$ can be shown to possess only the quantized values $j(j+1)\hbar^2$ for its square, where $j = l \oplus s = l \pm \frac{1}{2}$. Only positive values of j are allowed, so that for $l = 0$, $j = \frac{1}{2}$ only. The symbol $\oplus$ is used to represent a quantized vector addition such that $a \oplus b$ has the values $a + b,\ a + b - 1, \ldots, |a - b|$. Thus the resultant (spin + orbital) angular momentum for the electron has the magnitude

$$A_j = \sqrt{j(j+1)}\,\hbar \qquad \text{with } j = |l \pm \tfrac{1}{2}| \qquad (14.20)$$

The z component of $\mathbf{A}_j$ must have one of the quantized values

$$(A_j)_z = m_j\hbar \qquad -j \le m_j \le j \qquad (14.21)$$

where m_j takes on the integrally spaced values $j,\ j - 1, \ldots,\ -j$. Thus for $j = \frac{3}{2}$, $m_j = \frac{3}{2}, \frac{1}{2}, -\frac{1}{2}, -\frac{3}{2}$.

Before the spin-orbit interaction was included in the hamiltonian, the quantum states of the single-electron atom could be specified in terms of the four quantum numbers n, l, m_l, and m_s; for these states fixed values are assigned to the z components of $\mathbf{A}_s$ and $\mathbf{A}_l$ separately. With the spin-orbit interaction taken into account the true quantum states are characterized by a fixed value of $\mathbf{A}_j$ and of $(A_j)_z$. The appropriate four quantum numbers are then n, l, j, and m_j. The indefiniteness of all components of $\mathbf{A}_l$ and $\mathbf{A}_s$ when $\mathbf{A}_j$ is quantized is an example of a general principle. When the vector sum of two or more angular momenta is quantized, the individual momenta cease to have any definite components, although their squares retain definite magnitudes, for example, $l(l + 1)\hbar^2$. This corresponds to the fact that in classical mechanics the individual vectors may assume various directions without disturbing their vector sum.

The energy depends now on j, as well as on n and l. Thus each nl level with $l > 0$ is split by the spin-orbit effect into two. From the fact that $\mathbf{A}_j = \mathbf{A}_l + \mathbf{A}_s$, we have $\mathbf{A}_j \cdot \mathbf{A}_j = (\mathbf{A}_l + \mathbf{A}_s) \cdot (\mathbf{A}_l + \mathbf{A}_s) = A_l{}^2 + A_s{}^2 + 2\mathbf{A}_s \cdot \mathbf{A}_l$ or

$$\mathbf{A}_s \cdot \mathbf{A}_l = A_s A_l \cos (A_s, A_l) = \tfrac{1}{2}(A_j{}^2 - A_l{}^2 - A_s{}^2)$$
$$= \frac{j(j + 1) - l(l + 1) - s(s + 1)}{2} \hbar^2 \quad (14.22)$$

Substitution of Eq. (14.22) into Eq. (14.19) leads to the result that the shift ΔE_{ls} in energy level due to the spin-orbit interaction is $(l \neq 0)$

$$\Delta E_{ls} = \frac{e^2\hbar^2}{16\pi\epsilon_0 m^2 c^2 a_B{}^3} \frac{Z^4[j(j + 1) - l(l + 1) - s(s + 1)]}{n^3 l(l + \tfrac{1}{2})(l + 1)} \quad (14.23)$$

where the first fraction is $Rhc\alpha^2/2$, or 3.62×10^{-4} eV. For states with $l > 0$ the spin-orbit interaction splits each energy level into two, raising the energy for $j = l + \tfrac{1}{2}$ and lowering it for $l - \tfrac{1}{2}$. To distinguish between these levels, one commonly adds the value of j as a subscript to the letter used to denote l; thus the two levels arising from the splitting of the $3p$ level are written $3p_{\frac{1}{2}}$ and $3p_{\frac{3}{2}}$.

Since s states have no orbital angular momentum $(l = 0)$, there is no spin-orbit correction for an s level. However, in the relativistic theory the wave equation contains other small terms besides those giving rise to the spin-orbit effect. These terms cause the energy to vary somewhat with l; in particular, there is a so-called *special spin* correction for $l = 0$ states. As a consequence of the corrections neglected here, the fine structure of the hydrogen levels is complex. It has been the subject of intensive experimental and theoretical research which has been fruitful in elucidating several subtleties of nature.

14.9 Spatial Degeneracy of the Wave Function

For a particle in free space with no force field defining the z direction, the axis for the polar coordinates can be drawn in any desired direc-

374 *Introduction to Modern Physics*

tion.   Thus the set of quantum states and of associated wave functions is in some degree a matter of arbitrary choice.   This arbitrariness of the wave functions is another aspect of the general phenomenon of degeneracy, and its mathematical aspects deserve a short further discussion.

Whenever several wave functions are associated with the same value of the energy E, any linear combination of them is another possible wave function for the same energy.   Thus, let ψ_1, ψ_2, ψ_3, . . . , ψ_k be solutions of Eq. (14.3a) for $E = E_n$.   Then the linear combination ψ' where

$$\psi' = c_1\psi_1 + c_2\psi_2 + c_3\psi_3 + \cdot \cdot \cdot + c_k\psi_k \qquad (14.24)$$

c_1, c_2, c_3, . . . , c_k being any constants, is also a solution.   Thus ψ' is as valid a wave function to represent a quantum state for the energy E_n, as is any one of the functions ψ_1, ψ_2, . . . , ψ_k.   The possible wave functions associated with E_n are thus infinite in number.   It is possible, however, to find a number k of them in terms of which all others can be written as linear combinations.   These k functions can be chosen in infinitely many ways, but their number is always k.

Degeneracy of the quantum states has an exact analog in classical theory.   For example, a drumhead can vibrate at a certain one of its higher frequencies with any diameter as a nodal line, such as A_1, B_1, B_2, etc., in Fig. 14.7.   The mode of vibration at this frequency is thus not unique.   It is possible, however, to choose any two modes as fundamental modes, for example, A_1 and A_2 or B_1 and B_2, and to write the displacement during any vibration at the frequency in question as resulting from the superposition of vibrations of suitable

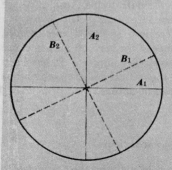

Fig. 14.7 Possible nodal lines of a vibrating drumhead.

amplitudes in the two chosen fundamental modes. The type of vibration that actually occurs in a given case depends upon the manner in which the drumhead is set into vibration. Similarly, when an atom is in a degenerate state characterized by a definite energy, the exact quantum state or mixture of states that it occupies depends upon its past history.

In the case of the one-electron atom, if the axis of polar coordinates is taken in a different direction, the factor $R_{nl}(r)$ in the expression for ψ_{nlm_l} is the same as before, but the geometrical significance of the directional factor $e^{im_l\phi}P_l^m(\cos\theta)$ is different. It can be shown, however, that the new directional factor can be written as a linear combination of the directional factors relative to the original axis and, in fact, in terms of those factors, $2l + 1$ in number, which are labeled with the same value of l.

The infinite variety of the quantum states for given n and l corresponds to the infinity of different possible orientations of a Bohr orbit, and it is closely related to the isotropy of space.

Appendix 14A - Solutions of the Radial Equation

To solve Eq. (14.8) for $R(r)$, it is convenient to set

$$\alpha = \frac{\sqrt{-8m_rE}}{\hbar} \qquad \lambda = \frac{Ze^2}{4\pi\epsilon_0\hbar}\sqrt{\frac{m_r}{-2E}} \qquad \text{and} \qquad \rho = \alpha r$$

Then Eq. (14.8) becomes

$$\frac{d^2R}{d\rho^2} + \frac{2}{\rho}\frac{dR}{d\rho} + \left[\frac{\lambda}{\rho} - \frac{1}{4} - \frac{l(l+1)}{\rho^2}\right]R = 0 \qquad (14A.1)$$

First we seek an asymptotic solution for $\rho \to \infty$. In this case

$$\frac{d^2R}{d\rho^2} - \frac{R}{4} = 0 \qquad \text{and} \qquad R(\rho) = B_1e^{\rho/2} + Be^{-\rho/2}$$

Since the positive exponent leads to $R \to \infty$ as $\rho \to \infty$, we set $B_1 = 0$ and assume that $R(\rho)$ can be written in the form

$$R(\rho) = e^{-\rho/2}F(\rho) \qquad (14A.2)$$

where $F(\rho)$ must satisfy the equation

$$\frac{d^2F}{d\rho^2} + \left(\frac{2}{\rho} - 1\right)\frac{dF}{d\rho} + \left[\frac{\lambda - 1}{\rho} - \frac{l(l + 1)}{\rho^2}\right]F = 0 \qquad (14A.3)$$

One can solve this equation by the power-series method, assuming that

$$F(\rho) = \rho^s(a_0 + a_1\rho + a_2\rho^2 + \cdots) = \rho^s \sum_{j=0}^{\infty} a_j\rho^j \qquad (14A.4)$$

where ρ^s appears ahead of the summation to take care of the singularity at $\rho = 0$. Substitution of Eq. (14A.4) into Eq. (14A.3) yields

$$\sum_{j=0}^{\infty} [(s + j)(s + j - 1)a_j\rho^{s+j-2} + 2(s + j)a_j\rho^{s+j-2}$$
$$- (s + j)a_j\rho^{s+j-1} - a_j\rho^{s+j-1} + \lambda a_j\rho^{s+j-1} - l(l + 1)a_j\rho^{s+j-2}] = 0$$

or

$$[s(s + 1) - l(l + 1)]a_0\rho^{s-2} + \sum_{j=1}^{\infty} \{[(s + j)(s + j + 1) - l(l + 1)]a_j$$
$$- (s + j - \lambda)a_{j-1}\}\rho^{s+j-2} = 0$$

Since the coefficient of each power of ρ must vanish separately, it follows that

$$s(s + 1) - l(l + 1) = 0 \qquad \textit{indicial equation} \qquad (14A.5a)$$
$$(s + j - \lambda)a_{j-1} = [(s + j)(s + j + 1) - l(l + 1)]a_j \qquad (14A.5b)$$

The indicial equation requires that $s = l$; the solution $s = -(l + 1)$ is excluded since s must be positive if R is not to have a singularity at $r = 0$. Therefore the *recurrence relation* (14A.5b) becomes

$$a_j = \frac{l + j - \lambda}{(l + j)(l + j + 1) - l(l + 1)} a_{j-1}$$
$$j = 1, 2, 3, \ldots \quad (14A.6)$$

If R is to go to zero at $r = \infty$, the series for $F(\rho)$ must terminate, which occurs if $\lambda = l + j$. Thus, if $\lambda = n$ is an integer greater than l, the series ceases after $n - l$ terms and R goes to zero at infinity. But λ is defined to be

$$\lambda = \frac{Ze^2}{4\pi\epsilon_0\hbar}\sqrt{\frac{m_r}{-2E}}$$

Therefore, the Schrödinger equation is satisfied only for those values of E given by

$$E_n = -\frac{m_r e^4 Z^2}{32\pi^2 \epsilon_0^2 \hbar^2 n^2} = -\frac{m_r e^4 Z^2}{8\epsilon_0^2 h^2 n^2} \tag{14A.7}$$

Insertion of this value of E_n into the definition of α yields

$$\alpha_n = \sqrt{\frac{8 m_r^2 e^4 Z^2}{32\pi^2 \epsilon_0^2 \hbar^2 n^2}} \frac{1}{\hbar} = \frac{2 m_r e^2 Z}{4\pi\epsilon_0 \hbar^2 n} = \frac{2Z}{n a_B} \tag{14A.8}$$

where a_B is the Bohr radius ($= 4\pi\epsilon_0 \hbar^2/m_r e^2 = 0.529 \times 10^{-10}$ m).

The solutions $R_{nl}(r)$ of Eq. (14.8) consist of the product of (1) a normalization constant C_{nl}, (2) $e^{-\alpha_n r}$, and (3) a polynomial in $\alpha_n r$ of $n - l$ terms. This polynomial in turn is the product of $(\alpha_n r)^l$ and the *associated Laguerre polynomial* $L_{n+l}^{2l+1}(\alpha_n r)$.

Problems

1. Show that the most probable value of r for the ground state of hydrogen is the Bohr radius a_B and that the expectation value of r is $3a_B/2$.

2. Calculate the expectation values of the potential energy P and the kinetic energy K for a hydrogen atom in the ground state. Compare these with the values of P and K in the Bohr model.

3. Show that the sum of the values of $\psi^*\psi$ for the six ($m_l = 1, 0, -1$; $m_s = +\frac{1}{2}, -\frac{1}{2}$) $2p$ states of hydrogen is spherically symmetric, i.e., is a function of r alone.

4. Classically how far could a hydrogen electron with the energy characteristic of the ground state get from the proton? Show that the quantum-mechanical prediction of finding the electron of a hydrogen atom in the ground state at this classical limit or farther from the proton is almost 25 percent (actually $13e^{-4}$).

5. Show that the operator for the square of the orbital angular momentum, which is given in Eq. (13.17) in cartesian coordinates, is given by Eq. (14.12) in spherical polar coordinates.

6. Show that the commutator operator $\hat{A}_x \hat{A}_y - \hat{A}_y \hat{A}_x$ is equal to $i\hbar \hat{A}_z$.

7. Show that the hamiltonian operator for the one-electron atom

$$-\frac{\hbar^2}{2m}\nabla^2 - \frac{Ze^2}{4\pi\epsilon_0 r}$$

commutes with the operator for the z component of the angular momentum; i.e., show that $\hat{H}(\hat{A}_l)_z - (\hat{A}_l)_z \hat{H} = 0$.

8. Show that for a Bohr orbit the ratio of the spin-orbit interaction energy to the electrostatic potential energy P is of the order of $|P/2mc^2| = |E_n/mc^2|$.

9. A solid cylinder of mass m and radius R has a charge q spread uniformly over the curved sides of the cylinder. If this *checker model* of the electron rotates about its axis with angular velocity ω, calculate the gyromagnetic ratio.

10. By operating directly on the function $f(\theta) = 5 \cos^3 \theta - 3 \cos \theta$ with operator $\hat{A}_l{}^2$ and $(\widehat{A}_l)_z$ in spherical coordinates, show that $f(\theta)$ is an eigenfunction of both operators. Find the corresponding eigenvalues.

11. Show that $15 \sin^2 \theta \cos \theta\, e^{2i\varphi}$ is an eigenfunction of the angular momentum operators $\hat{A}_l{}^2$ and $(\widehat{A}_l)_z$ by direct substitution in Eqs. (13.16) and (14.12). What are the corresponding eigenvalues?

12. After a series of measurements on the $n = 1$ state of hydrogen the angular part of the wave function has the form $\sqrt{3/4\pi} \sin \theta \sin \varphi$.

  (a) Compute $\langle (A_l)_z \rangle$.

  (b) Find the probability that a measurement of $(A_l)_z$ yields the value $\hbar$.

13. Show that if $pc \ll m_0 c^2$, the relativistic correction [Eq. (14.16)] is given by

$$\Delta E_r = - \int \Psi_n^* \frac{\hbar^4}{8m^2c^2} \nabla^4 \Psi_n \, d\mathcal{V}$$

Hint: Write the hamiltonian as $H = \sqrt{p^2c^2 + m_0{}^2c^4} - m_0c^2 + P$, expand as series in p, and apply first-order perturbation theory.

14. Use first-order perturbation theory to calculate the shift in the ground state of the hydrogen atom due to the finite radius of the proton if it is assumed that the proton charge is distributed uniformly throughout a sphere of radius $a = 1.2 \times 10^{-15}$ m.

  Ans: $\Delta E \approx 7 \times 10^{-9}$ eV

15. An electron is confined inside a sphere of radius a by a potential $P(r) = 0$ for $r \leq a$ and $P(r) = \infty$ for $r > a$. Show that the eigenfunctions for states with zero angular momentum are given by $\psi(r) = \text{const} (\sin kr)/kr$. Determine the relation between k and the energy, and show that the energy eigenvalues are $n^2h^2/8ma^2$.

16. If $P(r) = 0$ for $r \leq a$ and $P(r) = P_0$ for $r > a$, find the smallest value of P_0 for which there is a bound state with no angular momentum and energy P_0.

  Ans: $h^2/32ma^2$

17. The three-dimensional-oscillator problem (Chap. 13, Prob. 10) can also be solved in spherical polar coordinates. Take $P = \frac{1}{2}m\omega^2 r^2$ and solve the Schrödinger equation in these coordinates. What are the eigenvalues for angular momentum?

18. The wave function ψ_{211} represents a circulating current. Making use of Eq. (12.41), calculate the magnetic moment associated with this current and show it is 1 Bohr magneton.

19. Set up and solve the Schrödinger equation in terms of r and φ for the two-dimensional analog of the hydrogen atom, taking the potential energy to be $P = -e^2/4\pi\epsilon_0 r$. Show that the radial distribution function has its maximum at one-fourth of the Bohr radius and that the ground-state energy of the planar atom is 4 times that of the hydrogen atom. [See Zaslow and Zandler, *Am. J. Phys.*, **35**: 1118 (1967).]

20. When an electron moves in an "orbit" around a nucleus, the axes of a frame in which the electron is instantaneously at rest appear to precess relative to a frame fixed to the nucleus with an angular frequency $\omega_r = -\mathbf{v} \times \mathbf{a}/2c^2$, where $\mathbf{v}$ is the velocity and $\mathbf{a}$ the acceleration of the electron relative to the nucleus. [This is the Thomas precession (Sec. 14.7).] Derive this equation. *Hint:* Consider three inertial frames: S, fixed on the nucleus, S', fixed on the electron and moving with speed v along the x axis of S, and S'', moving with speed $a\,dt$ along the y axis of S'.

chapter fifteen

Atomic Structure

In Chap. 14 we considered atoms containing a single electron. We now discuss properties of atoms (and molecules) that contain two or more electrons. The Rutherford-Bohr model of the atom was built out of raw material from the field of physics. Originally, however, the atom was the child of chemistry. The ultimate theory of atomic structure must explain the facts of chemistry, which are just as cogent as the facts of physics. In the present chapter, we deal with the periodic table of elements; in later chapters we return to the study of spectra. The two topics are closely related, for wave mechanics furnishes the key to the theoretical understanding of both.

15.1 The Pauli Exclusion Principle

If there were no repulsion between the electrons, the theory of complex atoms would be very simple. Each electron, being subject only to the field of the nucleus, might be in the same sort of electronic state as if it were alone in the atom. Actually, however, the effect of electronic repulsion must be large, particularly for an electron relatively far from the nucleus. However, as a first approximation we may assume that the effect on any one electron of all the other electrons is to reduce the *effective* charge below the nuclear charge Ze. The total negative charge on all electrons in a neutral atom is numerically equal to the positive charge on the nucleus, so that at points far outside the atom the electrons screen off the nuclear field and the net field is zero. Inside the atom, as the nucleus is approached, the screening becomes less and less, until near the nucleus the field of force becomes essentially that due to the nucleus alone. Even the innermost electrons, however, are affected by the presence of the others, for the work required to remove any electron to infinity is considerably diminished by the repulsion of the others.

Suppose, for example, that the total negative charge on the electrons were spread out continuously with spherical symmetry, its (negative) density ρ being a function only of the distance r from the nucleus. Then, by a familiar result in electrostatics, the charge in a shell of radius r_1 and thickness dr_1 (Fig. 15.1), of magnitude $4\pi\rho r_1^2 \, dr_1$, would produce a potential of magnitude $(\rho r_1^2 \, dr_1)/\epsilon_0 r$ at any external point r and a corresponding electric field. Inside the shell it would produce no electric vector but only a uniform potential $(\rho r_1^2 . dr_1)/r_1\epsilon_0$. The total electrostatic potential at a distance r from the nucleus would thus be

$$\rho(r) = \frac{Ze}{4\pi\epsilon_0 r} + \frac{1}{r}\int_0^r \frac{\rho r_1^2 \, dr_1}{\epsilon_0} + \int_r^\infty \frac{\rho r_1 \, dr_1}{\epsilon_0} \tag{15.1}$$

Here e is the numerical electronic charge; Ze is the positive charge on the nucleus and is related to ρ by the equation $\int_0^\infty 4\pi\rho r_1^2 \, dr_1 = -Ze$.

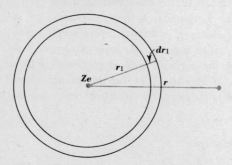

Fig. 15.1 A shell of radius r_1 and thickness dr_1 centered on the nucleus.

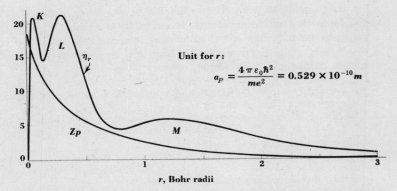

Fig. 15.2 The Hartree field and charge-cloud density for argon. The potential energy of an additional electron inside the argon atom would be $-eZ_p/4\pi\epsilon_0 r$, where Z_p is shown in the lower curve. K, L, M indicate the approximate locations of the $n = 1$, 2, and 3 shells, respectively.

This classical picture suggested the following simplified wave-mechanical approach to the problem, which has been very useful. The total electronic charge is imagined to be spread out in a spherically symmetrical charge cloud analogous to that described for one electron in Sec. 14.4, and the combined field of potential $P(r)$ due to the nucleus and to this electronic charge is calculated by Eq. (15.1). As an example, the calculated radial density η_r of the cloud charge for a neutral argon atom is plotted in Fig. 15.2; P is also plotted in a certain manner, as explained under the figure.

The field as thus obtained may be used as it stands in calculating the deflection of an external electron such as a cathode ray passing through the atom. An internal electron, however, will in reality be repelled only by the other electrons, not by itself. To correct for this fact, we omit the part of the charge cloud that is due to the internal electron under consideration.

For a first approximation to serve as the starting point for perturbation theory, the electrons are then assumed to move *without repelling each other* in this modified central field, the *average* effect of their repulsion having been included in the field. Because of the spherical symmetry of the field, the electronic wave functions are almost as simple as in the hydrogen atom. In classical mechanics, the vector angular momentum would remain constant. Analogously, in wave mechanics the electronic states are characterized by the same angular-momentum properties as in a one-electron atom; the wave functions contain, in fact, the same mathematical functions of θ and ϕ that are represented in Eqs. (14.11) for a single electron, only the radial factor R_{nl} being different. The radial factor is still characterized, however, by a number n, and, as in a coulomb field, it has $n - l$ numerical maxima, separated by zeros if $n - l > 1$. The electronic states may thus be characterized by four quantum numbers n, l, m_l, and m_s, where:

1. n is the *principal* (or radial) *quantum number* which is restricted to positive integer values 1, 2, 3, 4, etc.
2. l is the *orbital* (or azimuthal) quantum number, which may take on the values 0, 1, 2, . . . , $n - 1$. The orbital angular momentum of the electron has the magnitude $A_l = \sqrt{l(l + 1)}\,\hbar$.
3. m_l is the *magnetic quantum number* and takes on the positive and negative integer values l, $l - 1$, . . . , 0, . . . , $-l$. The z component of the orbital angular momentum is given by $(A_l)_z = m_l \hbar$.
4. m_s is the spin quantum number, which is $+\frac{1}{2}$ or $-\frac{1}{2}$. The z component of the spin angular momentum of the electron is $m_s \hbar$.

An assignment of all electrons to electronic quantum states will then define, in the zero order of perturbation theory, a quantum state for the atom as a whole. In assigning the electrons, however, account must be taken of a most important principle enunciated by W. Pauli in 1925 before the development of wave mechanics. In its early formulation the *Pauli exclusion principle* is:

No two electrons in any one atom can be in the same quantum state.

Since each electronic quantum state may be characterized by four quantum numbers n, l, m_l, and m_s (or n, l, j, and m_j), we may say:

No two electrons in any one atom may have the same four quantum numbers.

For discussing atomic structure it is convenient to use the restrictive form of exclusion principle presented above. However, it is possible to formulate the Pauli principle in a much more general form. In quantum mechanics identical quantum particles, such as electrons, cannot be distinguished from one another if they occupy a region such that their wave functions overlap. For example, within a helium atom we cannot say "electron 1 is here, and electron 2 is there." The wave function for the system predicts the probability of finding *an* electron here and *another* one there, but it cannot predict which electron is which. This wave function depends on the general coordinates x_i and x_j of the electrons, where x_i includes the x, y, z, and *spin* coordinates. Since the probability distribution is unchanged by exchanging the two electrons, we have

$$|\Psi(x_i, x_j)|^2 = |\Psi(x_j, x_i)|^2$$

from which we have

$$\Psi(x_i,x_j) = \pm \Psi(x_j,x_i)$$

It is an experimental fact that for all spin-$\frac{1}{2}$ particles (electrons, protons, neutrons, muons, etc.) the negative sign must be chosen, while for particles with zero or integer spin the positive sign must be chosen. In wave-mechanical language the Pauli exclusion principle may be stated in the form:

The wave function for the state of a system containing two or more electrons must be completely antisymmetric in the generalized coordinates of the electrons; i.e., the sign of the wave function must change if any two electrons are interchanged.

The wave-mechanical treatment of complex atoms is beyond the scope of this book, but in Appendix 15A some important features of that topic are illustrated, and a way of introducing the spin coordinate is presented.

15.2 Shells and Subshells

In a neutral atom in its ground state the Z electrons occupy the Z lowest energy states available. The energy of an electron in the modified central field depends, as in a coulomb field, chiefly upon n, but in this more general case it varies also somewhat with l, increasing (becoming less negative) as l is increased because of the penetrating orbit effect (Sec. 15.6). For the moment we shall ignore the fine-structure splitting of the (n,l) energy levels; in this approximation the energy does not depend on m_l and m_s.

The electrons in an atom normally fall into distinct groups differing in the value of the quantum number n. Electrons having the same n are said to belong to the same *shell*. Each shell is divided into sub-shells according to the values of l. Any subshell may contain up to $2(2l + 1)$ electrons, but no more, since m_l can range from the value l down through zero to $-l$, making $2l + 1$ different values; m_s may be either $\frac{1}{2}$ or $-\frac{1}{2}$. The numbers of electrons in the various types of closed (or full) subshells, together with their symbols, are:

l	0	1	2	3	4	5	6	7
Symbol	s	p	d	f	g	h	i	k
Number	2	6	10	14	18	22	26	30

Thus the full $4d$ subshell in tungsten (with $n = 4$, $l = 2$) contains 10 electrons.   The maximum possible number of electrons in a whole shell of any atom is $2n^2$.   To specify a zero-order quantum state for the atom, it suffices to specify a sufficient number of sets of the quantum numbers n, l, m_l, m_s, with no two sets alike in all four numbers, to accommodate all the electrons in the atom.   For the normal, or lowest-energy state of the atom, the electrons must be assigned to states of the lowest energy possible.   The first two electrons go into $1s$ states with $n = 1$, $l = 0$, $m_l = 0$, and $m_s = \pm\frac{1}{2}$.   Since no other states are possible with $n = 1$, the next two electrons go into $2s$ states, with much higher energy, and the next six into $2p$ states, with slightly higher energy still.   The next two go into $3s$ states with considerably higher energy, and so on.   A specification of n and l for all the electrons bound to a nucleus is called the *electronic configuration* of the atom (or ion).

The electrons in any completely filled shell or subshell contribute *zero* resultant angular momentum, both orbital and spin, to the angular momentum of the atom as a whole.   In such a closed subshell, for every electron with z component of spin angular momentum $+\frac{1}{2}\hbar$ ($m_s = +\frac{1}{2}$) there is another electron with z component $-\frac{1}{2}\hbar$.   Thus the resultant spin angular momentum for the closed subshell is zero.   Similarly for each electron with z component of orbital angular momentum $m_l\hbar$ there is one with z component $-m_l\hbar$.   The resulting charge distribution for a closed subshell is spherically symmetric as well.   It is these facts which make hydrogen-like wave functions a reasonable first approximation for the many-electron atom.   Any resultant orbital angular momentum and spin angular momentum and any asymmetry in charge distribution from the electron distribution in an atom arises from electrons which are in partially filled subshells.   The vast majority of the electrons in many-electron atoms are in closed subshells.

The shells are often referred to by the capital letters which became attached to them in the early days of x-ray study.   These designations are shown below and also, in part, in Table 15.1, in which the numbers and notation for the first three shells are summarized:

n	1	2	3	4	5	6	7
X-ray designation	K	L	M	N	O	P	Q

Table 15.1   Electron shells and subshells

Shell	K	L		M		
n	1	2		3		
Subshell, l	0	0	1	0	1	2
Letter designation	s	s	p	s	p	d
Number of electrons in subshell or shell $\Big\{$	2	2	6	2	6	10
	2	8		18		

15.3 The Periodic Table

The concept of atomic number was firmly established by the work of Moseley and others on x-rays. An atomic number Z may be unambiguously assigned to each element. Thus the periodic table, drawn up by the chemists primarily on the basis of chemical facts, is established with the assurance that there are no *unknown gaps* in it. The value of Ze represents the charge on the nucleus; it is also equal to the number of electrons surrounding the nucleus when the atom is neutral. The chemical properties of an element depend upon the number of the circumnuclear electrons in a neutral atom. Some understanding of the periodic table may be obtained by determining the arrangement of the electrons in the atom and the physical and chemical properties that follow from this configuration.

A bird's-eye view of the chief features to be explained is best gained by studying the arrangement of atoms in the periodic table inside the back cover. A striking feature is the recurrence of a *noble*, or "inert," *gas*, forming a series of turning points in the progression of the elements. It was pointed out by Rydberg that the values of Z for the noble gases could be expressed by a simple numerical series:

$$Z = 2(1^2 + 2^2 + 2^2 + 3^2 + 3^2 + 4^2 + \cdots)$$

viz., helium, $2 \times 1^2 = 2$; neon, 10; argon, 18; krypton, 36; and xenon, 54.

The noble gases show little tendency for their atoms to join with other atoms in chemical combination. The elements standing in the table on either side of a noble gas, on the other hand, are strongly active chemically and have contrasting properties. Those closely *following* a noble gas, like lithium, beryllium, potassium, calcium, and so on, are metallic and strongly electropositive; they readily form *positive* electrolytic ions. Furthermore, their maximum valence in chemical compounds is equal to the number of steps by which they lie beyond the noble gas, e.g., 1 for sodium, 2 for magnesium, 3 for aluminum; sodium forms univalent ions in solution; magnesium, bivalent; aluminum, trivalent. The elements closely *preceding* a noble gas, on the contrary, are electrically nonconducting, perhaps even gaseous, and strongly electronegative; they tend to form *negative* electrolytic ions, alone or in combination with other atoms, and they exhibit chemical valence equal to the number of steps by which they precede the noble gas in the table. An element of either of these two kinds scarcely combines with another element of the same kind; but an element closely preceding a noble gas in the table combines readily with any element closely following a noble gas.

These facts invite the conclusion that there is something very peculiar about the arrangement of the electrons of a noble gas. If we make the reasonable assumption that the chemical activity of an atom is conditioned somehow upon the magnitude of its external electric field, we may conclude that the atoms of the noble gases must be surrounded by very weak fields. If so, there should be little tendency for the atoms of these gases to combine into molecules or to condense into liquid or solid form; it is a fact that the noble gases are composed of monatomic molecules and also have very low boiling and freezing points.

The properties of the elements adjacent to the noble gases are then accounted for if we suppose that the arrangement of the outer electrons in an atom of a noble gas is an especially *stable* one, i.e., an arrangement of especially low energy. An atom of an element following a noble gas in the table will then contain one or two extra electrons outside a noble-gas core; e.g., sodium has one and magnesium two electrons outside of a neon core; and these extra electrons may well be comparatively easy to detach. The tendency of such atoms to form positive ions would thus be explained. Furthermore, in the solid state these extra *valence electrons* may easily come loose under the attraction of neighboring atoms, functioning, therefore, as free electrons, so that the elements in question ought to be good conductors of electricity, which they are.

An element such as chlorine, on the other hand, could arrange its electrons as they are arranged in argon if it had one more electron. Elements closely preceding a noble gas in the periodic table might exhibit a tendency to pick up an extra electron, thereby forming a negative ion. Some of these elements do, in fact, form negative ions that are more stable, i.e., have lower energy, than the neutral atom; this is true of the halogens and of oxygen and sulfur. In the solid state, such atoms would contain no electrons with a tendency to become free; thus, the absence of electrical conductivity in these elements would be accounted for.

These ideas furnish an explanation for the formation of a compound such as sodium chloride. In combining, the sodium atom loses 1 electron, its remaining 10 electrons then forming the stable configuration that is characteristic of neon (but, doubtless, somewhat more compressed because the nuclear charge of sodium is higher). The chlorine atom adds the electron lost by the sodium to its own 17, making 18 electrons arranged in the stable argon configuration (but slightly expanded). The electrostatic attraction of the two ions thus formed then binds them tightly together into a molecule. When the molecule of sodium chloride thus formed is put into water, the attraction of the ions is weakened and the molecules fall apart into the constituent ions, each with its outer group of electrons in the arrangement characteristic of an inert gas. In a crystal of sodium chloride, also, the grouping into molecules disappears; a crystal of this type is composed of ions but not of molecules. Thus many chemical and physical facts can be correlated if we make the assumption that the arrangement of the electrons occurring in a noble gas is a peculiarly stable one of low energy.

One of the principal problems of an explanatory theory of the atom is to account for the high stability of certain electronic configurations. Then an explanation must be found for the systematic sequences that occur in the periodic table between the noble gases. As the atomic weight increases, these sequences become longer and the elements of a sequence become more similar in chemical properties; many of the *rare earths* are even difficult to separate chemically. A successful theory should lead automatically to all these relations between chemical properties and atomic number on the basis of as few assumptions as possible.

15.4 The First Two Periods

The key to the periodic table was finally furnished by wave mechanics, with the help of the Pauli exclusion principle and the principle of electron spin. It will now be shown, taking the elements in the order of their atomic numbers, that the electronic groupings as inferred from wave mechanics enable us to understand the sequence of the elements. The *neutral atom* will be under discussion.

a. $Z = 1$: *Hydrogen* A single electron in an atom is normally in a $1s$ electronic quantum state. Degeneracy exists because of the two possible values of m_s.

b. $Z = 2$: *Helium* Two electrons combined with a nucleus for which $Z = 2$ can both go into $1s$ states, with quantum numbers 1, 0, 0, $\frac{1}{2}$ and 1, 0, 0, $-\frac{1}{2}$. Because the electronic charge cloud in helium is spherically symmetrical, there is almost no electric field outside the atom. Atoms of helium should therefore exhibit comparatively little tendency to associate themselves into a liquid or solid phase, in agreement with observation. In helium the K shell is complete; the electronic configuration is $1s^2$. All heavier atoms will be expected to contain, next to the nucleus, a complete K shell of this sort.

c. $Z = 3$: *Lithium* The third electron in lithium outside the helium-like core occupies a state with $n = 2$. The energy associated with a $2s$ wave function is somewhat less than that of a $2p$ function for an electron in the atomic central field (see Sec. 15.6); hence, the electronic constitution of normal lithium is $1s^2 2s$. Since s means $l = 0$, we have $m_l = 0$, but $m_s = \pm \frac{1}{2}$, so that again there is a twofold degeneracy in the ground state of lithium.

The energy for a $2s$ wave function lies much higher than that for a $1s$ function. If the field were a coulomb field, as for a one-electron atom, we could use Eq. (14.10), which shows that in such a field the $2s$ state lies only a quarter as far below the ionization level as the $1s$ state does. Hence, the $2s$ electron should be comparatively easy to remove. Lithium forms positive ions easily, and it conducts electricity when in the solid state, the $2s$ electrons functioning as "free" electrons. Lithium combines chemically with a valence of 1, as shown by such compounds as Li_2O, $LiOH$, $LiCl$.

To remove also a $1s$ electron from a lithium atom requires much more energy than to remove the $2s$, or valence, electron. In harmony with this conclusion, the first ionization potential of lithium is observed to be 5.39 V, as against 75.6 V for the second. Furthermore, since little instantaneous symmetry is expected from a single electron in the second shell, the lithium atoms should be surrounded by stray electric fields and should readily group themselves into a condensed phase. The melting point of lithium is 454°K, the boiling point 1599°K.

For lithium the *arc* spectrum, which is that from the *neutral atom*, is predominantly that to be expected from a single electron. The *spark* spectrum, which is ascribed to emission by *singly ionized atoms*, contains singlets and triplets of lines like the *arc* spectrum of *neutral helium*. This is what we should expect on the basis of the theory, for a singly ionized lithium atom contains the same number of electrons as neutral helium. The frequencies of corresponding lines are much higher than in helium, however, because the stronger nuclear charge causes all energy levels to lie much lower.

Throughout atomic spectra there is a striking similarity of the patterns of spectral lines emitted by neutral atoms and by ions with the same number of electrons. A sequence of atoms and ions having the same number of electrons, such as He, Li^+, Be^{++}, B^{3+}, C^{3+}, etc., is called an *isoelectronic sequence*. The patterns of spectral lines from members of an isoelectronic sequence are almost the same but shift to higher frequencies as the state of ionization increases.

d. $Z = 4$: Beryllium   Two electrons outside the K shell can both be in $2s$ states but with opposite spins. The resulting element should be bivalent, since both $2s$ electrons should come off relatively easily. Such is the case. The oxide, hydroxide, and chloride have the formulas BeO, $Be(OH)_2$, $BeCl_2$. The first two ionization potentials of beryllium are 9.3 and 18.2 V. To remove a third electron out of the $1s$ shell requires 154 eV, and to remove the last one, 218 eV.

Again the spectral evidence confirms the theory. The arc spectrum of beryllium is a two-electron spectrum of singlets and triplets like that of neutral helium. In the spark spectrum, on the other hand, doublet lines like those from neutral lithium are found; these are emitted by singly ionized beryllium·atoms. A few singlet lines are also known in the spark spectrum; they are part of a two-electron spectrum emitted by doubly ionized beryllium atoms. Lines have also been found which are ascribed to triply ionized atoms.

e. $Z = 5$: *Boron*   With five electrons in the atom, one goes into a $2p$ state. Boron is a trivalent element, witness B_2O_3, $B(OH)_3$, BCl_3. It is not metallic, however; crystals of boron are good insulators. Evidently, with three electrons present in the L shell, conditions are not favorable to the formation of free electrons in the solid state. To remove the $2p$ electron from a boron atom requires only 8.3 eV; to remove the two $2s$ electrons in succession requires 25 and 38 eV whereas to remove one of the $1s$ electrons as well requires an additional 259 eV.

The next elements in order are most easily understood if we pass them by for the moment and consider neon next.

f. $Z = 10$: *Neon*   It is possible to put into an atom two $2s$ electrons and six $2p$ electrons, or eight in all, with $n = 2$. The L shell is then filled. With every possible value of m_l and m_s represented, the electronic charge cloud is symmetrical about the nucleus, as it is in helium; and now the symmetry is sufficiently complete to result in a gas having a low boiling point. Neutral neon in its normal state has the electronic constitution $1s^2 2s^2 2p^6$.

g. $Z = 9$: *Fluorine* $(1s^2 2s^2 2p^5)$ If $Z = 9$, the neutral atom contains seven electrons in the L shell, or one less than enough to fill it. If one more electron were added, we should have a negative ion the exterior of which would be a closed shell, as in neutral neon. It cannot be expected that the same loss of energy would occur when an electron is added to a *neutral* atom as when it is added to a *positive ion;* but, on the other hand, the electron is added in the shell for $n = 2$ in an atom having a fairly strong nuclear field $(Z = 9)$. We can thus understand the fact that fluorine forms univalent negative ions which are stable, i.e., have lower energy than the neutral atom, and that it exhibits a negative valence of 1 in chemical combination. There is little tendency for a *second* extra electron to be bound by the fluorine atom, for it would have to occupy a state with $n = 3$, the energy of which would lie considerably higher.

h. $Z = 8, 7$: Oxygen $(Z = 8, 1s^22s^22p^4)$ commonly exhibits a negative valence of 2 in chemical combination, as in lithium oxide, Li_2O; and *nitrogen* $(Z = 7, 1s^22s^22p^3)$ is commonly trivalent, as in lithium nitride, Li_3N. Often nitrogen is united with oxygen into a compound as in $LiNO_3$.

i. $Z = 6$: Carbon $(1s^22s^22p^2)$ With four electrons in the L shell, which can accommodate eight, carbon is an element for which it becomes questionable whether it is energetically favorable to lose or gain electrons in interacting with other elements. A frequent compromise is to share electrons in a *covalent bond* (Sec. 21.2). In the amorphous form (graphite) carbon exhibits fair metallic conductivity, but in diamond it is an excellent insulator.

15.5   *Remainder of the Periodic Table*

The second octet of elements, from $Z = 11$ to $Z = 18$, parallels closely the first octet. *Sodium* $(Z = 11, 1s^22s^22p^63s)$ contains a single valence electron outside the neon core. Sodium is univalent, as in NaOH and the familiar NaCl. Its arc spectrum is a typical one-electron spectrum. In general, it resembles lithium closely. *Magnesium* $(Z = 12)$ is a bivalent element similar to beryllium. It burns with a brilliant white flame to form the oxide MgO. It has an arc spectrum of singlets and triplets resembling that of helium. *Aluminum* $(Z = 13)$ is trivalent, like boron, but it is metallic and an excellent conductor of electricity. The sesquioxide, Al_2O_3, occurs in crystalline form as sapphire and ruby. *Silicon* $(Z = 14)$ is a good deal like carbon. The dioxide SiO_2, however, which occurs in quartz, is a substance of extremely high melting point, whereas the analogous compound, CO_2, is a gas! Only an elaborately refined application of wave mechanics can explain contrasts such as these. *Phosphorus* $(Z = 15)$, although chemically much more active, forms compounds analogous to those of nitrogen. In the poisonous gas phosphine, or PH_3, it is trivalent, just as nitrogen is in gaseous ammonia, NH_3. *Sulfur* $(Z = 16)$ corresponds to oxygen. In H_2S, it is bivalent, just as oxygen is in H_2O. *Chlorine* $(Z = 17)$ is univalent and easily forms negative ions in solution; in general, it resembles fluorine closely but is less active. Finally, in *argon* $(Z = 18)$, we reach again an atom composed of complete subshells, with electronic formula $1s^22s^22p^63s^23p^6$. Thus, argon has completed K and L shells and, outside these, two more completed subshells. The M shell is not yet full, since there are no $3d$ electrons. Nevertheless, the symmetry of the $3p^6$ configuration is evidently enough to make the external field around argon very weak, so that it acts as a noble gas with a very low boiling point ($-186°$C).

At $Z = 19$, we might expect the addition of an electron in a $3d$ state ($n = 3, l = 2$). But the next element, *potassium*, closely resembles sodium and lithium, not only in chemical properties but also in its spectra even to the finest details. Since atomic spectra in the visible region are emitted by electrons in the periphery of the atom, they furnish valuable information concerning the state of the outermost electrons. A careful wave-mechanical calculation confirms the conclusions from chemistry and spectroscopy that the valence electron in potassium is in a $4s$ rather than a $3d$ state. A qualitative discussion of the reason appears in Sec. 15.6.

The next element, *calcium* ($Z = 20$), corresponds rather well to magnesium in the preceding octet, but beyond this point the sequence becomes quite different. Beginning with *scandium* ($Z = 21, 1s^2 2s^2 2p^6 3s^2 3p^6 4s^2 3d$) the $3d$ subshell begins to fill, expanding the $n = 3$ shell from the provisionally stable $s^2 p^6$ octet into the complete group of 18 electrons. Atoms with an incompletely filled d shell belong to the group known as *transition elements*, of which several are of great economic and scientific importance. They are somewhat alike in chemical properties, which are largely determined by the outer $4s$ electrons. For all $3d$ transition elements the $3d$ and $4s$ energy states are very close together. Indeed the outer electron configuration for *chromium* ($Z = 24$) is $3d^5 4s$. For *nickel* ($Z = 28$) the configuration is $3d^8 4s^2$, but from *copper* ($Z = 29$) onward the $3d$ level remains below $4s$.

Between *gallium* ($Z = 31$) and the noble gas *krypton* ($Z = 36$) the $4p$ subshell is filled. As is the case after every noble gas, the next element starts a new shell; *rubidium* ($Z = 37$) has an electron configuration consisting of a krypton core and a $5s$ electron. After the $5s$ subshell is filled, the $4d$, $5p$, and $6s$ levels are filled. Beginning at *cerium* ($Z = 58$) the 14 possible $4f$ electrons are added in turn, inside already completed $5s$ and $5p$ subshells. The exteriors of all these atoms are closely similar, so that we have a group of chemically similar elements, the *rare earths*.

The electronic configurations of the elements in their normal states are given in the Appendix. Also tabulated are the ionization energies for the neutral atoms of each element, the ionization energy being the energy required to remove the least tightly bound electrons from the atom. These ionization energies are plotted in Fig. 15.3. A systematic relation is evident between the quantum numbers of the outermost electrons and the ionization potentials, with the p^6 configuration of the

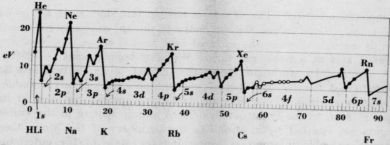

Fig. 15.3 The ionization energy in electron volts as a function of atomic number. *(Reprinted from H. E. White, "Introduction to Atomic Spectra." Copyright 1934. McGraw-Hill Book Company. Used by permission.)*

noble gases giving a particularly high ionization potential and the single *s* electron in a new shell, characteristic of the alkalis, giving particularly low ionization potential.

15.6 Penetrating Orbits

We now turn to the question of why, for a given principal quantum number *n*, the states of higher angular momentum (higher *l*) lie above those associated with those of smaller angular momentum in a multi-electron atom. To be specific, let us discuss the atom sodium $(Z = 11)$ for which 10 electrons fill all possible states with $n = 1$ and 2. These electrons produce a spherically symmetrical charge cloud about the nucleus, the net effect of which is to reduce the effective charge providing the field in which the eleventh electron moves. The probability of finding one of the inner 10 electrons between r and $r + dr$ is shown by the shaded region in Fig. 15.4. On the same figure are shown the quan-

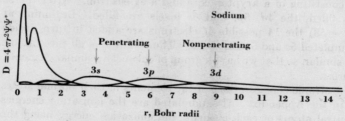

Fig. 15.4 Calculated charge density in the neutral sodium atom for three alternative states (3s, 3p, 3d) of the valence electron; the density due to the core is shown shaded. *(Reprinted from H. E. White, "Intro-duction to Atomic Spectra." Copyright 1934. McGraw-Hill Book Company. Used by permission.)*

tum-mechanical predictions for the radial charge density $4\pi r^2\psi^*\psi$ for the eleventh electron in a $3s$, a $3p$, and a $3d$ state. The probability of finding a $3d$ electron inside the core is extremely small; for a $3p$ electron the probability is larger; for a $3s$ electron it is much larger still. Clearly, for an electron inside the core the effective nuclear charge is substantially greater than for an electron outside (for which $Z_{eff} = 1$). If the electron lies almost entirely within the stronger field, its associated energy is expected to be lower. Such an argument is not entirely conclusive, however; for in the mathematical theory the energies are determined by the boundary condition at infinity. We must remember that as the electron penetrates the core, its total energy is not decreased, but instead there is a conversion of potential energy into kinetic. However, in a coulomb field of strength $Ze^2/4\pi\epsilon_0 r^2$ the energies of the corresponding states are all lower by a factor of Z^2 than for the coulomb field from a unit charge. It is thus not unreasonable that the energy level should sink as the degree of penetration into the core increases. That this is true is shown by detailed quantum-mechanical calculations; further, the fact is thoroughly supported by all the evidence of atomic spectroscopy.

It is customary to speak of the s and p states in terms of *penetrating orbits*, a term which remains from the days of Sommerfeld's expansion of the Bohr theory to elliptic orbits. While the idea of sharp orbits is unacceptable in modern wave mechanics, the language of Sommerfeld's model continues to be widely used. Not only is the penetrating-orbit effect enough to explain why for a given n states of smallest l lie lowest, but it is also adequate to explain why low-angular-momentum states of a higher shell may have a smaller energy than the high-angular-momentum states of a shell of lower n. For example, we have seen that for potassium the $4s$ state lies below the $3d$. Similarly, for rubidium the $5s$ state lies below the $4d$; for cesium the $6s$ lies below the $5d$ and the $4f$ as well.

As we have seen, the $4s$ level lies below the $3d$ for potassium ($Z = 19$). For hydrogen, on the other hand, the $n = 4$ levels all lie higher than the $n = 3$ levels; the dependence of the energy on l is very small and arises from entirely different causes. For elements of $Z = 2$ to 5 the $3d$ level remains below the $4s$, but at carbon ($Z = 6$) the electron core has grown sufficiently for the $4s$ level to fall below the $3d$. From there to chromium the $4s$ level lies lower, but at chromium ($Z = 24$) the two are extremely close together, and one of the two $4s$ electrons goes to the $3d$ state, giving a $3d^5 4s$ configuration. For manganese ($Z = 25$) through nickel ($Z = 28$) the $4s$ level falls below the $3d$, but at copper ($Z = 29$) the $3d$ state sinks below the $4s$ and remains there throughout the rest of the periodic table. Similar crossovers occur for the $5s$ and $4d$ levels and for the $6s$, $5d$, and $4f$ levels.

15.7 Low-lying Energy States

In discussing the periodic table, we have assumed that in a neutral heavy atom there are 2 electrons characterized by $n = 1$, 8 with $n = 2$, 18 with $n = 3$, and so forth.   Further, we have assumed that each occupies an energy state and satisfies a wave function which may be crudely approximated by the corresponding hydrogenic energy and ψ_{nlm_l}. The most direct experimental evidence for the validity of these assumptions lies in x-ray phenomena.   We begin by considering the absorption of x-rays in matter.   In contrast with the apparently chaotic state of affairs in regard to the absorption of light in the visible or near-visible portions of the spectrum, we find comparative simplicity in the laws for x-ray attenuation.   Measurements of the attenuation coefficient for a monochromatic beam of parallel rays are readily made by use of the ionization spectrometer (Fig. 7.6) with a tube voltage low enough to eliminate second-order reflections.   For a given crystal angle θ and wavelength λ, the ionization current is measured both with and without a sheet of absorbing material of known density ρ and thickness x placed in the path of the beam, say between the two slits S_1 and S_2.   These measurements give, respectively, I and I_0 in the equation

$$I = I_0 e^{-\mu x} = I_0 e^{-(\mu/\rho)\rho x} \tag{15.2}$$

from which either the *linear attenuation coefficient* or the *mass attenuation coefficient* μ/ρ can be computed.   Figure 15.5 shows the mass attenuation coefficient of lead in the wavelength range $0.1 < \lambda < 1.2$ Å.   Beginning at point o, μ/ρ rises rapidly with increasing wavelength, until point a, corresponding to $\lambda = 0.1405$ Å and $\mu/\rho = 8$ (about), is reached, at which the value of μ/ρ suddenly drops to point a'.   This is the *K absorption limit*, which we associate with the photoelectric ejection of electrons from the $n = 1$, or K, shell.   Photons with wavelength shorter than the K absorption limit can and do eject electrons from the K shell, while photons with longer wavelengths do not have enough energy to eject a K electron.   Thus the K absorption limit gives a measure of the energy with which a K electron is bound to the atom.   For lead λ_K, the wavelength of the K absorption edge is 0.1405 Å, and the corresponding energy is $E_K = hc/\lambda = 88{,}000$ eV.

With further increase in wavelength, the absorption again increases rapidly, being mostly due to the ejection of electrons from the L ($n = 2$) shell, until at $\lambda = 0.780$ Å there is another sharp drop.   After a brief rise there is another absorption edge at 0.813 Å, and a third at 0.950 Å. These three discontinuities are known respectively as the L_I, L_{II}, and L_{III} absorption edges.   They reveal that there are three distinct energy

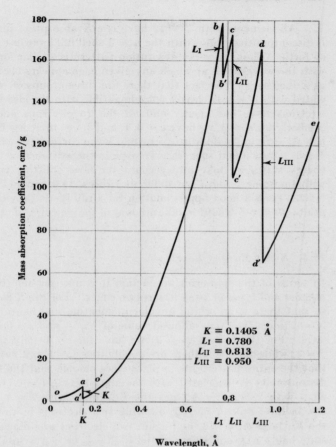

$$K = 0.1405 \text{ Å}$$
$$L_{\text{I}} = 0.780$$
$$L_{\text{II}} = 0.813$$
$$L_{\text{III}} = 0.950$$

Fig. 15.5   *K* and *L* absorption edges of lead.

states for *L* electrons; we shall see in the following section that there are two kinds of 2*p* states which have different energies.

Beyond the L_{III} discontinuity the absorption again increases rapidly. Extension of Fig. 15.5 beyond point *e* would reveal that in the region $3.2 < \lambda < 5.0$ Å, a group of five "breaks" occurs, representing the M_{I}, M_{II}, M_{III}, M_{IV}, and M_{V} absorption limits. The wavelength of these breaks gives us the five characteristic energies allowed for *M* (*n* = 3) electrons. Beginning near 14 Å comes seven *N* limits associated with the *n* = 4 shell, and at still longer wavelengths come limits due to *O* (*n* = 5) and *P* (*n* = 6) levels.

All elements with $Z > 10$ have one K absorption jump and three L discontinuities; those with the $n = 3$ shell full have five M edges, and so forth. Thus the x-ray absorption pattern is similar for all elements, with the wavelength at which any given edge appears becoming shorter as Z increases. The fact that there are three groups of electrons associated with $n = 2$ is clearly not compatible with the idea that all the $2s$ electrons have one energy and all the $2p$ electrons another energy. Similarly for $n = 3$, we have $l = 0, 1,$ and 2, but there are five characteristic M energies, and for $n = 4$ we find seven energies, which is more than $l = 0, 1, 2, 3$ might suggest. To explain the existence of the additional energy states we take into account the level splitting between states with the same n and l but different j (Secs. 14.6 and 14.7). Equation (14.23) gives a good approximation for x-ray levels provided an appropriate *effective* Z is used to take account of the shielding of other electrons.

15.8 X-ray Energy Levels

In terms of the spin-orbit interaction it is now possible to explain the number and types of x-ray absorption edges. For the K electrons $l = 0$, so that there is no orbital angular momentum and hence no spin-orbit interaction. The only allowed value of j is $\frac{1}{2}$, and we find a single K level. For the L shell ($n = 2$) we have for $l = 0$ only the possibility $j = \frac{1}{2}$, while for $l = 1$ there are possibilities $j = \frac{1}{2}$ and $j = \frac{3}{2}$. We find that the entire pattern fits in place if we assume that the L_I absorption discontinuity is associated with the state having $l = 0$, $j = \frac{1}{2}$; the L_{II} edge with $l = 1$, $j = \frac{1}{2}$; and the L_{III} limit with $l = 1$, $j = \frac{3}{2}$. For the L_I level m_j can be $\pm\frac{1}{2}$, allowing two electrons in this state. For the L_{II} level $m_j = \pm\frac{1}{2}$, and again two electrons are allowed. The L_{III} state has $j = \frac{3}{2}$, and so m_j can be $\frac{3}{2}, \frac{1}{2}, -\frac{1}{2}$, or $-\frac{3}{2}$, a total of four possibilities. Thus in the $nljm_j$ description, there are in a filled L subshell twice as many electrons in the L_{III} level as in the L_I or L_{II} states. One would expect that there would be a bigger drop in the absorption coefficient when four electrons are involved than two; this is indeed the case in Fig. 15.5.

When the spin-orbit interaction is included, it becomes convenient to think of each subshell as defined by a given nl being split into two nlj subshells, each composed of all electrons having the same values of n, l, and j. In the absence of electric or magnetic fields all electrons in a given nlj subshell have the same energy. Each closed subshell contains $2j + 1$ electrons, one for each value of m_j. Table 15.2 shows the symbols for the various inner subshells and the maximum number of electrons allowed in each.

Table 15.2 Symbols for electron shells, subshells, and x-ray levels

| Name | Shell | | Subshell | | | Symbol | |
	n	Maximum No. of Electrons	l	j	Maximum No. of Electrons	Spectroscopic	X-ray
K	1	2	0	$\frac{1}{2}$	2	$1s$	K
L	2	8	0	$\frac{1}{2}$	2	$2s$	L_{I}
			1	$\frac{1}{2}$	2	$2p_{\frac{1}{2}}$	L_{II}
			1	$\frac{3}{2}$	4	$2p_{\frac{3}{2}}$	L_{III}
M	3	18	0	$\frac{1}{2}$	2	$3s$	M_{I}
			1	$\frac{1}{2}$	2	$3p_{\frac{1}{2}}$	M_{II}
			1	$\frac{3}{2}$	4	$3p_{\frac{3}{2}}$	M_{III}
			2	$\frac{3}{2}$	4	$3d_{\frac{3}{2}}$	M_{IV}
			2	$\frac{5}{2}$	6	$3d_{\frac{5}{2}}$	M_{V}
N	4	32	0	$\frac{1}{2}$	2	$4s$	N_{I}
			1	$\frac{1}{2}$	2	$4p_{\frac{1}{2}}$	N_{II}
			1	$\frac{3}{2}$	4	$4p_{\frac{3}{2}}$	N_{III}
			2	$\frac{3}{2}$	4	$4d_{\frac{3}{2}}$	N_{IV}
			2	$\frac{5}{2}$	6	$4d_{\frac{5}{2}}$	N_{V}
			3	$\frac{5}{2}$	6	$4f_{\frac{5}{2}}$	N_{VI}
			3	$\frac{7}{2}$	8	$4f_{\frac{7}{2}}$	N_{VII}
O	5	Never full	0–4	$\frac{1}{2}-\frac{9}{2}$			$O_{I}-O_{IX}$
P	6	Never full	0–5	$\frac{1}{2}-\frac{11}{2}$			$P_{I} \cdots$

15.9 Photoelectrons Ejected by X-rays

When an electron is ejected from the K shell by x-ray photons of energy $h\nu$, the electron emerges from the atom with kinetic energy $h\nu - h\nu_K$, where ν_K is the frequency of the K absorption limit. If the atom lies on the surface of the absorbing material, the electron may escape into the surrounding space with this kinetic energy. Similar statements hold for the photoelectrons ejected from the L shell, which consist of three slightly different groups with maximum energies corresponding to the three L absorption limits, ν_{L_1}, $\nu_{L_{II}}$, $\nu_{L_{III}}$. Likewise, from the M shell there are five groups; and so on. If we start with x-rays of very high frequency, photoelectrons of all kinds are produced. If the frequency is decreased, as it passes ν_K, the K photoelectrons disappear; as it passes the L limits, the three groups of L photoelectrons disappear in turn; and so on.

Among the experiments demonstrating these facts are those by

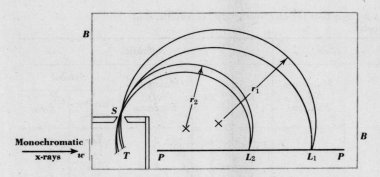

Fig. 15.6 Robinson's magnetic spectrograph for studying the photo-electric action of x-rays.

Robinson and his collaborators, whose apparatus is shown diagrammatically in Fig. 15.6. A beam of x-rays of frequency ν enters through a thin window W a highly evacuated brass box BB and falls upon a target T of the material under investigation. Photoelectrons are expelled from the surface of T in all directions and with various velocities. The whole apparatus is placed in a known magnetic field B, at right angles to the plane of the paper, which can be varied at will; the photoelectrons describe circles in this field. Some of them pass through the narrow slit S and eventually strike the photographic plate PP. If the electrons leaving T have velocities $v_1, v_2, \ldots$, they will move in circles of radii $r_1, r_2, \ldots$ and will strike the plate at points $L_1, L_2, \ldots$. As shown in the figure, the arrangement is such as to "focus" electrons leaving the different parts of the target with the same velocity onto the plate at such positions as L_1 and L_2, the diameter of the circle being the distance between S and L_1 or L_2.

Values of the kinetic energy K of the photoelectrons can be calculated from measured values of r, for comparison with $h(\nu - \nu_A)$, where ν_A is any absorption limit. Robinson found on his plates a number of "lines" representing groups of photoelectrons. The corresponding absorption limits were in complete agreement with values determined spectroscopically. He found one K level, three L levels, five M levels, and, in the case of U^{92}, five of the seven N levels, the pairs of levels $N_{\text{IV,V}}$ and $N_{\text{VI,VII}}$ being too close together to be resolved in his apparatus.

The technique of measuring the velocities of photoelectrons ejected by x-rays has been developed to the point where this method is now one of the most reliable for the precise determination of physical constants. By using double-focusing electron spectrometers and other improvements it is now possible to resolve lines differing in energy by only one one-hundredth of the energy difference formerly required. For example, Hagström, Nordling, and Siegbahn have found that the K electron of sulfur is bound 5.5 eV more tightly in the $6+$ state of ionization (Na_2SO_4) and 4.3 eV more tightly in the $4+$ valence state (Na_2SO_3) than in pure sublimed sulfur.

Appendix 15A - Wave Mechanics for Two Identical Particles with Spin One-half

Suppose that there are two identical particles in the square well of Sec. 13.3. Assume for the moment that they do not interact with each other and that no such phenomenon as spin need be considered. Let the coordinates of the particles be x_1 and x_2. Then, in classical mechanics, the combined (kinetic) energy K of the particles equals $(p_1{}^2 + p_2{}^2)/2m$, where p_1 and p_2 are their respective momenta. Replacing the momenta by operators as in Eq. (13.12a), we have as the time-free wave equation for the two particles

$$-\frac{h^2}{8\pi^2 m}\left(\frac{\partial^2\psi}{\partial x_1{}^2} + \frac{\partial^2\psi}{\partial x_2{}^2}\right) = E\psi \tag{15A.1}$$

Here ψ is a function of both x_1 and x_2; it has the significance that

$$|\psi|^2\, dx_1\, dx_2$$

is the probability of finding particle 1 in dx_1 and, simultaneously, particle 2 in dx_2. Our present problem is simple, however, in that solutions of (15A.1) can be found which are products of functions of x_1 and x_2 separately. Such a solution, satisfying the boundary conditions and normalized so that $\int_0^L dx_1 \int_0^L dx_2\, |\psi_{nk}|^2 = 1$, is

$$\psi_{nk} = \psi_n(x_1)\psi_k(x_2) = \frac{2}{L}\sin n\pi\,\frac{x_1}{L}\sin k\pi\,\frac{x_2}{L} \qquad 0 \le x \le L$$

where ψ_n and ψ_k are one-particle functions and n and k are any two positive integers. The associated energy of the system is

$$E_{nk} = \frac{h^2}{8m}\,\frac{n^2 + k^2}{L^2} \tag{15A.2}$$

This solution represents particle 1 as being in state n and particle 2 as in state k. *A new and very important form of degeneracy* now appears, however, which may be called *exchange degeneracy*.   For, if x_1 and x_2 are interchanged in ψ_{nk}, provided $n \neq k$, another wave function corresponding to the same value of the energy is obtained, viz.,

$$\psi_{kn} = \frac{2}{L} \sin k\pi \frac{x_1}{L} \sin n\pi \frac{x_2}{L}$$

15A.1   Interacting Particles

These functions as written are not suitable for use in a perturbation calculation.   For, if an attempt is made to correct for a small perturbing term $f(x_1,x_2)$ in the energy operator, a difficulty is encountered arising from the fact that ψ_{nk} and ψ_{kn} are associated with the same energy. Equations analogous to (13.42) are obtained, and, among them, one in which n is replaced by nk and k by kn; after removing the prime from E'_{nk} in this equation, as was done in proceeding to (13.42a), the equation takes the form

$$(E_{kn} - E_{nk})b_{nk} + \int_0^L dx_1 \int_0^L f(x_1,x_2)\psi_{kn}\psi_{nk} \, dx_2 = 0$$

Here, however, $E_{kn} - E_{nk} = 0$; whereas the integral may not vanish.
The simplest and usual procedure in such cases is to substitute suitable *new combinations* of the degenerate wave functions.   Instead of ψ_{nk} and ψ_{kn}, let us employ as zero-order functions two combinations of these functions so chosen that the matrix component of the given perturbation energy $f(x_1,x_2)$ between the new functions vanishes.   The perturbation matrix is then said to be *diagonalized* with respect to these particular zero-order functions, a completely diagonal matrix being one that has zeros for all its elements except those on the leading diagonal (top left to bottom right).   When such functions are used, the difficulty disappears.
The necessity of choosing suitable zero-order functions in cases of degeneracy has an exact classical analog.   In the case of the vibrating drumhead described in Sec. 14.8, if the symmetry is destroyed by the addition of a small eccentrically placed weight w, the effect on the frequency can be calculated by means of perturbation theory; but it is now necessary to start from a *particular choice* of the zero-order modes of vibration (see Fig. 15A.1).   One of the modified modes of vibration will closely resemble a zero-order mode in which the nodal line passes through the position of the added weight; the other new mode will have its nodal line in a perpendicular direction.   One effect of the weight is thus to fix the nodal lines.   Another effect is to replace the single zero-order

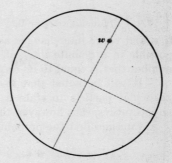

Fig. 15A.1 Nodal lines on a loaded vibrating drumhead.

frequency by two slightly unequal frequencies associated with two alternative patterns of vibration. Thus the degeneracy is removed by the added weight.

In the present case, the necessary combinations of ψ_{nk} and ψ_{kn} to serve as zero-order starting points for perturbation theory are found to be a *symmetric* one and an *antisymmetric* one defined as follows:

$$\psi_{nkS} = \frac{1}{\sqrt{2}} \, (\psi_{nk} + \psi_{kn}) \qquad\qquad (15A.3a)$$

$$\psi_{nkA} = \frac{1}{\sqrt{2}} \, (\psi_{nk} - \psi_{kn}) \qquad\qquad (15A.3b)$$

If x_1 and x_2 are interchanged in these functions, it is obvious that ψ_{nkS} is unaltered, so that this function is symmetric for exchange of the two particles; whereas ψ_{nkA} just changes sign and is, therefore, antisymmetric for such exchange. The factor $1/\sqrt{2}$ is inserted to preserve the normalization. [Symmetry for particle exchange should not be confused with symmetry in the coordinates of one particle; an example of the first would be $\psi = (x_1 - x_2)^2$, of the second, $\psi = x_1{}^2 x_2{}^4$.]

The new functions are orthogonal to all independent wave functions, and also to each other; for a short calculation shows that

$$\int_0^L dx_1 \int_0^L \psi_{nkS}\psi_{nkA} \, dx_2 = 0$$

Furthermore, these functions have the property that the matrix component with respect to them of any operator that is *symmetric* in x_1 and x_2 vanishes. For example, suppose the complete wave equation contained a small potential term of the form $P = f[(x_1 - x_2)^2]$. Then

$\int_0^L dx_1 \int_0^L f\psi_{nkS}\psi_{nkA}\, dx_2 = 0$. This is most quickly seen by noting that if x_1 and x_2 are interchanged as variables of integration, which is always possible in a definite integral, the integral changes sign; but the only number that is equal to its own negative is 0.

It will be noted that in both ψ_{nkS} and ψ_{nkA} the assignment of a particular particle to state n or to state k has disappeared. In two important respects, however, these functions have contrasting properties.

Consider first the probability densities

$$\psi_{nkS}^2 = \tfrac{1}{2}(\psi_{nk}^2 + \psi_{kn}^2) + \psi_{nk}\psi_{kn}$$

$$\psi_{nkA}^2 = \tfrac{1}{2}(\psi_{nk}^2 + \psi_{kn}^2) - \psi_{nk}\psi_{kn}$$

In the right-hand members of these two expressions, the *first* term represents the probability density on the assumption that the particles move independently. For, if particle 1 is in state n, the chance of finding x_1 in dx_1 is $\psi_n^2\, dx_1$, and similarly it is $\psi_k^2\, dx_2$ for x_2 in dx_2, so that the combined probability for both as independent events is the product of these two expressions, or $\psi_{nk}^2\, dx_1\, dx_2$; and similarly for the alternative arrangement represented by ψ_{kn}. The *second* term on the right is a wave-mechanical novelty. Wherever x_1 and x_2 have nearly equal values, $\psi_{nk}\psi_{kn} = \psi_{nk}^2 = \psi_{kn}^2$, approximately, so that the effect of the last term is almost to double ψ_{nkS}^2 but to reduce ψ_{nkA}^2 almost to zero. Thus there exists an *exchange effect* that causes the particles to favor close proximity when the wave function describing their state is symmetric under exchange of their coordinates but to avoid each other's company when the wave function is antisymmetric. This exchange effect is quite independent of any force-action between the particles. It has no analog in classical theory.

In the second place, consider again the first-order energy corrections due to a perturbation that is itself symmetric in the particles. For concreteness, suppose that the particles carry equal electric charges q and that their mutual electrostatic energy is P, where $P = q^2/4\pi\epsilon_0 r$. Then, in analogy with Eq. (13.41)

$$\Delta E_S = \frac{1}{2} \iint \frac{q^2}{4\pi\epsilon_0 r} (\psi_{nk}^2 + \psi_{kn}^2)\, dx_1\, dx_2$$
$$+ \iint \frac{q^2}{4\pi\epsilon_0 r} \psi_{nk}\psi_{kn}\, dx_1\, dx_2 \quad (15\text{A.}4a)$$

$$\Delta E_A = \frac{1}{2} \iint \frac{q^2}{4\pi\epsilon_0 r} (\psi_{nk}^2 + \psi_{kn}^2)\, dx_1\, dx_2$$
$$- \iint \frac{q^2}{4\pi\epsilon_0 r} \psi_{nk}\psi_{kn}\, dx_1\, dx_2 \quad (15\text{A.}4b)$$

Here it is possible to give to the first term in ΔE_S and ΔE_A a semi-classical interpretation. We can write

$$\iint \frac{q^2}{4\pi\epsilon_0 r} \psi_{nk}^2 \, dx_1 \, dx_2 = \int q[\psi_k(x_2)]^2 \, dx_2 \int \frac{q}{4\pi\epsilon_0 r} [\psi_n(x_1)]^2 \, dx_1$$

and the last integral, in dx_1, can be regarded as an expression for the potential at the point x_2 due to the cloud charge representing the first particle; whereas $q[\psi_k(x_2)]^2 \, dx_2$ is the density of the second charge cloud at this point. A similar interpretation can be put upon the contribution from ψ_{kn}^2. Thus the first term in ΔE_S and ΔE_A can be interpreted as the mutual potential energy of the two charge clouds due to classical coulomb repulsion of the particles. The second term on the right in (15A.4b), on the other hand, positive in ΔE_S and negative in ΔE_A, contains an integral, known as the *exchange integral*, which has no direct analog in classical theory. Its opposite sign in the two cases arises from the wave-mechanical tendency already noted for the particles to cluster together in a symmetric state but to avoid each other in an antisymmetric state, thereby resulting in different mean values of the perturbing potential in the two cases. This effect occurs in addition to other possible effects caused by mutual attraction or repulsion of the particles, which may result in further corrections to the energies.

It may be remarked that the exchange effect gives rise here to a physical distinction that does not exist in classical theory. Even in the presence of the perturbing potential, there would be two possible classical motions differing only in that the two similar particles are interchanged, and, once started, these motions could not be distinguished from each other by observation. In wave mechanics, on the other hand, the two corresponding quantum states differ in energy.

Similar features occur in three-dimensional cases also. The exchange effect is limited, however, to *exactly similar* particles.

15A.2 Electron Spin. The Exclusion Principle

In the example above, let the particles now be electrons, but let the box be large enough so that for a first approximation effects of electrostatic repulsion may be neglected. To be complete, the treatment must now be extended to include electron spin.

In dealing with many electrons, the handling of two separate wave functions for each electron becomes very intricate. It is more convenient to introduce some sort of symbolism by which these separate functions can be combined symbolically into a single, composite wave function. The following concise notation, although not in general use, appears to be both convenient and adequate.

Let S_{m_s} denote an electronic spin state characterized by the quantum number m_s; to avoid writing fractions continually, however, it will be convenient to write S_α for $S_{\frac{1}{2}}$ and S_β for $S_{-\frac{1}{2}}$. A wave function for one electron may then be written $\psi = S_\alpha u_1 + S_\beta u_2$, where u_1 and u_2 are functions, real or complex, of the spatial coordinates (and, in general, also of the time). To make things come out right in products of wave functions, let us assume by definition that $S_\alpha{}^2 = S_\beta{}^2 = 1$ but

$$S_\alpha S_\beta = S_\beta S_\alpha = 0$$

Otherwise, S_{m_s} is to behave like a real number. Then the probability density comes out correctly; thus

$$|\psi|^2 = \psi^*\psi = (S_\alpha u_1^* + S_\beta u_2^*)(S_\alpha u_1 + S_\beta u_2)$$
$$= u_1^* u_1 + u_2^* u_2 = |u_1|^2 + |u_2|^2$$

In dealing with more than one electron, we might now write a separate symbol S_{m_s} for each. It is simpler, however, to write the sets of coordinates for the various electrons always in a standard order, and to attach subscripts to a single S in the same standard order to denote their spins. Then the general expression for ψ for two electrons will be

$$\psi = S_{\alpha\alpha} u_1(x_1, x_2, t) + S_{\alpha\beta} u_2(x_1, x_2, t) + S_{\beta\alpha} u_3(x_1, x_2, t)$$
$$+ S_{\beta\beta} u_4(x_1, x_2, t)$$

We shall then postulate that $S_{(m_s)_1(m_s)_2} S_{(m_s)_1{}'(m_s)_2{}'}$ equals unity if $(m_s)_1' = (m_s)_1$ and $(m_s)_2' = (m_s)_2$ but otherwise equals zero.

We return now to the case of *two electrons in a box*. When their mutual repulsion is neglected, possible alternative wave functions corresponding to the same energy E_{nk} would seem to be the following four:

$$S_{\alpha\alpha}\psi_{nk} \qquad S_{\alpha\beta}\psi_{nk} \qquad S_{\beta\alpha}\psi_{nk} \qquad S_{\beta\beta}\psi_{nk}$$

Here, for example, $S_{\alpha\beta}$ indicates that the first electron has spin $\frac{1}{2}$ and the second $-\frac{1}{2}$, whereas $S_{\beta\alpha}$ indicates values $-\frac{1}{2}$ and $\frac{1}{2}$, respectively. If $n \neq k$, another set corresponding to the same energy E_{nk} would be obtained similarly from ψ_{kn}, making eight functions in all. At this point, however, account must be taken of the *exclusion principle*. *The wave function must be antisymmetric in the coordinates and spins of identical particles:* i.e., if the coordinates and spins of one particle are interchanged as a group with those of another, the wave function must change sign.

This requirement is an independent fundamental postulate of wave mechanics.

None of the wave functions so far written, however, has the required antisymmetry. It is necessary, therefore, again to substitute appropriate linear combinations of functions belonging to the same energy. Out of the eight product functions associated with energy E_{nk}, when $n \neq k$, the following four antisymmetric combinations can be made, and any other antisymmetric combination can be expressed in terms of these:

$$\psi_1 = \sqrt{\tfrac{1}{2}}\, S_{\alpha\alpha}(\psi_{nk} - \psi_{kn}) \qquad \psi_2 = \sqrt{\tfrac{1}{2}}\, S_{\beta\beta}(\psi_{nk} - \psi_{kn})$$

$$\psi' = \sqrt{\tfrac{1}{2}}\,(S_{\alpha\beta}\psi_{nk} - S_{\beta\alpha}\psi_{kn}) \qquad \psi'' = \sqrt{\tfrac{1}{2}}\,(S_{\beta\alpha}\psi_{nk} - S_{\alpha\beta}\psi_{kn})$$

Here, for example, simultaneous interchange of spatial coordinates and spins converts $S_{\alpha\beta}\psi_{nk}$ into $S_{\beta\alpha}\psi_{kn}$ and vice versa, so that the signs of both ψ' and ψ'' are reversed. The necessary reversals of ψ_1 and ψ_2 are produced by the spatial functions alone.

Our troubles are not yet over, however. The functions ψ' and ψ'' will probably not do as zero-order functions for use in perturbation theory! For, if f denotes a symmetric perturbation, we have

$$\iint f\psi'\psi''\, dx_1\, dx_2 = -\iint f\psi_{nk}\psi_{kn}\, dx_1\, dx_2$$

(Here $S_{\alpha\beta}S_{\beta\alpha} = S_{\beta\alpha}S_{\alpha\beta} = 0$, but $S_{\alpha\beta}S_{\alpha\beta} = S_{\beta\alpha}S_{\beta\alpha} = 1$.) The last integral is unlikely to vanish. For general use, therefore, it is preferable to substitute for ψ' and ψ'' the following new combinations proportional to their sum and difference:

$$\psi_3 = \tfrac{1}{2}(S_{\alpha\beta} + S_{\beta\alpha})(\psi_{nk} - \psi_{kn}) \qquad \psi_4 = \tfrac{1}{2}(S_{\alpha\beta} - S_{\beta\alpha})(\psi_{nk} + \psi_{kn})$$

It is now easy to show that if f is symmetric for interchange of x_1 and x_2, then $\iint f\psi_3\psi_4\, dx_1\, dx_2 = 0$.

It is an interesting fact that, from the properties of the spin momentum as an operator, it can be shown that ψ_1, ψ_2, and ψ_3 all give to the square of the resultant spin momentum a magnitude $S(S + 1)\hbar^2$ with $S = 1$, and, furthermore, that the component of spin momentum in the direction of the axis that was used in defining S_α and S_β has the value $M_S\hbar$ where $M_S = 1, 0, -1$, for ψ_1, ψ_3, ψ_2, respectively. These values of M_S correspond to $\alpha + \alpha = 1$, $\alpha + \beta = 0$, $\beta + \beta = -1$. For ψ_4, on the other hand, $S = 0$ and $M_S = 0$.

The two alternative values of S can be regarded as resulting from the addition of electronic spin vectors of magnitude $\tfrac{1}{2}$ in either parallel or antiparallel positions. Sometimes the states with $S = 1$ are described as states in which the electrons have "parallel spins," but this description seems questionable in the case of ψ_3.

To assist the memory, the four properly symmetrized wave-functions thus obtained for the two electrons may be repeated in symbolic array; thus

$$
\begin{array}{ll}
& \quad S = 1 \qquad\qquad\qquad\qquad\qquad S = 0 \\
M_S = 1 & \\
& \quad \sqrt{\tfrac{1}{2}}\, S_{\alpha\alpha} \\
M_S = 0 & \\
& \quad \tfrac{1}{2}(S_{\alpha\beta} + S_{\beta\alpha}) \quad (\psi_{nk} - \psi_{kn}) \qquad \tfrac{1}{2}(S_{\alpha\beta} - S_{\beta\alpha})(\psi_{nk} + \psi_{kn}) \\
M_S = -1 & \\
& \quad \sqrt{\tfrac{1}{2}}\, S_{\beta\beta}
\end{array}
\qquad (15A.5)
$$

Thus, for $n \neq k$, the occurrence of spin with $s = \tfrac{1}{2}$, in combination with the exclusion principle, merely doubles the number of degenerate wave functions associated with the same energy E_{nk}.

It will be noted that all three functions for $S = 1$ are antisymmetric in the space coordinates alone, whereas the single function for $S = 0$ is symmetric. This feature gives to the functions *exactly the same contrasting exchange properties that were found to occur when spin was neglected.* Many quantitative results are the same, since the spin symbols disappear in all products of the wave functions. In particular, Eq. (15A.4b) for ΔE_A due to a slight electrostatic interaction now holds without change for the three zero-order states with $S = 1$, whereas (15A.4a) for ΔE_S holds for $S = 0$. It will be observed that ΔE_A is the same for all three values of M_S. Thus degeneracy of the energy levels due to spin is not completely removed by electrostatic interaction, or, in fact, by any perturbation that is symmetric in the coordinates of the two particles.

In conclusion, the important case $k = n$ remains to be noticed. In this case $\psi_{nk} - \psi_{kn} = 0$, so that only *one* antisymmetric wave function is obtained, with $S = M_S = 0$; it can be written

$$
\psi = \sqrt{\tfrac{1}{2}}\,(S_{\alpha\beta} - S_{\beta\alpha})\psi_{nn} \tag{15A.6}
$$

If we try to start from $S_{\alpha\alpha}\psi_{nn}$ or $S_{\beta\beta}\psi_{nn}$, no usable wave function is obtained, after allowing for the exclusion principle.

Problems

1. Write the electron configuration for the ground state of each of the following elements: $_6$C, $_{17}$Cl, $_{50}$Sn, $_{74}$W, and $_{86}$Rn.

2. An excited K atom has an Ar core and the valence electron in a $5d$ state. What are the possible values of j? For each of these j values, find the angle between $\mathbf{A}_l$ and $\mathbf{A}_s$ and the angle between $\mathbf{A}_l$ and $\mathbf{A}_j$.

3. (a) Show that the separation between the fine-structure levels of a given nl state is

$$\Delta E_{l+\frac{1}{2},l-\frac{1}{2}} = \frac{\alpha^2 E_1 Z_{\text{eff}}^4}{n^3 l(l+1)}$$

where α is the fine-structure constant and E_1 ($= 13.6$ eV) is the magnitude of the energy of an electron in the first Bohr orbit of hydrogen.

(b) Prove that the separation is such that the average energy of the fine-structure levels is just the energy of the nl state before the spin-orbit interaction is taken into account.

4. The critical absorption wavelengths of gold are at 0.1532, 0.8622, 0.9009, and 1.038 Å for the K, L_{I}, L_{II}, and L_{III} levels, respectively. Find the energy in keV required to remove an electron from each of these levels in gold. From which of these levels will electrons be ejected by a photon of 14 keV energy?

5. Jönsson found that the ratio r_K of the attenuation coefficient on the short-wavelength side of the K absorption edge to that on the long-wavelength side is given approximately by $r_K = E_K/E_{L_{\text{I}}}$, where E is the energy required to remove an electron from the level indicated by the subscript. Using the data of the preceding problem and Jönsson's relation, calculate r_K for gold. (The observed value is 5.65.) Assuming that for all electrons other than the K ones the small wavelength shift from one side of the absorption limit to the other makes no difference, estimate the probability that a photoelectron ejected from a Au atom by a photon with $\lambda = 0.152$ Å came from the K shell. (Almost all the attenuation at this wavelength in Au may be assumed to come from photoelectric absorption.)

Ans: 5.63; about 0.9

6. The mass-attenuation coefficient of nitrogen is 0.745 m^2/kg for x-rays with $\lambda = 1.54$ Å. Find the fraction of a 1.54-Å beam which is lost in traversing 0.85 m of nitrogen at 300°K and 1 atm pressure.

7. The K absorption edges of silver, cadmium, indium, and tin lie at 0.4845, 0.4631, 0.4430, and 0.4239 Å, respectively. A beam of monochromatic γ rays is weakly absorbed in thin films of Ag, Cd, and In, but much more strongly absorbed in a thin Sn film of comparable thickness. Between what two energies, in keV, does this experiment bracket the energy of the γ ray?

8. A beam of Sr^{38} $K\alpha_1$ photons ($\lambda = 0.8735$ Å) falls on thin gold foil. Using the data of Prob. 4, find the kinetic energies of the photoelectrons ejected from the L_{II} and L_{III} levels.

9. The mass-attenuation coefficients of aluminum, copper, and lead for x-rays with $\lambda = 0.880$ Å are respectively 0.975, 9.12, and 13.5 m²/kg. Find the mass per unit area of each of these metals required to reduce the intensity of a beam of 0.880-Å photons by a factor of $\frac{1}{2}$. What are the thicknesses of the half-value layers if the densities are respectively 2700, 8940, and 11,350 kg/m³?

10. Consider the matrix operator $\hat{s} = 1_x\hat{s}_x + 1_y\hat{s}_y + 1_z\hat{s}_z$ with

$$\hat{s}_x = \frac{\hbar}{2}\begin{bmatrix} 0 & 1 \\ 1 & 0 \end{bmatrix} \qquad \hat{s}_y = \frac{\hbar}{2}\begin{bmatrix} 0 & -i \\ i & 0 \end{bmatrix} \qquad \hat{s}_z = \frac{\hbar}{2}\begin{bmatrix} 1 & 0 \\ 0 & -1 \end{bmatrix}$$

Show by direct calculation that $\hat{s}_x$, $\hat{s}_y$, $\hat{s}_z$ satisfy the commutation rules for angular momentum, that is, $[\hat{s}_x, \hat{s}_y] = i\hbar\hat{s}_z$. Show that $\alpha = \begin{bmatrix} 1 \\ 0 \end{bmatrix}$ and $\beta = \begin{bmatrix} 0 \\ 1 \end{bmatrix}$ are simultaneously eigenfunctions of $\hat{s}^2$ and $\hat{s}_z$ and find the corresponding eigenvalues.

chapter sixteen

X-ray Spectra

In the preceding chapter we discussed the atom as a nucleus surrounded by a cloud of electrons. To a reasonable approximation we regard each electron as associated with a hydrogen-like orbital determined by the field produced by the nucleus and by the average charge cloud of all the other electrons. From quantum mechanics and the Pauli exclusion principle we assume that any heavy atom in its ground state has two K electrons (with $n = 1$), eight L electrons ($n = 2$), and so forth. Data from photoelectric absorption are compatible with this idealization. We now consider various transitions involving electrons in the inner shells.

16.1 Low-lying Energy Levels

In treating atomic states thus far, the energies of levels have been referred to a zero corresponding to a free electron at rest at infinity; thus all bound states involve negative energy for this electron. This is the usual approach for treating optical spectra. Actually, the energy levels are characteristic of the atom as a whole. *In x-rays it is customary to choose as the zero point for energy measurements the neutral atom in its ground state.* Then all other states have positive energy. If a K electron is missing, the energy of the atom is positive and corresponds to the energy required to remove this electron from the atom.

Further, in optical spectroscopy one ordinarily reports transitions between electronic states, while in x-rays the atomic states are considered and attention is usually focussed on the hole (or absence of electron in a normally filled shell). For example, a $K \rightarrow L_{III}$ transition is one in which there is an initial vacancy in the K shell and a final vacancy in the L shell; the electron goes from the L_{III} subshell to the K shell, or just opposite to the transition for the hole. Clearly, it is not rigorous to think of this as a single-electron transition, because the energy states of the other electrons are also changed; the electronic charge cloud is different for the two states.

In spite of the fact that the energy levels belong to the **atom** as a whole, the one-electron quantum numbers and modified hydrogen orbitals are useful. For every filled shell and subshell, the resultants of orbital, spin, and total angular momenta are zero. Therefore, if one electron is missing from a closed subshell, the allowed values of the orbital, spin, and total angular momenta for the rest of the electrons in the subshell are identical with those for the single electron which can complete the subshell. Thus, n, l, and j for a subshell with one electron missing are the same as for an electron which can fill the vacancy.

From data on photoelectric absorption and from other measurements it is possible to determine the energies associated with atomic

Table 16.1   *X-ray energy levels of seven elements,* keV†

Level	Al¹³	Cu²⁹	Mo⁴²	Ag⁴⁷	W⁷⁴	Pb⁸²	U⁹²
K	1.562	8.996	20.036	25.556	69.637	88.163	115.80
L_I	0.1154	1.104	2.872	3.811	12.115	15.892	21.795
L_{II}	0.00730	0.955	2.633	3.529	11.559	15.231	20.974
L_{III}	0.00727	0.935	2.528	3.356	10.219	13.061	17.193
M_I		0.122	0.509	0.718	2.821	3.860	5.556
M_{II}		0.078	0.413	0.603	2.574	3.566	5.187
M_{III}		0.076	0.396	0.572	2.279	3.076	4.306
M_{IV}		⌠0.003	0.234	0.373	1.870	2.591	3.725
M_V		⌡	0.231	0.567	1.807	2.482	3.556

† λ in angstrom units is equal to $12.398/E$ in keV.

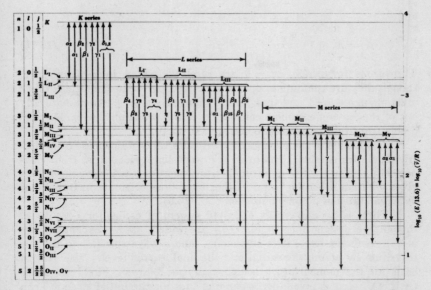

Fig. 16.1 X-ray energy-level diagram for U^{92}, showing the transitions allowed by the selection rules $\Delta l = \pm 1$, $\Delta j = 0, \pm 1$.

states characterized by small values of the principal quantum number n. Values for seven elements are quoted in keV in Table 16.1, while Fig. 16.1 shows the energy of these x-ray levels [in terms of log $\bar{\nu}/R$, which is the same as log $(E$ in eV/13.6$)$] for uranium. Also shown are the most prominent radiative transitions.

16.2 Characteristic X-ray Lines

On the basis of the energy-level diagram in Fig. 16.1 we might expect that there would be three lines associated with transitions from K to L, five lines associated with transitions from K to M, and so forth. However, a detailed study of x-ray spectra reveals that there are only two $K \rightarrow L$ transitions and two $K \rightarrow M$ transitions. Transitions between levels can occur only if certain selection rules are satisfied. For x-rays the selection rules for electric-dipole radiation (all the strong lines are of this type) are

$$\Delta l = \pm 1 \tag{16.1a}$$

and

$$\Delta j = 0, \pm 1 \tag{16.1b}$$

Transitions for which $\Delta n = 0$ are rarely observed in x-ray spectra, though they are common in optical spectroscopy. The conditions that l must change by ± 1 and that j must change by 0 or ± 1 come naturally out of the wave-mechanical calculation of the probability of an electric-dipole transition between two states (Sec. 13.10).

In addition to the prominent x-ray lines, all of which arise from electric-dipole transitions, there are a few weak lines which arise from magnetic-dipole and electric-quadrupole transitions for which wave mechanics gives different selection rules. Other faint lines originate in transitions between states of double ionization (Sec. 16.6).

There is no universally accepted system of names for x-ray lines, though the designations in Fig. 16.1 are widely used. Except for the most prominent lines, it is customary (and always specific) to list the transition by giving the initial and the final atomic level. The general basis for the nomenclature is essentially as follows. First, the series (K, L, M, etc.) is given, corresponding to n for the initial state. Then the strongest line in the series when observed under low resolution is named α, the next strongest β, and so forth. Under higher resolution the α line is in general a doublet; the more intense component is called α_1 and the weaker α_2. In the L series the β_1 line is next strongest. There are many weaker lines in the same wavelength region; these are called β_2, β_3, . . . , β_{15}. A similar system is used for the γ lines. The relative positions of the less prominent lines vary from element to element, and a numbering system which appears entirely logical for one element may seem chaotic for an element of substantially different atomic number.

The $K\alpha_1$ line arises from a $K \rightarrow L_{III}$ transition for the atom, corresponding to an L_{III} electron jumping to a vacancy in the K shell. The $K\alpha_2$ line is associated with a L_{II} electron filling the K vacancy. Since there are four $(2j + 1)$ L_{III} electrons and two L_{II} electrons, one might guess that the $K\alpha_1$ line is twice as intense as the $K\alpha_2$ line, an inference well substantiated by experiment. Similarly the line $K\beta_1$ ($K \rightarrow M_{III}$) is twice as strong as $K\beta_2$ ($K \rightarrow M_{II}$), again because the statistical weights $(2j + 1)$ are in the ratio of 4:2. However, this simple proportionality to the statistical weight is good only for the intensities of two lines of very nearly the same frequency. Thus $K\alpha_1$ is much stronger than $K\beta_1$ although the statistical weights of L_{III} and M_{III} are both 4; the probability that an L electron will fill a K vacancy is very much greater than that for an M electron. The relative intensities of certain L and M x-ray lines can be calculated by applying the Burger-Dorgello-Ornstein sum rule (Sec. 17.3).

16.3 Electron Excitation of X-ray Levels

One of the first experimental tests of the theory of x-ray spectra was made by Webster in 1916. Using a rhodium target, he increased the energy of the bombarding electrons and observed the intensity of the rhodium $K\alpha_1$ line at 0.613 Å. A plot of the intensity of x-radiation at this wavelength as a function of tube voltage is shown in Fig. 16.2. Such a plot is called an *isochromat*. The curve shows that the intensity increases gradually as the voltage across the x-ray tube increases, until at 23.2 kV it rises abruptly. This can be explained as follows. Below 23.2 kV the intensity at 0.613 Å is due to the continuous x-radiation, which increases as the voltage is raised, as discussed in Sec. 7.8. None of this radiation is characteristic of the rhodium; a similar intensity would be expected for any target of comparable atomic number. At 23.2 kV the incident electrons have sufficient energy to eject K electrons from rhodium atoms. As soon as vacancies in the K shell are produced, transitions from other states occur. In particular, electrons from the L_{III} level undergo transitions to the K shell and give rise to the rhodium $K\alpha_1$ line. As the energy of the bombarding electrons rises, the number

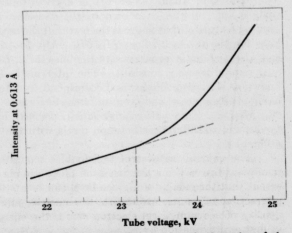

Fig. 16.2 An isochromat showing the intensity of the rhodium $K\alpha$ line as a function of the potential difference applied to the x-ray tube.

of K electrons ejected increases rapidly and the intensity of the $K\alpha_1$ line rises steeply.

Photons of the rhodium $K\alpha_1$ line have an energy of 20.1 keV, rather than the 23.2 keV. The difference of 3.1 keV corresponds to the energy of the L_{III} level. In order to excite the K lines it is necessary to eject an electron from the K shell. These electrons cannot go to an occupied level; they must either be ejected from the atom or else go to an allowed optical level. The excitation potential for all Rh K lines is exactly the same, since in every case it is necessary to eject an electron from the K shell before any K line can be radiated. In the case of x-rays elements do not strongly absorb the frequencies which they emit. This is in sharp contrast to the optical case. The reason is simple: in optical spectra electrons are raised to vacancies in higher levels and return to the ground states; in x-ray spectra electrons must be raised to an *unoccupied* level. When there is a vacancy in an x-ray level, it is most likely to be filled by one of the electrons from an intermediate level.

Many isochromats have been run for other x-ray lines. The results are in excellent agreement with data from other experiments designed to measure the energies of low-lying atomic energy levels.

16.4   X-ray Doublets and Screening Constants

The manner in which the energy of a given inner electron level varies with atomic number is shown in the Moseley plot of Fig. 16.3. A striking feature of that figure is the fact that the separation of the Moseley lines for the levels L_{II}-L_{III}, M_{II}-M_{III}, M_{IV}-M_V, etc., increase steadily as Z rises, while the separation of the lines for L_I-L_{II}, M_I-M_{II}, M_{III}-M_{IV}, etc., remain roughly constant. The intervals L_I-L_{II}, M_I-M_{II}, . . . are known as *screening doublets* and correspond to differences between levels having the same n, s, and j, but different l values. The intervals L_{II}-L_{III}, M_{II}-M_{III}, . . . are called *spin-relativity doublets*, or *regular doublets*, and correspond to differences between levels with the same n, l, and s, but different j.

The general features of the doublet separations can be explained semiquantitatively on the basis of fairly simple arguments. To the extent that the energy of a given level can be attributed to the coulomb interaction of an electron with the nucleus and the average charge density of a symmetrical electron cloud, the energy of a level is given in first approximation by $E_n = -Rhc(Z - \sigma_1)^2/n^2$. Here σ_1 is a *screening constant* which takes into account the effect of the electron cloud, so that $Z - \sigma_1$ is an *effective* nuclear charge.

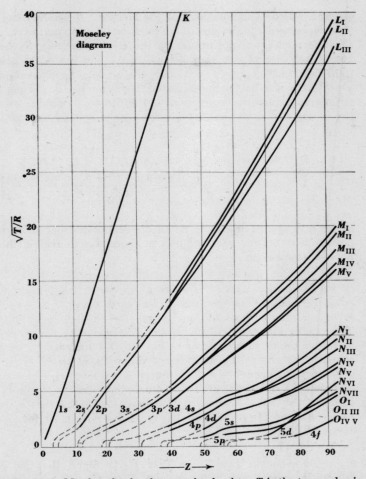

Fig. 16.3 Moseley plot for the x-ray levels where T is the term value in wavenumber units; $\sqrt{T/R} = \sqrt{E/13.6}$, where E is the energy of the level in electron volts. (*After H. E. White, "Introduction to Atomic Spectra." Copyright 1934. McGraw-Hill Book Company. Used by permission.*)

The splitting of an n,l level into two fine-structure levels for $j = l + \frac{1}{2}$ and $j = l - \frac{1}{2}$ is due to the spin-orbit interaction, for which Eq. (14.23) is appropriate provided we replace Z by $Z - \sigma_2$, where σ_2 is a screening constant which is less than σ_1. The reason σ_2 is smaller is that for the spin-orbit interaction it is the net electric intensity experienced by the

electron which is important, not the energy. A symmetrical shell of electrons external to the electron in question takes no part in reducing the effective nuclear charge insofar as the electric intensity is concerned, but it does insofar as energy is concerned. For the relativistic correction, Eq. (14.16), it turns out that $Z - \sigma_2$ should replace Z, and then Eq. (14.16) is a valid approximation, even for $l = 0$. When the relativistic and spin-orbit corrections are added to E_n above, we have for the energy E_{nlj} of an x-ray level characterized by n,l,j the result

$$E_{nlj} = - Rhc \left\{ \frac{(Z - \sigma_1)^2}{n^2} - \frac{\alpha^2 (Z - \sigma_2)^4}{n^3} \left[\frac{3}{4n} - \frac{1}{l + \frac{1}{2}} \right. \right. $$
$$\left. \left. + \frac{j(j + 1) - l(l + 1) - s(s + 1)}{2l(l + \frac{1}{2})(l + 1)} \right] \right\} \quad (16.2)$$

where $\alpha = \frac{1}{137}$. (In a more detailed calculation other terms in higher powers of α^2 occur.) We are now in a position to explain the various doublet separations.

a. Regular-doublet Law   *The energy difference ΔE between two spin-relativity doublets is proportional to the fourth power of $Z - \sigma_2$, where σ_2 is the screening constant appropriate for the spin-orbit interaction.* Two energy states differing only in j correspond to the same "orbits" and have the same first terms in Eq. (16.2). Increasing j by 1 clearly changes the energy by an amount proportional to $(Z - \sigma_2)^4$. Since only electrons inside the orbit in question influence σ_2, one expects that σ_2 should be independent of Z. But σ_2 increases as one goes to subshells further from the nucleus. Sommerfeld found that σ_2 is 3.5 for $L_{II,III}$; 8.5 for $M_{II,III}$, 13 for $M_{IV,V}$, 17 for $N_{II,III}$, 24 for $N_{IV,V}$, and 34 for $N_{VI,VIII}$.

b. Screening-doublet Law   In 1920 Hertz found that *the difference between the square roots of the energies of two screening-doublet levels is a constant essentially independent of Z.* This statement, sometimes called the *irregular-doublet law*, is only approximately true; the differences increase slowly with Z. To obtain this law from Eq. (16.2) we neglect all corrections involving α^2 and have immediately for two levels a and b of the doublet

$$\sqrt{-E_a} = \sqrt{Rhc} \, \frac{(Z - \sigma_1)_a}{n} \qquad \sqrt{-E_b} = \sqrt{Rhc} \, \frac{(Z - \sigma_1)_b}{n}$$

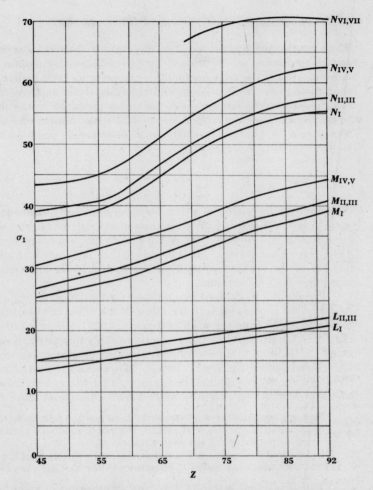

Fig. 16.4 The screening constant σ_1 as a function of atomic number. (*After Sommerfeld.*)

Since a and b involve different orbits, σ_1 is different in the two cases and we have $\Delta \sqrt{E} = \sqrt{-E_a} - \sqrt{E_b} = \sqrt{Rhc}\ \Delta\sigma_1/n$, where $\Delta\sigma_1$ is the difference in screening constants. The screening constant σ_1 increases with Z (Fig. 16.4), as we expect. The changes in slope of the N curves are associated with the filling of the $4f$ shell for $57 < Z < 72$.

16.5   The Fluorescence Yield and the Auger Effect

Thus far the one process we have discussed by which an excited atom can go to a lower energy state is by the emission of a photon, but this is by no means the only possibility.   As early as 1909 Sadler observed that the number of K photons emitted from metal appeared to be less than one-third the number of K vacancies produced in the metal.   If n_e electrons are ejected from the K shell of some element and as a result there are n_p K photons emitted, the K fluorescence yield w_K is defined to be

$$w_K = \left(\frac{n_p}{n_e}\right)_K \tag{16.3}$$

The fluorescence yields of other levels are similarly defined; for any state w is the *fraction of the excitations of the state which lead directly to the emission of a photon*.   Several methods for measuring fluorescence yields have been devised.   It is found that w_K increases with the atomic number of the emitter, going from less than 0.10 for light elements to more than 0.95 for uranium.

If only 40 K photons are emitted for 100 K electrons ejected in copper, what process leads to the filling of the other 60 vacancies?   The answer emerged from the work of Auger in 1925.   Using a Wilson cloud chamber, he studied the x-ray ejection of photoelectrons from argon atoms.   He found that over 90 percent of the long photoelectron tracks were accompanied by a short track.   To make a quantitative study of the latter, Auger diluted the argon with hydrogen to increase the length of what we now call the *Auger tracks*.   He found that:

1.   The Auger track always originates at the same point as a photoelectron track.
2.   The direction of ejection of the Auger electron is random and independent of the direction taken by the photoelectron.
3.   The length of the Auger track is independent of its direction, of the direction of the photoelectrons, and of the frequency of the incident x-ray photons.

The first point suggests that the photoelectron and the Auger electron come from the same atom.   The other two are consistent with the hypothesis that the ejections of the photoelectron and the Auger electron are independent events, occurring sequentially rather than simultaneously.   In Auger's experiments the ejection of the photoelectron left the argon atom with a vacancy in the K shell.   When one L electron filled this vacancy, a second L electron was ejected as the Auger electron, thus leaving the atom doubly ionized.   Some persons have called this process the *internal photoelectric effect*, imagining that a $K\alpha$ photon was absorbed on its way out of the atom by an L electron, which was thus internally ejected as a photoelectron. There exists an abundance of evidence that this is not the case. If this explanation were correct, one would find Auger electrons associated not only with atoms from which the photoelectron was ejected but also with other atoms. However, we do not find Auger tracks originating at other atoms. Also, from the absorption coefficient of argon for its own $K\alpha$ radiation, one expects that perhaps one photon in a million might be absorbed in the same atom from which it is emitted. But Auger found that 93 percent of the argon atoms had Auger electrons associated with the photoelectrons. It is generally agreed that no photon is emitted in the Auger process; rather the excited atom undergoes a transition in which one electron moves to a lower level and another electron is simultaneously ejected. An adequate wave-mechanical description of Auger transitions has been developed; it predicts transition probabilities of the observed magnitude.

In a $K \rightarrow L_{III}L_{III}$ transition the energy of the Auger electron is the energy of the K level less the energy of the doubly ionized atom described by $L_{III}L_{III}$. The latter is slightly more than twice the L_{III} ionization energy because it requires more energy to remove an electron from an ionized atom of atomic number Z than from the corresponding state in a neutral atom. (In many cases it is a good approximation to assume this is the energy needed to remove an electron from the same state in an atom of atomic number $Z + 1$.) The energy of the Auger electron depends on the properties of the ejecting atom only. Auger transitions need not leave the atom with two vacancies in the same shell. For example, one Auger transition which is of importance for the $L\alpha$ satellites is $L_I \rightarrow L_{III}M_V$; in general, Auger transitions from the L_I state are so probable that the L-series lines originating from L_I vacancies are very much less intense than would otherwise be the case.

It is the Auger effect that competes with radiation as processes by which an excited atom can lose energy. *An Auger transition is one in which an electron is emitted rather than a photon when an excited atom goes to a lower energy state.* When both electrons of a helium atom are simultaneously excited from their $1s$ orbitals, Auger transitions occur in which one electron is ejected and the other returns to the $1s$ state. Such *autoionization* is not uncommon in atomic spectroscopy.

16.6 *Multiple Ionization of Inner Shells*

It was perhaps fortunate that the spectral apparatus available to Moseley and the early workers did not have the sensitivity and resolving power of present-day spectrometers. The lines which they observed were the more intense and more easily resolvable lines of x-ray spectra, due to transitions between states of *single ionization* and called *first-order* lines.

With improvements in technique, other lines were discovered which did not fit into the conventional diagram. The majority of these lines were rather faint, were usually found close to, and on the short-wavelength side of, the more intense diagram lines, and hence were called *satellite lines*. A typical spectral curve of the satellite structure is reproduced in Fig. 16.5. The satellite structure is observed to be complex, containing numerous component lines of various intensities. Most (if not all) first-order lines are accompanied by such satellites; the total number of satellite lines now known far exceeds the number of diagram lines.

Because of the low intensity of the satellites, reliable experimental information on their characteristics is extremely difficult to obtain. The excitation voltage of certain satellites, however, has been definitely shown to be somewhat greater than the excitation voltage of the accompanying first-order, or parent, line. In the case of the type of K satellites that is illustrated in Fig. 16.5 the energy of excitation is found to be equal to the energy required to eject a K electron and *in addition* an L electron from the atom. Hence, we may

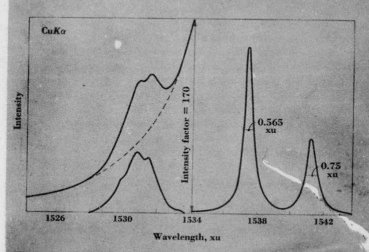

Fig. 16.5 The $K\alpha$ doublet of Cu^{29} with accompanying satellites. The satellite intensity, lower left, is obtained by subtracting the estimated intensity due to other lines. (*Parratt, Phys. Rev., vol. 50, p. 1, 1936.*)

assume that the initial state for the emission of these satellite lines is a state of *double ionization*, in which the atom has an electronic vacancy in both the K shell and the L shell. Such a state of the atom may be called a KL *atomic state*. In a similar way, other states of double ionization, such as KM, LM, etc., should be possible.

An atom in a KL state may undergo a radiative transition into any one of a number of other states of double ionization, for example, $KL \rightarrow KM$ (an electron dropping from the M shell into the L shell) or $KL \rightarrow LL$ (an electron dropping from the L shell into the K shell). Estimates of the atomic energy indicate that the loss of energy should be slightly greater in the transition $KL \rightarrow LL$ than in the diagram transition $K \rightarrow L$, which gives rise to the $K\alpha$ lines; hence, the former transition should give rise to satellites close on the short-wavelength side of the $K\alpha$ lines. Similarly, the transition $KL \rightarrow LM$ should give rise to satellites on the short-wavelength side of the $K\beta$ lines. Presumably, a cathode-ray electron can eject two electrons at once from an atom. If this is the origin of the doubly ionized atoms, theoretical estimates indicate that the intensity of satellites relative to the parent lines should decrease in a continuous manner with increasing atomic number. Such a variation with atomic number is found by experiment to hold for satellites accompanying lines of the K series but not for L or M satellites.

The intensity of the satellites accompanying the $L\alpha$ line is observed to decrease rather abruptly as the atomic number increases from 47 to 50 and to increase again rather abruptly at about 75; between atomic numbers 50 and 75, $L\alpha$ satellites are practically unobservable. This anomalous behavior was resolved in 1935, when Coster and Kronig pointed out the importance of the Auger effect in this connection. They suggested that the $L\alpha$ satellites came primarily from $L_{III}M_{IV}$ and $L_{III}M_V$ initial states and that these in turn were produced by Auger transitions from L_I states. The transitions $L_I \rightarrow L_{III}M_{IV}$ and $L_I \rightarrow L_{III}M_V$ can occur only if the energy of the L_I state is greater than that of the doubly ionized states $L_{III}M_{IV}$ and $L_{III}M_V$. This condition is satisfied only for elements with $Z < 50$ and $Z > 75$. Hence, if we suppose that few atoms in the states $L_{III}M_{IV,V}$ are produced directly by the cathode-ray bombardment, so that such states are produced chiefly by Auger transitions, the absence of $L\alpha$ satellites from $Z = 50$ to $Z = 75$ is explained.

The occurrence of Auger transitions should also have an effect upon the intensities of certain *diagram* lines. Thus, transitions for which L_I is the initial state should be weakened by removal of atoms from the L_I state through Auger transitions and hence should undergo a rather abrupt change in intensity at $Z = 50$ and $Z = 75$ (about), in agreement with experiment.

16.7   X-ray Spectra and the Outer Part of the Atom

X-ray spectra are related primarily to the *inner* atomic structure. Concurrent changes that may occur elsewhere in the atom can largely be ignored because these changes have relatively little effect upon the energy levels arising from vacancies in the inner shells. Certain finer features of x-ray spectra, however, involve in their explanation a consideration of the outer part of the atom or even, in a solid or liquid, of the surrounding material.

We have assumed that when an x-ray photon is absorbed, an electron is removed entirely from the absorbing atom. If the atom is isolated, however, as in a monatomic gas, it is possible for the electron to stop in some outer vacancy in the atom; the absorbed energy $h\nu$ would then be less than if the electron were removed to infinity. The K absorption edge of argon, studied with high resolving power, is shown in Fig. 16.6. In addition to the photoelectric absorption extending toward higher frequencies from the K absorption edge, the curve shows a series of absorption lines on the long-wavelength side of this edge, resulting from atomic transitions into one of the K "resonance" levels. These lines should be closely spaced, however; for the energy differences between the levels should be of the same order of magnitude as the differences between the ordinary optical levels, i.e., a few electron volts or less. The electronic configuration for argon ($Z = 18$), $1s^2 2s^2 2p^6 3s^2 3p^6$, is altered in nonionizing K absorption to $1s2s^2 2p^6 3s^2 3p^6 np$, a p state being the only kind to which an electron can be excited from a $1s$ state by photon absorption (since the selection rule allows l to change only by 1). The spacing of the atomic levels under discussion will thus be almost the same as that of the optical P levels of potassium ($Z = 19$). Hence, if we ascribe the most intense line in Fig. 16.5 to the electronic transition $1s \rightarrow 4p$, we can locate the positions of the other resonance lines and of their series limit by using the known optical P terms for potassium. The photoelectric absorption edge should have a finite "width" when observed with such high resolving power; theory indicates that the shape should be given by an arctangent curve, whose width is due to the same factors that cause the width of each of the absorption lines and also of the emission lines. The theoretical absorption edge is

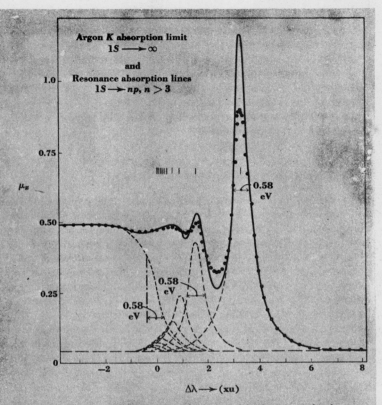

Argon *K* absorption limit
$$1S \longrightarrow \infty$$

and

Resonance absorption lines
$$1S \longrightarrow np, n > 3$$

Fig. 16.6 The *K* absorption edge of argon and its interpretation in terms of a series of resonance lines at wavelengths greater than the continuous *K* absorption band, which is centered at *O*. (*Courtesy of L. G. Parratt.*)

drawn in the figure as the left-hand dotted curve, centered at the calculated position of the *P*-series limit.

When an atom is brought into close proximity to other atoms, x-ray phenomena involving only ionization in an inner shell are little affected, but those involving also the atomic exterior are considerably modified. In a solid, the outermost atomic electrons come to be associated with the entire solid rather than with individual atoms, and essentially continuous bands of electronic states replace the discrete atomic states so far as these electrons are concerned (Sec. 21.5). The simplest case is that of the alkali metals. Here the set of electronic states for the valence electrons in their individual atoms is replaced in the crystal by a band of electronic states several electron volts wide in energy. Transition of electrons from this band to vacancies in an x-ray level leads to a broad radiation band rather than a sharp line. For example, the L emission from solid sodium, corresponding to transitions from the $3s$ band to the unresolved $2p$ level, is observed as a wedge-shaped band beginning at about 450 Å and rising to a maximum near 390 Å.

16.8   Refraction and Reflection of X-rays

a. Refraction   The first positive evidence that x-rays are measurably refracted came from the work of Stenström, who showed from accurate measurements of wavelength that Bragg's law for the reflection of x-rays from crystals does not yield identical values when the wavelength of a given line is computed from different orders of reflection. Hjalmar found, for example, that the apparent wavelength of the Fe $K\alpha_1$ line as measured in the first order by reflection from a gypsum crystal ($2d = 15.155$ Å) was 1.9341 Å; while measurements in the sixth order gave 1.9306 Å, nearly 0.2 percent less.

This apparent failure of the Bragg formula was shown to be due to refraction of the rays on entering the crystal, the index of refraction being *less than unity*. In Fig. 16.7 is shown the path of a ray entering at glancing angle θ and incident on the Bragg plane PP at an angle θ'. Bragg's law in the form

$$n\lambda' = 2d \sin \theta' \tag{16.4}$$

where λ' is the wavelength *in the crystal,* gives the true law of reflection *at the crystal plane.* To obtain the relation between θ and λ, the wavelength in air, we use the law of optics, $\mu = \lambda/\lambda' = \cos \theta/\cos \theta'$, μ being the refractive index from air to crystal. Substituting values of λ' and

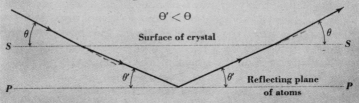

Fig. 16.7   Refraction of x-rays entering the surface of a crystal.

of $\sin \theta' = (1 - \cos^2 \theta')^{\frac{1}{2}}$ in Eq. (16.4), expanding in powers of the very small quantity $1 - \mu$, and keeping only the first power of this quantity, it is found that

$$n\lambda = 2d \sin \theta \left(1 - \frac{1 - \mu}{\sin^2 \theta} \right) \qquad (16.4a)$$

A few values of $1 - \mu$ or δ obtained with use of this formula by Larsson, from observations in which different orders of diffraction were compared, are as follows:

λ, Å	$\delta \times 10^6$	
	Mica	Calcite
1.537	8.94	8.8
2.499	24.6	22.4
3.447	49.1	41.9
5.166	103	
7.111	182	
8.320	262	

Good values of $1 - \mu$ have also been obtained by refraction through a prism and by measuring the critical angle for total reflection.

b. Reflection from Ruled Gratings   Once it was established that the index of refraction of materials is less than unity, it was clear that x-rays could be totally reflected in going from air to some medium, such as glass or a metal. As a consequence, the possibility arose that a ruled grating might be used to measure x-ray wavelengths in exactly the same way that a grating is used in the optical region, provided that the glancing angle between the x-ray beam and the ruled surface is less than the critical angle for total reflection. Compton and Doan in 1925 were the first to make measurements of this kind. Using a grating of speculum metal with 50 lines per millimeter, they found the wavelength of the $K\alpha_1$ line of molybdenum to be $\lambda = 0.707 \pm 0.003$ Å. Ten years later Bearden, using gratings having 100 or 300 lines per millimeter ruled on glass sputtered with gold, measured the wavelength of the Cu $K\alpha_1$ line (1.5406 Å) to an accuracy better than 0.01 percent.

Measurements made with ruled gratings afford the most reliable means of determining the absolute magnitudes of x-ray wavelengths, since they involve no assumptions as to the homogeneity of a crystal; in fact, the only elements that enter into the determination by this method are the wave theory of light as propagated in a vacuum and such well-tested operations as the measurement of angles and the counting of lines under a micrometer microscope.

Problems

1. Indicate all allowed (electric-dipole) transitions between the N and M x-ray levels.

2. If the K, L, and M energy levels of platinum lie at roughly 78, 12, and 3 keV respectively, compute the approximate wavelengths of the $K\alpha$ and $K\beta$ lines. What minimum potential difference across an x-ray tube is required to excite these lines? At approximately what wavelength is the K absorption edge? At approximately what wavelength are the three L edges?

Ans: 0.187 Å, 0.165 Å, 78 kV, 0.159 Å, 1 Å

3. The $K\beta_1$ line ($K \rightarrow M_{\text{III}}$) of rhodium ($Z = 45$) has a wavelength of 0.5445 Å. Combining this with Webster's data from the K isochromat (Sec. 16.3), calculate the energy of the M_{III} level of Rh. Check the reasonableness of your answer by interpolation, using Table 16.1.

4. (a) If silver $K\alpha$ photons strike a molybdenum absorber, find the kinetic energy with which K electrons are ejected.

(b) If the molybdenum atom undergoes an Auger transition, $K \rightarrow L_I L_{\text{III}}$, calculate the initial kinetic energy of the Auger electron, assuming that the excitation energy of the state of double ionization is just the sum of the individual energies. Does this assumption lead to too high or too low an energy for the Auger electron?

5. (a) In 1925 to 1926 Auger found that of 223 instances in which an incident photon ejected a photoelectron from the K shell of Kr^{36}, 109 of the photoelectron tracks were accompanied by an Auger electron track. Find the K fluorescence yield in krypton.

(b) Calculate the K fluorescence yield for argon ($Z = 18$) if out of 165 K photoelectron tracks, 153 are accompanied by Auger tracks.

Ans: (a) 0.51; (b) 0.07

6. Auger's experiments on the photoelectric absorption of photons in xenon ($Z = 54$) lead to the results that $w_K = 0.71$ while $w_L \approx 0.25$. (His data did not permit him to distinguish from which L subshell a photoelectron arose.) If he studied 270 ejections from the K shell and

175 from the L shell, how many photoelectrons from each shell were accompanied by Auger electrons?

7. Use the regular doublet law, Sommerfeld's value for σ_2 (Sec. 16.4), and the $L_{II} - L_{III}$ separation in Ag^{47} from Table 16.1 to calculate the $L_{II} - L_{III}$ separation in In^{49}. (The measured value is 227.5 eV.)

8. From the screening-doublet law and the data for Pb^{82} in Table 16.1, estimate the energy of the L_{II} level in Hg^{80} if the L_I level of mercury has an energy of 14.873 keV. (The observed energy for L_{II} is 14.238 keV.)

9. Negative muons captured by protons make transitions from $n = 2$ levels to the ground state. The photons emitted fall on a copper foil. From what Cu level is a photoelectron emitted with the smallest energy? What is the approximate kinetic energy of such a photoelectron?

10. The critical angle for x-ray reflection from glass is 15 minutes of arc for photons with $\lambda = 1.54$ Å. What is the index of refraction of glass at this wavelength?

Ans: 0.999906

11. The phase velocity of x-rays in a crystal is slightly greater than c, and so the index of refraction of the crystal may be written as $1 - \delta$, where $\delta \lesssim 10^{-5}$.

(*a*) Show that when refraction at the crystal surface is taken into account, Bragg's law is replaced by

$$n\lambda = 2d \left(1 - \frac{2\delta - \delta^2}{\sin^2 \theta}\right)^{\frac{1}{2}} \sin \theta$$

(*b*) Show that if λ_1 and λ_2 are the apparent wavelengths obtained for a given line by Bragg's law in orders n_1 and n_2, δ can be calculated from

$$\delta = \frac{2(\lambda_1 - \lambda_2)}{\lambda_1 + \lambda_2} \frac{n_2^2}{n_2^2 - n_1^2} \sin^2 \theta_1$$

which is known as *Stenström's formula*.

12. Calculate the percentage error introduced in $n\lambda$ by the use of the simple form of Bragg's law in place of the more exact form (given in the preceding problem) for the case in which $\delta = 10^{-5}$ and $\theta = 15°$.

Ans: 0.015 percent

13. (*a*) Show that when x-rays are totally reflected at almost grazing incidence from a ruled grating, the condition for an interference maximum is $n\lambda = a[\cos \theta - \cos(\theta + \alpha)]$, where a is the grating space, θ the glancing angle, and $\theta + \alpha$ the angle the reflected beam makes with the reflecting face of the grating.   What is the physical significance of negative values of n?

(b) If 2.5-Å x-rays are incident at a glancing angle of 40 minutes of arc on a grating ruled with 50 lines per millimeter, find the glancing angle $\theta + \alpha$ for each of the two first-order maxima.

Ans: 43.5 and 36.1 minutes of arc

chapter seventeen

Atomic Spectra

In the preceding chapters the outstanding success of one-electron quantum mechanics and the Pauli exclusion principle in explaining the periodic table and x-ray spectra has been described. We now turn to the broad diversity of optical spectra exhibited by the elements. These spectra ordinarily arise from the excitation of a single outer electron; other electrons may be regarded as remaining in their original states (to a good approximation). To explain spectra of the more complicated atoms we introduce other interactions in addition to the electric field of the nucleus, the spherically symmetric part of the average field of the other electrons, and the spin-orbit interaction.

17.1 Angular Momentum and Selection Rules

As the starting point for discussing the optical spectra of multielectron atoms, it is convenient to utilize a *vector model* in which the electron cloud about a nucleus of charge Ze arises from Z individual electrons, each in a single-electron state of the kind discussed in Sec. 15.1 and each characterized by four quantum numbers n, l, m_l, and m_s. Although the electrons lose their identity in the electron cloud, we can imagine each as contributing spin angular momentum, orbital angular momentum, and the associated magnetic moments to the atom as a whole. The resultant angular momentum of the electron system is designated by $\mathbf{A}_J$.

In classical mechanics the vector angular momentum of any isolated system remains constant in time. Analogously, *the quantum states of any atom not subject to external forces can be so defined that when the atom is in one of them, the total angular momentum* $\mathbf{A}_J$ *of its electron system has the fixed magnitude*

$$A_J = \sqrt{J(J + 1)}\, \hbar \tag{17.1}$$

and the component of $\mathbf{A}_J$ *about any chosen Z axis drawn through its center of mass has the fixed value*

$$(\mathbf{A}_J)_z = M_J \hbar \tag{17.2}$$

where J and M_J are two quantum numbers characteristic of the quantum state. The number J is a positive integer or half integer or zero, and M_J has one of the $2J + 1$ integrally spaced values between $M_J = -J$ and $M_J = J$, inclusively. Thus, M_J is integral or half integral according as J is. For example, if $J = 0$, $M_J = 0$; if $J = \frac{1}{2}$, $M_J = \frac{1}{2}$ or $-\frac{1}{2}$; if $J = 1$, $M_J = 1$ or 0 or -1; etc.

The rule for determining whether J and M_J are integral or half integral can be discovered by considering the electronic quantum states described in Sec. 15.1. In those states there is, for each electron, an integral value of the quantum number m_l, and also a value of m_s which is either $\frac{1}{2}$ or $-\frac{1}{2}$; and the total angular momentum of the electrons about the chosen axis is $[\Sigma(m_l + m_s)]\hbar$, where the sum Σ is to be taken for all electrons in the atom. Thus,

$$M_J = \Sigma(m_l + m_s) \tag{17.3}$$

From this equality and the character of the values of m_l and m_s it is at once obvious that M_J and hence also J are integral when the atom contains an even number of electrons and half integral when it contains an odd number. It can also be seen that the maximum possible value of M_J, and hence also of J, is $\Sigma l + N/2$, where N is the total number of electrons in the atom; for the maximum value of any individual m_l is l, and for each electron m_s cannot exceed $\frac{1}{2}$.

The relationship of the quantum numbers J and M_J can be visualized with the help of a vector diagram, just as the relation of l and m_l was visualized for a single electron (Sec. 14.5). As in the case of the one-electron atom, the vector angular momentum has no definite azimuth around the axis, and its components perpendicular to the axis have no definite values.

a. Selection Rules   Wave mechanics leads to the following selection rules for J and M_J, which are of great importance in spectroscopy. In any atomic jump associated with the dipole emission or absorption of a photon, *J may remain unchanged or it may increase or decrease by unity* but not by any larger amount. That is, if J_1 and J_2 are the values of J for the two states between which the jump occurs, then either $J_1 = J_2$ or $J_1 - J_2 = 1$ or $J_1 - J_2 = -1$. The same rule holds also for M_J. *No jump can occur, moreover, from a state with $J = 0$ to another state with $J = 0$.* In symbols,

$$\Delta J = 0 \text{ or } \pm 1 \text{ (and not } 0 \to 0) \qquad \Delta M_J = 0 \text{ or } \pm 1 \qquad (17.4)$$

These selection rules are not always strictly obeyed, however; sometimes jumps occur by quadrupole emission for which $\Delta J = \pm 2$ or $\Delta M_J = \pm 2$. Spectral lines due to such jumps are mostly very weak.

These selection rules have analogies in classical theory (and were first proposed on this basis). Orbital angular momentum, in classical mechanics, is associated with rotation. Suppose that a body capable of rotation contains an electrical vibrator of some sort; and let this vibrator emit electromagnetic radiation of frequency ν when the body is at rest. Then, if the body is set rotating with angular velocity ω, the radiation emitted will contain just the three frequencies ν, $\nu + \omega$, $\nu - \omega$. The reason is similar to that for the analogous effect of a magnetic field upon a classical vibrating electron, which was described in Sec. 6.3 in connection with the classical explanation of the Zeeman effect. The three classical frequencies $\nu + \omega$, ν, $\nu - \omega$ correspond (in emission) to the three possible changes in M_J. This is an example of Bohr's correspondence principle.

b. Angular Momenta of Closed Subshells.   From the description of atomic subshells in Sec. 15.2 it is clear that in a closed subshell every electron is matched by another whose quantum state differs only in that the sign of m_l is reversed; and the values of m_s are similarly paired off. Thus $M_J = \Sigma m_l + \Sigma m_s = 0$. This fact suggests that $J = 0$ also, and this surmise can be verified from the theory. Furthermore, $\Sigma m_l = 0$ and

$\Sigma m_s = 0$, separately, and, correspondingly, it can be shown that the total orbital angular momentum of the electrons in a full subshell likewise vanishes, and so does the resultant angular momentum of spin.

Thus, in considering the total angular momentum of an atom, all closed subshells can be ignored. An atom of a noble gas in its normal state, in which the electrons are all grouped into closed subshells, is thus necessarily in a state with $J = 0$. An atom containing one or more peripheral electrons outside the inner closed subshells will have angular momenta, orbital, spin, or total, that are determined entirely by the peripheral electrons, with no contributions from the closed subshells. The consideration of atomic angular momenta is thereby greatly simplified.

17.2   Alkali-type Spectra

At this point we may to advantage describe the gross features of the simplest type of spectra, that emitted by atoms containing a single valence electron outside one or more closed subshells. Among neutral atoms of this sort may be mentioned those of the alkali metals, lithium, sodium, potassium, rubidium, and cesium. Other examples are singly ionized atoms of beryllium, magnesium, calcium, strontium, or barium; doubly ionized atoms of boron or aluminum; and so on.

In an atom of this type only the valence electron is active, and only its quantum numbers vary. For the moment, let us disregard electron spin. The energy of an atom containing a single outer electron depends on the quantum numbers n and l. The latter specifies the orbital angular momentum not only of the valence electrons but, in this case, also of the entire electron system of the atom, since contributions from the inner closed subshells vanish. However, when there are two or more electrons outside of closed subshells, this total orbital angular momentum is the resultant of that associated with the individual electrons. The total orbital angular momentum of the electron system is $\sqrt{L(L + 1)}\, \hbar$ (Sec. 17.6), where L is zero or integer; for the alkalis $L = l$, of course. Energy levels of atoms are commonly labeled with capital letters according to the following scheme:

L	0	1	2	3	4	5	6	7
	S	P	D	F	G	H	I	K

In general, it is customary to use for the electron system of an atom capital letters L, S, J to characterize the resultants of the angular momenta represented by l, s, and j for the individual electrons.

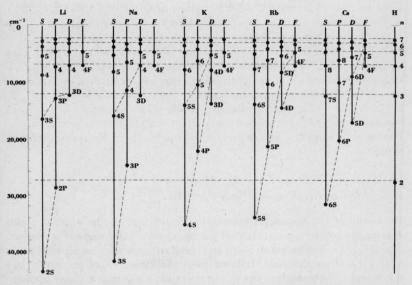

Fig. 17.1 The energy levels between 50,000 and 2000 cm⁻¹ of the five alkali metals and of hydrogen. All fine structure is ignored.

The gross features of the energy levels of the alkalis are shown in Fig. 17.1; the doublet fine structure, characteristic of all save the S levels and discussed in the following section, is not shown. In this figure the energies are given in terms of wavenumbers, i.e., the number of waves per unit length, $1/\lambda$, since it is wavelength which the optical spectroscopist measures. For the greatest L the energy is close to that of the corresponding hydrogen level. For smaller L the energy levels lie lower, by virtue of the penetrating-orbit effect (Sec. 15.6).

Spectral lines are emitted when the valence electron makes a transition from one level to another, the energy of the photon being equal to the energy difference between the two states. Only certain transitions compatible with the selection rules are allowed. The quantum number L is usually subject to the selection rule

$$\Delta L = 0,\ \pm 1 \qquad \text{for jumping electron } \Delta l = \pm 1 \qquad (17.5)$$

Hence, no dipole jumps occur, for example, between S and F levels. In the case of a single active electron, where $L = l$, $\Delta L = \pm 1$ only.

For the alkalis the most prominent lines arise from transitions from P to S levels. They are called *principal series* lines. A line arising from an $S \rightarrow P$ transition belongs to one of the *sharp series* (Sec. 9.2), while transitions from D to P levels lead to *diffuse series* lines. Bergmann found a series of sodium lines in the infrared which originated in $F \rightarrow D$ transitions and for which the energies were close to those of some hydrogen lines. This was called a *fundamental series*. The identification thus made of the terms in the alkali spectra constitutes the historical reason for the use by spectroscopists of the mysterious letters S, P, D, F (from the words sharp, principal, diffuse, fundamental) to represent various values of L (or the corresponding small letters for one electron). For higher values of L it was agreed later to continue down the alphabet, skipping J.

17.3 Fine Structure in Alkali Spectra

By virtue of the spin-orbit interaction the energy of the atom depends also on the relative orientation of the spin and orbital angular momenta, Eq. (14.19). For an alkali atom the resultant spin angular momentum is $\sqrt{(\frac{1}{2})(\frac{3}{2})}\,\hbar$ since the contributions from the closed cores are zero; only the valence electron gives the electron system a net spin angular momentum. Once again it is customary to use capital letters for quantum numbers for the complete electron system, and S is used to designate the total spin quantum number for the atom. In this case $S = \frac{1}{2}$.

For a sodium atom in its ground state the valence electron has $n = 3$, $l = 0$, and $s = \frac{1}{2}$; for the electron system $L = 0$, $S = \frac{1}{2}$, $J = \frac{1}{2}$. It is customary to write this state $3s\ ^2S_{\frac{1}{2}}$ or $3^2S_{\frac{1}{2}}$ (read "three doublet ess one-half"), where the first number stands for n, the superscript is $2S + 1$, the letter gives L, and the subscript J. Other states are specified in the same way; for example, $4^2D_{\frac{5}{2}}$ conveys the information that the outer electron has $n = 4$, the spin quantum number S is $\frac{1}{2}$, $L = 2$, and $J = \frac{5}{2}$.

The effect of spin-orbit interaction is to split the energy levels or terms into two, one with $J = L + \frac{1}{2}$ and the other with $J = L - \frac{1}{2}$. Only the S terms ($L = 0$) are single, with $J = \frac{1}{2}$. Almost always the level with the larger J is found experimentally to have the higher energy. As an example, a diagram of the energy levels for neutral sodium is shown in Fig. 17.2. Each level is represented in this diagram by a short horizontal line. The diagram shows the splitting of the P levels; the subscripts on S, P, etc., denote J values distinguishing fine-structure sublevels, which are not shown separately for D and F. Plotted are binding energies measured downward from zero for the state of ionization (with no kinetic energy in the removed electron); these *terms* are expressed in wavenumber units. The numbers written opposite the

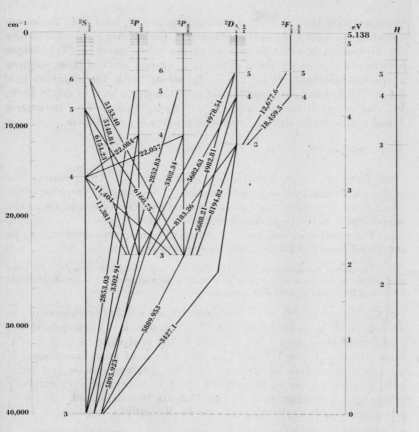

Fig. 17.2 Energy-level diagram for neutral sodium, showing the origin of the stronger spectral lines. Hydrogen levels are shown for comparison. (Wavelengths in angstrom units; the numbers opposite the levels are values of *n*.)

levels are the values of *n*. The most important spectral lines are indicated by oblique lines with wavelengths in Å. One weak line is shown $(3S \leftarrow 3D)$[1] that violates the $\Delta L = 0, \pm 1$ selection rule.

By far the most prominent features of the sodium spectrum are the *D* lines in the yellow region of the spectrum. They come from $3\,^2S_{\frac{1}{2}} \leftarrow 3\,^2P_{\frac{3}{2}}$ and $3\,^2S_{\frac{1}{2}} \leftarrow 3\,^2P_{\frac{1}{2}}$ transitions. There are four quantum states ($M_J = \frac{3}{2}$, $\frac{1}{2}$, $-\frac{1}{2}$, $-\frac{3}{2}$) associated with the $P_{\frac{3}{2}}$ level and two with $P_{\frac{1}{2}}$. It is essentially

---

[1] It is customary for spectroscopists to write the final state first and then the initial one. We shall follow this custom, using an arrow between level designations as a reminder.

twice as likely that an excited Na atom be in a $P_{\frac{3}{2}}$ level as in the corresponding $P_{\frac{1}{2}}$ level and, as a consequence, the shorter λ member of the Na D doublet is twice as intense as the other member. Transitions between two given nL terms give rise to several spectral lines, forming a spectral multiplet (Fig. 17.3). In accord with the selection rule $\Delta J = 0, \pm 1$, jumps are possible between the single level of any S term, $^2S_{\frac{1}{2}}$, and either of the two levels in a P term, $^2P_{\frac{1}{2}}$ and $^2P_{\frac{3}{2}}$. Between a P and a D, with levels $^2D_{\frac{3}{2}}$, $^2D_{\frac{5}{2}}$, three jumps are possible, viz.,

$$^2P_{\frac{1}{2}} \leftarrow {}^2D_{\frac{3}{2}} \qquad {}^2P_{\frac{3}{2}} \leftarrow {}^2D_{\frac{3}{2}} \qquad {}^2P_{\frac{3}{2}} \leftarrow {}^2D_{\frac{5}{2}}$$

(The jump $^2P_{\frac{1}{2}} \leftarrow {}^2D_{\frac{5}{2}}$ is forbidden, since $\Delta J = 2$.) The resulting trio of lines is known as a *compound doublet*. Thus *the chief spectrum of the alkali metals consists of spectral doublets and compound doublets*, hence the superscript 2 in the level designation.

The relative intensities of the lines of a compound doublet may be calculated from the *Burger-Dorgello-Ornstein sum rule.*

The sum of the intensities of all the lines of a multiplet which arise from the same initial state or terminate at the same final state is proportional to the statistical weight $2J + 1$ of the initial or final state respectively.

For the case of the $^2P \leftarrow {}^2D$ multiplet, let the relative intensities of the lines be α, β, γ, as shown in Fig. 17.3. Applying the sum rule to the initial state gives $\beta/(\alpha + \gamma) = \frac{6}{4}$ and to the final state $(\alpha + \beta)/\gamma = \frac{4}{2}$, from which $\alpha:\beta:\gamma = 1:9:5$. This sum rule is of great importance in the practical analysis of spectra, since the spectroscopist is first faced with a large number of lines which must be organized very carefully before a term diagram such as Fig. 17.2 can be established. The sum rule gives reliable predictions only for small multiplet splitting.

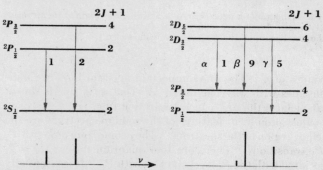

Fig. 17.3 Transitions between doublet levels and (*below*) relative frequencies of the lines. If S lies above P, the order of the lines is reversed; similarly if P lies above D.

As confirmation of the association of multiplet structure and the term diagram, the wavenumbers and separations of the first four lines of each of the chief series in the spectrum of neutral sodium are listed in Table 17.1.   The theoretical conclusions are seen to be confirmed by the observations.   The *sharp* series, $3P \leftarrow mS$, consists of spectral doublets, and, as shown in column 7 of Table 17.1, the frequency difference is constant within the experimental errors, representing the difference between the $3^2P_{\frac{1}{2}}$ and $3^2P_{\frac{3}{2}}$ levels.   This fact helps also to confirm the identification of the S levels as made above.

The *principal* series of lines, $3S \leftarrow nP$, also consists of doublets but with frequency differences that represent the spacings between the two levels in various P terms and hence decrease rapidly with increasing n. The separation of the two levels in a term with quantum numbers n and l, due to the spin-orbit effect, is calculated from Eq. (14.23) to be

$$\Delta\bar{\nu} = \frac{\Delta E_{l+\frac{1}{2}} - \Delta E_{l-\frac{1}{2}}}{hc} = \frac{Z_{\text{eff}}^4 e^2\hbar^2[(l + \frac{1}{2})(l + \frac{3}{2}) - (l - \frac{1}{2})(l + \frac{1}{2})]}{(hc)16\pi\epsilon_0 m^2 c^2 a_B{}^3 n^3 l(l + \frac{1}{2})(l + 1)}$$

$$= \frac{\alpha^2 R Z_{\text{eff}}^4}{n^3 l(l + 1)} = 5.84\,\frac{Z_{\text{eff}}^4}{n^3 l(l + 1)} \qquad \text{cm}^{-1} \qquad (17.6)$$

where Z_{eff} = effective atomic number

$\quad\quad R$ = Rydberg constant

$\quad\quad \alpha$ = fine-structure constant, Eq. (9.10)

Thus in a one-electron atom the spin-orbit separation decreases as $1/n^3$. The observed decrease in the spectrum of sodium is even more rapid, presumably because the effective value of Z decreases as n increases.

The lines of the *diffuse* series, $3P \leftarrow mD$ are compound doublets. One line should be very weak, however.   The two brightest lines result from the jumps $3^2P_{\frac{1}{2}} \leftarrow m\,^2D_{\frac{3}{2}}$ and $3^2P_{\frac{3}{2}} \leftarrow n\,^2D_{\frac{5}{2}}$.   If Eq. (17.6) holds roughly for sodium, because of the factor $l(l + 1)$ in the denominator, the separation $nD_{\frac{5}{2}} - nD_{\frac{3}{2}}$ should stand to the separation $nP_{\frac{3}{2}} - nP_{\frac{1}{2}}$ in the ratio $1/(2 \times 3):1/(1 \times 2)$ or as only 1:3; and, as n increases, the D separation should rapidly diminish further.   With ordinary resolving power the lines of the diffuse series are observed as apparent doublets with a frequency difference that is nearly constant and equal to the difference $3P_{\frac{3}{2}} - 3P_{\frac{1}{2}}$ (Table 17.1).   Doublets of *exactly* this separation are formed by the faint line $3P_{\frac{3}{2}} \leftarrow n_1 D_{\frac{3}{2}}$ and the brighter line $3P_{\frac{1}{2}} \leftarrow n_1 D_{\frac{3}{2}}$.

Thus, the theory accounts very well for the principal features of the spectrum emitted by neutral sodium atoms.   It is equally successful with the other alkali metals, the spectra of which are qualitatively very similar to that of sodium.

Table 17.1 Doublet intervals in the NaI spectrum

Series	n	J	Initial Term Value, cm^{-1}	Line		Interval $\Delta\bar{\nu}$, cm^{-1}
				λ, Å	$\bar{\nu}$, cm^{-1}	
Principal; $3^2S_{\frac{1}{2}} \leftarrow n^2P_J$ 3S term value, 41,449.65 cm^{-1}	3	$\frac{1}{2}$ $\frac{3}{2}$	24,493.47 24,476.37	5,895.92 5,889.95	16,956.18 16,973.38	17.20
	4	$\frac{1}{2}$ $\frac{3}{2}$	11,182.77 11,177.14	3,303.32 3,302.99	30,266.88 30,272.51	5.63
	5	$\frac{1}{2}$ $\frac{3}{2}$	6,409.38 6,406.86	2,853.03 2,852.83	35,040.27 35,042.79	2.52
	6	$\frac{1}{2}$ $\frac{3}{2}$	4,153.14 4,151.89	2,680.44 2,680.34	37,296.51 37,297.76	1.25
Sharp; $3^2P_J \leftarrow n^2S_{\frac{1}{2}}$ Term values, cm^{-1} $3^2P_{\frac{1}{2}} - 24,493.47$ $3^2P_{\frac{3}{2}} - 24,476.27$	4	$\frac{1}{2}$ $\frac{3}{2}$	15,709.8	11,381.2 11,403.6	8,783.7 8,766.5	17.2
	5	$\frac{1}{2}$ $\frac{3}{2}$	8,249.0	6,154.23 6,160.75	16,244.5 16,227.3	17.2
	6	$\frac{1}{2}$ $\frac{3}{2}$	5,077.0	5,148.84 5,153.40	19,416.5 19,399.3	17.2
	7	$\frac{1}{2}$ $\frac{3}{2}$	3,437.6	4,747.94 4,751.82	21,055.9 21,038.7	17.2
Diffuse; $3^2P_J \leftarrow n^2D$†	3	$\frac{1}{2}$ $\frac{3}{2}$	12,276.8	8,194.82 8,183.26	12,216.7 12,199.5	17.2
	4	$\frac{1}{2}$ $\frac{3}{2}$	6,900.9	5,682.63 5,688.20	17,592.6 17,575.4	17.2
	5	$\frac{1}{2}$ $\frac{3}{2}$	4,412.9	4,978.54 4,982.81	20,080.6 20,063.4	17.2
	6	$\frac{1}{2}$ $\frac{3}{2}$	3,062.4	4,664.82 4,668.56	21,431.1 21,413.9	17.2

† Only the mean value of the $D_{\frac{3}{2}}$ and $D_{\frac{5}{2}}$ level is shown and only the two strong lines.

The separation of the doublet levels increases rapidly with increasing Z. In Table 17.2 are shown the wavelengths and the wavenumbers

Table 17.2 *First lines of the principal series for the alkali metals*

	Li	Na	K	Rb	Cs
Z	3	11	19	37	55
n	2	3	4	5	6
$\lambda(^2S_{\frac{1}{2}} \leftarrow ^2P_{\frac{1}{2}})$, Å	6,707.8	5,895.9	7,699.0	7,947.6	8,943.5
$\lambda(^2S_{\frac{1}{2}} \leftarrow ^2P_{\frac{3}{2}})$, Å		5,890.0	7,664.9	7,800.2	8,521.1
$\bar{\nu}(^2S_{\frac{1}{2}} \leftarrow ^2P_{\frac{1}{2}})$, cm^{-1}	14,904	16,956	12,985	12,579	11,178
$\bar{\nu}(^2S_{\frac{1}{2}} \leftarrow ^2P_{\frac{3}{2}})$, cm^{-1}		16,973	13,043	12,817	11,732
$\Delta\bar{\nu}$, cm^{-1}	0.34	17	58	238	554

of the D lines or their analogs, i.e., the first lines of the principal series, for all the alkali metals and also the doublet differences for these lines. The enormous increase in the doublet separation with Z for alkali metals, in spite of the progressive increase in n, is in strong contrast with the more moderate variation in the wavelength of the lines in question. This results from great sensitiveness of the spin-orbit effect to the character of the central field near the nucleus.

17.4 The Spectrum of Helium

During the eclipse of 1868 Janssen detected helium lines in the solar spectrum. Lockyer suggested the name *helium* (*helios* is the Greek word for sun) for this element, spectral lines from which were observed many years before Ramsey first isolated helium on the earth in 1895. Long before the helium spectrum was understood, measurements of the wavelengths of the spectral lines had led to the energy-level diagram shown in Fig. 17.4, in which there are two sets of levels. No transitions were observed from a level in one set to a level in the other. At first it was believed that there might be two kinds of helium, called *orthohelium* and *parahelium*, one associated with each of the term systems. The parahelium terms led to series of singlet lines, while the orthohelium terms led to series of triplets. When efforts to separate these two types of helium proved futile, the two systems of terms were eventually explained in an entirely different way, viz., *an orthohelium atom has two electrons*

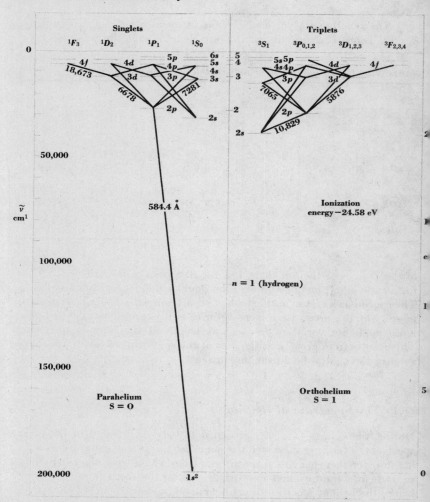

Fig. 17.4 Term diagram for neutral helium showing the division of terms into noncombining singlet and triplet terms.

with spin parallel and a parahelium atom has two electrons with spins antiparallel (see Sec. 15A.2).

The resultant spin angular momentum for two electrons is given by $\sqrt{S(S+1)}\,\hbar$, where

$$S = s_1 \oplus s_2 = \tfrac{1}{2} \oplus \tfrac{1}{2} \tag{17.7}$$

which has possible values 1 or 0. Here $\oplus$ implies quantized vector addition, as in Sec. 14.8. For orthohelium $S = 1$; for parahelium $S = 0$.

An outstanding feature of the triplet terms of orthohelium is that there is no ground state corresponding to that of the singlet system. According to the Pauli exclusion principle, two electrons in the $n = 1$, $l = 0$, $m_l = 0$ state cannot have the same value of m_s. Thus only for parahelium can the ground state involve a $1s^2$ configuration; for orthohelium the lowest configuration is $1s2s$.

For parahelium $S = 0$, so $J = L \oplus S = L$. Since for each value of L there is a single value of J, the energy levels are singlets with no fine structure and are designated by $^1S_0$, $^1P_1$, $^1D_2$, etc. For orthohelium $S = 1$, and so $J = L \oplus S$ has the values $L + 1$, L, $L - 1$ (but never minus values). Thus, except for $L = 0$ levels, each nL state has three fine-structure levels. Orthohelium levels are triplets, designated by such symbols as $^3S_1$, $^3P_0$, $^3P_1$, $^3P_2$, $^3D_1$, and so forth.

17.5   Many-electron Wave Theory

When an atom contains two or more electrons outside closed subshells, it is generally true that in a radiative transition only one electron is active but the atomic energy levels between which the jump occurs are themselves influenced by the presence of the other electrons. As a basis for the discussion of such cases, we describe briefly the various interactions which must appear in the complete hamiltonian function required to calculate the wave function for a neutral atom. Although the exact relativistic wave equation for a multielectron atom is not known, an approximate equation can be written in which the following terms may occur:

1. Terms representing the kinetic energy of the electrons; for the jth electron this takes the form $-(\hbar^2/2m)\nabla_j^2$.
2. Potential-energy terms representing the electrostatic interaction energy of each electron with the nucleus.
3. Potential-energy terms representing the mutual electrostatic energy of the electrons.
4. Spin-orbit terms of the form of Eq. (14.23) for electrons with $l > 0$.
5. Mixed spin-orbit terms representing interaction between each spin magnetic moment and the orbital motions of other electrons.
6. Spin-spin interactions between the spin magnetic moments of the electrons.
7. Interactions between the magnetic moments arising from the "orbital" motion of the electrons.
8. Relativistic corrections, presently only partly known.
9. Terms allowing for nuclear motion, for interaction between nuclear spin and the electron magnetic moments, interactions of electrons with the nuclear quadrupole moments, and so forth.

Fortunately, interactions associated with classes 5 to 9 are ordinarily small compared with the first four. The largest term in the potential energy is that of the nuclear electrostatic interaction (class 2) followed by that of 3. The latter may be split into two parts: a spherically symmetrical one and a nonspherically symmetric one. The first part we have taken into account in terms of shielding constants, which have the effect of reducing the nuclear charge from Z to Z_{eff}. The second part is associated with electrons outside of closed subshells and has not been treated as yet. Particularly for light elements, the effect of the nonspherically symmetric electrostatic field is the largest of the remaining terms, but for heavy elements the spin-orbit interaction is greatest. In developing the theory one considers the largest terms first and then treats smaller terms as perturbations. We shall discuss first the situation in which the residual electrostatic interaction dominates (LS coupling) and later that in which the spin-orbit interaction dominates (jj coupling).

a. Electronic Configurations and Coupling Schemes   The possible states of a single electron in the modified central field are characterized by quantum numbers n and l. As assignment of n's and l's for all Z electrons in the atom defines an *electronic configuration*. For example, a C^6 atom in a $1s^2 2s^2 2p 3d$ configuration has a $1s^2$ core, two electrons in $2s$ states, one in a $2p$ and another in a $3d$ state. Within this configuration a number of different energy states (or energy levels) are possible, each with its own total energy. Often the words *term* or *term value* are used in place of *energy level*, a hangover from the work of Rydberg and Ritz (Sec. 9.3). When a radiative transition occurs, the photon has energy $h\nu$ equal to the energy difference between initial and final states. For optical spectra, only the electrons outside of closed subshells undergo transitions.

An important *selection rule* holds for configuration changes. In general, radiative transitions occur only between configurations that differ in just one of the n,l electronic states; furthermore, the values of l in the initial and final states must differ by unity. Thus transitions may occur between levels belonging to a $2s3p$ configuration and a $2s3d$ configuration, but not between $2s3p$ and $2p3d$ because both l's are different, and not between $2s3p$ and $2s4f$ because $\Delta l = 2$. The selection rule $\Delta l = \pm 1$ diminishes enormously the number of transitions that have to be considered in spectroscopy.[1]

---

[1] The rule holds strictly so long as the wave function corresponds exactly to a single configuration. In some cases, ψ contains components of appreciable magnitude belonging to other configurations, for which the sum $l_1 \oplus l_2 \oplus \cdots \oplus l_n$ differs by an even integer. In such cases transitions, called two-electron jumps, may occur in which two electronic l's change, with $\Delta l = 2$ for the second.

17.6 LS, or Russell-Saunders, Coupling

In atoms that are not too heavy, the residual electrostatic interaction has a relatively large effect. For simplicity of discussion, let it be assumed for the present that spin-orbit interactions are initially negligible. Consider two or more electrons outside of closed subshells. By virtue of the residual electrostatic repulsions, which are not directed toward (or away from) the nucleus, the individual orbital angular momenta of these electrons are no longer constant, but their resultant is unaffected by these forces. This resultant angular momentum $\mathbf{A}_L$ has a magnitude $\sqrt{L(L+1)}\hbar$, where

$$L = l_1 \oplus l_2 \oplus l_3 \oplus \cdots \tag{17.8a}$$

Here $\oplus$ means quantized vectorial addition (Fig. 17.5) such that if $l_1 = 2$ and $l_2 = 3$, $L = l_1 \oplus l_2 = 5, 4, 3, 2$, or 1. If $l_3 = 1, L = l_1 \oplus l_2 \oplus l_3 = 6, 5, 4, 3, 2, 1$, or 0. The (degenerate) quantum states can also be chosen so that the component of $\mathbf{A}_L$ in the direction of any chosen axis is $M_L\hbar$, where the $2L + 1$ alternative values of M_L are integrally spaced from $-L$ to L. Thus $L\hbar$ is the maximum observable value of the component of $\mathbf{A}_L$ in any direction.

In addition to orbital angular momentum $\sqrt{l(l+1)}\hbar$, each electron has spin angular momentum $\sqrt{(\frac{1}{2})(\frac{3}{2})}\hbar$. As the result of a strong quantum-mechanical exchange correlation between the spins of electrons, there is an effect which tends to align electron spins parallel to each other and which further tends to keep electrons with parallel spin away from each other. The spin angular momenta have a resultant $\mathbf{A}_S$ of magnitude $\sqrt{S(S+1)}\hbar$, where

$$S = s_1 \oplus s_2 \oplus s_3 \oplus \cdots = \tfrac{1}{2} \oplus \tfrac{1}{2} \oplus \tfrac{1}{2} \oplus \cdots \tag{17.8b}$$

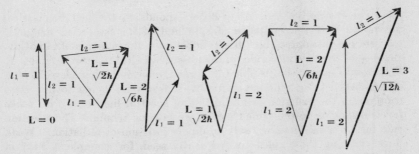

Fig. 17.5 Schematic diagram suggesting how $l_1 \oplus l_2$ gives $L = 0$, 1, or 2 when $l_1 = 1$, $l_2 = 1$ and gives $L = 1$, 2, or 3 when $l_1 = 2$, $l_2 = 1$.

The component of $\mathbf{A}_S$ in the z-direction has a value $M_S\hbar$, where M_S is any one of the integrally spaced numbers S to $-S$. Since each electronic spin can contribute at most $\hbar/2$ to the axial component, the maximum possible value of M_S, and hence the value of S, is $N/2$ for an atom containing N electrons outside of closed subshells. Both S and M_S are thus integers or zero, or half integral, according as N is even or odd.

In some configurations, two or more independent sets of LS quantum states occur having the same values of L and S.

Zero-order L,S,M_L,M_S wave functions are easily formed as linear combinations of the n,l,m_l,m_s functions. The L,S,M_L,M_S quantum states are not usually characterized by definite values of the individual m_l's and m_s's; the individual axial components of electronic angular momentum are not definite (except in a few cases). This corresponds to the fact in classical mechanics that fixing the vector sums $\mathbf{A}_L$ and $\mathbf{A}_S$ does not fix the individual vector momenta $\mathbf{A}_l$ and $\mathbf{A}_s$. Repulsion between electrons would usually have little average effect on the *magnitude* of any individual A_l; it would merely speed the electron up and then slow it down again during its orbital motion. A component of this repulsion normal to the orbital plane, on the other hand, would cause a precession of the individual A_l vectors about the fixed direction of their resultant $\mathbf{A}_L$, in analogy with the effect of a torque applied to a gyrostat. Such precessional motions symbolize vividly the necessity of abandoning the m_l's and m_s's and quantizing only the resultant momentum when full account is taken of electronic repulsion. The motions themselves, however, have no analog in the wave-mechanical description of the quantum states.

Simple *selection rules* hold for L and S; this is the most important property of such quantum numbers. The rules are

$$\Delta L = 0 \text{ or } \pm 1 \text{ (and not } 0 \to 0) \tag{17.9a}$$
$$\Delta S = 0 \tag{17.9b}$$

The restriction that $\Delta S = 0$ corresponds to the fact in classical mechanics that the existence of spin and of its associated magnetic moment has no appreciable (direct) effect upon the emission of radiation. Because of this selection rule, in the absence of all spin-orbit interaction, the atomic energy levels fall into broad, noncombining classes, each distinguished by a different value of S; levels belonging to one class combine in the emission or absorption of spectral lines only with other levels belonging to the same class, as we saw for helium. The selection rule for L refers, however, only to dipole emission of radiation. Weak lines violating these rules are frequently seen, for example, λ 3427 in Fig. 17.2.

The occurrence of atomic states characterized by quantum numbers L and S obeying the selection rules just stated is called LS or *Russell-Saunders coupling* (of the electronic momenta).

The LS Terms   In zero order, only a single energy level belongs to a given configuration such as $3p4d$. The residual electrostatic interaction then splits this single level into several levels belonging to different pairs of values of L and S. That this interaction should separate levels differing in L is reasonable. In classical theory, different values of L would imply different electronic orbits, and the average values of the repulsive potential energy of the electrons should therefore also be different. It may seem strange, however, that states differing only in S should likewise be separated in energy by electronic repulsions. The reason for such an effect lies in the different spatial symmetries of the wave functions associated with different values of S. The separation has little to do with the mutual energy of the spin magnets, although this would cause a slight separation if other causes were not active. In the mathematics, the separation of levels having different S results from exchange integrals of the electronic mutual energies. These integrals have no classical analog. Wave mechanics thus introduces a distinction that is absent in classical theory.

17.7 *LS Multiplets of Levels*

To explain the observed fine structure of the spectral lines, account must be taken of the *spin-orbit interactions* as a second perturbing energy much smaller (in light atoms) than the residual electrostatic energy. The spin-orbit interactions disturb both the orbital and the spin momenta but not the grand vector sum of all angular momenta. In the classical analog, the vectors A_L and A_S, unless actually parallel, would precess steadily about their resultant A_J, which has the magnitude $\sqrt{J(J+1)}\hbar$. Here

$$J = L \oplus S$$

so that the possible values of J that can occur in a given LS term are all those integrally spaced values that satisfy the inequality $|L - S| \leq J \leq L + S$. As usual, if the z direction is established by some field, $(A_J)_z = M_J\hbar$, where $M_J = J, J - 1, \ldots, -J$. Thus, if $L = 0$, it follows that $J = S$; if $S = 0$, $J = L$; in either case the level is single. If $L \geq S$, J takes on all the $2S + 1$ integrally spaced values from $L + S$ down to $L - S$, inclusive; whereas if $L < S$, J takes on the $2L + 1$

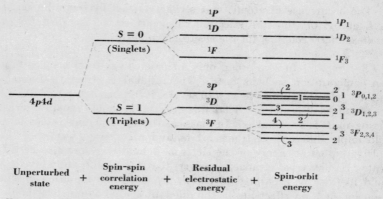

Fig. 17.6 The fine-structure splitting arising from a $4p4d$ configuration. The Landé interval rule is illustrated by the spacings of the triplet levels. (*From R. B. Leighton, "Principles of Modern Physics." Copyright 1959. McGraw-Hill Book Company. Used by permission.*)

integrally spaced values from $S + L$ to $S - L$. Because of these relations, the *observed* number of levels composing a given multiplet obviously furnishes important information about the values of L or of S that should be assigned to the term. The observed number of levels τ must equal either $2L + 1$ or (more commonly) $2S + 1$; hence, either $L = (\tau - 1)/2$ or, more likely, $S = (\tau - 1)/2$. A further test must then be made by noting whether or not the value of L or S so inferred is consistent with the selection rules as applied to transitions between the given multiplet of levels and other multiplets for which L or S may be assumed to be known.

Unless L or S is zero, the single energy level belonging to a given LS term is thus split by the spin-orbit interactions into sublevels characterized by different values of J. The various L,S,J levels arising in this way from the $4p4d$ configuration are shown in Fig. 17.6. This configuration gives rise, by electrostatic interaction and the exchange correlation between spins, to singlet and triplet P, D, and F terms. The triplet levels in turn are split by the spin-orbit interaction as shown.

All allowed transitions between two multiplets of levels, taken together, give rise to a group of spectral lines which may be called a *spectral multiplet.*

17.8   Spacing of the LS Multiplet Levels

For the spacing of the levels in an LS multiplet, wave mechanics furnishes a simple and useful formula known as the *Landé interval rule.* The increase in the atomic energy due to the spin-orbit effect, for any one of the J levels of a given LS level multiplet, is provided by extending Eq. (14.23) to the many-electron system

$$\Delta E = \tfrac{1}{2}B[J(J+1) - L(L+1) - S(S+1)] \tag{17.10}$$

Here B is a constant that varies from one level multiplet to another. The difference between the energy of a level for J and that for $J+1$ is the difference in the corresponding values of ΔE, or

$$E_{J+1} - E_J = \tfrac{1}{2}B[(J+1)(J+2) - J(J+1)] = B(J+1) \tag{17.11}$$

This equation expresses Landé's interval rule:

The energy differences between two successive J levels are proportional, in a given LS term, to the larger of the two values of J.

The rule is of great help in determining the values of J that are to be assigned to various observed levels.

According to Eq. (17.10), ΔE is positive for some values of J and negative for others. The weighted average of ΔE vanishes, provided each level is weighted in proportion to the number, equal to $2J+1$, of the M_J states composing it. The total energy of the weighted-average level is given by the formula

$$\bar{E}_{LS} = \frac{\displaystyle\sum_J (2J+1)E_J}{\displaystyle\sum_J (2J+1)} \tag{17.12}$$

It is to such a weighted-average level that a Rydberg formula really refers when it is written without regard to the fine structure of the terms.

17.9   The Arc Spectrum of Mercury

The familiar arc spectrum of mercury presents spectroscopic features of great interest. The principal levels and many of the lines are shown in the usual way in Fig. 17.7, wavelengths being given in Angstrom units. It is instructive to note how a configuration such as $6s6p$ gives rise, by electrostatic interaction, to singlet and triplet P terms, with the latter split into $^3P_0$, $^3P_1$, and $^3P_2$ levels by the spin-orbit interaction. The mercury energy-level diagram presents a fair example of LS coupling for two electrons. A number of intercombination lines (between singlet and triplet terms) are found, however, so that the LS coupling is not

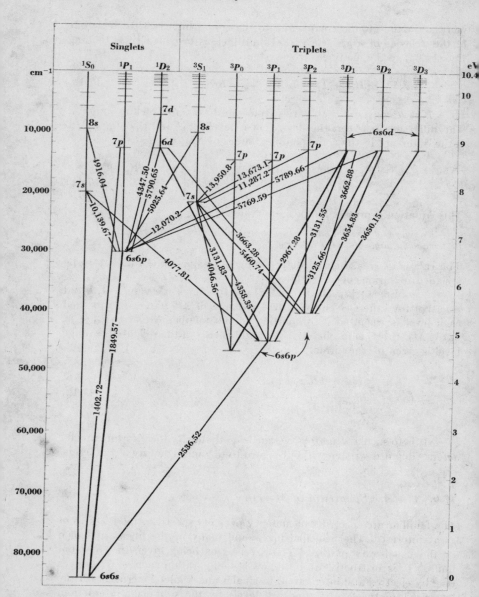

Fig. 17.7 The most important energy levels and spectral lines for neutral mercury.

perfect in mercury. The configuration for the ground state is $6s^2$; thus the normal state of a mercury atom is $^1S_0$. In four of the most important terms, viz., $6s6p$ ^1P, $6s6p$ ^3P, $6s6d$ ^1D, and $6s6d$ ^3D, the mere change of an electronic l with no change in n results in a comparatively large increment of atomic energy.

The intercombination line, $6s^2$ $^1S_0 \leftarrow 6s6p$ $^3P_1$, $\lambda = 2536.52$, is one of the principal ultraviolet lines in the mercury spectrum. Its strength is associated with the fact that the LS coupling approximation becomes less good as Z increases. For Hg ($Z = 80$) the spin-orbit interaction has become large, and the selection rules for L and S begin to fail (see Sec. 17.11). The transition probability between the levels is high, as is shown by the intensity of the line in absorption. In addition, the $6s6p$ $^3P_1$ state becomes densely populated as electrons drop into it from higher triplet states.

Atoms accumulate also in the $^3P_2$ and $^3P_0$ levels of the $6s6p$ configuration. From these levels they cannot pass by a radiative jump to the normal state, for $^1S_0 \leftarrow$ $^3P_2$ would mean $\Delta J = 2$, and $^1S_0 \rightarrow$ $^3P_0$ would mean a jump from $J = 0$ to $J = 0$, both of which are forbidden by the selection rules for J. Levels lying above the normal state out of which radiative transitions are extremely improbable are called *metastable* levels. An atom can stay in such a level for a long time if it is not disturbed by outside influences. It may eventually be brought back into the normal state, however, by a *collision of the second kind*,[1] and it is presumably in this manner that the atoms in the $^3P_2$ and $^3P_0$ levels are eventually returned to their normal state.

The selection rule for L is pretty well obeyed. With values of L assigned as in the diagram, S terms combine only with P terms, P with S and D, and so on. However, a few weak lines, not shown on the diagram, have been observed corresponding to $\Delta L = 2$. All the transitions shown in Fig. 17.7 are in harmony with the selection rule for configurations. Only one electron changes its n and l, and always $\Delta l = \pm 1$.

The spectral lines can be grouped into *series*, if desired. Thus, within the singlet system, all lines ending on the lowest $^1S_0$ level form the singlet principal series. Only these and the line $\lambda = 2536$ Å can be observed in absorption in mercury vapor. Of the lines ending on the lowest ^1P level, those originating from ^1S terms form a sharp series, those originating from ^1D terms, a diffuse series, just as in sodium; and so on. Similar series can be picked out within the triplet system. It

---

[1] Collisions resulting in the excitation of a previously unexcited atom or molecule are called "collisions of the first kind," while collisions in which an excited atom or molecule is returned to its ground state, the energy of excitation appearing as increased kinetic energy of the colliding particles, are known as "collisions of the second kind."

is really not very interesting to group the lines of such a complex spectrum into series, however, especially when the fine structure is as coarse as it is in the mercury spectrum. Thus the great spectral sextet of ultraviolet lines, $\lambda = 2967$ to $\lambda = 3663$, from the lowest ^3D to the lowest ^3P term, would constitute together the first "line" of the triplet diffuse series. The lines $^1P_1 \leftarrow {}^1D_2$ (λ 5791), $^3P_2 \leftarrow {}^3S_1$ (λ 5461), and $^3P_1 \leftarrow {}^3S_1$ (λ 4358) are the familiar yellow, green, and blue lines emitted from the mercury arc.

It may also happen that *both* valence electrons are displaced into higher electronic states. A few levels ascribed to the $6p^2$ configuration have been discovered.

Among other neutral atoms which have spectra similar to that of mercury may be mentioned helium; then the alkaline earths, beryllium, magnesium, calcium, strontium, and barium; and the close relatives of mercury, zinc, and cadmium. Ions with similar electronic exteriors are C^{++}, Al^+, Si^{++}, Pb^{++} (the number of plus signs indicating the number of positive charges on the ion).

A second subclass of atoms with two optically active electrons is formed by those which, in their normal states, contain two s and two p valence electrons. In such cases the two s electrons usually (but not always) stay put, only the two p electrons being active. Examples of such atoms are neutral carbon, silicon, germanium, tin, and lead. Furthermore, certain observed spectra of the same type have been ascribed to singly (positively) ionized atoms of nitrogen, phosphorus, and bismuth.

17.10   Equivalent Electrons

When two electrons have the same n and also the same l, they are called *equivalent electrons*. In configurations containing equivalent electrons, such as $5s^2$ or $5s7p^3$, certain LS terms that might otherwise occur are excluded through the operation of the exclusion principle.

Consider, for example, the configuration $6s^2$ in mercury. Both electrons have the same value of n, l, and m_l, and so they must have opposite spin quantum numbers m_s. The resultant S must be zero. Indeed for any closed shell or subshell, $L = 0$ and $S = 0$; a $^1S_0$ state results.

As an example of one way we may formally find the possible states for equivalent electrons. i.e., electrons in the same nl subshell, consider a p^2 configuration. If we represent $m_s = \frac{1}{2}$ by $\uparrow$ and $m_s = -\frac{1}{2}$ by $\downarrow$, we begin by producing an array like that of Table 17.3, in which we show all possible combinations of spin and angular-momentum components consistent with the Pauli exclusion principle. The resultant

Table 17.3   Terms for two equivalent p electrons

m_l			$M_L = \Sigma m_l$	$M_S = \Sigma m_s$	Term
1	0	−1			
↑	↑		1	1	*³P
↑		↑	0	1	*
↑↓			2	0	Δ¹D
↑	↓		1	0	Δ
↓	↑		1	0	*
↑		↓	0	0	*
↓		↑	0	0	Δ
	↑↓		0	0	¹S
↓	↓		1	−1	*
↓		↓	0	−1	*
	↑	↓	−1	0	*
	↓	↓	−1	−1	*
	↓	↑	−1	0	Δ
	↑	↑	−1	1	*
		↑↓	−2	0	Δ

M_S is Σm_s, while M_L is Σm_l. For this case the largest value of M_S is 1, and the largest M_L associated with $M_S = 1$ is also 1. This suggests a ³P term for which M_L can be 1, 0, −1, and M_S 1, 0, −1. The nine entries marked by asterisks satisfy these requirements. (All nine must appear in the array; which of the terms with $M_L = 1$, $M_S = 0$ is chosen is of no importance here.) The highest remaining M_S is 0, and the highest associated M_L is 2, so that there is a ¹D term. This requires $M_L = 2, 1, 0, -1, -2$; five appropriate combinations are marked by Δ. When this is done, there remains unassigned a single $M_L = 0$, $M_S = 0$, which corresponds to a ¹S state. Thus the possible terms for a p^2 configuration are ³P₀,₁,₂, ¹D₂, and ¹S₀, many fewer than a pair of non-equivalent p electrons $(np, n'p)$ which has the possibilities ³D₁,₂,₃, ¹D₂, ³P₀,₁,₂, ¹P₁, ³S₁, and ¹S₀.

The possible terms for other configurations of equivalent electrons can be obtained by a construction similar to that of Table 17.3. Actually it is unnecessary to concern oneself with any term in which M_L or M_S is negative, since there is a one-to-one correspondence between the negative and positive values. Thus all the entries below the line in Table 17.3 could have been omitted. Table 17.4 shows the possible terms for various equivalent electron configurations and the total number of independent states arising from each configuration.

Since the contribution of any filled subshell to L and S is always zero, it is clear that if one electron is missing from a closed subshell,

Table 17.4   *LS* terms for equivalent electrons

Configuration	Terms	Ind. States
s^2	^1S	1
p^2, p^4	^3P, ^1D, ^1S	15
p^3	^4S, ^2D, ^2P	20
d^2, d^8	^3F, ^3P, ^1G, ^1D, ^1S	45
d^3, d^7	^4F, ^4P, ^2H, ^2G, ^2F, $^2D(2)$, ^2P	120
d^4, d^6	^5D, ^3H, ^3G, $^3F(2)$, ^3D, $^3P(2)$, ^1I, $^1G(2)$, ^1F, $^1D(2)$, $^1S(2)$	210
d^5	^6S, ^4G, ^4F, ^4D, ^4P, ^2I, ^2H, $^2G(2)$, $^2F(2)$, $^2D(3)$, ^2P, ^2S	252
f^2, f^{12}	^3H, ^3F, ^3P, ^1I, ^1G, ^1D, ^1S	91

the resultant L and S values for the closed-shell-minus-one-electron system must be just l and $\frac{1}{2}$ respectively. Thus a p^5 configuration leads to a ^2P term, just as a p^1 configuration does. Similarly a p^4 configuration has the same terms as a p^2 one. In general, *when a subshell is more than half filled, the allowed terms are identical to those for a number of electrons equal to the number of vacancies.* Table 17.4 shows several examples.

The spectroscopist is interested not only in what terms are possible from a given configuration but also in how the energies of the terms are related. This is of particular importance for determining the ground state of atoms. For LS coupling there are two rules which apply to the ground configuration but not always to excited ones:

Hund's Rule. *Of terms for equivalent electrons* (1) *the one with greatest multiplicity* $2S + 1$ *lies lowest, and the term energies increase as the multiplicity decreases, and* (2) *of terms of a given multiplicity, the lowest is that with greatest L.*

The Multiplet Rule. *For multiplets formed from equivalent electrons, the smallest value of J lies lowest if the subshell is half filled or less; the highest value of J is lowest if the shell is more than half full.*

For example, tungsten, $_{74}W$, has a d^4 configuration. Since the d subshell can hold 10 electrons of which 5 can have the same spin, the greatest S for the ground state is $(4 \times \frac{1}{2} =) 2$. Since all 4 electrons have the same spin, they must have different m_l, and the biggest possible M_L is then $(2 + 1 + 0 + -1 =) 2$. With $L = 2$ and $S = 2$, the smallest J is 0 and hence the ground state for $_{74}W$ is $^5D_0$. Osmium ($Z = 76$) has a d^6 configuration, four electrons short of a closed subshell, and so again the ground state has $S = 2$ and $L = 2$. However, this time the greatest J lies lowest, the multiplet is "inverted," and the ground state is $^5D_4$.

If a given configuration contains both equivalent and nonequivalent electrons, the LS fine structure can be obtained by beginning with the equivalent electron terms and then considering the contributions of the other electrons one by one. For example, an excited state of nitrogen involves a p^2d configuration. Adding a d electron to the p^2 configuration yields for the $^3P_{0,1,2}$ term alone the possibilities, $^4F_{\frac{3}{2},\frac{5}{2},\frac{7}{2},\frac{9}{2}}$, $^4D_{\frac{1}{2},\frac{3}{2},\frac{5}{2},\frac{7}{2}}$, $^4P_{\frac{1}{2},\frac{3}{2},\frac{5}{2}}$, $^2F_{\frac{5}{2},\frac{7}{2}}$, $^2D_{\frac{3}{2},\frac{5}{2}}$, and $^2P_{\frac{1}{2},\frac{3}{2}}$; many other levels come from combining the d electron with the $^1D_2$ and $^1S_0$ terms.

17.11 Coupling of the jj Type

As the nuclear charge increases, the spin-orbit effects become rapidly larger; as a consequence, the J levels tend less and less to group themselves into recognizable LS multiplets and the selection rules for L and S fail more and more. Finally, in very heavy atoms the spin-orbit effects may predominate over the residual electrostatic interaction to such an extent that an approximation occurs to the other type of coupling mentioned in Sec. 17.5, known as *jj coupling*.

In the zero-order stage of perturbation theory, let the electrostatic interaction of the electrons be ignored, except insofar as allowance is made for its average effect as represented in the central field. Each electron can be assumed to occupy one of the n,l,j,m_j states. A quantum state for the atom is then specified by assigning a set of such quantum numbers for each of the N electrons in the atom:

$$n_1,l_1,j_1, \ n_2,l_2,j_2, \ \ldots \ , \ n_N,l_N,j_N$$

Electrons occupying closed shells may be ignored provided these shells remain closed in all the radiative transitions that are considered. (The electronic energy is independent of m_j.)

In considering only those energy levels belonging to a particular configuration it suffices to write down only a set of j's: $j_1, j_2, \ldots, j_N$, with the understanding that j_1 goes with n_1l_1, and so on. An energy level so specified may be called a jj level or term. The number of jj levels belonging to a given configuration cannot exceed 2^N; for each j can assume at most two values, $l + \frac{1}{2}$ and $l - \frac{1}{2}$.

In addition to the selection rule for configurations, there is now a selection rule for j: $\Delta j = 0$ or ± 1. The special rules for jj coupling can thus be summarized in the following form:

1. *Only one n,l,j set of quantum numbers can change in a radiative transition—"only one electron jumps at a time."*
2. *For the jumping electron,*

$$\Delta l = \pm 1 \qquad \Delta j = 0 \text{ or } \pm 1$$

Allowance may then be made for the residual electrostatic interaction as a second smaller perturbation. To prepare for this, we make combinations of functions containing different electronic m_j's so as to quantize the resultant angular momentum $\mathbf{A}_J$ with the introduction of the usual quantum numbers J and M_J. This may be visualized as the addition of the electronic j vectors in various ways into a J vector. In classical mechanics, the electrostatic forces would cause precessions of the electronic resultant angular momenta $\mathbf{A}_j$ about their vector sum $\mathbf{A}_J$.

The residual electrostatic interaction then has the effect of separating slightly the J levels belonging to each jj term, forming a jj multiplet of levels. A notation for these levels is easily invented. Thus, analogous designations of a J level with $J = 3$ might be, in jj or LS coupling, respectively,

$$
\begin{array}{cc}
jj & LS \\
(n_1l_1, \ldots, n_Nl_N)(j_1 \cdots j_N)_3 & n_1l_1 \cdots n_Nl_N \; {}^5D_3
\end{array}
$$

As the residual electrostatic interaction grows stronger, however, it tends to mix the various electronic zero-order functions together in the wave function; then it becomes less useful to attach the numbers $(n_1l_1j_1, \ldots)$ to the energy levels, and the associated selection rules tend to fail. All degrees of coupling intermediate between the ideal LS and jj types occur. However, the quantum numbers J and M_J and their selection rules persist so long as the atom is free from external fields.

An example showing the transition from LS to jj coupling is shown in Fig. 17.8. The relative positions of certain levels are shown for carbon $(Z = 6)$, germanium $(Z = 32)$ and lead $(Z = 82)$, corresponding levels being connected by dotted lines. In silicon $(Z = 14)$, the corresponding set of levels is observed to be arranged much as in carbon, whereas tin $(Z = 50)$ resembles lead in this respect. The number of closed subshells underlying the active electrons is different in the two cases, but this difference is immaterial to the relative arrangement of the levels.

In Fig. 17.8 we note that the J value of any level remains the same in all three spectra; in carbon the levels group themselves by their energy values into good LS multiplets, whereas in lead they form jj groups, for which the j values are given in parentheses. The observed radiative transitions as indicated by arrows in the figure for carbon and lead include all that are allowed by the selection rules.

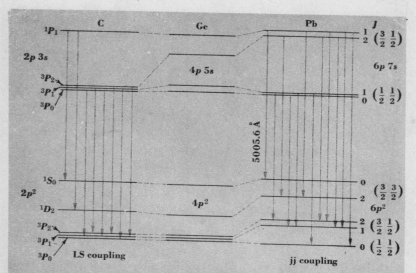

Fig. 17.8 A transition from LS to jj coupling. Within each multiplet of levels the spacing is drawn to scale, except that the splittings within each ^3P multiplet are exaggerated; however, the scale is different for each multiplet. For carbon the $2p^2$ multiplet has terms from 90,878 to 69,231 cm^{-1} and the $2p3s$ from 30,547 to 28,898 cm^{-1}; for lead the $6p^2$ range is 59,821 to 30,365 and the $6p7s$ from 24,863 to 10,383.

The student will find it instructive to note which lines occur with one form of coupling and not with the other.

In the course of transition from LS coupling, in which the electrostatic and spin interactions dominate, to jj coupling, where the spin-orbit interactions are predominant, two-electron configurations may sometimes be advantageously described by other coupling schemes. For example, when one of the $3p$ electrons of silicon is excited to the $4f$ state, the configuration is highly asymmetric in the two electrons, which now exist in quite different charge-cloud environments. Because of this we may speak of these indistinguishable particles as though we could tell them apart. In this case the spin-orbit interaction l_1s_1 for the $3p$ electron exceeds the l_1l_2 interaction. The latter in turn is greater than the s_1s_2 and l_2s_2 interactions. For this case we may think of $l_1 \oplus s_1 = j_1$; $j_1 \oplus l_2 = K$, and finally $K \oplus s_2 = J$ (Fig. 17.9c). This leads to jK coupling, described by $\{[(l_1s_1)j_1, l_2]K, s_2\}J$ just as LS coupling corresponds to $[(l_1l_2)L, (s_1s_2)S]J$ or jj coupling to $[(l_1s_1)j_1, (l_2s_2)j_2]J$. In the jK notation a term is

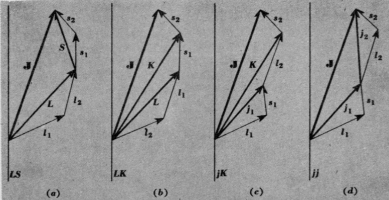

Fig. 17.9 Schematic diagram showing *LS*, *LK*, *jK*, and *jj* coupling for a pair of electrons.

written symbolically as $j_1[K]_J$; for the Si $3p4f$ example the term written in *LS* notation as $3p4f$ $^1F_3$ becomes in the *jK* description $\frac{1}{2}[\frac{5}{2}]_3$.

When the l_1l_2 interaction is stronger than l_1s_1, the term may be best described by $l_1 \oplus l_2 = L$, $L \oplus s_1 = K$, $K \oplus s_2 = J$ (Fig. 17.9*b*). This scheme $\{(l_1l_2)L, s_1]K, s_2\}J$ is called *LK* coupling and may be written symbolically as $L[K]_J$.

17.12   The Breadth of Spectral Lines

No spectral line as observed is perfectly sharp, no matter how great the resolving power of the spectrometer. A line devoid of structure appears densest in the center and fades out on the edges. The *full width at half intensity* Δ is defined as the separation (in λ or ν) between two points, one on each side of the center, at which the intensity is half as great as it is at the maximum (Fig. 17.10). (For convenience,

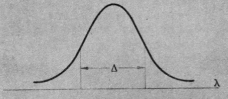

Fig. 17.10 The width Δ of a spectral line.

Δ is often called simply *the width*.) The principal causes of line broadening are the following.

a. Doppler Effect　The *observed* frequency of a spectral line is changed by motion of the radiating atom in the line of sight, owing to Doppler's principle, the apparent frequency *increasing* if the motion is *toward* the observer and *decreasing* if the motion is *away* from the observer.

In a luminous gas, such as the mercury vapor in the mercury-arc lamp, the atoms are moving with a maxwellian distribution of velocity. Therefore, a spectral line emitted by a gas must comprise, as observed, a range of frequencies symmetrically distributed about the frequency emitted by the atom when at rest; this range increases with increasing temperature. The distribution of intensity throughout the line is determined by Maxwell's distribution of velocities. According to Rayleigh, the brightness of the line at a distance $\lambda' - \lambda$ from the center is proportional to $e^{-b(\lambda'-\lambda)^2}$, where b is a constant depending on the temperature and on the mass of the atom. From this formula and the value of b it follows that the width Δ defined as above, if it is due entirely to the doppler effect, should be

$$\Delta = 0.72 \times 10^{-6}\lambda \sqrt{\frac{T}{M}} \qquad \text{wavelength units}$$

where T is the absolute temperature and M is the atomic or molecular weight of the radiating atom or molecule.

b. Pressure　Using the light from a single unresolved spectral line, it is possible, under favorable conditions, to produce interference fringes when the difference in the path of the two beams is as much as several hundred thousand wavelengths. In terms of classical theory, this fact was interpreted to mean that the wave train sent out by any particular atom is continuous, i.e., without change of phase, for at least that number of vibrations. In order that the atom may emit wave trains of this length, it must be "free from interruptions" for a corresponding period of time. This means that the *mean free time* between collisions with other atoms must, on the average, exceed the time required to emit a complete wave train, since it may be assumed that a collision would cause either a change of phase or excessive damping. Analogous conclusions are deduced from wave mechanics.

Collisions between atoms become more frequent the higher the pressure of the gas for a given temperature. The higher the pressure, the shorter the wave trains and the more frequent the abrupt changes of phase. An increase of temperature, at given density, increases both the collision rate and the pressure. Thus in general, at high gas pressure there is a broadening due to the increasing frequency of phase changes resulting from collisions. Michelson confirmed this by showing, from measurements with the interferometer, that below a pressure of the order of a millimeter, the breadth of the hydrogen line $\lambda = 6563$ Å is almost entirely due to the doppler effect but that at higher pressures the line becomes considerably broader.

The observed broadening is enhanced further by direct disturbances of the energy levels, or of the radiation process itself, when another molecule comes close to the radiating atom.

c. Natural Line Breadth   According to wave-mechanical theory, a line ought also to exhibit a small "natural" width even when emitted by an atom at rest. As an analogy, a classical oscillator, radiating energy, would decrease continually in amplitude; it would emit, therefore, a damped wave train of finite effective length. It can be shown that a damped train of sine waves is equivalent to the superposition of a large number of perfectly regular trains of great length, with slightly differing frequencies. Such radiation, observed as a spectral line, would, therefore, be broadened slightly. In the visible region, the natural line breadth is mostly much less than 0.001 Å and so is not detectable. It is easily observable, on the other hand, in the case of x-rays.

Natural line breadth is associated with an indeterminateness in the atomic energies. The principle of the conservation of energy is always found to hold insofar as it can be tested; but measurement of the energy takes time, and the precision of the result is thereby limited. If Δt denotes the time available for measurement or establishment of an energy, and if ΔE denotes the indefiniteness in the energy, by the indetermination principle

$$\Delta E \Delta t \gtrsim \hbar/2$$

For an atomic or molecular energy level, we may take $\Delta t = \tau$, where τ is the mean life of the level, whether limited by radiative processes or otherwise, as by Auger transitions. Thus

$$\Delta E \approx \frac{\hbar}{\tau}$$

The quantity ΔE is sometimes regarded as a natural level width. For a spectral transition into or out of the normal state, for which $\tau = \infty$ and $\Delta E = 0$, theory gives for the natural line breadth

$$\Delta \nu = \frac{\Delta E}{h} \approx \frac{1}{2\pi\tau}$$

where τ refers to the other state that is involved in the transition; thus $\Delta = \Delta\lambda = \Delta(c/\nu) \approx \lambda^2/2\pi c\tau$. In general, if τ_1 and τ_2 refer to the two levels, respectively,

$$\Delta \approx \frac{\lambda^2}{2\pi c}\left(\frac{1}{\tau_1} + \frac{1}{\tau_2}\right)$$

The formula $\nu = (E_1 - E_2)/h$ refers to the center of the line.

d. Incipient Stark Effect In the spectra emitted by discharge tubes, a common cause of broadening is the production of a small Stark effect by strong electric fields, which are not great enough to produce an observable splitting of the line but are sufficient to make it appear perceptibly broader than it otherwise would.

Problems

1. List the term designations ($^S L_J$) of all possible terms which can result from the following values of L and S: (a) $L = 2$, $S = \frac{1}{2}$; (b) $L = 4$, $S = 1$; (c) $L = 3$, $S = \frac{5}{2}$; (d) $L = 0$, $S = 2$; (e) $L = 1$, $S = \frac{3}{2}$.

2. Indicate the possible values of J associated with each of the following incomplete term designations: ^1P, ^2D, ^1D, ^3S, ^3P, ^4P, ^6H, ^5P. Under what circumstance is the actual number of possible J values less than the multiplicity? For each of the terms above, indicate whether it arises from a configuration with an even or an odd number of electrons.

3. Sodium atoms in the ground state $3s\ ^2S_{\frac{1}{2}}$ are bombarded by electrons with 3.5 eV energy. With the help of Fig. 17.2 list the wavelengths of all the spectral lines which can be excited. (Assume no excited atoms are struck by the 3.5-eV electrons.)

4. Estimate the relative intensities of the lines of compound doublets arising from transitions $^2D \leftarrow {}^2F$ and $^2F \leftarrow {}^2G$.

5. From the wavelengths of the sodium D lines compute the effective magnetic induction experienced by the Na valence electron as a result of its "orbital" motion.

Ans: 20 Wb/m²

6. From the wavelengths of the sodium D lines calculate Z_{eff} ($= Z - \sigma_2$) in Eq. (17.6) for the $3p$ state.

Ans: 3.55

7. Using the intervals of Table 17.1 and Eq. (17.6), compute the screening constant for the spin-orbit interaction of the $4p$, $5p$, and $6p$ levels. Compare the results with the $3p$ value of 7.45 from the preceding problem. How does the value of the screening constant for the spin-orbit effect compare with that for the energy of the n,l state determined by $E = Rhc(Z - \sigma)^2/n^2$?

8. A line arising from a transition to the ground state of an atom has a width Δ of 0.5 cm^{-1}. If the width is almost entirely due to the width of the initial state, calculate the approximate lifetime of this state.

9. Write the ground electron configuration for $_{14}$Si. What is the ground state? List all the energy states which can arise from this configuration in order of increasing energy, assuming LS coupling.

10. Write the ground electron configuration for $_7N$, determine the ground state, and list all energy levels which arise from the electron configuration in order of increasing energy (with the aid of Table 17.4).

11. Write the ground-state configuration and determine the ground-state term for $_{22}$Ti, $_{23}$V, $_{27}$Co, and $_{28}$Ni, given that in each case there are two electrons in the $4s$ level.

12. A level multiplet is observed with spacings in the ratio of $1:2:3:4$. What are possible values of L and S? Repeat for spacings of $3:5:7$ and $3:5$.

Partial Ans: ^5D

13. Prepare a diagram similar to Fig. 17.6 to show the relative locations of the levels of all terms arising in LS coupling for the configurations (a) $3p4p$; (b) $4p4f$; and (c) $2p^23s$.

14. On a figure similar to 17.3 indicate all six transitions allowed between ^3D and ^3P levels. Calculate the relative intensities of the lines given that the line $^3P_2 \leftarrow {}^3D_3$ is 84 times as intense as the line $^3P_2 \leftarrow {}^3D_1$.

Ans: 84, 15, 1, 45, 15, 20

15. If the constant B in Eq. (17.10) is 5 times as great for the ^3P level multiplet as for the ^3D, show the relative frequencies of six lines of the preceding problem on a sketch similar to the lower part of Fig. 17.3.

16. Find the possible terms for (a) a p^3 configuration and (b) a p^4 configuration.

17. Find the possible terms for a d^2 configuration.

chapter eighteen

Atoms in a Magnetic Field

In preceding chapters we have asserted without proof that the angular momentum of the electron system for an atom is quantized in such a way that when the z direction is fixed by a suitable field, the z component of the angular momentum can have only values which differ by an integer times ℏ. Further we have introduced relations between spin and orbital angular momentum and the corresponding magnetic moments. In the present chapter we treat the Zeeman effect and related phenomena which provided many of the clues from which the vector model of the atom was inferred.

18.1   Effects of a Magnetic Field on an Atom

The quantum states of an atom are modified when the atom is subjected to external magnetic or electric fields. States which are formerly degenerate, i.e., associated with the same energy, may become separated, and

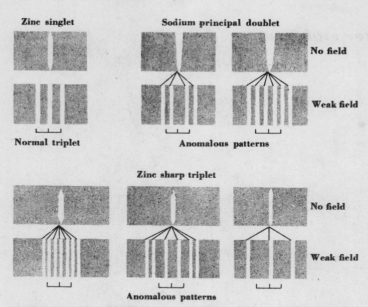

Zinc singlet

Sodium principal doublet

No field

Weak field

Normal triplet

Anomalous patterns

Zinc sharp triplet

No field

Weak field

Anomalous patterns

Fig. 18.1 Photographs of some Zeeman patterns in a weak magnetic field, viewed perpendicular to **B**. The brackets show the position of a normal triplet in the same field. (*From H. E. White, "Introduction to Atomic Spectra." Copyright 1934. McGraw-Hill Book Company. Used by permission.*)

spectral lines arising from transitions between these states may split into several components. An interesting and important case is the Zeeman effect, discussed in terms of classical theory in Sec. 6.3. A few of the observed Zeeman patterns are illustrated in Fig. 18.1; only the *normal* Zeeman pattern at the upper left was compatible with classical theory. No explanation of the other patterns, described as *anomalous*, was found until electron spin was introduced. Although the classical theory yielded correct separations for the components of the normal triplet, it used a model for the radiation which bears little resemblance to the *vector model* of the atom introduced in preceding chapters.

The splitting produced in energy levels by a magnetic field results from the interaction of the field with the *magnetic moment* of the electron system. This magnetic moment may arise from the orbital motion of charged particles and from the spin. A bar magnet with magnetic moment **M** experiences a torque **M × B** in a magnetic field of induction **B**, which we choose to establish the z direction. Referred to a zero of potential energy when **M** and **B** are perpendicular, the potential energy of the magnetic dipole is $-\mathbf{M} \cdot \mathbf{B}$. Similarly, provided the field is not too strong, an atom of magnetic moment $\mathbf{\mu}_a$ has potential energy of orientation given by $-\mathbf{\mu}_a \cdot \mathbf{B}$. If E_0 represents the energy of any atomic level in the absence of a field, application of a magnetic field removes the $(2J + 1)$-fold degeneracy of the level and the energy associated with each value of M_J becomes different. That is, the energy of the atom can be written

$$E = E_0 - \mathbf{\mu}_a \cdot \mathbf{B} = E_0 - \mu_z B \tag{18.1}$$

where μ_z is the z component of the magnetic moment $\mathbf{\mu}_a$.

18.2   The Normal Zeeman Effect

For all singlet states $S = 0$, and so there is no net spin angular momentum and no net magnetic moment associated with spin. The magnetic moment of the electron system is due to orbital motion, and the quantum-mechanical value is

$$\mathbf{\mu}_L = - \frac{e}{2m} \mathbf{A}_L = - \frac{e}{2m} \mathbf{A}_J \tag{18.2}$$

which comes directly by finding the resultant of the contributions of the individual electrons given by Eq. (9.16) and then making use of the fact that since $S = 0$, $L = J$ and $\mathbf{A}_L = \mathbf{A}_J$. Since the allowed values of $(\mathbf{A}_J)_z$ are $M_J \hbar$, we have immediately that, for the case of singlet levels, the z component of the magnetic moment can have only one of the values $-(e/2m)M_J \hbar$. Consequently, a singlet level of initial energy E_0 in the absence of a magnetic field is split by an applied $\mathbf{B}$ into $2J + 1$ equally spaced magnetic substates given by

$$E = E_0 + \frac{e\hbar}{2m} B M_J = E_0 + \mu_B B M_J \tag{18.3}$$

where μ_B $(= e\hbar/2m)$ is the Bohr magneton.

Figure 18.2 shows the splitting for $^1D_2$ and $^1P_1$ levels; the energy spacing between adjacent sublevels is $(e\hbar/2m)B$. Transitions occur only between magnetic sublevels for which the selection rule $\Delta M_J = 0$, ± 1 is satisfied. There are three transitions in Fig. 18.2 for which $\Delta M_J = 0$, but all three lead to the emission of the same frequency ν_0 as that of the line from the $^1P_1 \leftarrow {}^1D_2$ transition in the absence of a field. Similarly, the three $\Delta M_J = +1$ transitions lead to a single line of frequency $\nu_0 - eB/4\pi m$, while the $\Delta M_J = -1$ transitions give a line

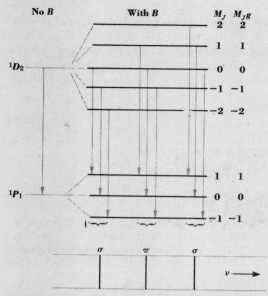

Fig. 18.2   Normal Zeeman pattern arising from a
$^1P_1 \leftarrow {}^1D_2$ transition.   Only three lines result from
the nine allowed transitions.   The displacement of
the σ lines is $\pm e\hbar B/2m$ from the π line.

of frequency $\nu_0 + eB/4\pi m$.   These are the observed frequencies, derived
on an entirely different model from that of Lorentz and Zeeman, who
nevertheless predicted the same frequencies, Eq. (6.3).

The selection rule for M_J comes directly from a quantum-mechanical calculation of transition probabilities.   Quantum mechanics further
predicts that the lines of the Zeeman pattern should be polarized in
accord with the rules (valid for LS coupling in a weak field whether
$S = 0$ or not):

For σ polarization
$$\Delta M_J = \pm 1 \tag{18.4a}$$

For π polarization
$$\Delta M_J = 0 \text{ (but no } 0 \to 0 \text{ for } \Delta J = 0) \tag{18.4b}$$

where π polarization means the electric intensity of the radiation is
parallel to **B** and σ means that the electric vector is perpendicular to
B when the radiation is observed at right angles to **B** (Sec. 6.3).

18.3 *Electron Spin and the Landé g Factor*

When Goudsmit and Uhlenbeck introduced the profoundly important concept of electron spin in 1925, they found that assigning a magnetic moment of 1 Bohr magneton for spin led to an explanation of several features of atomic spectra. In terms of quantum mechanics this is equivalent to writing spin magnetic moment $\mathbf{\mu}_s$ as $-2(e/2m)$ times the spin angular momentum $\mathbf{A}_s$, while for the orbital case $\mathbf{\mu}_l = -(e/2m)\mathbf{A}_l$. The question promptly arose whether one unit of spin angular momentum involves *exactly* or *approximately* twice as great a magnetic moment as one unit of orbital angular momentum.   Let us write

$$\mathbf{\mu}_s = -g_s \frac{e}{2m} \mathbf{A}_s \tag{18.5}$$

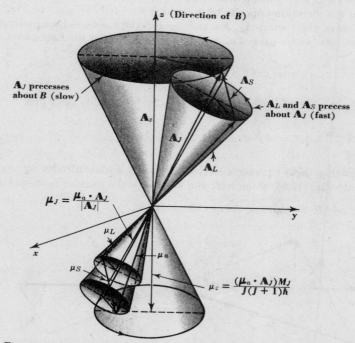

Fig. 18.3 Illustrating the approximations used in treating the weak field Zeeman effect. (*Modified from R. B. Leighton, "Principles of Modern Physics." Copyright 1959. McGraw-Hill Book Company. Used by permission.*)

where g_s, known as the *spin g factor*, is possibly exactly 2 and surely approximately 2.

In 1934 Kinsler produced the first evidence that g_s is slightly greater than 2. From careful measurements of the Zeeman effect in $^3P_1$ and $^1P_1$ states of neon he found that if Eq. (18.5) is accepted, $g_s = 2.0034 \pm 0.0032$. Subsequent measurements by a variety of techniques have led to the value[1] $g_s = 2.002319$. For our purposes it will suffice to take $g_s = 2$.

We now calculate the Landé g factor for LS coupling when B is low enough for Zeeman splitting to be small compared with the spin-orbit fine structure. Then the magnetic energy can be considered as a small perturbation. By virtue of the torque on the magnetic moment, $\mathbf{A}_J$ now precesses around $\mathbf{B}$ at a rate small compared with the precession rate of spin and orbital angular momentum about $\mathbf{A}_J$ (Fig. 18.3). Since $\mathbf{\mu}_a$ precesses about $\mathbf{A}_J$ rapidly compared with the precession rate of $\mathbf{A}_J$ about $\mathbf{B}$, the *time average* of $\mathbf{\mu}_a$ is approximately equal to its component in the direction opposite to $\mathbf{A}_J$, which we shall call μ_J. In Fig. 18.4 we redraw the vectors $\mathbf{A}_L$, $\mathbf{A}_S$, and $\mathbf{A}_J$ and the associated magnetic moment vectors, where $\mathbf{\mu}_S = \Sigma\mathbf{\mu}_s = -2(e/2m)\mathbf{A}_S$. From the figure we have

$$\mu_J = \mu_L \cos\theta + \mu_S \cos\phi \tag{18.6}$$

By the law of cosines $A_S{}^2 = A_J{}^2 + A_L{}^2 + 2A_L A_J \cos\theta$ and

$$A_L{}^2 = A_J{}^2 + A_S{}^2 + 2A_J A_S \cos\phi$$

Solving these equations for $\cos\theta$ and $\cos\phi$ and introducing the values into Eq. (18.6), along with the values for the angular momenta and the

[1] P. Kusch, *Phys. Today*, **19**(2):23 (1966).

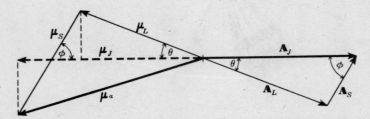

Fig. 18.4 Because each unit of spin angular momentum contributes twice as much magnetic moment as a unit of orbital angular momentum, $\mathbf{\mu}_a$ is not parallel to $\mathbf{A}_J$.

magnetic moments, give

$$\mu_J = -\frac{e\hbar}{2m}\left\{\frac{\sqrt{L(L+1)}\,[J(J+1)+L(L+1)-S(S+1)]}{2\,\sqrt{L(L+1)}\,\sqrt{J(J+1)}}\right.$$

$$\left.+\frac{2\,\sqrt{S(S+1)}\,[J(J+1)+S(S+1)-L(L+1)]}{2\,\sqrt{S(S+1)}\,\sqrt{J(J+1)}}\right\}$$

$$= -\frac{e}{2m}\,\sqrt{J(J+1)}\,\hbar\,\frac{3J(J+1)+S(S+1)-L(L+1)}{2J(J+1)}$$

$$= -g\,\frac{e}{2m}\,A_J \quad (18.7)$$

where

$$g = 1 + \frac{J(J+1)+S(S+1)-L(L+1)}{2J(J+1)} \quad (18.8)$$

is the Landé g factor for LS coupling. In the presence of a "weak" magnetic field which establishes the z direction, the allowed values of the z component of $\mathbf{A}_J$ are $M_J\hbar$ and the corresponding possible values of $(\mu_J)_z$ are

$$(\mu_J)_z = -g\,\frac{e}{2m}\,A_J\,\frac{M_J\hbar}{A_J} = -\frac{e\hbar}{2m}\,M_J$$

$$\times\left[1+\frac{J(J+1)+S(S+1)-L(L+1)}{2J(J+1)}\right] = -\mu_B M_J g \quad (18.9)$$

where μ_B is the Bohr magneton $(e\hbar/2m)$. By Eq. (18.1), in a magnetic field each energy level is split into $2J+1$ "magnetic" sublevels with energy

$$E = E_0 + \mu_B B M_J g \quad (18.10)$$

Equation (18.10) reduces to Eq. (18.3) when g is set equal to unity, the Landé g factor for singlet states.

18.4 *Zeeman Patterns of LS Multiplets in a Weak Field*

As examples of anomalous Zeeman patterns, consider those of the sodium D lines, which arise from the transitions $^2S_{\frac{1}{2}} \leftarrow {}^2P_{\frac{1}{2}}$ and $^2S_{\frac{1}{2}} \leftarrow {}^2P_{\frac{3}{2}}$. The splitting of the levels and the allowed transitions are shown in Fig. 18.5, with the resulting line patterns appearing below.

In discussing Zeeman patterns it is convenient to specify the displacement of each line of the Zeeman pattern from the original (zero-

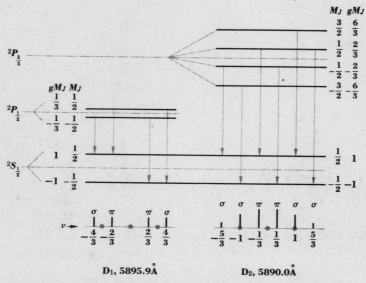

Fig. 18.5 Weak-field Zeeman splitting for the sodium *D* lines. The position of the normal triplet is shown by three dots. The same scale is used for all Zeeman separations. The relative heights of the lower lines suggest the theoretical relative intensities.

field) line in terms of the normal Zeeman displacement, which is $\mu_B B$ in energy units and $\mu_B B/h$ in frequency units. The displacement in wavenumbers of each outer line of a normal Zeeman triplet from the central line is sometimes called a *Lorentz unit* for the field *B*; its value $\tilde{L} = \mu_B B/hc = 0.467B$ cm^{-1} when *B* is measured in webers per square meter.

For the sodium *D* doublet the line D_1 ($^2S_{\frac{1}{2}} \leftarrow {}^2P_{\frac{1}{2}}$) is split into four components, two π components with displacements of $\pm\frac{2}{3}\tilde{L}$ and two σ with $\pm\frac{4}{3}\tilde{L}$ displacement. D_2 ($^2S_{\frac{1}{2}} \leftarrow {}^2P_{\frac{3}{2}}$) is split into six components, two π with displacements of $\pm\frac{1}{3}\tilde{L}$ and four σ with displacements of ± 1 and $\pm\frac{5}{3}\tilde{L}$. The selection rule $\Delta M_J = 0$ leads to π polarization and $\Delta M_J = \pm 1$ to σ polarization, just as for the normal Zeeman effect, which we now recognize as simply the special case in which $S = 0$.

It is evident that a wide variety of patterns is possible; a few are diagrammed in Fig. 18.6. *Corresponding lines* in the spectral multiplets of a given *series* exhibit the *same* type of pattern; such lines have a common final level, and their initial levels have the same *J*, *S*, and *L* and hence the same value of *g*. This conclusion is in agreement with

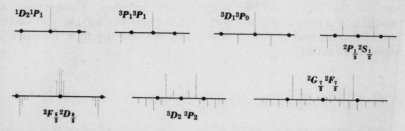

Fig. 18.6 Some Zeeman patterns for *LS* coupling in a weak field. Heights of lines indicate rough relative intensities; π lines are drawn above, σ below. The dots show the position of the normal triplet in the same field.

observation; it is known as *Preston's rule*, discovered empirically in 1898. Preston's rule is often of use in deciding what lines belong together in a series. From Zeeman patterns it is possible in many cases to infer values of both g and J for the two LSJ levels involved.

Before the development of wave mechanics a mathematical description of the anomalous Zeeman effect, including the g factor, was worked out by Landé, who showed the necessity of using $J(J + 1)$ rather than J^2. A few special cases may be noted which we discuss in terms of the Landé factors g' for the initial state and g'' for the final state. It can happen that a spectral line is not split at all in a magnetic field. An example is the $^5D_0 \leftarrow {}^5F_1$ transition, in which $g' = 0$ for $^5F_1$ while $M_J = 0$ for $^5D_0$; this case is illustrated by the line λ 5713 Å in the arc spectrum of titanium. If $g' = g''$, the pattern may be a simple triplet, but of nonnormal separation. Similarly, a transition from a level with $J = 1$ to one with $J = 0$ leads to a triplet line pattern with anomalous separation.

Intensity rules for Zeeman patterns can be derived from the knowledge that the statistical weights of all M_J levels are equal and from the facts that:

1. *The sum of the intensities for all transitions from any initial Zeeman level is equal to the sum of the intensities for all transitions from any other initial level.*
2. *The sum of the intensities of all transitions arriving at any final Zeeman level is the same as for any other final level.*

For the normal Zeeman pattern the two σ components are each half as intense as the central π line as seen transversely to the magnetic field; the σ components are twice as intense when viewed along the field as they are when viewed transversely. This latter result is in agreement with the classical theory of the Zeeman effect; each circular motion is equivalent to two linear vibrations at right angles to each other, but in the transverse direction, radiation is received from only one of these, the one that lies in the line of sight being invisible.   Furthermore, as much light is polarized in one way as in the opposite way, so that on the whole the emitted radiation is unpolarized; and the total intensity is equal in all directions.

18.5   The Paschen-Back Effect

The formulas of Sec. 18.3 are valid only for weak magnetic fields. The term "weak" is, of course, relative, since fields in excess of 1000 gauss are usually required to produce good Zeeman patterns.   A field is weak when the total spread of the Zeeman pattern of each line is small relative to the spacing of the lines themselves.   Figure 18.7, for example, shows diagrammatically to scale the D lines of sodium with their respective Zeeman patterns in a field of 30,000 gauss.   The Zeeman separations are seen to be small compared with the separation between D_1 and D_2.   For these lines, therefore, 3 Wb/m^2 is a weak field.   Consider, however, the Zeeman effect in lithium.   The zero-field separation of the two components of the first line of the principal series is of the order of 0.3 cm^{-1}.   A field of 3 Wb/m^2 would produce Zeeman separations of 1.4 cm^{-1}.   For lithium therefore, 3 Wb/m^2 is a strong field.

As the magnetic field is increased from weak to strong, the approximations of Sec. 18.3 are no longer reasonable, since as **B** increases, the precession rate due to **B** is no longer less than the precession rate of $\mathbf{A}_L$ and $\mathbf{A}_S$ about $\mathbf{A}_J$.   At strong fields $\mathbf{A}_L$ and $\mathbf{A}_S$ may precess separately about **B**.   As we have seen in Sec. 9.8, the orbital angular momentum vector $\mathbf{A}_l$ for a single electron precesses about **B** at the Larmor precessional rate $\omega_l = eB/2m$.   Since all electronic A_l's precess at the same rate, the magnetic field does not disturb the resultant orbital momentum, which in the absence of other influences

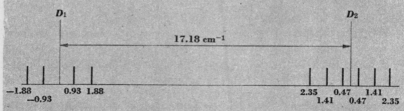

Fig. 18.7   Relative separation of the sodium D lines and their Zeeman components in a field of 3 Wb/m^2.

would itself merely precess at velocity ω_l. Since the gyromagnetic ratio for spin is twice that for orbital angular momentum, the resultant spin momentum would precess undisturbed at velocity $2\omega_l$. Since these rates are not equal, in general, the vector sum $\mathbf{A}_J$ of the two momenta is not constant. In wave mechanics a magnetic field has no disturbing effect upon LS coupling, but it does prevent the quantizing of $\mathbf{A}_J$. In a weak field, J continues to be a fairly good quantum number, but as the field becomes strong, J disappears and its selection rules cease to hold.

In classical theory the vector moment of force due to a uniform magnetic field $\mathbf{B}$ is perpendicular to $\mathbf{B}$; hence it has no tendency to alter the component of the vector angular momentum in the direction of $\mathbf{B}$. Analogously, in wave mechanics, in the presence of a uniform magnetic field each quantum state of an otherwise isolated atom is characterized by a fixed component of angular momentum parallel to the field; or, in cases of degeneracy, the quantum states can be so defined that this is true. The principal term in the expression for this angular momentum is $M_J h$; the quantum number M_J is integral or half integral according to the rule stated in Sec. 17.1 and is subject to the selection rule stated there. There is also a much smaller component of momentum proportional to the magnetic field itself, but this is negligible in spectroscopy. Although the total angular momentum may not be fixed (so that J is no longer a valid quantum number), M_J as expressed in Eq. (17.3) continues to be significant. Since $M_J h = \Sigma(A_l)_z + \Sigma(A_s)_z = (M_L + M_S)h$, we may write for the z component of the magnetic moment of the electron system

$$\mu_z = -\frac{e}{2m}[\Sigma(A_l)_z + 2\Sigma(A_s)_z] \times \frac{M_J h}{\Sigma(A_l)_z + \Sigma(A_s)_z} = -\mu_B M_J g \tag{18.11a}$$

where the g factor is now given by

$$g = \frac{\Sigma(A_l)_z + 2\Sigma(A_s)_z}{\Sigma(A_l)_z + \Sigma(A_s)_z} \tag{18.11b}$$

The view was expressed by Paschen and Back in 1912 that if the magnetic field could be made sufficiently great, the (then) unknown cause of multiplet splitting would be overpowered and all patterns would revert to the normal triplet. This reduction might occur either through the coalescence of lines or through the disappearance of certain lines. In their observations, Paschen and Back were able to demonstrate such an approximation to a triplet in two cases. Later they showed experimentally that in a strong field the J selection rule ceases to hold and new lines may appear, only to disappear again as the field becomes still stronger.

This phenomenon, now called the Paschen-Back effect, is readily inferred from wave-mechanical theory or from the vector modéi. The magnetic field **B** has no tendency to destroy the coupling of l vectors into L, or of s vectors into S, but it does destroy the further coupling of L and S into J. Let us consider a field so strong that the Zeeman splitting of the levels is much larger than the splitting due to the spin-orbit effect. Then the latter effect can be ignored to a first approximation and the atomic states may be the L,S,M_L,M_S states described in Sec. 17.6, with the axis drawn parallel to **B**. In a classical motion, the l vectors would now all precess about **B** at the same angular velocity ω_l and their resultant would precess with them. Similarly, the s vectors and their resultant would precess at the rate $2\omega_l$. Since these rates are unequal, the *resultant* angular momentum varies in magnitude, although its component in the direction of **B** remains constant.

In wave mechanics, the total component in the **B** direction is $M_J\hbar$, and here $M_J = M_L + M_S$; furthermore, $\Sigma(A_l)_z = M_L\hbar$ and $\Sigma(A_s)_z = M_S\hbar$. Thus, as in a weak field, $\Sigma(A_l)_z$ and $\Sigma(A_s)_z$ are independent of **B**, and from Eqs. (18.4), (18.6), and (18.11a,b) we have, as an approximation valid in a sufficiently strong field,

$$g = 1 + \frac{M_S}{M_J} \tag{18.11c}$$

$$\mu_z = -(M_J + M_S)\mu_B \tag{18.11d}$$

$$E = E_0 + (M_J + M_S)\mu_B B \tag{18.12}$$

Because of the selection rules, $\Delta M_J = 0$ or ± 1, $\Delta M_S = 0$, the normal triplet now arises just as in the case of one electron in a huge field.

When account is taken of the spin-orbit interactions, however, it is found that the spacing of the levels is not quite constant, and levels having the same value of $M_J + M_S$ but different M_J may be slightly separated. A fine structure is thereby introduced into the strong-field pattern, of the same absolute order of magnitude as the original fine structure in the zero field. This is shown by $E = E_0 + (M_S + M_S)\mu_B B + KM_L M_S$.

The transition from a weak to a strong field is illustrated in Fig. 18.8 for a $^2S \leftarrow {}^2P$ transition such as that which gives rise to the D lines of sodium. The plots are constructed relative to the spacing

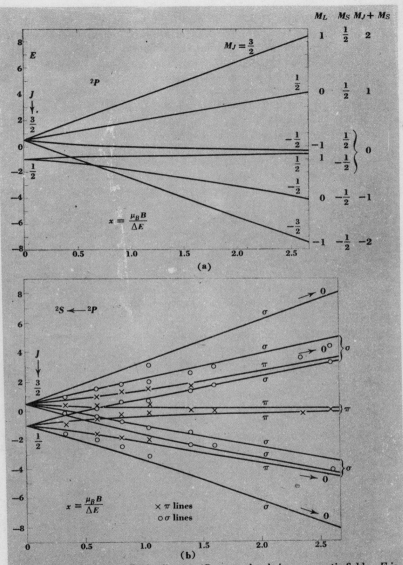

Fig. 18.8 (a) Paschen-Back effect on ^2P energy levels in a magnetic field. E is energy in arbitrary units; ΔE is the $\frac{3}{2} - \frac{1}{2}$ separation in zero field, taken as 1.5 on the ordinate. In x, ΔE is in joules and B in webers per square meter. (b) Calculated Paschen-Back effect on the $^2S \leftarrow {}^2P$ Zeeman pattern with a few observations by Kent on lithium.

between the $P_{\frac{3}{2}}$ and $P_{\frac{1}{2}}$ levels in zero field, so that they are valid for any case of this type; the abscissa is proportional to field strength. In Fig. 18.8a are shown the calculated positions of the ^2P levels, labeled with values of $M_J = M_L + M_S$, and with quantum numbers for no field and strong field shown at the sides of the plot. The two ^2S levels simply separate in proportion to B and are not shown. In Fig. 18.8b the calculated positions of the 10 lines for the transition $^2S \leftarrow {}^2P$ are similarly shown, π and σ lines being so labeled. Lines which become weak and would ultimately vanish in an infinite field are indicated by the symbol "$\rightarrow 0$." It will be noted that as the field becomes strong, two of the ^2P levels coalesce, and of the 10 weak-field lines, 4 disappear while the others tend to coalesce in pairs.

A corresponding theory can be developed for jj coupling. A help in tracing the levels is provided by the rule that levels having the same M_J do not cross as the field increases in strength.

It should be added that in the discussion so far the magnetic field has been tacitly assumed not to be so extremely strong as to bring together levels belonging to different terms, such as a P and a D. If this limit could be exceeded, the normal triplet should disappear again, to reappear only in a field so huge as to swamp all other influences except that of the central field, thus producing the case described below.

18.6 Zeeman Effect in a Huge Field

The simplest type of Zeeman effect should be produced by a field so strong that *complications due to all other sources can be ignored*. Such a field we shall call *huge*.

Consider first a single electron in a central field, placed in a uniform magnetic field **B**. Since we are assuming that the spin-orbit effect is swamped by the huge field, we can use the n, l, m_l, m_s quantum states with the axis parallel to **B**. When the atom is in one of these states, $(A_l)_z$ and $(A_s)_z$ have the magnitudes

$$(A_l)_z = m_l \hbar \qquad (A_s)_z = m_s \hbar$$

Therefore,

$$\mu_z = -\frac{e\hbar}{2m}(m_l + 2m_s) = -\mu_B(m_j + m_s)$$

where $m_j = m_l + m_s$. Since μ_s does not change (appreciably) with further increase in B, Eq. (18.1) gives for the energy

$$E = E_0 + \mu_B B (m_j + m_s) \tag{18.13}$$

Here $m_j + m_s$ is always an integer or zero.

From Eq. (18.13) we see that the magnetic field splits the single original level with energy E_0 into several magnetic levels, each characterized by a value of $m_j + m_s$. An example is illustrated in Fig. 18.9. Since $m_j + m_s$ may have any integral value from $l + 1$, when $m_l = l$ and $m_s = \frac{1}{2}$, down to $-(l + 1)$, there are in general $2l + 3$ magnetic levels in all; if, however, $l = 0$, there are only two levels, with $m_j + m_s = \pm 1$. All the levels except the upper two and the lower two are double. The levels are equally spaced, the difference in energy being $\mu_B B$.

Consider now transitions between two such sets of energy levels. As selection rules we have, besides $\Delta l = \pm 1$,

$$\Delta m_j = 0, \pm 1 \qquad \Delta m_s = 0$$

The allowed transitions, shown in Fig. 18.9, lead to the normal Zeeman

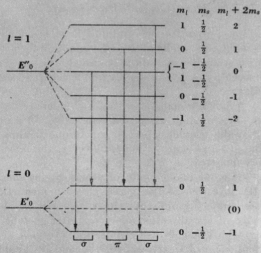

Fig. 18.9 Energy levels for one electron in a huge magnetic field. Transitions giving rise to the same ν are bracketed.

triplet with:

For π polarization

$$\Delta m_l = 0 \qquad \text{giving} \qquad \nu = \nu_0$$

For σ polarization

$$\Delta m_l = \pm 1 \qquad \text{giving} \qquad \nu = \nu_0 \pm \frac{\mu_B B}{h}$$

Finally, it may be remarked that in a huge field the magnetic moment of the atom, in general, becomes variable, and $E = E_0 - \mu_z B$ no longer holds. The atom then behaves rather like a set of magnets connected by springs. To move such a compound magnet to a position where the field B is greater by dB with no change in its direction requires work $dE = -\mu_z \, dB$. Thus the general relation between μ_a, B, and the energy E is

$$\mu_{eff} = -\frac{dE}{dB} \tag{18.14}$$

18.7   The Stern-Gerlach Experiment

The space quantization of atomic angular momenta constitutes a characteristic feature both of the older quantum theory and of wave mechanics. An experiment which directly reveals the space quantization itself was proposed by O. Stern in 1921 and was carried out in collaboration with Gerlach.

A magnet tends to move so as to increase the magnetic flux through it in the direction of its magnetic axis. In a uniform field, the result is that the magnet experiences a torque tending to line it up with the field. In a *nonuniform* field, however, the magnet experiences a *translatory force* $\mathbf{F} = \mathbf{\mu} \cdot \nabla \mathbf{B}$.

Suppose that a beam of atoms having magnetic moments travels in the direction of the x axis across a magnetic field whose lines are approximately parallel to the y axis but whose magnitude decreases rapidly in the direction of $+y$. After passing through the field, let these atoms be collected on a suitable target. Then atoms which have a component of their magnetic moment in the direction of the field are deflected; if the component of the moment has the same direction as the field, they will be deflected toward $-y$; if it has the opposite direction, toward $+y$.

According to *classical* theory each atom will enter the field with its magnetic axis inclined at some angle θ to the field, and the axis will then execute a Larmor precession about the field at the fixed angle θ. Since all values of θ occur among the atoms, their deflections will be distributed in continuous fashion, and the atoms, instead of forming a small spot on the target, will be drawn out into a continuous band. According to *quantum* theory, on the other hand, each atom enters the field in a certain quantum state, defined with the direction of the field as an axis. Its magnetic moment in the direction of the field is gM_J Bohr magnetons (if the field is not too strong). The beam is broken up, therefore, into separate beams and forms on the target a series of distinct spots, one for each possible value of M_J.

The arrangement used in the experiment of Stern and Gerlach is shown diagrammatically in Fig. 18.10. The nonhomogeneous field was produced between pole pieces of which one had a sharp edge, so that near it the field was much stronger than elsewhere. A strap-shaped beam of silver atoms was formed by evaporating silver in a heated oven O and allowing atoms from the vapor to stream out through collimating slits; the beam (shown by the small rectangle in Fig. 18.10a) traveled closely past the sharp edge of pole piece P_2 and was condensed on a plate at T. With no field, the beam formed a narrow line on the plate (Fig. 18.10c). When the magnetizing current was turned on, the line was not widened continuously but was divided into two lines, except at the ends, which were produced by atoms passing at some distance from the sharp edge. Space quantization of the silver atoms was thus clearly revealed. From the separation of the two lines and the gradient of the magnetic field strength, it was calculated that each silver atom had a magnetic moment in the direction of the field of 1 Bohr magneton.

These results, and others obtained subsequently, are in agreement with the predictions of wave mechanics. The silver atom is normally in a $^2S_{\frac{1}{2}}$ state, for which $g = 2$; thus half the atoms should have $M_J = \frac{1}{2}$ and a moment in the direction of the field of 1 Bohr magneton; for the

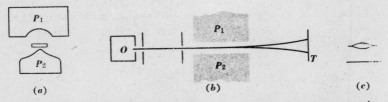

Fig. 18.10 The Stern-Gerlach experiment. (*a*) Approximate cross section of poles; (*b*) path of ions; and (*c*) the trace on the target with (*above*) and without (*below*) magnetic field.

other half, $M_J = -\frac{1}{2}$, and the moment is -1. Similar results were obtained by Taylor and by Leu for sodium and potassium, which are likewise in $^2S_{\frac{1}{2}}$ states. Atoms of zinc and cadmium, in Leu's experiments, were unaffected by the magnetic field. Atoms of thallium gave a double line corresponding to a moment of $\frac{1}{3}$ magneton. Zinc and cadmium are normally in $^1S_0$ states, which have $M = 0$ and no magnetic moment. For thallium the normal state is, from spectroscopic evidence, $^2P_{\frac{1}{2}}$; thus $M_J = \frac{1}{2}$ or $-\frac{1}{2}$ again, but $g = \frac{2}{3}$.

18.8   Isotope Structure and Hyperfine Structure

The ordinary fine structure due to spin-orbit interaction does not exhaust the possibilities of fine details in spectral lines. Even before 1900, Michelson and others had shown, by means of the interferometer, that many spectral lines possess a further structure much finer still; this came to be known as *hyperfine structure*.

After the discovery of isotopes, it was believed for a time that each component of a hyperfine pattern of lines was emitted by a different isotope. Later, hyperfine structure was discovered in the spectra of some elements which consist of only one isotope. An example is bismuth; the line λ 3596 Å contains six hyperfine components spread over a range of 0.3 Å, or 2.3 cm⁻¹. As suggested by Pauli in 1924, this effect is due to the fact that the nucleus has angular momentum and an associated magnetic moment. There is a tendency now to limit the expression *hyperfine structure* to the spectrum from a given isotope and to refer to the other effect as *isotope structure*. In the spectrum from a mixture of isotopes, both types may occur superposed.

In light atoms the *isotope shift* may arise from simple differences in the nuclear motion. The simplest case is that of hydrogen, in which the isotope shift led Urey and his collaborators to the discovery of heavy hydrogen, or deuterium. The Rydberg constants for the two kinds of hydrogen atoms can be found by substituting $M = M_1$ and $M \approx 2M_1$ successively in Eq. (9.12) and making use of $R = m_r e^4 / 8\epsilon_0{}^2 h^3 c$

$$R_1 = \frac{M_1 R_\infty}{M_1 + m} \qquad R_2 = \frac{2M_1 R_\infty}{2M_1 + m}$$

M_1 being the mass of a proton and m the mass of an electron. The difference in wavelength for a given line of wavelength λ is then

$$\Delta\lambda \approx -\frac{\lambda(R_2 - R_1)}{R_\infty} \approx -\frac{\lambda m}{2M_1}$$

Since $M_1/m = 1,836$, it follows that the Hβ line for deuterium should lie $4861/3.672 = 1.32$ Å on the violet side of that for ordinary hydrogen. This line was observed to be faintly visible, in the expected position, in the spectrum from a sample of common hydrogen; it increased in strength as the relative concentration of deuterium was increased.

In the spectrum of a *heavy* atom, isotope shifts are found, in general, to be proportional to the differences in atomic mass. A photograph illustrating such structure in the spectrum of tungsten is shown in Fig. 18.11. The spectrum was formed with a Fabry-Perot etalon; hence the same pattern appears repeated many times in different orders. Tungsten consists chiefly of four isotopes, three having mass numbers 182, 184, and 186 about equally abundant, and 183 about half as abundant as the others. The three observed lines are ascribed to the more abundant isotopes, any effect of 183 being assumed to be masked by the others.

In such heavy atoms, the isotope shift cannot be explained as a simple difference in the effects of nuclear motion, which are far too small. The cause lies in a departure from simple coulomb interaction between nucleus and electrons, associated with the finite size and shape of the nucleus, the effect varying from isotope to isotope. Many nuclei have significant electric quadrupole and magnetic dipole moments.

The theoretical treatment of *true hyperfine structure due to a nuclear magnetic moment* resembles closely the treatment of *LS* fine

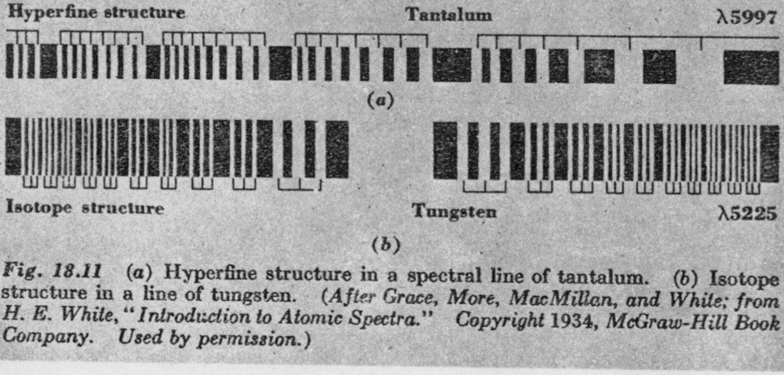

Fig. 18.11 (a) Hyperfine structure in a spectral line of tantalum. (b) Isotope structure in a line of tungsten. (*After Grace, More, MacMillan, and White; from H. E. White, "Introduction to Atomic Spectra." Copyright 1934, McGraw-Hill Book Company. Used by permission.*)

structure caused by the spin-orbit effect. The angular momentum of the nucleus is associated with spin and orbital angular momenta of the protons and neutrons of which the nucleus is composed. The maximum possible value of the component of the nuclear angular momentum in any direction is denoted by $I\hbar$; the square of the total is $I(I + 1)\hbar^2$. The spin number I is an integer or 0 whenever the mass number (number of protons and neutrons in the nucleus) is even, and half integral when the mass number is odd. Moreover, $I = 0$ whenever both the mass number and the atomic number Z are even. If $I > 0$, there exists in the usual way a second quantum number M_I restricted to integrally spaced values such that $|M_I| \leq I$; thus the nuclear state is $(2I + 1)$-fold degenerate. Protons and neutrons have $I = \frac{1}{2}$; some other values of I are listed in Table 18.1.

Table 18.1   Nuclear Spin and Magnetic Moment

Atomic Number Z	Element	Mass Number A	Nuclear Spin I	Magnetic Moment μ_I Nuclear Magnetons
0	n	1	$\frac{1}{2}$	-1.9131
1	H	1	$\frac{1}{2}$	2.7928
		2	1	0.8574
2	He	3	$\frac{1}{2}$	-2.1276
		4	0	
3	Li	6	1	0.8220
		7	$\frac{3}{2}$	3.2564
4	Be	9	$\frac{3}{2}$	-1.1776
5	B	10	3	1.8007
		11	$\frac{3}{2}$	2.6885
7	N	14	1	0.4036
		15	$\frac{1}{2}$	-0.2831
9	F	19	$\frac{1}{2}$	2.6287
11	Na	23	$\frac{3}{2}$	2.2176
13	Al	27	$\frac{5}{2}$	3.6414
21	Sc	45	$\frac{7}{2}$	4.7564
41	Nb	93	$\frac{9}{2}$	6.1670
47	Ag	107	$\frac{1}{2}$	-0.1135
81	Tl	203	$\frac{1}{2}$	1.6115
		205	$\frac{1}{2}$	1.6274

If the Dirac relativistic theory for charged particles held for protons of mass M_p as well as for electrons, the effective magnetic moment of the proton would be, in analogy with Eq. (9.21a) for electrons, $e\hbar/2M_p = \mu_B/1836$, with a sign corresponding to rotation of positive charge. As we have seen in Sec. 10.5, the magnetic moment of the proton is 2.793 of these *nuclear magnetons*, while the neutron has a moment of -1.91 nuclear magnetons. In general, the moment μ_I for any nucleus represents the maximum possible value for the component of the nuclear moment in any given direction, measured in nuclear magnetons. In a nuclear orientation state in which the nuclear angular momentum about a given axis is $M_I \hbar$, its component of magnetic moment in the direction of this axis is $(M_I/I)\,\mu_I$. Values of μ_I are also listed in Table 18.1.

The nuclear angular momentum may then be added vectorially to the resultant angular momentum of the atomic electrons by a process of IJ coupling exactly similar to the LS coupling of orbital and spin momenta. Such coupling is necessary whenever weak noncoulomb interaction between the nucleus and the atomic electrons occurs. For the grand resultant momentum there are then two new quantum numbers $F = J \oplus I$ and M_F, with $|J - I| \leq F \leq J + I$ and $|M_F| \leq F$.

Interactions between the nucleus and the electronic motions and spins may then slightly separate levels having different F, forming a hyperfine multiplet of levels. The number of these levels is $2I + 1$ if $J \geq I$ but $2J + 1$ if $J < I$. If the interaction is magnetic in nature, equations are obtained of the same form as Eqs. (17.10) and (17.11) for LS multiplets except that here L, S, and J are replaced, respectively, by J, I, and F. As selection rules, in addition to others, we have

$$\Delta F = 0 \text{ or } \pm 1 \qquad \Delta M_F = 0 \text{ or } \pm 1$$

As an alternative, however, the interaction may arise from an asymmetry in the electrostatic field around the nucleus, usually of the quadrupole type, and in this case the formulas are more complicated.

A *hyperfine spectral multiplet* results from transitions between the hyperfine levels composing two ordinary J levels. Usually it happens that the spacing of the hyperfine sublevels in one of the two J levels is much larger than that in the other, so that the former spacing stands out in the multiplet as observed, the finer structure due to the other J level being frequently not resolved at all. The result is then an easily recognizable *flag* type of pattern, a good example of which is shown in Fig. 18.11.

As an illustration of the theory, consider the following doublet in the spectrum of thallium ($_{81}$Tl):

$$6s^2 7s \ ^2S_{\frac{1}{2}} \leftarrow 6s^2 6p \ ^2P_{\frac{1}{2}} \qquad \lambda = 3777 \text{ Å}$$
$$\bar{\nu} = 26{,}478 \text{ cm}^{-1}$$
$$6s^2 7s \ ^2S_{\frac{1}{2}} \leftarrow 6s^2 6p \ ^2P_{\frac{3}{2}} \qquad \lambda = 5352 \text{ Å}$$
$$\bar{\nu} = 18{,}684 \text{ cm}^{-1}$$

Under high resolution each line of this doublet is seen to be made up of three components. Their wavelengths, estimated intensities, and relative positions on a frequency scale are shown in Fig. 18.12a.

To produce three component lines, as observed, there must be two hyperfine sublevels in each J level. Here $2 = 2I + 1$, and so $I = \frac{1}{2}$ for the thallium nucleus. Then for $J = \frac{1}{2}$, $F = I \pm \frac{1}{2}$ or $F = 1$ or 0; for $J = \frac{3}{2}$, $F = 2$ or 1. The spacing of the hyperfine levels is easily determined from the observed separations in the spectrum. The level diagram and spectral fine structure for one line of the doublet are shown in Fig. 18.12b. The same pattern would result if the doublet separations were interchanged, but the Zeeman pattern for the hyperfine components unambiguously assigns the energy differences.

Hyperfine spectra are an important source of information concerning the values of I and μ_I. To find μ_I, however, requires a wave-mechanical calculation, which is easiest for light atoms. The sign of μ_I affects the order of the F levels; very commonly, when $\mu_I > 0$, levels with larger F lie above those with smaller F, and vice versa for $\mu_I < 0$.

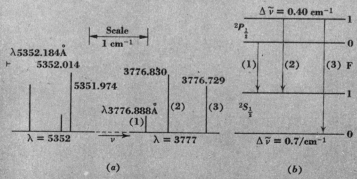

Fig. 18.12   (a) Hyperfine structure of the thallium doublet; $\lambda = 5352$ and 3777 Å. (b) Energy levels and transitions for the 3777 Å thallium line.

Furthermore, complications frequently arise which have been ascribed to slight departures from spherical symmetry in the electrostatic field of the nucleus, especially such a departure as can be described by assuming an electric quadrupole moment in the nucleus. If $I > \frac{1}{2}$, such a moment causes perturbations, large or small, in the hyperfine pattern; it may even be the sole cause of a hyperfine splitting.

18.9 The Stark Effect

An applied electric field can also modify the quantum states of an atom, separating states which were formerly degenerate. This effect was discovered in 1913 by Stark, who observed splitting of the Balmer lines of hydrogen. When a hydrogen atom is in a homogeneous electric field E which establishes the z direction, a perturbation energy $-eEz$ is added to the coulomb energy in the potential-energy term of the Schrödinger equation. This results in a first-order change in energy of $-eE\langle z \rangle$, where $\langle z \rangle$ depends on the electron "orbit" and its eccentricity. The details of the calculation are complicated, but they lead to the prediction that when E is weak enough for its effect to be smaller than the spin-orbit interaction, each level is split into $2n - 1$ components with different energies, where n is the principal quantum number. The electric field does not interact through the magnetic moment, and each pair of Zeeman levels m_j and $-m_j$ have the same energy. States of $+m_j$ and $-m_j$ correspond to the same charge distributions and therefore have the same shift in energy in the electric field.

For weak fields the Stark splitting for hydrogen lines leads to symmetrical patterns, and the line spacings are proportional to E. This situation is described as the *linear* or *first-order* Stark effect. When E exceeds 10^7 V/m, there are shifts in the line patterns which are proportional to E^2 and we speak of the *second-order* Stark effect. A qualitative explanation is that under the influence of the strong field the atom becomes polarized so it has an induced electric dipole moment proportional to E. The torque on such a dipole is proportional to E times the dipole moment, or to E^2.

The Stark effect becomes still more complicated for multielectron orbits. The selection rule $\Delta l = \pm 1$ is no longer required, and many new lines arise, corresponding to $\Delta l = 0, \pm 2, \pm 3$, and so forth. For penetrating orbits no linear effect is observed, but there is a *quadratic* Stark effect due to the interaction of the field with the induced

---

Problems

1. Compute the displacement in Å of the two outer Zeeman components from the mercury 1849.6 Å line ($6s^2$ $^1S_0 \leftarrow 6s6p$ $^1P_1$) in a magnetic field of 2 Wb/m².

Ans: 0.032 Å

2. (a) Calculate the magnetic moment of an isolated aluminum atom in its ground state by use of the vector model of the atom. What is the ratio of this value to μ_J?

   (b) Do the same for a fluorine atom.

3. Determine the weak-field Zeeman splitting for a line arising from a $^3S_1 \leftarrow {}^3P_1$ transition in terms of the normal Zeeman splitting.

Ans: π lines at ± 0.5; σ lines at ± 1.5, ± 2

4. Find the weak-field Zeeman pattern for a $^2P_{\frac{1}{2}} \leftarrow {}^2D_{\frac{3}{2}}$ transition in terms of the normal Zeeman separation. Make a sketch similar to those of Fig. 18.5, showing the positions and polarizations (but not the relative intensities) of the lines.

5. In the triplet spectrum of zinc there are sharp series lines arising from $^3P_0 \leftarrow {}^3S_1$ and diffuse series lines from $^3P_0 \leftarrow {}^3D_1$. Calculate the Zeeman pattern for both kinds of lines. To which series did the line at the lower right of Fig. 18.1 belong?

6. (a) Show that $g = 1.50$ for ^3P, ^5D, and ^7F levels and, in general, for $S = L$.

   (b) Under what conditions is $g = 0$? Give an example for both even and odd multiplicity.

   (c) Show that in addition to the singlet levels, $g = 1$ for $^5F_2$ and $^7I_5$ states. What is the general condition on S, L, and J if $g = 1$ for LS coupling?

7. Determine the weak-field Zeeman pattern for a line arising from a $^4P_{\frac{5}{2}} \leftarrow {}^4D_{\frac{5}{2}}$ transition and show the displacements of all lines of the pattern in a sketch similar to those of Fig. 18.6. (Do not concern yourself with relative intensities.)

8. Calculate the Landé g factor for the $^{10}H_{\frac{5}{2}}$ and $^{10}G_{\frac{5}{2}}$ levels and find the Zeeman pattern for lines arising from transitions between the two states.

9. (a) Show that any object with magnetic moment μ experiences a force in the z direction given by $\mu_z \, \partial B_z/\partial z$ when it moves through a non-uniform magnetic field.

   (b) In a Stern-Gerlach experiment silver atoms with average kinetic energy of 4×10^{-20} J pass through an inhomogeneous field 0.75 m long with a $\partial B_z/\partial_z$ equal to 2500 Wb/m³ along the atom path. Find the maximum separation of the "lips" of Fig. 18.10c at the end of the magnet.

10. A spectral line arises from a transition $^3P_1 \leftarrow {}^3D_2$ for an atom for which $I = \frac{1}{2}$. Draw a diagram showing the hyperfine splitting of these levels and indicate all allowed transitions by arrows.

11. (a) Show that the classical interaction energy of two magnetic dipoles μ_1 and μ_2 separated by r is given by

$$E_{\mu_1\mu_2} = \frac{\mu_0}{4\pi} \left[\frac{\mu_1 \cdot \mu_2}{r^3} - \frac{3(\mu_1 \cdot r)(\mu_2 \cdot r)}{r^5} \right]$$

(b) Use this relation to find the maximum interaction energy of one nuclear magneton and 1 Bohr magneton separated by 1 Bohr radius.

chapter nineteen

Molecules and Molecular Spectra

That atoms attract one another is clear from the facts that (1) two or more atoms may join together to form a stable gaseous molecule and (2) larger aggregates of atoms cling together to form liquids and solids. A satisfactory theory of the atom should explain how and why certain combinations of atoms form stable molecules. The ability of modern quantum theory to supply a highly successful description of interatomic forces is of fundamental importance no less to the chemist than to the physicist.

19.1 Interatomic Forces

To explain the binding between atoms, we consider first the interaction of two neutral atoms as they approach one another from a distance. If the electrons of each atom formed a spherically symmetrical charge cloud about the nucleus, neither atom would be surrounded by an electric field and hence there would be little force action between them. However, the picture of electrons in orbitals suggests that although on the average the field due to electrons might just cancel the electric field of the nucleus, there should be also a rapidly fluctuating residual field. This field might polarize a neighboring atom and consequently exert a fluctuating attractive force upon it. When wave-mechanical perturbation theory is applied to the problem, an attraction of one atom for another, called the *van der Waals attraction*, is found to exist. This attraction is important for the phenomenon of cohesion, but it does not play a major role in the binding of the atoms into chemical compounds.

Chemical forces come into play when atoms approach so closely that their electronic charge clouds begin to overlap. New effects then occur. Of great importance is the *electron-exchange effect*, described for one particular case in Appendix 15A and discussed qualitatively in Sec. 19.4. Atoms held together by means of a pair of exchanged electrons are said to share a *covalent bond;* binding of this type accounts for the compounds called *homopolar* by the chemist. Long before quantum mechanics led to an understanding of the covalent bond, classical ideas were invoked to explain the attraction between a strongly electropositive atom (such as sodium) and a strongly electronegative atom (such as bromine). It is well known that a compound such as NaBr dissociates into Na^+ and Br^- ions when it is dissolved in water. This suggested that there might be a transfer of an electron from a Na atom to a Br atom, thereby forming a pair of ions which might be held together by electrostatic attraction. Such a bond, called *ionic* or *heteropolar*, was the first to be understood and offers a reasonable starting point for discussing gaseous molecules.

Although we find it convenient to discuss the limiting cases of purely ionic and purely covalent bonds, in reality all grades of transition occur between the two types. Thus the bonds in molecules of CO_2 or CdS are intermediate between the two idealized cases. The mathematical treatment is best described by saying that such chemical bonds are partly of one type and partly of the other.

19.2   The Ionic Bond

As a specific example of an ionic bond, we shall discuss potassium chloride. When a K atom with a single electron outside of a closed electron core interacts with a Cl atom, which requires one more electron to complete the p^6 subshell to give the most stable of electron configurations, a primarily ionic bond is formed. We may think of the K atom as giving up its outer electron which joins the Cl atom, thereby leaving

K+ and Cl⁻ ions, bound together by electrostatic interaction. The best support for this picture is its success in predicting the observed binding energy and the correct general properties of the molecule. Clearly, a molecule formed by a positive ion and a negative ion should be polar in nature and have a permanent electric dipole moment. Quantitative measurements of the observed dipole moment, of the binding energy, and of the molecular spectra are all in excellent agreement with predictions based on the assumption that the KCl molecule is a union of the K+ ion and Cl⁻ ion with their nuclei separated by a distance of 2.79 Å.

When an atom of chlorine and an atom of potassium interact to form a molecule of potassium chloride, an energy of 4.42 eV/molecule is released. This is called the *binding energy* of the molecule. As expected, 4.42 eV can also dissociate the KCl molecule into an atom of potassium and an atom of chlorine. For a conservative system, the binding energy is numerically equal to the *dissociation energy*. If we select as the zero for our energy measurement the situation in which a Cl and a K atom are infinitely far apart, the KCl molecule in its ground state will have a negative energy equal to the dissociation energy. To obtain an estimate of this energy, imagine that we form a molecule by a series of idealized steps, which are not those by which an actual molecule is formed. But since the processes involved conserve energy, the dissociation energy is independent of the manner in which the molecule is formed. We begin with a K atom and a Cl atom infinitely separated. Let us remove an electron from the K and give it to the Cl atom, thus producing the corresponding ions. To remove the outer electron from the K atom requires its ionization energy 4.34 eV. The *electron affinity* of Cl is 3.80 eV; it represents the energy released when an electron becomes attached to a neutral Cl atom. The process of forming the ions has required from us a net energy of 0.54 eV. We now regain this energy and more as we allow the K+ and Cl⁻ ions to approach one another under their electrostatic attraction. The potential energy P of charges e and $-e$ at a distance r apart is given by

$$P = -\frac{1}{4\pi\epsilon_0}\frac{e^2}{r} = -\frac{14.4}{r}\text{ eV-Å} \qquad (19.1)$$

Here we have evaluated the constant so that the potential energy is in electron volts when the separation is measured in angstrom units. At the equilibrium separation 2.79 Å, the potential energy P is −5.16 eV.

Clearly, the potential energy decreases without limit as r decreases (Fig. 19.1). There can be no equilibrium separation unless some repelling force is brought into play. The primary factor which limits the approach

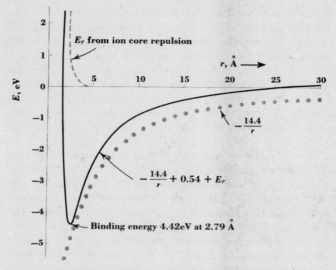

Fig. 19.1 The heavy curve is the energy of a KCl molecule as a function of the distance r between the nuclei of the K$^+$ and Cl$^-$ ions. The zero-energy configuration consists of a neutral K atom an infinite distance from a neutral Cl atom.

of the ions is associated with the charge cloud surrounding each nucleus. In this case both ions have a $1s^2 2s^2 2p^6 3s^2 3p^6$ core. Once the K and Cl ions are close enough for the electron clouds to begin overlapping, the Pauli exclusion principle requires that one or more of the electrons move to a higher energy state. This brings into play a strong repulsive force which rises rapidly as the separation is decreased, as shown in Fig. 19.1. The result is that there is a minimum in the potential at $r = 2.79$ Å, which represents the equilibrium separation of the ions. In terms of quantum mechanics, as the two ions approach, the wave functions for the individual electrons become strongly distorted and the quantum states are no longer the simple hydrogenic ones which we have assumed. As the two ion cores interpenetrate, the Pauli principle requires that electrons move to states of greater energy. There is also repulsion between the positively charged nuclei which may be lumped with the ion-core repulsion to give the net dashed curve in Fig. 19.1. One can estimate the repulsive energy from the fact that the observed dissociation energy for KCl is 4.42 eV, while the ion potential energy less the cost of forming the ion pair is $5.16 - 0.54 = 4.62$ eV. This suggests that the repulsive energy is approximately 0.20 eV at the equilibrium separation.

There are two other small contributions to the energy which roughly cancel each other. The first is the van der Waals attraction between the electron clouds, and the second is the small vibrational energy required by the uncertainty principle. The latter arises because the potential energy near the minimum in Fig. 19.1 is essentially the harmonic-oscillator potential; the zero-point energy of $\frac{1}{2}h\nu$ is a quantum requirement for such a system. As we see in Sec. 19.8, the harmonic vibrations of the atoms are a major consideration in explaining the spectra of molecules.

19.3 The Hydrogen Molecule Ion

There are many diatomic molecules of elements, e.g., hydrogen, oxygen, nitrogen, and chlorine, in which atoms of the same type are bound together. To gain an understanding of such covalent bindings it is desirable to begin to study a system composed of two protons and a single electron, which is an ionized H_2 molecule. This is a three-body system for which no general solution exists. However, since the mass of a proton is 1836 times that of the electron, the speeds of the protons are much less than that of the electron. Therefore it is a reasonable first approximation to assume that the protons are at rest while the electron is in motion. We then treat the wave equation for the electron for some particular separation R of the two protons by letting the x axis lie along the line connecting the protons and putting the origin midway between them. The Schrödinger equation for the electron is

$$-\frac{\hbar^2}{2m}\nabla^2\psi - \frac{e^2}{4\pi\epsilon_0}\left\{\frac{1}{[(x-R/2)^2+y^2+z^2]^{\frac{1}{2}}} + \frac{1}{[(x+R/2)^2+y^2+z^2]^{\frac{1}{2}}}\right\}\psi = E\psi \quad (19.2)$$

When R is large, the solution of (19.2) leads to the eigenvalues of the hydrogen atom, corresponding to the electron being bound to one of the protons; thus we have a neutral hydrogen atom and a proton off at distance R.

If we repeat the problem for R a few angstrom units, we find that because of the symmetry of the potential-energy function, any nondegenerate eigenfunction must have either even or odd parity. Figure 19.2 shows approximate ground-state eigenfunctions for four values of

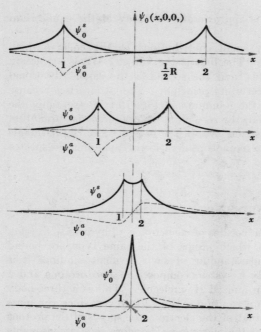

Fig. 19.2 Schematic illustration of the symmetric function ψ_0^s and the antisymmetric ψ_0^a for four values of the proton separation R. The ordinate is the value of the wave function along the x axis. (*Modified from R. B. Leighton, "Principles of Modern Physics." Copyright 1959. McGraw-Hill Book Company. Used by permission.*)

R ranging from very large to zero. The latter case corresponds to that of the helium ion, for which we know that the ground-state energy is $4 \times -13.6 = -54.4$ eV, while the first corresponds to -13.6 eV binding energy for the electron.

The equilibrium separation for the protons of the hydrogen molecule-ion is that for which the energy of the *system* (not of the electron alone) is minimium. As the protons approach one another, the energy of the electron decreases, but the energy due to coulomb repulsion of the protons increases. These two energy terms are plotted in Fig. 19.3, where the zero of energy is chosen to be that with the two protons infinitely far apart and the electron bound to one of them. The energy of the system is the difference between the previous two; it has a minimium of -2.65 eV at $R_0 = 1.06$ Å. The electron distribution along the x axis for the equilibrium separation is shown at the top of Fig. 19.4, while below are plotted contour lines of equal $\psi^*\psi$ for values 0.9, 0.8, . . . , 0.1 times the maximum value as computed by Burrau. The probability of finding the electron between the two protons is very great, as one would expect; binding is achieved through the *sharing* of the electron between the two nuclei.

It should be pointed out that only the even-parity wave function gives binding in this case; antisymmetric wave functions such as the dashed one in Fig. 19.2 correspond to increasing energy as R decreases and hence to a net repulsion between the protons. The one-electron bond can be reasonably strong for two protons, but it is very weak if the two nuclei do not have the same charge.

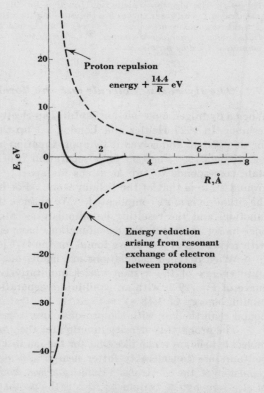

Fig. 19.3 The energy of the hydrogen molecule-ion as a function of the interproton separation R referred to a zero-energy configuration consisting of a proton and a hydrogen atom infinitely far apart.

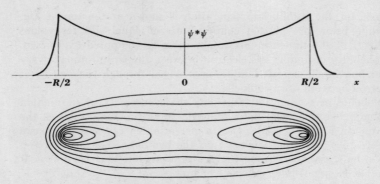

Fig. 19.4 The electron distribution function for the normal hydrogen molecule-ion. The lower curves are contour lines for value 0.9, 0.8, . . . , 0.1 times the maximum value. *(From L. Pauling and E. B. Wilson, Jr., "Introduction to Quantum Mechanics." Copyright 1935. McGraw-Hill Book Company. Used by permission.)*

19.4   The Hydrogen Molecule and the Covalent Bond

When a hydrogen molecule-ion captures an electron, a hydrogen molecule results.   In 1927 Heitler and London set up the Schrödinger equation for this system and solved it by approximation methods similar to those we used above.   For very large separation R of the protons the ground state corresponds to two H atoms far apart.   As R goes to zero, the ground state is that of the helium atom.   For intermediate separations the situation is more complicated.   We observe that the potential-energy function, and the resulting hamiltonian operator, are symmetrical in x once more; as a result the eigenfunctions have either even or odd parity with respect to x, just as we found for the H_2^+ ion.

When detailed calculations are made, one finds that the ground-state energy of the system varies qualitatively with R like the solid curve of Fig. 19.3, with an equilibrium separation $R_0 = 0.742$ Å and a binding energy of 4.48 eV.   Thus the neutral molecule is more tightly bound than the ion, with the protons closer together.

The probability-density function for the ground state of the neutral molecule behaves much like that for the ion in Fig. 19.4, except that the contours are considerably fatter near $x = 0$ as a result of the mutual repulsion of the electrons.   Binding arises from the sharing of a pair of electrons by the two nuclei; this is the prototype of *covalent* (or *homopolar*) *binding*.   But a very important feature of this binding should again be mentioned: *the two electrons involved in a covalent bond must have opposite spin.*

The ground-state spatial wave function for the electron pair has even parity. But the Pauli exclusion principle permits two electrons to have this spatial wave function only if they differ in spin. Any interchange of space and spin coordinates of two electrons must be antisymmetric. If the spins are parallel, the second electron must go into a higher energy state; for the H_2 molecule this yields no bonding. Hence two H atoms can form an H_2 molecule only when the electrons have opposite spin.

Electron-pair bonds between dissimilar atoms are roughly of the same strength as between similar atoms, although one-electron bonds are weak unless the atoms are the same. A single electron between dissimilar atoms is almost always found nearer the one which gives the electron the lowest energy. However, for the two-electron bond, the mutual repulsion of the electrons keeps the electrons well separated unless one of the atoms has a strong electron affinity. (The latter leads to *ionic* binding.) Only two electrons participate in an ordinary covalent bond. If a third electron is present, it must have an eigenfunction corresponding to higher energy and this results in either a much weaker bond (as in H_2^-) or in no binding at all (HeH). It is possible to have more than one covalent bond between two atoms; e.g., in O_2 we have two covalent bonds and in N_2 three. In methane, CH_4, the carbon atom has four covalent bonds, one to each H atom.

19.5   Binding between Square Wells

The mathematical complexities of dealing with atoms in three dimensions are considerable, but many of the most important features of atomic binding are shown by the mathematically simple problem of binding between one-dimensional square wells. We take two identical square wells of depth P_0, each containing one electron, initially far apart and study what happens as we bring them together (Fig. 19.5). On each plot the wave function ψ for the ground state is shown.

Now ψ and $-\psi$ are equally good eigenfunctions. When the wells are far apart (no interaction), the energy of the system is the same regardless of what signs are chosen for ψ at the two wells. As the wells begin to interact, the sign of the total ψ_T remains unimportant, but we get different eigenvalues for the energy, depending on whether the sign of ψ is the same in both wells (symmetric ψ_T) or opposite (antisymmetric ψ_T). That this is true comes from the fact that the symmetric ψ_T is associated with lower kinetic energy. The expectation value for K, given by Eqs. (13.14) and (13.15), depends on $|d\psi/dx|^2$, and clearly $|d\psi/dx|$ is smaller for the symmetric case. By the time the wells meet

Symmetric

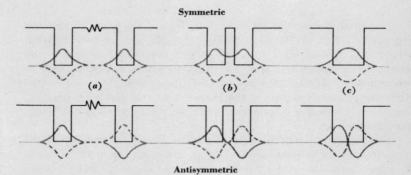

Antisymmetric

Fig. 19.5 Symmetric and antisymmetric wave functions for two square wells (a) far apart, (b) close together, and (c) so close that the barrier disappears.

(Fig. 19.5c) the symmetric ψ_T corresponds to the ground state for a well of double width, while the antisymmetric ψ_T corresponds to the first excited state.

The energy of the system corresponding to the lowest symmetric and antisymmetric states is plotted in Fig. 19.6. For the symmetric case it decreases steadily as R decreases, until the barrier disappears, but for the antisymmetric it rises slightly (the effective wavelength becomes slightly smaller). As the wells approach, the ground-state probability of finding one of the electrons between the wells increases rapidly as though the electrons were moving back and forth between the wells. It is often said that the electrons *resonate* between the

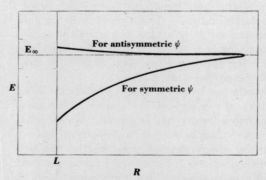

Fig. 19.6 Energy of symmetric and antisymmetric states as a function of separation for the two square wells of Fig. 19.5.

two wells. In terms of the resonance concept the attractive force
$(F = -dE/dR)$ between the two wells is due to the resonant exchange
of electrons between them. Again the two electrons can both have the
ground-state spatial eigenfunction only if they have opposite spin.

The crucial feature of this model, which applies equally to bringing
two atoms together, is this: *When two identical systems approach one
another and interact, each nondegenerate energy level is split into two levels.*
The greater the interaction, the greater the splitting becomes. There
are many analogous situations in physics. If two identical resonant
LC circuits are brought together so that they interact through mutual
inductance, the loosely coupled circuits have two resonant frequencies,
one slightly greater and one slightly less than the resonant frequency
of either circuit when it is isolated. Similarly, when two identical
pendulums are loosely coupled, the interaction leads to two normal
modes with frequencies above and below the natural frequency of each
pendulum.

When three identical square wells (or atoms, or *LC* circuits, or
pendulums) interact, a single energy level (or frequency) is split into
three. In general, for N interacting particles there is an N-fold splitting
of states. When N is very large, as it is when many atoms interact
to form a solid, what was for each atom a sharp energy level may be
spread out into a broad band of energies, as we shall see in Chap. 21.

19.6 Molecular Spectra

In addition to spectra emitted by *atoms*, there are other spectra in vast
variety emitted by *molecules* containing two or more atoms. Thus, in
the visible part of the spectrum emitted by a discharge tube containing
hydrogen, only three or four lines belonging to the Balmer series are
emitted by free atoms, which have been produced by the dissociation
of molecules. Many other lines are observed, mostly fainter, which
are ascribed to emission by the undissociated molecules. Again, if one
looks at the spectrum of the carbon arc with a spectroscope of moderate
resolving power, one observes, at the extreme (violet) edge of the visible
part of the spectrum, *bands*, which are sharply defined and brightest
on the long-wavelength edge and which fade out gradually toward shorter
wavelengths. With higher resolving power, these bands are seen to be
composed of a large number of lines which are crowded together at the
long-wavelength edge, called the *head* of the band, and are separated
farther and farther toward the short-wavelength side. The lines are so
close together as to appear, under low resolving power, like a *continuous*
spectrum. These bands are ascribed to molecules of cyanogen, CN.

There are also many groups or bands of lines in the infrared which have a molecular origin.

In general, the spectrum emitted by any kind of molecule can be divided into *three spectral ranges* which correspond to different types of transition between molecular quantum states. (The spectrum of some molecules is confined to only one of these ranges.) Simple reasoning in terms of classical assumptions concerning molecular structure leads us to expect such a feature in molecular spectra, and the reasoning needs only to be translated into wave-mechanical terms in order to constitute a correct theoretical approach to the subject.

a. Rotation Spectra   Suppose a molecule were a rigid structure but contained electric charges so disposed that the molecule possessed an electric moment. If such a molecule were to rotate, according to classical theory, it would emit radiation for essentially the same reason that an electron revolving in a circle would radiate. The radiation would consist of sine waves having the frequency of rotation. Conversely, radiation falling upon such a molecule could set it into rotation, energy of the radiation being at the same time absorbed.

Spectral lines corresponding to this simple picture are observed in the far infrared and constitute *rotation spectra*. Typical photon energies involved are 10^{-5} to 10^{-3} eV.

b. Vibration-Rotation Spectra   If the molecule were not rigid but contained atoms capable of vibration under elastic forces about equilibrium positions, and if the binding were ionic so that some atoms contained an excess of positive charge and others an excess of negative, then according to classical theory, radiation would be emitted by the vibrating atoms as they move back and forth. Unless the molecule were at the same time rotating, the frequency emitted would be that of the atomic vibration. If the molecule were rotating, however, the emitted line would be divided into two lines having frequencies respectively greater or less than the frequency of the atomic vibration, in essentially the same way as, in the classical theory of the Zeeman effect, a magnetic field modifies the frequencies emitted by a vibrating electron.

Furthermore, it would be anticipated that if the amplitude of vibration became large, the atomic vibrations, although still periodic, would no longer be simple harmonic. This is true even in the familiar example of the vibrations of a pendulum. The radiation emitted could then be resolved by Fourier analysis into wave trains with frequencies representing the fundamental and the harmonic overtones of the atomic vibrations. Each of these separate frequencies would then be split up further by rotation of the molecule.

Many spectra corresponding roughly to this classical picture are known in the infrared and are called *vibration-rotation spectra*. Photon energies are often in the range 0.2 to 2 eV.

c. Electronic Spectra   Finally, according to classical ideas, an electron in the molecule might vibrate by itself and so radiate. The emitted radiation would be affected, however, both by the vibrations of the atoms in the molecule and by the rotation of the molecule as a whole. It would probably be one of the outer electrons that radiated in the optical region of the spectrum, and its frequency would be much affected by the instantaneous position and motion of the nuclei. The rotation of the molecule would tend to split up the emitted lines as in the emission of the vibration-rotation spectrum. Molecular bands in the visible and ultraviolet, such as the cyanogen bands described above, correspond roughly to this third classical picture.

The three types of molecular spectra thus characterized will be taken up in succession for a brief discussion.

19.7 Rotation Spectra

As a simple model to illustrate certain features of the behavior of actual molecules, we may imagine a molecule to consist of several mass points held rigidly at fixed distances from each other. The quantum states for such a molecule, according to wave mechanics, would be characterized by fixed values for the angular momentum, in the same way as the states of an atom. The corresponding quantum number J, however, is confined here to integral values (zero included). The discussion will be restricted hereafter almost entirely to *diatomic molecules*. If there are only two mass points in the molecule, the line joining them is an axis of symmetry, and only rotation about an axis perpendicular to this line has significance; furthermore, the moment of inertia about all such perpendicular axes will have the same value.

The relation between angular momentum and energy is found to be the same according to wave mechanics as in classical theory. For the angular momentum A and energy E, we shall have, therefore, in terms of the angular velocity ω and moment of inertia I,

$$A = I\omega \qquad E = \tfrac{1}{2} I\omega^2$$

Therefore

$$E = \frac{A^2}{2I}$$

Inserting here the wave-mechanical value of A^2, we have

$$A^2 = J(J + 1)\hbar^2$$

$$E = J(J + 1)\frac{\hbar^2}{2I} = \frac{J(J + 1)h^2}{8\pi^2 I} \tag{19.3}$$

Such a molecule can radiate by dipole emission only if it possesses an electric moment, which will be the case, for example, if one mass point has associated with it a positive charge and the other an equal negative charge. The selection rule for J is then found to be the same as that for the quantum number l of a single electron in a central field:

$$\Delta J = \pm 1$$

Since in the present case E and J increase or decrease together, $\Delta J = -1$ corresponds to emission of energy and $\Delta J = +1$ to absorption. In a transition from state J to state $J - 1$, the emitted frequency is

$$\nu = \frac{\Delta E}{h} = 2BJ \tag{19.4a}$$

$$B = \frac{h}{8\pi^2 I} \tag{19.4b}$$

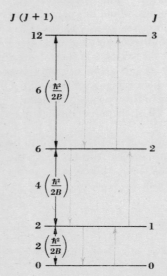

Fig. 19.7 Energy-level diagram for a rotation spectrum.

since $J(J + 1) - (J - 1)(J - 1 + 1) = 2J$. Thus the molecule emits a spectrum consisting of *equally spaced lines, with frequencies equal to a multiple of a fixed number, B.* For an emission line, J refers to the initial state for the molecular transition; for an absorption line, to the final state. The corresponding type of energy-level diagram is illustrated in Fig. 19.7.

All the hydrogen halides in the gaseous state show broad absorption lines in the far infrared which are nearly equally spaced and whose wavenumbers are nearly multiples of a constant quantity. These lines are believed to result from transitions in which the rotational state of the molecule alone changes, almost according to the simple theory just described. Thus for HCl, Czerny found a series of absorption maxima at wavelengths ranging from 120 to 44 μ. The corresponding wave numbers are listed as $\bar{\nu}$ in Table 19.1. Under the heading J is given the

Table 19.1   *Absorption spectrum of* HCl *in the far infrared*

J	$\bar{\nu}$, cm^{-1}	$\Delta\bar{\nu}$, cm^{-1}
4	83.03	20.70
5	103.73	20.57
6	124.30	20.73
7	145.03	20.48
8	165.51	20.35
9	185.86	20.52
10	206.38	20.12
11	226.50	

larger of the two values of J assigned to each transition. We note that the spacing of the maxima is almost uniform but shows a slight trend. A decrease in $\Delta\bar{\nu}$ as J increases, and hence also in the apparent value of B, implying an increase in I, is what would be expected if the atoms are not tightly bound together but become slightly pulled apart by centrifugal action as the speed of rotation increases. Substituting in (19.4*a, b*) $\nu = 83.03 \times 3 \times 10^{10}$ and $J = 4$, we find for HCl, $I = 2.7 \times 10^{-47}$ kg-m^2.

19.8   *Vibration-Rotation Spectra*

a. Approximate Theory of a Vibrating Diatomic Molecule   A simple model of a diatomic molecule in which atomic vibrations can occur supposes the atoms themselves to be point masses that are held by a force that varies with the distance between them. Let this force correspond to a potential energy P, which, plotted as a function of the distance r between atomic centers, is represented by the curve marked P in Fig. 19.8. The potential has been arbitrarily taken to be zero at $r = \infty$. The force exerted by either atom on the other is proportional to the slope of this curve. From the point r_0 at which P has its minimum value outward, the force is attractive; for $r < r_0$, it is repulsive, rising rapidly to high values. Under the influence of such a force, according to classical theory, the atoms could be at rest and in equilibrium at the distance r_0. If disturbed moderately, they would vibrate about this point; if given kinetic energy exceeding $-P_0$, however, where P_0 is the value of P at $r = r_0$, they would fly apart; i.e., dissociation of the molecule would ensue. Such a picture corresponds to the observed properties of molecules, and it is also suggested by wave-mechanical theory.

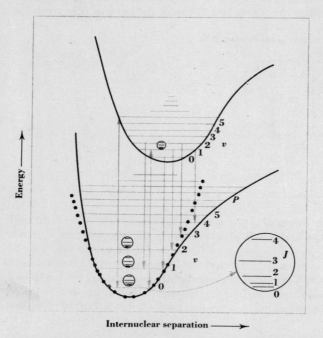

Fig. 19.8 The potential energy P and vibrational levels of a diatomic molecule in its electronic ground state; associated with each vibrational state are many rotational levels. The dotted curve represents the potential function for a simple harmonic oscillator. Also shown is an excited electronic state with a few of its vibrational levels and some electronic-band transitions (Sec. 19.10).

According to wave mechanics, however, there would be a set of *discrete quantum states* for the positions of the atoms relative to each other, with energies as suggested by the horizontal lines in Fig. 19.8. The lowest of these quantum states would correspond to an energy a little greater than P_0; even in their lowest state the atoms have a certain amount of kinetic energy, analogous to the zero-point energy $h\nu/2$ of a harmonic oscillator. Thus the interatomic distance is not quite fixed even when the molecule is in this state, although the most probable value of the interatomic distance will be close to r_0. Let us number the quantum states in the order of increasing energy, denoting the number of a state by $v \geq 0$. Let the corresponding energy be denoted by E_v. The values of E_v are all negative, zero energy belonging to a state in which the atoms are at rest at infinity. Thus, $-E_v$ represents the energy of dissociation of the molecule when it is in state number v and not rotating. The total number of the discrete states may be finite or infinite, depending upon the form of the potential curve.

The wave functions and energies for the first few states should resemble those for a harmonic oscillator. For, if we expand P in a Taylor series about $r = r_0$ and note that $dP/dr = 0$ at $r = r_0$, we obtain

$$P = P_0 + \frac{1}{2}\left(\frac{d^2P}{dr^2}\right)_{r=r_0} (r - r_0)^2 + \frac{1}{6}\left(\frac{d^3P}{dr^3}\right)_{r=r_0} (r - r_0)^3 + \cdots \tag{19.5}$$

The first two terms of this series represent a potential function of the same type as that for a harmonic oscillator with $(d^2P/dr^2)_{r=r_0}$ playing the role of spring constant β. Near $r = r_0$ the potential curve follows the harmonic-oscillator curve (dotted line) closely, and the lowest energy states correspond to those for the harmonic oscillator with v playing the role of n in Eq. (13.28). As long as the harmonic-oscillator curve is followed, the vibrational energy levels are given by

$$E_v = (v + \tfrac{1}{2})h\nu_0 \tag{19.5a}$$

where $\nu_0 = (d^2P/dr^2)^{\frac{1}{2}}/2\pi m_r^{\frac{1}{2}}$ (m_r is the reduced mass of the vibrating system), and the energy is now measured relative to the minimum of the potential energy curve at r_0.

As v increases, the potential energy departs from the parabolic approximation, and the vibrational levels get closer together. In the higher vibrational states the average separation of the atoms is greater than for the lower states. As the temperature is increased, the molecule is likely to be found in a higher vibrational level and the average interatomic spacing increases. The wave functions and energies for the higher states depart considerably from those for an oscillator, owing to the influence of the remaining terms of the series. Thus the selection rule for the harmonic oscillator, $\Delta v = \pm 1$, cannot be expected to hold, in general, for a diatomic molecule, even $\Delta v = 0$ being possible. The probability of a jump may be expected, however, to fall off rather rapidly as $|\Delta v|$ increases above 1.

To obtain quantum states for the whole molecule, allowance must then be made for rotation. If we suppose that, to a sufficient approximation, the energies of rotation and of vibration are additive, we may write for the total energy, by Eq. (19.3),

$$E = E_v + J(J + 1)Bh \tag{19.6a}$$

$$B = \frac{h}{8\pi^2 I} \tag{19.6b}$$

The various values of E_v are called *vibrational levels* of the molecule; the energies represented by both terms on the right in Eq. (19.60) are called *vibration-rotation levels*, or simply *rotational levels*. In actual cases Bh is usually very small relative to the difference between successive values of E_v; hence, the rotational levels belonging to each vibrational level form a closely spaced group. The general arrangement of the rotation-vibration levels is illustrated in Fig. 19.9 in which, however, the relative spacing of the rotational levels is enormously exaggerated and only a few of these levels are shown.

Because of the relative smallness of B, all lines arising from transitions between two given vibrational levels will lie close together; they are said to constitute a *band*, because with low resolving power they appear continuous. Bands arising from transitions in which only the vibrational and perhaps rotational energies of the molecule change are called *vibration-rotation* bands.

For a molecule composed of two point masses, the selection rule for J is found to be, as for the pure rotation lines,

$$\Delta J = \pm 1$$

For many molecules of more complicated structure, $\Delta J = 0$ is also allowed. A few transitions allowed by the rule $\Delta J = \pm 1$ are indicated in Fig. 19.9.

From the approximate expression for the energy in Eq. (19.6a), we find for the frequencies emitted in a transition between levels v',

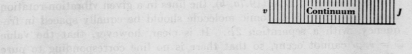

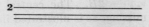

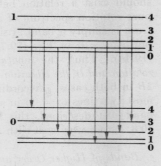

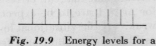

Fig. 19.9 Energy levels for a diatomic molecule with the separations of the rotational states enormously exaggerated relative to the vibrational separations. A few radiative transitions for the first vibration-rotation band are shown.

$J - 1$ and v'', J, or between v', J and v'', $J - 1$, respectively,

v', $J - 1$; v'', J

$$\nu = \nu_{v'v''} - 2BJ \qquad J = 1, 2, 3, \ldots \qquad (19.7a)$$

v', J; v'', $J - 1$

$$\nu = \nu_{v'v''} + 2BJ \qquad J = 1, 2, 3, \ldots \qquad (19.7b)$$

$$\nu_{v'v''} = \frac{E_{v'} - E_{v''}}{h} \qquad (19.7c)$$

It is assumed here that $v' > v''$ and $E_{v'} > E_{v''}$.

According to Eqs. (19.7a, b), the lines in a given vibration-rotation band emitted by a diatomic molecule should be equally spaced in frequency, with a separation $2B$. It is clear, however, that the value $\nu = \nu_{v'v''}$ cannot occur, so that there is no line corresponding to pure vibration. The nearest frequencies to this are $\nu = \nu_{v'v''} + 2B$ and $\nu_{v'v''} - 2B$. The central line of the band should thus appear to be missing, as is illustrated in Figs. 19.9 and 19.10.

b. Relation between Rotation and Vibration-Rotation Bands   There should exist a relation between the vibration-rotation bands and the pure rotation spectrum of a given substance; indeed, the rotation spectrum is simply a band arising from transitions in which $\Delta v = 0$. If we put $\nu_{v'v''} = 0$ in Eq. (19.7b), we obtain $\nu = 2BJ$, in agreement with (19.4a). Thus, the *separations of the lines should be the same in the rotation and in the vibration-rotation parts* of the spectrum, being equal to $2B$ in both cases. According to (19.7a, b), the separation of the innermost two lines ($J = 1$) in a vibration-rotation band is $4B$. In the HCl band just described, this separation is 41.60 cm^{-1}. Half of this, or 20.8, agrees very well with $\Delta\bar{\nu}$ as shown in Table 19.1.

c. Bands of Higher Order and the Heat of Dissociation   Another theoretical point that can be tested is the prediction that although transitions for $\Delta v > 1$ are to be expected, the resulting bands should be relatively weak. If the vibrational levels were equally spaced, as for the harmonic oscillator, the vibrational frequencies of the bands for $\Delta v \geq 1$ would be proportional to 1, 2, 3, . . . , just as they are for the harmonic overtones of a classical vibrating system. If the HCl band at 3.46 μ arises from

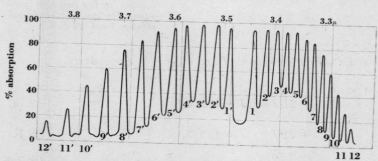

Fig. 19.10   The principal absorption band of HCl in the near infrared.   *(After E. S. Imes.)*   The numbers on the lines give the larger J for the transition; those for which ΔJ is opposite in sign to Δv are primed.   *(Reprinted from G. Herzberg, "Spectra of Diatomic Molecules." Copyright 1950.   D. Van Nostrand Co., Inc. Used by permission.)*

a transition between the lowest two vibrational levels ($v'' = 0$, $v' = 1$), we expect to find bands diminishing progressively in intensity at about 1.73, 1.15, 0.865 μ, etc. Actually, by observing the absorption through very thick layers of HCl gas, bands have been observed at 1.76, 1.20, and 0.916 μ, the last mentioned being 10,000 times weaker than the 3.46 band.

A further check is furnished by a connection with the *heat of dissociation* of the molecule, which is the energy that must be added to a molecule in its lowest quantum state in order to separate the atoms and leave them at rest an infinite distance apart. Obviously, the energy of the molecule in any bound state must be less than the energy of the separated atoms, else the molecule would dissociate spontaneously. Hence, the quantity $h\nu$ for any molecular line, representing the difference in energy between two molecular states, must be less than the heat of dissociation. For the HCl band at $\lambda = 0.916$ μ, $h\nu = 1.4$ eV. The heat of dissociation of 1 g molecule of HCl, for dissociation into H_2 and Cl_2, is 92 kJ, whereas the heats of combination of 1 g atom of H or Cl into H_2 or Cl_2 are, respectively, 211 and 120.3 kJ. Hence to dissociate HCl into H and Cl requires $92 + 211 + 120.3$ kJ/g molecule, or 4.4 eV/molecule. This is more than 3 times the value of $h\nu$ as just calculated for the highest-frequency vibration band of HCl that has been observed.

d. Effect of Thermal Agitation upon Infrared Bands

No absorption bands are·observed for which $v'' > 0$ in the spectrum of HCl. This is explained as a consequence of the wide spacing of the vibrational levels relative to the quantity kT. Molecules are seldom thrown by thermal agitation into a vibrational level which can serve as the initial level for an absorption band with $v'' > 0$. For $\lambda = 3.5$ μ, we find $h\nu = 0.35$ eV; whereas, at $T = 290°K$, $kT = 0.025$ eV. If E_0, E_1 are the energies of the two states involved in the production of $\lambda = 3.5$ μ, the ratio of the number of atoms in the upper of these two states to that in the lower will be the ratio of their Boltzmann factors, Eq. (4.20),

$$\frac{n_1}{n_0} = e^{-(E_1 - E_0)/kT} = e^{-h\nu/kT} = e^{-35/2.5} < 10^{-6}$$

since the statistical weights g_1 and g_0 are equal. This is so small that practically all molecules of HCl are in their lowest vibrational states ($v = 0$) at room temperature. Hence only the absorption band for which $v'' = 0$ can be observed; and the observed rotation spectrum is the zero-zero vibration-rotation band ($v = 0$ to $v = 0$).

Among the various rotational levels a wide distribution of the molecules occurs even at room temperature. The experimental value of B from Table 19.1 is 20.6/2 cm^{-1}, making $Bhc = 1.3 \times 10^{-3}$ eV. Thus relative to the state with $J = 0$, the energy of any other rotational state is

$$E_J = J(J + 1) \times 1.3 \times 10^{-3} \quad \text{eV}$$

compared to room-temperature kT of 0.025 eV. The statistical weight of each J state is $2J + 1$, corresponding to $M_J = J, \ldots, -J$. To find the ratio of the number of HCl molecules with $J = 10$ to the number with $J = 0$ at room temperature, we have, by Eq. (4.20),

$$\frac{n_{10}}{n_0} = \frac{21}{1} e^{-\frac{1}{2}x^2} = 21e^{-5.7} = 0.07$$

so that even states with $J = 10$ occur 0.07 times as often as states with $J = 0$, a result in agreement with observed intensities.

In some other substances, such as I_2, an appreciable fraction of the molecules are in higher vibrational states at room temperature, and other bands can occur in absorption or in the rotation spectrum.

e. Infrared Bands of Other Types of Molecules Vibration-rotation spectra are known for many molecules. Their structure is often more complex than that described above; in many cases transitions for $\Delta J = 0$ are permitted, so that an additional sequence of lines occurs and there is no gap in the center of the band. Thus the absorption s̲ ̲ctrum of CO_2 shows many vibration-rotation bands in the region from 1.46 to 15.50 μ, and that of water vapor shows many from 0.69 to 6.26 μ; these bands, together with the rotation lines of water vapor, are responsible for the marked infrared absorption of the earth's atmosphere.

In order to exhibit vibration-rotation and rotation spectra of appreciable intensity, a molecule must possess an electric moment. *Homonuclear* diatomic molecules, such as O_2, N_2, H_2, Cl_2, possess no moments and, hence, have no spectra of these two types. Gases composed of such molecules are transparent in the infrared.

19.9 Molecular Quantum States

Up to this point, the *electrons* in the molecule have been ignored. Actually, the wave function for a molecule must contain the coordinates and spins of all the electrons as well as the coordinates of the nuclei.

When atoms unite into a molecule, the inner electrons in each atom can be regarded as remaining associated with the nucleus, but the outer electrons come to belong to the molecule as a whole rather than to any individual nucleus.

a. Approximate Separation of Electronic and Nuclear Motions   In many cases, especially for the diatomic molecules, the wave function can be separated approximately into two factors, of which one has reference to the electrons whereas the second factor represents vibrations of the nuclei and rotation of the molecule as a whole. The first factor then represents the electrons as being in a certain *electronic quantum state*. Both the energy and the wave function of the electronic state depend upon the relative positions of the nuclei; in this respect the situation in a molecule is quite different from that in an atom. The mutual electrostatic energy of the nuclei is commonly included in the energy of the electronic state. The electronic energy as so defined possesses a minimum value for certain relative positions of the nuclei; if they move closer together or farther apart, the electronic energy rises, and so does the energy of the molecule. Thus the energy of the electronic state functions as a potential energy tending to hold the atoms together in definite relative positions of equilibrium. It is in this way that a potential energy such as was sketched in Fig. 19.8 comes to exist in a diatomic molecule.

In this same approximation, the energy E of the molecule can be written as the sum of two parts, a negative part E_e, representing roughly the average of the electronic energy but including also the electrostatic energy due to the mutual repulsion of the nuclei, and a much smaller positive part E_{rv}, which is associated with vibration of the atoms relative to each other and with rotation of the molecule as a whole. Thus $E = E_e + E_{rv}$. For a diatomic molecule, the energy E_{rv} is what was denoted by E in Sec. 19.8.

For a detailed discussion of electronic states we must refer the student to other books, but something may be said here by way of explanation of the notation that the student will meet. Only *diatomic* molecules will be discussed.

b. ΛS Coupling   As in the theory of atomic states, electronic spin may be treated in different ways according to circumstances. If the *electronic spin-orbit effects are small*, an analog of *LS* coupling occurs. It is possible, as a first approximation, to assign a fixed value to the component of the orbital angular momentum of the electrons about the nuclear line, or line joining the two nuclei, which is an axis of symmetry. The magnitude of this momentum is denoted by $\Lambda\lambda$, where Λ is a positive integer or zero: the molecular quantum number Λ corresponds to M_L in atomic theory. In a diatomic molecule a unique choice of axis is supplied by the nuclear line. The *total* orbital momentum of the electrons cannot have a fixed value in a molecule, however; hence there can be no quantum number L. Spectroscopically, Λ plays a role analogous to L, since states corresponding to different values of Λ have different energies; in imitation of the atomic notation for LS coupling, states with $\Lambda = 0, 1, 2, 3, \ldots$ are indicated by the letters $\Sigma, \Pi, \Delta, \Phi, \ldots$, respectively.

The electronic spins are then coupled with the introduction of a quantum number S. The value of $2S + 1$ is written as a superscript. Thus, we obtain a set of ΛS electronic states represented by such symbols as

$$^1\Sigma, \ ^3\Sigma, \ ^1\Pi, \ ^3\Pi, \ \ldots \qquad ^2\Sigma, \ ^4\Sigma, \ ^2\Pi, \ ^4\Pi, \ \ldots$$

As with atoms, an important significance of the quantum numbers lies in the associated selection rules. There is a strong tendency for transitions to be limited to those for which

$$\Delta\Lambda = 0 \text{ or } \pm 1 \qquad \Delta S = 0$$

Instead of introducing next the electronic spin-orbit effects, we turn to the consideration of the *nuclear* motions, the effect of which upon the energy we are assuming to be much larger than the electronic spin-orbit effects. Each electronic state of the molecule may be combined with any one of the vibrational states for the nuclei. The orbital angular momentum of the electrons is then added vectorially to the angular momentum of rotation of the nuclei about axes perpendicular to the nuclear line. This results in the introduction of a rotational quantum number K, which may have any integral value such that

$$K \geq \Lambda$$

It is not possible for K to be less than Λ because in a diatomic molecule the two angular momenta are necessarily perpendicular to each other. The general selection rule for K is that

$$\Delta K = 0 \text{ or } \pm 1$$

We obtain in this way a set of *electronic-vibration-rotation states* for the molecule, numbered with quantum numbers Λ, S, v, K.

The *electronic spin-orbit effect* must then be considered. We add the total orbital angular momentum of electrons and nuclei, represented by K, to the total angular momentum due to electronic spin, represented by S, thus obtaining a grand resultant angular momentum, for which we introduce the usual quantum numbers J and M_J. The possible values of J are integrally spaced from $K + S$ down to $|K - S|$. The effect of the electronic spin-orbit interaction is as in atomic LS coupling, to separate states having different values of J. Thus a *fine structure*, analogous to the LS multiplet structure, is introduced into the electronic-vibration-rotation levels. The superscript in such a symbol as $^3\Pi$ refers to the (normal) number of J levels in this fine structure.

In singlet levels, with $S = 0$, and $J = K$, there is no fine structure, just as in atoms. The HCl molecule, for example, is normally in a $^1\Sigma$ electronic state; hence, the discussion of the vibrational and rotational states of this molecule as given above was adequate.

c. Ω *Coupling*   When the electronic spin-orbit effects are not smaller than the effects of molecular rotation, other modes of approach by means of perturbation theory become appropriate. If the *spin-orbit* effect is actually *large*, we have the analog of jj coupling in atoms; the electronic orbital and spin momenta about the nuclear line are first added together, their magnitude being represented by a quantum number Ω. The values of Ω are integral or half integral according as the number of electrons in the molecule is even or odd. Each electronic state, characterized by a value of Ω, is then combined with a vibrational state represented by quantum numbers v and J. Thus in this form of coupling, as when $S = 0$, the rotational levels correspond to the value of J.

The *energy of a molecular state* thus depends in the case of ΛS coupling upon five quantum numbers, Λ, S, v, K, J, or in the case of Ω coupling upon three, Ω, v, J. Its variation with the various numbers presents several different orders of magnitude. The variation with K or J is comparatively small. Spectral lines that differ only in the K values for the initial and final states lie close together and form a band, similar to the vibration-rotation bands that have been described. In such cases, if $J \neq K$, the energy varies still less with J, the effect of this variation being only to introduce a fine structure into the lines of the band. In the case of Ω coupling, on the other hand, the lines of a band arise from differences in the values of J. The energy varies

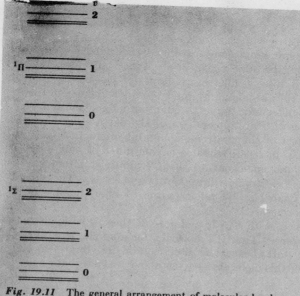

Fig. 19.11 The general arrangement of molecular levels (spacings exaggerated between rotational levels and grossly contracted between $^1\Sigma$ and $^1\Pi$ groups).

much more rapidly with v than it does with K or J.   Each pair of values of v, one for the initial and one for the final state, gives rise to a possible band.   The electronic states, finally, characterized by various values of Λ and S, or of Ω, are separated in energy by differences of the order of those between atomic LS terms.   These general features of the array of molecular levels are illustrated for a simple case in Fig. 19.11.

19.10   Electronic Bands

The most general type of transition between molecular states is one in which changes occur in the electronic state of the molecule as well as in its nuclear vibration-rotation state.   The spectra hitherto discussed, of rotational or vibrational type, represent special cases in which the electronic state does not change.   Such spectra constitute, however, only a small fraction of all known band spectra.

When the electronic state does change in a transition, the resulting change in energy is usually so large that the band lies in the visible or ultraviolet region of the spectrum. Such bands may be called *electronic* bands.

In transitions characterized by a given pair of electronic states and by given values v' and v'', representing a fixed pair of vibrational states, various changes of J (and of K, if $K \neq J$) may occur. The resulting lines form a single *band*. All the bands due to transitions between a given pair of electronic states, for all possible values of v' and v'', are said to form a *band system*. Because various electronic jumps are possible, the band spectrum of any molecule consists of many band systems.

The lines in a given electronic band are limited by the selection rules

$$\Delta K = 0 \text{ or } \pm 1 \qquad \Delta J = 0 \text{ or } \pm 1$$

Even aside from the fine structure that exists when the rotational levels are numbered by a quantum number K which is different from J, electronic bands are commonly more complicated than the simple type of vibration-rotation band described above because transitions for $\Delta K = 0$ (or $\Delta J = 0$) are allowed. Lines for which ΔK has the opposite sign to Δv constitute the *P branch* of the band; those for $\Delta K = 0$ (or $\Delta J = 0$) constitute the *Q branch;* those for which ΔK (or ΔJ) is in the same direction as Δv form the *R branch*. The Q branch is frequently missing, e.g., in all bands arising from a $\Sigma \rightarrow \Sigma$ electronic transition. When $K \neq J$, what is regarded as a single band sometimes contains more than one branch of each type.

A good approximate expression for the energies of the molecular levels is often obtained if we employ for the vibration-rotation part an expression like that in Eq. (19.6a) but allow B to vary with v. We have then for the molecular energy, in cm^{-1},

$$\bar{E} = \bar{E}_e + \bar{E}_v + B_v K(K + 1) \cdots \tag{19.8}$$

where B_v is positive. Thus, for the levels belonging to two different electronic states, we may write

$$\bar{E}' = \bar{E}'_e + \bar{E}'_{v'} + B'_{v'} K(K + 1)$$
$$\bar{E}'' = \bar{E}''_e + \bar{E}''_{v''} + B''_{v''} K(K + 1)$$

and we then find for the three branches of the band:

$P: v', K - 1; v'', K$

$$\bar{\nu} = \bar{\nu}_e + \bar{\nu}_{v'v''} - (B'_{v'} + B''_{v''})K + (B'_{v'} - B''_{v''})K^2 \tag{19.9a}$$

$Q: v', K; v'', K$

$$\bar{\nu} = \bar{\nu}_e + \bar{\nu}_{v'v''} + (B'_{v'} - B''_{v''})K(K + 1) \tag{19.9b}$$

$R: v', K; v'', K - 1$

$$\bar{\nu} = \bar{\nu}_e + \bar{\nu}_{v'v''} + (B'_{v'} + B''_{v''})K + (B'_{v'} - B''_{v''})K^2 \tag{19.9c}$$

Here K refers in each case to the larger of the two values of K concerned in the transition. If the quantum number K does not exist for the levels in question, K is to be replaced by J in all equations from (19.8) to (19.9c). The symbol $\bar{\nu}_e = (\tilde{E}'_e - \tilde{E}''_e)/h$ and represents the frequency that would arise from the electronic transition alone; whereas $\bar{\nu}_{v'v''} = (\tilde{E}'_{v'} - \tilde{E}''_{v''})/h$, and it is assumed that $v' \geqq v''$.

Since the quantities $B'_{v'}$ and $B''_{v''}$ in Eqs. (19.9a, b, c) refer to different electronic states, they may differ considerably. We may suppose the forces between the atoms to be quite different in the two cases; hence, their positions of equilibrium and the values of the moment of inertia of the molecule are also different. This is in contrast to the case of the vibration-rotation bands, where B_v varies only a little from one level to another. The quadratic terms in (19.9) soon make themselves felt. As K increases, the trend of $\bar{\nu}$ in one branch is soon reversed, in the P branch if $B'_{v'} > B''_{v''}$, in the R branch if $B'_{v'} < B''_{v''}$. It is thus a general characteristic of electronic bands that one branch is folded back on itself and on top of the others. At the point in the spectrum where a branch turns back, the lines are crowded together, forming a *band head*. The band appears to shade away from the head; some bands are shaded toward the red, some toward the violet. When a Q branch is present, it may form a second head, although such behavior is not obviously predicted by Eq. (19.9b).

A band of this type is often represented graphically by means of a Fortrat diagram, on which each line is represented by a point or circle, the ordinate representing K (or J) and the abscissa, $\bar{\nu}$. Such a diagram is shown in Fig. 19.12a for the CuH band λ 4280 Å, which has no Q branch; a spectrogram of this band is shown in Fig. 19.12b. The band is shaded toward the red from a head which is plainly shown in the spectrogram. The electronic transition is $^1\Sigma \rightarrow {}^1\Sigma$.

The *number of different bands in a band system*, all arising from the same electronic transition but with various values of v' and v'', may be very large. The change in v, $v' - v''$, however, although not confined to ±1 as for a harmonic oscillator, tends to be restricted to moderate values. A spectrogram showing parts of three such band systems is reproduced in Fig. 19.13. Extremely low dispersion was used in order to include a wide spectral range; hence each *band* appears as a single *line* in the figure; the groups that seem to the eye to stand out are groups of bands having in each case the same value of the difference $v' - v''$. The values of v', v'', for example, $6 - 3$, are shown

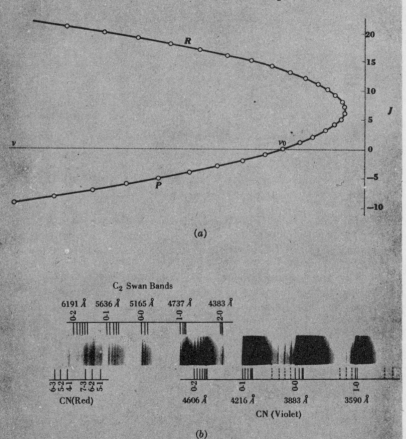

(a)

(b)

Fig. 19.12 (a) Fortrat diagram of the CuH band λ 4280 Å. The lines are represented by circles; the ordinate represents the larger J for the transition; $v_0 = v_{v'v''}$; (b) spectrogram of this band with J value of upper energy level. (*Due to R. Mecke; reprinted from G. Herzberg, "Spectra of Diatomic Molecules." Copyright 1950. D. Van Nostrand Co., Inc. Used by permission.*)

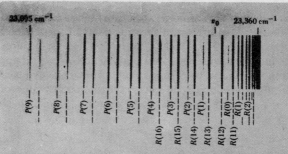

Fig. 19.13 Three band systems of CN and C_2 (carbon arc in air). (*Reprinted from G. Herzberg, "Spectra of Diatomic Molecules." Copyright 1950. D. Van Nostrand Co., Inc. Used by permission.*)

for a few bands. The Swan bands of C_2 result from a $^3\Pi \rightarrow {}^3\Pi$ electronic transition; the violet CN bands, from $^2\Sigma \rightarrow {}^2\Sigma$; the red CN bands, from a $^2\Pi \rightarrow {}^2\Sigma$ transition.

Because electrons move much faster than nuclei, *an electronic transition occurs so rapidly compared with the vibrational motion of the nuclei that the internuclear separation remains essentially constant during the transition.* This fact, known as the *Franck-Condon principle*, is illustrated by the vertical lines of Fig. 19.8. Note that the average internuclear separation is greater in the higher electronic state.

The band spectra of molecules containing *different isotopes* of the same element are slightly different, because of the difference in nuclear mass. Extra lines due to this cause in band spectra as observed have sometimes led to the discovery of rare isotopes, such as the oxygen isotope of atomic weight 18, present in only about one five-hundredth as great concentration as O^{16}.

19.11 The Raman Effect

At this point a slight digression may conveniently be made to describe the Raman effect, which can occur in atomic spectra but is of greatest interest in relation to band spectra.

When light passes through a "transparent" substance, solid, liquid, or gaseous, a certain part of the light is scattered in all directions (the Tyndall effect). The most familiar example is the light from a clear sky. Rayleigh ascribed such effects to scattering by the individual molecules or by groups of molecules much smaller in linear dimension than the wavelength of the light. If the incident light is monochromatic, the scattered light is ordinarily observed to be unchanged in frequency, in accordance with Rayleigh's theory.

It was shown in 1925 by Kramers and Heisenberg, however, that, according to classical electromagnetic theory, if the scattering electrons in an atom or molecule are in motion, the scattered light should contain other frequencies in addition to that of the incident light. Failure to notice this implication of classical theory had been due to the common assumption that the radiating particles were at rest except as disturbed by radiation. In 1928, independently of the theoretical prediction, Raman and Krishnan discovered the phenomenon experimentally, in the course of an extensive study of scattering by liquids and solids.

The scattering of light with a change in its frequency has been studied extensively since then and is commonly called the *Raman effect*. To observe it, the incident light should be monochromatic and very intense. The scattered light is then seen to contain, besides a line of the same frequency as the incident light, several weak lines of other frequencies. If the incident frequency ν is varied, these other lines move along the frequency axis at the same rate, maintaining constant frequency differences from ν and not changing greatly in intensity. In these respects the Raman lines differ sharply in behavior from fluorescent lines, the frequencies of which are fixed by the scattering substance and which flash out only when the incident frequency falls upon an absorption line of the substance. In the Raman effect it is *frequency shifts* in the scattered spectrum that are determined by the nature of the scatterer rather than the frequencies themselves.

When a photon of frequency ν is scattered by an atom or molecule whose quantum state is not altered in the process, the scattered photon has the same frequency as had the incident photon. But it may happen that the atom or molecule is changed in the process from a state in which its energy is E_1 to a state of different energy E_2. Conservation of energy then requires that the frequency ν' of the scattered photon be modified so that $h\nu' + E_2 = h\nu + E_1$; hence,

$$\nu' = \nu + \frac{E_1 - E_2}{h}$$

In terms of such ideas the Raman effect had been predicted at a much earlier date by Smekal.

In the expression just written, the term $(E_1 - E_2)/h$ can be interpreted as a frequency ν_{12} that the atom might conceivably emit or absorb in the usual way, in jumping from the first state to the second. Thus for the Raman line we may write

$$\nu' = \nu + \nu_{12}$$

The difference between the frequency of each Raman line and the frequency of the incident light is thus equal to the frequency of some conceivable emission or absorption line of the scattering atom or molecule.

The *intensity* of a Raman line has nothing to do with the intensity of the emission or absorption line that is thus correlated with it. The selection rules for the two are quite different; transitions that are forbidden in ordinary spectra occur freely in the Raman effect. This is one reason for its great theoretical interest. According to wave mechanics, *a Raman jump is possible between two atomic or molecular levels A and B only when there exists at least one third level, C, such that ordinary radiative transitions are allowed between A and C and between B and C.* It is almost as if the atom or molecule actually jumped first from A to C and then from C to B. The relative probabilities of the various processes do not correspond to this simple picture, however, nor can it be said that the atom or molecule exists for any definite time in state C, as it does in the production of fluorescence.

In light scattered by *polyatomic molecules* there may be Raman lines correlated in this manner with energetically possible lines in either the rotational or the vibration-rotation or the electronic spectra of the molecule. It does not matter whether these spectra can actually be observed or are prevented from direct occurrence by a selection rule.

In the common type of rotation spectrum, the selection rule is $\Delta J = \pm 1$. For the Raman lines associated with this spectrum, therefore, the selection rule will be either $\Delta J = 0$ (that is, $J \rightarrow J \pm 1$ for a jump from A to C and $J \pm 1 \rightarrow J$, from B to C) or $\Delta J = \pm 2$. The case $\Delta J = 0$ involves no change in the molecular energy and hence merely contributes to the ordinary, or Rayleigh, scattering. Thus, effectively, for Raman lines of purely rotational origin, we must have

$$\Delta J = \pm 2$$

The incident line, as seen in the spectrum of the scattered light, should be accompanied on each side by several lines spaced twice as far apart on the frequency scale as are the lines of the rotational spectrum itself. From Eq. (19.3) it is seen that the displacements of the Raman lines from the incident line will be proportional to $J(J + 1) -$ $(J - 2)(J - 1)$, or to $2(2J - 1)$, J referring to the upper rotational level. Hence, since $J \geq 2$, the displacements are proportional to 6, 10, 14, . . . ; thus the distance from the unmodified line to the first Raman line is 1.5 times the mutual spacing of the Raman lines themselves. The origin of the lines is illustrated diagrammatically in Fig. 19.14.

Besides these *rotational* Raman lines, there may also appear, at a much greater distance and usually only on the long-wavelength side, a band corresponding to the ordinary *vibration-rotation* spectrum. For this band the selection rule is easily seen to be

$$\Delta J = 0 \text{ or } \pm 2$$

Lines for which $\Delta J = 0$, involving almost no change in rotational energy, coalesce into an intense line in the approximate position of the missing central line of frequency $\nu_{v'v''}$ or ν_{10} in the vibration-rotation band. On each side of this line there are much fainter lines for $\Delta J = \pm 2$. In this band the vibrational change is from $v = 0$ to $v = 1$, as for absorption. In case the vibrational level for $v = 1$ lies so low that a considerable fraction of the molecules are maintained in this state by thermal agitation, a similar weaker Raman band should be observed on the short-wavelength side of the exciting line as well, associated with the transition from $v = 1$ to $v = 0$.

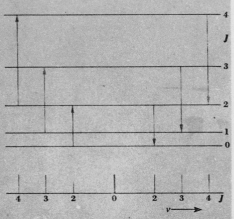

Fig. 19.14 Diagram for a rotational Raman spectrum.

Many Raman spectra have been observed. They constitute a valuable source of information concerning molecular quantum states. The light scattered by HCl gas when strongly illuminated by light from a mercury arc in glass was studied by Wood and Dieke. They found a line at 4581.8 Å which they interpreted as the single intense line in a vibration-rotation Raman spectrum excited by the Hg line at 4047 Å. The frequency difference between these two lines is 2886.0 cm^{-1}, which is in excellent agreement with the frequency of the missing central line in the 3.5-μ band from HCl, viz., 2885.4 cm^{-1}. Much closer to the exciting line and on both sides of it there were also a number of lines interpreted as a rotational Raman spectrum. In the corresponding spectrum adjacent to the mercury line at 4358 Å, the measured spacings between the lines lay mostly between 41 and 42 cm^{-1}, in good agreement with double the spacing in the ordinary rotation spectrum of HCl, as shown in Table 19.1.

Even homonuclear molecules such as O_2 and N_2 give Raman spectra corresponding to vibration-rotation and rotational spectra, although the latter cannot be observed directly. Such molecules receive special discussion in Sec. 20.9.

19.12   *The Ammonia Inversion Spectrum*

The band spectra of *polyatomic* molecules are complicated. Several different modes of vibration of the molecule are possible, and if these modes involve oscillating electric moments, they give rise to observable vibration-rotation bands. The qualitative features to be expected in simple cases were worked out in 1927 by Hund. One feature of special interest is represented in the spectrum of gaseous ammonia.

In the ammonia molecule, NH_3, the nuclei may be thought of as lying at the corners of a low pyramid, N at the apex and H at the corners of an equilateral triangular base (Fig. 19.15). It is calculated that the N nucleus is about 0.36 Å distant from the H_3 plane, the NH distance being about 1.01 Å and HH, 1.63 Å; the HNH angles are about 108°. In a vibration of the molecule, the nuclei move only short distances from their positions of equilibrium. A system of four particles will have a total of 12 degrees of freedom, but 3 of these belong to their common center of gravity and 3 more correspond to rigid rotations, leaving 6 degrees of vibrational freedom. Because of symmetry, however, degeneracies occur, so that only four distinct vibrational frequencies are observed in the ammonia spectrum, namely: $\bar{\nu}_1 = 3336$ cm^{-1} and $\bar{\nu}_2 = 3414$ cm^{-1}, the latter weak, and both near $\lambda = 3\ \mu$; $\bar{\nu}_3 = 949.9$ cm^{-1} (10.5 μ) and $\bar{\nu}_4 = 1627.5$ cm^{-1} (6.2 μ), the latter very intense. In $\bar{\nu}_1$ and $\bar{\nu}_3$ the nuclei move symmetrically relative to the axis of the pyramid; for $\bar{\nu}_2$ and also for $\bar{\nu}_4$, two independent modes exist, so that there are six modes in all.

It will be noted that for given positions of the protons, the N nucleus may lie either on one side of the H$_3$ plane or on the other, as indicated in Fig. (19.15). These arrangements are related to each other as if by reflection in a mirror placed parallel to the H$_3$ plane; one of them cannot be converted into the other by a pure rotation. They *can* be interconverted by a rotation plus interchange of two of the protons. Since all protons have the same properties, it follows *in classical mechanics* that these two forms of NH$_3$ are physically indistinguishable. In such situations, however, wave mechanics introduces a nonclassical distinction.

Consider the variation of the potential energy P of the molecule as N moves along a line perpendicular to the H$_3$ plane, x being its

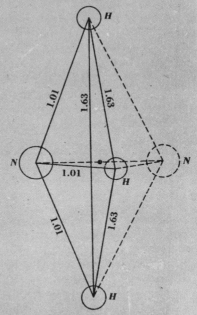

Fig. 19.15 The ammonia molecule NH$_3$ with numbers between atoms giving separations in angstrom units.

distance from this plane. The curve for P is symmetrical in x with two similar minima representing two possible positions of stable equilibrium for N relative to the H atoms, separated by a potential barrier, a situation qualitatively resembling that of Fig. 19.5*b*. If the protons stood still, the barrier would remain high. However, as the protons move out and in, the barrier becomes low, perhaps less than 3000 cm^{-1}. In classical mechanics the N atom would simply be located in one or the other of the two "wells." The wave function for a quantum state, on the other hand, is in such cases either symmetric or antisymmetric relative to the center of symmetry and thus represents simultaneous *equal probabilities* of finding N in one well or the other. If the potential barrier is very low or very narrow, the energies of the quantum states are widely separated, but when the barrier is high and wide, they come almost together in pairs, a symmetric and an antisymmetric state close together. That they do not quite coincide is connected with the fact that, contrary to the classical case, $|\psi|^2$ represents a small probability, different for the two functions, that N might be found *inside the barrier*, perhaps even in the same plane with H$_3$.

In ammonia the result of this inversion effect is that each of the vibration-rotation states of the molecule is doubled. The lowest two levels, one of them being the ground state, lie only about 0.66 cm^{-1} apart. (The next lowest pair of excited states are 35.7 cm^{-1} apart.) The effect on the band spectra is to convert all vibration-rotation or rotational lines into doublets, the selection rules being such that the doublet separation $\Delta \nu$ is the sum of the inversion separations in the initial and final states. This effect was found in 1932 by Dennison and Hardy in the "ν_1 band"; the vibration-rotation lines were observed as doublets spaced 1.3 cm^{-1} apart. Soon thereafter Wright and Randall found a split averaging 1.3 cm^{-1} in the purely *rotational* lines at 71, 83, and 100μ, observed in absorption.

Radiative transitions between the lowest two molecular levels themselves are possible, corresponding to a calculated frequency $\bar{\nu} = 0.66$ cm^{-1} or a wavelength $\lambda = 1.5$ cm. This, the NH$_3$ *inversion line*, was found by Cleeton and Williams in 1934. They observed a clear-cut peak in the absorption spectrum of NH$_3$ at $\lambda = 1.25$ cm (0.8 cm^{-1}). A predicted *fine structure* in the inversion line, caused by differences in the various rotational states of the molecule, was observed in 1946 by W. E. Good, who found 30 inversion-rotation lines spread over a range of 6000 MHz, or 0.20 cm^{-1}.

Finally, even a *hyperfine* structure was noted by Good in certain lines. For example, at very low gas pressures two faint lines could be seen on each side of the main peak of the (3,3) line, whose half-intensity width was about 7×10^5 Hz. This fine structure, studied by Coles and Good using separated isotopes, appeared in lines due to $N^{14}H_3$ but not in those due to $N^{15}H_3$. The splitting is ascribed to a slight local departure of the nuclear field of N^{14} from the coulomb form, representable by an electric quadrupole moment. The effect of this moment on the energy levels varies for different coupling states of the nuclear spin with other spins. This interpretation was confirmed by later theoretical analysis. Quadrupole splitting occurs only with a nucleus having $I > \frac{1}{2}$; for N^{14}, $I = 1$, for N^{15}, $I = \frac{1}{2}$. Although with $I = \frac{1}{2}$ two opposite orientations are possible, a quadrupole looks the same when turned end for end.

Problems

1. (a) Show that if point masses m_1 and m_2 are attached to the ends of a linear spring with spring constant k, the system vibrates at a frequency $\nu = (1/2\pi)\sqrt{k/m_r}$, where $m_r = m_1 m_2/(m_1 + m_2)$ is the reduced mass of the system.

(b) Show that the moment of inertia of these two masses about an axis through the center of mass and perpendicular to the line of centers is given by $m_r R^2$, where R is the distance between the masses.

2. From the data of Table 19.1 calculate the equilibrium separation of the H^+ and Cl^- ions, assuming that the H^+ ions are protons and that the chlorine ions are of mass number 35. The handbook value of the equilibrium separation in the ground state is 1.2746 Å. Why is your answer a little larger? Why does $\Delta\bar{\nu}$ decrease as J increases?

3. From Raman spectra of hydrogen gas it is found that the fundamental vibrational transition $v = 0$ to $v = 1$ occurs at $\bar{\nu} = 4159.2$ cm^{-1}.

(a) Find the effective "spring constant" $(d^2 E/dR^2)_{R_0}$ for H^2.

(b) Calculate the wavenumber of the corresponding transition for D_2 (a hydrogen molecule composed of two deuterium atoms), for which the experimental value is 2990.3 cm^{-1}.

Ans: 520 N/m

4. Calculate the moment of inertia and the energies of the first five rotational levels for the KCl molecule. (Assume the Cl atom has $A = 35$.) What are the wavelengths of rotational lines arising from transitions between these levels?

5. Estimate the repulsive energy term for RbCl, given that the equilibrium separation is 2.89 Å and the dissociation energy is 3.96 eV. (Assume that the van der Waals attractive interaction is approximately equal to the vibrational zero-point energy.)

6. (a) Assume that the ion-core repulsion energy has the form a/R^n for KCl, where a and n are numerical constants. From the data of Sec. 19.3, obtain values for a and n.

(b) Using the crude values from part (a), estimate the vibrational frequency for small displacements. (The experimental value is 8.34×10^{12} Hz.)

7. Show that the frequency of the rotational line arising from a transition $J + 1$ to J for a diatomic molecule lies between the classical rotational frequencies of the molecule in the upper and lower states, assuming the rotational kinetic energy of state J is given by Eq. (19.3).

8. The force constant for the HCl bond is 470 N/m. Find the probability that an HCl molecule is in its lowest excited vibrational state at 290°K.

9. If the hydrogen molecule has a force constant of 573 N/m, assuming there were no anharmonic terms in the expansion for the potential energy, find the vibrational quantum number which corresponds to the dissociation energy.

10. Show that both the P and R branches of the Fortrat diagram are given by the same formula if K is allowed to take on negative values in Eq. (19.9).

11. A molecule of LiH (lithium of mass 7) has an equilibrium separation of 1.595 Å and a vibrational frequency with a wavenumber of 1406 cm^{-1}.

(a) Find the energy of the first excited rotational state and the ratio of this energy to that required to excite the first vibrational state.

(b) Compute the force constant for the vibrational frequency of the molecule.

12. Of HCl molecules in the ground vibrational state ($v = 0$) compute the ratio of the number in the ground rotational state ($J = 0$) to the number in the first excited state ($J = 1$) at $T = 290°K$. (See Sec. 19.7 for moment of inertia of HCl molecule; note that the statistical weight of each J state is $2J + 1$.)

Ans: about 0.4

13. (a) Show that when $kT = h\nu$ for vibrational energy states, roughly one-half of the molecules are in excited vibrational states.

(b) Show similarly that when kT is equal to the minimum rotational energy, roughly half the molecules are rotating. (Do not forget the $2J + 1$ statistical weight.)

14. (*a*) Find the temperature at which kT is equal to the energy of the first excited rotational state for H_2.

(*b*) If the force constant for H_2 is 573 N/m, find the temperature at which kT is equal to the excitation energy for the first excited vibrational level.

(*c*) What rotational level has energy nearest to that required to raise the molecule to the $v = 1$ vibrational state?

(*d*) If $J = 1$ and $v = 1$, compute the number of rotations made by the H_2 molecule during one vibration.

15. In 1929 Morse suggested an equation giving the energy of a diatomic molecule as a function of the internuclear spacing R in terms of the dissociation energy (or binding energy) E_b, the equilibrium separation, and one arbitrary constant a. The Morse function may be written

$$E(R) = E_b \left(e^{-2a(R-R_0)} - 2e^{-a(R-R_0)} \right)$$

(*a*) Show that the Morse function gives $E(R)_0 = E_b$ and $E(\infty) = 0$.

(*b*) Show that the vibration frequency for small displacements is

$$\frac{a}{2\pi} \sqrt{\frac{2E_b}{m_r}}$$

where m_r represents the reduced mass.

(*c*) Calculate a for the Morse curve for Na_2, given that $E_b = 0.75$ eV, $R_0 = 3.07$ Å, and the wavenumber for the fundamental vibration transition is 157.8 cm^{-1}.

Ans: (*c*) 0.84 Å^{-1}

16. The muonic hydrogen molecule-ion $(p - \mu^- - p)$ is known to exist. Assuming that the μ^- acts as a massive electron, estimate the interproton separation and the binding energy of the molecule-ion.

chapter twenty

Introduction to Quantum Statistics

Just as the failure of classical mechanics made it necessary to establish quantum mechanics for the description of nature on the atomic scale, the failure of classical Maxwell-Boltzmann statistics to provide correct solutions for many problems involving fundamental particles led to the development of new systems of quantum statistics. In the present chapter we reexamine the statistical bases underlying the Maxwell-Boltzmann statistics and introduce modifications which lead to the Fermi-Dirac statistics, applicable to all fundamental particles with spin one-half, and to Bose-Einstein statistics, applicable to particles which have zero or integer spin.

20.1 Statistics of Distinguishable Objects

Suppose we have N distinguishable objects which we wish to place in order along the x axis. In how many ways can this be done? Clearly there are N possible choices for the object nearest the origin, $N - 1$ choices for the next position, $N - 2$ for the third, etc. The number of possible orders is

$$N(N - 1)(N - 2) \cdots (1) = N!$$

If we have s boxes, in how many different ways can these N objects be placed in these boxes with n_1 in the first box, . . . and n_s in the sth box, assuming that we are not concerned with the order of the objects in the boxes? For the first box there are $[N(N - 1) \cdots (N - n_1 + 1)]/n_1!$, where the denominator is the number of ways the n_1 objects could be arranged in the box. Similarly, for the second box the n_2 objects could then be selected in $[(N - n_1)(N - n_1 - 1) \cdots (N - n_1 - n_2 + 1)]/n_2!$ ways, etc. For all s boxes the total number of different ways becomes

$$\frac{N(N - 1) \cdots (N - n_1 + 1)}{n_1!}$$
$$\frac{(N - n_1) \cdots (N - n_1 - n_2 + 1)}{n_2!} \cdots$$
$$\frac{n_s(n_s - 1) \cdots 1}{n_s!} = \frac{N!}{n_1! n_2! \cdots n_s!} \quad (20.1)$$

Consider an evacuated box of volume V which is arbitrarily divided into s subvolumes (Fig. 20.1). Let N idealized, distinguishable point "molecules" be injected. Here and throughout our development of distribution laws we assume that the molecules interact only very weakly with one another, so that only during brief collisions do they exert

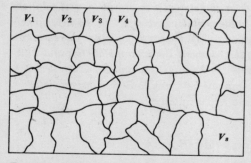

Fig. 20.1 A box of volume V arbitrarily divided into s subvolumes.

substantial forces on each other.   Thus we are dealing with an essentially ideal gas.   The probability that any particular molecule is in the ith subvolume at any instant is given by V_i/V, that any preselected pair is in this subvolume is $(V_i/V)^2$, and that any *preselected* n_i molecules are in V_i is $(V_i/V)^{n_i}$.   At any instant the probability that there are n_1 molecules in V_1, n_2 in V_2, . . . , and n_s in V_s is therefore

$$\mathcal{P}(n_1,n_2,\ldots,n_s) = \frac{N!}{n_1!n_2!\cdots n_s!}\left(\frac{V_1}{V}\right)^{n_1}\left(\frac{V_2}{V}\right)^{n_2}\cdots\left(\frac{V_s}{V}\right)^{n_s}$$

$$= N!\prod_{i=1}^{s}\frac{1}{n_i!}\left(\frac{V_i}{V}\right)^{n_i} \tag{20.2}$$

If we sum $\mathcal{P}$ over all possible sets of values of n_1, n_2, etc., subject to the condition that $\Sigma n_i = N$, we must, of course, get 1 as a result, since we sum over the probabilities of all possible distributions.   Indeed by the multinomial theorem

$$\left(\frac{V_1}{V} + \frac{V_2}{V} + \frac{V_3}{V} + \cdots + \frac{V_s}{V}\right)^N$$

$$= \sum_{\substack{\text{all possible sets} \\ \text{of values } n_i \ldots n_s}} \frac{N!}{n_1!\cdots n_s!}\left(\frac{V_1}{V}\right)^{n_1}\cdots\left(\frac{V_s}{V}\right)^{n_s} = 1 \tag{20.3}$$

(If $n_i = 0$, it is omitted; this can be taken care of by setting $0! = 1$.)

If N distinguishable but otherwise identical monatomic molecules were suddenly injected into an evacuated, perfectly reflecting container with a known arbitrary distribution of velocities, knowledge of the positions and collision characteristics of the molecules would permit us *in principle* to apply the laws of classical mechanics to predict the future motion of each molecule.   However, experiments with real molecules and containers show that, regardless of the injection mechanism, the memory of the initial state is completely obliterated in a very short time.   Collisions soon produce the same equilibrium state for a given N and average energy, quite independent of the injection distribution.   *This equilibrium state is the state of maximum probability corresponding to the number of molecules N, total energy E, and volume V chosen.*   Every measurement of a parameter of the gas made over a finite time interval leads to a time-average-over-the-interval value of the parameter in agreement with predictions based on the laws of probability.

To find the equilibrium distribution of the energies we imagine a one-dimensional space in which the energy of each molecule is represented as an arrow with its tail at the origin and its length proportional to the energy (Fig. 20.2).   We divide this energy space into a large number s of contiguous cells of sizes $\Delta\epsilon_1$, $\Delta\epsilon_2$, . . . such that all arrows which end in the first cell have average energy ϵ_1 and so forth.   Let n_1 arrows end in the first cell and n_i in the ith cell.

If E represents the total energy of the molecules, which are assumed to have negligible interaction energies, it is clear that the numbers n_i

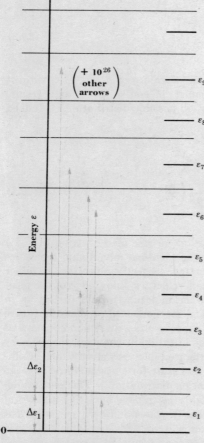

Fig. 20.2 At some instant the energy of every molecule is represented by an arrow in a one-dimensional energy space.

must satisfy the two equations

$$\sum_{i=1}^{s} n_i = N = n_1 + n_2 + n_3 + \cdots + n_s \tag{20.4}$$

$$\sum_{i=1}^{s} n_i\epsilon_i = E = n_1\epsilon_1 + n_2\epsilon_2 + \cdots + n_s\epsilon_s \tag{20.5}$$

At this point in a classical treatment we introduce a result of quantum mechanics which proves useful. Specifically, we accept the facts (1) that the allowed energy states of a bound system are discrete and (2) that if any degeneracy exists, it can be removed by application of a magnetic field or in some other way. Assuming then a system of nondegenerate discrete energy states, we let g_1 represent the number of allowed energy states in cell 1 and g_i the number in the ith cell. Then the total number of possible energies is

$$G = g_1 + g_2 + \cdots + g_s = \sum_{i=1}^{s} g_i \tag{20.6}$$

In the absence of any exclusion principle or constant associated with the conservation of energy, each molecule might be in any one of the G states and the total number of ways N distinguishable molecules could be distributed among the states is G^N. Each of these possibilities specifies a particular *microscopic state* or *microstate* of the system. However, if the molecules are all alike, we are ordinarily not concerned with which particular molecules have energies that place them in cell i but only with how many are in each energy cell. A statement that there are n_1 molecules in cell 1, n_2 in cell 2, and n_i in cell i serves to specify a *macroscopic state* of the system. If we now postulate that *all microstates have the same a priori probability*, each g_i plays the role of V_i in Eqs. (20.2) and (20.3) with G corresponding to V. Then the probability of each distinct macroscopic distribution becomes

$$\mathcal{P}(n_1, \ldots, n_s) = \frac{N!}{n_1! \cdots n_s!} \left(\frac{g_1}{G}\right)^{n_1} \cdots \left(\frac{g_s}{G}\right)^{n_s} = \frac{N!}{G^N} \prod_{i=1}^{s} \frac{g_i^{n_i}}{n_i!} \tag{20.7}$$

In the limit of continuous energy states, Eq. (20.7) is still valid if we take g_i/G to represent the probability that an arbitrary energy arrow terminates in the ith cell.

Boltzmann proposed that the equilibrium distribution of energies of the molecules in a gas is the one for which $\mathcal{P}(n_i, \ldots, n_s)$ is maximum, subject to the restrictions of Eqs. (20.4) and (20.5), which respectively require the conservation of molecules and the conservation of energy. He elected to work with $\ln \mathcal{P}$, which obviously is maximum when $\mathcal{P}$ is maximum. Boltzmann proved that there is a logarithmic relation between entropy and probability; subsequently Planck proposed that the entropy S could be written as $S = K \ln \mathcal{P}$, where K is a constant. Thus, as required by thermodynamics, the equilibrium configuration is that of maximum entropy. Clearly

$$\ln \mathcal{P} = \ln N! - N \ln G - \ln n_1!$$
$$- \cdots - \ln n_s! + n_1 \ln g_1 + \cdots + n_s \ln n_s \quad (20.8)$$

According to Stirling's theorem, when N is large,

$$\ln N! = N \ln N - N + \ln \sqrt{2\pi N} + \text{terms of the order of } \frac{1}{N}$$
$$(20.9)$$

If all n's are large compared to 1, we need take only the first two terms of the Stirling formula and write

$$\ln \mathcal{P} = N \ln N - N - N \ln G + n_1 \ln \frac{g_1}{n_1} + n_1 + \cdots$$
$$+ n_s \ln \frac{g_s}{n_s} + n_s \quad (20.8a)$$

To find the maximum of $\ln \mathcal{P}$ subject to the restrictions imposed by Eqs. (20.4) and (20.5) we use Lagrange's method of undetermined multipliers. Let α and β be two as yet undetermined constants (Lagrange multipliers). Equations (20.4) and (20.5) can be written

$$\alpha \left[\left(\sum_{i=1}^{s} n_i \right) - N \right] = 0 \quad \text{and} \quad \beta \left[\left(\sum_{i=1}^{s} n_i \epsilon_i \right) - E \right] = 0$$

Combining these with Eq. (20.8a) leads to the result that

$$\ln \mathcal{P} = N \ln N - N - N \ln G + \sum_{i=1}^{s} n_i \ln \frac{g_i}{n_i} + \sum_{i=1}^{s} n_i$$
$$- \alpha \left(\sum_{i=1}^{s} n_i - N \right) - \beta \left(\sum_{i=1}^{s} n_i \epsilon_i - E \right) \quad (20.10)$$

if and only if Eqs. (20.4), (20.5), and (20.8a) are all satisfied.

For ln $\mathcal{P}$ to be maximum, it is necessary that for each n_i

$$\frac{\partial \ln \mathcal{P}}{\partial n_i} = \ln \frac{g_i}{n_i^*} - 1 + 1 - \alpha - \beta \epsilon_i = 0 \tag{20.11}$$

where n_i^* is the value of n_i for which $\mathcal{P}$ is maximum. The distribution numbers n_1^*, n_2^*, etc., are of great importance because the number of distinct macroscopic states for them is vastly greater than the number for any other possible set of n_i's. Indeed, it is overwhelmingly likely that the gas will be found in a state characterized by n_i's virtually equal to the n_i^*'s satisfying Eq. (20.11). Any other state is so unlikely that it does not persist throughout the time of a measurement. Solution of Eq. (20.11) for n_i^*/g_i yields

$$f_{MB}(\epsilon_i) = \frac{n_i^*}{g_i} = \frac{1}{e^\alpha e^{\beta \epsilon_i}} = e^{-\alpha} e^{-\beta \epsilon_i} \tag{20.12}$$

According to Eq. (4.19), the number of molecules n_i in a small volume $g_i = dp_x \, dp_y \, dp_z \, dx \, dy \, dz$ of phase space is given by

$$n_i = g_i \left[\frac{N}{(2\pi mkT)^{\frac{3}{2}} CV} \right] e^{-\epsilon_i/kT} \tag{20.12a}$$

which is Eq. (20.12) if we identify β with $1/kT$ and $e^{-\alpha}$ with the constant in the brackets. That these identifications are proper can be shown by evaluating the constants α and β by use of Eqs. (20.4) and (20.5). Thus the Maxwell-Boltzmann (M-B) distribution arises from the statistics of distinguishable particles.

In classical statistical mechanics these ideas are applied to the molecules of real gases as though these molecules were distinguishable. The resulting M-B distribution function is applicable in many situations, but in spite of its successes, the M-B statistics failed to yield acceptable solutions to many problems involving electrons in atoms and in solids.

20.2   Indistinguishable and Exclusive Particles

In deriving the M-B distribution function above, it was assumed that the molecules are distinguishable. When two billiard balls collide, we can follow each ball and state with confidence, for example, that ball A came from $-x$ direction, struck ball B, and then moved off in the y direction. However, when two electrons (or protons or identical atoms) collide, we cannot in general follow the motions of each electron. If electron A comes from the $-x$ direction and after collision with B, an electron moves off in the y direction, we cannot tell whether it is A or B. Electrons are indistinguishable. Each electron has an associated wave packet, and once the wave packets of the two electrons overlap, we can no longer keep track of which electron is which. What happens while the wave functions overlap strongly is hidden in the fog of uncertainty. Thus billiard balls are distinguishable, but electrons, and indeed all identical fundamental particles, are indistinguishable.

Further, electrons obey the Pauli exclusion principle (as do other particles of half-integer spin); on the other hand, photons, pions, and other particles of zero or integer spin do not. We shall call particles obeying an exclusion principle *exclusive* and the others *nonexclusive*. To see how indistinguishability and exclusiveness may influence the statistics of particles, consider the number of possible distinct ways in which two particles ($n = 2$) can be placed in three energy states ($g = 3$).

CASE 1. DISTINGUISHABLE, NONEXCLUSIVE PARTICLES   Let the two particles be represented by x and o. The distinct ways in which these particles can be arranged in three energy states are shown below (without regard to order).

```
- - - - -   - - - - -   - o - x -   - - - - -   - - x - -   - - - - -   - - o - -   - - x - -   - - o - -
- - - - -   - o - x -   - - - - -   - - x - -   - - - - -   - - o - -   - - - - -   - - o - -   - - x - -
- o - x -   - - - - -   - - - - -   - - o - -   - - o - -   - - x - -   - - x - -   - - - - -   - - - - -
```

The nine possibilities above are the g^n expectations which have been used in deriving the M-B results.

CASE 2. INDISTINGUISHABLE, NONEXCLUSIVE PARTICLES   Since the particles are indistinguishable, we use an x to represent either one of them. There are now six distinct ways of arranging two particles in three energy levels:

```
- - - - -   - - - - -   - x - x -   - - - - -   - - x - -   - - x - -
- - - - -   - x - x -   - - - - -   - - x - -   - - - - -   - - x - -
- x - x -   - - - - -   - - - - -   - - x - -   - - x - -   - - - - -
```

Indistinguishable, nonexclusive particles clearly obey different statistical rules than those of Maxwell and Boltzmann. Bose and Einstein introduced statistics applicable for such particles, which are known as *bosons*. A glance at the example above suggests that particles which obey B-E statistics are more likely to be bunched together in the same energy states than are *boltzons;* in the M-B case one-third of the possible arrangements involve two particles in the same level, while in the B-E case it is one-half.

CASE 3. INDISTINGUISHABLE, EXCLUSIVE PARTICLES Let us represent each particle by an x and assume the particular exclusion rule that only one particle can occupy any one energy state. Then the possible arrangements are

$$
\begin{array}{ccc}
- - - - - & - - x - - & - - x - - \\
- - x - - & - - - - - & - - x - - \\
- - x - - & - - x - - & - - - - -
\end{array}
$$

The statistics for indistinguishable exclusive particles was developed by Fermi and Dirac. Such particles are often called *fermions.* Clearly, the introduction of exclusiveness prevents particles from bunching together in a common energy state. Thus B-E statistics leads to more bunching and F-D statistics to less bunching than that given by M-B statistics.

Before deriving the B-E and F-D distribution functions it is desirable to introduce the following *postulate* which applies to all three of the statistics under consideration.

The a priori probability of every physically distinct microscopic arrangement of N particles among the allowed energy states is the same, provided the arrangement is consistent with the principle of conservation of energy and any exclusion principle which applies.

For the case of M-B statistics we have found that Eq. (20.7) gives the thermodynamic probability of finding n_1 particles in the energy cell characterized by energy ϵ_1, and so forth. We turn now to finding the corresponding probabilities for the B-E and F-D cases.

20.3 The Bose-Einstein Distribution

Consider the number of ways in which n indistinguishable balls can be apportioned among g boxes, allowing any number from 0 to n to be placed in any one box. This problem can be solved by a simple artifice. Imagine a linear array of $g + n - 1$ holes into which either black pegs (balls) or white pegs (partitions) can be inserted. Let there be n black pegs and $g - 1$ white pegs, which are just enough to provide g "boxes." Then the possible distinguishable permutations of the pegs correspond to the number of ways in which the n balls could be placed in g boxes. Below is one way in which 10 balls could be put in 6 boxes—with 2 in the first, 3 in the second, 0 in the third, 1 in the fourth, etc.

$$|\bullet\bullet\circ\bullet\bullet\bullet\circ\circ\bullet\circ\bullet\bullet\bullet\circ\bullet|$$
$$\cdot\;2\qquad\quad 3\qquad\quad 0\;\;1\qquad\quad 3\qquad\quad 1$$

The number of permutations of $g + n - 1$ distinguishable pegs is $(g + n - 1)!$, but neither the n black pegs nor the $g - 1$ white ones are distinguishable, so the number of distinct arrangements is

$$\frac{(g + n - 1)!}{n!(g - 1)!}$$

Now we return to the energy diagram of Fig. 20.2 and ask for the most probable distribution of N indistinguishable particles among the energy cells. The probability that n_1 particles have their energy arrow terminating in the first cell, n_2 in the second, etc., is given by

$$\mathcal{P}(n_1, \ldots, n_s) = \left[\frac{(g_1 + n_1 - 1)!}{n_1!(g_1 - 1)!} \cdots \frac{(g_s + n_s - 1)!}{n_s!(g_s - 1)!}\right]\frac{1}{C_B}$$
$$= \frac{1}{C_B}\prod_{i=1}^{s}\frac{(g_i + n_i - 1)!}{n_i!(g_i - 1)!} \qquad (20.13)$$

where

$$C_B = \sum_{\substack{\text{all possible}\\ \text{values of } n_1 \ldots n_s}}\;\prod_{i=1}^{s}\frac{(g_i + n_i - 1)!}{n_i!(g_i - 1)!}$$

$$\ln \mathcal{P}(n_1, \ldots, n_s) = \sum_{i=1}^{s}\left[\ln (g_i + n_i - 1)! - \ln n_i! - \ln (g_i - 1)!\right]$$
$$- \ln C_B$$

Once again we apply Stirling's theorem and reimpose the lagrangian conditions $\alpha\left[\left(\sum_{i=1}^{s} n_i\right) - N\right] = 0$ and $\beta\left[\sum_{i=1}^{s} (n_i\epsilon_i) - E\right] = 0$ to obtain

$$\ln \mathcal{P}(n_1, \ldots, n_s) = \sum_{i=1}^{s}\left[(g_i + n_i - 1)\ln (g_i + n_i - 1)\right.$$
$$- (g_i + n_i - 1) - n_i\ln n_i + n_i - (g_i - 1)\ln (g_i - 1) + (g_i - 1)]$$
$$- \ln C_B - \alpha\left[\left(\sum_{i=1}^{s} n_i\right) - N\right] - \beta\left[\left(\sum_{i=1}^{s} n_i\epsilon_i\right) - E\right] \qquad (20.14)$$

Again for the overwhelming most probable distribution $\partial(\ln \mathcal{P})/\partial n_i$ must be zero for each n_i, from which

$$\ln (g_i + n_i^* - 1) + 1 - 1 - \ln n_i^* - 1 + 1 - \alpha - \beta\epsilon_i = 0$$

and

$$g_i - 1 = n_i^*(e^\alpha e^{\beta\epsilon_i} - 1)$$

Now if $g_i \gg 1$, we have

$$f_{BE}(\epsilon_i) = \frac{n_i^*}{g_i} = \frac{1}{e^\alpha e^{\beta\epsilon_i} - 1} = \frac{1}{e^\alpha e^{\epsilon_i/kT} - 1} \tag{20.15}$$

since again $\beta = 1/kT$.

Comparison of experimental data with theoretical predictions reveals that B-E statistics are obeyed by photons, pions, α particles, and in general by all particles which have zero or integer spin. In quantum mechanics interchange of two bosons leads to no change in the wave function; in other words, the wave function is symmetric to the interchange of bosons.

20.4 The Fermi-Dirac Distribution Law

The number of ways in which n indistinguishable particles can be distributed among g energy states, subject to the exclusion condition that no more than one particle occupies any single state, is

$$\frac{g(g - 1)(g - 2) \cdots (g - n + 1)}{n!} = \frac{g!}{n!(g - n)!}$$

The number of possible ways in which n_1 particles can be placed in g_1 states, n_2 in g_2 states, etc., subject to the exclusion principle, is then the product of s terms like that above; and the probability of having this situation is

$$\mathcal{P}(n_1, \ldots, n_s) = \frac{1}{C_F} \prod_{i=1}^{s} \frac{g_i!}{n_i!(g_i - n_i)!}$$

where

$$C_F = \sum_{\substack{\text{all possible} \\ \text{values of } n_1 \ldots n_s}} \prod_{i=1}^{s} \frac{g_i!}{n_i!(g_i - n_i)!}$$

and

$$\ln \mathcal{P}(n_1, \ldots, n_s) = \sum_{i=1}^{s} [\ln g_i! - \ln n_i! - \ln (g_i - n_i)!] - \ln C_F$$

Use of Stirling's theorem and the lagrangian multiplier equations leads directly to

$$\ln \mathcal{P} = \sum_{i=1}^{s} [g_i \ln g_i - g_i - (g_i - n_i) \ln (g_i - n_i) + (g_i - n_i)$$
$$- n_i \ln n_i + n_i] - \ln C_F - \alpha \left[\left(\sum_{i=1}^{s} n_i \right) - N \right]$$
$$- \beta \left[\left(\sum_{i=1}^{s} n_i \epsilon_i \right) - E \right]$$

from which

$$\left(\frac{\partial \ln \mathcal{P}}{\partial n_i} \right)_{n_i^*} = \ln (g_i - n_i^*) - 1 + 1 - \ln n_i^* - 1 + 1$$
$$- \alpha - \beta \epsilon_i = 0$$

or

$$g_i - n_i^* = n_i^* e^{\alpha} e^{\beta \epsilon i}$$

and if we make use of $\beta = 1/kT$,

$$f_{FD}(\epsilon_i) = \frac{n_i^*}{g_i} = \frac{1}{e^{\alpha} e^{\epsilon_i/kT} + 1} \qquad (20.16)$$

If we let $\alpha = -\epsilon_F/kT$, we obtain

$$f_{FD}(\epsilon_i) = \frac{n_i^*}{g_i} = \frac{1}{e^{(\epsilon_i - \epsilon_F)/kT} + 1} \qquad (20.16a)$$

where the quantity ϵ_F is called the *Fermi energy*.

Since n_i^*/g_i is simply the fraction of the g_i states which are filled, it is clear that $f_{FD}(\epsilon_i)$ is the probability that a state of energy ϵ_i is filled and that for a state with $\epsilon_i = \epsilon_F$ the probability of occupancy is $\frac{1}{2}$. Throughout our discussion of solids in Chaps. 21 to 23, we shall find the Fermi energy to be a topic of major importance.

It is found that F-D statistics are followed by electrons, muons, protons, neutrons, and, in general, by all particles of half-integer spin. Particles which obey F-D statistics are called *fermions*. Exchange of two fermions in the wave function ψ for a system leads to a reversal of the sign of ψ; thus the wave function for a system of fermions is antisymmetric in the exchange of a pair of particles.

20.5 Comparison of the Distribution Laws

In each of the distributions of Eqs. (20.12), (20.15), and (20.16) we have the ratio n_i^*/g_i given as a function of energy ϵ_i. Here n_i^* is the expected number of particles, and g_i the number of allowed energy states in the ith energy range of Fig. 20.2. The ratio n_i^*/g_i is called the *occupation index* for the states of energy ϵ_i. As we saw above, *for F-D statistics* $f_{FD}(\epsilon_i) = n_i^*/g_i$ *represents the probability that the state is occupied.* No such simple interpretation is possible for M-B and B-E distributions because many nonexclusive particles can occupy the same state.

The M-B distribution is plotted in Fig. 20.3 for three different values of temperature and of α. The M-B occupation index is a pure exponential decreasing by a factor $1/e$ for every increase in ϵ by kT. Although $f_{MB}(\epsilon_i)$ itself is a function of T and α (where the latter depends on temperature, the number of molecules, and the distribution of possible energy levels), the ratio between $f_{MB}(\epsilon_1)$ and $f_{MB}(\epsilon_2)$ depends only on the difference in the energies divided by kT according to the relation

$$\frac{f(\epsilon_1)}{f(\epsilon_2)} = e^{-(\epsilon_1-\epsilon_2)/kT} = e^{-\Delta\epsilon/kT} \tag{20.17}$$

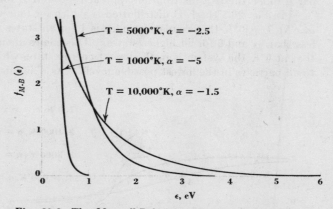

Fig. 20.3 The Maxwell-Boltzmann distribution function as a function of ϵ for three values of T and α.

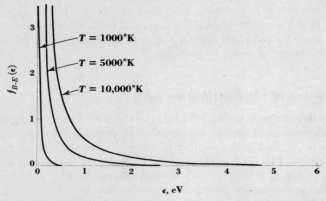

Fig. 20.4 The Bose-Einstein occupation index for $\alpha = 0$ at three temperatures.

The B-E distribution function is plotted in Fig. 20.4 for a gas of photons. In this case $\alpha = 0$, because the number of photons within a cavity is not constant; photons are being emitted and absorbed at the walls of an isothermal cavity at all times. (The fact that N is not constant removes the restriction that $\alpha[(\Sigma n_i) - N] = 0$, through which α was introduced. Removal of this condition is equivalent to setting $\alpha = 0$.) For small energy ϵ the occupation index is larger for the B-E distribution than for the M-B distribution, while at energies large compared to kT the B-E distribution approaches the M-B form.

In Fig. 20.5 the F-D distribution is shown for four values of T and α. At $T = 0°K$ the occupation index is 1 for all states with an energy less than ϵ_F and 0 for all higher states. This comes about from the fact that at $0°K$ the system is in the lowest energy state, which corresponds to all particles in the lowest possible levels. As T is raised, the occupa-

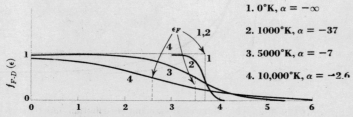

1. $0°K, \alpha = -\infty$
2. $1000°K, \alpha = -37$
3. $5000°K, \alpha = -7$
4. $10,000°K, \alpha = -2.6$

Fig. 20.5 The Fermi-Dirac distribution function for four values of T and α. At the Fermi energy ϵ_F the occupation index is $\frac{1}{2}$.

tion index remains nearly 1 for energies well below the Fermi energy and almost zero for energies far above. A number of particles are excited to energy states a little above the Fermi level from states a little below this level. As T is raised to very high values, the Fermi energy decreases somewhat (this energy is that for which the occupation index is $\frac{1}{2}$) and the distribution $f_{FD}(\epsilon)$ gradually becomes more and more like the M-B distribution.

Both the F-D and the B-E distributions approach the M-B distribution when $e^{\alpha}e^{\epsilon/kT} \gg 1$. When this inequality is satisfied, Eq. (20.17) is valid for both B-E and F-D distributions. In the sections which follow we shall cite examples of the applications of M-B and B-E statistics; numerous applications of F-D distributions appear in later chapters.

20.6 The Specific Heats of Gases

In Sec. 4.3 we discussed the failure of the classical equipartition of energy to explain the specific heats of gases. All major difficulties disappear when the internal energy of the molecule is described in terms of the quantum states that play such an important role in the theory of molecular spectra. For a gas in equilibrium the occupation index of the various levels is given by the M-B distribution law (20.12), according to which the number of molecules in states with energy ϵ_i is given by

$$n_i = Cg_i e^{-\epsilon_i/kT} \tag{20.18}$$

where C is a constant such that $N = \Sigma n_i$. By eliminating C we may write

$$n_i = \frac{Ng_i e^{-\epsilon_i/kT}}{\sum_{i=1}^{s} g_i e^{-\epsilon_i/kT}} \tag{20.19}$$

These equations are applicable to the states of both atoms and molecules. We consider first an example from the realm of atomic physics. The sodium D lines result from transitions between $3^2P_{\frac{3}{2}}$ and $3^2P_{\frac{1}{2}}$ levels to the $3^2S_{\frac{1}{2}}$ level (Sec. 17.3). Let us calculate the fraction of the sodium atoms in a bunsen flame at 1800°C which are in the 3^2P levels, remembering that the fine structure involves a very small energy difference compared with the energy separation of 3^2P from 3^2S, which is 3.36×10^{-19} J. The statistical weight of 3^2P is 6 (4 for $P_{\frac{3}{2}}$ and 2 for $P_{\frac{1}{2}}$) while that for $3^2S_{\frac{1}{2}}$ is 2. If we let n_0 represent the number of atoms in the ground state and n_1 the number in the 3^2P excited levels, we have for the ratio

$$\frac{n_1}{n_0} = \frac{6}{2} e^{-(\epsilon_1-\epsilon_0)/kT} = 3e^{-(3.36\times10^{-19}/1.38\times10^{-23}\times2073)} = 2.3 \times 10^{-5}$$

Thus only a small fraction of the sodium atoms at 2073°K are excited thermally at any given time. They suffice, however, to cause a considerable emission of sodium light.

In a gas the molecules are distributed statistically among the quantum states according to Eq. (20.19), where ϵ refers to the internal energy of the molecules. This internal energy may be associated with (1) rotation, (2) vibration of the atoms, or (3) electronic excitation. The total internal energy for 1 kmole containing N_A molecules is

$$E_i = \sum_{j=1}^{s} n_j\epsilon_j = N_A \frac{\Sigma\epsilon_j g_j e^{-\epsilon/kT}}{\Sigma g_j e^{-\epsilon/kT}} \tag{20.20}$$

The translational kinetic energy E_t is $N_A \times \frac{3}{2}kT$ [by Eq. (4.4)] or $3RT/2$. The total energy of the molecules is $E = E_i + E_t$ and, since the specific heat per kilomole c_V is $(\partial E/\partial T)_V$, we have

$$c_V = \frac{3}{2}R + N_A \frac{d}{dT} \frac{\Sigma\epsilon_j g_j e^{-\epsilon j/kT}}{\Sigma g_j e^{-\epsilon j/kT}} \tag{20.21}$$

It is now possible, in principle at least, to take the values of ϵ_j derived from a study of band spectra and to calculate the specific heat at constant volume from spectroscopic data by means of Eq. (20.21). This has actually been done with complete success in a number of cases.

Equation (20.21) predicts the variation of specific heat with temperature. When T is very small, $c_V = 1.5R$, the classical value for a monatomic gas. As the temperature rises, polyatomic molecules pass by thermal agitation into higher quantum states, beginning with the rotational levels associated with the lowest vibrational and electronic states. In general, the second rotational level lies from 0.0025 to 0.0075 eV in comparison with $kT = 0.025$ eV at room temperature (17°C), whereas the first step upward in vibration requires 0.025 to 0.25 eV and the first electronic excitation requires at least 2 eV. It is clear, therefore, that at ordinary temperatures the molecules are widely distributed among the first set of rotational levels. Under such conditions the average rotational energy per molecule approximates the classical value, which is kT for a diatomic molecule or $\frac{3}{2}kT$ for a molecule composed of three or more noncollinear atoms. Thus we understand the approximate success of classical theory in dealing with the simpler types of molecules.

At higher temperatures, vibrational energy contributes appreciably to c_V, eventually to the extent of kT for each vibrational level; at temperatures well above 10,000°C contributions may be expected from the higher electronic levels as well.

Comparison with Observation: Data on the specific heat of hydrogen, plotted as circles in Fig. 20.6, are in good agreement with the theoretical predictions. Below 60°K, $c_V = 1.5R$, whereas from 300 to 500°K, $c_V \approx 2.5R$ very closely, but from 600° upward vibration plays a part. The complete theory of a molecule composed of two *similar* atoms is more complicated than as here described. In the case of hydrogen it has been possible to verify experimentally the remarkable conclusions that follow from the theory outlined briefly in Sec. 20.9.

In further illustration of the use of quantum theory, we consider chlorine. Whereas the second vibrational state ($v = 1$) for gases such as N_2 or H_2 lies far above the first ($v = 0$), for chlorine it lies only 560 cm^{+1} above the first, as may be inferred from a study of the electronic band system that extends in the absorption spectrum from 4800 to 5800 Å.

The difference in energy between these two states is thus

$$\epsilon_1 - \epsilon_0 = h\nu = 6.62 \times 10^{-34} \times 560 \times 3 \times 10^{10} = 1.110 \times 10^{-20} \text{ J}$$

At $T = 288°K$ the ratio of the Boltzmann factors for the two states is $e^{-(\epsilon_1-\epsilon_0)/kT} = e^{-2.794} = 0.0611$. Thus, 0.061 times as many molecules will be in the second vibrational state as in the first. The additional energy

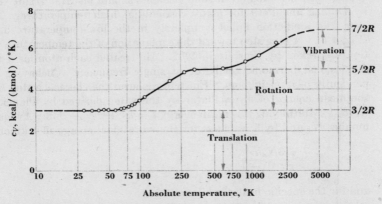

Fig. 20.6 Specific heat of hydrogen at constant volume as a function of absolute temperature.

due to this vibration in 1 kmole of Cl_2 is $N_A(\epsilon_1 - \epsilon_0)e^{-(\epsilon_1-\epsilon_0)/kT}$. The contribution of this energy to c_V is

$$N_A \frac{(\epsilon_1 - \epsilon_0)^2}{kT^2} e^{-(\epsilon_1-\epsilon_0)/kT} = R\left(\frac{\epsilon_1 - \epsilon_0}{kT}\right)^2 e^{-(\epsilon_1-\epsilon_0)/kT} = 0.48R$$

Thus $c_V = (2.5 + 0.48)R = 2.98R$, in fair agreement with the observed value, $3.02R$. A more exact calculation using Eq. (20.21) gives

$$c_V = 3.06R$$

When a similar calculation is made for HCl, whose vibration-rotation band at 3.4 μ or 2886 cm^{-1} (Sec. 19.8) indicates that

$$\epsilon_1 - \epsilon_0 = 5.73 \times 10^{-20} \text{ J}$$

the ratio of the Boltzmann factors is found to be only 5×10^{-7}. Thus molecular vibration can make no appreciable contribution to the specific heat of HCl. For O_2, $\epsilon_1 - \epsilon_0 = 1556$ cm^{-1}, and vibration may contribute about $0.026R$ to c_V. Other variations in c_V may be due to quantum corrections on the rotational energy.

20.7 Specific Heats of Solids

Classical equipartition of energy led (Sec. 4.3) to the prediction that the specific heat per kilomole of solid elements should be $3R$, a result verified by Dulong and Petit for many elements at high temperatures. However, the prediction failed completely in the low-temperature domain (Fig. 4.1). A long step toward the explanation of the temperature variation was taken by Einstein in 1907. He treated each atom as a three-dimensional harmonic oscillator of single frequency ν behaving independently of its neighbors. Einstein replaced the classical value kT for the mean energy of an oscillator by the value derived by Planck, Eq. (5.18). Multiplying this mean energy $\bar{\epsilon} = h\nu/(e^{h\nu/kT} - 1)$ by Avogadro's number N_A and by 3, we have for the vibrational energy of 1 kmole

$$E_0 = \frac{3N_A h\nu}{e^{h\nu/kT} - 1}$$

and for the specific heat per kilomole

$$c_V = \frac{dE_0}{dT} = 3R\left[\frac{e^{h\nu/kT}}{(e^{h\nu/kT} - 1)^2}\left(\frac{h\nu}{kT}\right)^2\right] \tag{20.22}$$

At sufficiently high temperatures this expression reduces to the classical value, $3R$; at low temperatures the exponential becomes large and c_V decreases rapidly. The formula was found to fit the data fairly well, as is illustrated by the lower curve in Fig. 20.7. At *very* low temperatures, however, the roughly exponential shape of the curve does not agree with experimental points. Careful measurements show that c_V varies as T^3 near 0°K.

An improved approach was introduced in 1912 by Debye, who took into account the fact that the displacement of any one atom is intimately correlated with the displacement of its neighbors at any instant. Each atom exchanges energy with its neighbors by means of elastic waves propagated through the crystal. These traveling waves can be treated as particles known as *phonons*. Phonon energies are quantized in units of $h\nu$, just as photons are. Thus the vibrational energy of the crystal can be associated with phonons traveling through the lattice.

Any solid is capable of vibrating elastically in many modes of different frequency. Debye calculated the number of independent modes of vibration in essentially the same manner in which standing waves

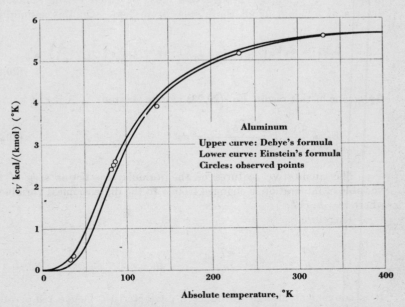

Fig. 20.7 Comparison of the specific-heat formulas of Einstein and Debye with experimental data for aluminum with Θ taken to be 398°K.

of an electromagnetic field within an enclosure were treated in Sec. 5.3. The principal differences are that (1) the number of modes of vibration in a solid are limited in number and (2) there are two kinds of elastic vibrations. There are transverse vibrations with two independent polarizations, in analogy with those for light; the number dn_t of these modes per unit volume of the solid, given by Eq. (5.9), is $dn_t = g(\nu)\, d\nu = (8\pi\nu^2\, d\nu)/v_t^3$, where v_t is the speed of transverse waves. Then there are longitudinal, or compressional, waves of a single type traveling with speed v_l and having a density $dn_l = (4\pi\nu^2\, d\nu)/v_l^3$.

Now let us assume that the energy associated with each of these modes of vibration has the average quantum value as stated in Eq. (5.18). Integrating over all frequencies from the lowest, sensibly zero, to the highest possible ν_m and multiplying by the volume V_0 occupied by 1 kmole, we find for the thermal energy of 1 kmole

$$E_0 = 4\pi V_0 \left(\frac{1}{v_l^3} + \frac{2}{v_t^3}\right) \int_0^{\nu_m} \frac{h\nu}{e^{h\nu/kT} - 1}\, \nu^2\, d\nu \qquad (20.23)$$

The true value of ν_m is difficult to determine; hence Debye used an approximate value derived from the condition that the total number of modes of vibration of a kg atom must be equal to the number of degrees of freedom of all the atoms, or to $3N_A$. On this assumption

$$3N_A = 4\pi V_0 \left(\frac{1}{v_l^3} + \frac{2}{v_t^3}\right) \int_0^{\nu_m} \nu^2\, d\nu = \tfrac{4}{3}\pi V_0 \left(\frac{1}{v_l^3} + \frac{2}{v_t^3}\right) \nu_m^3 \qquad (20.24)$$

By means of this result, Eq. (20.23) can be written

$$E_0 = \frac{9N_A}{\nu_m^3} \int_0^{\nu_m} \frac{h\nu}{e^{h\nu/kT} - 1}\, \nu^2\, d\nu \qquad (20.25)$$

The qualitative features of the formula are better seen if the variable of integration is changed from ν to the dimensionless variable x, where $x = h\nu/kT$.

Therefore

$$\nu = \frac{kT}{h}\, x \qquad d\nu = \frac{kT}{h}\, dx$$

It is convenient also to introduce a characteristic Debye temperature Θ defined as $\Theta = h\nu_m/k$, so that $\nu_m = k\Theta/h$. With these substitutions,

Eq. (20.25) becomes

$$E_0 = 9R \frac{T^4}{\Theta^3} \int_0^{\Theta/T} \frac{x^3}{e^x - 1} \, dx \tag{20.26}$$

and the specific heat is

$$c_V = \frac{dE_0}{dT} = 9R \left[4 \left(\frac{T}{\Theta} \right)^3 \int_0^{\Theta/T} \frac{x^3 \, dx}{e^x - 1} - \frac{\Theta}{T} \frac{1}{e^{\Theta/T} - 1} \right] \tag{20.27}$$

At *high temperatures*, c_V as given by Eq. (20.27) approaches the classical value $3R$. At *very low temperatures*, on the other hand, the exponentials become large, so that the last term can be neglected; the upper limit in the integral can also be replaced by ∞ with little error. It is known that

$$\int_0^\infty \frac{x^3 \, dx}{e^x - 1} = \frac{\pi^4}{15}$$

Hence Eq. (20.27) becomes, approximately,

$$c_V = \tfrac{12}{5} \pi^4 R \left(\frac{T}{\Theta} \right)^3 = 234R \left(\frac{T}{\Theta} \right)^3 \tag{20.28}$$

Thus, at temperatures much below the Debye temperature Θ, the atomic heat of a simple solid is *proportional to the cube* of the absolute temperature as confirmed by experiment.

Variation of the specific heat as T^3 implies a finite value of the entropy $(\int c_V \, dT)/T$ at absolute zero—an example of the Nernst heat theorem or the *third law of thermodynamics*. According to classical theory the entropy of a solid at $0°K$ would be negatively infinite.

For intermediate temperatures, it is necessary to evaluate the integral in (20.26) numerically. Good agreement with observation has been found for many substances when Θ is chosen for best fit with the data and even when Θ is calculated from the elastic constants with the use of Eq. (20.24). The data for many crystalline compounds can be fitted by adding one or more terms of the Einstein type to represent, in effect, internal vibrations of the atoms within each crystal cell.

20.8 The Photon Gas

Photons follow the B-E distribution law, and by applying the B-E distribution to blackbody radiation we are led directly to the Planck radiation law. As we have seen in Sec. 20.5, α is zero for a gas of photons, and so we have by Eq. (20.15)

$$\frac{n_i^*}{g_i} = \frac{1}{e^{\epsilon_i/kT} - 1} \tag{20.15a}$$

In Sec. 5.3 we have followed Rayleigh in finding the number of degrees of freedom per unit volume for standing waves in an enclosure with frequency between ν and $\nu + d\nu$ to be $(8\pi\nu^2\, d\nu)/c^3$, Eq. (5.9). Since the energy of a photon is given by $\epsilon = h\nu$, we have for the number of allowed energies per unit volume between ϵ_i and $\epsilon_i + d\epsilon_i$ the result that

$$g_i = \frac{8\pi\epsilon_i^2\, d\epsilon_i}{h^3 c^3} \tag{20.29}$$

Consequently the most probable number of photons per unit volume with energy between ϵ_i and $\epsilon_i + d\epsilon_i$ is

$$n_i = \frac{8\pi\epsilon_i^2\, d\epsilon_i}{h^3 c^3 (e^{\epsilon_i/kT} - 1)} \tag{20.30}$$

Each photon has energy $h\nu$, and so the energy per unit volume associated with frequencies between ν and $\nu + d\nu$ is given by

$$U_\nu\, d\nu = \frac{n_i h\nu_i}{V} = \frac{8\pi\epsilon_i^3\, d\epsilon_i}{c^3 h^3 (e^{\epsilon_i/kT} - 1)} = \frac{8\pi h\nu^3\, d\nu}{c^3 (e^{h\nu/kT} - 1)} \tag{20.31}$$

This is precisely Eq. (5.21) giving the Planck radiation law in terms of frequency.

20.9 Homonuclear Molecules

Molecules that contain at least two *similar* nuclei have certain peculiar and sometimes astonishing properties. Thus, it was early observed that in certain bands from such molecules alternate lines were weaker or perhaps missing entirely. After the advent of wave mechanics, it was shown by Hund that this phenomenon could be explained as an effect of nuclear spin. The effect is essentially statistical in nature, and in discussing it no account need be taken of the nuclear magnetic moments or of other minor forms of nuclear interaction, which serve only to produce, as in atomic spectra, a minute hyperfine structure. Diatomic molecules only will be considered here.

a. Ortho and Para States   The wave function ψ for a molecule must contain as variables the coordinates of the nuclei and, in general, an allowance for nuclear spin. The spin numbers I and M_I were described in Sec. 18.8. All the known facts are consistent with the assumption that ψ is always *symmetric* in the coordinates and spins of any two similar nuclei for which I is *integral or zero* (the B-E type), whereas if I is *half-integral* (F-D type) the wave function is *antisymmetric*. That is, interchange of the spatial coordinates and of the values of M_I for any two similar nuclei leaves ψ unchanged if I is integral or zero but reverses the sign of ψ, as in the case of electrons, if I is half-integral.

As with electrons, so long as spin energies can be ignored, *spatial* symmetry and *spin* symmetry can be treated separately. States for a pair of similar nuclei that are *symmetric* in the nuclear spins have been called *ortho* states, those antisymmetric in the spins, *para* states. (Ortho means right or proper, i.e., ordinary; para, alongside of, i.e., a variant, the less common variety.) Since M_I has $2I + 1$ possible different values, there are, for two similar nuclei, $(2I + 1)^2$ possible spin states in all. Of these, the states in which M_I has the same value for both nuclei are symmetric in the nuclear spins; there are $2I + 1$ such states. Then, each combination of two different values of M_I yields both a symmetric and an antisymmetric state, in analogy with the $\alpha\beta$ combination for two electrons as described in Appendix 15A. There are $(2I + 1)I$ such possible combinations. The number N_p of antisymmetric, or para, states will thus have the value $(2I + 1)I$, whereas the total number of symmetric, or ortho, states is

$$N_0 = (2I + 1)I + (2I + 1) = (2I + 1)(I + 1)$$

Thus we have the statistical ratio

$$\frac{N_p}{N_0} = \frac{I}{I + 1} \tag{20.32}$$

The ortho states have always the greater statistical weight; if $I = 0$, the para states are missing altogether.

The *spatial* symmetry of ψ relative to nuclear exchange, i.e., for exchange of only the spatial coordinates of the particles, must then be considered. Sometimes this is referred to simply as the *symmetry* of ψ. A wave function will have the required grand symmetry when I is *integral* provided the spin factor by itself is symmetric and the spatial factor is also symmetric, i.e., remains unaltered when the coordinates of the nuclei are interchanged; or, both factors may be antisymmetric. Thus, for integral or zero I, the ortho states have spatially symmetric wave functions, whereas the para states have spatially antisymmetric functions. Conversely, for *half-integral* I, the ortho states are those with spatially antisymmetric wave functions, while the para states are symmetrical in the spatial coordinates.

No general rules can be given for predicting with certainty the spatial symmetry of ψ. Usually a good approximate form for ψ, with nuclear spin omitted, is

$$\psi = \psi_e \psi_v \psi_J$$

where ψ_e, ψ_v, and ψ_J stand for electronic, vibrational, and rotational factors corresponding to the division of the molecular energy into three parts as described in Secs. 19.8 and 19.9. The vibrational factor ψ_v is always symmetric for nuclear exchange, essentially because a vibration of similar nuclei looks the same seen from either end. The rotational factor ψ_J is even for even values of J and odd for odd J. But ψ_e, although often symmetric in states with $\Lambda = 0$, may be either symmetric or antisymmetric for nuclear exchange. Thus the ortho states are the even-J rotational states in some cases but the odd-J states in others.

An important *selection rule*, $\Delta I = 0$, exists in this connection, paralleling that for the different S states of atoms. If the nuclei had no spin-dependent form of interaction in addition to their coulomb fields, the ortho and para states would never combine at all; a molecule in one type of state would remain in states of that type forever. Conversion from ortho to para or vice versa can take place, as a result of slight noncoulomb nuclear forces, but this process is very slow, requiring perhaps months or years.

b. Effects on Band Spectra   The selection rule $\Delta I = 0$ effectively limits spectral transitions to those between two ortho states or between two para states. Since a certain type of spatial symmetry accompanies each type of spin state, dependent upon the parity of I, it follows also that the *spatial* symmetry of ψ for nuclear exchange cannot change in a radiative transition. These restrictions result in two major peculiarities in the bands from homonuclear diatomic molecules.

In the first place, since in a purely rotational or vibration-rotational transition the electronic state of the molecule does not change, to preserve the symmetry of ψ the spatial symmetry (for exchange) of the rotational factor ψ_J cannot change. Thus transitions of this kind can occur only between two even values of J or between two odd values. But for such bands there is also a spectral selection rule that $\Delta J = \pm 1$. It follows that so far as dipole radiation is concerned, *homonuclear diatomic molecules cannot have any purely rotational or vibration-rotational spectra at all!* As was noted at the end of Sec. 19.8 the absence of such spectra may also be regarded as due to the absence of a molecular electric moment in such molecules. In Raman spectra, on the other hand, frequency shifts corresponding to such transitions may occur, since for Raman lines the usual selection rule is that

$$\Delta J = \pm 2$$

The second major consequence of exchange symmetry concerns the *relative intensity* of the lines composing a given band. Ordinarily, some of the molecules will be in ortho states while others are in para states, and in one of these states transitions will occur only between even values of J while in the other they occur only between odd values. The relative intensity of the even-J and odd-J lines will thus depend upon the relative population of the ortho and para states. If a substance has been for a long time at a given temperature, it may be assumed that all *fundamental* states will be populated in proportion to a Boltzmann factor $e^{-E/kT}$. Thus the intensity of a spectral line as observed will be proportional to (1) the Boltzmann factor for the initial state, (2) the transition probability, and (3) the nuclear statistical weight of the initial state, which is proportional, for successive values of J, alternately to I or to $I + 1$. Since the first two factors do not usually vary rapidly from line to line, the observed lines alternate in intensity. If quantitative allowance can be made from theory for the first two factors, the magnitude of I for the nucleus can be inferred with certainty from measurements of the relative intensities of the lines. Furthermore, if $I = 0$, there are no para states and no para lines at all, so that either the even-J or the odd-J lines are missing altogether.

Many examples of these phenomena are known, and they all agree with the theoretical predictions provided the right value of I is assumed. Such observations are regarded as a reliable source of information concerning I. Thus, O_2 has no rotation or vibration-rotation bands, and in its Raman spectra the odd-J lines are missing. It is concluded that $I = 0$ for the O nucleus.

Perhaps the most famous case is that of N_2. A photograph showing alternating intensities in a Raman spectrum from nitrogen is shown in Fig. 20.8. The spectrum is a rotational one excited by

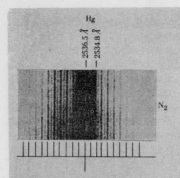

Fig. 20.8 Alternating intensities in a rotational Raman spectrum of N_2. (*From G. Herzberg, "Molecular Spectra and Molecular Structure," vol. 1, "Spectra of Diatomic Molecules," 2d ed., D. Van Nostrand Company, Inc., 1950.*)

the mercury line at 2536.5 Å, photographed by F. Rasetti. Careful measurements had also been made much earlier by Ornstein and van Wijk on four electronic bands of nitrogen (at 3984, 4278, 3884, and 4237 Å). After correcting for the Boltzmann factor and the transition probabilities, they found a ratio of statistical weights for even-J states to odd-J states very close to $\frac{1}{2}$. Equating $\frac{1}{2}$ to $I/(I+1)$, we find $I = 1$ for the nitrogen nucleus.

These observations were published in 1928, and at that time a value $I = 1$ for nitrogen was "extraordinarily surprising." It was supposed then that the nitrogen nucleus must consist of 14 protons and 7 electrons and that an odd number of fundamental particles should require a half-integral value of I. The difficulty was resolved after the discovery of the neutron had led to the view that the nitrogen nucleus is composed of 7 protons and 7 neutrons.

c. Effects on Other Physical Properties   The different statistical weights of the ortho and para states must be taken account of in dealing with any physical property, such as specific heat, that depends upon the distribution of the molecules among their quantum states. The question may also be raised as to the possibility of a mass separation of molecules in the two states.

These features have been studied extensively in hydrogen. This gas is a mixture of orthohydrogen with molecules rotating in odd-J states and parahydrogen in even-J states, the latter including the state of no rotation with $J = 0$. Here $I = \frac{1}{2}$ for the protons, and the electronic state is symmetric for nuclear exchange; hence odd J's must go with the ortho spin. If all fundamental I, M_I states occurred with equal frequency, the ratio of para to ortho would be $I/(I + 1) = \frac{1}{3}$. At room temperature, it is calculated that ordinary hydrogen should be about 25.1 percent para. In ordinary experimentation there is not time for appreciable change to occur in the para-ortho ratio; nevertheless, the difference in statistical weights should have some effect on the observed specific heat at low temperatures because of the unequal representation of the even-J and odd-J rotational states.

A more remarkable fact is that it was found possible to accelerate the rate of interconversion between the two forms by adsorbing the gas onto charcoal, so that at 20°K it becomes almost pure parahydrogen, in the state $J = 0$, within a few hours. In this way the special properties of parahydrogen could be studied, especially its specific heat as a function of temperature, and agreement with theoretical predictions was found. The properties of pure orthohydrogen were then inferred by comparison with those of the normal mixture (Fig. 20.9). Orthodeuterium (H_2^2) has also been prepared; here

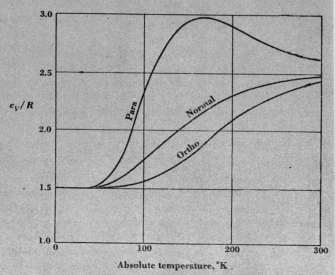

Fig. 20.9 Experimental values of the specific heat at constant volume for parahydrogen and normal hydrogen and the inferred values for orthohydrogen.

$I = 1$, $N_p/N_0 = \frac{1}{3}$, and it is the ortho states that have even values of J.

---

1. (a) In how many ways can four persons occupy seven seats in a room?

(b) A bridal party consists of six persons. In how many ways can a photographer arrange them in a straight line if the bride and groom always occupy the middle two places?

Ans: (a) 840; (b) 48

2. (a) In how many ways can a committee of 5 persons be selected from a department of 11 members?

(b) A city council has 15 members, of which 9 belong to party A and 6 to party B. In how many ways can a 7-man committee be selected if the majority party must have 4 representatives?

Ans: (a) 464; (b) 2520

3. A manufacturer of blocks for children makes initially identical wooden cubes and then paints each face one of six bright colors. Calculate how many distinguishable blocks can be produced in this way.

Ans: 30

4. Find the number of ways in which two particles can be distributed in five states if (a) the particles are distinguishable, (b) the particles are indistinguishable and obey Bose-Einstein statistics, and (c) the particles are indistinguishable and only one particle can occupy any one state. Show the possibilities in a diagram for each case.

Ans: 25, 15, 10

5. Show that the Einstein relation for the heat capacity per kilomole of a solid reduces to the classical value of $3R$ when $kT \gg h\nu$.

6. Assume that there exists an infinity of energy levels equally spaced 1 eV apart starting at 0. Assume further that there are three particles which share a total energy of 6 eV. Finally, assume that all possible arrangements are equally likely and that the system passes through all possible arrangements, spending an equal time in each.

(a) If the particles are indistinguishable and exclusive (one per level), show all possible arrangements of particles in levels. What's the probability of finding a particle in level 0? in level 2? in level 4? in level 6?

(b) If the particles are indistinguishable (but not exclusive), show all possible arrangements. From your sketch determine (i) the probability of finding all three particles in the same level, (ii) the probability of finding two particles in the same level, and (iii) the relative populations of the various levels.

(c) Repeat (b) for the case of distinguishable particles.

7. (a) Calculate the probability that an allowed state is occupied if it lies above the Fermi level by kT, by $2kT$, by $4kT$, by $10kT$, and by $100kT$.

(b) Prove that for a system obeying Fermi-Dirac statistics the probability that a level lying ΔE below the Fermi level is *not* occupied is the same as the probability that a level ΔE above the Fermi level *is* occupied.

8. The dark absorption lines in the solar spectrum which were called C and F by Fraunhofer correspond to the first two members of the Balmer series of hydrogen. If the surface temperature of the sun is 6000°K, find the ratio of the number of hydrogen atoms in states with $n = 2$ to the number in the ground state for hydrogen at the solar surface.

9. The Debye temperature for carbon crystallized as diamond is 2230°K. Calculate the heat capacity per kilomole for diamond at 20°K. Compute the highest lattice frequency ν_m involved in the Debye theory.

Ans: 1.4 J/kmole-K°, 4.7 × 10¹³ Hz

10. At low temperature the specific heat of graphite is proportional to T^2 and is thus an exception to the Debye T^3 law. This has been attributed to the fact that graphite has a layer structure with feeble forces between layers. Assume a two-dimensional plane lattice for graphite, show that the distribution of low-frequency modes goes as $\nu \, d\nu$, and derive the T^2 law.

chapter twenty-one

Solids—Insulators and Metals

The conditions of temperature and pressure at the surface of the earth are such that a large number of materials exist in the solid state, in which they maintain both shape and volume. Ionic and covalent binding are dominant for some solids, but a different variation of the coulomb interaction leads to metallic crystals with high strength, ductility, and conductivities, both electrical and thermal.

21.1 Crystals

Some scientists prefer to reserve the term *solid* for what we shall call a *crystalline solid*. Of a piece of glass, a bar of silver, and a strip of wood, only the silver qualifies as crystalline. The distinction is made on the basis of the kind of order which exists in the material, i.e., on the type of geometrical arrangement in which the atoms fall. By means of x-rays it is possible to determine the relative positions of various kinds of atoms in a material. Suppose for a moment that we could actually see the individual atoms in a piece of glass. If we locate a silicon atom and examine its neighbors, we find that they are not distributed in a random fashion. There is a definite geometrical order. However, if we look at other silicon atoms, we find that their neighbors form a somewhat different pattern. For glass there is no orderly arrangement which persists throughout the material; there is a *short-range* order but not *long-range* order. Such a material which retains its shape is sometimes called an *amorphous solid* or a *supercooled liquid.* *Crystalline solids exhibit long-range order;* in a large single crystal the ordering extends uninterrupted over distances of many million atomic diameters.

For discussing crystals it is convenient to introduce the concept of a *space lattice, a network of straight lines constructed in such a way that it divides space into identical volumes with no space excluded.* The *lattice points* are at the intersections of these lines. Associated with each lattice point there is either a single atom or an identical complex of atoms. For every crystal there is thus a network of lattice points for each of which there is either a single atom or a group of atoms called the *basis* of the crystal.

In 1848 Bravais showed that there are precisely 14 possible space lattices (Fig. 21.1); of these we shall be concerned primarily with the three cubic systems, which fortunately are the simplest ones. For any type of lattice there exist three *fundamental translation vectors* $\mathbf{a}$, $\mathbf{b}$, and $\mathbf{c}$; these are the shortest vectors which satisfy the condition that if we go from any initial point in a crystal any integer $n_1\mathbf{a}$ distances, $n_2\mathbf{b}$ distances, and $n_3\mathbf{c}$ distances, we arrive at a point which has identically the same environment as the initial point. Thus any two lattice points can be connected by a *translation vector* $\mathbf{T} = n_1\mathbf{a} + n_2\mathbf{b} + n_3\mathbf{c}$. The simplicity of the cubic systems is associated with the fact that for them $\mathbf{a}$, $\mathbf{b}$, and $\mathbf{c}$ are equal in magnitude and mutually perpendicular.

Once $\mathbf{a}$, $\mathbf{b}$, and $\mathbf{c}$ are established, it is simple to use them to specify (1) the locations of any points within a crystal cell, (2) the direction of any vector, and (3) the orientation of any plane in the crystal. Thus $(1\frac{1}{2}\frac{1}{2})$ represents the point which is $1\mathbf{a}$, $\frac{1}{2}\mathbf{b}$, and $\frac{1}{2}\mathbf{c}$ from the origin. For the body-centered cell of Fig. 21.1, atoms are located at (000), (100), (010), (001), (110), (101), (011), (111), and $(\frac{1}{2}\frac{1}{2}\frac{1}{2})$.

To specify the direction of a vector we translate the vector, placing its tail at the origin, and give the coordinates of the first lattice point reached by the line determined by the vector. Thus the y axis lies in the [010] direction, while the [210] direction is in the plane formed by $\mathbf{a}$ and $\mathbf{b}$, and corresponds to two $\mathbf{a}$'s and one $\mathbf{b}$. The body atom in a

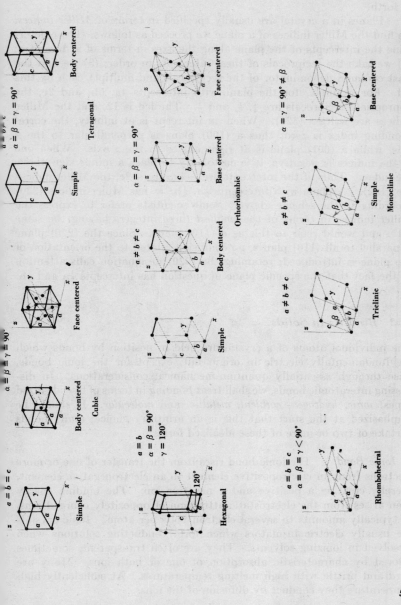

Fig. 21.1 The 14 Bravais, or space, lattices. The cells outlined in heavier lines are the conventional *unit cells* which can be used to build an infinite lattice by suitable translations. (The angle between **b** and **c** is α, etc.)

body-centered cube is in the [111] direction from the cube corner, and so forth.

Planes in a crystal are usually specified in terms of *Miller indices*. To find the Miller indices of a plane we proceed as follows: (1) we determine the intercepts of the plane along the axes in terms of **a**, **b**, and **c**; (2) we take the reciprocals of these intercepts in order; (3) we find the least common denominator of the reciprocals and multiply each by this lcd. For example, for the plane with intercepts 4**a**, 6**b**, and 2**c**, the appropriate reciprocals are $\frac{1}{4}$, $\frac{1}{6}$, and $\frac{1}{2}$. The lcd is 12, and the Miller indices are written (326). When an intercept is at infinity, the corresponding index is zero; thus a (100) plane is perpendicular to the x axis, while a (001) plane is at right angles to the z axis. When one of the indices is negative, it is customary to place a minus sign above the index. Thus, if the intercepts are -4**a**, 2**b**, -2**c**, the Miller indices are ($\overline{1}2\overline{2}$). A plane with intercepts $\frac{1}{2}$**a**, $\frac{1}{2}$**b**, ∞ has Miller indices (220) according to the scheme above. Some scientists prefer to express the Miller indices in terms of the *smallest* three integers having the same ratio and would refer to this as a (110) plane. Since the (220) plane is parallel to all (110) planes, no confusion related to the orientation of the plane is introduced; retaining the (220) designation calls attention to the fact that the specific plane in question has intercepts $\frac{1}{2}$**a** and $\frac{1}{2}$**b**.

21.2 Binding in Solids

The individual atoms of a crystal are held in position by bonds which are fundamentally electric in origin but, except for the ionic bonds, arise through essentially quantum-mechanical considerations. In discussing interatomic bonds, we shall treat bonding in terms of five idealized types: *ionic*, *hydrogen*, *covalent*, *metallic*, and *molecular*. It should be emphasized at the start that this is an arbitrary choice; many bonds partake of two or more of these idealized forms.

a. Ionic Bond   The ionic bond rises from the transfer of one or more electrons from an electropositive element to an electronegative element, thereby creating a positive and a negative ion. The binding energy then arises from the electrostatic attraction of oppositely charged ions. It typically amounts to several electron volts per atom. Ionic crystals are usually electric insulators which form conducting solutions when dissolved in ionizing solvents. They are often transparent, sometimes colored by characteristic absorption of one or both ions. Many are hard and brittle with high melting temperatures. At sufficiently high temperature they conduct by diffusion of the ions.

. *Hydrogen Bond*   The hydrogen bond is a special type of ionic bond in which a hydrogen nucleus (H^+ or D^+) holds together two electronegative atoms with a binding energy of the order of 0.1 eV. Unlike most ions, H^+ can have only two near-neighbor negative ions, because he proton has a radius of the order of only 10^{-15} m (all other ions have adii thousands of times this great). In the crude approximation in which each ion is regarded as a sphere and the solid is composed of spheres in contact, the proton can have only two negative ions touching t, while a typical positive ion has six or eight nearest neighbors. The hydrogen bond is important in ice, in hydrogen fluoride, and in many organic molecules, particularly the proteins.

. *Covalent Bond*   The covalent bond in its idealized sense involves the sharing between two identical atoms of a pair of electrons of opposite spin. In a diamond crystal each carbon atom forms covalent bonds with four nearest neighbors. Covalent binding energies of a few electron volts per atom are common. The crystals are insulators or semiconductors, having high melting points. They are extremely hard and break by cleavage. Group IV elements in the periodic table come closest to forming ideal covalent bonds; III-V and II-VI compounds have bonds which are a mixture of covalent and ionic.

d. Metallic Bond   The metallic bond arises from the sharing of outer electrons by all the atoms of a crystal. In copper, for example, each atom contributes roughly one electron to the crystal as a whole. These *free electrons*, leaving the metal atoms in place as ions, serve as a pervasive "glue," moving among the ions and binding them together. Because of the free electrons, metallic crystals are opaque and strongly reflecting; they are excellent conductors with resistance decreasing as the temperature is lowered. Excellent thermal conductivity comes also from the free electrons. Typical metals have moderately high melting points; they are mechanically tough and ductile rather than brittle. The binding energy is usually a few electron volts per atom.

e. Molecular Bond   The molecular bond forms solids from noble-gas atoms and between many organic molecules, such as methane. Unlike the three preceding classes of binding, in which electrons are either exchanged or shared, molecular bonds involve no transfer or exchange of charge. Rather the bond arises from the van der Waals interaction between electrically neutral molecules. Even an atom, such as neon or argon, with a spherically symmetric electron distribution has a small instantaneous electric dipole moment as a result of motions of the electrons. This produces at the location of a neighboring molecule a small

electric field which polarizes the latter. A weak attractive interactio
results. The potential energy associated with the dipole moments i
proportional to the square of these moments and to the inverse sixt
power of the separation. Because the dipole moments are very sma
and ion-core repulsion prevents the molecules from coming very clos
molecular binding is weak—of the order of a few hundredths of an eleo
tron volt per atom. This small binding leads to very low melting an
boiling points. The crystals are electrically insulating, insoluble, sof
and readily deformable.

21.3   The Ionic Crystal

Probably the simplest of all solids to understand are those composed o
positive and negative ions, such as lithium fluoride and sodium chloride
The primary interaction between the ions is electrostatic and involve
spherical charge distributions. As a specific example, we shall conside
potassium chloride, since we can use information given in the descriptio
of the KCl molecule. In the molecule each ion is bound to a single io
of the opposite sign, whereas in the solid each ion has a number of neares
neighbors and it is no longer meaningful to pair each positive ion with a
single negative ion.

The crystal lattice for potassium chloride is face-centered cubio
with one K^+ and one Cl^- ion associated with each lattice point (Fig. 21.2).

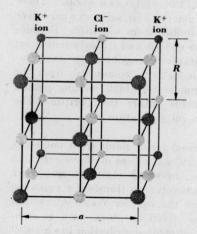

Fig. 21.2 The KCl crystal involves
interleaved face-centered arrays of
K^+ and Cl^- ions.

Alternatively we may think of it as the interleaving of face-centered arrays of K$^+$ ions and of Cl$^-$ ions. We begin by calculating the *binding energy* of KCl, which we define as the energy released when K and Cl atoms combine to form the solid. As we saw in Sec. 19.2, it requires 4.34 eV to remove the valence electron from a K atom to produce a K$^+$ ion. Similarly, 3.80 eV is released when an electron joins a Cl atom to form a Cl$^-$ ion. Thus the process of forming an ion pair from neutral atoms requires 0.54 eV. The energy released when these ions are assembled to form a crystal may be calculated as follows. Consider the central Cl$^-$ ion of Fig. 21.2. If R represents the distance to the nearest K$^+$ ion in the crystal, there is an electrostatic-interaction energy released of $-e^2/4\pi\epsilon_0 R$ for each nearest neighbor, of which there are six. The next nearest neighbors are Cl$^-$ ions at distance $\sqrt{2}\,R$; there are 12 of them. Next there are eight K$^+$ ions at a distance of $\sqrt{3}\,R$. Then come six Cl$^-$ ions at a distance of $2R$, etc. The total lattice energy E_L is obtained by summing the potential-energy contribution from all the ions in the crystal and is given by

$$E_L = -\frac{e^2}{4\pi\epsilon_0 R}\left(\frac{6}{1} - \frac{12}{\sqrt{2}} + \frac{8}{\sqrt{3}} - \frac{6}{2} + \cdots\right) = -\frac{\alpha e^2}{4\pi\epsilon_0 R}$$

$$(21.1)$$

where α is the value to which the series in the parentheses converges. (This series converges very slowly; however, schemes have been developed which give much more rapid convergence.) For the ionic face-centered crystal $\alpha = 1.747558$; α is known as the *Madelung constant* in honor of the man who computed it in 1912.

The appropriate Madelung constant can be calculated for any ionic crystal; e.g., its value is 1.762760 for the CsCl structure (Fig. 21.3), which is a simple cubic array of Cs$^+$ ions interleaved with a cubic array of Cl$^-$ ions. (Note that if all the atoms were the same in Fig. 21.3, the crystal would be body-centered.)

When the ions are sufficiently close for their cores to begin to overlap, the Pauli exclusion principle again leads to a repulsive force which rises rapidly as R decreases. We may write the repulsive-energy term in the form $E_R = \Lambda/R^n$, where Λ is an appropriate constant and n is typically of the order of 10. We can then write the approximate binding energy per ion pair as a function of R for the crystal by combining the cost of producing an ion pair, the net lattice attractive energy, and the repulsive energy term to obtain the relation

$$E_b(R) = (4.34 - 3.80) - 1.748 \times \frac{14.40}{R} + \frac{\Lambda}{R^n} \qquad \text{eV} \qquad (21.2)$$

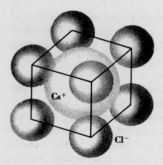

Fig. 21.3 The CsCl crystal involves interleaved simple cubic arrays of Cs⁺ and Cl⁻ ions.

The sums of these terms for the solid behave very much as they did for the molecule and, except for scale factors, the curves of Fig. 19.1 are roughly applicable. For the KCl crystal the equilibrium separation is 3.14 Å, and the binding energy is −6.67 eV/ion pair. It should be noted that in this approximate treatment the vibrational energies of the ions have been neglected.

Because the exclusion-principle repulsion is small until the ion cores begin to overlap and then rises extremely rapidly, the radii of ions in alkali halide crystals are reasonably well defined and one may approximate each ion by a sphere of fixed radius. For example, Pauling has shown that the grating spaces of many alkali halide crystals can be estimated by treating the ions as spheres in contact with the following assignments to radii in angstrom units:

Ion	Li⁺	Na⁺	K⁺	Rb⁺	F⁻	Cl⁻	Br⁻	I⁻
Radius, Å	0.60	0.95	1.33	1.48	1.36	1.81	1.95	2.16

Evidence supporting the ionic nature of KCl comes from measuring the dielectric constant as a function of frequency. For static and low-frequency electric fields the dielectric constant of ionic crystals is substantially larger than it is for optical-frequency fields. This arises from the fact that the ions are too massive to respond significantly to high frequencies. For low-frequency fields the ions have time to move; ionic polarization results when the applied frequency is less than the natural frequency of ion vibration. This frequency is typically of the order of 10^{13} Hz. As examples, the ratio of the static dielectric constant to that at optical frequencies for some familiar ionic crystals are LiF, 9.27/1.92; NaCl, 5.62/2.25; KCl, 4.68/2.13.

21.4   Covalent Binding

The covalent bond in solids arises, just as it did for molecules, from the sharing of a pair of electrons of antiparallel spin between neighboring atoms. The prototype of covalent bonding is that of the group IV elements: carbon, germanium, silicon, and gray tin. These all crystallize in the diamond lattice (Fig. 21.4). Each atom has four nearest neighbors with each of which it shares a covalent bond (Fig. 21.5a). The space lattice is face-centered cubic with a basis of two atoms associated with each lattice point, one at (000) and the other at $(\frac{1}{4}\frac{1}{4}\frac{1}{4})$. The ground-state configuration for group IV atoms is s^2p^2. However, the p level lies very near to the s, and in the diamond lattice the configuration may be better described as sp^3. Actually the four bonds in the diamond structure are of equal strength, and the angles between bonds are the same. This is perhaps best explained by assuming that the proximity of other atoms modifies the character of the s and p orbitals in such a way that their energies become degenerate. Then any linear combination of the wave functions is also a possible orbital; by a technique known as

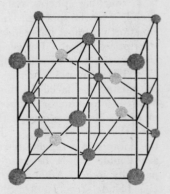

Fig. 21.4 If the sites marked by black and gray circles are occupied by the same kinds of atoms, the resulting crystal cell is typical of diamond, silicon, and germanium; if the sites are occupied by different kinds of atoms, e.g., zinc and sulfur, the structure is that of zinc blendes.

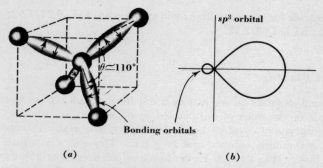

Fig. 21.5 (a) Covalent bonds typical of atoms which crystallize in the diamond structure. (b) A hybridized sp^3 orbital.

sp^3 hybridization it is possible to make combinations of the wave functions which result in four orbitals, each of which is directed toward a corner of a tetrahedron occupied by a nearest neighbor. An example of such an orbital is shown in Fig. 21.5b.

The diamondlike covalent bond is a strong one and highly directional. This leads to hard but brittle materials since the effective sharing of electrons can occur only in a few directions. It leads to a relatively loosely packed crystal in the sense that if a model is assembled from touching spheres, only 34 percent of the volume of the model is occupied by the spheres. In crystals with covalent binding the dielectric constant is almost independent of frequency from zero frequency through the optical region. For germanium, for example, the dielectric constant is approximately 16 while for silicon it is about 12 over this entire range.

The diamond structure consists of two face-centered cubic lattices, the second displaced by one-quarter of the (111) diagonal of the first. Many semiconductor compounds crystallize in the zinc blende structure in which Zn atoms occupy the lattice points of one lattice and S atoms the other (see Fig. 21.4). Thus each Zn atom has four S neighbors at the corners of a regular tetrahedron and vice versa. In this case there are double bonds between atoms which are partially covalent and partially ionic. Among the many compounds which crystallize in the zinc blende structure are SiC (IV, IV); AlP, InAs, and InSb (III, V); ZnSe, ZnS, and CdS (II, VI); CuF, CuCl, and AgI (I, VII).

21.5   *Metallic Binding*

In metals the atoms no longer hold their outermost electrons, which then form a *Fermi gas* of electrons, held within the metal, but otherwise essentially free. It is these electrons which are responsible for the excellent electrical and thermal conductivity of metals. As an example, we discuss sodium, which crystallizes in the body-centered cubic lattice with a nearest-neighbor distance of 3.71 Å and a lattice constant of 4.28 Å. The probability of finding the 3s electron of sodium more than 4 Bohr radii from an isolated atom is substantial, as Fig. 15.4 shows. But an electron that far away from its atom in the solid would be closer to some neighboring atom. As a result of the interaction due to near neighbors, the valence electron of sodium becomes almost free within a crystal; to a reasonably good approximation each atom of sodium contributes one electron to the crystal as a whole. A sodium crystal is composed of positive ion cores and free electrons. The energy is minimized for this arrangement, since the electron can be in the immediate vicinity of positive ions at all times and therefore have low potential energy, while at the same time its wave function is spread out, and this corresponds to relatively low kinetic energy.

As we saw in Sec. 19.5, the effect of bringing one atom close to another is to split what was a single sharp energy level into a pair of levels. If there are many interacting atoms, each outer level is spread into a continuous band of levels, some of which lie lower and some higher than the original level. When sodium atoms form a crystal, the sharp atomic levels are spread into bands, as shown in Fig. 21.6. At the equilibrium separation the 3s level is spread out into a broad band with

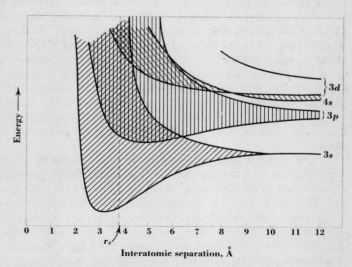

Fig. 21.6 As sodium atoms are brought together to form a metal, the higher-lying sharp atomic energy levels are spread into bands, which overlap strongly at the interatomic spacing of solid Na. (*After Slater.*)

the lowest energy states lying lower than the atomic 3s level. Meantime, the 3p band overlaps the 3s, the 3d and 4s bands overlap 3p, and so forth. As a result, a free electron in the Na crystal can have any energy from that at the bottom of the 3s band upward. An extended range of levels from which electrons can gain energy from an applied electric field and move through the crystal is known as a *conduction band*.

All the alkali metals crystallize in the body-centered cubic lattice in which each ion has eight nearest neighbors. The *Pauling radius* of sodium ions is 0.95 Å, but the equilibrium separation of nearest neighbors is 3.71 Å. This had led Hume-Rothery to describe metallic sodium as "small Na$^+$ ions swimming in regular array in the electron cloud of the valency electrons." In part it is because the ions occupy so small a fraction of the volume of the crystal that the electron-gas model (Sec. 21.8) is particularly appropriate for the alkalis.

Most other metals form face-centered cubic or hexagonal close-packed (Fig. 21.7a) crystals. These arrays give 12 nearest neighbors per atom, the largest possible *coordination number* of identical atoms. The face-centered cubic and hexagonal close-packed structures are intimately related, as one can see by study of Fig. 21.7b. If one starts with a layer A of identical spheres having six touching neighbors and a second layer B is added so that the spheres fit into the depressions, each sphere in the second layer is in contact with nine others, three below and six in the second layer. When a third layer is added, there is a choice which of two sets of depressions the spheres occupy. If they lie at the A sites directly above those of the first layer and the crystal is built up of layers ABABA · · · , hexagonal close packing results; if they lie at the C sites and the crystal is built of layers ABCABC · · · , a face-centered cubic crystal is modeled. These two arrays are the closest possible packings for equal spheres; in each case the spheres occupy 74 percent of the total volume. For metals like copper and silver, a face-centered cubic array of touching spheres is a fair model for the ion cores.

The metallic bond arises from the interactions of the ubiquitous free electrons and the ion cores. It is directionally insensitive, being almost equal in all directions. Further, as long as the ion cores are not too different in size, the electron glue serves equally well for various kinds of ion cores, which suggests why metals can be alloyed or joined together. The close packing favored by the metallic bond leads to structures which are dense and strong. The high ductility of metals can be related to the fact that it is relatively easy to make one set of planes slip relative to another. Close-packed planes can slip past each other with a minimum of interference and reach another stable position

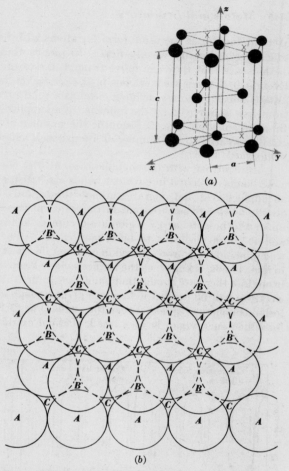

Fig. 21.7 (a) The hexagonal close-packed structure. (b) When close-packed layers are built up in the order *ABABA* · · · , hexagonal close packing results; stacking in the order *ABCABC* · · · leads to a face-centered cubic array.

after a minimum displacement. Thanks to the nondirectional nature of the bond and its sameness for all the ion cores, the atoms of a layer displaced from one equilibrium position to the next can join up with a new set of neighbors as perfectly as before, thereby restoring the original structure of the crystal.

21.6   *Metals and Insulators*

Metallic binding is the usual form for atoms with few valence electrons, which are ordinarily weakly held. In metals these electrons are free in the sense that they can move through the crystal with relative ease. Even at 0°K there are electrons in the *conduction band* of a metal (Fig. 21.8a); indeed, we may define a *metal* as a solid in which there are electrons free to move under the influence of an applied electric field at 0°K. In some substances, e.g., bismuth, the number of free carriers is very small compared to the number of atoms; such materials are often called *semimetals*.

In contrast with the metallic case, at 0°K electrons transferred in ionic bonds or shared in covalent bonds are not free to gain energy from an applied field and migrate through the crystal. Although the sharp atomic levels are smeared out in the solid, all the allowed energy states up to a forbidden energy region are occupied (at least at 0°K) and there are no more electrons to be accounted for. The highest filled band, which includes electrons shared in covalent bonds or electrons transferred in ionic bonds, is known as the *valence band*. Even in excellent crystalline insulators there are conduction bands above the valence band into which electrons can be excited if they are given enough energy, but there is a forbidden energy region between the bands (Fig. 21.8b and c). When the forbidden energy gap is very small, thermal excitation may excite elec-

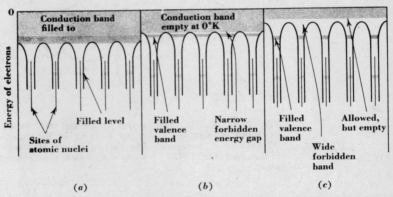

Fig. 21.8 Comparison of energy bands of (a) a metal, (b) a semiconductor, and (c) an insulator. At 0°K electrons occupy all states indicated in black; energy states in gray are allowed but unoccupied at 0°K. Low-lying occupied states (such as *K* and *L* levels) are not shown. (*From A. S. Smith and J. N. Cooper, "Elements of Physics," 7th ed. Copyright 1964. McGraw-Hill Book Company. Used by permission.*)

trons from the valence band to the conduction band, so that the material may be a conductor at elevated temperatures although a good insulator at 0°K. A material with a small gap between the valence and conduction bands is known as a *semiconductor*.

21.7 Metallic Conduction

Although metallic binding is a quantum-mechanical phenomenon, as early as 1900 Drude proposed that in metals there are one or more free electrons per atom plus a crystal of positive ions. On this basis he was able to explain electrical and thermal conductivity semiquantitatively. In 1909 Lorentz applied Maxwell-Boltzmann statistics to this electron gas and obtained reasonable behavior so far as the optical properties of metals are concerned. However, his theory led to unsatisfactory predictions for the temperature coefficient of resistance, for specific heats, and for the paramagnetic susceptibility.

A primitive model of metallic conduction begins with the assumption that the conduction electrons behave in an identical manner on the average. When an electric field E is applied, the electrons are accelerated by the driving force $-eE$. The fact that an equilibrium current is quickly achieved can be explained by assuming that there arises from collisions with the ion core a net damping force which equals the driving force. Let this damping force be represented by $-mv_D/\tau$, where m is the electron mass, v_D is the average drift velocity, and τ is a constant we shall call the *relaxation time*. For the steady state the driving and damping forces are equal, and so we have

$$v_D = \frac{e\tau E}{m} \tag{21.3}$$

If E is in the x direction, we may write $J_x = nev_D$, where J_x is the current density and n is the number of conduction electrons per unit volume. If we substitute v_D from Eq. (21.3), we have

$$J_x = \frac{ne^2\tau}{m} E_x \tag{21.4}$$

Thus the assumption of a damping force proportional to the average drift velocity leads directly to Ohm's law, which is $J = \sigma E$ in its microscopic form. We see that $\sigma = ne^2\tau/m$ for this model, in which we attribute the joule heating to the work done against the damping force.

Suppose that a steady electric field E_x has established a steady current density J_x corresponding to an average drift velocity $v_D(0)$. At $t = 0$ let us switch off the field. Then the electrons are decelerated according to $d(mv_D)/dt = -mv_D/\tau$, from which

$$v_D(t) = v_D(0)e^{-t/\tau} \tag{21.5}$$

The constant τ is called the relaxation time by virtue of its dimension of time and its role in Eq. (21.5); τ is very small for metals, since in a noninductive circuit equilibrium is achieved very quickly. For example, if we take the experimental value $\sigma = 6 \times 10^7$ mhos/m for copper and assume that each atom contributes one free electron, we find that $\tau \approx 2 \times 10^{-14}$ s.

In discussing electrical conductivity it is convenient to introduce the *mobility* of charge carriers. The *mobility μ is the magnitude of the ratio of drift velocity to applied electric field.* Thus

$$\mu = \left| \frac{v_D}{\mathsf{E}} \right| = \frac{e\tau}{m} = \frac{\sigma}{ne} \tag{21.6}$$

The mobility of charge carriers can be directly measured as we shall see in Sec. 22.1.

For any material the resistivity ρ is the reciprocal of the conductivity, and by Eq. (21.4)

$$\rho = \frac{1}{\sigma} = \frac{m}{ne^2\tau} \tag{21.7}$$

In terms of the particle model of an electron, τ is due to collisions with ion cores; from the wave point of view we think of the electron wave packet scattered by phonons or by imperfections and impurities as the electrons pass through the crystal. In terms of either model one can express the resistivity for a typical metal approximately as the sum of two terms

$$\rho = \rho_i + \rho_L \tag{21.8}$$

where ρ_i is the contribution due to interactions with crystal imperfections and impurities and ρ_L is due to the phonons (lattice vibrations associated with the thermal motions of the atoms). Equation (21.8) is known as *Matthiessen rule* and expresses the experimental fact that contributions to the resistivity from several sources are additive. (For example, for very thin films another term ρ_s must be added for the contribution of surface scattering.)   The first term ρ_i is essentially independent of temperature T, while ρ_L is temperature-dependent.

For many metals ρ varies with temperature roughly as curve A of Fig. 21.9.   Near $T = 0$, ρ_L is negligible.   As T is raised, the periodicity of the lattice is disturbed by phonons passing through the crystal. Conduction electrons interact with these photons, which serve to remove energy from the electrons.   For T small compared with the Debye temperature Θ, ρ_L varies roughly as the fifth power of absolute temperature; at temperature high compared with Θ, the lattice contribution is essentially proportional to T.   For reasonably pure metals ρ_i is small compared with ρ_L at room temperature, and consequently ρ is roughly a linear function of T.   For many alloys ρ_i is very great and completely dominant over ρ_L.

For more than 20 metals the resistivity falls suddenly to zero at some critical temperature, as shown by curve B of Fig. 21.9.   This phenomenon of *superconductivity* was discovered in 1911 by H. Kamerlingh Onnes in experiments with mercury wires.   For mercury the critical temperature T_c is 4.15°K; technetium has the highest critical temperature (11.2°K) of any element.   However many alloys have higher critical temperatures.   For Nb$_3$Sn, widely used for winding superconductive magnet coils, T_c is slightly over 18°K.

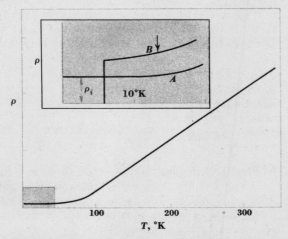

Fig. 21.9   The resistivity of a metal as a function of temperature.   In the insert, curve B is for a metal which becomes superconducting and curve A for one which does not.

21.8   *The Sommerfeld Free-electron Model*

Although the considerations of the preceding section predict many reasonable results, they lead to entirely unsatisfactory predictions of such quantities as metallic specific heats.   In 1928 Sommerfeld applied quantum theory and Fermi-Dirac statistics to the electron-gas model with great success.   In this model one assumes that the sole effect of the ion cores is to neutralize the electric charge on the electrons themselves so that they behave as though they were noninteracting.   In spite of the artificiality of this assumption, the Sommerfeld theory predicts correctly many features of metallic behavior.

Following Sommerfeld, let us think of the electrons as noninteracting particles in a box, a problem which is solved in Sec. 13.7 for a cubical box of side L with the result that the allowed energies are, by Eq. (13.26a),

$$E = \frac{p^2}{2m} = \frac{h^2}{8mL^2}(n_x{}^2 + n_y{}^2 + n_z{}^2) = \frac{h^2 n^2}{8mL^2} \tag{21.9}$$

where n_x, n_y, and n_z are positive integers.   We wish to calculate the density of states $g(E)$ for the box, where $g(E)\,dE$ is the number of allowed electron states per unit volume with energy between E and $E + dE$. The number of values of n_x, n_y, and n_z for which n lies between n and $n + dn$ is $\frac{1}{8}(4)\pi n^2/dn$, as one can see by referring to Fig. 5.4.   Since there are two spin states for each value of n_x, n_y, n_z, we have for the total number of states per unit volume with n between n and $n + dn$

$$g(n)\,dn = \frac{2(4)\pi n^2\,dn}{8V} \tag{21.10}$$

and, by use of Eq. (21.9),

$$g(E)\,dE = \frac{\pi}{V}\frac{8mL^2E}{h^2}\frac{\sqrt{8mL^2}}{h}\frac{dE}{2\sqrt{E}} = \frac{2^{\frac{1}{2}}\pi m^{\frac{3}{2}}}{h^3}\sqrt{E}\,dE \tag{21.11}$$

In Fig. 21.10 the dashed curve shows $g(E)$ as a function of E.

The number $N(E)\,dE$ of electrons per unit volume with energy between E and $E + dE$ is the product of $g(E)\,dE$ and the probability that a state of energy E is occupied; the latter is the Fermi function $f(E)$ given by Eq. (20.16a); thus

$$N(E)\,dE = g(E)f(E)\,dE = \frac{2^{\frac{1}{2}}\pi m^{\frac{3}{2}}\sqrt{E}\,dE}{h^3(e^{(E-E_F)/kT} + 1)} \tag{21.12}$$

where E_F is the Fermi energy.

To calculate the Fermi energy E_F for a metal in the Sommerfeld theory, we consider the metal at $0°K$ when the electrons occupy the lowest possible energy states and $f(E)$ is 1 for $E < E_F$ and 0 for $E > E_F$. Thus all states below E_F are filled and all above E_F are empty at $0°K$; $N(E)$ at $0°K$ is shown by the light solid line of Fig. 21.10. Now if n represents the number of free electrons per unit volume in the solid, it is clear that

$$n = \int_0^{E_F} N(E)\, dE = \frac{2^{\frac{3}{2}}\pi m^{\frac{3}{2}}}{h^3} \int_0^{E_F} \sqrt{E}\, dE = \frac{2^{\frac{5}{2}}\pi m^{\frac{3}{2}}}{3h^3} E_F^{\frac{3}{2}}$$

from which

$$E_F = \frac{3^{\frac{2}{3}}h^2}{8\pi^{\frac{2}{3}}m} n^{\frac{2}{3}} = 3.65 \times 10^{-19} n^{\frac{2}{3}} \qquad \text{eV} \tag{21.13}$$

For sodium we can readily find n if we assume one free electron per atom. The density is 971 kg/m³, and the atomic weight is 22.99, so that $n = 6.02 \times 10^{26} \times 971/22.99$ m⁻³ and $E_F = 3.16$ eV. Similar

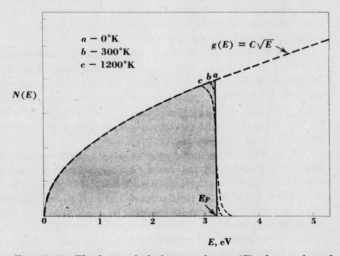

Fig. 21.10 The heavy dashed curve shows $g(E)$, the number of allowed energy states per unit volume per unit energy range at energy E. The light solid line designates the number $N(E) = g(E)f(E)$ of occupied energy states per unit volume per unit energy range at $0°K$; light dashed lines indicate $N(E)$ at higher temperatures.

calculations for other monovalent elements yield the results:

Metal	Li	K	Rb	Cs	Cu	Ag
E_F, eV	4.72	2.14	1.82	1.53	7.04	5.51

Even at 0°K many sodium conduction electrons have a kinetic energy more than 3 eV since the potential energy is zero in this model. For Na E_F corresponds to an electron speed in excess of 1.05×10^6 m/s. The average kinetic energy of the free electrons at 0°K is given by

$$\bar{E} = \frac{\int_0^{E_F} E N(E)\, dE}{\int_0^{E_F} N(E)\, dE} = \frac{\int_0^{E_F} E^{\frac{3}{2}}\, dE}{\int_0^{E_F} E^{\frac{1}{2}}\, dE} = \frac{\frac{2}{5} E_F^{\frac{5}{2}}}{\frac{2}{3} E_F^{\frac{3}{2}}} = \frac{3}{5}\, E_F \qquad (21.14)$$

When the temperature is raised, some electrons with energy near E_F are raised to higher energy states; the $N(E)$ curve then is shown by the dashed curves (b) and (c) of Fig. 21.10. At 290°K kT is only 0.025 eV. For an energy 1.1 kT greater than E_F, $f(E)$ is $\frac{1}{4}$; for 1.1 kT less than E_F, $f(E) = \frac{3}{4}$. [E_F continues to represent the value of E for which $f(E) = \frac{1}{2}$.] Since only a small fraction of the free electrons gain energy as T is increased, and since those which do gain pick up only about kT, it is clear why the free electrons make only a small contribution to the specific heat at room temperature.

We are now in a position to contrast the Sommerfeld model with that used by Lorentz, who assumed the average kinetic energy of the electrons was given by $K = \frac{3}{2}kT$. At 300°K this is less than 0.04 eV, while the Sommerfeld theory predicts a kinetic energy of several electron volts. On the other hand, the older theory predicts a specific heat of $\frac{3}{2}R$ per kilomole for the electrons, while the newer model calls for two orders of magnitude less.

Consider a copper conductor bearing a current. The average drift velocity is typically of the order of millimeters per second, while the average speed of the electrons is about 1.2×10^6 m/s. In the absence of an applied field, the average velocity is zero even though the average speed is great. A finite drift velocity in a metal is analogous to a wind in the atmosphere; air molecules and electrons in metals are in rapid motion at all times. A wind and a current result when a small average drift velocity is superimposed on the random velocity distribution in each case. The mean free path λ of an electron in copper is approximately τ times the Fermi velocity v_F (the velocity corresponding to the Fermi kinetic energy $K_F = E_F$, since $P' = 0$). Using the value of τ from

Sec. 21.7, we have

$$\lambda_{cu} \approx 2 \times 10^{-14} \text{ s} \times 1.6 \times 10^6 \text{ m/s} \approx 3 \times 10^{-8} \text{ m} = 300 \text{ Å}$$

Thus a typical electron may travel 100 atom layers between collisions.

For a metal, or any other kind of Fermi gas, the Fermi energy depends somewhat on the temperature (see Fig. 20.5). However, as long as the Fermi energy at 0°K is large compared to kT, the change is very small; $E_F(T)$ is given by

$$E_F(T) = E_F(0°\text{K}) \left[1 - \frac{\pi^2}{12} \left(\frac{kT}{E_F(0)} \right)^2 \cdots \right] \tag{21.15}$$

Since E_F is several electron volts for metals, one must go to very high temperatures before the Fermi energy differs significantly from its 0°K value.

Throughout this discussion of the free-electron model, we have assumed the potential energy to be zero; thus we have placed the energy zero at the bottom of the conduction band (Fig. 21.11). For sodium the Fermi energy lies at about 3.2 eV. It requires about 2.27 eV to take an electron at the top of the conduction band and move it an infinite distance from the crystal. This energy may be supplied by photons (photoelectric effect), by heat (thermionic emission), or by bombarding the crystal with electrons (secondary emission) or other particles. Raising the temperature from 0°K to room temperature decreases the work function ϕ very

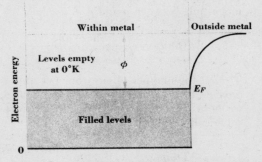

Fig. 21.11 In the Sommerfeld free-electron model the potential energy is taken as 0 at the bottom of the conduction band and electrons have kinetic energies ranging from 0 to E_F at 0°K. To escape from the metal an electron at the Fermi level needs an additional energy ϕ, the work function of the metal.

little, since kT is small compared to ϕ. However, at sufficiently high temperatures a significant number of electrons have enough energy to escape.

21.9   Thermionic Emission

As the temperature of a metal is increased, electrons near the Fermi surface are excited to higher energy states; eventually some have enough energy to surmount the surface barrier and escape. From Fig. 21.11 the energy required is $E_F + \phi$, but for an electron to escape it must not only have at least this energy but must also arrive at the surface with its momentum **p** suitably directed. If we consider escape through a surface defined by the plane $x = 0$, the x component of the momentum must be equal to or greater than the critical value

$$p_{xc} = + \sqrt{2m(E_F + \phi)} \qquad (21.16)$$

since the kinetic energy associated with the y and z components of the momentum does not aid in the escape.

Let us now assume that all electrons which reach the surface with $p_x > p_{xc}$ are thermionically emitted. Then the emission current density J_x has the magnitude ev_x times the number of electrons per unit volume with $p_x > p_{xc}$. To calculate J_x we convert the energy distribution of Eq. (21.12) to a momentum distribution by using $E = p^2/2m$, thereby obtaining for the number of electrons per unit volume with momentum magnitude between p and $p + dp$,

$$N(p)\, dp = \frac{8\pi p^2\, dp}{h^3(e^{(p^2/2m - E_F)/kT} + 1)} \qquad (21.17)$$

To find the number of electrons per unit volume with momentum components between p_x and $p_x + dp_x$, p_y and $p_y + dp_y$, and p_z and $p_z + dp_z$, we note from Fig. 21.12 that, because the momentum distribution is spherically symmetrical,

$$N(p_x,p_y,p_z)\, dp_x\, dp_y\, dp_z = \frac{dp_x\, dp_y\, dp_z}{4\pi p^2\, dp} N(p)\, dp$$

$$= \frac{2 dp_x\, dp_y\, dp_z}{h^3(e^{(p^2/2m - E_F)/kT} + 1)} \qquad (21.18)$$

Only electrons with momenta to the right of the plane $p_x = p_{xc}$ in Fig. 21.12 can escape. We now find the emission current density by

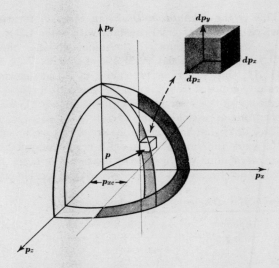

Fig. 21.12 Only electrons with $p_x > p_{xc}$ can escape from the metal. Because the momentum distribution is spherically symmetric, the number of electrons with a momentum terminating in $dp_x \, dp_y \, dp_z$ is $N(p) \, dp$ multiplied by the ratio of the volume $dp_x \, dp_y \, dp_z$ to the volume of the spherical shell $4\pi p^2 \, dp$.

evaluating

$$J_x = \int_{p_x = p_{xc}}^{\infty} \int_{p_y = -\infty}^{\infty} \int_{p_z = -\infty}^{\infty} e \frac{p_x}{m} \frac{2dp_x \, dp_y \, dp_z}{h^3 (e^{[(p_x{}^2 + p_y{}^2 + p_z{}^2)/2m - E_F]/kT} + 1)} \tag{21.19}$$

Since only electrons with energy greater than $E_F + \phi$ can be emitted and ϕ is many times kT, we may neglect the unity in the Fermi factor and write

$$J_x = \frac{2ee^{E_F/kT}}{h^3 m} \int_{p_{xc}}^{\infty} p_x e^{-p_x{}^2/2mkT} \, dp_x \int_{-\infty}^{\infty} e^{-p_y{}^2/2mkT} \, dp_y \int_{-\infty}^{\infty} e^{-p_z{}^2/2mkT} \, dp_z \tag{21.19a}$$

These three integrals are standard ones evaluated in Appendix 4A; upon substituting the appropriate values, using $p_{xc} = \sqrt{2m(E_F + \phi)}$, one obtains

$$J = \frac{4\pi mek^2}{h^3} T^2 e^{-\phi/kT}$$

$$= A_0 T^2 e^{-\phi/kT} \tag{21.20}$$

This is the *Richardson-Dushman equation* for thermionic emission, in which A_0 is a universal constant with value 1.2×10^6 A/m²-°K².

Some representative thermionic-emission data are given in Table 21.1. The functional variation of J with T is well described by Eq. (21.14), but experimental values of A_0 are considerably less than the value computed from the atomic constants for most metals. Of the many reasons for this discrepancy we mention two: (1) we have assumed that all electrons approaching the surface with $p_x > p_{xc}$ escape, but we know (Sec. 12.2) that some electrons are always reflected at a step barrier; (2) the work function ϕ varies with crystallographic direction and for the polycrystalline wire ordinarily studied, the most copious emission comes from that fraction of the surface with favorable crystal orientation.

Table 21.1   Thermionic-emission constants

Emitter	A_0, A/m²-°K²	ϕ, eV
Cr	0.48×10^6	4.60
Cs	$1.6 \ \times 10^6$	1.8
Mo	0.55×10^6	4.3
Ni	0.30×10^6	4.61
Ta	0.55×10^6	4.19
W	0.60×10^6	4.52
Ba on W	0.15×10^6	1.56
Cs on W	0.32×10^6	1.36

21.10   Metals in Contact

When two dissimilar conductors are placed in contact, a *contact potential difference* appears between them. Typical values range from a fraction of a volt to several volts. Consider two metals, initially far apart, which have work functions ϕ_1 and ϕ_2 and Fermi kinetic energies shown in Fig. 21.13a. When they are brought together, electrons can go to lower energy states by transferring from metal 1 to metal 2. Adding electrons to a material raises its Fermi level; removing them lowers the Fermi level. Electrons continue to move preferentially from metal 1 until the Fermi levels coincide (Fig. 21.13b), at which point equilibrium is achieved with equal numbers of electrons crossing the surface in each direction per unit time. (Equal Fermi levels correspond to equal chemical potentials from the thermodynamical point of view.)

The transfer of electrons leaves metal 1 positively charged and metal 2 negative, thus establishing an electric-dipole layer at the sur-

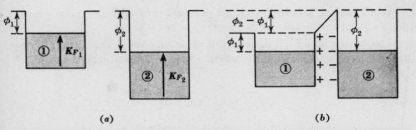

Fig. 21.13 Two metals with different work functions (*a*) far apart and (*b*) in contact. When any two materials are in contact and in equilibrium, their Fermi levels coincide.

face. The added energy an electron needs to cross this layer is the *contact potential energy;* from Fig. 21.13*b* this energy is $\phi_2 - \phi_1$, and the corresponding *contact potential difference* ΔV is $(\phi_2 - \phi_1)/e$. An electron crossing from metal 1 to 2 gains kinetic energy at the expense of potential energy and vice versa.

One way to measure the contact potential difference is to arrange two insulated parallel plates, one of each metal, connected to the terminals of an electrometer, with a bit of radioactive material so placed as to ionize the air between the plates. A zero electrometer reading is taken with the terminals connected together; then the terminals are disconnected, and a second reading is taken after the accumulation of charge on the plates has destroyed the electrostatic field between them and so stopped the flow of ions. The difference of the two readings is the contact potential difference, the plate showing a negative charge in the second reading being the one that is positive when the plates are in contact. The difference of surface potential thus defined depends greatly, however, upon the state of the surfaces. The potential difference can be regarded as the algebraic sum of a potential jump between the metals themselves and other jumps, perhaps sometimes voltaic in origin, between each metal and whatever layer of oxide or other contaminating substance is present on its surface.

Problems

1. Find the Miller indices for planes with each of the following sets of intercepts: (*a*) 6a, 2b, 3c; (*b*) a, 2b, ∞ ; (*c*) 2a, −b, 2c.

2. Imagine a crystal formed of identical hard spheres in contact with nearest neighbors. Show that the ratio of the volume of the spheres to the volume of the crystal is 0.74 for face-centered cubic and hexagonal close-packed, 0.68 for body-centered cubic, 0.52 for simple cubic, and 0.34 for the diamond packing.

3.   Show that the number of atoms per unit area of any crystal plane is proportional to the spacing between adjacent planes.

4.   Most of the alkali halides crystallize in the NaCl structure. Some of them have the following internuclear separations of nearest neighbors:

Halide	LiF	NaF	KF	LiCl	NaCl	KCl
Internuclear separation, Å	2.01	2.31	2.67	2.57	2.81	3.14

Assuming a hard-sphere model for the ions and an ionic radius of 1.33 Å for the F^- ion, compute radii for the other ions of these crystals. Why does the hard-sphere model give reasonably consistent values when we know the solids are compressible and that the ions are far from hard?

5.   Find the radius of the largest sphere which can fit into the largest interstice for a simple cubic structure composed of identical hard spheres in contact. Do the same for face-centered and body-centered crystals.

6.   Show that in a cubic crystal the distance between adjacent planes with Miller indices (hkl) is given by $d_{hkl} = a/(h^2 + k^2 + l^2)^{\frac{1}{2}}$, where a is the lattice constant.

7.   Show that in the diamond structure, with atoms at the corners of a regular tetrahedron, the tetrahedral bonds make an angle of 109.5° with each other.

8.   At about 1180°K iron transforms into a face-centered cubic structure from the body-centered cubic form normal at room temperature. Assuming there is no change in density, calculate the ratio of the nearest neighbors distance in the face-centered modification to that in the body-centered form.

   Ans: 1.029

9.   Calculate a rough approximation to the Madelung constant for KCl by considering only the first cube of ions surrounding a central K^+ ion and assuming that for an ion on any face one-half the charge belongs to the cube, for an ion on an edge one-fourth, and for a corner ion one-eighth.

   Ans: 1.45

10. Show that the Madelung constant for a one-dimensional infinite line of alternating positive and negative ions is 2 ln 2, or 1.386.

11. If the cohesive energy of NaCl is 6.61 eV/ion pair and the interatomic spacing is 2.82 Å, find the electrostatic-attractive energy and

the ion-core repulsive energy per ion pair and calculate an approximate value of n.

12. From the fact that the density of copper is 8940 kg/m³, calculate the number of Cu atoms per cubic meter, the Fermi kinetic energy on the basis of the Sommerfeld model (one free electron per atom), the average kinetic energy of the free electrons, and the Fermi velocity (speed of electron having the Fermi kinetic energy).

  Ans: 8.47×10^{28} m⁻³; 7.02 eV; 4.2 eV; 1.6×10^6 m/s

13. A piece of No. 12 copper wire carrying its rated maximum current of 25 A is at 50°C. At that temperature the resistance of a piece 1 m long is 0.0058 Ω; its cross-sectional area is 3.3×10^{-6} m². Assuming that there is one conduction electron per atom, find (*a*) the conductivity σ, (*b*) the average drift velocity of the electrons, (*c*) the mobility μ, and (*d*) the mean free path. (Use the results of the preceding problem.)

14. The Fermi temperature for an electron gas is defined as the ratio of the Fermi energy E_F to the Boltzmann constant k. Calculate the Fermi temperature of the electrons for each of the metals for which an approximate Fermi energy is tabulated in Sec. 21.8.

15. Compute the Fermi kinetic energy for aluminum assuming that there are three free electrons per atom. (The density of Al is 2750 kg/m³.) Over what temperature range is this energy greater than $10kT$?

16. The potential energy of a pair of atoms in a solid can be written $P(x) \approx Ax^2 - Bx^3 - Cx^4$, where the atomic separation is $R_0 + x$ and the zero of potential energy is taken to be the equilibrium configuration at 0°K. Show that $\langle x \rangle = 3BkT/4A^2$ when $|Bx^3 + Cx^4|$ is small compared to Ax^2 and to kT. *Hint:* Take

$$\langle x \rangle = \frac{\int_{-\infty}^{\infty} x e^{-P(x)/kT}\, dx}{\int_{-\infty}^{\infty} e^{-P(x)/kT}\, dx}$$

expand small terms in series, and integrate.

17. Show that the velocity distribution of electrons perpendicular to the surface of a thermionic emitter is maxwellian in form, i.e., is proportional to the number which would escape if electrons in the metal obeyed Maxwell-Boltzmann statistics and had no barrier to overcome. Show that for the escaping electrons in the vacuum the average kinetic energy associated with the x component of the momentum is kT while for the y and z components it is $\frac{1}{2}kT$.

18. Show that as long as the Fermi kinetic energy K_F of the free electrons in a metal is very large compared to kT, the molar heat capacity of these electrons is

$$(c_V)_{el} = \frac{\pi^2 R k T}{2 K_F}$$

where R is the general gas constant and it is assumed that there is one free electron per atom.

chapter twenty-two

The Band Model for Metals ————————————

In many respects the Sommerfeld model of a metal as a Fermi gas of free electrons provides amazingly good predictions of the properties of the alkali metals and reasonable ones for copper, silver, and gold. It leads directly to acceptable values for the specific heats of these metals and to the Richardson-Dushman equation for thermionic emission. But the interactions between electrons and the ion cores are of transcendent importance for a quantitative understanding of properties of other metals and of alloys. A dramatic example of the failure of the free-electron model to account for the properties of certain metals is found in the Hall effect, the subject with which we begin this chapter. When the electron-lattice interaction is taken into account, we are led to several ideas which have proved to be of great utility.

22.1 The Hall Effect

Half a century before Sommerfeld's free-electron theory and more than 15 years before the electron was discovered, Hall performed experiments designed to tell whether metallic conduction is due to positive or negative carriers. In 1879 he passed a current through a thin sheet of metal with a pair of contacts C and D attached at opposite sides of the central equipotential surface (Fig. 22.1). When a magnetic field B_z is applied normal to the sheet, the equipotentials are shifted and a potential difference develops between C and D. Suppose for the moment that the carriers are positive and move in the x direction. The Lorentz force $q(\mathbf{v} \times \mathbf{B})$ is then in the $-y$ direction, so that the carriers are initially deflected downward. This makes the bottom of the sheet (and D) more positive, while the top (and C) becomes more negative. The surface charge builds up until the resulting electric field E_y exerts an upward force equal on the average to the downward Lorentz force. Thus at equilibrium

$$1_y q \mathsf{E}_y + q(\mathbf{v} \times \mathbf{B}) = 0 \tag{22.1a}$$

or

$$\mathsf{E}_y = v_x B_z \tag{22.1b}$$

where v_x is the average speed of the carrier (and hence is to be identified with the drift velocity rather than the Fermi velocity).

Now let us suppose that the carriers are negative. For a conventional current in the x direction, negative carriers move in the $-x$ direc-

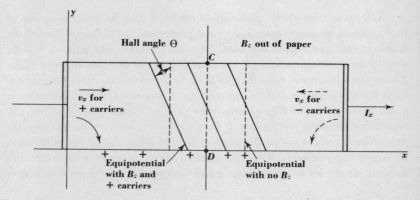

Fig. 22.1 Positive carriers moving in the x direction are deflected downward by a magnetic induction out of the plane of the paper. When equilibrium is established, there is an induced electric intensity in the y direction. If the current is due to negative carriers, the carriers are also deflected downward (dashed lines at right) and E_y is in the $-y$ direction.

tion, but the Lorentz force is still downward. In this case negative charge is deflected downward, D becomes negative relative to C, and E_y is negative. Thus Hall reasoned that if the charge carriers in a metal are positive, D should become positive relative to C when the magnetic field is applied; if the carriers are negative, C should become positive relative to D.

We now define the *Hall coefficient* to be

$$R_H = \frac{E_y}{J_x B_z} \tag{22.2}$$

where J_x is the current density: R_H is positive for positive carriers, negative for negative carriers. The current density J_x is given by qnv_x, where n is the number of carriers per unit volume. If we substitute this in Eq. (22.2) and make use of (22.1b), we have

$$R_H = \frac{1}{nq} \tag{22.2a}$$

Thus a measurement of the potential difference between C and D in the presence of a known field B_z not only tells us the sign of the carrier but also predicts the number of carriers per unit volume, since $|q| = e$, the fundamental unit of charge.

The Hall coefficients of several metals are listed in Table 22.1. Also tabulated is the number of free electrons per atom computed from R_H and the density of the metal. For the alkalis, and other valence-1

Table 22.1 Hall coefficients and corresponding number of free electrons per atom at room temperature

Metal	R_H, m³/C	Free Electrons per Atom
Na	-2.5×10^{-10}	0.99
K	-4.2×10^{-10}	1.1
Cu	-0.55×10^{-10}	1.3
Ag	-0.84×10^{-10}	1.3
Al	-0.30×10^{-10}	3.5
Be	$+2.4 \times 10^{-10}$	-2.2
Zn	$+0.33 \times 10^{-10}$	-2.9
Cd	$+0.60 \times 10^{-10}$	-2.5
Sb	230×10^{-10}	-0.008
Bi	-7500×10^{-10}	3×10^{-4}

metals, Hall measurements are reasonably compatible with the idea of one conduction electron per atom.   However, the Hall coefficient is positive for beryllium, zinc, cadmium, lead, molybdenum, and several other metals.   More than half a century had to elapse before an explanation was available in terms of the band theory of metals (Sec. 22.7).

From Hall measurements we can obtain not only the sign of the carrier and the number per unit volume but also the mobility.   One way to determine μ is to measure the *Hall angle* θ through which the equipotential lines are rotated (Fig. 22.1) by the magnetic field; this angle is given by $\tan \theta = \mathsf{E}_y/\mathsf{E}_x$.   If we make use of Eq. (22.2) and recall that $J_x = (nev_x) = ne\mu\mathsf{E}_x = \sigma\mathsf{E}_x$, we obtain

$$\tan \theta = \frac{\mathsf{E}_y}{\mathsf{E}_x} = R_H\sigma B_z = \mu B_z \tag{22.3}$$

The Hall angle is positive for positive carriers and negative for electrons. From measured mobilities one can calculate relaxation times and mean free paths.

The Hall effect was first used to study metals, but it is an even more valuable tool for the study of semiconductors.   Since the number of carriers per unit volume for semiconductors is small compared with the number for metals, Hall coefficients and Hall angles are correspondingly larger.

Throughout this discussion of the Hall effect it has been assumed that only one kind of carrier is present, although for the metals with positive Hall coefficients and for many semiconductors this is not the case.   When more than one kind of carrier is present, a more elaborate theory is required and Eqs. (22.2a) and (22.3) are no longer applicable.

22.2   *Electrons in a Periodic Lattice*

To gain further insight into the nature of metals and to understand how some can have positive Hall coefficients, we must take into account the periodic fields which arise from the arrangement of ion cores in a lattice. While free electrons in a metal tend to smooth out the fluctuations in potential due to the ions, the assumption that the potential is constant throughout the metal is a gross oversimplification.   Even though the free-electron model successfully predicts many metallic properties, the lattice periodicity has far-reaching consequences.

The solutions of the Schrödinger equation for a free electron in one dimension are of the form (Sec. 12.2)

$$\psi = C e^{\pm ikx} \tag{22.4}$$

where k is the wavenumber given by $k = 2\pi/\lambda = p/\hbar$. However, if the electron is subject to a potential $P(x)$ which has the spatial periodicity a of the lattice, Bloch has shown that the solutions are of the form

$$\psi = u_k(x) e^{\pm ikx} \tag{22.4a}$$

where $u_k(x)$ has the periodicity of the lattice, so that

$$u_k(x) = u_k(x + a) = u_k(x + na) \qquad n \text{ an integer} \tag{22.5}$$

The *Bloch function* of Eq. (22.4a) differs only by a periodic function from the free-electron function of Eq. (22.4). The effect of the lattice periodicity is to modulate the free-electron solution, since

$$\psi(x + a) = u_k(x + a) e^{\pm ik(x+a)} = \psi(x) e^{\pm ika} \tag{22.6}$$

The specific form of $u_k(x)$ depends on the particular $P(x)$ involved and on the value of the wavenumber k. For example, Kronig and Penney have solved the Schrödinger equation for a one-dimensional array of rectangular potential wells (Fig. 22.2). They found that for an electron in this periodic field, certain bands of energy are allowed and certain other bands forbidden. Without carrying through the calculations, we can see how such a result develops. For widely spaced levels of considerable depth, an electron is virtually bound in a well, and the eigenenergies are those for a single well. For this case we found

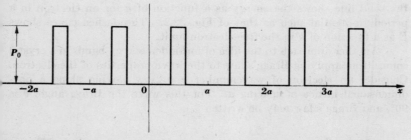

Fig. 22.2 The potential energy for a one-dimensional array of identical square wells a distance a apart.

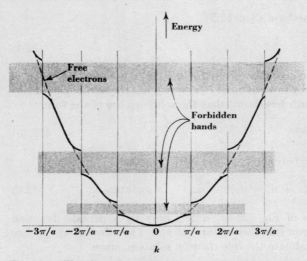

Fig. 22.3 The allowed energies for a one-dimensional array of square wells (Fig. 22.2) as a function of the wave vector k. The discontinuities occur at values of k for which Bragg's law is satisfied.

sharp energy levels (Secs. 13.3 and 13.4). As the wells are brought closer together, the wave function penetrates the barrier more and more. For two wells this gave rise (Sec. 19.5) to the splitting of a single energy state into two; for a large number of equally spaced wells it leads to the spreading of a single level into a continuous band, which becomes wider and wider as the spacing of the levels is reduced. Finally, when the barrier thickness is reduced to zero, we have a free electron in an infinitely wide well; all kinetic energies are allowed. In Fig. 22.3 the solid line shows the energy as a function of k for an electron in a periodic potential such as that of Fig. 22.2. The dashed curve shows E as a function of k in the free-electron limit.

Another approach to the idea of forbidden energy bands in a crystal comes from applying Bragg's law to the wave properties of the electron. Consider an electron of wavenumber $k = 2\pi/\lambda$ moving along a one-dimensional lattice of spacing a. For this wave the Bragg angle θ is 90°, and Bragg's law may be written

$$n\lambda = \frac{2\pi n}{k} = 2a \sin 90° \tag{22.7a}$$

or

$$k = \frac{n\pi}{a} \tag{22.7b}$$

Thus an electron with wavenumber $k = n\pi/a$ experiences strong Bragg reflection as it moves through the one-dimensional crystal. An electron with wavenumber $k = n\pi/a$ does not propagate as a traveling wave moving progressively in one direction but is represented by a set of standing waves arising from Bragg reflections. If we increase k gradually starting at $k = 0$, we find that electrons can pass quite freely through a periodic one-dimensional crystal until k approaches π/a, at which point strong Bragg reflection occurs. As we increase k further, traveling waves are allowed again, until at $2\pi/a$ they experience second-order Bragg reflection.

22.3 Effective Mass

For a free electron the kinetic energy $K = p^2/2m$, and since $p = h/\lambda = \hbar k$, the E-vs.-k curve is the parabola $E = \hbar^2 k^2/2m$. Obviously the latter equation does not apply to the heavy curve of Fig. 22.3, which may be discussed in terms of an *effective mass* m^*. The concept of effective mass is widely used in solid-state physics. When a free electron is subjected to an electric field E it experiences an acceleration $\mathbf{a} = -e\mathsf{E}/m$. However, when the electric field is applied to a crystal, few electrons, if any, have the acceleration $-e\mathsf{E}/m$. For example a copper K electron is so tightly bound to its atom that it is not accelerated at all; its "effective mass" is infinite. For an electron which is not bound to any single atom, Newton's second law gives

$$m\mathbf{a} = -e\mathsf{E} + \text{forces due to neighboring ion cores and electrons}$$

These latter forces we do not know quantitatively, but we can transfer our ignorance to the left side of the equation by writing

$$m^*\mathbf{a} = -e\mathsf{E} \tag{22.8}$$

Let us estimate m^* for an electron in a one-dimensional lattice, for which the E-vs.-k curve is shown in Fig. 22.3. To do this, we invoke wave properties of the electron, which we assume moves with a group velocity v_g given by [see Eq. (11.5)]

$$v_g = \frac{d\omega}{dk} = \frac{2\pi \, d\nu}{dk} = \frac{2\pi}{h} \frac{dE}{dk} = \frac{1}{\hbar} \frac{dE}{dk} \tag{22.9}$$

where we have made use of the fact that $E = h\nu$.   In a time δt the electric field $\mathbf{E}$ does work on the electron, increasing its energy by

$$\delta E = -e\mathbf{E}v_g \, \delta t \tag{22.10}$$

while by Eq. (22.9), $\delta E = \hbar v_g \, \delta k$.   Therefore,

$$\hbar v_g \, \delta k = -e\mathbf{E}v_g \, \delta t$$

or

$$-e\mathbf{E} = \hbar \frac{dk}{dt} \tag{22.11}$$

Thus the external applied force on the electron is equal to $\hbar \, dk/dt$ when we are describing the electron in terms of the wave formulation.   The acceleration of the electron is, by Eqs. (22.9) and (22.11),

$$a = \frac{dv_g}{dt} = \frac{1}{\hbar} \frac{d^2E}{dk \, dt} = \frac{1}{\hbar} \frac{d^2E}{dk^2} \frac{dk}{dt} = \frac{1}{\hbar^2} \frac{d^2E}{dk^2} (-e\mathbf{E})$$

Comparison with Eq. (22.8) leads immediately to

$$m^* = \frac{\hbar^2}{d^2E/dk^2} \tag{22.12}$$

For a three-dimensional crystal the effective mass is represented by a tensor of the form

$$m_{ij}^* = \frac{\hbar^2}{\partial^2E/(\partial k_i \, \partial k_j)} \tag{22.12a}$$

We now consider the behavior of the effective mass for the range $0 \le k \le \pi/a$ in Fig. 22.4a.   For small k the curve corresponds closely to the free-electron curve, so that $m^* \approx m$.   At point A the slope, and hence v_g, reaches its maximum value; d^2E/dk^2 is zero, and m^* becomes infinite.   Between A and B, d^2E/dk^2 is negative, corresponding to negative values of m^*.   (As we shall see in Sec. 22.7, it is desirable to discuss this region of k space in terms of holes with positive charge and positive mass.)   As k approaches π/a, v_g approaches 0 and m^* is $-m$; one interpre-

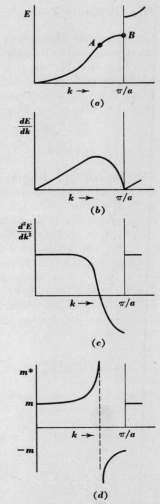

Fig. 22.4 (a) E, (b) dE/dk, (c) d^2E/dk^2, and (d) m^* plotted as functions of k for a periodic potential function like that of Fig. 22.2.

tation is that when k is essentially π/a, applying an accelerating field leads to Bragg reflection. Thus the force on the electron in one direction leads to the electron's gaining momentum in the opposite direction; by virtue of its interaction with the crystal, the electron behaves as though it had negative mass.

22.4 Brillouin Zones

Consider a two-dimensional square lattice such as that of Fig. 22.5. An electron moving in the [10] direction experiences first-order Bragg reflection from the (10) planes if its k vector has magnitude π/a. However, an electron moving at an angle θ with the (10) planes undergoes first-order Bragg reflection from these planes when k reaches the critical value k_c, given by $2\pi/k_c = \lambda = 2a \sin \theta$, or

$$k_c = \frac{\pi}{a \sin \theta} \tag{22.13}$$

Thus the critical k depends on the propagation direction of the electron. For example, an electron moving in the [31] direction has $\sin \theta = 3/\sqrt{10}$, so that $k_{c[31]} = \sqrt{10}\,\pi/3a$. When θ is less than 45°, Bragg reflection occurs from the (01) planes for a smaller k_c than that given by Eq. (22.13), viz., for $k_c = \pi/(a \sin \phi) = \pi/(a \cos \theta)$.

In a two-dimensional k space let us represent **k** for each of many electrons in our lattice by an arrow with its tail at the origin (Fig. 22.6). Let us further draw the square bounded by $k_x = \pm \pi/a$, $k_y = \pm \pi/a$ and thereby outline what is known as the *first Brillouin zone* for our two-dimensional crystal. Clearly, the condition for Bragg reflection for the [10] direction is that the appropriate arrow terminate on the zone

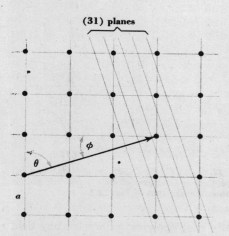

Fig. 22.5 A two-dimensional square lattice with a vector in the [31] direction which is perpendicular to the (31) planes.

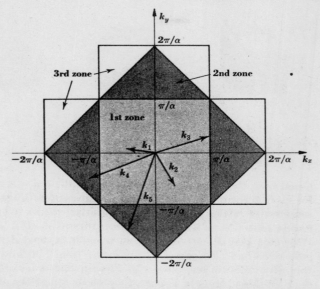

Fig. 22.6 First, second, and third Brillouin zones for a two-dimensional square lattice. Any *k* vector which terminates on a zone boundary satisfies Bragg's law for reflection from some set of lattice planes.

boundary; similarly, a $k_{[31]}$ vector of length $\sqrt{10}\,\pi/3a$ (k_3 in Fig. 22.6) just reaches the boundary. In general, *any k vector which terminates on the boundary of the first Brillouin zone satisfies the condition for Bragg reflection.*

An electron with a *k* vector such as k_1 or k_2 has too long a wavelength to suffer Bragg reflection from any set of planes. It can move almost freely through the crystal. An electron associated with k_4 has too short a wavelength for first-order Bragg reflection from (10) planes and too long for reflection from (11) or (01) planes. However, k_5 terminates on the boundary of the second Brillouin zone, satisfying the Bragg condition for the (11) planes, for which the spacing is

$$a/\sqrt{1^2 + 1^2} = a/\sqrt{2}$$

in the square lattice. This corresponds to a spacing $\sqrt{2}/a$ in *k* space, since distances there are reciprocal to distances in the crystal; for this reason *k* space is sometimes called a *reciprocal space.*

For two- and three-dimensional lattices, just as for the one-dimen-

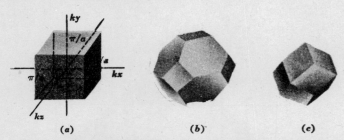

Fig. 22.7 The first Brillouin zone in k space for (a) simple cubic, (b) face-centered cubic, and (c) body-centered cubic crystal.

sional case, there are discontinuities in the allowed energies for electrons for each k vector that satisfies the condition for Bragg reflection from some set of lattice planes. In the positive quadrant of Fig. 22.6, this means reflection of any order from (10), (01), (11), (21), (12), (31), (13), etc., planes. [It is common to refer to second- and third-order reflection from (10) planes as first-order from (20) and (30) planes; in this nomenclature one always deals with first-order reflection.] If we draw all these planes in k space, they serve as portions of the boundaries of the various Brillouin zones. If we imagine a short k vector in any direction starting from the origin and growing steadily longer in time, this vector terminates in the first Brillouin zone until it reaches a plane. When it crosses this plane, it is in the second Brillouin zone until it reaches the next plane, and so forth. In Fig. 22.6 a k vector growing in the [11] direction would cross the (10), (01), and (11) planes at the point $k_x = \pi/a$, $k_y = \pi/a$. Having crossed three sets of planes at this point, it finds itself in the *fourth* Brillouin zone. Figure 22.6 shows the first three Brillouin zones for the square lattice; all the zones have the same area. We shall be concerned primarily with the first two Brillouin zones.

Application of these ideas to three dimensions leads to the first Brillouin zones for simple, face-centered, and body-centered cubic lattices shown in Fig. 22.7a, b, and c, respectively. *The first Brillouin zone is the region in k space bounded by planes such that any electron with a* **k** *vector terminating inside the zone has too long a wavelength to suffer Bragg reflection from any set of lattice planes, while one with a* **k** *terminating on the zone boundary has a wavelength satisfying the Bragg condition.* An important conclusion brought out by the Brillouin zones is that the magnitude of **k** for which Bragg reflection occurs is not the same in all directions. Thus the energy at which forbidden energy gaps occur is a function of the direction in which an electron moves through the crystal.

22.5 The Fermi Surface

It is possible to show the allowed energy values for electrons in a crystal in k space, since we have, through Fig. 22.3 (and the equations from which it was calculated), a relationship between k and the kinetic energy. For k vectors which are small enough so that they do not come close to Brillouin zone boundaries, the allowed energies are very similar to those for free electrons. If we join all points in k space which correspond to the same energy, we obtain a set of energy contours (Fig. 22.8). (Once again we discuss a two-dimensional cubic array, but the ideas are readily extended to three dimensions.) For small k's, the constant-energy contours are circles in two dimensions (spheres in three). However, as k approaches a zone boundary, the corresponding energy increases very slowly (Fig. 22.3); consequently, the energy contours bulge toward the boundaries. Once the k vectors reach the zone boundary in the [10] and [01] directions, the corresponding energy contours terminate on the zone boundaries, which they approach at right angles. In Fig. 22.8 the largest value of k and the largest energy allowed in the first zone are associated with the [11] direction and the corners of the lattice.

In a crystal at 0°K electrons are all in the lowest possible energy states; they fill all states with energy less than or equal to the Fermi energy $E_F(0)$, while all levels with $E > E_F(0)$ are empty. In terms

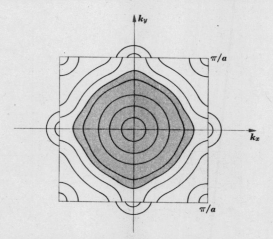

Fig. 22.8 Equal-energy surfaces in k space. At 0°K all electrons are in the lowest possible energy states; all levels with energy less than the Fermi energy E_F are filled. The energy surface associated with E_F is called the Fermi surface.

of the Brillouin zone of Fig. 22.8, this corresponds to occupation of all k states which lie inside the equienergy contour corresponding to $E_F(0)$. This latter contour in k space is known as the *Fermi surface*. In Fig. 22.8 the boundary of the shaded region pictures the Fermi surface in the case in which electrons occupy somewhat less than half the states in the Brillouin zone. For small k's, where the free-electron approximation is reasonable, the Fermi surface is spherical. As soon as any k vectors approach the Brillouin zone boundaries, the shape of the Fermi surface is modified.

Also shown in Fig. 22.8 are a few energy contours in the second Brillouin zone. Because of the energy gap at the zone boundary, these contours are not continuous with contours in the first zone. In some cases the lowest allowed energies in the second zone lie below the highest energies in the first zone; in other cases the first zone is completely filled before any states in the second zone are occupied. When the lowest energy in the second zone lies below the highest in the first zone, the two zones are said to *overlap*. In Fig. 22.9 are shown E-vs.-k curves which lead to overlapping and to separate zone patterns. To see what this implies for the Brillouin zones, imagine a crystal at 0°K from which all conduction electrons have been removed so that the first Brillouin zone is empty. Now adding a group of electrons would lead to a Fermi surface like that in Fig. 22.10a. If we add more, the Fermi surface

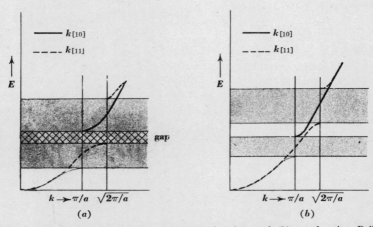

Fig. 22.9 E-vs.-k curves for (a) nonoverlapping and (b) overlapping Brillouin zones for a two-dimensional square lattice. The E-vs.-k curves differ in different directions, with the maximum variation between the $k_{[10]}$ and $k_{[11]}$ curves. In (a) the crosshatched range of energies is forbidden for every **k** direction, while in (b) every energy is allowed for at least some **k** direction.

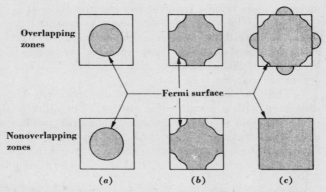

Fig. 22.10 When just enough electrons are present to fill the first Brillouin zone, in the case of   overlapping zones some of the electrons spill over into an overlapping zone where lower energy states are available.

becomes that in Fig. 22.10*b*.   When we have introduced enough electrons to fill the first Brillouin zone, the resulting Fermi surface depends on whether the zones are overlapping or not (Fig. 22.10*c*).   If they overlap, some of the electrons occupy states in the second zone, because they lie below the highest levels of the first zone.

22.6   Density of States

We have approached Brillouin zones through the Kronig-Penney model of a periodic array of square wells.   In each allowed state of a square well there can be at most two electrons with antiparallel spins.   As we bring square wells together, each energy level spreads into a band of energies; with each band we associate a Brillouin zone.   Thus, for each zone we may have a maximum of two electrons per square well; when we extend the theory to atoms, we associate each atom with a well and find that *two electrons per atom fill each Brillouin zone*.   For example, in a sodium crystal of 1 cm³ volume there are 2.4×10^{22} atoms, each of which contributes óne electron to the first Brillouin zone for conduction electrons. Since the zone can hold two electrons per atom, it is half filled.

For electrons with k small enough to lie far from the zone boundary in Fig. 22.8, the electrons behave almost as if they were free, and Eq. (21.11) represents a good approximation to the density-of-states function. As k approaches π/a, a small increase in E leads to a large increase in the number of k states (see Fig. 22.3) and the density-of-states function increases more rapidly than the free-electron curve (Fig. 22.11a), leading to a cusp. Once the zone boundary is reached, the only available states lie toward the corners of the Brillouin zone and the density of states decreases above the cusp, reaching zero when the zone is filled. When two Brillouin zones overlap as in Fig. 22.9, the density-of-states curves also overlap (Fig. 22.11b).

Thus far we have been discussing primarily the Brillouin zone associated with a simple cubic crystal. For other lattices the shapes of the Brillouin zones in k space are considerably different in detail, but most gross features are retained. In Fig. 22.12 the density-of-states function is shown for the body-centered cube and the face-centered cube. Of particular interest is the fact that the cusps lie at significantly different energies.

Density-of-state curves like those of Fig. 22.12 play an important role in understanding alloy systems. An example is found in the copper-zinc alloys (brass). Copper crystallizes in the face-centered cubic structure with one free electron per atom. When a small amount of zinc is added, the zinc atoms occupy Cu sites in the face-centered cubic structure; however, each Zn atom brings two electrons to the Brillouin zone, thus raising the Fermi level. If one assumes that the cusp for the face-centered cubic structure occurs when the Fermi surface (based on a free-electron model) first touches the Brillouin boundary, one finds that the cusp is associated with 1.36 electrons per atom. This suggests that

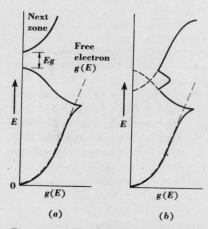

Fig. 22.11 The density of states $g(E)$ (abscissa) as a function of energy E (ordinate) for the cases of (a) nonoverlapping bands and (b) overlapping bands.

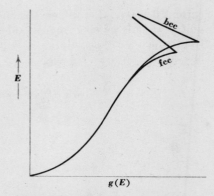

Fig. 22.12 The density-of-states function $g(E)$ for face-centered and body-centered cubic crystals.

when more than 36 percent of the atoms in the alloy are Zn, the Fermi energy should rise rapidly as more Zn is added. A similar calculation for the body-centered structure shows that the cusp occurs at 1.48 electrons per atom. Over the range from 1.36 to 1.48 the density of states is falling for the face-centered cubic but rising for the body-centered cubic structure. Thus, so far as electronic energy is concerned, the body-centered cubic structure becomes increasingly stable. Experimentally it is observed that brass changes from the face-centered cubic α phase to the body-centered cubic β phase at an electron concentration near 1.38. At higher electron concentrations the copper-zinc alloy system exhibits still other phases.

22.7 Filled Bands and Holes

If an electric field is applied to a crystal, electrons can achieve a net drift velocity with resulting current only if there are vacant allowed energy states available to them. For excellent conductors like copper, with one valence electron per atom, the first Brillouin zone is half filled, and there is a continuum of allowed energy states just above the Fermi surface to which electrons may be excited by the field.

On the other hand, when a nonoverlapping band is filled, no allowed energy states are accessible immediately above the Fermi surface; therefore, electrons cannot gain energy from the applied field. As a result, there is no net drift velocity and no current. If the band is not entirely full, a few electrons may be able to gain energy and produce a small current density. Let us consider the case in which a single electron per unit volume is missing from an otherwise filled band. When this electron, which we designate by i, is present, there is no current density and so

$$J = -e \sum_{j=1}^{n} v_j = -e \sum_{j \neq i}^{n} v_j - ev_i = 0 \qquad (22.14)$$

or

$$-e \sum_{j \neq i}^{n} v_j = +ev_i \qquad (22.14a)$$

In the absence of electron i, the summation on the left is the current density due to all the electrons present. It is equal to the right side, which we may interpret as a current density due to the motion of a single *positive* charge, which we call a *hole*. *A vacant quantum state in an almost filled band acts like a positively charged particle*, which is directly analogous to an electron. It is a great simplification to be able to consider a current to be carried by holes rather than focusing attention on the complex motion of a host of electrons.

The hole concept presents us with a way of explaining the positive Hall coefficients observed for many metals. On the basis of band theory one might at first expect that elements with s^2 outer electron configurations, for example, Be, Mg, Ca, Zn, might be electrical insulators since they have just enough electrons to fill the s^2 atomic level and the corresponding band. However, in all these materials the p band overlaps the s band considerably, so that there are p electrons as well as an equal number of holes in the s band. To see that we may get a positive Hall coefficient when we have equal numbers of positive and negative carriers, we derive a relation for the Hall coefficient for the case in which we have n electrons and p holes per unit volume. In this case we have for the current density J_x (Fig. 22.1)

$$J_x = -ne(v_n)_x + pe(v_p)_x = (ne\mu_n + pe\mu_p)\mathsf{E}_x \qquad (22.15)$$

where the subscripts p and n refer to holes and electrons respectively. If E_y represents the y component of the transverse Hall field due to B_z, holes carry net y current density $(J_p)_y = pe\mu_p[\mathsf{E}_y - (v_p)_x B_z]$, while for electrons $(J_n)_y = ne\mu_n[\mathsf{E}_y - (v_n)_x B_z]$. However, the equilibrium

$$J_y = (J_p)_y + (J_n)_y$$

must be zero, and therefore

$$pe\mu_p[E_y - (v_p)_x B_z] + ne\mu_n[E_y - (v_n)_x B_z] = 0 \qquad (22.16)$$

from which

$$E_y = \frac{[p\mu_p(v_p)_x + n\mu_n(v_n)_x]B_z}{p\mu_p + n\mu_n} = \frac{(p\mu_p{}^2 - n\mu_n{}^2)E_x B_z}{p\mu_p + n\mu_n} \qquad (22.16a)$$

By definition, $R_H = E_y/J_x B_z$; use of Eqs. (22.16a) and (22.15) yields

$$R_H = \frac{1}{e}\frac{p\mu_p{}^2 - n\mu_n{}^2}{(p\mu_p + n\mu_n)^2} \qquad (22.17)$$

In the case of metals with an s^2 atomic configuration the mobility of the s-band holes is considerably greater than that of the p-band electrons. Since the number of holes and the number of electrons are equal, R_H is positive for these metals.

The solid elements Bi, As, Sb, Se, and Te have primarily covalent binding, but the conduction band overlaps the valence band slightly, so that a small number of electrons go to the conduction band, leaving an equal number of holes in the valence band. Because of the small number of carriers, these solids are known as *semimetals*. Even at 0°K they are conductors, but not good ones. The sign of the Hall coefficient depends [by Eq. (22.17)] on which type of carrier has the higher mobility; for Bi it is electrons, but for As and Sb it is holes.

22.8 Transition Metals

Several members of the group of metals with incompletely filled d subshells are of particular economic and scientific interest. Among these *transition elements* are iron, cobalt, and nickel, which are ferromagnetic; their outer electron configurations are respectively $4s^23d^6$, $4s^23d^7$, and $4s^23d^8$ for isolated atoms. However, in the solid state the $4s$ level is spread into a very broad band (Fig. 22.13), which strongly overlaps the much narrower $3d$ band. Many of the unique properties of transition metals are associated with the overlap of the s and d bands.

In metallic nickel, on the average, 0.6 electron per atom is in the s band and 9.4 in the d band. Most of the conductivity of Ni is due to the s-band electrons: in atomic language the d electrons are

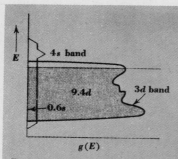

Fig. 22.13 Schematic diagram showing the overlap of the 3d and 4s bands in metallic nickel.

involved in bonds between atoms, while in free-electron language the d electrons have a high effective mass.   Further, the s electrons accelerated by an electric field undergo lattice collisions which may scatter them into vacant states in the d band.   As a consequence of the ease with which s electrons are scattered out of the band, nickel (and the other transition metals) has higher resistivity than elements with filled d shells; for Ni the resistivity is more than 4 times that for Cu, with its filled 3d band and one 4s electron.

Of the average 9.4 electrons per nickel atom in the 3d band, roughly 5 have one spin and 4.4 the antiparallel one.   As a consequence of the 0.6 unpaired spin per atom, Ni atoms in a crystal have a net magnetic moment of approximately 0.6 Bohr magneton.   Iron and cobalt atoms have higher average magnetic moments, viz., 1.7 μ_B for Co and 2.2μ_B for Fe.   Heisenberg has explained the ferromagnetic properties of these elements in terms of an exchange interaction between electrons which favors parallel orientations of the unpaired electron spins of neighboring atoms.   Zener has proposed a different theory with an interaction similar to the one that favors unpaired electrons in atoms having parallel spins (Hund's rule).

Problems

1.   In a Hall experiment on silver a current of 25 A is passed through a long foil which is 0.1 mm thick (in the direction of B) and 3 cm wide. Find the Hall voltage produced across the width by a magnetic induction

of 1.4 Wb/m². If the conductivity of silver is 6.8×10^{-3} mho/m, estimate the Hall angle and the mobility of electrons in silver.

Ans: 29 μV; 8×10^{-3} rad; 5.7×10^{-3} m²/V-s

2. A conductor has 5×10^{19} electrons and 8×10^{20} holes per cubic meter. If $\mu_p = 0.05$ and $\mu_n = 0.09$ m²/V-s, calculate the conductivity and the Hall coefficient.

3. From the Hall coefficient (Table 22.1) calculate the mobility of electrons in copper. From the mobility compute the relaxation time τ by use of Eq. (21.6) and show that the resulting value is compatible with the value of 2×10^{-14} s quoted in Sec. 21.7. The latter was calculated by assuming 1 electron per atom, while the Hall coefficient suggests 1.3. How can one explain such good agreement in τ under the circumstances? (The conductivity of Cu is 6×10^7 mho/m.)

4. Draw a diagram in k space showing the first seven Brillouin zones for a two-dimensional square lattice.

5. Draw a diagram showing the first three Brillouin zones for a two-dimensional rectangular lattice with $b = 2a$.

6. Show that in a two-dimensional square lattice the free-electron model predicts that an electron with its k vector terminating at the corner of the first Brillouin zone has twice the energy of an electron with a $k_{[10]}$ vector terminating on the zone boundary. In a simple cubic lattice what is the ratio of the lengths of the longest and shortest k vectors which reach the boundary of the first Brillouin zone? What is the corresponding energy ratio for free electrons?

7. An electron moves in the [43] direction in a two-dimensional square lattice with $a = 3.6$ Å. For this direction what is the distance in k space to the boundary of the first Brillouin zone? The second zone? The third zone? Calculate the energy of a free electron with each of these values of k.

8. Assume that the constant-energy surfaces for a simple cubic lattice are spheres until they first reach the boundary of the first Brillouin zone and that the cusp of the density-of-states function occurs at the energy corresponding to the first contact. Show that under these conditions the cusp comes when the zone holds 1.047 electrons per atom.

9. Copper crystallizes in a face-centered cubic structure. Calculate the lattice constant a for the Cu crystal. Find the shortest k vector which reaches the boundary of the first Brillouin zone. To what free-electron kinetic energy does this correspond? (There are 8.47×10^{28} atoms/m³ in Cu.)

Ans: 3.61 Å; 1.61 Å^{-1}; 9.9 eV

10. Find the volume of the first Brillouin zone for a face-centered cubic crystal of lattice vector a. Calculate the length of the shortest k vector which reaches the zone boundary. If the full zone holds 2 electrons and the cusp in the density-of-states function occurs for the shortest k to reach the zone boundary, show that the cusp comes at 1.36 electrons per atom.

Ans: $32\pi^3/a^3$; $\sqrt{3}\,\pi/a$

chapter twenty-three

Semiconductors

Metallic conductors have free carriers responsive to an applied electric field, even at $0°K$.   Indeed, for pure metals the conductivity is greatest at low temperatures, since the mean free path of electrons is least reduced by lattice collisions.   On the other hand, in good electrical insulators virtually all electrons are firmly held in atomic or valence bonds so that the conductivity is extremely low.   Between these extremes there is an increasingly important group of substances, the semiconductors, which have almost no free carriers at low temperature but which have, at room temperatures, a modest conductivity that increases with rising temperature.

23.1   Semiconducting Materials

A semiconductor is a material with a filled valence band and an empty conduction band at $0°K$ but with the energy gap between the bands small enough so that thermal excitation can create usable carrier densities at operating temperature.   There are many semiconducting crystals, but silicon and germanium are of central importance in practice.   Many semiconductors are binary chemical compounds.   Some are composed of one element from the third and one from the fifth column of the periodic table.   Such materials are called III-V compounds; examples are InAs, InSb, and GaP.   Still others are II-VI compounds such as CdS, CdTe, and ZnS.   These compounds frequently crystallize in the zinc blende structure (Fig. 21.4), which is identical with the diamond structure except that the two face-centered sublattices are occupied by different elements. One of the earliest practical uses of semiconductors was in rectifiers involving oxides of copper and aluminum.

The valence and conduction bands of the pure covalent group IV

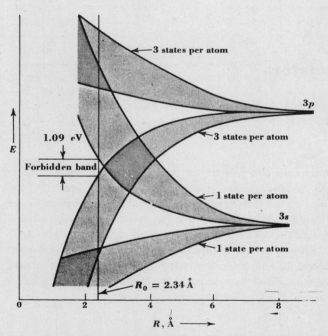

Fig. 23.1   Band structure of silicon as a function of nearest-neighbor separation. At normal spacing R_0 the 3s and 3p bands have crossed, and there is an energy gap of 1.1 eV.   The structure is similar for the 4s and 4p bands of germanium.   (*From H. D. Young, "Fundamentals of Optics and Modern Physics." Copyright 1968. McGraw-Hill Book Company.   Used by permission.*)

materials result from the hybridization of the valence orbitals of the free atoms (Sec. 21.4). In the crystalline state there is a mixing of the broadened *s* and *p* atomic levels in such a way that the eight (2s and 6p) states fall into two bands (each involving four original states), separated by a forbidden energy region (Fig. 23.1). In a crystal the lower (valence) band is filled at 0°K, while the upper (conduction) band is empty. The energy gap decreases with atomic number; for C (diamond) it is 5.3 eV, for Si 1.10 eV, for Ge 0.72 eV, and for gray tin 0.01 eV. (The precise width of the gap is different for *k* vectors in various directions; the values above are approximate minimum gaps.)

23.2 Intrinsic Semiconductors

A chemically pure specimen of any semiconducting material has properties which are characteristic of the material alone. Such a material is an *intrinsic semiconductor* in contrast to an *extrinsic* semiconductor, the conducting properties of which are intimately related to "impurities" deliberately introduced. At 0°K the energy bands of an intrinsic semiconductor, shown descriptively in Fig. 23.2a, consist of a filled valence band and an empty conduction band, separated by an energy gap E_g.

At room temperature a small fraction of the valence electrons are thermally excited to the conduction band. The conductivity of the semiconductor is due to these electrons and to the equal number of holes left in the valence band. To find the number of electrons per unit volume

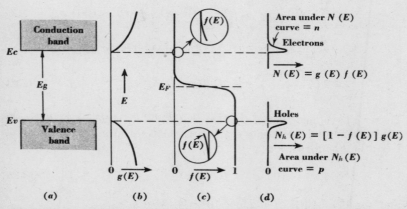

Fig. 23.2 (a) Schematic diagram of band structure in a particular direction for a semiconductor; (b) the density of states $g(E)$; (c) the Fermi function $f(E)$; and (d) the functions $N(E)$ and $N_h(E)$ for the semiconductor of (a).

in the conduction band at temperature T, we need to know the density-of-states function for the band and the Fermi energy E_F. Since we shall always be dealing with the situation in which the conduction band has few electrons compared to its capacity, it is a reasonable approximation to assume the free-electron functions of Eq. (21.11) (Fig. 23.2b)

$$g(E) \, dE = \frac{2^{\frac{3}{2}} \pi m^{*\frac{3}{2}}}{h^3} \sqrt{E - E_c} \, dE \qquad (23.1)$$

where we have taken cognizance of the fact that the bottom of the conduction band is now at E_c rather than at 0 and we have replaced the free-electron mass m by the effective mass. The probability that any state in the conduction band is occupied is given by the Fermi function [Eq. (20.16a) and Fig. 23.2c], and the number of electrons per unit volume in the conduction band with energy between E and $E + dE$ is (Fig. 23.2d)

$$N(E) \, dE = g(E) f(E) \, dE = \frac{2^{\frac{3}{2}} \pi m^{*\frac{3}{2}}}{h^3} \frac{\sqrt{E - E_c}}{e^{(E - E_F)/kT} + 1} \, dE \qquad (23.2)$$

where the Fermi energy E_F remains to be found.

To obtain the number n of electrons per unit volume in a conduction band, we must integrate $N(E) \, dE$ from the bottom of the conduction band E_c up to the highest occupied energy. Since $f(E)$ approaches zero for $E \gg E_F$, we may extend the integral to infinity without changing n significantly, so we write

$$\begin{aligned} n &= \int_{E_c}^{\infty} N(E) \, dE = \int_{E_c}^{\infty} \frac{2^{\frac{3}{2}} \pi m^{*\frac{3}{2}}}{h^3} (E - E_c)^{\frac{1}{2}} e^{-(E - E_F)/kT} \, dE \\ &= \frac{2^{\frac{3}{2}} \pi m^{*\frac{3}{2}}}{h^3} e^{-(E_c - E_F)/kT} \int_{E_c}^{\infty} (E - E_c)^{\frac{1}{2}} e^{-(E - E_c)/kT} \, dE \\ &= 2 \left(\frac{2\pi m^* kT}{h^2} \right)^{\frac{3}{2}} e^{-(E_c - E_F)/kT} \end{aligned} \qquad (23.3)$$

Here we have made use of the fact that $E_c - E_F$ is many kT for useful semiconductors in dropping the 1 in the denominator. If $m^* \approx m$, we have the useful relation

$$n = 4.8 \times 10^{21} T^{\frac{3}{2}} e^{-(E_c - E_F)/kT} \text{ m}^{-3} \qquad (23.3a)$$

and, for $T = 300°\text{K}$, $n = 2.5 \times 10^{25} e^{-(E_c - E_F)/kT} \text{ m}^{-3}$.

To find the Fermi energy E_F, we compute the number of holes per cubic meter and equate it to n. The probability that an allowed state at an energy E below E_F is unoccupied is given by

$$1 - f(E) = 1/(e^{(E_F-E)/kT} + 1)$$

We now assert, appealing to Fig. 22.10, that the density-of-states function at the top of the valence band is of shape similar to that at the bottom of the conduction band (Fig. 23.2b), described analytically by substituting $\sqrt{E_v - E}$ for $\sqrt{E - E_c}$ in Eq. (23.1) and the appropriate effective mass m_h for m^*. If $N_h(E)\, dE$ represents the number of holes per unit volume with energy between E and $E + dE$ and p is the number of holes per unit volume, we have, in close analogy with Eq. (23.3),

$$p = \int_{-\infty}^{E_v} N_h(E)\, dE = \int_{-\infty}^{E_v} \frac{2^{\frac{1}{2}}\pi m_h^{\frac{3}{2}}}{h^3} \sqrt{E_v - E}\, e^{-(E_F-E)/kT}\, dE$$

$$= 2\left(\frac{2\pi m_h kT}{h^2}\right)^{\frac{3}{2}} e^{-(E_F-E_v)/kT} \tag{23.4}$$

From $p = n$ for an intrinsic semiconductor,

$$m_h^{\frac{3}{2}} e^{-(E_F-E_v)/kT} = m^{*\frac{3}{2}} e^{-(E_c-E_F)/kT}$$

or

$$E_F = \frac{E_v + E_c}{2} + \frac{3kT}{4} \ln \frac{m_h}{m^*} \tag{23.5}$$

For silicon, germanium, and many other semiconductors the last term of Eq. (23.5) is small compared with the first, and an error of only a few hundredths of an electron volt is introduced by neglecting it. For such a semiconductor *the Fermi level lies at the middle of the energy gap between the top of the valence band and the bottom of the conduction band.* Therefore, if E_g is as much as 0.5 eV, $e^{(E_c-E_F)/kT}$ and $e^{(E_F-E_v)/kT}$ are at least e^9; since $e^9 \gg 1$, dropping the 1 in the Fermi functions above is well justified.

23.3   Conductivity

For germanium $E_g \approx 0.72$ eV, and at room temperature, where $kT = \frac{1}{40}$ eV, n and p are approximately $2.5 \times 10^{25} e^{-14.4} \approx 3 \times 10^{19}$ m^{-3}. The number of atoms per unit volume is 4.45×10^{28} m^{-3}, so that there is one electron in the conduction band and one hole in the valence band for every 1.5×10^9 atoms (compared with one electron per atom in copper). In spite of the small number of carriers the conductivity of Ge is 2 mhos/m compared with 6×10^7 for Cu. The conductivity is given by [see Eq. (22.4)]

$$\sigma = ne\mu_n + pe\mu_p \tag{23.6}$$

That the ratio of the conductivity of Cu to that of Ge is 3×10^7 while the corresponding ratio of conduction electron density is 2×10^9 comes primarily from the fact that the mobility of electrons in Ge ($\sim$0.4 m²/V-s) is about 100 times that in Cu. The mobility of holes in Ge is roughly 0.2 m²/V-s, so that electrons carry about two-thirds of the current while holes carry the remainder.

One method of determining the energy gap E_g in an intrinsic semiconductor is to measure σ as a function of T over a broad range. From Eqs. (23.3a), (23.4), and (23.5) we have the approximate result that

$$n = p = 4.8 \times 10^{21} T^{\frac{3}{2}} e^{-E_g/2kT} \tag{23.7}$$

Combining this with Eq. (23.6) yields

$$\sigma = 4.8 \times 10^{21} e(\mu_n + \mu_p) T^{\frac{3}{2}} e^{-E_g/2kT} \tag{23.7a}$$

A plot of $\ln(\sigma/T^{\frac{3}{2}})$ as a function of $1/T$ leads to an approximately straight line of slope $-E_g/2k$. Thus E_g can be computed directly from the measured slope. The deviation from linearity of the curve is due, in part at least, to the fact that the mobilities are not independent of temperature.

An optical method for determining E_g involves measurement of the linear absorption coefficient of a thin film of the semiconductor as a function of the wavelength of incident radiation. For such a film the intensity I which penetrates a thickness x of the material is given by $I = I_0 e^{-\alpha x}$, where I_0 is the intensity at $x = 0$ and α is the linear absorption coefficient. A plot of α as a function of λ for a particular germanium specimen is shown in Fig. 23.3. Here the radiation is in the infrared region. As λ decreases toward 2 μ, the absorption begins to increase, but at $\lambda = 1.7$ μ it rises abruptly. The energy of a 1.7 μ photon is 0.72 eV, and the sharp rise at this wavelength is due to the fact that 0.72-eV photons can excite electrons from the valence band to the conduction band while 0.6-eV photons cannot. That there is some absorption at low photon energies is associated with other possible processes. Because of the presence of free electrons in the conduction band, most semiconductors exhibit a metallic luster, but

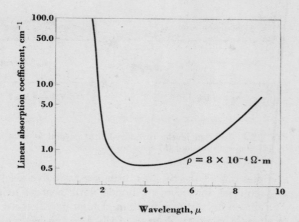

Fig. 23.3 Linear absorption coefficient of a lightly doped *n*-type germanium crystal as a function of wavelength of infrared radiation.

the number per unit volume is insufficient to make the semiconductor an excellent reflector.

23.4 *Extrinsic Semiconductors*

In practice the most important semiconductors are those in which certain atomic impurities have been incorporated to introduce new allowed energy levels. To be specific, consider the result of adding a few parts per million of valence-5 arsenic to a silicon crystal. Each As atom

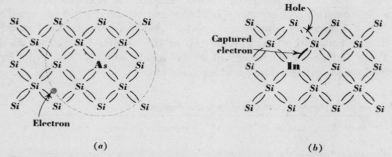

(a)   *(b)*

Fig. 23.4 *(a)* An arsenic atom brings five electrons to a lattice site for which four are desired for covalent bonds; the fifth is readily excited into the conduction band. *(b)* An indium atom brings three electrons; a fourth is readily captured to complete the fourth covalent bond.

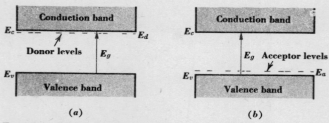

Fig. 23.5 Isolated levels introduced into crystal at sites occupied by (*a*) donor atoms and (*b*) acceptor atoms.

occupies one of the Si sites, bringing to it five electrons when only four are required for covalent bonds with nearest neighbors (Fig. 23.4*a*). The silicon crystal is a medium of high (≈ 12) dielectric constant; partly as a result, the fifth electron of the As atom has a binding energy of only 0.049 eV. By virtue of thermal excitation, this electron is rarely held by the impurity atom and is in the conduction band most of the time. For this reason the As atom is referred to as a *donor*. A semiconductor doped by adding donor atoms is said to be of *n* type, because the conduction is primarily by electrons from the donors. In the energy diagram (Fig. 23.5*a*) donor atoms introduce isolated donor levels near the bottom of the conduction band.

When a semiconductor is doped by adding a few parts per million of a valence-3 element such as indium, the impurity atom brings three electrons but occupies a site calling for four covalent bonds. Such an atom can readily capture an electron (Fig. 23.4*b*), thereby establishing the fourth bond; the atom serves as an *acceptor*. The captured electron is not available for conduction, but there remains a hole in the valence band from which the electron was excited. A material doped with acceptors conducts primarily by holes and is called a *p*-type semiconductor. Acceptors introduce isolated energy levels in the forbidden band about 0.01 eV above the top of the valence band (Fig. 23.5*b*). Table 23.1 shows the energy of levels introduced in Ge and Si by light dopings of common donor and acceptor atoms.

Table 23.1 *Levels introduced by light dopings*

Host Crystal	Donor Ionization Energy $E_c - E_d$, eV			Acceptor Ionization Energy $E_a - E_v$, eV			
	P	As	Sb	B	Al	Ga	In
Ge	0.0120	0.0127	0.0096	0.0104	0.0102	0.0108	0.0112
Si	0.045	0.049	0.039	0.045	0.057	0.065	0.16

Addition of either donors or acceptors increases the conductivity above the intrinsic level. In the n-type material, electrons are *majority carriers*, while holes are *minority carriers*. The situation is reversed in p-type conductors, where the majority carriers are holes.

23.5 The Fermi Level in Extrinsic Semiconductors

Adding donors to an intrinsic material raises the Fermi level above the level of the gap by an amount which depends both on the doping and on the temperature. To see how and why the Fermi level shifts with temperature, consider an n-type material at $0°K$, when the crystal is in its ground state. All donor levels are occupied, and there are no electrons in the conduction band. Since the occupation index $f(E)$ is 1 up to the donor levels and 0 at the conduction band, the Fermi level must lie somewhere in the range $E_d \leq E_F \leq E_c$. It is not typically midway between E_d and E_c, because the density-of-states functions lack the symmetry shown by the intrinsic semiconductor. As T is increased, the donors are rapidly excited, and soon half of the donor states are emptied, at which temperature the Fermi level coincides with the donor levels. As the temperature rises further, electrons are excited from the valence band. Eventually the number of electrons from donors is a trivial fraction of the electrons in the conduction band. At this point the Fermi level has descended to almost the center of the energy gap (Fig. 23.6) and the semiconductor behaves essentially as though it were intrinsic. Similarly, in a p-type material the Fermi level moves from between E_v and E_a at $0°K$ to $(E_v + E_c)/2$ at high temperature.

As an example, let us compute the Fermi level for a Ge crystal doped with gallium at the rate of 3.7×10^{22} atoms/m^3 at $300°K$, where $kT = 0.026$ eV. By conservation of charge, the sum of the number of electrons in the conduction band and the number of ionized acceptors is equal to the number of holes in the valence band; therefore, if we let $E_v = 0$,

$$2.5 \times 10^{25} e^{-(0.72 - E_F)/0.026} + \frac{3.7 \times 10^{22}}{e^{(0.0108 - E_F)/0.026} + 1} = 2.5 \times 10^{25} e^{-E_F/0.026}$$

$$2.5 \times 10^{25} \times 2 \times 10^{-12} e^{E_F/kT} + \frac{3.7 \times 10^{22}}{1.5 e^{-E_F/kT} + 1} = 2.5 \times 10^{25} e^{-E_F/kT}$$

$$1.48 \times 10^{-3} = e^{-E_F/kT} \qquad \text{or} \qquad E_F = 0.17 \text{ eV}$$

In this case the Fermi level lies 0.17 eV above the top of the valence

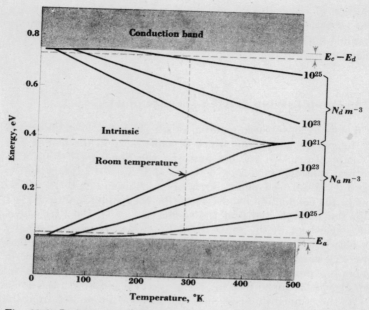

Fig. 23.6 Position of the Fermi level as a function of temperature for both *n*- and *p*-type germanium of various impurity concentrations. (*From L. V. Azároff and J. J. Brophy, "Electronic Processes in Materials." Copyright 1963. McGraw-Hill Book Company. Used by permission.*)

band and roughly one-fourth the distance to the bottom of the conduction band.

For a doped (or impurity) semiconductor, the carrier densities n and p are not equal, but their product is the same as the corresponding product for an intrinsic semiconductor so long as E_F is several kT from both conduction and valence bands.   Then

$$np = (4.83 \times 10^{21})^2 T^3 e^{-E_g/kT} = (np)_{\text{intrinsic}} \tag{23.8}$$

Even when a crystal is doped by adding acceptors, there are always some donor levels present.   However, as long as the density of acceptors exceeds that of donors, the donor electrons fill acceptor levels and the result is a *p*-type material with an effective acceptor density given by the difference in densities of acceptor and donor levels.   Thus donors and acceptors can "compensate" each other.   In fact, it is easier to fabricate specimens with equal densities of donors and acceptors than to produce them free from impurities; a doped crystal with intrinsic behavior is said to be *compensated.* If N_d^+ and N_a^- are the densities of ionized donors and acceptors respectively, the condition for charge neutrality requires that

$$n + N_a^- = p + N_d^+ \tag{23.9}$$

The temperature variation of n and p for an n-type semiconductor is shown in Fig. 23.7. Once the Fermi level lies a few kT below the donor levels, essentially all the donors are ionized, but over a considerable temperature range few electrons are excited from the valence band. Over this *exhaustion range* the conductivity σ varies much less rapidly with temperature than it does in the *impurity range* (low temperatures) or the *intrinsic range* (high temperatures). For Ge the exhaustion region commonly ends about 400°K, while for Si it is about 500°K. Above these temperatures the semiconductors behave essentially intrinsically, and the purposes of doping are no longer achieved.

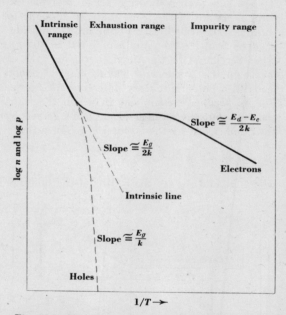

Fig. 23.7 Variation of the electron and hole concentrations in an n-type semiconductor with the reciprocal of the absolute temperature. (*From L. V. Azároff and J. J. Brophy, "Electronic Processes in Materials." Copyright 1963. McGraw-Hill Book Company. Used by permission.*)

23.6   The p-n Junction

The interface between a *p*-type region and an *n*-type region is called a *p-n* junction.   The transition region is typically about 1 μ in width and is produced in a single host crystal.   (If one attempts to make a *p-n* junction by pressing a piece of *p* material against one of *n* material, such a plethora of imperfections and associated energy levels in the forbidden region occur that the desired electrical properties of the junction are very unlikely.)   For the junctions we discuss below we shall assume that the junction width is small compared with the average distance a minority carrier (either electron or hole) diffuses in the crystal before it recombines with a majority carrier.

The equilibrium energy diagram for a *p-n* junction is shown in Fig. 23.8, where for convenience we place the arbitrary zero for energy at the top of the valence band in the *n* region, and where ΔE is then the energy corresponding to the top of the valence band of the *p* region. To see how such a diagram arises, imagine an ideal *n* crystal and an ideal *p* crystal of the same material brought together.   When equilibrium is established, the Fermi levels coincide.   This is achieved by electron transfer from the *n* material to the *p*, giving rise to a double layer of charge in the junction region with electrons filling acceptor levels on the *p* side and leaving ionized donors on the *n* side.   This leads to a potential-energy difference ΔE between the *n* and *p* regions. For an *abrupt junction*, one in which the transition region is narrow, the fact that the number of electrons gained by the *p* side is equal to the number lost by the *n* material requires that

$$N_d X_n = N_a X_p \qquad (23.10)$$

where X_n and X_p are the widths of the junction in the *n* and *p* region.

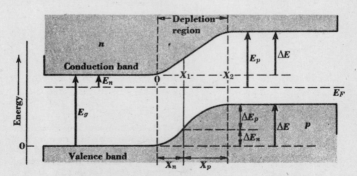

Fig. 23.8   Energy diagram for an unbiased *n-p* junction.

respectively, while N_d and N_a are the donor and acceptor densities. In writing Eq. (23.10) we are assuming that all donors in X_n have contributed electrons to fill all acceptor states in X_p, an approximation which is good for an abrupt junction. Over the entire width of the junction there are a reduced number of carriers, and we speak of it as a *depletion layer*.

The resultant charge density ρ in the region 0 to X_n is $N_d e$; associated with it is an electric field E and displacement D. By $\nabla \cdot D = \nabla \cdot \kappa\epsilon_0 E = \rho$, we have for $0 \leq x \leq X_n$

$$\kappa\epsilon_0 \frac{dE}{dx} = N_d e \tag{23.11a}$$

or

$$E = \frac{N_d e x}{\kappa\epsilon_0} \tag{23.11b}$$

since $E = 0$ at $x = 0$ (see upper part, Fig. 23.8). For $0 < x < X_1$ there is an electric intensity $E_n = N_d e x / \kappa\epsilon_0$ to the right. A similar treatment for the region $X_1 < x < X_2$ leads to electric intensity $E_p = N_a e(X_2 - x)/\kappa\epsilon_0$, also to the right. (In graphs like Fig. 23.8 the ordinate is always *electron energy*.) In the range $0 < x < X_1$ the electron energy rises by $\int_0^{X_1} (e^2 N_d x \, dx)/\kappa\epsilon_0$ and for $X_1 < x < X_2$ by $\int_{X_1}^{X_2} (e^2 N_a x \, dx)/\kappa\epsilon_0$, so that

$$\Delta E_n = \frac{N_d e^2 X_n^2}{2\kappa\epsilon_0} \tag{23.12a}$$

and

$$\Delta E_p = \frac{N_a e^2 X_p^2}{2\kappa\epsilon_0} \tag{23.12b}$$

From Eqs. (23.10) and (23.12) one sees that heavy dopings (great N_d and N_a) lead to narrow junctions and light dopings to broad ones.

23.7 The p-n Rectifier

When equilibrium is established at a *p-n* junction, electrons migrate at equal rates in both directions and so also do holes. While there is a vastly greater electron density in the conduction band of the *n* region, only electrons with kinetic energy greater than ΔE in Fig. 23.8 can surmount the potential barrier.   On the other hand, any electron reaching the junction of the p side finds the junction field exerting a force toward the n side.   The hole density in the p region exceeds that in the n region, but again the barrier layer equalizes the flow of holes. (Remember that for holes lower energy states lie upward, just opposite to those for electrons.)

When the Fermi levels on both sides of the junction coincide, the current I_{0e} carried by electrons moving from the p to the n region has a magnitude proportional to the electron density in the p region and may be written by Eq. (23.3a) as $C_1 e^{-E_p/kT}$.   This is balanced by an equal flow of electrons from the n to the p region which is proportional to the number of electrons in the n region with energy great enough to surmount the barrier ΔE.   By Eq. (21.14), this current has the magnitude $C_2 e^{-\Delta E/kT}$ and we have

$$I_{0e} = C_1 e^{-E_p/kT} = C_2 e^{-\Delta E/kT} \tag{23.13a}$$

Similar relations for holes can be found (with the aid of Fig. 23.8) to be

$$I_{0h} = C_3 e^{-\Delta E/kT} = C_4 e^{-(E_g - E_n)/kT} \tag{23.13b}$$

Let us now consider what happens when we apply an external bias voltage at the junction.   At this point we follow convention by defining the junction current to be positive when the p region is biased positive and the n region negative.   Then the conventional current is in the $-x$ direction of Fig. 23.9a and one speaks of the *forward current* associated with the forward bias.   For this case V is positive.   When the n region is biased positive, V is taken as negative and the junction is

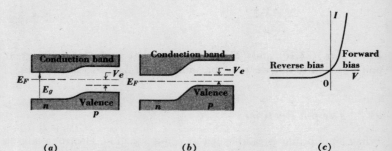

Fig. 23.9 An n-p junction (a) forward-biased and (b) reverse-biased;   (c) the current as a function of bias voltage.

reverse-biased (Fig. 23.9*b*); the resulting flow of charge is termed the *reverse current.*

When a bias voltage is applied to a *p-n* rectifier, most of it ordinarily appears across the junction region, which is depleted of carriers and therefore has high resistivity. We shall let V represent the bias across the junction region proper; the applied voltage must then be V plus the IR drops in the p and n regions on each side of the junction.

Whether the junction is biased or not, the currents $C_1 e^{-E_p/kT}$ and $C_4 e^{-(E_g - E_n)/kT}$ are unaffected, since they depend only on the appropriate carrier densities. However, the barrier for electrons going from n to p and for holes going from p to n material becomes $\Delta E - Ve$ (recall that V is positive for forward bias, negative for reverse bias). The resultant current at the junction is then given by

$$I = C_2 e^{-(\Delta E - Ve)/kT} - C_1 e^{-E_p/kT} + C_3 e^{-(\Delta E - Ve)/kT} - C_4 e^{-(E_g - E_n)/kT}$$
$$= (I_{0e} + I_{0h})(e^{Ve/kT} - 1) = I_0(e^{Ve/kT} - 1) \qquad (23.14)$$

where $I_0 = I_{0e} + I_{0h}$. The current-voltage characteristics of the junction are shown in Fig. 23.9*c*, which illustrates the strong rectifying property of the *p-n* junction. The current rises sharply for increasing forward bias but approaches $-I_0$ for increasing negative bias. (At high negative bias *avalanche* breakdown occurs when the carriers gain enough energy to create additional carriers in collisions.)

Not only does the *p-n* junction have an interesting current-voltage relationship, but it also has in some cases important capacitance characteristics. As we saw in Sec. 23.6, the junction region is depleted in carriers over a small distance and thus has the properties of an electrical capacitance per unit area of magnitude $C_A = \kappa\epsilon_0/w$, where κ is the dielectric constant of the crystal and w the junction width. Under reverse bias the junction width is increased, so that the junction capacitance decreases.

23.8   The Photovoltaic Effect

When photons fall on a *p-n* junction, electrons are lifted from the valence band to the conduction band provided the photon energy exceeds E_g, the gap energy. Thus additional carriers, both electrons and holes, are provided by the photons; in Fig. 23.10*a* only these additional carriers generated by the photons are indicated. Few of the additional electrons in the n region have enough energy to diffuse to the p region, but electrons at the p side of the junction are swept by the junction electric field to the n side; similarly holes are swept to the p side. Thus photons

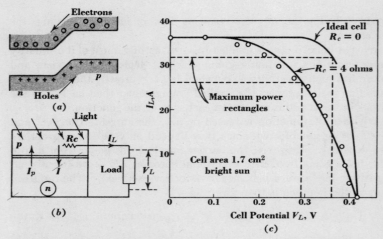

Fig. 23.10 (a) Flow of minority carriers driven by absorption of photons near a p-n junction. (b) Schematic diagram of a circuit with a photovoltaic emf. (c) Load current I_L in circuit of (b) as a function of the terminal potential difference V_L from the photovoltaic cell.

increase the minority-carrier densities substantially in both p and n regions and an appreciably greater minority-carrier transfer is produced across the junction.

When a p-n junction is reverse-biased with $|Ve| \gg kT$ and kept in the dark, the resulting reverse current as given by Eq. (23.14) is $I_R = I_0(1 - e^{-|Ve|/kT}) \approx I_0$, where I_0 is called the *dark current* in this application. When the junction is illuminated, photons generate new carriers which enhance the reverse current by an amount I_p, so that in the presence of the light the reverse current is $I_0 + I_p$. By cooling the junction, one can make $I_0 \ll I_p$. The resulting *photodiode* is a sensitive device for detecting and measuring beams of photons. Properly designed photodiodes are useful detectors of high-energy charged particles, x-rays, and γ rays as well as of infrared and visible radiation. For detecting in the far-infrared region a base material such as InSb with a small energy gap is required, since the gap energy must be less than the photon energy.

When no external potential difference is applied to an illuminated p-n junction, the photon-induced charge transfer results in producing a photovoltage V_p across the barrier layer with the n region negative and the p region positive. We can compute the photovoltaic emf in terms of I_p and I_0 as follows. Once equilibrium is established in the crystal in the absence of an external circuit, the forward current due to V_p just balances the photon-induced I_p, so that $0 = I_0(e^{V_p e/kT} - 1) - I_p$, from which

$$V_p = \frac{kT}{e} \ln\left(\frac{I_p}{I_0} + 1\right) \tag{23.15}$$

When $I_p \gg I_0$, the photovoltaic emf is proportional to the logarithm of the photocurrent I_p, a desirable response for a cell to be used as a light meter.

An important application of the photovoltaic effect is found in the *solar cell*, which converts radiant energy from the sun to electrical energy. The solar radiation curve is continuous with a broad maximum in the green region of the visible spectrum, corresponding to a photon energy of 2.5 eV. For conversion of solar radiation by a *p-n* junction an energy gap of 1.0 to 1.5 eV is desirable. Silicon with $E_g = 1.1$ eV has been widely used for solar cells. Since electrons and holes produced far from the junction ordinarily recombine, it is desirable that the junction of a silicon solar cell be very close to the surface; typically a *p* layer about 1 μ in thickness is formed at the surface which is subsequently covered with an antireflection coating similar to that used on lenses.

When a solar cell is used to supply power to a load (Fig. 23.10*b*), the photon-generated current I_p is equal to the load current I_L plus the forward junction current due to the induced potential difference V.

$$I_p = I_L + I_0(e^{Ve/kT} - 1) \tag{23.16}$$

The voltage V_L appearing across the load is the generated potential difference minus the potential drop $I_L R_c$, where R_c is the solar cell resistance, due almost entirely to the very thin *p* layer. Combining $V_L = V - I_L R_c$ with Eq. (23.16) leads to

$$V_L = \frac{kT}{e} \ln\left(\frac{I_p - I_L}{I_0} + 1\right) - I_L R_c \tag{23.17}$$

The current-voltage curve for a 1.7-cm² solar cell with $R_c = 4\ \Omega$ is shown in Fig. 23.10*c*. A good silicon solar cell may have an efficiency (electrical power delivered to load divided by total incident solar power) of about 15 percent.

23.9 The Tunnel (or Esaki) Diode

When junctions are very heavily doped ($\geq 10^{25}$ atoms/m³), the donor and acceptor levels are no longer sharp and isolated but are spread into bands, which overlap the valence and conduction bands (Fig. 23.11*a*). The Fermi level on the *n* side moves into the conduction band, and that in the *p* region moves into the valence band. Further, the junction width becomes very narrow (≈ 100 Å), so that quantum-mechanical tunneling may occur through the barrier at the interface. The tunnel current depends sensitively on the junction width.

At equilibrium the number of electrons tunneling in both directions through the barrier is the same. For a small forward bias (Fig. 23.11*b*) the number of electrons tunneling from the *n* to the *p* side is greatly increased because the number of empty states to which they can transfer is large; at the same time tunneling in the opposite direction is reduced. However, as the forward bias is increased, the states into which electrons may transfer by tunneling decrease and soon disappear (Fig. 23.11*c*). As *V* is further increased, the forward current is due to electrons which have enough energy to surmount the potential barrier for which Eq. (23.14) is applicable.

The resulting current-voltage curve for the tunnel diode is shown in Fig. 23.12. In the region between the peak and the minimum, the slope of the curve is negative, and the diode has a *negative dynamic resistance dI/dV*. Such a negative resistance is a very useful property. By suitable choice of components it is possible to produce a circuit with no dynamic resistance. A circuit containing an inductor, capacitor, and a biased tunnel diode can form a simple oscillator capable of generating frequencies as high as 10^{11} Hz.

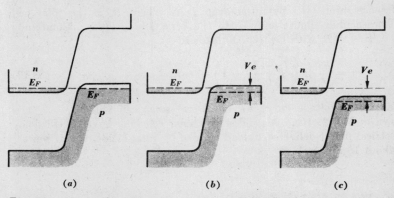

Fig. 23.11 A tunnel, or Esaki, diode (*a*) in equilibrium with no bias, (*b*) with a small forward bias (≈ 0.05 V in Fig. 23.12), and (*c*) with a large forward bias (≈ 0.2 V in Fig. 23.12).

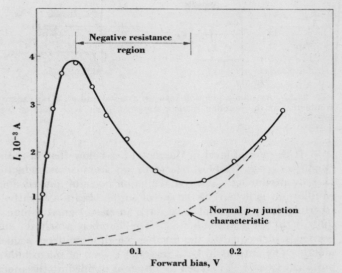

Fig. 23.12   The current as a function of applied potential bias for a tunnel diode.   (*After L. Esaki.*)

23.10   *Metal-semiconductor Junctions*

When one makes an electrical contact to a semiconductor crystal, the contact may act as a rectifier or as an ohmic connection, depending on the Fermi levels of the materials involved.   If the connection is to an *n*-type crystal, the contact is ohmic when the Fermi level of the metal lies above that of the semiconductor and rectifying when it lies below.   For a connection to a *p*-type semiconductor, the contact is ohmic when the Fermi level of the metal lies lower than that of the semiconductor and rectifying if it lies above.

Consider a junction formed between the metal and *n*-type semiconductor of Fig. 23.13.   When contact is made, electrons flow from the metal until the Fermi levels coincide, thereby charging the semiconductor negative and leaving additional electrons on the semiconductor side of the junction.   When a potential difference is applied in either direction, there are abundant carriers at the junction to give current in either direction and the current is proportional to the applied potential difference.   The junction is ohmic.

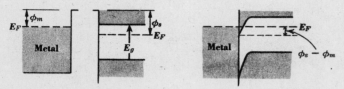

Fig. 23.13 An ohmic junction between a metal and an n-type semiconductor; at the junction carriers are plentiful on both sides of the junction.

If the Fermi level in the metal lies below that in the n-type semiconductor (Fig. 23.14a), electrons go from semiconductor to metal, thereby producing a positive depletion layer of ionized donors. When equilibrium is achieved, the metal surface bears a negative charge and electrons from the metal must have an energy greater than $E_F + \Delta E$ to diffuse to the semiconductor. The number which do surmount the barrier is just equal to the number which move from semiconductor to metal. The current in both directions, each of magnitude I_0, are proportional to $e^{-\Delta E/kT}$. When a voltage is applied at the junction, it does nothing about the height of the barrier on the metal side; that is determined entirely by the nature of the metal and of the semiconducting crystal. Because the depletion layer has reduced conductivity, most of the potential difference appears across it. For forward bias (n material negative) the Fermi level is raised relative to that in the metal (Fig. 23.14b), and the current associated with electrons going from semiconductor to metal increases by a factor $e^{Ve/kT}$. As a result, the forward current becomes

$$I = I_0(e^{Ve/kT} - 1) \tag{23.18}$$

For reverse bias, Eq. (23.18) is still applicable; V is then negative. Thus we see the junction current follows the same equation as that of a p-n junction.

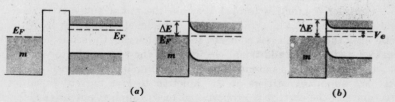

(a) (b)

Fig. 23.14 (a) A rectifying junction between a metal and an n-type semiconductor; there is a depletion layer at the junction. (b) The junction of (a) subject to forward bias.

When metal contacts are made to a *p*-type crystal, conduction in the semiconductor is primarily by holes. If the Fermi level of the metal lies above that of the semiconductor, electrons from the metal fill the holes at the junction and a rectifying depletion layer forms. When the Fermi level lies lower in the metal, electrons from the semiconductor go to the metal, giving additional carriers at the junction, which is then ohmic.

23.11 Thermoelectricity

When a current passes across the junction between dissimilar conductors, heat is either generated or absorbed at the junction (in addition to the joule heat liberated). This phenomenon is called the *Peltier effect*. The thermal energy H removed from the junction when a charge q passes from material a to material b is

$$H = \Pi_{ab}q \qquad (23.19)$$

where Π_{ab} is called the *Peltier emf* (or *coefficient*).

Consider the circuit of Fig. 23.15, in which a current is passed through an *n*-type semiconductor with ohmic contacts to the same metal at each end. (The double layer implied in Fig. 23.13 exists over such a short distance at each junction that it does not appear on this scale.) For the current direction shown, the junction at the right is cooled (H positive), while that at the left is heated (H negative). At the left contact, electrons going from semiconductor to metal enter the metal with energy greater than E_F in the metal. This energy is rapidly transferred to the metal lattice and appears as heat. At the right contact only electrons with energy greater than $E_F + \Delta E$ can enter the semi-

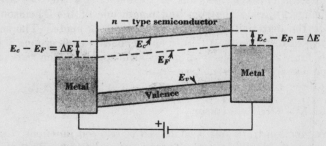

Fig. 23.15 When a current is passed through the semiconductor in the circuit, the junction at the right is cooled and that at the left is heated.

conductor. The metal loses electrons occupying higher energy states but not those of lower energy. This leads to a lowering of the average electron energy and to a cooling of the junction.

At the left junction, which is heated, H is negative for conventional current going from metal to semiconductor. On the average electrons fall from a level $E_c + 2kT$ in the semiconductor to energy E_F in the metal, with this difference appearing as heat. By (Eq. 23.19),

$$\Pi_{m \to n} = -\Pi_{n \to m} = -\frac{E_c + 2kT - E_F}{e} \tag{23.20}$$

where $e = +1.6 \times 10^{-19}$ C. The $2kT$ in this expression is the average kinetic energy of the electrons crossing the junction. The average kinetic energy of electrons in the conduction band is about $\frac{3}{2}kT$ (if $E_c - E_F > 5kT$), but electrons of higher speed are more likely to cross the junction. For p-type material Π is positive for current from metal to semiconductor and $\Pi_{m \to p} = -\Pi_{p \to m} = (E_F + 2kT - E_c)/e$.

Peltier emfs for metal-semiconductor junctions are typically much larger than for metal-metal junctions because the average potential energy of the carriers is larger than the Fermi energy in semiconductors, whereas the reverse is true for metals. The Peltier emf for doped semiconductors depends on the junction temperature, not only by virtue of the term $2kT$ but also because the position of the Fermi level in the energy gap varies with temperature. By careful choice of p and n semiconductors, it is possible to pass a conventional current from the n specimen to metal and then to the p specimen, thus cooling the metal at both contacts. With such an arrangement the metal can be cooled as much as 80°C below the temperature of the remainder of the circuit.

In addition to the Peltier emf, which appears at any junction between dissimilar conductors, there is also a *Thomson emf*, which appears between the ends of a conductor when they are at different temperatures. The sign and magnitude of the Thomson emf depend on the material of which the conductor is made. The sum (with due regard to sign) of all Peltier and Thomson emfs around a closed circuit is known as the *Seebeck emf*, which appears when two junctions in a circuit are at different temperatures.

23.12   The Transistor

A junction transistor is composed of two p-n junctions close together in the same single crystal. There are two basic types, the p-n-p and the n-p-n transistor, depending on the conduction type of the crystal between the two junctions. The operation of both types is fundamentally similar, but holes are the majority carriers for the p-n-p transistors and electrons for the *n-p-n*. Just as with vacuum tubes, a broad range of operating characteristics can be achieved by various choices of the many controllable variables in transistor design. We describe below some of the physics underlying the operation of an *n-p-n* transistor.

A common type of *n-p-n* transistor consists of a strongly doped *n* region separated from a moderately doped *n* region by a thin ($\approx 20\ \mu$) lightly doped *p* region. The energy-level diagram for equilibrium in the absence of applied voltages is shown in Fig. 23.16*a*. In operation, the stronger *n* region, known as the *emitter*, is forward-biased, so that it injects electrons into the *p* region, or *base*. It is very important that the diffusion length (average distance a minority carrier travels before recombining with a majority carrier) for electrons be very large compared to the thickness of the base, so that a very small fraction of the electrons entering the base from the emitter combine with holes. Almost all the electrons pass through the *p-n* junction into the *collector*, which is reverse-biased. The energy-level diagram for a biased transistor appears in Fig. 23.16*b*. It should be noted that the equilibrium concentration of electrons is not achieved in the base during normal operation, when most of the electrons injected from the emitter simply diffuse through the base. At the base-collector junction the reverse bias sweeps these electrons into the collector.

The ratio of the collector current I_c to the emitter current I_e is known as the *current gain factor* α, which approaches 1 in many well-designed transistors. This factor is the product of the *emitter efficiency* γ and the *base transport efficiency* ϵ. The emitter efficiency is the fraction of the emitter-junction current carried by majority carriers—in this case, electrons. It is to achieve a γ near unity that the emitter is strongly doped. For the *n-p-n* transistor the base transport efficiency is the ratio of the collector current to the electron current injected at

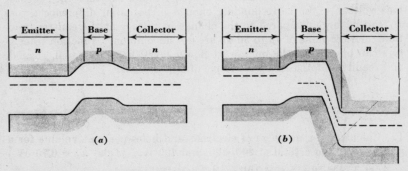

Fig. 23.16 The energy bands for *n-p-n* junction transistor (*a*) in equilibrium with no applied potential differences and (*b*) biased for operation as an amplifier.

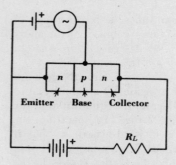

Fig. 23.17 A grounded-emitter circuit in which an *n-p-n* transistor serves as an amplifier.

the emitter junction; ϵ approaches unity when the base is very narrow and the diffusion length for electrons in the *p* material is large. It is advantageous to have modest doping for the base region, since heavier doping results in a shorter diffusion length for minority carriers. With a collector potential of only a few volts, the collector current becomes almost equal to the emitter current.

Transistors can perform many functions in electronic circuits. For example, the transistor of Fig. 23.17 is connected to serve as a grounded-emitter amplifier. The term *grounded-emitter* implies that the emitter is common to both input and output. (For specific purposes *grounded-base* and *grounded-collector* circuits are also used.) While a full analysis of an amplifier circuit requires knowledge of all the circuit components, it is easy to see qualitatively how amplification arises. A small change in the forward bias at the emitter junction produces a large gain in emitter current and a corresponding change in collector current if $\alpha \approx 1$. If R_L is large (but $I_c R_L$ remains less than the collector bias voltage!), the corresponding change in potential difference across R_L can be large compared to the change in voltage at the emitter junction. Power gains of 40 to 50 db are not uncommon. Junction transistors can operate at the microwatt power level; they are small, rugged, and readily adapted to a wide range of applications.

Problems

1. Calculate the numbers of electrons and holes per unit volume for a pure germanium crystal at 200, 300, and 400°K. (Take $E_g = 0.70$ eV.)

Ans: 3.6×10^{16}, 5.1×10^{19}, and 2.0×10^{21} m⁻³

2. Silicon is fairly transparent to infrared radiation of wavelength greater than 1.1 μ but absorbs shorter wavelengths strongly. From this information estimate the energy gap in silicon.

3. (*a*) Calculate the number of electrons and holes in intrinsic silicon at 300°K.

(*b*) If the mobilities of electrons and holes are 0.12 and 0.05 m²/V-s respectively, find the conductivity of Si at 300°K.

(*c*) Compute the expected Hall coefficient at 300°K.

Ans: (*a*) 1.5×10^{16} m⁻³; (*b*) 4.1×10^{-4} mho/m; (*c*) 170 m³/C

4. Assume that four electrons of an isolated As donor atom in a Ge crystal form valence bonds and that the fifth exists in the first Bohr orbit about an effective nuclear charge of $+e$ in a medium of dielectric constant 16. Find the binding energy and the radius of this orbit. Compare the radius with the nearest-neighbor separation in the Ge crystal.

Ans: -0.053 eV; 8.5 Å; 3.5 times nearest-neighbor distance

5. (*a*) Estimate the energy of the lowest bound state of the fifth valence electron of an arsenic atom in a silicon lattice (Fig. 23.4*a*), assuming that the specific inductive capacitance of Si is 12.

(*b*) Calculate the radius of the first Bohr orbit.

(*c*) Find the ratio of this radius to the interatomic spacing between silicon atoms if $a = 5.4$ Å.

6. Find the position of the Fermi level for a germanium semiconductor doped with phosporus at the rate of 5×10^{22} atoms/m³ at temperatures of 200, 300, and 400°K.

Ans: 0.10, 0.16, 0.23 eV below E_c

7. A germanium crystal is doped with phosphorus at 10^{23} atoms/m³ and with aluminum at 5×10^{22} atoms/m³. How does the conductivity of this crystal compare with that of the crystal of Prob. 6 when both crystals are at 300°K? Calculate the conductivity if the mobilities of holes and electrons are respectively 0.20 and 0.40 m²/V-s. How many times is the conductivity greater than that of pure Ge?

Ans: 3.2×10^3 mhos/m; 1600

8. Show that if a semiconductor has a zero Hall coefficient, the fraction f of the current carried by electrons is given by $f = \mu_p/(\mu_n + \mu_p)$. Find f for a specimen of silicon for which the mobility is 0.12 m²/V-s for electrons and 0.05 for holes.

9. An alternative way to explain the rectifying characteristics of a *p-n* junction is based on changes in the width of the depletion region. Discuss the rectification on that basis.

10. Estimate with the aid of Fig. 23.6 the internal potential difference $\Delta V \ (= \Delta E/e)$ in a germanium p-n junction at room temperature for an acceptor concentration of 10^{21} atoms/m^3 and a donor concentration of 10^{25} atoms/m^3. Calculate the width of the junction.

11. An intrinsic semiconductor has a band gap of 0.2 eV as determined by optical absorption. If it has a conductivity of 3 mhos/m at 290°K, calculate the conductivity at 400°K, assuming that the mobilities of both electrons and holes remain constant and that the effective masses of holes and electrons are equal.

12. (a) Show that the width w of an abrupt junction (given by $w = X_n + X_p$) is equal to

$$\frac{1}{e} \left[(2\epsilon_0 K \,\Delta E) \left(\frac{1}{N_a} + \frac{1}{N_d} \right) \right]^{\frac{1}{2}}$$

where the quantities are defined in Sec. 23.6.

(b) Show that when $N_a \ll N_d$,

$$w = \frac{1}{e} \left(\frac{2\epsilon_0 K \,\Delta E}{N_a} \right)^{\frac{1}{2}}$$

13. A p-n junction involves a double layer of opposite charges separated by a narrow depletion layer and thus acts as a capacitor. From the fact that a parallel-plate capacitor has a capacitance $C = \epsilon_0 K A/d$, where A is the area and d the separation, show that a p-n junction for which $N_d \gg N_a$ has a capacitance per unit area given approximately by

$$C_{\text{area}} = \left[\frac{\epsilon_0 K N_a e^2}{2(\Delta E + Ve)} \right]^{\frac{1}{2}}$$

where V is the potential difference applied at the junction. Thus the capacitance varies with applied voltage. (Use the results of the preceding problem.)

14. A silicon p-n junction has gallium at 10^{24} atoms/m^3 for N_a and arsenic at 2×10^{22} atoms/m^3 for N_d. What is the approximate value of ΔE for this junction? What is the approximate junction thickness?

15. Show that the average kinetic energy of electrons going "over the falls" from the n material of Fig. 23.15 to the metal is $2kT$.

chapter twenty-four

Interactions of High-energy
Particles with Matter

When high-energy charged particles and photons pass through a sub-stance, they interact with the atoms of the material in a variety of ways. Some understanding of the processes involved is a valuable preface to a discussion of nuclear physics, since in nuclear reactions the target nuclei, the bombarding particles, and the resulting products are far too small to see and too light to weigh on the most sensitive balance. Consequently, the unraveling of a nuclear reaction is often based on information inferred from indirect evidence involving interactions of the particles with matter through which they pass. We discuss first the processes by which high-energy photons are removed from a beam.

24.1   Attenuation of a Photon Beam

When photons have their origin in nuclear transitions, they are called γ *rays*. If they result from the acceleration of free electrons or other

635

charged particles, they are known as *bremsstrahlung* (braking radiation) or as x-rays. If they arise from interactions of primary cosmic-ray particles in the upper atmosphere, they are called *secondary cosmic rays*. If their origin is in the combination of a positron and an electron, they are described as *annihilation radiation*. The properties of the photons depend on the frequency and not at all on their origin.

When a narrow collimated beam of monochromatic high-energy photons passes through a thin film of thickness x, the intensity I transmitted in the same direction and at the same wavelength is given by

$$I = I_0 e^{-\mu x} \tag{24.1}$$

where I_0 is the incident intensity and μ is the *linear attenuation coefficient*. (It is sometimes referred to in the literature as the linear absorption coefficient; we prefer to reserve the term absorption coefficient to refer to that part of the energy actually converted to internal energy within the absorber.) An exponential attenuation law is characteristic of processes in which a particle is removed from a beam in a single event. In contrast, when α particles are absorbed in air, each one loses a small fraction of its energy in each of many encounters with air molecules; when all the kinetic energy has been dissipated, the α particle is said to have been stopped.

The ratio of the linear attenuation coefficient to the density of the absorber is the *mass attenuation coefficient* μ_m, which has the useful property that it is independent of the particular phase in which a given absorbing material may be. For example, it is the same for water vapor, water, and ice. If the mass attenuation coefficient for any element is multiplied by the atomic weight A and divided by Avogadro's number N_A, the result is the *atomic attenuation coefficient* μ_a

$$\mu_a = \frac{\mu_m A}{N_A} = \frac{\mu A}{\rho N_A} \tag{24.2}$$

The atomic attenuation coefficient is a useful concept in that it enables one to calculate the molecular attenuation coefficient of any type of molecule simply by adding the contributions of each of the atoms composing the molecule. For example, the molecular attenuation coefficient of copper sulfate is given by

$$\mu_{CuSO_4} = \mu_{Cu} + \mu_S + 4\mu_O \tag{24.3}$$

Thus one can calculate the molecular attenuation coefficient for any chemical compound for which one knows the chemical formula and the atomic coefficients of the atoms involved.

24.2 Attenuation Processes

X-ray photons can be removed from a well-collimated beam by a number of processes, of which photoelectric absorption is usually most important for energies in the lower range, scattering for intermediate energies, and pair production at high energies. In addition, photons of sufficiently high energy can be absorbed through nuclear reactions resulting in the ejection of a particle from a nucleus or in the fission of an atom. For example, 2.2-MeV photons can split a deuteron into a proton and a neutron, a process called *photodisintegration*. Ordinarily the cross section for such reactions is relatively small, and so we can write the atomic attenuation coefficient as

$$\mu_a = \tau_a + \sigma_a + \pi_a \tag{24.4}$$

where τ_a, σ_a, and π_a are respectively the atomic cross sections for photoelectric absorption, scattering, and pair production respectively. Which of these processes is dominant depends on the energy of the photons and on the atomic number of the absorber (Fig. 24.1).

a. Photoelectric Absorption   X-ray photons with energies in the keV range are absorbed primarily through the photoelectric process. For photon energies greater than the K binding energy, the atomic photo-

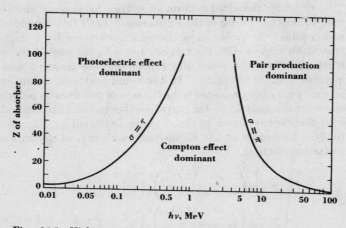

Fig. 24.1 High-energy photons are removed from a collimated beam by photoelectric absorption, scattering, and pair production. The lines show the values of Z and $h\nu$ for which the two neighboring effects are equal. (*From R. D. Evans, "The Atomic Nucleus." Copyright 1955. McGraw-Hill Book Company. Used by permission.*)

electric cross section of an atom decreases rapidly as λ decreases. A crude approximation is

$$\tau_a = CZ^4\lambda^3 \tag{24.5}$$

where C is a constant equal roughly to 2.6×10^{-30} m^2. For λ well below the K edge, τ_a varies more nearly as $\lambda^{3.5}$ for a modest range of wavelengths; the exponent then decreases so that for photon energy $E > 0.5$ MeV, τ_a is roughly proportional to λ. As photon energies increase, the exponent of Z also increases; for $E = 3$ MeV, τ_a goes as $Z^{4.6}$, and for very high energies Z^5 is probably a reasonable estimate. However, for energies in the MeV range, τ_a is less than the scattering cross section.

b. Scattering   Photons are also removed from a well-collimated beam by scattering processes, both of the Compton type (Sec. 7.10), in which part of the photon energy is transferred to an electron, and of the Thomson-Rayleigh type (Sec. 7.9). The latter is more probable for long wavelengths, while for high-energy photons Compton scattering is the more important type. In 1929 Klein and Nishina applied Dirac's relativistic formulation of quantum mechanics to achieve a unified theory of scattering which correctly predicts both the angular intensities and the polarization of scattered radiation.

Consider polarized quanta of energy $h\nu$ moving along the z axis with the electric vector in the x direction. Let I be the incident intensity and consider the radiation of energy $h\nu_\theta$ scattered by a single free electron at an angle θ with the z axis. Let ψ be the angle between plane of scattering (i.e., the plane determined by the incident and scattered quanta) and the xz plane. In the beam scattered at an angle θ, the intensity may be considered to be the sum of two linearly polarized beams, one with the electric vector perpendicular to the xz plane and the other with the electric vector in the xz plane, indicated by $I_{\theta\perp}$ and $I_{\theta\parallel}$ respectively. According to the Klein-Nishina theory, at distance r from the scattering electron,

$$I_{\theta\perp} = \frac{I}{4r^2}\left(\frac{e^2}{4\pi\epsilon_0 mc^2}\right)^2\left(\frac{h\nu_\theta}{h\nu}\right)^3\left(\frac{h\nu_\theta}{h\nu} + \frac{h\nu}{h\nu_\theta} - 2\right) \tag{24.6a}$$

$$I_{\theta\parallel} = \frac{I}{4r^2}\left(\frac{e^2}{4\pi\epsilon_0 mc^2}\right)^2\left(\frac{h\nu_\theta}{h\nu}\right)^3\left(\frac{h\nu_\theta}{h\nu} + \frac{h\nu}{h\nu_\theta} + 2 - 4\sin^2\theta\cos^2\psi\right) \tag{24.6b}$$

$^2/4\pi\epsilon_0 mc^2$ is the classical radius of the electron, Eq. (2.16). The total intensity I_θ scattered at angle θ from an unpolarized primary beam can be computed by averaging Eqs. (24.6a, b) over ψ and adding, which gives the general relation

$$I_\theta = \frac{I}{2r^2}\left(\frac{e^2}{4\pi\epsilon_0 mc^2}\right)^2\left(\frac{h\nu_\theta}{h\nu}\right)^3\left(\frac{h\nu_\theta}{h\nu} + \frac{h\nu}{h\nu_\theta} - \sin^2\theta\right) \qquad (24.7)$$

From (Eq. 7.16a) it follows directly that

$$\nu_\theta = \frac{\nu}{1 + \epsilon(1 - \cos\theta)} \qquad (24.8)$$

where $\epsilon = h\nu/mc^2$. Substitution of Eq. (24.8) into (24.7) yields

$$I_\theta = \frac{I}{2r^2}\left(\frac{e^2}{4\pi\epsilon_0 mc^2}\right)^2 (1 + \cos^2\theta)\left[\frac{1}{1 + \epsilon(1 - \cos\theta)}\right]^3$$
$$\left\{1 + \frac{\epsilon^2(1 - \cos\theta)^2}{(1 + \cos^2\theta)[1 + \epsilon(1 - \cos\theta)]}\right\} \qquad (24.9)$$

The intensity distribution for Klein-Nishina scattering as a function of θ is shown in Fig. 24.2 for several values of ϵ. The experimental points are those of Friedrich and Goldhaber for a wavelength of 0.14 Å scattered by carbon.

When $\epsilon(1 - \cos\theta) \gg 1$, the scattered radiation is unpolarized, even

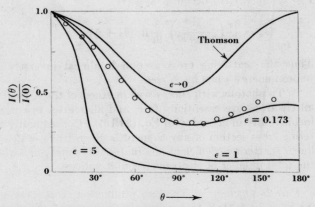

Fig. 24.2 Ratio of the intensity of the beam scattered at angle θ to the scattered intensity at $\theta = 0°$ for four values of $\epsilon = h\nu/mc^2$.

when the incident photons are polarized. In this case

$$I_\theta = I_{\theta\perp} + I_{\theta\parallel} = 2I_{\theta\perp} = \frac{I}{2r^2}\left(\frac{e^2}{4\pi\epsilon_0 mc^2}\right)^2 \frac{1}{\epsilon^2(1 - \cos\theta)^2} \quad (24.9a)$$

To obtain the total power P_s removed from the beam by scattering by an electron, it is not sufficient to integrate I_θ over a sphere surrounding the electron because each photon removed has energy $h\nu$, while the scattered photon has energy $h\nu_\theta$. Rather

$$P_s = \int_0^\pi I_\theta \frac{\nu}{\nu_\theta} 2\pi r^2 \sin\theta\, d\theta$$

By use of Eqs. (24.8) and (24.9) P_s can be found and from it the free-electron scattering cross section σ_e.

$$\sigma_e = \frac{P_s}{I} = 2\pi\left(\frac{e^2}{4\pi\epsilon_0 mc^2}\right)^2 \left\{\frac{1 + \epsilon}{\epsilon^2}\left[\frac{2(1 + \epsilon)}{1 + 2\epsilon} - \frac{1}{\epsilon}\ln(1 + 2\epsilon)\right]\right.$$
$$\left. + \frac{1}{2\epsilon}\ln(1 + 2\epsilon) - \frac{1 + 3\epsilon}{(1 + 2\epsilon)^2}\right\} \quad (24.10)$$

For $\epsilon \ll 1$, Eq. (24.10) may be expanded in a power series in ϵ.

$$\sigma_e = \frac{8\pi}{3}\left(\frac{e^2}{4\pi\epsilon_0 mc^2}\right)^2 (1 - 2\epsilon + 5.2\epsilon^2 - 13.3\epsilon^3 + \cdots) \quad (24.10a)$$

This reduces to the Thomson classical cross section as $\epsilon \to 0$. For $\epsilon \gg 1$, Eq. (24.10) becomes

$$\sigma_e = \frac{8\pi}{3}\left(\frac{e^2}{4\pi\epsilon_0 mc^2}\right)^2 \frac{3}{8\epsilon}\ln(2\epsilon + \tfrac{1}{2}) \quad (24.10b)$$

Thus the scattering cross section is almost inversely proportional to photon energy when $h\nu \gg mc^2$.

To photons with energy well in excess of the K-shell binding energy, all electrons are essentially free, and all scatter essentially equally and incoherently in accord with Eq. (24.10). Consequently the atomic scattering cross section of any element is given by $\sigma_a = Z\sigma_e$. Similarly, the linear scattering coefficient of a material is simply $n\sigma_e$, where n represents the total number of electrons per unit volume.

c. Pair Production   The pair-production cross section rises from zero for $h\nu < 1.02$ MeV to become the dominant factor in the attenuation coefficient for very high energies. For $h\nu < 50$ MeV the atomic pair-

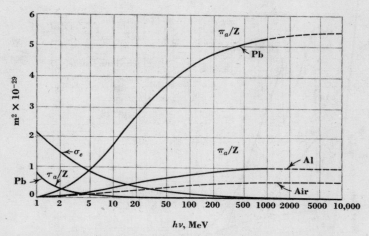

Fig. 24.3 Theoretical cross sections for the interaction of high-energy photons with atoms of lead, aluminum, and air (Z taken as 7.26 for air). For all three, $\sigma_a = Z\sigma_e$. The photoelectric cross section τ_a is shown only for Pb; it is inappreciable for Al and air for $h\nu > 1$ MeV. (*From data in W. Heitler*, "*The Quantum Theory of Radiation*," 2d ed., 1944, Oxford University Press.)

production cross section is proportional to Z^2. As the photon energy increases above the threshold value, π_a increases slowly at first and then more rapidly as the energy increases, finally reaching an essentially constant value at some hundreds of MeV. For $h\nu \gg 137m_0c^2/Z^{\frac{1}{3}}$, the cross section is given by

$$\pi_a = \frac{Z^2}{137}\left(\frac{e^2}{4\pi\epsilon_0 mc^2}\right)^2 \left(\frac{28}{9}\ln\frac{183}{Z^{\frac{1}{3}}} - \frac{2}{27}\right) \tag{24.11}$$

Theoretical values of π_a divided by Z are shown for three materials, in Fig. 24.3.

The mean distance L that a high-energy photon travels before producing a pair is $L = 1/\pi_a N$, where N represents the number of atomic nuclei per unit volume. Some values of L are as follows:

Energy, $h\nu$, MeV	Mean Distance for Pair Production L, cm		
	In Standard Air	*In* Al	*In* Pb
25	9.8×10^4	27.4	1.25
100	5.9×10^4	17.0	0.86
1000	4.5×10^4	13.2	0.70

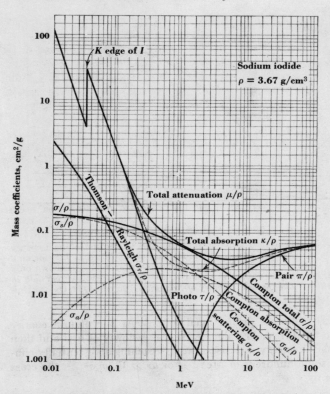

Fig. 24.4 Mass attenuation coefficients for sodium iodide. The scattering and absorption components of the Compton attenuation coefficient and the total mass absorption coefficient are shown by dashed lines. (*Adapted from R. D. Evans, "The Atomic Nucleus." Copyright 1955. McGraw-Hill Book Company. Used by permission.*)

For all elements a plot of the attenuation coefficient as a function of photon energy is decreasing for $h\nu = 1$ MeV. As the photon energy increases, the contribution of pair production grows rapidly. This leads to a minimum of μ_a at a photon energy which depends on Z. Minimum attenuation occurs for lead at $h\nu \approx 3.5$ MeV, for copper about 10 MeV, for aluminum about 25 MeV, and for air near 50 MeV. To summarize the processes of photon attenuation, Fig. 24.4 shows by solid lines the mass attenuation coefficients for sodium iodide, a material of great importance for scintillation detectors. The dashed lines are discussed in the next section.

24.3 Absorption vs. Attenuation

When a physicist measures attenuation coefficients, ideally he has a fine beam of single-energy photons (Fig. 24.5a) and considers as being removed from the beam every photon which fails to emerge with the same energy and same direction as it entered. In this case one speaks of a *narrow-beam* or *good-geometry* experiment. On the other hand, an experiment in which a significant fraction of the scattered or secondary particles can continue in the beam (Fig. 24.5b) is described as *broad beam* or *bad geometry*. In a broad-beam arrangement much of the incident intensity may leave the absorber with wavelength and direction modified. The photons which escape are not absorbed, since their energy does not remain in the absorbing material. If we consider the processes which go into making up the total attenuation coefficient, we observe that in the photoelectric absorption a photon is completely removed from the beam and its energy transformed into kinetic energy of the ejected photo-electron plus potential energy associated with the binding energy of that electron. Even so, some atom in the absorber is left in an excited state and may emit a photon of longer wavelength, which rejoins a broad beam. Similarly, a photon scattered through a small angle such as 10° can escape from the absorber in the broad-beam experiment, while it is counted as removed in a good-geometry experiment. A high-energy photon which creates pairs is removed from the beam, but the high-energy electron and positron may produce bremsstrahlung and annihilation radiation, which augment the broad beam. A dramatic example of this effect is to be found in the cosmic radiation. High in the upper atmosphere there is a flux of extremely high-energy photons, or secondary cosmic rays. Through pair production, followed by bremsstrahlung and annihilation radiation, each very high-energy photon on the average becomes a large number of lower-energy photons and associated secondary electron pairs. These offspring are much more strongly absorbed in air; as a consequence, the energy absorbed per gram of air increases over a substantial region as the cosmic rays descend through the atmosphere, even though

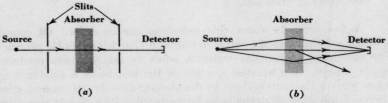

Fig. 24.5 An example of (a) narrow-beam and (b) broad-beam geometry.

the total intensity (energy incident per unit area per second) is actually decreasing. In both Compton scattering and pair production, a portion of the energy of the incident beam is absorbed and appears as heat in the absorbing layer. Another substantial portion may pass through with reduced photon energy.

In a layer of thickness dx the energy dI_{abs} *absorbed* per second per unit area is proportional to the intensity I which reaches the layer; hence

$$dI_{abs} = \kappa I \, dx \tag{24.12}$$

where κ is the linear absorption coefficient. Clearly κ is smaller than the linear attenuation coefficient, since the latter takes into account the energy content of all scattered photons as well as secondary photons from excited atoms and from pair bremsstrahlung and pair annihilation. For photons with energy of the order of 1.5 McV, about half the energy of a Compton scattered photon is absorbed on the average and half remains in the beam. In Fig. 24.4 the dashed curves show κ/ρ 'and how the Compton coefficient is divided between absorption and scattered photons.

The damage produced in a biological specimen or an inanimate object such as a semiconductor circuit depends primarily on the amount of energy per unit volume which is *dissipated* in the absorber rather than on the total energy *incident*. Indeed, photons which pass completely through without interacting with the specimen cause no damage. For discussing radiation damage it is convenient to define the *dose rate* for a sample exposed to a photon beam as the *energy absorbed per unit mass per unit time*.

In medical work the total x-ray energy falling on the subject is of minor interest, but the amount of energy absorbed per unit volume of the subject is of paramount importance. Therefore a unit of x-ray dosage, called the *roentgen*, was defined officially by the Fifth International Congress of Radiology in 1937 as follows:

> The roentgen shall be that quantity of x- or gamma-radiation such that the associated corpuscular emission per 0.001293 g of dry air produces, in air, ions carrying one electrostatic unit of quantity of electricity of either sign.

A few remarks about this definition are in order. Of course, 0.001293 g is the mass of 1 cm³ of dry air at 0°C and 760 mm Hg pressure. The associated corpuscular emission refers to the electrons produced by the x-rays. The greatest fraction of the ionization is produced by the secondary electrons and not by the photons directly; all the ionization produced by these electrons is to be collected. In a simple ionization chamber some of the photoelectrons produced in 1 cm³ of gas will leave this volume and produce part of their ionization in other portions of the chamber. These ions are to be collected and counted. The task of doing this cannot be solved by simply making an ionization chamber with a volume of 1 cm³ of dry air and collecting the resulting ionization, because in this case some of the photoelectrons would strike the wall and produce secondary electrons which would be counted. These secondary electrons are not ions produced by the associated corpuscular emission. The design of a reliable dosage meter involves careful consideration and balance among many factors.

Because photons of 100 keV energy are much more readily absorbed than 100-MeV photons, beams of equal intensity of monochromatic 100-keV and 100-MeV photons do not yield equal dosage rates; the 100-keV beam corresponds to many more roentgens (r) per unit time. A dosage of approximately 500 r from a 200-kV x-ray tube is sufficient to produce reddening in a typical human skin and is taken as the *threshold erythema dose*. A dosage of 700 r over the entire body is sufficient to produce death in approximately half of the human beings subjected to it; for this reason it is known as the *median lethal dose*.

Since the roentgen is defined for the passage of photons through air, it does not specifically involve biological effects; in this sense the roentgen is a measure of *exposure dose* to high-energy photons. The effect on a biological specimen is properly expressed by introducing an *absorbed dose* for which the rad[1] is the common unit. *The rad is the absorbed dose of any high-energy radiation which is accompanied by the liberation of 100 ergs of energy per gram of absorbing material* (or 0.01 J/kg). For photons with energies between 0.3 and 3 MeV 1 r of exposure dose corresponds closely to 1 rad of absorbed dose in tissue.

Although biological effects are produced by all ionizing radiations, the absorbed dose in rads required for a certain effect may be very different for different kinds of radiation. For example, 1 rad of neutrons is far more effective in producing cataracts in the eye than 1 rad of x-rays. This difference in response is taken into account by introducing the *relative biological effectiveness* (RBE), defined for any radiation as the ratio of the absorbed γ-ray dose in rads to the absorbed dose of the specified radiation which produces the same biological effect. The dose unit for biological effects is called the rem for *roentgen equivalent mammal*, where

$$\text{Dose in rems} = \text{RBE} \times \text{dose in rads}$$

[1] An older unit, the rep or *roentgen equivalent physical*, corresponds to the liberation of 97 ergs/g of body tissue.

24.4 *Energy Loss of Charged Particles*

When a charged particle with mass great compared with the mass of the electron moves through matter, it loses energy through electromagnetic interactions with electrons which are raised to excited states or torn away from atoms. The radii of atomic nuclei are so small compared with atomic dimensions that nuclear scattering and interactions are rare compared to interactions with electrons; therefore, nuclear reactions may be neglected in a first approximation.

In 1913 Bohr derived an expression for the space rate of energy loss for a charged particle on the basis of classical considerations. He considered a heavy particle, such as an α particle or a proton, of charge ze, mass M, and velocity V passing an atomic electron of mass m at a distance b (Fig. 24.6a). Bohr assumed that the electron was moving so slowly that it could be regarded as remaining at essentially the same point during the passing of the heavy particle. (Clearly this assumption is not valid if the electron acquires a speed comparable to that of the incident particle.) As the heavy particle passes, the electrostatic force acting on the electron changes direction continuously. However, if the electron moves negligibly during the passage of the heavy particle, the impulse $\int F_x\, dt$ parallel to the path is zero by symmetry, since for each position of the incident particle in the $-x$ direction there is a corresponding position in the $+x$ direction which makes an equal and opposite contribution to the x component of the momentum. However, throughout the passage, there is a force in the y direction, and the associated impulse I_y is given by

$$I_y = \int_{-\infty}^{\infty} F_y\, dt = \int_0^{\pi} \frac{ze^2 \sin \theta}{4\pi\epsilon_0 r^2} \frac{dx}{V} = \int_0^{\pi} \frac{ze^2 \sin \theta\, d\theta}{4\pi\epsilon_0 b V}$$

since $x = -b \cot \theta$ and $dx = (b\, d\theta)/(\sin^2 \theta)$.

The momentum transferred to the electron during the full passage is therefore

$$p = I_y = \frac{ze^2}{2\pi\epsilon_0 b V} \tag{24.13}$$

If the electron has not achieved a relativistic velocity, its kinetic energy is given by

$$K = \frac{p^2}{2m} = \frac{z^2 e^4}{8\pi^2 \epsilon_0^2 b^2 m V^2} \tag{24.14}$$

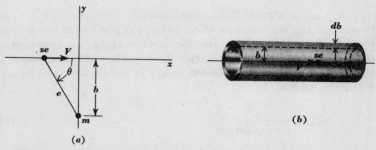

Fig. 24.6 (a) A heavy particle of charge ze passing an electron at distance b. (b) A shell of radius b and thickness db with its axis the path of the heavy charged particle.

To calculate the energy lost per unit path length by the heavy particle, we observe that the number of collisions per unit length for which b lies between b and $b + db$ is equal to the number of electrons per unit length in a shell bounded by cylinders of radii b and $b + db$ (Fig. 24.6b). If n represents the number of electrons per unit volume, the number of electrons per unit length in the shell is $2\pi bn\, db$ and the energy loss due to these electrons is

$$- \frac{dE(b)}{dx}\, db = \frac{z^2 e^4 n}{4\pi\epsilon_0^2 bm V^2}\, db$$

while the total energy per unit path length lost by the heavy charged particle to electrons in all shells in the range between b_{max} and b_{min} is

$$- \frac{dE}{dx} = \int_{b_{min}}^{b_{max}} - \frac{dE(b)}{dx}\, db = \frac{z^2 e^4 n}{4\pi\epsilon_0^2 m V^2} \ln \frac{b_{max}}{b_{min}} \qquad (24.15)$$

A glance at Eq. (24.15) shows why the integral over db is not taken from zero to infinity; if it were, the energy loss per unit path length would be infinite. It now becomes a challenge to select reasonable values for b_{min} and b_{max}. To choose a meaningful value for b_{min}, we observe that if the heavy particle collided head on with the electron, the maximum speed the electron could have is $2V$, since in an elastic collision the velocity of separation is equal to the velocity of approach. The corresponding maximum kinetic energy (for a nonrelativistic V) is $K_{max} = \frac{1}{2}m(2V)^2 = 2mV^2$. If this value of K_{max} is inserted in Eq. (24.14), the corresponding b_{min} becomes

$$b_{min} = \frac{ze^2}{4\pi\epsilon_0 m V^2} \qquad (24.16)$$

If b_{max} is allowed to become infinite, $-dE/dx$ goes to infinity because of the contribution of an unlimited number of small energies given to distant electrons. But the smallest energy an atomic electron can accept must be sufficient to raise it to an allowed excited state. If I represents the average excitation energy of an electron and we choose $K_{min} = I$, we find

$$b_{max} = \frac{ze^2}{2\pi\epsilon_0 \sqrt{2mV^2I}} \tag{24.17}$$

When Eqs. (24.16) and (24.17) are substituted in Eq. (24.15), we obtain

$$-\frac{dE}{dx} = \frac{z^2e^4n}{8\pi\epsilon_0^2mV^2} \ln \frac{2mV^2}{I} \tag{24.18}$$

This result is, of course, only a crude approximation. The value for b_{min} [Eq. (24.16)] is reasonable only in the nonrelativistic limit and only when the coulomb field of the heavy particle is essentially constant over the wave packet which represents the electron. This packet has a size about $\lambda/2\pi$, where λ is the de Broglie wavelength of the electron. In the center-of-mass frame the electron has a speed of almost V, and its de Broglie wavelength is $\lambda = h\sqrt{1 - V^2/c^2}/mV$, where m is the rest mass of the electron. Quantum-mechanically, only values of b larger than $\lambda/2\pi$ are meaningful, and this leads to the prediction that

$$b_{min} \approx \frac{\hbar \sqrt{1 - V^2/c^2}}{mV} \tag{24.19}$$

Relativistic quantum-mechanical treatments have been given by Bethe, by Bloch, and by others. According to Bethe, if $V < 0.95c$,

$$-\frac{dE}{dx} = \frac{z^2e^4n}{4\pi\epsilon_0^2mV^2} \left[\ln \frac{2mV^2}{I(1 - V^2/c^2)} - \frac{V^2}{c^2} \right] \tag{24.20}$$

The average excitation energy I is difficult to compute and is ordinarily treated as an adjustable parameter selected to fit experimental data. For most elements I in electron volts is roughly $13Z$, where Z is the atomic number of the stopping atoms; for the very light elements hydrogen, helium, and beryllium the experimental values of I are 19, 44, and 64 eV, respectively, for incident particles with energies in the MeV range.

The energy loss per unit path length has a minimum when $V \approx 0.96c$.

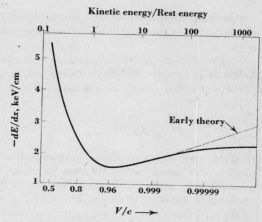

Fig. 24.7 Energy loss per unit path length for a heavy singly ionized particle passing through oxygen (760 mm, 0°C) as a function of particle speed (*lower scale*) and of the ratio of kinetic energy to rest energy (*upper scale*).

For higher speeds the early theories predicted (Fig. 24.7) that $-dE/dx$ should increase as the logarithm of the energy of the heavy particle because of a relativistic increase in the radius of action of the electric field of the penetrating particle. However, in 1938 Swann observed that polarization effects would screen distant electrons, and in 1940 Fermi developed a theory showing that the polarization effects depend primarily on the electron density n. As a result of the polarization and the *density effect*, $-dE/dx$ is almost constant in the extreme relativistic range.

The quantity $-dE/dx$ is called the *linear stopping coefficient* or the *stopping power* for the material through which the heavy charged particle passes. For a particle with $V < 0.95c$, this space rate of energy loss is (1) proportional to the number of electrons per unit volume in the stopping material and (2) proportional to the square of the charge and roughly inversely proportional to the square of the velocity of the incident particle.

The stopping coefficient depends on the charge and the speed (but not on the mass) of the incident particle. Hence a measurement of the stopping power of a material for protons as a function of speed permits one to calculate the stopping power of a material for other charged particles, such as α particles, muons, or deuterons, provided $V > \sqrt{I/2m}$, and provided the penetrating particle is moving so rapidly that it does not pick up or lose electrons. Since $|dE/dx|$ increases as V decreases, slower particles produce more ion pairs per unit path length than faster ones.

The distance a charged particle moves through a stopping material before it is brought to rest is called the *range* of the particle. For a particle of initial energy E_i, the range R is

$$R = \int_0^{E_i} \frac{dE}{-dE/dx} \tag{24.20a}$$

In early experiments on nuclear transmutations, the energy of an emitted particle was usually deduced from its range in air at 15°C and 1 atm pressure (Fig. 24.8). In 1910 Geiger found that the range of natural α particles in air is given in centimeters by $R \approx 0.32E^{\frac{3}{2}}$ (E is the energy in MeV); *Geiger's rule* is a good approximation for 4 MeV $< E <$ 10 MeV.

As an α particle approaches the end of the range, it may capture

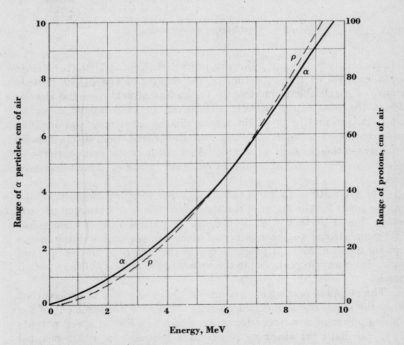

Fig. 24.8 Range-energy relations for α particles and protons in standard air (760 mm, 0°C). (*Data from E. Segrè, "Experimental Nuclear Physics," vol. 1, John Wiley & Sons, Inc., New York, 1953.*)

and lose electrons very rapidly; thus the effective charge varies between 0 and 2. An α particle in air loses energy at a maximum rate when its residual range is about 0.4 cm; then it produces 6600 ion pairs per millimeter of path length, compared to about 2000 when it has many MeV of energy. The number of ion pairs per unit path length is called the *specific ionization* of the particle. For protons the specific ionization has a maximum at 2750 ion pairs per millimeter when it is about 1 mm from the end of its range in air.

Although magnetic and electrostatic deflection methods have long since displaced range measurements for accurate energy determinations, the energy loss of charged particles as they pass through matter continues to be a topic of considerable importance. In using the better techniques, it is still necessary to make corrections for the energy loss of particles in layers of material on the source or in layers through which the particles pass. To make such corrections it is necessary to know the differential rate of energy loss with path.

24.5 *The Stopping of Electrons*

A high-energy electron passing through matter loses energy both by ionization and excitation of atoms and by the emission of electromagnetic radiation. For energies up to a few MeV the former processes are dominant, but at very high energy radiation losses are the more important.

There are two principal reasons why Eq. (24.20) is not applicable to electrons: (1) in the derivation it was assumed that the incident particle was undeflected, while when an incident electron interacts with another or with a nucleus, the transverse momentum given to the incident particle is far from negligible; there is considerable scattering of the electrons in encounters with electrons and nuclei; (2) for collisions between identical, and hence indistinguishable, particles one cannot tell which of the two final electrons was the incident one; in the quantum-mechanical theory there exist important exchange phenomena which alter the results substantially. The ionization-excitation loss for electrons is given by a formula due to Bethe:

$$
-\left.\frac{dE}{dx}\right|_{\text{ion}} = \frac{ne^4}{8\pi\epsilon_0^2 mV^2}\left[\ln\frac{mV^2K}{2I^2(1-\beta^2)}\right.
$$
$$
\left. - (2\sqrt{1-\beta^2}-1+\beta^2)\ln 2 + 1 - \beta^2 + \cdots\right] \quad (24.21)
$$

where K is the kinetic energy, V the speed of the electron, and $\beta = V/c$; all other symbols have the same meaning as in Eq. (24.10).

Bethe and Heitler have shown that when $K \gg mc^2$, the differential energy loss due to bremsstrahlung (continuous x-radiation primarily from acceleration of the electrons in the electric fields of nuclei) is given by

$$-\left.\frac{dE}{dx}\right|_{\text{rad}} \approx \frac{4Ze^6NK(Z + 1.3)[\ln{(183Z^{-\frac{1}{3}})} + \frac{1}{8}]}{4\pi\epsilon_0{}^3\hbar m^2c^5} \tag{24.22}$$

where N is the number of atoms per unit volume of atomic number Z. The radiation loss is proportional to the kinetic energy K of the electron (or positron), whereas the ionization loss is proportional to $\ln K$ and inversely to the square of the velocity. Thus the radiation loss increases very much more rapidly with K than the ionization loss. For lead the radiation loss is equal to the ionization loss at 7 MeV, while for hydrogen the equality comes at 340 MeV.

In the extremely high energy range the energy loss per unit path length is proportional to the kinetic energy K. This leads to an exponential loss of energy as the electron or positron passes through a material.

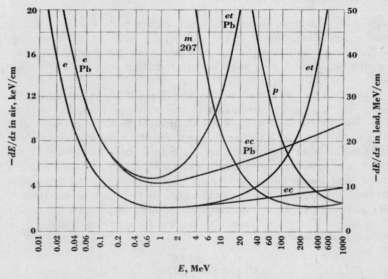

Fig. 24.9 Theoretical curves for energy loss per unit path length by charged particles as a function of the kinetic energy E of the particle. Curves *e-ec* and *e-et*: collision loss and total (including radiation) energy loss for electrons in standard air. Curves *e*Pb-*ec*Pb and *e*Pb-*et*Pb: ionization-excitation loss and total loss for electrons in lead. Curves *m*-207 and *p*: total loss in air by a muon and by a proton.

In this case $-dE/dx = -dK/dx = K/\Lambda$, where Λ is called the *radiation length*. In the region where this relation holds, $K = K_0 e^{-x/\Lambda}$. For various materials it is customary to tabulate $\Lambda\rho$ in grams per square centimeter, where ρ is the density of the stopping material. For air and water $\Lambda\rho$ is about 36 g/cm² when $K > 85$ MeV; for iron $\Lambda\rho = 14$ g/cm² when $K > 24$ MeV; for lead $\Lambda\rho = 6$ g/cm² when $K > 7$ MeV. Figure 24.9 shows curves for the energy loss of electrons in air and in lead along with air data for a muon and a proton for comparison.

24.6 Čerenkov Radiation

When a charged particle moves through a medium with refractive index n with a speed greater than the speed c/n of light in the medium, coherent light is emitted which is analogous to an acoustic shock wave and to the bow wave of a ship. This phenomenon was discovered and investigated quantitatively by Čerenkov in 1934. Three years later Frank and Tamm explained the Čerenkov radiation on the basis of classical electromagnetic theory. From the early days of radioactivity it was known that various substances, such as solutions of mineral salts, emitted feeble light when exposed to the radiations from radioactive materials. In some cases the light was fluorescent in origin. In 1929 Mallet recorded ultraviolet spectra from water irradiated by γ rays, which came from Čerenkov radiation although the latter had not yet been identified.

A particle of charge ze moving through a dielectric medium has an accompanying electric field that temporarily polarizes the atoms. They follow the waveform of the pulse and, in so doing, reradiate. If the speed V of the particle is less than the local speed of light, the scattered radiations from all segments of the track interfere destructively. However, when V exceeds c/n, scattered radiation from all points on the track can be in phase in certain directions. A coherent wavefront (Fig. 24.10) is propagated through the medium in the directions AC which satisfy the condition

$$\cos \theta = \frac{c/n}{V} = \frac{c}{nV} \tag{24.23}$$

The electric vector of the Čerenkov radiation is perpendicular to the surface of the cone.

According to the theory of Frank and Tamm, the energy loss per unit path length due to Čerenkov radiation is

$$-\frac{dE}{dx} = \frac{\pi z^2 e^2}{\epsilon_0 c^2} \int \left(1 - \frac{c^2}{V^2 n^2}\right) \nu \, d\nu \tag{24.24}$$

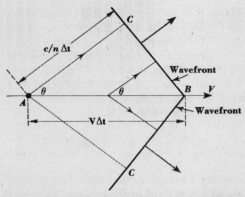

Fig. 24.10 A charged particle moving through a medium at a speed V greater than the local speed c/n of light emits Čerenkov radiation at angle θ such that $\cos \theta = c/nV$.

where the integral extends over all frequencies for which $Vn/c > 1$. This limits the radiation to frequencies less than those of the ultraviolet absorption bands of the medium. Below this limit the energy loss goes as $\nu \, d\nu$ (or $d\lambda/\lambda^3$), which reveals that the light is concentrated in the violet end of the spectrum. From Eq. (24.24) one finds that a fast ($V \approx c$) electron passing through water emits about 23 visible photons per millimeter of path length.

Čerenkov radiation is not a major cause of energy loss from the point of view of stopping the particle. For example, a 100-MeV electron in water loses about 2 MeV/cm in ionization, an equal amount in bremsstrahlung, and only 2.7 keV/cm in Čerenkov radiation. For a 1-GeV proton ($V/c \approx 0.87$) the ionization loss is again 2 MeV/cm, while the Čerenkov loss is about 1.7 keV/cm and the bremsstrahlung loss is only 10 eV/cm. Although the Čerenkov loss is small, it is extremely important because (1) the light is concentrated in the visible and near ultraviolet, (2) it appears at a well-defined angle and with well-defined polarization, and (3) the Čerenkov pulse is sharp ($\ll 10^{-10}$ s). These features make it possible to develop energy-selective directional counters, capable of detecting and of measuring the speed of high-energy charged particles.

24.7 Detection of Charged Particles

In most detecting devices, the passage of a charged particle is manifested by its ionization of the medium through which it passes. An

α particle, for example, in passing through a gas will both ionize and excite the atoms along its path by the action of its electric field on the atomic electrons, producing free electrons and positive ions.   In many gases—oxygen and chlorine are examples—the electrons so released immediately attach themselves to neutral atoms and form negative ions; in others, like the noble gases, the electrons may travel by themselves.   On the average, a charged particle loses about 34 eV of energy for each ion pair that is formed.   Both γ rays and x-rays can be detected through their interaction with atoms; in the photoelectric or Compton processes, for example, part of the energy of the radiation is converted into kinetic energy of an electron, and the electron is then detected through its ionizing power.

a. Ionization Chambers and Counters   In electrical detection methods, the ions produced by the passage of a charged particle are separated by means of an electrostatic field.   The ions move toward the electrodes and on arrival are registered by an electrometer or some type of vacuum-tube amplifier.   A typical arrangement is shown in Fig. 7.2.   Particles may enter through a thin window or, in the case of γ rays, secondary electrons may be ejected from the wall.   If a 5-MeV α particle, for example, is stopped in the gas of the chamber, it will produce about $5 \times 10^6/34 = 150,000$ ion pairs.   Supposing the wire and measuring system to have a capacity of 10 pF, the potential of the wire will be changed by

$$\frac{1.5 \times 10^5 \times 1.6 \times 10^{-19}}{10^{-11}} = 0.0024 \text{ V}.$$

With a reasonably good amplifier, a pulse of this size is easily detectable. It is ordinarily preferable to use a noble gas since electrons move much more rapidly in the collecting field than negative ions do, and recombination of ions in the gas is minimized.

With a sufficiently fine wire and high positive voltage on the wire, it is possible to obtain so strong a field that electrons nearing the wire are accelerated to the point where they may produce further ions.   The new electrons so formed are again accelerated and produce still more ions, resulting in a multiplication of the original charge by a factor which may in practical cases be as much as 1000.   Under these conditions the device is called a *proportional counter*, since the pulse size is proportional to the original ionization.   The counter used by Rutherford and Geiger in their determination of the number of α particles from radium (Fig. 8.2) was of this type.   At still higher voltages, the so-called *Geiger counting region* is reached; in this condition the rate of multiplication is so great that the discharge spreads through the whole of the counter, giving rise to a current which is limited only by the power supply. In practice, a limiting resistance is inserted in the circuit to lower the voltage and quench the discharge. The pulse output represents then merely a count, and its magnitude bears no relation to the number of original ions. A Geiger counter will easily respond to individual β particles.

b. Scintillation Counters   Many of Rutherford's early experiments on charged particles were made by counting scintillations in a zinc sulfide screen. When an ionizing particle passes through a crystal lattice such as that of zinc sulfide, it excites electrons in the lattice, and in the course of rearrangement the lattice may reemit some of the absorbed energy in the form of light. It is important that the emitted radiation be of a wavelength which is not strongly absorbed by the lattice. The action in the case of zinc sulfide is dependent on a trace of impurity such as Cu or Ag. A number of other substances, both inorganic and organic, have the same property. In modern applications a photomultiplier tube is used to detect the scintillations, and a response which is directly proportional to the energy loss in the scintillator can be obtained. A typical

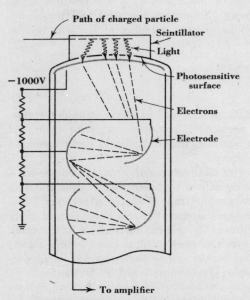

Fig. 24.11 Scintillation counter (schematic diagram).

arrangement is shown in Fig. 24.11.   The light emitted by the scintillator is collected on a photosensitive surface which comprises the cathode of the photomultiplier tube.   The photoelectrons are accelerated by an electric field to the first electrode; at this electrode each impinging electron produces two to six secondaries, and these in turn travel to the second electrode, there multiplying again.   The whole process, which may continue through 10 or more stages, requires less than 10^{-8} s, and the amplification is sufficient to allow direct display of the pulse on an oscilloscope screen.

c. Cloud Chamber   The path of an ionizing particle through a gas can be made visible by means of the cloud chamber, invented by C. T. R. Wilson in 1912.   This device consists essentially of a cylindrical or rectangular chamber provided with transparent windows and closed at the bottom by a movable piston or a flexible diaphragm.   In the chamber are contained a gas and a supply of some liquid which has a reasonable vapor pressure at room temperature; a mixture of ethyl alcohol and water is often used.   After the vapor has reached equilibrium, the chamber is suddenly expanded by pulling out the piston or diaphragm, and the gas is rapidly cooled, producing a condition of supersaturation of the vapor.   In the absence of any condensation nuclei, the supersaturation can persist for a long time, but if any ions are present, they act as centers on which droplets of liquid can grow.   Thus, with suitable illumination the trail of ions which mark the path can be seen and photographed.   If a magnetic field is provided, the tracks are curved, and from a measurement of the reprojected images the magnetic rigidity can be determined.   If the particle velocity is nonrelativistic, i.e., not too near the velocity of light, the density of droplets per unit path length can be used to estimate Z/v.

d. Photographic Plate   One of the earliest detection methods—and with some modifications, one of the most modern—makes use of the photographic plate.   Röntgen's and Becquerel's initial investigations were based on the blackening of the photographic plates by charged particles.   For reasons which are not yet completely understood, the production of a few hundred ions in the neighborhood of a silver halide grain in the emulsion renders the grain developable.   When a large number of such events have occurred in a small region of a plate, that region is blackened in the development process.   Ordinary emulsions contain only a small amount of silver compared to the other components present, and hence many ionizing events may be required to produce a detectable effect.   With recent development of emulsions containing over 80 percent silver halide by weight, however, the sensitivity is much increased, and under microscopic examination of the developed plate, the path of each ionizing particle appears as a track of silver grains. The photographic-plate method has had considerable application, particularly in the study of cosmic rays, where emulsions sufficiently thick to stop even very high-energy particles are used.

e. Bubble Chamber   In 1952 Glaser invented the bubble chamber, in which a charged particle moves through a superheated liquid rather than a supersaturated gas as in a cloud chamber. Bubble chambers containing liquid hydrogen have proved to be particularly valuable for studying many high-energy reactions for which the particle tracks could not be completely shown in a cloud chamber of reasonable size. The charged particle interacts with many more atoms per unit path length in a liquid than in a gas and loses its energy in a much shorter distance. In the bubble chamber the liquid is kept at an elevated pressure with its temperature just too low for boiling to occur. When a burst of charged particles enters the chamber, the pressure is suddenly reduced. Ions produced by the particles serve as nuclei on which bubbles form preferentially. The bubble tracks are photographed and the high pressure restored before general boiling begins.

f. Spark and Streamer Chambers   If a metal plate or wire is charged to a high potential and another conductor is placed nearby, a spark may jump the gap between the conductors when a "bridge" of ions is created across the gap by the passage of an ionizing particle. A common form of spark chamber is constructed from a series of closely spaced parallel metal plates with alternate plates charged to a high potential by a short pulse triggered by the arrival of an ionizing particle. Sparks leap through the noble gas (often a mixture of helium and neon) filling the region between the plates, revealing the path of the particle by the location of the sparks. The spark track is photographed and analyzed much as a cloud-chamber photograph is. Another form of spark chamber utilizes a two-dimensional array of parallel wires to locate the spark path by electrical signals which can be fed directly into a computer.

In the streamer chamber an extremely short pulse is applied between conductors. So short is the pulse that it produces very short flashes, called *streamers*, which are localized along the path of the particle. As photographed in the direction perpendicular to the field, the track consists of a series of short parallel streamers; photographed parallel to the field lines, the track appears as a series of dots. The streamers are localized and do not extend from conductor to conductor. Consequently the streamer chamber can delineate a track parallel to the conductors and can show simultaneously many tracks, a major advantage over the spark chamber.

1. Assuming that $\tau_a = CZ^4\lambda^3$, where $C = 2.6 \times 10^{-30}$ m^2/Å^3, and that $\sigma_a = Z\sigma_e$, calculate the atomic attenuation coefficient of copper for $\lambda = 0.2$ Å and $\lambda = 0.7$ Å. (Handbook values are 164 and 512 barns, respectively.)

2. Starting with Eqs. (24.6a, b), derive Eqs. (24.7), (24.9), and (24.9a).

3. Derive asymptotic expressions for Eq. (24.10) for high energies [Eq. (24.10b)] and for low energies through the term in to the first power [first two terms in Eq. (24.10a)].

4. Estimate the mass scattering coefficient of magnesium for 20-MeV bremsstrahlung.

5. Calculate the number of ion pairs produced per gram of dry air by an exposure dose of 1 r. Assuming that on the average it requires 33 eV to produce one ion pair, calculate the energy dissipated per gram of air by an exposure dose of 1 r.

6. A beam of 1.5-MeV photons passes through a sodium iodide crystal 0.8 cm thick. With the aid of Fig. 24.4 make a rough estimate of what fraction of the incident intensity appears beyond the crystal in the case of (a) a narrow-beam experiment, and (b) a broad-beam experiment.

7. Prove that the smallest energy a photon can have and still produce a pair in the field of a free electron is 2.04 MeV.

8. Find the maximum energy which a 10-MeV proton can impart to an electron initially at rest.

9. The *specific ionization* of a particle is the number of ion pairs produced per unit path length. If it requires 33 eV on the average to produce an ion pair in air, estimate the stopping power of standard air and the specific ionization for 9-, 4-, and 2-MeV α particles from the curves of Fig. 24.8.

10. An experimental value for the atomic stopping cross section of copper for 1-MeV protons is 1.3×10^{-18} eV-m^2/atom. From this value estimate the atomic stopping cross section of Cu for 4-MeV α particles for which the experimental value is 5.1×10^{-18} eV-m^2/atom.

11. Calculate $-dE/dx$ for 5-MeV protons and 20-MeV α particles in air by use of Eq. (24.18). Compare your result with the values implicit in Fig. 24.8.

12. For protons in the visible region the index of refraction is about 1.5 for glass, lucite, or mica. Calculate the angle θ for Čerenkov radiation for (a) a 100-MeV electron and (b) a 1-GeV proton.

13. Prove that photons are concentrated in the short-wavelength region of the Čerenkov spectrum by showing that the number of photons with wavelength between λ and $\lambda + d\lambda$ is proportional to $1/\lambda^2$.

14.   (a) Show from Eq. (24.24) that the number N of Čerenkov photons emitted per length L with wavelength between λ_1 and λ_2 is

$$\frac{N}{L} = \frac{\pi Z^2 e^2}{\epsilon_0 hc} \left(1 - \frac{c^2}{V^2 n^2} \right) \left(\frac{1}{\lambda_1} - \frac{1}{\lambda_2} \right)$$

providing n is essentially constant over the wavelength range in question.

(b) When an electron with $V \approx c$ moves through water with $n = 1.33$, how many photons are emitted per centimeter of path with wavelength between 4000 and 6000 Å?

15.   A magnetic analyzer, used to select particles of a given momentum, passes both muons and charged pions. In certain energy ranges it is possible to tell which particles in the beam are muons by use of a Čerenkov counter. Find over what momentum range (in units of MeV/c) a Čerenkov counter with index of refraction 1.33 will give a pulse for a muon but none for a pion of the same momentum.

chapter twenty-five

The Nucleus

We turn now from a consideration of crystalline aggregates of atoms to a further study of atomic nuclei, their properties, and structure. It occurred to a number of physicists, after the high energy of the radiations from radioactive substances was discovered, that these radiations might be capable of inducing transmutations in other elements. Several experiments were tried in which radioactive materials were mixed with stable substances and the products of transmutation searched for by chemical analysis. Production in this way of neon and argon from water was reported as early as 1907, but later experiments failed to substantiate this claim and the effect was generally attributed to contamination. Actually the number of atoms which one could hope to transmute by such procedures is so small as to preclude detection by chemical means; it was not until methods had been developed whereby individual events could be detected that artificial transmutation could be demonstrated.

25.1 Discovery of Artificial Transmutation

The first conclusive evidence for artificially induced transmutation was obtained by Rutherford in 1919, in connection with some experiments on the scattering of α particles in various gases. The apparatus used in these studies is illustrated in Fig. 25.1. In a gastight chamber B is mounted a source D of Po^{214}, which emits α particles of about 7 cm range in standard air. A thin silver foil W covers a hole in the end of the chamber, and a zinc sulfide scintillation screen F, observed with a microscope M, permits counting of individual particles coming through the window. Provision is made for inserting absorbing foils between the window W and the screen F and for varying the distance x between the source and the window. The whole apparatus is placed in a transverse magnetic field to eliminate the β rays. In the first experiments, carried out by Marsden in 1914, hydrogen gas was used in the chamber, and scintillations due to the hydrogen recoils were observed with absorbers having thicknesses up to about 28 cm air equivalent. This range was approximately in agreement with expectations. If we consider the conservation of momentum and energy in an elastic collision between an incident particle of mass M_0 and the struck particle, of mass M_1, the energy transferred is

$$E = E_0 \frac{4M_0M_1}{(M_0 + M_1)^2} \sin^2 \frac{\theta}{2} \tag{25.1}$$

where θ is the angle of deflection of the incident particle and E_0 is the initial energy. For a Po^{214} α particle striking a proton head on, the maximum energy transfer is $\frac{16}{25} \times 7.680$ MeV = 4.9 MeV, and the range of such a proton in air is 32.5 cm (Fig. 24.8), in reasonable agreement with the value found by Marsden.

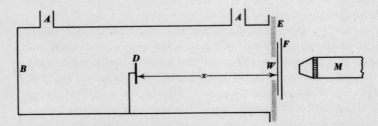

Fig. 25.1 Apparatus by which Rutherford observed production of protons in the disintegration of nitrogen by α particles. The movable source is at D; particles produced by transmutations in the gas strike the screen F, giving rise to scintillations observed by use of the microscope M.

In an investigation of certain peculiarities of the effects in hydrogen, Rutherford made some tests in 1919 with air in the chamber, with the unexpected result that long-range particles appeared again. Further work showed that the effect was due to nitrogen, that neither oxygen nor carbon dioxide produced similar particles, and that the particles were much too numerous to be ascribed to hydrogen contamination. Rough magnetic analysis indicated that the particles probably were protons, and Rutherford concluded that they must have been knocked out of the nitrogen nucleus by the impact of the α particles. The number observed was about 1 per million α particles. The work was extended by Rutherford and Chadwick in 1922, and protons with ranges as long as 40 cm were observed from nitrogen. Five other elements, boron, fluorine, sodium, aluminum, and phosphorus, were also found to yield long-range particles; with aluminum as the target, they observed protons with ranges up to 90 cm air equivalent. In a later version of the experiment (1924), the protons were observed at 90° to the α particles, where no elastically scattered protons from hydrogen impurities would be expected at all. With this apparatus, they could be sure that any particles they saw at ranges greater than 7 cm (the maximum range of the α particles) were due to processes other than elastic scattering. In testing a variety of target materials, they found that all the light elements up to and including potassium could be disintegrated, with the exception of hydrogen, helium, lithium, beryllium, carbon, and oxygen.

Rutherford's original surmise that the α particle knocked out a proton from the target nucleus turned out to be only partially correct. In a series of cloud-chamber pictures of α particles traversing a nitrogen atmosphere, Blackett photographed eight disintegrations (in 415,000 α tracks). In all eight cases, the α track terminated in a fork, with a lightly ionizing particle, identifiable as a proton, as one branch and a heavy recoil, similar in behavior to the nitrogen recoils observed in elastic scattering, as the other. In no case was a third track corresponding to the α particle itself visible after the collision. From this it was concluded that the α particle had been absorbed into the nucleus, releasing a proton and forming a residual nucleus 3 u heavier than nitrogen and with one more unit of charge. In the terminology of nuclear reactions, the events observed by Rutherford, Chadwick, and Blackett may be written

$$_7N^{14} + {_2}He^4 \rightarrow {_8}O^{17} + {_1}H^1$$

where the conservation of charge and mass number are explicitly exhibited. The existence of O^{17} in ordinary oxygen was unknown in 1925. From the range of the protons observed, the energy release could be calculated. In the case of nitrogen, this turned out to be negative; i.e., part of the kinetic energy of the α particle was used up in forming the two products, but in several other instances a positive energy release was observed. In aluminum, for example, a 90-cm proton was produced, corresponding to a kinetic energy of 8.6 MeV. Even ignoring the kinetic energy of the residual nucleus, this represents a gain of nearly 1 MeV over the initial kinetic energy of the α particle.

At the time of these experiments, the question arose whether the new nuclides produced were stable, and a systematic investigation was undertaken to detect radioactive effects, with negative results. Unfortunately the detection methods used were insensitive to β rays; otherwise the discovery of the activity induced in aluminum, among others, might have been anticipated by more than a decade.

25.2   *Discovery of the Neutron*

For nearly 12 years after Rutherford's discovery of the production of protons from α-particle bombardment of light elements, this was the only type of artificial disintegration known. By 1930 several other laboratories had entered the field, and several important new facts had been brought to light. Bothe and Fränz had shown, in 1927, that the less energetic α particles of Po^{210} (5.3 MeV) were also capable of producing transmutations, and Pose in 1929 was able to show that α-particle bombardment of aluminum (a single isotope) produced three distinct groups of protons. A particularly significant observation was made by Pose in 1930, when he found that the yield of protons exhibited maxima for certain α-particle energies; this is the phenomenon of *resonance*, predicted by Gurney in 1929, about which we shall have more to say in Sec. 25.13.

The first observation of nuclear radiations other than protons from the transmutation of the light elements was made by Bothe and Becker in 1930. Using a Geiger counter, they observed penetrating radiation from α bombardment of several elements, among them Be and B. The radiation was so weak that they could only make the roughest estimate of the absorption coefficient; they could establish, however, that the radiation they observed was more penetrating than any known γ ray.

In 1932, Irène and Frédéric Joliot-Curie discovered a remarkable property of these radiations. In an attempt to compare the absorption coefficients in various materials, they set up the arrangement shown schematically in Fig. 25.2. α particles from the polonium source P bombarded a piece of beryllium B, and the radiation was detected in the ionization chamber I. A 1.5-cm-thick Pb filter F reduced the effect

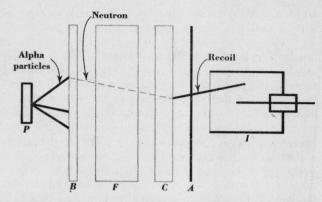

Fig. 25.2 Arrangement used by the Joliot-Curies to study the recoils ejected from various materials by radiation produced in the α-particle bombardment of beryllium and boron.

on the onization chamber of the soft γ radiation from the Po source. Inserting absorbers of such materials as Al, Cu, Ag, and Pb in front of the ionization chamber, at *C* in the figure, they found only a slight diminution in intensity, but with paraffin, water, and cellophane, they actually observed an *increase* in the ionization-chamber reading. The greatest effect was observed with paraffin, which produced an increase by a factor of 2. Acting on the conjecture that the effect was due to protons ejected from the paraffin by the radiation, the Joliot-Curies placed aluminum absorbers between the paraffin and the ionization chamber (at *A* in the figure) and found that 0.2 mm of Al was sufficient to destroy the effect. The increase in ion-chamber current was therefore not attributable to fast electrons, and tests with a magnetic field showed that no easily deflectable particles, like slow electrons, were responsible. This evidence and the association of the effect with hydrogenous substances made the identification with protons reasonably certain. Knowing the range-energy relation for protons in aluminum, they concluded that the protons had an energy of about 4.5 MeV, and concluded that they were recoils from a high-energy γ ray.

From Eq. (7.17a) for the Compton process, the maximum energy transferable to a proton by a γ ray of energy *hν* is

$$E_{\max} = h\nu \, \frac{2\epsilon}{1 + 2\epsilon} \qquad \epsilon = \frac{h\nu}{Mc^2} \tag{25.2}$$

where *M* is the mass of the proton. Using their value of 4.5 MeV for the proton energy, they arrived at a γ-ray energy of approximately

50 MeV.   A similar examination of the radiation from boron bombarded with α particles yielded a γ-ray energy for this reaction of 35 MeV.

Two serious difficulties involved in this interpretation were pointed out by Chadwick in his famous paper of 1932.   In the first place, the number of protons ejected in the Compton process could be calculated from the Klein-Nishina formula and was many thousand times too small to account for the observations.   In the second place, he showed that, even allowing rather generous limits for the errors in the masses involved, no conceivable reaction could produce the required energy.   The most energetic possible reaction would be given by the simple capture of the α particle by the Be^9, forming C^{13}; the difference in mass energy between $Be^9 + He^4$ and C^{13} could then be radiated as a γ quantum.   The mass of C^{13} and of the α particle were known, and an upper limit for the mass of Be^9 could be obtained from the fact that Be^9 is stable.[1]   The mass difference $Be^9 + He^4 - C^{13}$ added to the kinetic energy available from the α particle gave an upper limit for the energy of the γ ray of only 14 MeV—considerably less than the 50 MeV reported.

In extension of the work of the Joliot-Curies, Chadwick redetermined the range of the recoil protons from hydrogen, obtaining a range of 40 cm, corresponding to a velocity of 3.3×10^7 m/s and an energy of 5.7 MeV.   He also observed recoils from a number of other light substances, including nitrogen, and made rough measurements of the recoil energies.   From Eq. (25.2), the maximum energy transferred by a γ ray should be roughly inversely proportional to the mass of the struck nucleus [note that $2\epsilon \ll 1$, so that $E_{max} \sim 2(h\nu)^2/Mc^2$], and so it would be expected that nitrogen recoils would have an energy of 5.7/14 MeV.   The observed energy was more than 3 times this value.

Thus the assumption that the observed effects were produced by high-energy γ rays was untenable.   If, on the other hand, the radiation consisted of uncharged particles of about the same mass as the proton, all difficulties were resolved.   Such particles, having no interaction with the atomic electrons, would be able to pass relatively freely through matter, losing their energy only through comparatively rare nuclear collisions.   In such collisions they could transfer a large fraction of their energy and the observed recoils could be accounted for with only moderate primary energies.   The existence of such particles, called *neutrons*, had in fact been suggested by Rutherford some 12 years earlier; many attempts to observe them had been made at the Cavendish Laboratory before Chadwick's discovery.

Comparison of the recoil energies from hydrogen and nitrogen per-

---

[1] If the mass of Be^9 were greater than the sum $2He^4 + n^1$, it would disintegrate spontaneously.   Chadwick assumed, in his preliminary estimate, $Be^9 \leq 2He^4 + p + e$.

mitted a quantitative determination of the mass of the neutron. From the ordinary laws of elastic collisions, the maximum velocity which can be imparted to a proton by a particle of velocity V_0 having a mass equal to M_0 times the proton mass is

$$V_H = V_0 \frac{2M_0}{M_0 + 1} \qquad (25.3)$$

while the maximum velocity of a nitrogen recoil will be

$$V_N = V_0 \frac{2M_0}{M_0 + 14} \qquad (25.3a)$$

Dividing these two relations to eliminate V_0, we obtain

$$\frac{V_H}{V_N} = \frac{M_0 + 14}{M_0 + 1} \qquad (25.4)$$

For V_H, Chadwick used his value of 3.3×10^7 m/sec; V_N was obtained from cloud-chamber measurements by Feather as 4.7×10^6 m/s. Inserting these values in Eq. (25.4), Chadwick obtained $M_0 = 1.15$ with an estimated error of about 10 percent. The velocity of the neutrons was 3.2×10^7 m/s, corresponding to a kinetic energy of about 6 MeV.

Chadwick suggested that the reaction involved in the production of neutrons from α-particle bombardment of beryllium is

$$_4Be^9 + {}_2He^4 \rightarrow {}_6C^{12} + n^1$$

where n^1 refers to the neutron. From the known masses, again assuming an upper limit for Be^9, he concluded that a maximum of 8 MeV was available for the neutron when polonium α particles (5.3 MeV) were used. In the case of boron he assumed the reaction

$$_5B^{11} + {}_2He^4 \rightarrow N^{14} + n^1$$

Here the masses were all known, and from the energy balance he was able to obtain a mass for the neutron of 1.0067 u, with a probable error of about 0.1 percent.

Nuclear Binding Energies and Nuclear Forces

The discovery of the neutron revolutionized our ideas about the structure of nuclei; until that time it was generally believed that nuclei contained protons, electrons, and, possibly, α particles. The theory of nuclear structure based on this assumption was fraught with perplexing difficulties, particularly with respect to the properties of electrons in the nucleus. The introduction of the neutron as a nuclear constituent eliminated some of these problems and provided a much more satisfactory picture of the composition of nuclei. The theory has still many difficulties, it must be confessed, but at least they are now of another kind.

25.3   Properties of Nuclei

Before we embark on a discussion of the theory of nuclear structure, it will perhaps be appropriate to review some of the more important properties of nuclei as they were known in 1932. In the ensuing discussion we shall have the advantage of another 35-odd years of development, but most of the concepts we shall need for our preliminary discussion were at hand by 1932.

a. Charge   We begin with the nuclear charge. Since atoms are normally neutral, the nuclear charge is equal in magnitude and opposite in sign to that of the atomic electrons. Since the latter all have the same charge e, the nuclear charge is Ze, where Z is an integer. The value of Z is known from x-ray scattering experiments, from the nuclear scattering of α particles, and from the x-ray spectrum.

b. Mass   The masses of nuclei, determined by measurement of the ratios of charge to mass in the mass spectrometer (Sec. 10.4) and by nuclear reaction energies (Sec. 25.10), are all very close to integers when expressed in terms of $C^{12} = 12$ exactly. The mass number A is about $2Z$ for light elements and increases to about $2.5Z$ for heavy ones.

c. Size   The "size" of a nucleus, like that of an atom, is not a uniquely defined quantity, since the specification depends upon what property is under consideration. If two atoms approach one another, they may exhibit a weakly attractive force at larger distances, which changes at short distances to a strongly repulsive force. The distance at which this change occurs may be used as a measure of the sum of the radii of the two atoms. On the other hand, the "radius" may also be defined in terms of the distribution of electronic charge in the atom. The two definitions may give different results; both are subject to arbitrariness because the functions involved change slowly with distance. In the nuclear case, too, various definitions of the radius are used, but all agree fairly well; in each case the measurements suggest a rather sharp boundary.

One useful measurement of the size of nuclei comes from α-particle scattering experiments of the kind described in Sec. 8.4. In these experiments, a beam of α particles is directed onto a thin foil of the material under study and the number of α particles scattered at various angles is determined. For moderate α-particle energies, the experimental results are consistent with the assumption of a simple inverse-square force acting between the nucleus and the α particle. As the α-particle energy is increased, deviations from the Rutherford scattering law appear, particularly at large angles. These deviations are associated with the fact that the α-particle trajectory reaches closer to the nucleus at the higher energies, and can thus be interpreted as a failure of the inverse-square law at small distances. Analysis of the scattering experiments leads to the conclusion that the radius at which the force law changes varies from a few times 10^{-15} m in light nuclei to about 10^{-14} m in heavy nuclei.

Scattering of high-energy neutrons by nuclei (Sec. 25.15) also gives a measure of the extent of the force field around a nucleus. Since neutrons are unaffected by the electrostatic forces, their scattering must be interpreted in terms of some nonelectrical force. The experimental results indicate an attractive force which sets in rather abruptly at short distances. The radius at which the force begins to act strongly varies approximately as the cube root of the mass number and appears to be well represented by the expression

$$R = R_0 A^{\frac{1}{3}} \qquad R_0 = 1.1 \text{ to } 1.5 \times 10^{-15} \text{ m} \qquad (25.5)$$

We shall discuss in later sections still other determinations of the nuclear radius as defined in various ways and shall find that all are reasonably consistent with Eq. (25.5).

d. Spin The spin—better referred to as the total angular momentum—of the nucleus can be obtained from the hyperfine-structure pattern, from spectra of homonuclear molecules, or by atomic-beam methods. The magnitude of the angular momentum is specified, in units of $\hbar$, by the number I,[1] where I is the maximum possible projection of the angular-momentum vector on any arbitrary axis in space. In nuclei with even A, I is found to be an integer or zero; the ground state I is always zero if both Z and A are even. In nuclei with odd A, I is a half-integer. Protons, electrons, and neutrons have spin $I = \frac{1}{2}$.

e. Magnetic Moment A magnetic dipole moment is always associated with nuclear spin; the magnitudes of nuclear moments, obtained from the spacing in hyperfine structure and from atomic-beam experiments, are of the order of the nuclear magneton $e\hbar/2M$, where M is the proton

---

[1] The symbol J is sometimes used.

mass.   They are thus about one one-thousandth of the magnetic moment of the electron.   Nuclei with zero spin have no magnetic moment.

f. Statistics   Quantum-mechanical systems can be divided into two classes depending upon what kind of statistics are obeyed by the particles composing them (Secs. 20.3 and 20.4).   If the wave function describing two identical particles, e.g., two electrons with the same spin direction, changes its sign when the space coordinates of the two particles are interchanged, the wave function is antisymmetric with respect to the exchange and the particles obey Fermi-Dirac statistics.   For the other class of particles, the wave function is symmetric—does not change sign—and the particles obey Bose-Einstein statistics.   F-D particles obey the Pauli exclusion principle, while B-E particles do not.   Experimentally it is found that all particles with half-integral spins, e.g., protons, neutrons, and electrons, obey F-D statistics, and all those with integral or zero spin obey B-E statistics.   All nuclei with odd A obey F-D statistics, while those with even A obey B-E statistics.

g. Parity   The parity of a system refers to the behavior of the wave function under inversion of the coordinates through the origin, i.e., when x is replaced by $-x$, y by $-y$, and z by $-z$.   If the potential-energy function is unchanged by this process $[P(x,y,z) = P(-x,-y,-z)]$, then it can be shown that the wave function may either remain the same or it may change sign: it is said to have *even* parity if the sign remains the same; if the sign changes, the parity is *odd*.   The wave functions describing a particle in a box have alternately even and odd parities if the potential is symmetric in x.   The parity of the wave functions of the one-electron atom are even or odd as l is even or odd.   Nuclear states are characterized by a definite parity (which may be different for different states of the same nucleus), and the conservation of parity has an important bearing on nuclear reactions.

h. Electric Quadrupole Moment   A system of charges with nonzero spatial extension may produce, in addition to the ordinary radial electrostatic field of force, fields of higher complexity, characterized as dipole, quadrupole, and higher-multipole fields.   There is good evidence that nuclei do not have electric dipole moments, but electric quadrupole moments have been observed in many nuclei.

25.4   Constituents of Nuclei

The fact of the transmutability of nuclei argues that they are complex systems, composed of some common constituents.   The integral values of Z and the near-integral values of M suggest the proton as one candidate, while the occurrence of α particles and β particles as decay products early suggested electrons as fundamental constituents and helium nuclei either as constituents or as particularly stable subassemblies. Thus, in the early theories, a nucleus of charge number Z and mass number A was thought of as comprising A protons and $A - Z$ electrons, grouped as far as possible in units of four protons and two electrons. The hypothesis that electrons exist in the nucleus encounters a number of difficulties, however, several of which were already apparent long before the neutron was discovered. For one thing, there was the problem of the magnetic moment: a single odd electron in a nucleus would be expected to contribute, from its spin alone, a moment of 1 Bohr magneton; nuclear moments are very much smaller than this. More perplexing still was a problem arising in the light nuclides with even A and odd Z. N^{14}, for example, would have 14 protons and 7 electrons—all F-D particles with spin $\frac{1}{2}$. With an odd number of such particles, N^{14} should evidently obey F-D statistics and have a half-integral spin, but study of the band spectrum of nitrogen gas showed in fact that N^{14} obeys B-E statistics and has a spin of 1. Finally, according to the uncertainty principle, a particle with so small a mass as the electron would have an implausibly large kinetic energy if it were confined to nuclear dimensions.

With neutrons and protons as the building blocks of nuclei, on the other hand, these problems find a natural solution. On this picture a nucleus with charge number Z and mass number A contains Z protons and $A - Z$ neutrons, or A particles in all. Thus N^{14} has seven protons and seven neutrons; neutrons are Fermi particles and have half-integral spin, so that the spin and statistics of N^{14} are accounted for. The magnetic moments present less difficulty than before, since only heavy particles are involved. The problem of β decay is taken care of—or at least put in another category—by assuming that a neutron in the nucleus changes to a proton, *creating* the β particle (and an antineutrino) in the process. Two other types of decay, which were discovered later, also fit into this picture: a nuclear proton may be converted into a neutron, creating a positive electron and a neutrino (positron decay) or capturing an atomic electron (electron capture) and emitting a neutrino. The quantum-mechanical treatment of β decay bears a close resemblance to that describing the creation of light quanta in atomic transitions, and the formal similarity of the two processes emphasizes the point that the appearance of an electron coming out of a nucleus does not imply that it ever was in the nucleus. The assumption that neutrons and protons are constituents of nuclei and electrons are not is consistent with all known facts and may be regarded as a basic tenet of nuclear theory. The term *nucleon* is often used in referring to either protons or neutrons.

25.5   Masses and Binding Energies

Under the assumption that nuclei are composed of neutrons and protons, we can now discuss the various factors which enter into the masses of nuclei and gain some insight into their binding. The mass of any permanently stable nucleus is found to be less than the sum of the masses of the neutrons and protons which it contains. This fact is accounted for by the conversion of part of the mass energy of the particles into energy of binding, the relation between the change in mass and the binding energy being given by Einstein's equation, $E_{\text{binding}} = \Delta Mc^2$. The significance of the binding energy can perhaps be seen most easily in the more familiar atomic case. In the capture of a free electron by a proton to form a hydrogen atom, energy, 13.6 eV, is released (in the form of one or more light quanta); the mass of the hydrogen atom is therefore slightly less than the sum of the masses of a free electron and free proton. To ionize the hydrogen atom, an amount of energy at least equal to the binding energy of 13.6 eV must be added. In an atom containing more than one electron, the binding energy is the total energy required to remove all the electrons; it is thus to be distinguished from the ionization potential, which is the minimum energy required to remove one electron. In atoms, the ionization potential of the most loosely bound electron is a few electron volts; the total binding energy for all but the lightest atoms is given approximately by the empirical expression $E_B = 15.6Z^{\frac{5}{3}}$ eV. The change in mass corresponding to the binding energy in atoms is small—in the uranium atom it amounts to about 1 electron mass, or about 3×10^{-6} of the total—and is just at the limit of direct measurability; in the nucleus, on the other hand, the binding energy may represent an appreciable fraction of the total mass.

The mass of the nucleus with charge number Z and mass number A may be written

$$M_N(A,Z) = ZM_p + (A - Z)M_n - \frac{E_B}{c^2} \tag{25.6}$$

where M_p is the mass of the proton and M_n is the mass of the neutron. The term E_B/c^2 represents the mass equivalent of the total binding energy, the energy which must be added to the nucleus in order to break it up into Z protons and $A - Z$ neutrons. Because it is more nearly the atomic than the nuclear mass which is measured in the mass spectrograph, the values given in mass tables usually refer to the mass of the neutral atom, including the electrons. Expressed in terms of *atomic* rather than nuclear masses, Eq. (25.6) becomes

$$M(A,Z) = M_N(A,Z) + ZM_e = ZM_H + (A - Z)M_n - \frac{E_B}{c^2}$$
(25.7)

where now M_H is the mass of the hydrogen atom. Actually a very small correction for the binding energy of the atomic electrons should be made if the quantity E_B is to refer to the nuclear binding energy only, but the correction is negligible for our purposes, except perhaps in the heaviest nuclei.

With the help of Eq. (25.7) and the known masses of the neutron and hydrogen atom, we can now compute the binding energy of any nuclide for which the mass has been measured. As a simple example we take the deuterium atom (hydrogen isotope of mass number 2):

$$M(H^2) = M_H + M_n - \frac{E_B}{c^2}$$

From Table 10.2

$$2.014102 = 1.007825 + 1.008665 - \frac{E_B}{c^2}$$

$$\frac{E_B}{c^2} = 0.002388 \text{ u} \qquad E_B = 2.225 \text{ MeV}$$

We conclude that it requires 2.225 MeV of energy to dissociate a deuteron and that the capture of a free neutron by a proton releases 2.225 MeV of energy (in the form of a γ ray). In the same way, we find for the binding energy of the U^{238} nucleus

$$\frac{E_B}{c^2} = 92 \times 1.007825 + 146 \times 1.008665 - 238.050770$$

$$= 1.93422 \text{ u}$$

$$E_B = 1802 \text{ MeV}$$

Thus the binding energy in this nucleus amounts to nearly 1 percent of its mass.

One interesting result of the computation of nuclear binding energies is the observation that, except for the lightest nuclides, the binding energy varies practically linearly with A, or to put it another way, the binding

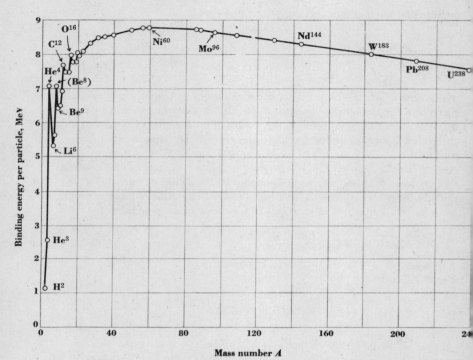

Fig. 25.3 Average binding energy per nucleon for the naturally occurring nuclides and the unstable nuclide Be⁸, which decays into two α particles.

energy per particle is very nearly constant over a large range of A. In Fig. 25.3 is shown a plot of the binding energy per particle as a function of A for the stable nuclides. Aside from certain irregularities among the light elements, the general course of the curve shows a gradual increase to a value of about 8.8 MeV/particle near the middle of the table with a falling off to about 7.6 MeV/particle in the region of the heavy elements. It is worth observing that the near constancy of the binding energy per particle is in striking contrast to the behavior of atoms, where the average binding energy per electron is a steadily increasing function of the number of electrons (proportional to $Z^{\frac{1}{3}}$).

a. Stability against Heavy-particle Emission   That the total binding energy of a nucleus is positive is not a sufficient condition for its stability; in order that a nucleus be stable, it is necessary that its mass be less than the sum of the masses of any other combinations of its constituent protons and neutrons, whether free or bound in small groups. The α-radioactive nuclei are unstable because the α particle itself is a tightly bound structure, and at the upper end of the periodic table, the binding energy per particle increases as the number of particles decreases. The energy required to remove any given particle from the nucleus is called the *separation energy*,[1] E_s. The separation energy for an α particle in the nucleus (Z, A) is

$$\frac{E_s(\alpha)}{c^2} = M(\text{He}^4) + M(Z - 2, A - 4) - M(Z, A) \tag{25.8}$$

If the α-particle separation energy is negative, the nucleus (Z, A) is unstable to α emission. For the light elements, the separation energy of an α particle is positive, e.g., it requires 4.730 MeV to remove an α particle from Ne^{20}. On the other hand, the α separation energy of Po^{210} is -5.4 MeV. There exist many isotopes with $A > 150$ which are energetically unstable with respect to α-particle emission. Because of the extremely long lifetimes expected for low-energy α emitters, these isotopes are nonetheless stable for all practical purposes. A still larger class are, in principle, unstable with respect to fission into two or even three roughly equal parts, but again the lifetimes are so long that the effect is generally unobservable. Stability against neutron or proton emission is no problem; the separation energy for these particles is generally around 7 to 8 MeV, or about the same as the average binding energy per particle, for all but the lightest nuclides.

b. Stability against Beta Decay Still a further condition for the stability of nuclei comes as a result of the phenomenon of β decay. It may occur, and frequently does, that although a nucleus has insufficient energy to emit a nuclear particle, a system of lower energy can be formed by transforming a neutron into a proton, or vice versa. Most often it is the requirement of stability against β decay which is the most stringent condition for the existence of a given nuclide in nature. The nuclide (Z, A) is stable against negative β decay if

$$M(Z, A) \leq M(Z + 1, A) \tag{25.9}$$

where the M's are atomic (not nuclear) masses. To see this, we note that $M(Z, A)$ is the mass of the nucleus plus Z electrons; if the nucleus is to decay, it must supply sufficient mass for the new nucleus plus the

---

[1] Sometimes also called the *binding energy* of the particle in the nucleus (Z, A).

β particle.   The mass condition then is

$$M(Z,A) = M_N(Z,A) + ZM(e) \leq M_N(Z + 1, A) \\ + ZM(e) + M(\beta)$$

where M_N refers to the mass of the nucleus alone, $M(e)$ to the mass of an atomic electron, and $M(\beta)$ to the β particle.   Since the negative β particle is an electron, the right-hand side adds up to $M(Z + 1, A)$, the atomic mass.

In the decay by *positron* emission, the initial nucleus must furnish mass for the new nucleus and a β^+ particle; the condition for positron stability of the nuclide $(A, Z + 1)$ is then

$$M_N(Z + 1, A) \leq M_N(Z,A) + M(\beta^+)$$

Adding $Z + 1$ electron masses to both sides and noting that the positron and electron masses are the same, we obtain the expression in terms of atomic masses:

$$M(Z + 1, A) \leq M(Z,A) + 2M(e) \tag{25.10}$$

A third type of β process which is of particular importance for stability considerations is decay by electron capture, in which an orbital electron—usually from the K shell—is absorbed into the nucleus and a neutrino is emitted.   Since the latter may have zero energy, the condition for stability becomes

$$M_N(Z + 1, A) + M(e) \leq M_N(Z,A)$$

or, in terms of atomic masses,

$$M(Z + 1, A) \leq M(Z,A) \tag{25.11}$$

We have in all three of these expressions for β stability ignored the change in the binding energy of the atomic electrons; this quantity is in almost all cases quite negligible.

The two conditions (25.9) and (25.11), for stability against negative β decay and electron capture, guarantee that of any two neighboring isobars one must be unstable, for the heavier will always[1] be able to transform into the lighter by one process or the other.   Thus if the masses

---

[1] Still ignoring the electron-binding-energy correction; no case is known where this prevents a decay which would otherwise be possible.   There exist some cases where K-electron capture is not possible and capture from L or higher shells occurs instead.

of isobars (nuclides with fixed A) increase uniformly as Z is varied away from some optimum value, only one will be stable. It is just this fact that gives the study of the naturally occurring nuclides its great importance in connection with the understanding of nuclear structure, for of any combination of A neutrons and protons, just the one which we find in nature will represent the configuration of minimum energy. There do occur cases among species with even A where the mass is not a uniform function of Z; we have then the situation where a nuclide may be unstable with respect to both its neighbors, and, for these values of A, two isobars and sometimes three, differing in charge by 2 units, may occur. The isobars Ca^{40}–A^{40} and Zn^{64}–Ni^{64} are examples.[1]

25.6 Nuclear Forces

We wish now to see what can be inferred from the material at hand about the character of the forces which hold nuclei together. It is in the first place clear that they are not ordinary electrostatic forces, if for no other reason than that all the particles involved either have positive charge or are neutral, and such a system could not be bound by classical forces. Even aside from the sign of the force, electrostatic forces are much too weak to account for the main effects; gravitational forces are weaker still and offer no assistance in the problem. It is therefore necessary to postulate an entirely new type of interaction, a type which never manifests itself in large-scale phenomena and which has no evident relation to either gravitational or electric forces. Since we have no previous experience with such forces, the best we can do is to make the simplest possible postulates about them consistent with the available data. With a finite number of facts available, we cannot be sure of a unique solution; it can only be hoped that the solution will contain sufficient elements of truth to permit further progress as experimental knowledge expands.

One characteristic feature of nuclear forces is their short range. Although the exact dependence on distance is not well established, there is good evidence that the force vanishes for all practical purposes at distances greater than a few times 10^{-15} m. There are many experiments which suggest that the force between two nucleons[2] sets in quite abruptly at a separation of 1 or 2 $\times$ 10^{-15} m. In the interest of simplicity, the force is usually regarded as derivable as the gradient of a potential

---

[1] There do occur a few apparent exceptions to the rule prohibiting neighboring stable isobars: for example, Cr^{50}, V^{50}, and Ti^{50} all exist in nature. Direct mass measurements show that V^{50} should be unstable. Its long life is attributed to the very high angular-momentum change of 6 units, which inhibits the decay.

[2] Aside from the electrostatic force, which acts only between protons.

(in the hope, among other things, that the force between nucleons is independent of their velocity), and a shape of the potential is chosen which is mathematically not intractable. It will be sufficient for our purposes to use as a convenient approximation the so-called square well. Conceptual difficulties with the infinite force at the boundary can be resolved by letting the transition take place over a small range in radius.

a. Theory of the Deuteron   In order to exhibit some of the concepts involved in discussing nuclear potentials and the quantum states of nuclei, we consider the deuteron. Although the deuteron is the only nucleus which admits of a reasonably exact solution, many features of the deuteron problem occur in more complicated nuclei and an understanding of this relatively simple case is useful in these applications. Our discussion will be based on the square-well potential illustrated in Fig. 25.4, where the heavy line indicates the potential energy of the neutron in the nuclear field of the proton (or vice versa). The potential energy has the value $P = -P_0$ out to the radius R and is zero for larger values. We write the wave function for the neutron, in coordinates relative to the proton, as $\psi(r,\theta,\phi)$, and look for a spherically symmetric[1] solution to the wave equation, of the form $\psi = u(r)/r$; the absolute square of $u(r)$ represents the probability of finding the neutron at distance r.

[1] With the assumed potential, the spherically symmetric solution will have the lowest energy.

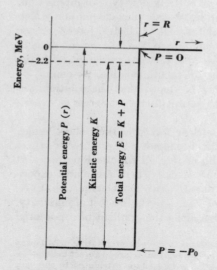

Fig. 25.4 Simplified square-well potential representing the interaction of a neutron and a proton.

The Schrödinger equation may then be written

$$\frac{d^2u(r)}{dr^2} + \frac{M}{\hbar^2}(E + P_0)u(r) = 0 \qquad r < R \tag{25.12a}$$

$$\frac{d^2u(r)}{dr^2} + \frac{M}{\hbar^2}Eu(r) = 0 \qquad r > R \tag{25.12b}$$

Because the neutron and proton have practically equal mass M, the reduced mass $M/2$ is used in these equations. We must now ascertain whether there exist any solutions to Eqs. (25.12) which correspond to stationary, bound states; evidently we require solutions for which E, the total energy, is negative (see Fig. 25.4). Since the deuteron is known to be a bound system, with $E = -2.225$ MeV, the question really amounts to choosing values of P_0 and R which will reproduce this fact. For $r < R$, the solution of (25.12a) is of the form

$$u(r) = A \sin \kappa r \qquad \kappa = \sqrt{\frac{M(E + P_0)}{\hbar^2}} \qquad r < R \tag{25.13a}$$

The cosine solution is inadmissible because it is finite at $r = 0$ and $\psi = u(r)/r$ would be infinite. The outside solution is

$$u(r) = Be^{-kr} \qquad k = \sqrt{\frac{M(-E)}{\hbar^2}} \qquad r > R \tag{25.13b}$$

where the positive-exponential solution is rejected because it produces an infinite ψ as r approaches infinity. The condition that these two solutions join smoothly at $r = R$ then determines whether a real eigenvalue for the energy exists. Equating the values and first derivatives of u at $r = R$, we obtain

$$A \sin \kappa R = Be^{-kR} \tag{25.14a}$$

$$A\kappa \cos \kappa R = -Bke^{-kR} \tag{25.14b}$$

Dividing (25.14b) by (25.14a)

$$\kappa \operatorname{ctn} \kappa R = -k \qquad \text{or} \qquad \operatorname{ctn} \kappa R = -\frac{k}{\kappa} \tag{25.15}$$

This transcendental equation can be solved numerically or graphically to produce the required condition on P_0 and R for any (negative) value of E. For the present purpose, it is sufficient to observe that if we choose $P_0 \gg E$, then k/κ is small and the cotangent of κR is small and negative; κR is then slightly greater than $\pi/2$. The situation is illustrated in Fig. 25.5, where $u(r)$ is indicated as a function of r. The inside (sine) function has reached a maximum and just started to decrease where it joins the outside exponential. This condition represents the first possible eigenvalue, since if the interior function had not yet passed its first maximum, it could not be made to fit a decreasing exponential outside. The condition $\kappa R \approx \pi/2$ gives

$$\frac{\pi}{2} \approx \kappa R \approx \frac{R}{\hbar} \sqrt{MP_0}$$

or

$$P_0 R^2 \approx \frac{\hbar^2}{16M} = 1.02 \times 10^{-28} \text{ MeV-m}^2 \tag{25.16}$$

using the definition of κ in (25.13a) and ignoring the small quantity E. If we choose for R the reasonable value of 2×10^{-15} m, we obtain $P_0 \approx 25$ MeV for the potential energy of the neutron-proton bond in the ground state of the deuteron. Taking $\kappa R = n\pi/2$, with $n = 3, 5, \ldots$, would yield excited states of the system, corresponding to the excited levels of an atom. It can be shown, however, that no bound excited states exist in the deuteron.

b. Scattering of Free Nucleons From the fact that the deuteron is a bound system, attractive forces exist between neutrons and protons. Further information on the forces can be obtained from a study of the scattering of free neutrons by protons. In such experiments a parallel beam of neutrons is allowed to impinge upon a target containing hydrogen atoms—paraffin, for example—and the number of neutrons deflected through various angles in elastic encounters is determined as a function of neutron energy (Sec. 25.15c). Since they have no charge, neutrons are unaffected by the electrostatic field and their scattering will directly reflect the operation of the nuclear forces.

The wave-mechanical theory of neutron-proton scattering is con-

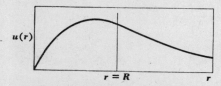

Fig. 25.5 Radial variation of the function $u(r) = r\psi(r)$ describing the deuteron in its ground state.

structed along much the same lines as is the theory of the deuteron, with the difference that the total energy E is positive for a free neutron, and hence all energy values are allowed. The wave function outside the radius R is no longer an exponential function but is a superposition of a plane wave, representing the stream of incoming particles, and expanding spherical waves, representing the scattered particles. The calculated scattering turns out to be relatively independent of the assumed shape of the potential in the case of low-energy neutrons, and the result is essentially fixed by the binding energy of the deuteron.

The amount of scattering observed for low-energy neutrons by protons turns out to be more than twice the maximum possible value predicted by this simple theory. So great a discrepancy clearly indicates that some factor enters the scattering problem which is not present in the deuteron. The missing factor was pointed out by E. P. Wigner in 1936, who observed that while they are in the ground state of the deuteron, the neutron and proton necessarily have parallel spin orientations (to produce the observed spin of 1 for the deuteron); no such restriction applies in the scattering of free neutrons, where both parallel and antiparallel spins occur. It is then possible to adjust the assumed interaction between neutrons and protons with *antiparallel* spins in such a way as to fit the scattering data. The results of this adjustment show that the forces depend strongly on the relative spins, the potential well in the antiparallel case being considerably less deep and having a larger effective radius than that in the parallel case. The situation is sometimes described in terms of an excited (singlet) state of the deuteron, with $I = 0$ and a negative binding energy of about 0.1 MeV.

c. Charge Independence of Forces   It was for some time generally believed that nuclear forces operated only between neutrons and protons, and a number of theories of nuclear structure were based on this assumption. In 1935, however, experiments on the scattering of high-velocity protons in hydrogen gas showed that nuclear forces also play a role in the interaction of protons with protons. Further work on the same line has led to the conclusion that the nuclear part of the force between two protons (p-p) is the *same* as that between a proton and a neutron (p-n) to a high degree of accuracy.[1] In addition, a comparison of energies of certain isobars in which neutrons are replaced by protons and protons by neutrons yields evidence for the equality of n-n, p-p, and p-n forces. The now generally accepted view that these three forces are essentially the same is often referred to as the assumption of *charge independence*

---

[1] This statement applies only to interactions which are allowed in both cases. The Pauli principle excludes certain interactions of two protons; e.g., two protons with otherwise identical quantum numbers must have antiparallel spins.

of nuclear forces. This close similarity between neutrons and protons in their behavior in the nucleus suggests that these two particles are fundamentally closely related, that they may be regarded as two states—differing as to charge—of the same particle.

d. Interaction of Nucleons in Nuclei: Exchange Forces   Thus far, we have discussed mainly the character of the forces acting between pairs of nucleons, as observed in simple two-body systems. We have found that an adequate account of these interactions can be given under the assumption that the forces are derivable from a potential which depends somewhat upon the relative spin orientations of the particles but apparently not upon whether they are protons or neutrons. When we come to consider systems comprising more than two particles, we must expect complications. Even if the forces between pairs of nucleons were known in detail, the mathematical complexity of the many-body problem would be formidable. Again, in order to obtain some insight into the binding of nuclei, we assume the simplest possible forces and test the consequences of these assumptions by experiment.

The most natural assumption about internucleon forces is that each nucleon is attracted by every other one in the nucleus with a force like that acting between two isolated nucleons. That this is not the case can be shown from the fact that the observed binding energy per particle is essentially constant in all but the lightest nuclei. In a nucleus containing A particles, $A(A-1)/2$ pairs can be formed; if attractive forces existed between all such pairs, the total binding energy would be roughly proportional to A^2 rather than to A. The observed radii of nuclei increase with A as if each nucleon occupied a constant volume, irrespective of how many others are present in the nucleus. It is evident then that the forces between nucleons "saturate" in some way; that once a nucleon has formed a few bonds, it ignores any further nucleons which may be added to the system.

A common hypothesis about internucleon forces in the nucleus is that they have at least in part an "exchange" character, i.e., that the law of force between two particles depends in a direct way upon the symmetry of the wave function with respect to their interchange. The binding of homonuclear molecules has something of this character, in that the effectiveness of the electrostatic force is determined by the spatial distribution of the charges, which in turn is governed by the symmetry requirements. In molecular binding, however—and, in fact, in all atomic phenomena—the forces are electrostatic or electromagnetic in origin and do not themselves contain any explicit dependence on symmetry properties. The energy difference between states of different symmetry character is then given by the *exchange integral*, which takes account f the tendency of particles to cluster together in some states and to void one another in other states. Such effects presumably operate within the nucleus, and a force depending only on distance, called an "ordinary" force, will have some saturation effects. Quantitative calcuations show that the effects so produced are insufficient, and it is necessary to introduce a force which depends *explicitly* on the symmetry of the wave function; such a force is called an *exchange force*. Several types of exchange forces have been proposed; perhaps the simplest of these is the so-called *Majorana interaction*, in which it is assumed that two particles attract one another if the wave function describing the entire system does not change sign, i.e., is symmetric, when the spatial coordinates of the two particles are interchanged, and that they repel f the wave function changes sign, i.e., is antisymmetric. For this case we assume that the wave function describing a nucleus can be written as a combination of one-particle wave functions, each characterized by a set of quantum numbers n, l, m_l, analogous to those used for atomic wave functions. Because of the Pauli principle, there can be at most four nucleons in any state with a given value of n, l, and m_l: two protons with opposite spins and two neutrons with opposite spins. The wave function is automatically symmetric with respect to the exchange of the spatial coordinates of any pairs among these four particles, and hence there are attractive forces between all pairs. Thus an α particle, comprising four nucleons in the lowest state, will be a tightly bound structure. If now a fifth nucleon is added, it must necessarily go into another state, and the wave function will be antisymmetric with respect to exchange of this particle with an identical particle in the first level. Exchanges with any of the other three lead to wave functions which are partly symmetric and partly antisymmetric. Since the antisymmetric pairs are assumed to lead to repulsive forces, the net effect is to make the five-nucleon system less strongly bound than the α particle (Li^5 and He^5 are in fact not bound, and do not occur in nature). As further nucleons are added, the number of both symmetric and antisymmetric pairs increases, and by a suitable adjustment of the assumed magnitudes of the resulting attractive and repulsive forces, the observed linear dependence on A of the binding energy and volume per nucleon can be accounted for. Detailed studies of the scattering of high energy—up to 350 MeV—neutrons and protons in hydrogen give strong evidence for the existence of exchange forces; however, the amount of repulsion observed in the antisymmetric interactions is too small to account for the saturation and other factors must be considered.

e. Nuclear Potential   Although the details of the interactions between nucleons within an actual nucleus are complicated, it is possible and useful to make a representation in terms of a relatively simple potential well for the interaction between an "average" nucleon and the rest of the nucleus.   The crude square well used in the deuteron problem (Fig. 25.4) predicts the situation of an average neutron; for a moderately heavy nucleus, the binding energy is about 8 MeV and the kinetic energy is 10 to 15 MeV.   The nuclear potential acting on the *most loosely* bound neutron fluctuates from one nucleus to the next; the binding energy (or separation energy) of the last neutron varies between 6 and 8 MeV in the stable nuclides.

The potential acting on a *proton* must include the electrostatic effect due to the repulsion of the other protons in the nucleus.   Outside the nuclear surface, the electrostatic potential has the form $(Z - 1)e^2/4\pi\epsilon_0 r$ as far as the effect on the last bound proton is concerned.   Inside, the electrostatic potential is modified because of the volume distribution of the charge, and there it may be approximated as a constant added (algebraically) to the nuclear potential.

The *total* electrostatic energy of a nucleus may be estimated from the work necessary to assemble a charge Ze uniformly distributed in a sphere of radius R.   This coulomb potential energy is

$$P_C = \frac{3}{5}\frac{(Ze)^2}{4\pi\epsilon_0 R} \tag{25.17}$$

An alternative calculation, in which only the mutual electrostatic energy of the Z protons is included, leads to

$$P_C = \frac{3}{5}\frac{Z(Z-1)e^2}{4\pi\epsilon_0 R} \approx 0.61\frac{Z(Z-1)}{A^{\frac{1}{3}}} \quad \text{MeV} \tag{25.18}$$

The situation of the last proton in a medium or heavy nucleus is shown in Fig. 25.6a, where E is again about 8 MeV.   It may be mentioned that even that part of the potential which is due to purely nuclear forces (P_N on the diagram) is not quite the same for protons and neutrons in a heavy nucleus which contains more neutrons than protons.   The protons, being in the minority, form more symmetric bonds per particle and hence are subject to a deeper potential, as far as nuclear forces are concerned, with the result that the net binding energy is about the same for protons and neutrons.

f. Nuclear Barrier   A diagram of the type shown in Fig. 25.6a can also be used to represent the binding of a combination of particles, such as an α particle, to the remainder of the nucleus.   In particular this representation can help us to understand the phenomenon of α decay.   In

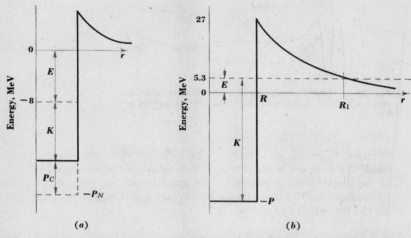

Fig. 25.6 Potential-energy functions for (a) the most loosely bound proton in a medium or heavy nucleus and (b) the most loosely bound α particle in Po^{210}.

Fig. 25.6b are shown the energy relations which might apply, for example, to $_{84}Po^{210}$. Since 5.3-MeV α particles are emitted, it is evident that the total energy, kinetic plus potential, of the α particles before ejection is 5.3 MeV, as represented by the dashed line (a small correction for the recoil energy is ignored here). Inside the nucleus, the potential energy is negative, and so we may presume that the kinetic energy is quite high—several times the value which it will ultimately have when the α particle escapes outside to the region of zero potential. In the intermediate region, however, the electrostatic potential is given by

$$P_C = \frac{Z_1 Z_2 e^2}{4\pi\epsilon_0 r}$$

where $Z_1 = 2$ and $Z_2 = 82$. If we use the value of the nuclear radius R given by Eq. (25.5), we find a potential of some 27 MeV at this point, just outside the range of the nuclear potential. Classically, then, the α particle is excluded from the whole region from R to some larger value R_1 where the potential energy has dropped to 5.3 MeV, since in this region the kinetic energy would be negative. Quantum-mechanically, however, there is a finite probability of penetrating this region. The wave function will have qualitatively the appearance of Fig. 25.7, with a rapidly oscillating character inside the nucleus, joining on to an exponentially falling function in the barrier region, and, for $r > R_1$, a sinusoidal

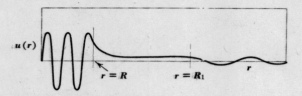

Fig. 25.7 Qualitative appearance of the wave function of an α particle escaping through the nuclear barrier of Fig. 25.6*b*.

function of small amplitude. The ratio of the flux outside to that inside is a measure of the probability of escape of the α particle and hence of the disintegration constant. An approximate calculation, first made by Gamow, and independently by Gurney and Condon, in 1929, yields for the probability of escape

$$\mathcal{P} \approx \exp\left[-\frac{2\sqrt{2M}}{\hbar} \int_R^{R_1} \sqrt{P(r) - E}\, dr \right] \tag{25.19}$$

This expression for the probability is a sensitive function of the radius R, and, by comparing lifetimes predicted by the formula with those observed for known α emitters, one can estimate R. The values so obtained are generally consistent with Eq. (25.5), with $R_0 = 1.2$ to 1.4×10^{-15} m. The probability of escape is also a most sensitive function of the energy E; for example, the energies of the α particles from Po^{214} and Th^{232} differ by less than a factor of 2, while the half-lives differ by a factor of about 10^{20}! It is evident that the electrostatic forces play a decided role in many nuclear phenomena and that to separate out the effects of the nuclear forces may be a matter of some difficulty.

g. Neutron-Proton Ratio in Stable Nuclides   Saturation of nuclear forces is at least partially accounted for by their exchange character. Certain other features of the distribution of stable nuclides also find a natural explanation under this hypothesis. Thus, the observed preference for near equality of N and Z follows immediately from the strong dependence of the nuclear force on the spatial symmetry of the wave function. Referring to our previous argument, we remember that four *different* nucleons (differing as to charge or spin orientation) can be placed in one n,l,m_l level (Fig. 25.8*a*), where they attract one another strongly. If we wish to construct a nucleus from four protons, on the other hand, two would have to be placed in another level (Fig. 25.8*b*) and the total number of attractive bonds would be smaller and the system less stable.

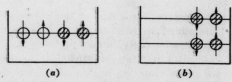

|(a)|(b)|

Fig. 25.8 Occupancy of nuclear quantum states by (a) two neutrons and two protons and (b) four protons.

The same argument can be extended to larger numbers of nucleons and shows that nuclei with $N = Z$ or $N = Z \pm 2$ are the most stable as far as the purely nuclear forces are concerned. As the number of particles increases, the mutual electrostatic repulsion of the protons will begin to make itself felt, effectively reducing the potential to which the protons are subject (Fig. 25.6a). In consequence of this effect, the proton levels will be displaced from the neutron levels, as indicated schematically in Fig. 25.9, and it will become advantageous to add a few extra neutrons. Thus $_{28}Ni^{60}$, with 28 protons and 32 neutrons, is more stable than $_{30}Zn^{60}$. Since the electrostatic energy increases roughly as Z^2, this effect becomes more pronounced as Z increases, and for high Z there are not only many more neutrons than protons, but the average binding energy per particle is effectively decreased.

A word of warning is perhaps in order at this point concerning such diagrams as Fig. 25.9. Since the potential is contributed by the nucleons

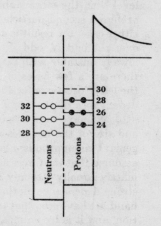

Fig. 25.9 Neutron and proton levels in $_{28}Ni^{60}$. Only the uppermost levels are shown.

themselves, the shape of the potential function will change as nucleons are added or removed, and the location of levels representing other than the most loosely bound nucleons has little significance. The diagram indicates qualitatively the effect of adding or moving one or two nucleons or of moving one nucleon from one state to another, but even so the magnitude of the total energy change may be quite different from the value the diagram suggests.

We observe a strong preference among the stable nuclides for even N ($N = A - Z$, the number of neutrons) and even Z, as opposed to odd N and odd Z, while nuclides with odd Z and even N and those with even Z and odd N occur in roughly equal numbers. These facts we can now understand in terms of the displacement of proton levels with respect to neutron levels. In $_{28}Ni^{60}$ (Fig. 25.9), for example, there are 14 filled proton levels, each with two protons (with spins opposed), and 16 occupied neutron levels. If another particle is to be added, it must go into an unoccupied level, and the chances are about even that the next available level will be a neutron level. Thus among the stable isotopes with an odd total number of particles A, it is to be expected that about half will have odd N and about half odd Z. In Ni^{60}, it happens that the next available level is a neutron level, corresponding to Ni^{61}; if now another particle is to be added, adding a second neutron will be more favorable than adding a proton, because the neutron level is lower and there is room for two particles in it. Thus $_{28}Ni^{62}$ is more stable than $_{29}Cu^{62}$; in fact Cu^{62} is radioactive and decays to Ni^{62}. Had the proton level been the more stable, i.e., if Cu^{61} were more stable than Ni^{61}, the preferred second particle would have been a proton, forming $_{30}Zn^{62}$. In either case, the result is a preference for the even-N, even-Z combination over the odd N, odd Z, and the latter, in medium and heavy nuclei, is always unstable with respect to the former. Among the light elements there are a few cases of stable odd-N, odd-Z nuclei; in these instances, the electrostatic effect is relatively unimportant, and the small spin dependence of the force gives rise to a preference for equal N and Z.

We have thus accounted for the main properties, the mass, charge, and size of the nuclides which are found in nature, making only rather general assumptions about the nature of nuclear forces. We have assumed that the forces act only over a short range, that they act approximately equally strongly on neutrons and protons, and that they have a strong dependence on the symmetry of the wave function. On the other hand, we have seen that there exist several alternatives to the last assumption, and it is by no means clear that this assumption is either adequate or essential.

About the *structure* of the nucleus, very little has been assumed in the foregoing. Whether the "quantum states" which we have used

really represent the independent motions of individual nucleons, or
whether the motions are so strongly perturbed that the "states" must
be taken to refer to some sort of collective motion of the nucleons as a
whole is not essential to these arguments. We discuss in later sections
experiments which have a bearing on a specific model for the nucleus.

Positrons, Artificial Radioactivity, Artificially Accelerated Particles

25.7 *Induced Radioactivity*

In 1930 Becker and Bothe bombarded Be with α particles and reported a
spectrum of electrons extending up to 4.5 MeV, which they attributed
to a γ ray of about this energy. To study the matter further, the Joliot-
Curies placed a lead foil and a polonium-beryllium source in a cloud
chamber, and with a magnetic field of about 1000 gauss, proceeded to
study the electrons ejected from the lead by the beryllium radiation. In
1932 they observed a number of "electrons" which seemed to be moving
toward the lead foil, since their tracks were curved in the wrong sense.
(The sense of deflection depends, of course, both on the sign of the charge
and the direction of motion.) It is now clear that the tracks were due
to positrons. After the announcement of the discovery of the positron,
the Joliot-Curies conducted further experiments on the radiation produced in Be by α-particle bombardment. They confirmed that positive
particles are produced when the hard γ radiation is converted in
lead. In a similar experiment in which aluminum, instead of beryllium, was bombarded by α particles, they found positrons again,
but this time no corresponding electrons. Further study showed that
the positrons were not produced in the converter but came from the
aluminum source itself. Finally, in the course of an investigation of
the effect of varying the α-particle energy, they noticed that the positrons
did not appear immediately when the polonium source was put in place but
built up gradually over a period of minutes. When the α particles were
removed, the effect died down gradually, again within a period of some
minutes. As a result of these experiments, the Joliot-Curies concluded
in 1934 that the effect must be ascribed to a radioactive material produced
by the α-particle bombardment. The reaction they suggested was

$$_{13}\text{Al}^{27} + {}_2\text{He}^4 \rightarrow {}_{15}\text{P}^{30} + {}_0n^1$$

The isotope P^{30} does not exist in nature, and they assumed that it decays spontaneously, with emission of positrons, to a known stable isotope of silicon by the reaction

$$_{15}P^{30} \rightarrow {}_{14}Si^{30} + \beta^+ + \nu$$

The half-life was 3.25 min. That the radioactive substance was an isotope of phosphorus they subsequently confirmed by showing that the active material had the same behavior under chemical analysis as elemental phosphorus.

In the same experiments, the Joliot-Curies found that boron bombarded by α particles also produced a positron activity with a half-life of about 14 min. This activity they attributed to another new isotope, N^{13}, formed in the reaction

$$_5B^{10} + {}_2He^4 \rightarrow {}_7N^{13} + {}_0n^1$$

and decaying to C^{13}:

$$_7N^{13} \rightarrow {}_6C^{13} + \beta^+ + \nu$$

They suggested that the same radioisotope might be produced by artificially accelerated deuterons bombarding carbon, through the reaction

$$_6C^{12} + {}_1H^2 \rightarrow {}_7N^{13} + {}_0n^1$$

This reaction was reported by Crane and C. C. Lauritsen in 1934.

Since the time of these early experiments, a great many radioactive isotopes have been produced artificially, with protons, deuterons, α particles, neutrons, and many other particles as bombarding agents; more than 1200 species are known. In addition to the fact that the study of these radioisotopes has yielded an enormous amount of information about the nucleus, they have had a considerable practical importance, particularly as tracers in chemical and biological investigations. In work of this kind, certain compounds are "tagged" by incorporating into them a radioactive isotope of one of their normal constituents, and the subsequent progress of this constituent through various chemical changes is traced by means of its easily detectable radioactivity. This technique permits many operations not amenable to chemical analysis, such as study of the rate of exchange of an element from one ionic form to another or the rate at which a given element is exhausted and replaced in a living organism. Radioisotopes of almost all the elements of biological interest are available: some of the most used are C^{14} (with a half-life of 5400 years), P^{32} (30 days), Na^{21} (14 hours), and S^{35} (87 days).

25.8 Nuclear Transformations with Artificially Accelerated Particles

From the time of the earliest observations of transmutations induced by α particles from radioactive sources, there was speculation as to whether it might not be possible to use artificially accelerated positive ions for the purpose. One obvious advantage would lie in the much greater number of particles available, for even a current as small as 1 μA represents some 6×10^{12} (singly charged) particles per second, or about the number of α particles emitted by the polonium derived from 160 g of radium. The prospect of inducing nuclear reactions with particles of practically attainable energies seemed rather hopeless, however, in view of the coulomb barrier. The barrier height of a nucleus of charge Z and mass number A for a singly charged particle is

$$P = \frac{Ze^2}{4\pi\epsilon_0 R} \qquad \text{or} \qquad P \approx \frac{Z}{A^{\frac{1}{3}}} \qquad \text{MeV}$$

so that something of the order of 1 or 2 MV would seem to be required even for the lightest elements.

With the development of the theory of barrier penetration, it appeared that nuclear processes might be observable even at energies considerably less than the barrier height, particularly in light nuclei. Accordingly, Cockroft and Walton at the Cavendish Laboratory set out to construct an accelerating system. Their first system, described in 1930, comprised a transformer-rectifier voltage source and a glass discharge tube into which ions were injected from a Wien-type canal-ray tube. With this installation they obtained a beam of 300 keV protons and looked briefly for reactions leading to γ-ray emission, with no definitive results. By 1932, they had extended and improved the apparatus, obtained proton beams up to 10 μA at energies up to 700 keV. The accelerating tube was provided with an extension into which various targets could be inserted. Thin mica windows in the side of the extension permitted the escape of disintegration products to the outside, where they could be detected by means of a zinc sulfide scintillation screen. The mica window was thick enough to stop elastically scattered protons. The first successful experiment was with lithium. With protons of 125 keV energy bombarding a target of lithium metal, they observed bright scintillations—about five per minute with a beam of 1 μA. As the proton energy was increased, many more scintillations were observed; at 500 keV, they estimated the efficiency to be about 10^{-8} disintegrations per incident proton. Absorption measurements gave a value of 8.4 cm for the maximum range of the particles producing the scintillations, and measurements of the ionization showed that they were α particles. It was concluded by Cockroft and Walton that the reaction involved was

$$_3\mathrm{Li}^7 + {_1}\mathrm{H}^1 \rightarrow {_2}\mathrm{He}^4 + {_2}\mathrm{He}^4$$

From the (poorly known) masses, they estimated that 14.3 ± 2.7 MeV should be available for kinetic energy of the α particles; from the observed range they calculated 17.2 MeV, in reasonable agreement. That two α particles were in fact produced simultaneously, in opposite directions, they checked with the arrangement shown in Fig. 25.10. The target consisted of a thin layer of lithium deposited on a thin sheet of mica, mounted at 45° to the incident proton beam. Two mica windows in opposite sides of the target tube allowed α particles produced in the target to escape and to be observed on the two scintillation screens S_1 and S_2. Two observers recorded the scintillations independently on a single moving tape: the existence of a large proportion of coincidences on the record was evidence that the process assumed was taking place. This conclusion was later confirmed by Dee and Walton with cloud-chamber photographs.

A number of other elements were investigated by Cockroft and Walton in 1932 to 1933. Both boron and fluorine were found to give comparatively large α-particle yields under proton bombardment; other elements, including Be, Na, and K, produced detectable numbers. The disintegration of Li, B, and F by 1.2-MeV protons from the cyclotron was reported by Lawrence, Livingston, and White in 1932, only a few months after Cockroft and Walton's first report. Oliphant, Kinsey, and Rutherford observed the disintegration of Li and B by deuteron bombardment and Crane, Lauritsen, and Soltan announced the artificial production of neutrons by helium-ion bombardment of beryllium in 1933.

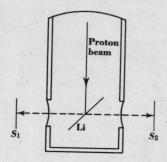

Fig. 25.10 Arrangement for detecting coincident particles produced in the proton bombardment of Li⁷.

Experiments on the disintegration of lithium with protons provided the first direct quantitative check of Einstein's equation, $E = mc^2$, relating mass and energy. The fact that all the masses involved were known from mass-spectrometer measurements and that the energy release was large compared to the errors of observation made it possible to check the agreement between the loss in mass and the observed kinetic energy within about 20 percent. Modern determinations from other nuclear reactions check this relation to high accuracy.

25.9 Accelerators

The years that have passed since the experiments of Cockroft and Walton have seen a considerable development in techniques and equipment for production of high-energy charged particles. We shall enumerate here some of the principal types which have found application in problems of nuclear physics.

a. High-tension Set   Voltages up to a few million have been produced by straightforward cascade connection of transformer-rectifier sets. Because of the difficulty of building rectifiers capable of withstanding high voltages, such sets are ordinarily built in units of about 200 kV, either with individual transformers, each excited by a special winding on the previous one, or with a single transformer and some form of voltage-multiplying circuit. The principal load on such a set is often corona loss (local ionization of the air near sharp points), which may amount to several milliamperes, and, unless large condensers are used, an undesirable ripple in voltage results. For example, if the capacitance is of the order of $10^{-2}\mu$F, a drain of 1 mA produces a voltage change of 1.6 kV in $\frac{1}{60}$ s. For many purposes, a ripple of this magnitude is acceptable, but situations do occur where a steadier voltage is desirable. For this reason, such installations are often designed to work at frequencies of 500 Hz or more.

A sketch of a typical accelerating tube is shown in Fig. 25.11. The tube is broken up into many sections, preferably with a controlled voltage for each section, partly to aid in focusing the ion beam and partly to distribute the voltage gradient as uniformly as possible along the insulating surfaces. The positive ions to be accelerated are generated in an *ion source*, located on the top of the tube. Gas—hydrogen, if protons are desired—is admitted into a discharge tube, in which the atoms are ionized by electron bombardment. An electrode at the bottom of the ion source, provided with a fine hole and held at a negative potential with respect to the discharge tube, extracts the ions and injects them into the main accelerating tube. By means of a series of electrostatic lenses, the ions

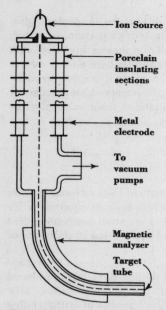

Ion Source

Porcelain
insulating
sections

Metal
electrode

To
vacuum
pumps

Magnetic
analyzer

Target
tube

Fig. 25.11 Schematic diagram
of a high-voltage accelerating
tube and a 90° magnetic analyzer.

are collected into a narrow bundle as they are accelerated down the tube. Since a certain amount of gas also escapes into the tube, it is necessary to evacuate the tube continuously with large diffusion pumps.

The ion source produces, besides protons (H⁺), also singly charged ions H_2^+ and H_3^+. In addition, incidental ions formed in the residual gas in the main tube are collected in the beam and accelerated. It is necessary, therefore, to provide a magnetic deflection or combined magnetic and electrostatic deflection at the lower end of the tube to reject the undesired components. With a properly designed magnetic or electrostatic analyzer, the measured field can be used to determine the particle energy with much higher accuracy than a direct voltage measurement can give.

b. Electrostatic Accelerator   A convenient way to produce a steady high voltage without the difficulties of ripple associated with transformer-rectifier sets is to use a continuously charged moving belt, as devised by Van de Graaff in 1931. The principle is shown by the schematic diagram of Fig. 25.12. Two rollers are provided, the lower driven by a motor, the upper located in the high-voltage terminal, well insulated from ground. Over the rollers passes an endless belt of insulating material, and, across the face of the belt, opposite the lower roller, is a "comb" consisting of a row of sharp pins with their points just clear of the moving belt. A supply of 10 to 30 kV between the comb and the belt produces a corona discharge, causing positive ions to flow from the comb to the belt and negative ions and electrons from the belt to the comb. The positive charges are then carried by the belt to the high-voltage terminal where a second comb connected to the terminal picks them off. The voltage to which the terminal can rise is determined by the balance between the current supplied by the belt and that lost by corona or drained down the accelerating tube. In a variation designed by Herb, Parkinson, and Kerst and now universally used for high-voltage applications, the machine, together with the accelerating tube, is mounted in a pressure tank so that it can be operated under several atmospheres' pressure of air or other insulating gas. Since the electrical breakdown strength of a gas is roughly proportional to pressure, it is possible to construct a machine with much smaller dimen-

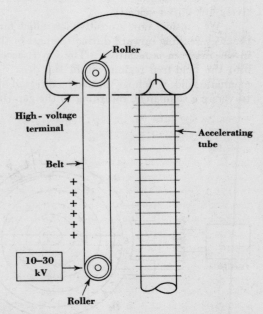

Fig. 25.12 An electrostatic accelerator in which charges are sprayed on a moving belt and carried up to an insulated high-voltage terminal.

sions when high pressure is used.   For example, clearances of the order
of 10 to 15 ft are required around a 1-MV installation at atmospheric
pressure, while 5- to 8-MV electrostatic accelerators can be housed in
8- or 10-ft-diameter pressure tanks.

c. Cyclotron   High particle energies can be obtained without the use
of high voltages by means of the *magnetic-resonance accelerator*, or
cyclotron, devised by Lawrence and Livingston in 1932.   In the cyclotron,
charged particles are given repeated accelerations in a radio-frequency
field.   Under the influence of a strong magnet, the particles move
alternately through field-free regions and regions of high field, in phase
with the changing electric field.   A diagram illustrating the principle
is shown in Fig. 25.13.   Within a flat, cylindrical vacuum chamber
B, placed between the poles of a magnet (not shown), are two *D*-shaped
electrodes D_1, D_2 (called dees) consisting of hollow, flat half-cylinders.
The dees are coupled to, or made part of, a resonating electric circuit
driven by a high-power radio-frequency oscillator, so that an alternating
voltage appears across the gap separating the dees.   An ion source *C*,
located at the center of the chamber, supplies positive ions with a relatively low initial velocity.

We consider now a positive ion which finds itself in the gap between
the dees, moving upward during the part of the cycle when the upper dee
in the diagram is negative.   The ion is accelerated in the electric field
into the field-free region within the dee.   Under the influence of the
magnetic field, which we take to be directed out of the paper, the particle
traverses a semicircle, returning to the gap (but now moving downward)

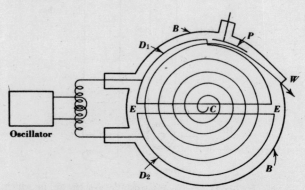

Fig. 25.13   Schematic diagram of a cyclotron operating in a
magnetic field out of the plane of the paper.

after a time[1] $t = \pi/\omega_B$, where ω_B is its angular frequency in the magnetic field:

$$\omega_B = \frac{v}{r} = B \frac{e}{M} \tag{25.20}$$

If the angular frequency of the radio-frequency field ω_E is so adjusted that the potential difference of the dees has reversed during the time that the particle was inside the dee, i.e., if $\omega_E = \omega_B$, the ion will again be accelerated. The crucial point is that the angular frequency ω_B is constant, independent of the velocity of the particle (insofar as the relativistic variation of mass can be neglected); as the velocity increases, the radius of curvature also increases and the time per half-turn remains constant. This process of acceleration every half period continues, with the particle moving in a series of connected semicircular arcs, until a limiting radius R_0, determined by the dimensions of the magnetic field, is reached. At this point, the ion can be extracted by means of a radial electrostatic field supplied by the deflector P, which provides enough deviation in the trajectory to allow the particle to escape through the fringing field of the magnet and through the window W to a target outside. Alternatively, targets may be inserted into the dee chamber and the circulating ion beam used without extraction. The ions are focused in the inhomogeneous electrostatic field between the dees and by a (deliberately introduced) slight decrease of B with radius so that a parallel beam is produced.

The maximum energy which can be achieved depends upon the available magnetic field, the limiting radius R_0, and the nature of the particle.

$$K_{max} = \frac{p^2}{2M} = \frac{B^2 e^2 R_0^2}{2M} \tag{25.21}$$

For example, a 60-in.-diameter cyclotron, operating at a magnetic field of 16,000 gauss, can produce 26-MeV deuterons with a radio frequency of 12 MHz. In principle, protons of twice this energy can be produced by doubling the frequency, but changing the frequency usually involves a major operation, so that cyclotrons are sometimes operated at reduced B for protons and produce only· half the possible energy.

---

[1] We have here ignored the motion during the electric acceleration; i.e., we assume that the distance over which the electric field operates is small compared to the path in the field-free space. This is not a good assumption near the beginning of the motion, and the trajectories during the first few turns are rather complicated.

d. Synchrocyclotron   The constancy of the frequency ω_B in Eq. (25.20) breaks down when the particles have relativistic velocity ($v \approx c$), so that M deviates appreciably from the rest mass. This change could, in principle, be allowed for by shaping the magnetic field in such a way as to increase B slightly as the radius increases, but such a variation would introduce serious defocusing of the beam. The alternative, suggested independently by McMillan and Veksler in 1945, is to allow the magnetic field or the radio frequency to vary with time as the particles are accelerated in such a way as just to compensate for the changing mass. It can be shown that in either case there exist stable orbits in which the particles tend to keep in step with the changing fields. A cyclotron making use of this principle is called a synchrocyclotron. Ordinarily, a synchrocyclotron is used to accelerate heavy particles—protons, deuterons, or α particles—and the radio frequency is varied while the magnetic field is held fixed. In the Berkeley 184-in. frequency-modulated synchrocyclotron, which produces 340-MeV protons, the required frequency change is about 35 percent. For electron acceleration, or for acceleration of heavy particles to energies exceeding a few hundred MeV, the *synchrotron* is generally preferred.

e. Synchrotron   The synchrotron operates on essentially the same principle as the synchrocyclotron except that the particles are injected at near-relativistic speeds and the magnetic field is varied in time, with the radio frequency held constant. The particles are injected at a relatively low value of the magnetic field and pick up energy from the radio-frequency accelerating gap as the magnetic field increases. With constant radio frequency ω_0 and essentially constant velocity (near c) the particles remain at nearly the same radius, $R = c/\omega_0$, throughout the acceleration, and only at this radius is a magnetic field needed. This feature means a great saving, since the entire center portion of the magnet is eliminated. A number of electron synchrotrons yielding energies of 1 GeV or more have been built.

For heavy particles, it is not so easy to inject with $v \approx c$. For this reason, one ordinarily varies both the radio frequency and the magnetic field in such a way as to keep the orbital radius approximately constant.

f. Betatron   The betatron or *induction accelerator*, invented by Kerst in 1940, is a device for accelerating electrons. Its operation depends simply on the fact that a changing magnetic flux induces an electromotive force in any enclosing circuit. From the law of Ampère, we have, for any closed path,

$$\oint \mathbf{E} \cdot d\mathbf{l} = \frac{d}{dt} \int \mathbf{B} \cdot d\mathbf{S}$$

An electron describing a circular path of radius R in a changing magnetic field is subject to a force $f = e\mathbf{E}$. The rate of change of momentum p is then

$$\frac{dp}{dt} = f = \frac{e}{2\pi R}\frac{d\phi}{dt}$$

where $\phi = \int \mathbf{B} \cdot d\mathbf{S}$ is the flux enclosed by the path. The momentum gained if the electron starts from rest when $\phi = 0$ is

$$p = \frac{e\phi}{2\pi R}$$

On the other hand, the value of the magnetic field at the orbit required to keep the electron at the constant radius R is

$$B_R = \frac{p}{eR}$$

Thus we have an equilibrium situation at all times if

$$B_R = \frac{\phi}{2\pi R^2} \tag{25.22}$$

i.e., if the value of B_R at the orbit is one-half the average of B over the area bounded by the orbit. The required variation of B can be obtained by choosing a suitable shape for the pole pieces of the magnet. The fact that the so-called *betatron condition* [Eq. (25.22)] can be satisfied independently of relativistic considerations means that the device can be used to accelerate electrons to very high energy. The maximum energy is determined by the available magnetic field and the radius.

g. Linear Accelerator   The linear accelerator also uses the principle of repeated accelerations, but the particles are allowed to travel in straight lines instead of being deflected in a magnetic field. The accelerator consists essentially of a long pipe, with coaxial cylindrical accelerating electrodes mounted inside, separated by short gaps. The whole chamber constitutes a resonant cavity, excited by external oscillators in such a mode that the time-varying electric field is directed along the axis. The lengths of the electrodes are so adjusted that the ion is "coasting" in the field-free interior during the time that the radio-frequency voltage has the wrong sign and traverses the gaps only when the field is so directed as to accelerate the ion. Linear accelerators can be designed either for electrons or for heavy particles.

Nuclear Reactions and Nuclear Models

With the availability of artificially accelerated particles, the number of nuclear reactions amenable to study has increased enormously. Thousands of reactions have been studied, many in considerable detail. In the following sections we shall consider some reactions which occur, the techniques used in studying them, and what can be said about nuclear models as a result of such studies.

25.10   General Features of Nuclear Reactions

a. Types   The most commonly used bombarding particles are protons, deuterons, and neutrons, the latter themselves produced in a nuclear reaction. Artificially accelerated α particles are also frequently used, as well as H^3 ions (tritons), He^3 ions, and heavier ions. γ rays and x-rays can also be used as disintegrating agents, and some reactions induced by electron bombardment have been observed. In general, at moderate bombarding energies—less than 10 MeV—two products appear as a result of a nuclear reaction: a light particle, usually one of the above mentioned, and a heavier one, referred to as the residual nucleus. Three-body production sometimes occurs, and at high energies even more products may be observed. As a convenient classification of the simpler types of nuclear reactions, we enumerate the following categories.

(I) ELASTIC SCATTERING   The incident particle strikes the target nucleus and undergoes an elastic collision. When this process is viewed in a coordinate system in which the target nucleus is initially at rest (the *laboratory system*), the scattered particle loses energy because the target nucleus recoils. In the center-of-mass system, in which the two particles approach one another in such a way as to keep the center of mass at rest, there is no energy transfer.

(II) INELASTIC SCATTERING   The incident particle reappears after the interaction with a lower energy; some of its energy has been taken up by the target nucleus, which is excited to a higher quantum state. An example is

$$Li^7 + H^1 \rightarrow Li^{7*} + H^1$$

The asterisk is used to indicate that the residual nucleus is in an excited state. In the present example the excess energy is later radiated away in the form of a γ quantum.

(III) SIMPLE CAPTURE The incident particle is captured by the target nucleus, and a new nucleus is formed. Nearly always the residual nucleus has a considerable excess of energy and radiates one or more γ quanta. An example is

$$C^{12} + H^1 \rightarrow (N^{13}) \rightarrow N^{13} + \gamma$$

Here the parentheses are used to indicate that the nucleus in question is in an excited state with an energy which is determined by the initial conditions of the reaction.

Particularly in elements of high atomic number, it occurs not infrequently that the energy released in the transition between two nuclear quantum states is taken up directly by an atomic electron, without the intervention of a γ ray. In this process, called *internal conversion*, the oscillating electromagnetic field associated with the nuclear transition interacts with the atomic electrons, resulting in the ionization of the atom. Just as in the ordinary photoelectric effect, the transition energy is given by the sum of the observed energy of the ejected electron plus its binding energy in the atom. The electron lines so produced are very sharp, and they have been much used in determining the energy levels of the heavy elements. By comparing two internal-conversion electron lines from the same transition, for example, lines due to conversion by K- and L-shell electrons, one can determine the difference in the atomic binding energies and by comparing with the known values from x-ray data, one can identify (as to Z) the radiating atom. This procedure has been used to ascertain whether the transition in question follows or precedes the disintegration. The ratio of the number of transitions of a given energy which result in internal conversion to the number which yield γ rays is called the internal conversion coefficient α. In the natural radioactive elements, α normally lies in the range 10^{-4} to 10^{-1}. In a few cases, however, where the γ radiation is forbidden or strongly discouraged, as for example when the angular momenta of the initial and final states differ by several units, the internal conversion coefficient may be much larger than 1. It is possible by means of the quantum theory to calculate the internal conversion coefficients to be expected for various values of the angular-momentum change in any given transition, so that the experimental determination of α can yield important information on the nuclear quantum states involved. A case of some interest occurs in Po^{214}, where an excited state, at 1.41 MeV, makes a transition to the ground state entirely through internal conversion: no corresponding γ ray is observed at all. There is reason to believe that this situation occurs because both the initial and final states have total angular momentum zero, single-quantum electromagnetic transitions between spin zero states being strictly forbidden.

(IV) DISINTEGRATION   On striking the target nucleus, the incident particle is absorbed and a different particle ejected.   An example is the disintegration of beryllium by α particles, producing neutrons:

$$Be^9 + He^4 \rightarrow C^{12} + n^1$$

(V) PHOTOEXCITATION AND PHOTODISINTEGRATION   A γ ray is absorbed by the target nucleus, exciting it to a higher quantum state; if the energy is sufficient, a particle may be ejected.   The photodisintegration of the deuteron, requiring 2.225 MeV, is an example:

$$H^2 + \gamma \rightarrow H^1 + n^1$$

A special case of photoexcitation of some interest is the so-called *coulomb excitation*, in which the varying electric field of a charged particle passing by a nucleus may induce transitions to higher states.

(VI) PRODUCTION OF "STRANGE" PARTICLES   In high-energy reactions mesons, hyperons, and antiparticles of the proton and neutron may be produced.   Use of these particles to bombard targets leads to additional reactions.   All the mesons and the unfamiliar members of the baryon group are brought forth in the laboratory through high-energy nuclear reactions.

(VII) SPONTANEOUS DECAY β- and α-decay processes may be regarded as nuclear reactions; they differ from those discussed above in that the total energy of the system is not under the experimenter's control.

b. Notation   It is convenient to have a shorthand notation for nuclear reactions; a commonly used form is the following:

$$X^A(a,b)Y^{A'}$$

where X^A and $Y^{A'}$ indicate the chemical symbol and mass number of the initial target and residual nuclei, respectively, and a and b the incoming (bombarding) and outgoing particles.   Thus, the reaction first studied by Rutherford (Sec. 25.1) would be written

$$N^{14}(\alpha,p)O^{17}$$

A radioactive disintegration is sometimes represented as

$$X^A(\beta^{\pm})Y^A$$

for example,

P$^{30}(\beta+)$Si^{30}

The concomitant emission of the neutrino is usually not indicated. A reaction leading to an excited state of the residual nucleus which subsequently decays by γ emission may be written in two stages,

N$^{14}(\alpha,p)$O^{17}*   O^{17}*(γ)O^{17}

or abbreviated,

N$^{14}(\alpha,p\gamma)$O^{17}

c. Conservation Laws   In any nuclear reaction, certain quantities must be conserved:

1. The *total energy* of the products, including both mass energy and kinetic energy of the particles, plus the energy of any γ rays or neutrinos which may be involved, must equal the mass energy of the initial ingredients plus the kinetic energy brought in by the bombarding particle (the target nucleus is ordinarily at rest).

2. The total *linear momentum* of the products must be equal to the momentum brought in by the bombarding particle.

3. The total *electric charge* is conserved.

4. The total *number of baryons* is constant.

5. The total *angular momentum* J, comprising the vector sum of the intrinsic angular momenta I (spins) and relative orbital momentum l of the particles, is conserved. If two particles collide head on, they have no relative orbital angular momentum and the value of J will be given by vector addition of the two intrinsic spins. In a glancing collision, orbital angular momentum is present and must be added (again vectorially) to the two I's to produce the total angular momentum J. In the general case, the beam of incident particles is represented as a plane wave which may be resolved into components representing various quantized values of the orbital angular momentum.

6. The *parity* (Sec. 25.3) of the system determined by the target nucleus and bombarding particle must be preserved throughout the reaction. The initial system comprises the target nucleus and the incident particle, approaching with a certain fixed orbital angular momentum. If the wave function describing this situation remains unchanged under inversion of the coordinates, the parity is even; if the wave function changes sign, the parity is odd. The parity for such a system is the product ($+1$ for even parity, -1 for odd) of the intrinsic parities of the target nucleus and bombarding particle and the contribution from the orbital angular momentum l; for even l, the contribution is $+1$, for odd l, it is -1.

d. Nuclear-reaction Kinetics   The energy balance in a typical nuclear reaction may be written

$$M_0c^2 + M_1c^2 + E_1 = M_2c^2 + M_3c^2 + E_2 + E_3 \qquad (25.23)$$

where M_0 = exact mass of target particle
$\quad\quad M_1$ = mass of bombarding particle
$\quad\quad M_2$ = mass of ejected particle
$\quad\quad M_3$ = residual nucleus
E_1, E_2, E_3 = kinetic energies of the corresponding particles
The *energy change* or Q of the reaction is defined by the mass difference

$$M_0 + M_1 = M_2 + M_3 + \frac{Q}{c^2} \tag{25.24}$$

Evidently

$$Q = E_2 + E_3 - E_1 \tag{25.25}$$

If Q is positive, the reaction is exoergic, i.e., energy is released in the process; if it is negative, the reaction is endoergic, and energy is absorbed. The determination of Q is of interest both from the standpoint of measuring masses—as a check on mass-spectrometer values for stable isotopes, and as a method of obtaining masses for unstable ones, e.g., the neutron— and also, where excited levels of the residual nucleus occur, for determining the energies of these levels.

In an experiment to measure a Q value, the bombarding energy E_1 and the energy of the ejected particles E_2 at some specified angle θ are measured. The Q value can then be obtained by applying the laws of conservation of momentum. The momentum diagram for a typical reaction is illustrated in Fig. 25.14. We have

$$p_{1x} = p_{2x} + p_{3x} \qquad 0 = p_{2y} + p_{3y} \qquad E_1 = E_2 + E_3 - Q$$

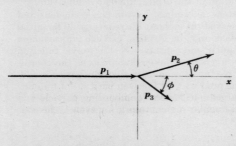

Fig. 25.14 Momentum diagram for a nuclear reaction. The incident particle has momentum p_1, the ejected particle momentum p_2, and the residual nucleus momentum p_3.

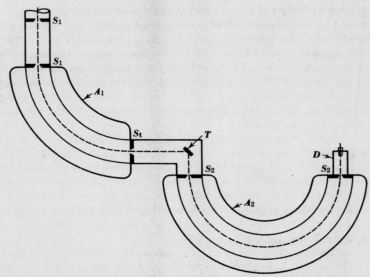

Fig. 25.15. Arrangement for a precise Q-value measurement. The incident particles are analyzed by the magnet A_1 and strike the target T; the ejected particles are analyzed by A_2.

or, since $p = \sqrt{2ME}$ in the nonrelativistic case,

$$\sqrt{2M_1E_1} = \sqrt{2M_2E_2}\cos\theta + \sqrt{2M_3E_3}\cos\phi$$
$$0 = \sqrt{2M_2E_2}\sin\theta + \sqrt{2M_3E_3}\sin\phi$$

We eliminate ϕ and E_3 to obtain Q in terms of E_1, E_2, and the masses

$$Q = E_1\frac{M_1 - M_3}{M_3} + E_2\frac{M_2 + M_3}{M_3} - \frac{2}{M_3}(M_1M_2E_1E_2)^{\frac{1}{2}}\cos\theta \tag{25.26}$$

Since the masses M rarely differ from the mass numbers A by more than a fraction of 1 percent, it is usually sufficiently accurate to use mass numbers in Eq. (25.26). Relativistic corrections are rarely necessary in work at moderate energies.

A typical experimental arrangement for measurements of this kind is illustrated schematically in Fig. 25.15. A beam of particles, produced in some type of accelerator, enters the magnetic analyzer A_1, where it suffers a 90° deflection before striking the target T. The path of the beam is defined by the set of slits S_1, and the energy of the particles is then determined by a measurement of the magnetic field in A_1. Of the particles produced in the target T, some of those which emerge at $\theta = 90°$ enter the magnetic spectrometer A_2. A measurement of the magnetic field in A_2 which is required to pass these particles through the slit system S_2 to the detector D then permits determination of $E_2(90°)$, the energy of the particles emitted at $\theta = 90°$. In principle, once the geometry is established, the Q-value determination requires only the measurement of the two magnetic fields.

With modern techniques, it is possible to measure Q values, for nuclear reactions producing charged particles, to an accuracy of 0.05 percent or better, and a great many Q values are now known to this precision. As an example of the close agreement obtained in different laboratories, the Q values of the reaction $Li^7(p\alpha)He^4$, are shown in Table 25.1. The mean deviation of these four values is less than 3 parts in 10,000.

Table 25.1 Q values for $Li^7(p,\alpha)$ He^4

Q, MeV	Source
17.338 ± 0.011	California Institute of Technology
17.340 ± 0.014	Massachusetts Institute of Technology
17.352 ± 0.009	University of Birmingham (England)
17.344 ± 0.013	Rice Institute

e. Mass Determinations   As one result of the growing collection of such precise Q values as the one just discussed, it has become possible to establish the masses of a large number of nuclides which do not exist in nature. The mass of the neutron, for example, can be determined relative to that of the hydrogen atom in the following set of reactions:

$$H^2 + H^2 \rightarrow H^3 + H^1 \qquad Q_1 = 4.0337 \pm 0.0017 \text{ MeV} \qquad (25.27a)$$

$$H^2 + H^2 \rightarrow He^3 + n \qquad Q_2 = 3.267 \pm 0.007 \qquad (25.27b)$$

$$H^3 \rightarrow He^3 + \beta^- \qquad Q_3 = 0.0186 \pm 0.0002 \qquad (25.27c)$$

If the second of these relations is subtracted from the sum of the first and third, there results $[M(n) - M(H^1)]c^2 = Q_1 + Q_3 - Q_2$. We obtain, for the neutron-hydrogen atom mass difference, $n - H^1 = 0.785$ MeV. A better value, based on many different measurements, is

$$n - H^1 = 0.78245 \pm 0.0001 \text{ MeV}$$

It is also possible, by combining reactions involving various nuclides, to derive independently of mass-spectrographic data a set of masses based on the standard C^{12}. For example, the Q value for the reaction $N^{14}(d,\alpha)C^{12}$ (the symbol d represents the deuteron) is accurately known. By combining this value with the Q values of several other reactions, the mass difference $N^{14} - C^{12}$ may be directly determined. In a similar way, other reactions can be found, connecting various of the light nuclei, and eventually a chain can be established leading directly to C^{12}. Such a chain necessarily contains many links, with errors which accumulate, but because it is mass *differences* which are determined, and these are small compared to the actual masses, the final accuracy is still good.

25.11 Masses of Mirror Nuclides

Among the many unstable nuclides whose masses are known from nuclear-reaction Q values, a set of particular importance is the so-called *mirror nuclides*. Two nuclei are said to be mirrors if one can be derived from the other by replacing all protons with neutrons and vice versa. Thus the mirror of $_5B^{11}$, with five protons and six neutrons, is $_6C^{11}$. Evidently it is not possible for both to exist in nature permanently, because one can be converted into the other by β decay; in the present instance, C^{11} is unstable to positron emission and has a half-life of 20.4 min. The mass difference $C^{11} - B^{11}$ can be found either from the observed end point of the positron spectrum,

$$M(C^{11}) - M(B^{11}) = E_{\max}(\beta^+) + 1.022 \text{ MeV}$$

(Sec. 25.5b), or from the reaction $B^{11}(p,n)C^{11}$. The Q value of the (p,n) reaction has been determined with great precision, and is

$$Q = -2.763 \pm 0.001 \text{ MeV}$$

Taking into account the 0.782-MeV $n - H^1$ difference, we obtain

$$M(C^{11}) - M(B^{11}) = 1.981 \text{ MeV}$$

The observed mass differences of mirror nuclides provide a sensitive comparison of the energies associated with p-p and n-n binding forces, for the only difference between two mirrors, as far as nuclear forces are concerned, is that some n-n interactions in one are replaced by p-p interactions in the other. This fact is most easily seen if we consider (for counting purposes only) a core, comprising equal numbers of protons and neutrons, which is the same for both nuclei, plus an odd neutron in the one case or an odd proton in the other. The extra neutron in B^{11}, for example, interacts with the protons and neutrons in the "core" and has a certain number a of n-n bonds and another number b of n-p bonds. In C^{11} the a bonds will be of the p-p character, while the b will remain of the n-p type. If now the p-p bonds formed in C^{11} differ in strength from the n-n bonds of B^{11}, there will be a difference in the masses of B^{11} and C^{11} from this effect. It is important to note that this result, though here derived with a crude model, is actually quite general and is independent of such considerations as whether, for example, all the various n-p bonds are equally effective.

In order to compare the contribution of the purely nuclear forces to the binding energies of B^{11} and C^{11} we write the masses in the following form:

$$M(B^{11}) = 5M(H^1) + 6M(n) - \frac{E_N}{c^2} + \frac{E_E}{c^2} \qquad (25.28a)$$

$$M(C^{11}) = 6M(H^1) + 5M(n) - \frac{E_N'}{c^2} + \frac{E_E'}{c^2} \qquad (25.28b)$$

where the terms E_N, E_N' represent the binding energy due to nuclear forces and the terms E_E and E_E' represent the electrostatic energy of the protons. An approximate expression for the latter was given in Eq. (25.17) (Sec. 25.6); we find, from this expression, $E_E(B^{11}) = 5.50$ MeV, $E_E'(C^{11}) = 8.25$ MeV. Subtracting the second of the relations above from the first and solving for the difference of the nuclear binding energies, we obtain

$$\frac{E_N}{c^2} - \frac{E_N'}{c^2} = M(C^{11}) - M(B^{11}) + M(n) - M(H^1) + \frac{E_E}{c^2} - \frac{E_E'}{c^2}$$
$$(25.29)$$

Inserting the measured values of the mass differences times c^2, measured in MeV,

$$E_N - E_N' = 1.981 + 0.782 + 5.50 - 8.25 = 0.01 \text{ MeV}$$

Since the average binding energy per particle is 5 or 10 MeV, it would appear that the assumption that p-p and n-n forces are equal is quite well supported by the observations. Of course the result depends to some extent on the validity of the electrostatic correction, but since this quantity occurs as a difference, even rather major changes in the assumed model could hardly alter the conclusion. Among the 20 or so mirror pairs for which the masses are known, the apparent differences in the nuclear contribution to the binding energies range from a few tens of keV up to a few hundred. The largest discrepancies occur among the lightest nuclei, where the assumption of a uniform charge distribution might be expected to be least accurate. We shall have more to say about the comparison of mirror nuclei in a later section.

25.12 Particle Groups

It is frequently found in the study of the products of nuclear reactions that particles of more than one energy (or Q value) are present. This multiplicity of groups was early observed in the natural α-radioactive elements (Sec. 10.1) and was there interpreted as indicating the existence of excited quantum states in the residual nucleus. In Fig. 25.16 is shown a part of the spectrum of proton groups observed in the bombardment of B^{10} by deuterons. The energy of the protons was measured in a semicircular magnetic spectrograph (see Fig. 25.15) placed at 90° to the incident beam, which in these observations had an energy of 1.510 MeV. The group labeled 1 on the figure corresponds to the transition to the ground state of B^{11}. From the figure, one can estimate the magnetic rigidity Br of these particles as 0.448 Wb/m. The energy is then

$$E_p = \frac{p^2}{2M} = (Br)^2 \frac{e^2}{2M} = 1.5 \times 10^{-12} \text{ J} = 9.6 \text{ MeV}$$

From Eq. (25.26), we obtain

$$Q = -\tfrac{9}{11}E_d + \tfrac{12}{11}E_p = -1.23 + 10.5 = 9.3 \text{ MeV} \qquad (25.30)$$

The 1967 value from this and other experiments is 9.2314 ± 0.0006 MeV. That the group is properly identified can be tested by measuring E_p at some other value of the incident deuteron energy; if the Q value is the same, the mass numbers appearing in Eq. (25.30) have been correctly assigned. By this test, it was ascertained that groups 1, 3, 6, and 7 pertain to the present reaction, while the remaining groups are associated with some contaminants in the target. The latter point was checked by separate bombardment of targets composed of the suspected contaminant materials.

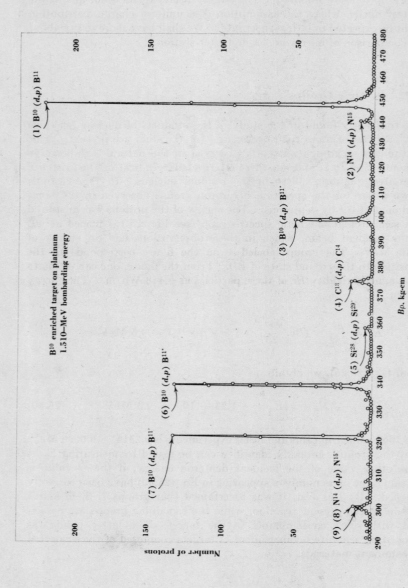

Fig. 25.16 Part of the spectrum of proton groups from the reaction $B^{10}(d,p)B^{11}$. [Van Patter, Buechner, and Sperduto, *Phys. Rev.*, 82:248 (1951).]

The Q values for the $B^{10}(d,p)B^{11}$ groups of Fig. 25.16 are listed in Table 25.2. In the right-hand column are indicated the differences

Table 25.2 *Proton groups[1] observed in the reaction* $B^{10}(d,p)B^{11}$

Proton group	Q, MeV	Excitation energy in B^{11}
1	9.2329 ± 0.0034	0
3	7.1083	2.1246 ± 0.0011
6	4.7871	4.4458 ± 0.0021
7	4.2137	5.0192 ± 0.0024

[1] Browne and O'Donnell, Phys. Rev., vol. 149, p. 767, 1966.

between each Q value and that corresponding to the ground state. These differences give directly the energies of excited states of B^{11}. In Fig. 25.17 is shown an energy-level diagram which illustrates the energetics involved. The difference in mass between the initial ingredients, $B^{10} + H^2$, and the final products, $B^{11} + H^1$, is Q/c^2, equivalent to 9.231 MeV. If the total energy is plotted on a vertical scale, the initial state (ignoring for a moment the bombarding energy) lies this amount higher than the final state. Since the zero of energy is arbitrary, we may for convenience subtract the mass of the proton ($\times c^2$) from both energies and represent the difference $M(B^{10} + H^2 - H^1) \times c^2 - M(B^{11}) \times c^2$ by a 9.231-MeV vertical displacement. All energies on the diagram are then referred to the ground state of B^{11}. Considering now the 1.510-MeV bombarding energy, we can see that a part of this energy goes into kinetic energy of the center of mass and is hence not available for internal rearrangements of the nuclei. A straightforward computation shows that a fraction $M_0/(M_1 + M_0)$, or $\frac{10}{12}$ in the present instance, of the bombarding energy is available in the system in which the center of mass is at rest. The total energy available in the center-of-mass system, relative to the ground state of B^{11}, is then $9.231 + \frac{10}{12} \times 1.510$, or 10.49 MeV. In the transition to the ground state, represented by the longest slant arrow on the diagram, the proton and B^{11} recoil share this energy in such a way as to conserve momentum. (Their velocities in the laboratory system are then obtained by adding vectorially the velocity of the center of mass.) If the B^{11} is left excited in the 2.1-MeV or higher levels, the kinetic energies will be correspondingly less, as indicated by the shorter arrows. Many other proton groups are known in addition to those in Table 25.2; the corresponding levels are shown in the diagram. The subsequent decay of most of the levels shown proceeds by γ radiation, either directly to the ground state or by cascades through lower states;

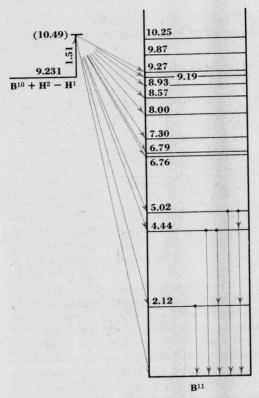

B¹¹

Fig. 25.17 Energy-level diagram for B¹¹. Slanted arrows represent proton groups; vertical arrows show γ-ray transitions.

some of the possible combinations are indicated by vertical arrows on the diagram. In many cases, it has been possible to check the level energies obtained from particle groups by direct measurements of the γ-ray energies.

The energy levels represented in Fig. 25.17 do not necessarily represent the energy states of any single nucleon. Just as in the similar atomic diagrams, a given level represents the energy of the system as a whole and need not refer to any specific model; thus the line at 2.12 MeV, for example, means only that at this energy above the ground state there exists a particularly long-lived configuration or mode of excitation, which may involve one or many particles.. For this reason also, it is convenient to plot energies from the ground state (as is done in x-ray diagrams) rather than from the ionization potential, which, in the nucleus, has various values for different particles.

25.13 Nuclear Resonances

The yield of a nuclear reaction, i.e., the number of processes observed per incident particle, usually depends strongly on the bombarding energy. A curve showing the yield as a function of energy is called an *excitation function*. With charged particles of low energy, the penetrability of the coulomb barrier increases rapidly with energy, and the yield will reflect this variation. In addition, sharp peaks, called *resonances*, may also be observed in the excitation function, and the properties of these resonances are of considerable interest to the nuclear physicist.

A striking example of a resonance effect occurs in the reaction $Li^7(p,\gamma)Be^8$. When Li^7 is bombarded by protons, there is produced, in addition to the α particles, which were discussed earlier, a γ radiation of more than 17-MeV energy. The process responsible for this γ ray is the simple capture of the proton by the Li^7 nucleus, forming Be^8, which then radiates the excess energy in 1 quantum. From the known masses, the Q of the reaction is

$$\frac{Q}{c^2} = M(Li^7) + M(H^1) - M(Be^8) = 0.018521 \text{ u}$$

$$Q = 17.252 \text{ MeV}$$

The energy available for the γ radiation is the Q value plus that part of the proton's kinetic energy which is available in the center-of-mass system

$$E_\gamma = 17.252 + \tfrac{7}{8}E_p \qquad \text{MeV} \tag{25.31}$$

where E_p is the bombarding energy measured in the laboratory coordinate system, i.e., the system in which the target nucleus is at rest.

If the yield of γ radiation from a very thin layer of lithium is measured as a function of bombarding energy, the curve shown in Fig. 25.18 results. The principal feature of this curve is a sharp maximum in the yield at a proton energy of 441 keV; the yield is reduced to one-half the maximum value by a change in either direction of only 6 keV. The observed variation of the yield is well approximated by the Breit-Wigner

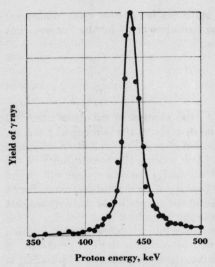

Fig. 25.18 Resonance in the yield of γ rays from the reaction $Li^7(p,\gamma)Be^8$. [*T. W. Bonner and J. E. Evans, Phys. Rev.,* **73**: 666 (1948).]

resonance formula

$$Y(E) = A(E)\, \frac{\Gamma^2/4}{(E - E_r)^2 + \Gamma^2/4} \tag{25.32}$$

where $Y(E)$ is the yield of γ rays at the proton energy E, and E_r is the value of the proton energy at the peak of the curve. The quantity Γ is the width of the curve in energy units, measured between the two points where the yield is one-half the maximum. The factor $A(E)$ is a function of energy which varies sufficiently slowly to be disregarded for our purpose.

The form of Eq. (25.32) suggests that at the energy corresponding to E_r, the system formed by $Li^7 + H^1$ has an approximately stationary quantum level. The remarkable feature of this level is that it lies in the region which the atomic spectroscopist would refer to as the *continuum*. Let us consider for a moment the analogous situation in the spectrum of the hydrogen atom. There the solutions of the Schrödinger equation are of two kinds, depending on the sign of the total energy measured with respect to the energy of the system comprising an ion with the electron at rest at infinity. If the total energy is negative, corresponding to a bound electron, the solution yields a set of discrete, nearly stationary[1] quantum states. If, on the other hand, the energy is positive, i.e., if the electron has kinetic energy at infinity, a continuum of states is allowed. When a hydrogen ion captures an electron in one of these positive-energy states, the energy available is equal to the ionization potential of 13.6 eV plus the kinetic energy of the electron, measured in the system in which the center of mass is at rest. If this energy is radiated in 1 quantum, the resultant radiation forms part of the continuum beyond the limit of the Lyman series. In Fig. 25.19a is shown a level diagram for the hydrogen atom illustrating the bound, discrete states below the ionization potential and the continuum above; a transition corresponding to capture of a positive-energy electron is indicated by the vertical arrow. Because there exist no discrete states of positive

[1] Nearly stationary in the sense that if the interaction with the radiation field is included, the quantum state decays in time, corresponding to a radiative transition to a lower level.

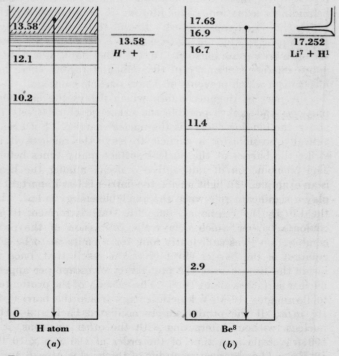

Fig. 25.19 Energy states measured from the ground state for (a) the hydrogen atom (in electron volts) and (b) the Be^8 nucleus (in MeV).

energy, the probability of capture of a free electron is a monotonic function of energy.

The nuclear case is illustrated in Fig. 29.19b. The ionization potential is now the separation energy of the proton, 17.252 MeV, measured with respect to the ground state of Be^8. Indicated on the right side of the diagram is the yield curve of Fig. 25.19b, with the yield now plotted horizontally, to the left, against energy on a vertical scale. The state formed by 441-keV protons lies at 17.64 MeV (Eq. 25.31), and the transition to the ground state results in a γ ray of this energy. Several other excited levels of Be^8 are also indicated.

The *width* of any quantum state is determined by its lifetime, for according to the uncertainty principle, if a system exists only for a time Δt, its energy can be defined only to an accuracy $\Delta E = \hbar \Delta t$. Actually all atomic and nuclear quantum states, except the ground states, have finite widths because they can decay to lower states by radiation; they represent, for this reason, only approximately stationary solutions of the Schrödinger equation. The lifetime is determined by the ease with which the decay can proceed. Bound states decay chiefly by γ radiation,[1] a process which is relatively slow on a nuclear time scale, and hence are ordinarily quite narrow. On the other hand, if a state is to be even approximately stationary in the unbound region, there must be some mechanism which prevents or delays particle emission.

A part of the mechanism which delays the emission of charged particles and makes possible the existence of relatively narrow levels above the binding energy is the nuclear barrier. For even if it is energetically possible for a particle to leave the nucleus, it may have to strike the barrier at the nuclear surface many times before it escapes. The phenomenon of radioactive α decay among the heavy elements is an example. In light nuclei, the barrier is less important but may still play a significant role when the available energy is low. In the case of the Li$^7(p,\gamma)$Be^8 resonance, once the state is formed, it is mainly the coulomb barrier which delays the reemission of the proton and the combination lives sufficiently long for γ emission to be possible. Calculation of the barrier effect gives the result that, once the proton is inside the Be^8 nucleus, the probability of escape per appearance at the nuclear surface is about 10^{-2}. The velocity of the proton (corresponding to its roughly 10-MeV kinetic energy inside the barrier) is about 5×10^7 m/s. If the proton simply oscillated back and forth across the nucleus, without interacting with the other nucleons, it would make 100 traversals in a time of the order of $100 \times 6 \times 10^{-15}/5 \times 10^7$, or 10^{-20} s. The observed mean life of the level is $\Delta t = \hbar/\Gamma = 6 \times 10^{-20}$ s.

[1] In some cases internal conversion, nuclear pair formation, or β decay occur instead of, or in addition to, γ radiation, but these are also slow processes.

In view of the uncertainties in the calculations we may regard this fair agreement as evidence that the long life of this level is mainly ascribable to the barrier effect.

The time required to radiate a γ quantum is estimated to be about 10^{-16} s, so that about one interaction in a thousand will result in γ radiation; in the remainder of the cases the proton reemerges seeming to have been elastically scattered. This process, called *resonance scattering*, results in a characteristic interference with the normal Rutherford scattering.

When several types of decay are possible for a given quantum state, the total decay rate is the sum of the decay rates for the individual processes, and the reciprocal of the mean life is the sum of the reciprocals of the individual mean lives

$$\frac{1}{t} = \frac{1}{t_a} + \frac{1}{t_b} + \cdots$$

where t_a may represent the mean time required for a proton to escape, t_b the mean life for γ radiation, and $t_c + \cdots$, other conceivable processes. It follows then that the observed width of a level Γ may be regarded as made up of *partial widths* $\Gamma_a = \hbar/t_a$ characteristic of the individual modes of decay.

In the present instance, it is energetically possible for the level to decay by α-particle emission; in fact the reaction $\mathrm{Li}^7(p,\alpha)\mathrm{He}^4$ has a Q value of 17.347 MeV. With this amount of energy available, the α particles might be expected to leave rapidly, contributing thereby an extremely large width to the level. That this does not occur is attributed to the operation of a rigid selection rule. The character, as to total angular momentum and parity, of the compound system formed by proton bombardment of Li^7 depends upon the orbital angular momentum brought in by the protons—different l values corresponding in the classical sense to different impact parameters—and at any given bombarding energy, several different kinds of compound systems may be formed. Thus protons with $l = 0$ (head-on collisions) will produce states with $I(\mathrm{Be})^8 = I(\mathrm{Li}^7) + I(\mathrm{H}^1) = \frac{3}{2} \oplus \frac{1}{2} = 2$ or 1, by vector addition. With $l = 1$ protons, states of $I(\mathrm{Be}^8) = 3$, 2, 1, or 0 may be formed. States with odd total angular momentum cannot possibly decay into two identical particles with zero spin,[1] and so if the Be^8 state in question has $I = 1$ or 3, it cannot produce α particles and its width will be determined entirely by the partial widths for proton or γ emission. It has

---

[1] The relative orbital angular momenta of two identical Bose-Einstein particles are restricted to even integral values by the requirement that the wave function be symmetric.

been confirmed that the state actually has $I = 1$ and is formed by $l = 1$ (*p*-wave) protons.

25.14   *Liquid-drop Model*

There occur many cases of resonances much too narrow, i.e., much too long-lived, to be accounted for by barrier effects.   Conspicuous among these are the extremely sharp resonances observed in capture of slow neutrons (Sec. 25.15).   It was in an attempt to explain these cases that N. Bohr proposed the *liquid-drop model* of the nucleus in 1936.   Prior to this time, most theories of the nucleus were based on analogy with the atom, in the sense that the nuclear particles were thought of as moving more or less independently of one another in a potential well which simply represented the *average* effect of the other nucleons and which took no other account of their existence.   The resonances were thought to arise mainly from the motion of a single nucleon, and interactions with other particles were treated by perturbation methods, just as in atomic theory.

In the liquid-drop model the interaction of nucleons is treated as a major effect, and any incoming particle is regarded as almost immediately losing its identity once it enters the nucleus.   The course of a nuclear reaction is thus divided into two distinct stages.   In the first stage, the incoming particle enters the target nucleus, forming a *compound nucleus*, and quickly shares its energy with nucleons already present, so that no single particle has sufficient energy to escape.   The second stage, the decay or disintegration of the compound system, occurs some time later, when the energy is again accidentally concentrated on some one particle or when the energy is lost by radiation.   In this view, the compound nucleus embodies more than a formal description of a transitory interaction between the incident particle and the target nucleus: it is a system which may have an independent existence for a time many orders of magnitude longer than the time required for its formation, and its intrinsic properties have a decisive influence on the outcome of the reaction.   There may be several ways of forming a given compound nucleus with a given energy, by various combinations of bombarding particles and target nuclei.   Subject to certain conservation laws relating to the angular momentum and parity, the character of the subsequent disintegration of the compound nucleus is considered to be independent of the details of its formation, the outcome in a particular instance being determined entirely by competition between the various possible modes.

As a simple analog for this picture of the course of nuclear reactions,

Bohr suggested the liquid drop. In such a drop, the binding forces between molecules hold the system together and prevent evaporation unless heat is added from outside. We suppose now that an additional molecule is introduced; since it was initially free, it must gain kinetic energy in going through the surface of the drop. In principle, it gains sufficient kinetic energy to permit its immediate reevaporation the next time it appears at the surface; it is likely, however, to collide with another molecule and lose some of its energy in the meantime. In the course of further collisions, the energy is divided among all the molecules present and no one has sufficient energy to evaporate. From the point of view of the drop as a whole, the temperature has been raised slightly. The condition of the drop is thus defined completely by the number of particles and the total energy available, and its future behavior, e.g., the eventual evaporation of some molecule through an accidental concentration of energy, will be quite the same as if the extra energy had been supplied to it in the form of heat.

Because of the many ways in which the extra energy made available by the capture of a particle can be shared among the other particles in the nucleus, a rather long time may elapse before the compound nucleus decays. Consequently the energy of the compound nucleus may be defined with relatively sharp limits. It is observed, e.g., in the capture of low-energy neutrons, that resonances only a fraction of an electron volt wide may occur. From the relation $\Delta E \, \Delta t \approx \hbar$, we find for the mean life of a level 1 eV wide, $\Delta t = 0.7 \times 10^{-15}$ s. Comparing this number with the characteristic time, of the order of 10^{-22} s, for a single traversal of the nucleus by an incoming neutron,[1] we find that the neutron travels some 10^6 or 10^7 nuclear diameters during the mean life of the compound nucleus.

If sufficient energy is available, there may be several ways for a given compound nucleus to decay. For example, bombardment of C^{12} with deuterons is observed to lead to the following reactions:

$$C^{12} + H^2 \rightarrow (N^{14}) \rightarrow \begin{cases} C^{13} + H^1 \\ N^{13} + n^1 \\ B^{10} + He^4 \\ C^{12} + H^2 \\ N^{14} + \gamma \end{cases}$$

The compound nucleus here (N^{14}) has an excitation energy of more than 10 MeV; with the extra energy brought in by deuterons of 2 MeV or more,

---

[1] Inside the nucleus, the kinetic energy is about 10 to 15 MeV, corresponding to a velocity of about 5×10^7 m/s. The diameter of a medium-weight nucleus is about 1 to 1.5×10^{-14} m.

it can eject a proton, a neutron, or an α particle, or it can reemit a deuteron. Resonances have been observed in the first two reactions and presumably occur for the others also. The same compound nucleus can be formed by bombarding C^{13} with protons. The disintegration of a compound nucleus of given energy is independent (subject to conservation of spin and parity) of the mode of formation and emphasizes the dominant role which the properties of the compound nucleus plays in determining the course of nuclear reactions.

25.15   Neutron Reactions

At the time of the discovery of the neutron, it was realized that such a particle might provide a powerful tool in inducing nuclear reactions, and the first cloud-chamber studies of neutron recoils in nitrogen gave evidence for a number of transmutations produced by neutrons. The efficacy of the neutron is attributable to the fact that it has no charge and hence has little or no interaction with the atomic electrons or with the coulomb field of the nucleus. When a beam of protons passes through matter, by far the greatest number are brought to rest through electronic interactions, and only a few produce a nuclear reaction. Neutrons travel through matter indifferent to the electric fields, and most of them end up by producing a nuclear reaction. The fact that they do not produce ions directly makes them somewhat more difficult to detect, but they can be counted with acceptable efficiency through their nuclear encounters.

At low bombarding energies (less than a few MeV) positively charged particles find it difficult to enter the target nucleus because of the coulomb barrier. In the classical theory, they could not enter unless they had sufficient energy to surmount the barrier, but wave mechanics admits a small penetrability. For a 200-keV proton striking an O^{16} nucleus, for example, the probability of penetrating is of the order of 10^{-3} as great as for a neutron of the same energy. At lower energies and for higher atomic numbers the disparity is greater.

a. Production of Neutrons   The bombardment of beryllium with α particles provides a convenient source of neutrons for experiments. When high intensity is not required, a radioactive material is simply mixed with beryllium powder and sealed in a small container. With polonium α particles (5.3 MeV), about 1 α particle in 10^4 produces a neutron, so that the number of neutrons from a source of, say, 10 mcuries strength is $3.7 \times 10^8 \times 10^{-4} \approx 3.7 \times 10^4$ neutrons/s. Stronger sources can be made with artificially accelerated particles. For example, in the reaction

$Be^9(d,n)B^{10}$ with 10 μA of 10-MeV deuterons, some 4×10^{11} neutrons are produced per second. For many purposes, it is desirable that the neutrons be homogeneous in energy. Monoenergetic neutrons can be produced in several reactions; a commonly used one is $Li^7(p,n)Be^7$, which has a negative Q value of 1.644 MeV. Neutrons first appear at a proton energy of 1.881 MeV and can be produced with any desired energy[1] by varying the bombarding voltage. A copious source of low-energy neutrons is the nuclear reactor.

b. Neutron Total Cross Section Because neutrons are essentially unaffected by the atomic electrons, a measurement of the attenuation of a neutron beam in passing through matter provides direct and useful information on the properties of nuclei. As a quantitative measure of the probability of an interaction between a moving particle and a nucleus, the term *cross section* is used (Sec. 7.9). To define this quantity in the present connection, we consider a parallel beam of I particles per second crossing an area A in which there is located a single target nucleus; we assume that of the I particles a certain number i suffer some kind of interaction with the target nucleus, for example an elastic scattering. The cross section for this process is then defined as

$$\sigma = \frac{\text{number of interactions/target nucleus}}{\text{number of incident particles/unit area}} = \frac{i}{I/A}$$

or, equivalently,

$$\sigma = \frac{\text{number of interactions/(target nucleus/unit area)}}{\text{number of incident particles}} = \frac{iA}{I}$$

Thus, if a target contains n nuclei per unit area perpendicular to an incident beam of I particles per second, the number of interactions per second is $I\sigma n$. In a layer of material of thickness dx containing N nuclei per unit volume, the probability of an interaction is

$$dP = \sigma N \, dx \tag{25.33}$$

The cross section has the dimensions of an area and is, in the classical sense, simply equal to that portion of the projected "area" of the target nucleus which is effective for the process in question.

The total cross section σ_t for interaction of neutrons with nuclei is made up of several terms, including the cross sections for elastic scatter-

---

[1] For proton energies greater than 2.37 MeV, an excited state of Be^7 may be formed, and a second group of neutrons appears.

ing, inelastic scattering, disintegration, and simple capture. For reasonably fast neutrons, the total cross section can be crudely estimated as equal to the projected area of the nucleus πR^2, where R is the radius of action of the nuclear force field. Actually, because even with high-energy neutrons wave-diffraction effects are important, an effective area of $2\pi R^2$ is a better estimate. Experimental values of σ_t can be obtained by measurement of the attenuation of a narrow beam of neutrons in a known thickness of material. If such a beam, of intensity I_0 neutrons per second, is incident on a slab of material of thickness x, the number of neutrons emerging unchanged in direction is

$$I = I_0 e^{-\sigma_t N x} \tag{25.34}$$

From a measurement of I/I_0, the cross section and hence the nuclear radius can be found. The values obtained generally agree with Eq. (25.5), with values of R_0 lying in the neighborhood of 1.4×10^{-15} m. For a medium-weight nucleus, πR^2 is an area of the order of 10^{-28} m^2, or 1 barn.

c. Neutron-Proton Scattering   Generally, the most probable result of a collision between a neutron of moderate energy—of the order of 1 to 5 MeV—and a light nucleus is a simple elastic scattering. This was the process observed by the Joliot-Curies and by Chadwick in their experiments on the recoils produced in hydrogen and in nitrogen. One type of neutron detector consists essentially of an ionization chamber lined with some hydrogenous material; the neutrons eject protons from the wall by elastic collision and the protons are counted through their ionizing ability. Measurements of the elastic scattering of neutrons in hydrogen are of considerable interest in that they lead to direct information on the character of nuclear forces (Sec. 25.6b).

An arrangement for a scattering experiment is shown in Fig. 25.20. A collimated monoenergetic beam falls on the scatterer S. Those neutrons which are scattered through the (variable) angle θ are counted

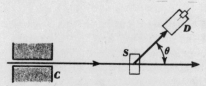

Fig. 25.20 Neutrons passing through the collimator C are scattered in the block of material S and counted at detector D.

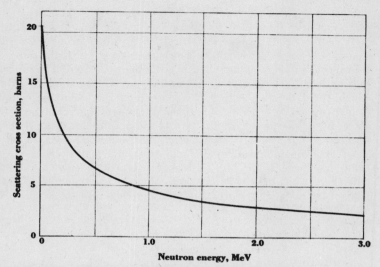

Fig. 25.21   Scattering cross section for neutrons in hydrogen (1 barn = 10^{-28} m²).

by the detector D.   If scattering in hydrogen is under study, the scatterer may be made of some compound of carbon and hydrogen, like paraffin or polyethylene.   The observed scattering is then the sum of the effects from the hydrogen and carbon nuclei, and the necessary correction is determined by an auxiliary experiment with a carbon scatterer.   The result of the experiment is expressed in terms of the number of neutrons scattered into the detector as a function of the angle θ and of the energy of the incident neutrons.   If the solid angle subtended by the detector and the number of scattering nuclei are known, the *differential cross section* $d\sigma/d\Omega$, that is, the cross section for scattering into unit solid angle, can be calculated.   For neutron energies above 1 eV and less than 10 MeV, the scattering in hydrogen turns out to be spherically symmetric when the results are expressed in the center-of-mass system.   The total cross section for elastic scattering, obtained by integrating the differential cross section over all values of θ, exhibits a smooth variation in energy, decreasing monotonically from the value 20.4 barns at a few electron volts to about 2 barns at 4 MeV (Fig. 25.21).

d. Neutron-induced Reactions   In addition to undergoing elastic scattering, neutrons produce nuclear reactions, of which many result in radioactive nuclides.   An extensive study of these effects made by

Fermi and his collaborators in the years 1934 to 1936 led to the identification of a number of reactions of this character. The experimental arrangement in these early investigations was simple. A neutron source (≈ 0.5 curie of radon mixed with beryllium powder) was used to irradiate a sample of some element for a short time, after which the sample was transferred to the neighborhood of a Geiger counter and examined for radioactivity. Generally, the samples were in the form of cylinders which could be slipped over the source and then over the cylindrical Geiger counter in order to obtain the most favorable geometry. When an activity was produced, the half-life was determined, and wherever possible a chemical identification of the radioisotope was made. Of some 60 elements investigated, about 40 showed radioactivity. Among the lighter elements—F, Mg, Al, Si, and several others—it was established that the activity resulted from (n,p) or (n,α) reactions. For example, bombardment of silicon produced radioactive aluminum with a half-life of about 3 min; the process is

$$\text{Si}^{28} + n^1 \rightarrow \text{Al}^{28} + \text{H}^1$$

Reactions of the character $\text{X}(n,p)\text{Y}$, where X is a stable nuclide, invariably lead to radioactive products, for Y is an isobar of X, differing in charge by one unit, and according to the rules developed in Sec. 25.5, it is not possible for both X and Y to be stable. For the same reason (p,n) reactions also lead to unstable products when the target is stable. Most (n,p) reactions are endoergic, because the product nucleus is usually unstable by more than the 0.78 MeV which the $n - \text{H}^1$ mass difference makes available. A few cases exist in which (n,p) reactions have positive Q values; $\text{He}^3(n,p)\text{H}^3$ and $\text{N}^{14}(n,p)\text{C}^{14}$ are examples. The (p,n) reactions are all endoergic by more than 0.78 MeV.

The (n,α) reactions are generally endoergic in light- and medium-weight nuclei and are relatively weak because the coulomb barrier discourages the escape of low-energy α-particles from the compound nucleus. Two important exoergic reactions are

$$\text{Li}^6(n,\alpha)\text{H}^3 \quad Q = 4.786 \text{ MeV} \qquad \text{B}^{10}(n,\alpha)\text{Li}^7 \quad Q = 2.792 \text{ MeV}$$

Both have high cross sections for low-energy neutrons and, because they produce charged particles, are much used for neutron detectors. If a proportional counter is filled with boron trifluoride gas—preferably enriched in B^{10}—neutrons entering the counter produce α particles which, because of their dense ionization, give rise to large pulses. The associated amplifier can then be set to reject small pulses due to γ-ray background, and the counter responds only to neutrons.

In nuclei with high Z, the coulomb barrier becomes so high that neutron reactions leading to charged-particle emission are improbable except with high (>10 MeV) bombarding energies. The important processes in heavy nuclei with moderate energies are elastic scattering, inelastic scattering, and simple capture followed by γ radiation (Sec. 25.10). Among the reactions of this type which were identified by Fermi's group are $Au^{197}(n,\gamma)Au^{198}$, in which a strong activity of 2.7-day half-life is produced, and $I^{127}(n,\gamma)I^{128}$, resulting in a radioisotope of iodine with a half-life of 25 min. As a rule, the cross section for radiative capture is small compared with that for elastic and inelastic scattering when the neutron energy exceeds a few keV.

e. Slow Neutrons   In the course of their investigations of radioactivity induced in silver by neutrons, Fermi and his collaborators observed that the amount of activation produced appeared to depend on the geometrical arrangement of the experiment. In a more careful study of this question, they found that the activation was affected by the presence of hydrogen-containing substances near the neutron source and silver sample and, finally, that surrounding the whole arrangement with a large quantity of water or paraffin enormously increased the yield of radioactivity. Further tests showed that the enhancement of the yield in the presence of hydrogen was exhibited by a number of elements in addition to silver; several had cross sections which were more than 100 times greater than the geometric cross section.

When neutrons pass through a substance containing hydrogen, they are strongly scattered and rapidly lose energy in elastic collisions. A neutron striking a proton transfers an amount of energy [see Eq. (25.1)]

$$\Delta E = E_0 \sin^2 \tfrac{1}{2}\theta$$

where θ is the angle of scattering in laboratory coordinates. For neutrons in the energy region of interest here, the scattering is isotropic in center-of-mass coordinates, and all possible values of the energy loss are equally probable. It follows that the energy of the neutrons is reduced on the average by one-half in a collision with a proton; the average energy of a group of neutrons which had initially the same energy E_0 will after n collisions per neutron be

$$E_{av} = \frac{E_0}{2^n}$$

However, in determining the average energy, the relatively few neutrons with high energy are given a disproportionate weight; a better index of the energy distribution for our purposes is the *median* energy. The median energy can be shown to be approximately, again for collisions in hydrogen,

$$E_{\text{med}} = \frac{E_0}{e^n}$$

From this we can compute that of a group of 1-MeV neutrons, half will have energies less than 1 eV after about 14 collisions.   The mean free path per collision, defined as the average distance traveled between collisions, is given by $\Lambda = 1/\sigma N$, where σ is the scattering cross section and N the number of protons per unit volume.   Inserting the measured values of σ (Fig. 25.21) we find for 1-MeV neutrons a mean free path of about 3.3 cm in water and for 1-keV neutrons, about 0.7 cm (we ignore the collisions with oxygen nuclei since they have relatively little effect on the energy). Thus a layer of water 10 cm thick around the neutron source reduces most of the neutrons to an energy where they are in equilibrium with the thermal motion of the water molecules (about $\frac{1}{40}$ eV at 290°K).   Neutrons reaching thermal energies are moving in random directions, and their motion is not unlike the diffusion of a gas.

The fact that slowing down the neutrons greatly increases their effectiveness in producing nuclear reactions is easily explained when the problem is examined from the point of view of wave mechanics.   In our earlier discussion, we estimated the maximum cross section for interaction of fast neutrons with nuclei by appealing to a simple geometrical picture of the projected area presented by a nucleus to an incoming particle.   A usual measure of the applicability of such geometrical pictures is the ratio of the *reduced wavelength* $\lambda/2\pi$ to the nuclear dimensions.   For a fast, say 5-MeV, neutron, the reduced wavelength is

$$\frac{\lambda}{2\pi} = \frac{\hbar}{\sqrt{2EM}} = 2 \times 10^{-15} \text{ m}$$

which is sufficiently smaller than the nuclear radius for visualization of the neutron's trajectory to present no difficulty.   For neutrons of 1 eV energy, on the other hand, $\lambda/2\pi \cong 0.5 \times 10^{-11}$ m, many times larger than the nucleus.   In this situation it becomes appropriate to think of the nucleus as a small diffracting center in a very long wave. Instead of trying to trace individual trajectories we regard the incident neutrons as a beam, with a certain average probability of having a neutron in a unit volume.   In such a beam, containing I_0 neutrons of velocity v crossing unit area per second, the number of neutrons per unit volume is $\rho = I_0/v$.   The probability per unit time that a neutron will be captured is proportional to the density of neutrons

$$N_{capt} \propto \rho = \frac{I_0}{v}$$

But the number of captures per unit time is, by definition, proportional to σI_0. Thus

$$\sigma_{capt} \propto \frac{N_{capt}}{I_0} \propto \frac{1}{v} \qquad (25.35)$$

Equation (25.35), often referred to as the $1/v$ *law*, is accurately confirmed in many neutron processes. The cross section for the reaction $B^{10}(n,\alpha)Li^7$, for example, shows no appreciable deviation from the $1/v$ dependence over the energy range from 10^{-2} to nearly 10^3 eV, and the activation of silver follows the law from 10^{-2} to about 1 eV. In some nuclides, the observed cross sections for thermal neutrons ($E \approx kT = \frac{1}{40}$ eV) are many thousand times greater than those for fast neutrons. One isotope of xenon has a thermal cross section of 3.5×10^6 barns!

f. Resonance Absorption   If the cross section for activation of, say, silver is very large for slow neutrons, the absorption coefficient in silver of such neutrons should also be high. That this is indeed the case was shown by Amaldi et al. by measuring the activation produced in a silver sample with various thicknesses of silver absorbers wrapped around it. For silver, the thickness required to reduce the activation to half value was about 1 mm, indicating an average cross section [Eq. (25.34)] of some hundreds of barns. It would thus appear that a relatively thin layer of silver should absorb many of the slow neutrons and greatly reduce the effect of the hydrogen thermalizer on all activations. Tests made with iodine as the detector showed, however, that this was not so: the activation of iodine was only moderately affected by the silver absorber, and although an iodine absorber would reduce the iodine activity greatly, it had little effect on silver samples. The implication is that neither silver nor iodine follows strictly the $1/v$ law but that both have extraordinarily high absorptions for certain different neutron energies. The activation of silver must then be mainly due to a narrow band of neutron energies—for which its absorption is correspondingly high—and that of iodine is due to another band of a different energy. It was shown by Amaldi and Fermi that these so-called *resonance-absorption bands* were in some cases only a fraction of an electron volt wide. These were the resonances whose discovery led to the formulation of the liquid-drop model of the nucleus.

Our information on these low-energy-neutron resonances has enormously increased with the development of more refined techniques.   One of these has been the neutron *velocity selector*, schematically illustrated in Fig. 25.22.   The neutrons are produced in the reaction $Be^9(d,n)B^{10}$, using deuterons accelerated by a cyclotron, and "thermalized" by a few centimeters of paraffin.   Slow neutrons pass through the material whose absorption is to be studied and to the detector some distance away. Velocity selection is accomplished by operating the cyclotron in such a way as to produce short bursts of neutrons and synchronizing the detector so that it counts only those neutrons whose time of flight lies in a selected narrow range.   In a typical arrangement, the distance to the detector is 6.4 m; the time of flight for a 1-eV neutron is then 470 μs.   With a time resolution of 5 μs, energies as high as 10,000 eV can be selected.

Another instrument, with particular applicability to the low-energy region, is the crystal spectrometer.   The de Broglie wavelengths of thermal neutrons are of the order of magnitude of the lattice spacings of crystals, and they are scattered by such a lattice according to the Bragg law.   With the large neutron fluxes available from nuclear reactors, it is possible to obtain scattered neutron beams of acceptable intensity and high homogeneity in energy.   The sample is placed in the beam, and its attenuation is measured as a function of the crystal angle.   A curve for silver, showing two strong resonances, is reproduced in Fig. 25.23.   The close agreement obtained in the location of resonance energies by the crystal and direct time-of-flight measurements provides an accurate check of the validity of the de Broglie relation for neutrons.

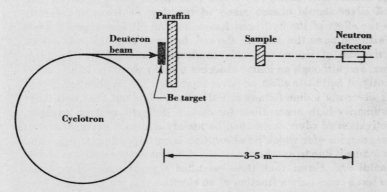

Fig. 25.22 Neutron velocity selector. Neutrons produced in the Be target are slowed down in the paraffin block and selected according to the time required to traverse the distance to the detector.

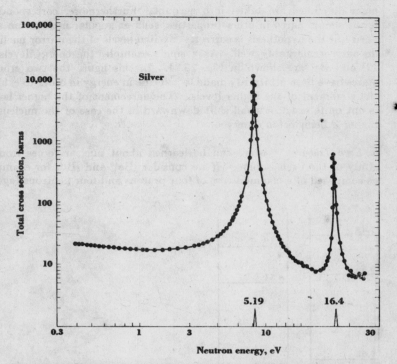

Fig. 25.23 Neutron capture resonances observed in silver with a crystal spectrometer. [*L. B. Borst and V. L. Sailor, Rev. Sci. Instr.*, **24**:141 (1953) *and R. E. Wood, Phys. Rev.*, **95**:644 (1954).]

25.16 Energy Levels of Nuclei

One important result of the study of nuclear reactions is the location and identification of the excited states of nuclei. In the atomic case, it was the study of the excited levels, as derived from spectra, which guided the theoretical development of our present model of the atom. Of all the various classes of nuclear energy levels we discuss here only a few which have a particularly straightforward interpretation.

a. Energy Levels of Mirror Nuclei   In Sec. 25.11, evidence that the nuclear parts of *p-p* and *n-n* forces are equal was adduced from comparison of mirror nuclei. From the arguments presented there, not only the ground states but also the excited states of mirror nuclei should agree in energy, once the electrostatic energy and the neutron-proton mass difference are taken into account. Furthermore, corresponding levels should agree in other properties, such as angular momentum and parity, if the hypothesis is correct. Excited levels of the mirror nuclides do agree remarkably well. As a single example, the energy levels of Li^7 and Be^7 are shown in Fig. 25.24. In this figure the two ground states have been arbitrarily made to coincide in energy in order to exhibit the alignment of the higher levels. The agreement of the higher levels is not quite exact, a small shift downward in the case of the nucleus of higher Z being often observed.

b. Even Isobars   More can be learned about nuclear forces from a study of the even isobars. If we consider $_4Be^{10}$ and $_5B^{10}$, for example, as composed of a common core of four protons and four neutrons (again,

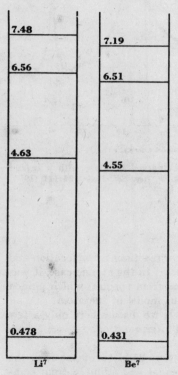

Fig. 25.24 Energy levels of Li^7 and Be^7. The ground states are arbitrarily adjusted to coincide.

only for the purpose of counting the number of interactions of various types) plus two neutrons or a neutron-proton pair, respectively, we find that some of the n-n bonds in Be^{10} are replaced in B^{10} not by p-p bonds, but by p-n bonds. If the nuclear binding energies of these two nuclides are found to be equal, we have evidence for the far-reaching assertion that, at least in light nuclei, the interaction between unlike nucleons is the same as that between like nucleons, i.e., that the forces are not only *symmetric* in the charge state (see Sec. 25.6c) of nucleons but are *charge-independent*.

A complication arises, however, in the comparison of Be^{10} and B^{10}, because certain p-n interactions in the latter are not permitted in the former when the proton is replaced by a neutron. A proton and neutron in the same quantum level may have either parallel or antiparallel spins; two neutrons, on the other hand, can only have antiparallel spins. If the nuclear force is greater when the spins are parallel, as would appear to be the case from the neutron-proton scattering experiments (Sec. 25.6b), the ground state of B^{10} may have a character which is forbidden in Be^{10}. On the other hand, there should exist an *excited* state of B^{10} in which the spins of the extra proton-neutron pair are antiparallel, and this state should have the same energy as the ground state of Be^{10}. The same argument applies to the mirror of Be^{10}, viz., $_6C^{10}$, which has two extra protons. All the levels of these two nuclides should have counterparts in B^{10}, but B^{10} may have levels which are not possible in either C^{10} or Be^{10}.

The known low-lying energy levels of Be^{10}, B^{10}, and C^{10} are shown in Fig. 25.25, in which a correction for the electrostatic energies and $n - H^1$ mass difference has been applied, so that the nuclear parts of the binding energies are shown relative to the ground state of B^{10}. Be^{10} has thus been shifted upward by 1.44 MeV and C^{10} downward by 2.05 MeV. The levels of B^{10}, C^{10}, and Be^{10}, which are connected by dashed lines, are found not only to agree well in energy but also to exhibit other properties which confirm their close genetic relation.

In general, the greater the difference between the numbers of protons and neutrons in a nucleus with a given total number of particles, the more restrictions placed by the Pauli principle on the number of possible interactions and the smaller the number of levels as compared with configurations comprising equal numbers of neutrons and protons. In a formal way, this property of a nuclear configuration may be expressed[1]

[1] E. P. Wigner, *Phys. Rev.*, 51:106 (1937). In the wave-mechanical treatment, the isotopic spin plays a role quite analogous to the ordinary spin. Just as the two spin states of a nucleon can be represented by components of a spin vector, the two charge states of a nucleon, i.e., neutron and proton, may be represented by components of the isotopic-spin vector.

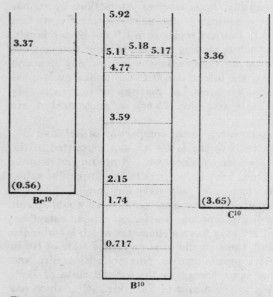

Fig. 25.25 Energy levels of Be¹⁰, B¹⁰, and C¹⁰. The ground states of Be¹⁰ amd C¹⁰ have been shifted by the electrostatic energy and the $n - H^1$ mass difference. Corresponding levels are connected by dashed lines.

by the so-called *isotopic-spin projection* $T_\zeta = \frac{1}{2}(N - Z)$. The greater the value of T_ζ, the fewer the permitted levels, and any level which is possible for a nuclide with a given T_ζ will also appear in all nuclides with lower absolute values of T_ζ. The various common levels are said to represent components of an *isotopic-spin multiplet* characterized by a quantum number T equal to the maximum value of T_ζ. Thus all the states of B¹⁰ have $T_\zeta = 0$; the ground state and first excited state do not occur in Be¹⁰ or C¹⁰ (for which $T_\zeta = 1$ and $- 1$) and are therefore isotopic-spin singlets: $T = 0$, $T_\zeta = 0$. The ground states of Be¹⁰ and C¹⁰ and the 1.74-MeV state of B¹⁰ comprise an isotopic-spin triplet: $T = 1$, $T_\zeta = 1, 0, -1$. In a sense the various isobars comprise a kind of fine structure in nuclear spectroscopy much as the different components of ordinary spin comprise a fine structure in atomic energy levels. It is thus again emphasized that it is the total number of particles, and not the somewhat incidental nuclear charge, which principally determines the behavior of a nucleus.

25.17 The Shell Model

We have discussed in previous sections the experiments which led to the development of the liquid-drop nucleus model for nuclear reactions. It is, as we have seen, an essential feature of this model that the nucleons are regarded as strongly interacting within the nucleus, and any picture envisaging separate particles moving independently of one another in an average central field of force would give results at variance with many of the facts of nuclear reactions. There has, however, been a simultaneous development in the opposite direction, in which it has been made increasingly clear that for some properties of nuclei the *independent-particle model* gives surprisingly good results. This development is closely connected with the discovery of the so-called *magic numbers*, configurations of neutrons and protons which exhibit certain rather special properties.

a. Periodicities in Nuclear Species   A conspicuous feature of atomic structure is the marked periodicity in chemical properties. We saw earlier how this phenomenon found a straightforward explanation in terms of the electronic shells and how the closing of a shell or subshell made itself evident in a strong increase in the ionization potential (Fig. 15.3). It is natural to search for analogous effects in nuclei in terms of specially stable nuclear species. Among the lightest nuclei, the peculiar stability of nuclei of the type $A = 2Z = 4n$, where n is an integer, was recognized at an early state. This stability is exhibited in the pronounced peaks of the binding-energy curve of Fig. 25.3. This observation, and the fact that α particles are emitted by many of the naturally radioactive substances, led for a time to the belief that the α particle is a fundamental building unit of nuclei. That this is not so in any literal sense is evident from the fact that heavier nuclei contain more neutrons than protons and the extra neutrons are just as tightly bound as the protons. Among light nuclei, the preference for equal, even numbers of protons and neutrons is accounted for by the high symmetry of such configurations (Sec. 25.6).

The suggestion that there are other shell-structure effects was first made in 1932 by Bartlett, who observed that the pattern of composition of the naturally occurring isotopes underwent a change at O^{16} and again at A^{36}. Between He^4 and O^{16}, all the stable nuclides belong to the sequence $He^4 + n + p + n + p + \cdots$. From O^{16} to A^{36}, the sequence becomes $O^{16} + n + n + p + p + n + n + \cdots$. It was suggested that these changes might be associated with the filling of shells of neutrons and protons of given orbital angular momentum. In the first *s* shell (orbital angular momentum = 0) there could be just two neutrons and two protons without violating the Pauli exclusion principle. The next, the "*p*" shell, with orbital angular momentum 1, has room for six neutrons and six protons and would be complete at O^{16}. The "*d*" shell, with 10 neutrons and 10 protons, would be filled at A^{36} ($Z = 18$). In general, as in atoms, the number of particles of each kind in a shell is $2(2l + 1)$. Evidence for the existence of other shells was brought forward in 1933 and 1934 by Elsasser and by Guggenheimer, who found indications of particular stability for several values of neutron and proton numbers, among them $Z = 20$, N or $Z = 50$, 82, and $N = 126$. Unfortunately the data available were insufficiently accurate, and the argument was weakened to some extent by the inclusion of several numbers which turned out not to be "magic" at all.

b. Central-field Approximation   Attempts were made to account for the existence of shells by calculations based on the Hartree central-field model, in which the mutual interactions of the nucleons are represented by an average central field. The solution takes essentially the same form as that for the atomic case, where, as a first approximation, the wave function for the system is written as a linear combination of single-particle wave functions and the main effect of the electrostatic repulsion is included in the potential function. Other interactions, such as the residual electrostatic repulsion or the spin-orbit effect, are taken into account by perturbation methods.

As a first approximation to the central field in the nucleus, a simple square-well potential is assumed. Since the potential is solely a function of radius, the Schrödinger equation may be separated into radial and angular parts, and the angular-dependent solutions are identical with those of the hydrogen atom. There exist a number of stationary solutions, depending upon the depth and width chosen for the potential, and these are characterized by the quantum numbers ν, l, m_l, where l and m_l refer to the orbital angular momentum and its projection on any chosen axis and ν is the radial quantum number, so defined that the number of nodes in the radial solution is given by[1] ν (excluding nodes at $r = 0$, but including $r = \infty$). The energy of a given stationary state depends explicitly upon both ν and l, and the level order is $1s$, $1p$, $1d$, $2s$, $1f$, $2p$, . . . , as shown in the left-hand side of Fig. 25.26. The actual energies may be different for neutrons and protons, particularly in heavier nuclei, where the electrostatic effect is important. In any

---

[1] This ν is not the same as the quantum number n used in atomic spectroscopy. In another commonly used notation, however, the solutions are designated by a total quantum number n (our $\nu + l$), and l; in this notation, the level order is $1s$, $2p$, $3d$, $2s$, ...

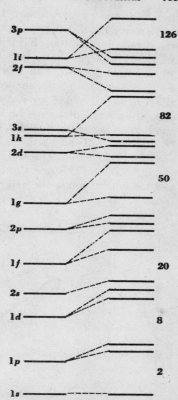

Fig. 25.26 Spacing and order of levels in an infinitely deep square potential well (*left*) and on the Mayer-Jensen potential (*right*). The total number of particles at each shell closing is indicated at the right.

case, with $2(2l + 1)$ particles in a "shell," closed shells should occur for 2, 8, 18, 20, 34, 40, 58, . . . neutrons, and similarly for protons. The numbers 50, 82, and 126 do not appear to occur in any natural way, although they can be produced by arbitrarily omitting certain terms, for example 2s and 2p.

c. Spin-orbit Coupling   At about this stage of development, the independent-particle model fell into disfavor, partly because the experimental evidence for shells among the heavier elements was something less than clear and partly because the conspicuous success of the liquid-drop model made it seem that any picture based on a fixed central potential was doomed to failure.   Evidence for periodicities in nuclear structure continued to accumulate, however, and the question was reopened by M. G. Mayer in 1948.   In a survey of existing information on the naturally occurring nuclides, she noted that species with neutron or proton numbers 50, 82, or 126 are distinctly more abundant in the earth's crust than their immediate neighbors, that there occur particularly large numbers of *isotones*—nuclides with the same neutron number N—for $N = 50$ and 82, and particularly large numbers of isotopes for $Z = 20$ and 50.   With these data, and certain others which we shall not enumerate here, she established the magic character of the numbers 50, 82, and 126 in addition to the well-recognized ones 2, 8, and 20.   Supporting

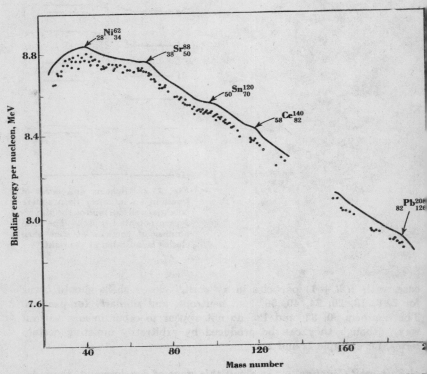

Fig. 25.27   Magnified portion of curve showing binding energy per nucleon.   [*H. E. Duckworth, Nature, 170*:158 (1952).]

evidence for the special stability of magic-number nuclides, obtained in direct mass-spectrograph measurements, is shown in Fig. 25.27, in which the binding energy per nucleon (see Fig. 25.3) is exhibited on a magnified scale for mass numbers 40 to 240. Changes in slope occur at Ni^{62} (28 protons), Sr^{88} (50 neutrons), Sn^{120} (50 protons), Ce^{140} (82 neutrons), and Pb^{208} (82 protons, 126 neutrons).

In 1950, Mayer, and independently Haxel, Jensen, and Suess, proposed a simple independent-particle model which led to the observed magic numbers in straightforward fashion. Starting with a potential differing only slightly from a square well, they postulated a strong spin-orbit coupling, proportional to $1 \cdot s$, which would split the zero-order states obtained from the central potential. Thus the $1p$ state would be split into two: $1p_{\frac{3}{2}}$ and $1p_{\frac{1}{2}}$ (the subscript gives the total angular momentum for a single particle); and $1d$ would become $1d_{\frac{3}{2}}$ and $1d_{\frac{5}{2}}$; the s state of course remains single. If the spin-orbit splitting is of the same order of magnitude as the separation between various l terms, the grouping of states may be considerably modified. The grouping which they derived is shown in Table 25.3, and the spacing of the levels is shown on the right-hand side of Fig. 25.26. On the basis of strong spin-orbit split-

Table 25.3 *Grouping of neutron or proton states in the shell model with spin-orbit coupling*

State designation	$1s$	$1p_{\frac{3}{2}}$	$1d_{\frac{5}{2}}$	$1f_{\frac{7}{2}}$	$1g_{\frac{9}{2}}$	$1h_{\frac{11}{2}}$
		$1p_{\frac{1}{2}}$	$1d_{\frac{3}{2}}$	$1f_{\frac{5}{2}}$	$2d_{\frac{5}{2}}$	$2f_{\frac{7}{2}}$
			$2s_{\frac{1}{2}}$	$2p_{\frac{3}{2}}$	$2d_{\frac{3}{2}}$	$2f_{\frac{5}{2}}$
				$2p_{\frac{1}{2}}$	$3s_{\frac{1}{2}}$	$3p_{\frac{3}{2}}$
				$1g_{\frac{9}{2}}$	$1h_{\frac{11}{2}}$	$3p_{\frac{1}{2}}$
						$1i_{\frac{13}{2}}$
Total neutrons or protons	2	8	20	50	82	126

ting for large l values, the numbers 50, 82, and 126 become magic in a straightforward manner. Among the light elements where the l values are low, in the $1s$ and $1p$ shells, it is assumed that the spin-orbit splitting is not so great, and the level order and magic numbers (2, 8) are those of the square well. In the next shell, the close proximity of the $1d$ and $2s$ states is a consequence of the assumed potential, and the spin-orbit effect again plays a minor role. It is possible that the spin-orbit effect may lower the $1f_{\frac{7}{2}}$ state in the next group sufficiently to make 28 a magic number; there is some indication for this in the relative stability of $_{28}Ni^{62}$. The next shell contains $1f$ and $2p$ states closely grouped, and the spin-orbit effect depresses the $1g_{\frac{9}{2}}$ state, so that this term appears in the shell which closes at 50 nucleons. By the same procedure, shells at 82 and 126 nucleons are accounted for.

d. Single-particle Model   To make more specific comparisons between the predictions of the independent-particle model and experiment, it is necessary to make some further assumptions. The most obvious of these, which is common to all models and which follows directly from the Pauli principle, is that the nucleons of a closed shell combine their spins and magnetic moments to give a resultant of zero. In the light nuclei He^4 and O^{16}, where $N = Z$, and in the *doubly magic* nucleus $_{82}Pb^{208}$, where both N and Z are magic, the ground-state angular momentum is zero, but, since this is apparently a general characteristic of nuclei with even N and even Z, the point is not strong. As regards the next assumption, two extremes are possible: in the so-called single-particle model, it is assumed that all like particles, including those outside closed shells, couple together in pairs to give angular momentum zero. The angular momentum and magnetic moment of all even-even nuclei would then be zero—as observed—and that of even-odd (Z even, N odd), or odd-even nuclei would always be that of the odd nucleon. The single-particle model works surprisingly well in accounting for the observed angular momenta. According to this model, in the $1p$ shell (He^4 to O^{16}), the total angular momenta expected are $\frac{3}{2}, \frac{3}{2}, \frac{3}{2}, \frac{3}{2}, \frac{1}{2}, \frac{1}{2}$ for A = 5, 7, 9, 11, 13, and 15, just as observed. Just above O^{16} there is some uncertainty as to whether the $s_{\frac{1}{2}}$ or $d_{\frac{5}{2}}$ "orbits" should fill first: for O^{17} and F^{19} the observed values of I are $\frac{5}{2}$ and $\frac{1}{2}$. For Ne^{21} and Na^{23}, on the other hand, $I = \frac{3}{2}$, which is a little difficult to account for. The remainder of the shell fills regularly to $_{19}K^{39}$ (odd protons) or $_{16}S^{35}$ (odd neutrons). In the next shell, the observed and predicted spins agree well in the main, but a few troublesome exceptions occur.

The single-particle model has also had some measure of success in accounting for the magnetic moments of odd-Z, even-N and even-Z, odd-N nuclei. It was in fact suggested as early as 1937, by Schmidt, that the magnetic moments of nuclei might be attributable to the motion of a single nucleon, with the contribution from its intrinsic spin moment and that from its orbital motion coupled in the same way as the corresponding electronic moments in an atom. Thus the total angular momentum **I** is considered to be the vector sum of the intrinsic spin of the nucleon **s** and an orbital angular momentum **l**. The magnetic moment so produced depends upon the value of l for the odd nucleon, and this value is so chosen as to give the closest possible agreement with the observed moment. Although the numerical agreement between predicted and observed moments is not very close, it is usually sufficiently good to permit determination of a unique l value and this value turns out in most cases to be just that predicted by the shell model.

25.18   The Collective Model

The curious dichotomy of the behavior of the nucleus as a liquid drop in some experiments and as an open structure with well-defined shells in others has occasioned no little perplexity.   In the *collective model* it is assumed that the "core" of the nucleus, comprising the particles in closed shells, is not completely inert but can be distorted by the extra nucleons.   For nuclei with closed shells there is no angular momentum, and the equilibrium shape is expected to be spherical.   However, if many nucleons exist outside closed shells, the interaction may be relatively strong, and the motion of the core so produced influences even the ground-state features.   The nucleus then exhibits a permanent deformation arising both from the "loose" nucleons and from the nuclear core distorted by the interaction with them.   As a result of the deformation, the nucleus may have a substantial quadrupole moment, a feature prominent in the rare-earth region.   There is every evidence that nuclear matter is easily deformed, vastly easier, for example, than the charge cloud of an atom.   This is not surprising when one remembers that the dominant force on an atomic electron is the coulomb field of

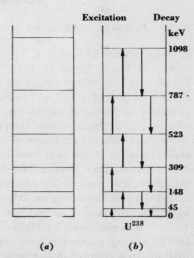

Fig. 25.28   (a) Theoretical levels for rotational band in a deformed even-*A*, even-*Z* nucleus.   (b) Experimental levels for U^{238} obtained by coulomb excitation.   [*Stephens, Diamond, and Perlman, Phys. Rev. Letters,* **3**:435 (1959).]

the nucleus, which is spherically symmetric.   For nucleons, on the other hand, the strongest interactions are with neighboring nucleons.

In molecular spectra one finds rotational, vibrational, and electronic excitation levels.   In distorted nuclei one finds analogous levels assigned to rotation, vibration, and nucleon excitation.   The nuclear rotational motion is not that of a rigid body but resembles a hydrodynamical wave moving around the nucleus; it is a rotation of the shape of the deformed surface which bounds the *A* nucleons.   The resulting energy-level pattern is very similar to that for the rotational states of a molecule (Fig. 25.28).   Not only does the deformed shape rotate, but there is also a vibration about the equilibrium shape.   By choosing a suitable interaction between the motion of a single particle and collective motions of the nucleus as a whole it is possible to obtain a description of nuclear reactions which combines the essential features of both the independent-particle and the liquid-drop model.   This approach has enjoyed a striking success in accounting for observed regularities in neutron-scattering cross sections at intermediate energies in medium and heavy nuclei.

Nuclear Fission and Nuclear Energy

25.19   Discovery of Fission

a. Transuranic Elements   The experiments which led to the discovery of fission began in 1934, in connection with investigations by Fermi and his collaborators of neutron-induced radioactivity.   The striking ability of neutrons to produce β^--unstable isotopes from many naturally occurring elements suggested to Fermi that it might be possible by this means to extend the periodic table to higher atomic numbers by neutron-induced processes in uranium.   It was very soon shown by experiments in Rome that uranium samples can in fact be made radioactive by neutron bombardment, and, furthermore, several different materials are produced.   Attempts at chemical identification soon revealed an embarrassing wealth of activities, which could be accounted for by known types of nuclear reactions only with considerable difficulty.   Thus, it was concluded by Hahn, Meitner, and Strassmann in 1937 that at least three separate chains of transitions must be assumed, all starting with neutron capture in the most abundant isotope of uranium, $_{92}U^{238}$, and involving several long-lived excited states of elements 93, 94, and perhaps 95.   More problems arose with the discovery by Irène Joliot-Curie and Savitch in 1938 of active materials which coprecipitated with lanthanum ($Z = 57$) and appeared to be "almost" inseparable from this element by chemical means. Hahn and Strassmann showed that there are several activities which coprecipitate with barium and whose daughter products behave like lanthanum. Since radium and barium have almost identical chemical properties and can be separated only with extreme difficulty, it was concluded that isotopes of radium ($Z = 88$) must be involved, formed by an $(n,2\alpha)$ process from uranium. However it was difficult to understand how an $(n,2\alpha)$ process could occur in such a heavy nuclide and especially in one so relatively stable as U^{238}. To establish the identification beyond doubt, Hahn and Strassmann set out to make the separation of the supposed radium isotopes from barium, using ThX and MsTh$_1$ (isotopes of radium) as carriers. To their surprise, they found that the activity followed the barium rather than the radium carriers and hence must be identified as barium itself! They soon found other activities which could be identified as isotopic with barium and lanthanum, and they were led to suggest that several of the supposed *transuranic elements* might actually be isotopes of other elements considerably lighter than uranium, possibly produced by an "explosion" of the uranium nucleus.

b. Physical Demonstration of Fission   A physical picture of the mechanism by which a uranium nucleus could break up into two halves was first proposed by Meitner and Frisch, within a few weeks after Hahn and Strassmann's publication of the chemical evidence. Basing their argument on the liquid-drop model of the nucleus, they pointed out that the binding forces acting on particles at the nuclear surface and producing an effective surface tension might be overcome by the electrostatic repulsion when the charge is sufficiently high. From a detailed model worked out by Bohr and Kalckar, they estimated that at $Z \approx 100$, only a small amount of extra energy—such as that brought in by a single neutron—might suffice to overcome the surface tension and cause the nucleus to break up into two roughly equal fragments. An estimate of the change in energy indicated that 100 to 200 MeV might be released in the process, an amount which roughly corresponds to the difference in (total) binding energy of U and of two elements near the middle of the atomic table. The elements thus formed would be quite unstable, because of their great neutron excess, and would decay through several stages before reaching a stable product.

Confirmation of this hypothesis came almost immediately in the demonstration by Frisch, in Copenhagen, that ionizing fragments having energies of over 70 MeV were produced in a uranium-lined ionization chamber under neutron bombardment, and by Joliot-Curie, in Paris, that the active materials had sufficient kinetic energy to travel nearly 3 cm in air; the energy he estimated to be of the order of 250 MeV.

The announcement of the results of Meitner and Frisch by Bohr at a conference in 1939 led to an intense activity among experimentalists in this country. Within a matter of days, the existence of heavily ionizing particles with enormous kinetic energies was confirmed in several laboratories, and many new and important discoveries were made in the course of the next few months. By mid-1939 the following facts were well established:

1. Fission occurs in both uranium and thorium under fast-neutron bombardment; in uranium, thermal neutrons are also effective, the cross section following the $1/v$ law.
2. U^{239}, with a half-life of 23 min, is formed in slow-neutron bombardment of uranium, the cross section exhibiting one or more strong resonances in the range 10 to 25 eV.
3. The fragments resulting from uranium fission occur in two broad groups, with energies around 60 and 100 MeV. The one group comprises fragments with mass numbers near 140, the other, near 95; symmetrical disintegrations, producing nuclides with mass numbers 115 to 120, seldom occur.
4. In about 1 percent of the fissions, delayed neutrons, presumably associated with the decay of one of the fragments, are emitted.
5. A small number of neutrons—between one and three—is emitted simultaneously with the fission.

25.20   Theory of Fission

a. Mechanism of Fission   The theory of the fission process was worked out in detail by Bohr and Wheeler. In accordance with the common treatment of nuclear reactions (see Sec. 25.14), they assumed the process to occur in two distinct stages: the formation of an excited compound nucleus by capture of a neutron, and the subsequent decay of the system in one of several possible ways. In the compound nucleus, the excess energy is distributed among the many possible modes of motion of the particles, sometimes being concentrated on a single particle and sometimes exciting much more complicated motions, involving many particles. The final outcome of the reaction is determined by the competition between the various modes of motion which can lead to disintegration and is independent of the way in which the compound nucleus was formed. In a heavy nucleus it may well occur that disintegrations resulting from these more general motions can compete favorably with single-particle emission or radiation.

In describing the complex motions of a system of many particles it is useful to deal with the changes in shape of the system as a whole, rather than to attempt to visualize the behavior of individual particles. This treatment is particularly appropriate when the particles are closely packed, so that the system is essentially incompressible. As a first approximation, then, the compound nucleus formed in the capture of a neutron by a heavy element is regarded as a drop of incompressible fluid. The simplest distortion possible for a fluid drop is that in which the drop elongates on some axis and assumes an ellipsoidal shape. The elongation results in an increase of surface area, and so the surface-tension forces tend to restore it to a spherical shape. Given a sufficient energy, it is possible to produce a distortion so extreme that the drop divides into two nearly equal spherical parts. The energy required for this can be shown to be 0.26 of the total surface energy. If the effect of a uniformly distributed electric charge on the drop is considered, it is evident that the elongation decreases the electrostatic energy (since the charges are farther apart), so that the electrostatic forces oppose the surface-tension forces and less energy is required to split the drop.

In the nuclear case, too, it is appropriate to think of a surface-tension effect, since particles near the surface have fewer neighbors than those in the interior and hence contribute less to the total binding. In fact, in the extreme liquid-drop model, it is just this effect which accounts for the general trend toward low binding energy per particle in the light nuclei (Fig. 25.3), where the surface-to-volume ratio is high. The surface area of a nucleus, and hence its surface energy, is proportional to $A^{\frac{2}{3}}$, where A is the mass number. The electrostatic energy, on the other hand, is proportional to $Z^2/A^{\frac{1}{3}}$ (Sec. 25.6), and the ratio Z^2/A is thus a measure of the relative importance of the electrostatic and surface effects. Above a certain critical value for this ratio, a nucleus is unstable with respect to arbitrarily small distortions and will elongate more and more until it breaks into two roughly equal fragments. The magnitude of the critical value can be estimated from known nuclear constants derived from the binding energies of stable nuclides and turns out to be $(Z^2/A)_{\text{limit}} = 45$, about 25 percent higher than the value for $_{92}U^{238}$. Thus it may be expected that elements of much higher atomic number than uranium are unstable to spontaneous fission.

b. Fission Threshold   For nuclei with less than the critical value of Z^2/A, a finite distortion is required to overcome the surface tension, and the amount of excess energy needed increases as the ratio Z^2/A decreases. The question whether fission can result from the capture of a neutron by a given nucleus then depends upon whether the energy brought in by the neutron (binding plus kinetic energy) is sufficient to produce the necessary distortion of the compound system. This energy turns out to be about 6 MeV for the nucleus U^{239}, formed by capture of neutrons in the most abundant isotope of uranium, U^{238}, and about 5.3 MeV for U^{236}, formed from U^{235}. The binding energy of a neutron in U^{239} is estimated to be 5.2 MeV, so that neutrons of about 1 MeV kinetic energy are required to induce fission of U^{238}. Because U^{236} has the favored even-Z, even-N character, the binding energy is nearly 1 MeV higher in this nucleus and exceeds the energy required for fission. Thus fission by thermal neutrons is possible in U^{235}. The observed thermal-neutron fission observed in natural uranium is due to this isotope, which has a relative abundance of about 1 part in 140.

c. Energy Release   The total energy released in fission of a uranium nucleus into two equal fragments can be estimated from the binding-energy curve of Fig. 25.3; near $A = 120$, the binding energy per particle is 8.5 MeV, while at $A = 240$, it is 7.6 MeV. Thus there is available 0.9 MeV/nucleon, or 216 MeV. Fission at low excitation energies is rarely symmetric. For the most common type of division, with mass ratio about 1:1.4, the energy release is about 200 MeV. Most of this energy appears immediately as kinetic energy, but about 20 MeV is released later in β decay of the fragments. A considerable adjustment—the transformation of some six or eight neutrons into protons—is needed to bring the fission products to the curve of stable nuclides, and radioactive chains of considerable complexity are involved. In a very few cases (less than 1 percent) neutrons are emitted in the course of the decay process; these are the so-called *delayed neutrons*, which follow the fission process with time delays varying from seconds to minutes. Altogether, more than 160 different radioactive nuclides have been found among the fission products: small wonder that early workers had difficulty in unscrambling their chemical properties!

25.21   *Prompt Neutrons—Chain Reactions*

Since the fission of uranium leaves the fragments with an excess of neutrons as compared with the final, stable products, it might be expected that some neutrons would be set free even in the first separation. If each fission results in the production of one or more neutrons, it may be possible, with a suitable geometrical arrangement, to use these neutrons to produce fissions in other uranium atoms and, in this way, to bring about a self-sustaining *chain reaction*. This possibility was recognized in the first discussions of fission and led to considerable speculation as to the practicability of producing nuclear energy on a large scale by this means. The magnitude of the energy available, about 200 MeV per event, is staggering compared with the few electron volts available from a chemical reaction, and it was easy to see that the development had potentially a considerable importance as a source of power and, as events progressed, as a military weapon. The crucial point in evaluating the feasibility of a self-sustaining reaction is the neutron economy. How many neutrons are emitted per fission, and how many of these can be expected to induce other fissions before being lost in some other reaction?

a. Chain Reaction in Uranium The number of *prompt neutrons*, i.e., neutrons emitted simultaneously with the fission, produced in the thermal-neutron-induced fission of U^{235} can be determined by bombarding a thin sample of separated U^{235} with a beam of thermal neutrons and observing the number of fast neutrons produced. The total number of fast neutrons is found to be 2.65 per fission. Their kinetic energies range up to 15 MeV (a few go still higher), with a most probable value near 1 MeV. If one of these 2.6 neutrons can be made to produce another fission, a chain reaction is possible; if more than 1.6 are lost or absorbed in other processes, the reaction will not be self-sustaining.

Let us consider the problem of producing a chain reaction in a block of ordinary uranium. For simplicity we take the block to be infinite in extent so that no neutrons are lost through the surface. We assume that 2.6 neutrons have been produced in a fission process at some point in the block, and we endeavor to estimate whether one of these will produce another fission. Some of the pertinent cross sections are listed in Table 25.4. For neutron energies about 1 MeV, fission of U^{238} is possible;

Table 25.4 *Neutron cross sections for some fissionable materials, in barns*

	U^{235}	U^{238}	U†	Pu^{239}
Average fission cross section, fast neutrons	1.3	0.6	0.6	2.0
Fission cross section, thermal neutrons	577	0	4.1	741
Radiative-capture cross section, thermal neutrons	101	2.7	3.4	274
Inelastic-scattering cross section, fast neutrons	3		3	2.5
Elastic-scattering cross section, fast neutrons	4		3	4

† Natural uranium; $U^{238}/U^{235} = 140$.

the average effective cross section is 0.6 barn. The cross section for U^{235} fission in this energy range is larger than that for U^{238}, but because of the small fraction ($\frac{1}{140}$) present, its effect here is negligible. Of the neutrons initially above the U^{238} threshold, few have an opportunity to cause fission; most of them will lose energy quickly through inelastic scattering processes, in which a sizable fraction of the neutron energy is used up in exciting the U^{238} nucleus. The cross section for inelastic scattering is 3 barns—5 times that for fission. In addition, neutrons may lose energy, albeit rather slowly, in elastic collisions ($\sigma = 3$ barns).

Thus of the roughly 30 percent of the initial neutrons which had energies above 1.5 MeV, about 10 percent—3 percent of the total—may produce fission in U^{238}, and the rest are slowed down to low energies.

As the neutrons approach thermal velocities, fission of U^{235} assumes importance: the fission cross section of U^{235} at thermal energy is 577 barns. In its natural dilution this represents an effective cross section of 4.1 barns. Not all the neutrons which reach thermal velocity produce fission, however, because the (n,γ) capture process occurs also, both in U^{235} and U^{238}, with an effective cross section of 3.4 barns in ordinary uranium. Thus only about half the neutrons which reach thermal energy produce fission in U^{235}. A still more serious difficulty appears in the course of the slowing-down process; this is the resonant absorption of neutrons by U^{238}, in which U^{239} is formed. The (n,γ) cross section exhibits many peaks in the region from 5 eV up to a few keV neutron energy, of which the most important lies in the neighborhood of 6.7 eV, with an effective width of 0.03 eV and a cross section of 20,000 barns. The scattering cross section at resonance is only about one twentieth of the capture cross section, and since elastic collisions reduce the energy in steps of only about one percent [see Eq. (25.1)], most of the neutrons will be captured in this resonance before reaching thermal energies. The actual situation is complicated, but the conclusion that a sustained reaction is impossible in ordinary uranium is probably correct.

b. Moderated Reactors   The problem of avoiding the resonant capture in U^{238} can be solved by the use of a *moderator*, i.e., a light material mixed with the uranium to slow the neutrons rapidly through the resonance region. In hydrogen, for example, the median energy is changed by a factor $1/e$ per collision, and the scattering cross section for slow neutrons is 20 barns (Fig. 25.21). With an admixture of, say, 100 atoms of hydrogen (in the form of water or paraffin) to one of uranium, most of the neutrons would pass through the resonance region without loss by capture. Unfortunately hydrogen has a thermal-neutron absorption cross section of 0.33 barns, so that it swallows up the neutrons as soon as they reach thermal energy and prevents a self-sustaining reaction. Of the other light substances, lithium and boron present the same difficulty. Helium, deuterium, beryllium, and carbon, having very small thermal-absorption cross sections, remain as possible moderators.

c. Critical Size   In a reactor of finite size, escape of neutrons may be a problem. Since the escape is a surface effect and all other interactions are simply proportional to the volume, the smaller the reactor, the more serious the escape. The minimum size of reactor which can just maintain itself is called the *critical size;* its magnitude can be roughly estimated from the following argument. In the course of being thermalized in a graphite moderator, a neutron makes about 100 collisions with carbon nuclei. The cross section for collision is about 4 barns; the mean free path between collisions is then

$$\lambda = \frac{1}{\sigma N} = \frac{12}{4 \times 10^{-24} \times 6 \times 10^{23} \times 1.6} \approx 3 \text{ cm}$$

where σ is the elastic-scattering cross section and N the number of nuclei per cubic centimeter. Since the neutron changes direction in a random way at each collision, the radial distance it has traveled from its point of origin after n_1 collisions is, on the average, $\sqrt{n_1} \times 3$, or 30 cm. After it reaches thermal energy, the neutron makes further collisions before it is finally captured, producing either fission or an (n,γ) process. The number of collisions before capture will be $n_2 = \Lambda/\lambda$, where Λ is the mean free path for absorption and λ the mean free path for scattering. The displacement is then $\sqrt{n_2}\,\lambda$, or $(\Lambda\lambda)^{\frac{1}{2}}$. In a pile containing 100 atoms of carbon to 1 of uranium, Λ will have roughly 100 times its value in uranium

$$\Lambda = \frac{100}{\sigma N} = \frac{100 \times 238}{7.5 \times 10^{-24} \times 6 \times 10^{23} \times 19} \approx 300 \text{ cm}$$

where σ is the combined cross section for fission and radiative capture. The scattering mean free path λ, determined by the graphite, is again about 3 cm. The combined distance for thermalizing and capture, $(30^2 + 300 \times 3)^{\frac{1}{2}} \approx 40$ cm, is then a rough measure of the average displacement of a neutron between production and capture. The actual paths of individual neutrons are subject to statistical fluctuations, and an appreciable fraction—roughly $1/e$—will suffer larger displacements than 40 cm. Also, since the displaced neutrons constitute the sources for further neutrons through the fission processes they induce, it is clear that the dimensions of the system must be several times the average path if the loss is to be kept below the few percent tolerance allowed. An actual calculation based on diffusion theory yields a value of about 300 cm for the critical radius of a spherical graphite-uranium reactor.

d. Reactor Control In practice, a reactor is built of such dimensions as to make it slightly above critical size, and the system is controlled by inserting rods or sheets of some material with a high neutron-absorption coefficient—cadmium is often used—into the reactor to reduce the multiplication factor to 1.00. The control of a reactor could be a rather ticklish problem; the lifetime of a neutron from production to ultimate absorption is only a few milliseconds, and even a small excess of the multiplication factor over 1.00 (*supercriticality*) might lead to catastrophic results in a short time.   Fortunately there are the delayed neutrons, with lifetimes ranging from fractions of a second to some minutes. These neutrons, comprising about 1 percent of the total, provide an adequate time delay so that small changes in criticality are reflected in changes in neutron flux only after a time of the order of minutes or more.

e. Power Production   As a practical matter, the most interesting characteristic of a pile is the number of fissions per unit time or, what amounts to the same thing, the power level.   Each fission releases about 200 MeV of energy, most of which appears eventually in the form of heat. A reaction rate of 3.1×10^{10} fissions per second corresponds to a power of 1 W.   The number of reactions which can be allowed to take place per second is limited by the rate at which this heat can be removed.   The first successful pile had no special cooling and could be operated at only about 2 kW.   Some reactors operate at power levels of thousands of megawatts. At a power level of 1 MW, the consumption of U^{235} is about 1.0 g/day. The attractiveness of a nuclear reactor as a power source is such that a major fraction of the large new electric power stations use nuclear energy.

25.22   Fusion

Although the phenomenon of fission is a most spectacular example of the conversion of mass energy into other, less passive forms, considerably more energy can be released by fusion of the lightest elements into elements lying near the middle of the atomic table.   From the observed variation of the average binding energy per nucleon (Fig. 25.3) it can be seen that the combination of free protons and neutrons to make $_{28}Ni^{60}$, for example, would release 8.8 MeV/nucleon—more than 10 times the energy released per nucleon in fission.   About 7 MeV/ nucleon of this energy is already released in the combination of two neutrons and two protons into an α particle: because the α particle is so tightly bound, further combinations of α particles release comparatively little energy.

Stellar Energy Sources   The conversion of hydrogen into helium is generally believed to be the principal source of energy in the sun and the stars.   Hydrogen appears to constitute more than 90 percent by mass of the total material in the universe, so that there is a comparatively large amount of energy available.   Two possible mechanisms of conversion have been given serious consideration: the so-called carbon-nitrogen cycle and the direct proton-proton chain.   Both processes depend for their operation upon extremely high temperatures. The reactions are brought about by the bombardment of one particle by another in the course of thermal agitation. Because the particles are charged, the probability of approach within the range of nuclear forces is very small indeed, even at the high temperatures and densities which obtain in stellar interiors, and, in fact, the reactions proceed mainly with particles whose energies lie high upon the tail of the Maxwell energy distribution. The carbon-nitrogen cycle requires the previous existence (formed in another process) of a small amount of carbon: the carbon is not depleted in the cycle but acts only as a kind of catalyst for conversion of four protons into an α particle. The reactions involved were proposed by Bethe and are

$$C^{12}(p,\gamma)N^{13} : N^{13}(\beta^+)C^{13}$$
$$C^{13}(p,\gamma)N^{14}$$
$$N^{14}(p,\gamma)O^{15} : O^{15}(\beta^+)N^{15}$$
$$N^{15}(p,\alpha)C^{12}$$

The net result is

$$4H^1 \rightarrow He^4 + 2\beta^+ + 2\bar{\nu} + 25 \text{ MeV}$$

Most of this energy will be absorbed locally and converted into thermal energy which maintains the reaction; the temperature will remain fixed at such a value that the rate of production just equals the rate of loss by thermal radiation.

The probability of any of these (p,γ) processes at low energies is governed by the Gamow penetration factor [Eq. (25.19)], which is a steep exponential function of the kinetic energy. At the estimated temperature of the sun's center the most probable energy kT of the protons is about 1 keV, but the rise of the cross section is so steep that the main contribution comes from the few protons with thermal-agitation energies some 10 times this value. The rate of reaction increases about as the twentieth power of the temperature.

The direct combination of protons may take place through the reaction

$$H^1(p,\beta^+)H^2$$

followed by such reactions as

$$H^2(p,\gamma)He^3 \qquad He^3(He^3,2p)He^4 \qquad He^3(d,p)He^4$$

The coulomb barrier is relatively less important here than in the reactions involving carbon and nitrogen, and the rate is less temperature-dependent. In the sun the (p,p) reactions appear to contribute most of the energy; in hotter stars, the C-N reactions dominate.

Problems

1. Compute the Q's for the following reactions (see Table 10.2 for masses): (a) $_7\text{N}^{14}(\alpha,d)_8\text{O}^{16}$; (b) $_4\text{Be}^9(d,n)_5\text{B}^{10}$; (c) $_4\text{Be}^9(\alpha,n)_6\text{C}^{12}$; (d) $_3\text{Li}^6(d,\alpha)_2\text{He}^4$; (e) $_9\text{F}^{19}(p,\alpha)_8\text{O}^{16}$.

2. An important reaction in fusion research is $\text{H}^2(d,p)\text{H}^3$. Find the Q of this reaction and the energies of the proton and triton if the energies of the deuterons can be neglected.

3. Calculate the de Broglie wavelength of a 30-GeV proton and of a 20-GeV electron. How do these values compare with the distance between nucleons in nuclei?

4. A neutron of mass m and initial speed v collides head on with a nucleus of mass M at rest. Show that if the collision is elastic, the kinetic energy K transferred to the nucleus is $2Mm^2v^2/(M + m)^2$. Find the fraction of the energy of the neutron transferred to (a) a proton of mass m, (b) an α particle of mass $4m$, and (c) a uranium atom of mass $238\ m$.

5. (a) Calculate the maximum energy of protons obtainable from a cyclotron of 1.2 m dee diameter and 1.5 Wb/m^2 magnetic induction. At what frequency must the cyclotron be operated? If the average energy gain per dee passage is 50 keV, how many revolutions do the protons make?

   (b) Calculate the maximum energy which the cyclotron can give α particles. At what frequency must the cyclotron be operated to achieve this energy? What is the α-particle energy which can be achieved using the frequency optimum for protons?

   Ans: (a) 39 MeV; 22.9 MHz; 390

6. Show that if the 4.5-MeV protons observed by the Joliot-Curies (Sec. 25.2) were indeed recoils from γ rays, the latter would have an energy of about 50 MeV.

7. Calculate the energy of the γ ray emitted by an excited B^{11} nucleus formed by the capture of a slow neutron by a B^{10} nucleus.

8. It is estimated that the energy released in the atomic-bomb explosion at Hiroshima was about 7.6×10^{13} J, equivalent to 20,000 tons of TNT. If an average of 200 MeV was released per fission, and if all fissions were by neutron capture in U^{235}, find the number of U^{235} atoms fissioned and the mass of U^{235} consumed. If 20 percent of the U^{235} atoms fissioned, what mass of U^{235} was needed for the bomb?

9. If a beam of thermal neutrons is passed through a cadmium foil 0.5 mm thick, what fraction of the beam is absorbed? All Cd isotopes have negligible absorption cross sections except for Cd^{113}, which has a capture cross section of 20,000 barns and a natural abundance of 12.26 percent. The density of Cd is 8650 kg/m^3.

10. When U^{235} captures a slow neutron, it fissions. If the fission products are Rb^{92} and Cs^{140}, how many neutrons are emitted? If the masses of U^{235}, Cs^{140}, and Rb^{92} are respectively 234.043915, 139.917110, and 91.919140 u, find the energy released in this particular fission.

11. A crude estimate of the reaction (or absorption) cross section of a nucleus for neutrons or protons of moderate energies can be obtained by assuming that any nucleon for which the center of mass reaches the edge of a nucleus is captured. Show that for neutrons this cross section is about πR^2, where R is the nuclear radius, while for protons it is $\pi R^2(1 - Ze^2/4\pi\epsilon_0 RK)$ for proton kinetic energy K greater than $Ze^2/4\pi\epsilon_0 R$ and zero for smaller values of K. *Hint:* Use the conservation of angular momentum.

12. Estimate the radius of the $_{14}Si^{27}$ nucleus from the fact that the maximum kinetic energy of the positron emitted in the reaction $_{14}Si^{27} \rightarrow {}_{13}Al^{27} + e^+ + \nu$ is 3.48 MeV. To what value of R_0 in Eq. (25.5) does this lead? (Assume that the charge is distributed uniformly throughout the nuclear volume.)

13. (*a*) Calculate the Q of the reaction $Li^7(p,\alpha)He^4$.

(*b*) If the radius of the Li nucleus is $1.4 \times 10^{-15}A^{\frac{1}{3}}$ m, compute the height of the coulomb barrier encountered by a proton reaching a Li^7 nucleus. How do you explain the fact that Cockroft and Walton excited this reaction with 125-keV protons?

(*c*) If the reaction is produced by 1.2-MeV protons striking a thin Li target, what is the maximum kinetic energy of an α particle ejected forward in the laboratory reference frame?

14. Find the Q value of the reaction $He^4(\alpha,p)Li^7$. If a beam of α particles is incident on a stationary helium target, calculate the minimum kinetic energy required of the incident α particles to excite this reaction.

15. Show that when a nuclear reaction has a negative Q value, the threshold bombarding energy for observation of particle 2 (Fig. 25.14) at angle θ is given by

$$(E_1)_\theta = \begin{cases} -Q \dfrac{M_2 + M_3}{M_2 + M_3 - M_1 - (M_1 M_2 / M_3)\sin^2\theta} & 0 \leq \theta \leq \dfrac{\pi}{2} \\[2ex] -Q \dfrac{M_3}{M_3 - M_1} & \dfrac{\pi}{2} < \theta \leq \pi \end{cases}$$

16.   Find the Q of the endoergic reaction $Li^7(p,n)Be^7$ if the mass of Be^7 is 7.016929 u.   What is the minimum bombarding energy required for protons striking stationary lithium nuclei?

Appendix I—Electronic Structure of Atoms†

Z	El.	Ground state	Ground state	IP	Ion ground state	IP	Resonance potentials		Resonance lines	
(1)	(2)	(3)	(4)	(5)	(6)	(7)	(8)		(9)	
1	H	$1s$	^2S	13.595	....	......	10.15		1,215.67(^2P)	
2	He	$1s^2$	^1S	24.581	^2S	54.403	20.96m	21.13	591.43($^3P_1$)	584.35(^1P)
3	Li	[He]$2s$	^2S	5.390	^1S	75.619	1.84		6,707.85($^2P_{\frac{3}{2}}$)	
4	Be	$-2s^2$	^1S	9.320	^2S	18.206	2.71	5.25	4,548.3($^3P_1$)	2,348.61($^1P_1$)
5	B	$-2s^22p$	$^2P_{\frac{1}{2}}$	8.296	^1S	25.149	3.57	4.94	3,470.6($^4P_{\frac{5}{2}}$)	2,497.72(^2S)
6	C	$-2s^22p^2$	$^3P_0$	11.256	$^2P_{\frac{1}{2}}$	24.376	C, 4.16	7.46	2,967.22(^5S)	1,656.998(^3P)
7	N	$-2s^22p^3$	^4S	14.53	$^3P_0$	29.593	C, 10.28		1,200.71(^4P)	
8	O	$-2s^22p^4$	$^3P_2$	13.614	^4S	35.108	C, 9.11	9.48	1,355.60(^5S)	1,302.17(^3S)
9	F	$-2s^22p^5$	$^2P_{\frac{3}{2}}$	17.418	$^3P_2$	34.98	12.69	12.98	976.50(^4P)	954.82(^2P)
10	Ne	$-2s^22p^6$	^1S	21.559	$^2P_{\frac{3}{2}}$	41.07	16.62m	16.84	743.71($^3P_1$)	735.89($^1P_1$)
11	Na	[Ne]$3s$	^2S	5.138	^1S	47.29	2.10		5,889.95($^2P_{\frac{3}{2}}$)	
12	Mg	$-3s^2$	^1S	7.644	^2S	15.031	2.71m	4.33	4,571.10($^3P_1$)	2,852.12($^1P_1$)
13	Al	$-3s^23p$	$^2P_{\frac{1}{2}}$	5.984	^1S	18.823	3.13		3,961.52(^2S)	
14	Si	$-3s^23p^2$	$^3P_0$	8.149	$^2P_{\frac{1}{2}}$	16.34	C,	4.93		2,516.11($^1P_1$)
15	P	$-3s^23p^3$	^4S	10.484	$^3P_0$	19.72	C, 6.93		1,787.65(^4P)	1,774.94
16	S	$-3s^23p^4$	$^3P_2$	10.357	^4S	23.4	C, 6.50	6.83	1,900.27(^5S)	1,807.31(^3S)
17	Cl	$-3s^23p^5$	$^2P_{\frac{3}{2}}$	13.01	$^3P_2$	23.80	8.88	9.16	1,389.78($^4P_{\frac{5}{2}}$)	1,347.32($^2P_{\frac{3}{2}}$)
18	A	$-3s^23p^6$	^1S	15.755	$^2P_{\frac{3}{2}}$	27.62	11.55m	11.83	1,066.66($^3P_1$)	1,049.22($^1P_1$)
19	K	[A]$4s$	^2S	4.339	^1S	31.81	1.61		7,664.91($^2P_{\frac{3}{2}}$)	
20	Ca	$-4s^2$	^1S	6.111	^2S	11.868	1.88	2.92	6,572.78($^3P_1$)	4,226.73($^1P_1$)
21	Sc	$-3d4s^2$	$^2D_{\frac{3}{2}}$	6.54	$^3D_1$	12.80	1.94m	1.98	6,378.82($^4F_{\frac{3}{2}}$)	6,305.67($^6D_{\frac{9}{2}}$)
22	Ti	$-3d^24s^2$	$^3F_2$	6.82	$^2F_{\frac{3}{2}}$	13.57	C, 1.96	2.39	6,296.65($^5G_2$)	5,173.74($^3D_1$)
23	V	$-3d^34s^2$	$^4F_{\frac{3}{2}}$	6.74^b	$^5D_0$	14.65	C, 2.23	2.54	5,527.72($^6G_{\frac{3}{2}}$)	4,851.48($^4D_{\frac{1}{2}}$)
24	Cr	$-3d^54s$	^7S	6.764	^6S	16.49	C, 2.90		4,289.72($^7P_2$)	
25	Mn	$-3d^54s^2$	^6S	7.432	^7S	15.636	C, 2.27	3.06	5,432.55($^8P_{\frac{7}{2}}$)	4,034.49($^6P_{\frac{7}{2}}$)
26	Fe	$-3d^64s^2$	$^5D_4$	7.87	$^6D_{\frac{9}{2}}$	16.18	C, 2.39	3.20	5,166.29($^7D_5$)	3,859.91($^5D_4$)
27	Co	$-3d^74s^2$	$^4F_{\frac{9}{2}}$	7.86^c	$^3F_4$	17.05	C, 2.91	3.50	4,233.99($^6F_{1\frac{1}{2}}$)	3,526.85($^4F_{\frac{9}{2}}$)
28	Ni	$-3d^84s^2$	$^3F_4$	7.633^d	$^2D_{\frac{5}{2}}$	18.15	C, 3.18	3.64	3,884.58($^5D_4$)	3,670.43($^3P_2$)
29	Cu	$-3d^{10}4s$	^2S	7.724	^1S	20.29	C, 3.79		3,273.96($^2P_{\frac{3}{2}}$)	
30	Zn	$-3d^{10}4s^2$	^1S	9.391	^2S	17.96	4.01m	5.77	3,075.90($^3P_1$)	2,138.56($^1P_1$)
31	Ga	$-3d^{10}4s^24p$	$^2P_{\frac{1}{2}}$	6.00	^1S	20.51	3.06		4,032.98(^2S)	
32	Ge	$-3d^{10}4s^24p^2$	$^3P_0$	7.88	$^2P_{\frac{1}{2}}$	15.93	C, 4.64m		2,651.58($^3P_1$)	
33	As	$-3d^{10}4s^24p^3$	^4S	9.81	$^3P_0$	18.63	C, 6.26		1,972.62($^4P_{\frac{5}{2}}$)	
34	Se	$-3d^{10}4s^24p^4$	$^3P_2$	9.75	^4S	21.5	C, 5.95	6.30	2,074.79(^5S)	1,960.90($^3S_4$)
35	Br	$-3d^{10}4s^24p^5$	$^2P_{\frac{3}{2}}$	11.84	$^3P_2$	21.6	7.83	8.29	1,576.5($^4P_{\frac{5}{2}}$)	1,488.6($^2P_{\frac{3}{2}}$)
36	Kr	$-3d^{10}4s^24p^6$	^1S	13.996	$^2P_{\frac{3}{2}}$	24.56	9.91m	9.99*	1,235.82($^3P_1$)	1,164.86(^1P)
37	Rb	[Kr]$5s$	^2S	4.176	^1S	27.5	1.56		7,947.64($^2P_{\frac{3}{2}}$)	
38	Sr	$-5s^2$	^1S	5.692	^2S	11.027	1.78m	2.68*	6,892.58($^3P_1$)	4,607.33*(^1P)
39	Y	$-4d5s^2$	$^2D_{\frac{3}{2}}$	6.377	^1S	12.233	1.31		9,494.81($^2P_{\frac{3}{2}}$)	
40	Zr	$-4d^25s^2$	$^3F_2$	6.835	$^4F_{\frac{3}{2}}$	13.13	C, 1.83	2.71	6,762.38($^5G_2$)	4,575.52($^3G_3$)
41	Nb	$-4d^45s$	$^6D_{\frac{1}{2}}$	6.881	$^5D_0$	14.32	C, 2.07m	2.97*	5,320.21($^6F_{\frac{1}{2}}$)	4,168.12*($^6F_{\frac{1}{2}}$)
42	Mo	$-4d^55s$	^7S	7.10	^6S	16.15	C, 3.18		3,902.96($^7P_2$)	
43	Tc	$-4d^55s^2$	^6S	7.28	^7S	15.26	2.09	2.88	5,924.57($^8P_{\frac{7}{2}}$)	4,297.06($^6P_{\frac{7}{2}}$)
44	Ru	$-4d^75s$	$^5F_5$	7.365	$^4F_{\frac{9}{2}}$	16.76	C, 3.13	3.26*	3,964.90($^7D_5$)	3,799.35*($^5D_4$)
45	Rh	$-4d^85s$	$^4F_{\frac{9}{2}}$	7.461	$^3F_4$	18.07	C, 3.36		3,692.36($^4D_{\frac{7}{2}}$)	
46	Pd	$-4d^{10}$	^1S	8.33	$^2D_{\frac{5}{2}}$	19.42	C, 4.22m	5.01*	2,763.09($^3P_1$)	2,447.91*($^1P_1$)
47	Ag	$-4d^{10}5s$	^2S	7.574	^1S	21.48	3.66		3,382.89($^2P_{\frac{1}{2}}$)	
48	Cd	$-4d^{10}5s^2$	^1S	8.991	^2S	16.904	3.73m	5.29	3,261.04($^3P_1$)	2,288.02($^1P_1$)
49	In	$-4d^{10}5s^25p$	$^2P_{\frac{1}{2}}$	5.785	^1S	18.828	3.02		4,101.76(^2S)	
50	Sn	$-4d^{10}5s^25p^2$	$^3P_0$	7.342	$^2P_{\frac{1}{2}}$	14.63	C, 4.29m	4.33*	2,863.32*($^3P_1$)	
51	Sb	$-4d^{10}5s^25p^3$	^4S	8.639	$^3P_0$	16.5	C, 5.36		2,311.47($^4P_{\frac{5}{2}}$)	
52	Te	$-4d^{10}5s^25p^4$	$^3P_2$	9.01	^4S	18.6	C, 5.49	5.78	2,259.02($^5S_2$)	2,142.75($^3S_1$)
53	I	$-4d^{10}5s^25p^5$	$^2P_{\frac{3}{2}}$	10.454	$^3P_2$	19.130	6.77	7.66	2,062.1($^4P_{\frac{5}{2}}$)	1,617.7($^2P_{\frac{3}{2}}$)
54	Xe	$-4d^{10}5s^25p^6$	^1S	12.127	$^2P_{\frac{3}{2}}$	21.21	8.31m	8.44*	1,469.62*($^3P_1$)	1,295.56($^1P_1$)
55	Cs	[Xe]$6s$	^2S	3.893	^1S	25.1	1.38		8,943.46($^2P_{\frac{3}{2}}$)	8,521.10($^2P_{\frac{1}{2}}$)

(Table continued on next page)

753

Z	El.	Ground state	Ground state	IP	Ion ground state	IP	Resonance potentials		Resonance lines	
(1)	(2)	(3)	(4)	(5)	(6)	(7)	(8)		(9)	
56	Ba	$-6s^2$	^1S	5.210	^2S	10.001	C, 1.52m	2.24*	7,911.36($^3P_1$)	5,535.53*(^1P)
57	La	$-5d6s^2$	$^2D_{\frac{3}{2}}$	5.61^c	$^3F_2$	11.43	C, 1.64	1.84*	7,539.24($^4F_{\frac{3}{2}}$)	6,753.05($^2D_{\frac{3}{2}}$)
58	Ce	$-4f5d6s^2$	$^3H_5$	f	$^4H_{\frac{7}{2}}$					
59	Pr	$-4f^36s^2$	$^4I_{\frac{9}{2}}$	5.48^i	$^5I^4$					
60	Nd	$-4f^46s^2$	$^5I_4$	5.51^{9i}	$^6I_{\frac{7}{2}}$					
61	Pm	$-4f^56s^2$	$^6H_{\frac{5}{2}}$							
62	Sm	$-4f^66s^2$	$^7S_0$	5.6	$^8F_{\frac{1}{2}}$	11.2	C, 1.71m	1.74*	7,141.13*($^9F_1$)	6,725.88($^9G_1$)
63	Eu	$-4f^76s^2$	^8S	5.67	^9S	11.24	C, 1.74	2.66*	7,106.48($^{10}P_{\frac{7}{2}}$)	4,661.88*($^8P_{\frac{7}{2}}$)
64	Gd	$-4f^75d6s^2$	$^9D_2$	6.16	$^{10}D_{\frac{5}{2}}$	12.+	....			4,225.85
65	Tb	$-4f^85d6s^2$								
66	Dy									
67	Ho									
68	Er									
69	Tm	$-4f^{13}6s^2$	$^2F_{\frac{7}{2}}$	......	$^3F_4$	......	....			5,675.83
70	Yb	$-4f^{14}6s^2$	^1S	6.22	^2S	12.10				3,987.99
71	Lu	$-4f^{14}5d6s^2$	$^2D_{\frac{3}{2}}$	6.15	^1S	14.7				
72	Hf	$-4f^{14}5d^26s^2$	$^3F_2$	7.0	$^2D_{\frac{3}{2}}$	14.9	C, 2.19			
73	Ta	$-4f^{14}5d^36s^2$	$^4F_{\frac{3}{2}}$	7.88	$^5F_1$	16.2	C, 2.90		4,280.47	
74	W	$-4f^{14}5d^46s^2$	$^5D_0$	7.98	$^6D_{\frac{1}{2}}$	17.7	C, 2.40m		4,982.16($^7F_1$)	4,008.75
75	Re	$-4f^{14}5d^56s^2$	^6S	7.87	^7S	16.6	C, 2.35	3.58	5,275.53($^8P_{\frac{5}{2}}$)	3,464.72($^6P_{\frac{7}{2}}$)
76	Os	$-4f^{14}5d^66s^2$	$^5D_4$	8.7	$^6D_{\frac{9}{2}}$	17	C, 2.80		4,420.67($^7D_4$)	
77	Ir	$-4f^{14}5d^76s^2$	$^4F_{\frac{9}{2}}$	9.2	....	......	C, 3.26		3,800.12($^6D_{\frac{9}{2}}$)	
78	Pt	$-4f^{14}5d^96s$	$^3D_3$	9.0	$^2D_{\frac{5}{2}}$	18.56	C, 3.74	4.04	3,315.05($^5D_4$)	3,064.71($^3P_2$)
79	Au	$[^h]6s$	^2S	9.22	^1S	20.5	C, 4.63		2,675.95($^2P_{\frac{3}{2}}$)	2,427.95($^2P_{\frac{3}{2}}$)
80	Hg	$-6s^2$	^1S	10.434	^2S	18.751	C, 4.67m	6.70	2,536.52($^3P_1$)	1,849.57($^1P_1$)
81	Tl	$-6s^26p$	$^2P_{\frac{1}{2}}$	6.106	^1S	20.42	3.29		3,775.72(^2S)	
82	Pb	$-6s^26p^2$	$^3P_0$	7.415	$^2P_{\frac{1}{2}}$	15.028	C, 4.33m	4.37	2,833.07*($^3P_1$)	
83	Bi	$-6s^26p^3$	^4S	7.287	$^3P_0$	16.68	C, 4.04		3,067.72($^4P_{\frac{3}{2}}$)	
84	Po	$-6s^26p^4$	$^3P_2$	8.43	....	......	....			2,449.99
85	At	$-6s^26p^5$	$^2P_{\frac{3}{2}}$*							
86	Rn	$-6s^26p^6$	^1S	10.745	$^2P_{\frac{3}{2}}$*	......	6.77m	6.94*	1,786.07*($^3P_1$)	1,451.56($^1P_1$)
87	Fr	$[Rn]7s$	^2S*	......	^1S*					
88	Ra	$-7s^2$	^1S	5.277	^2S	10.14	1.62m	2.57*	7,141.21	4,825.91*
89	Ac	$-6d7s^2$	$^2D_{\frac{3}{2}}$	6.9	^1S	12.1				
90	Th	$-6d^27s^2$	$^3F_2$	......	$^4F_{\frac{3}{2}}$					
91	Pa	$-6d^37s^2$*								
92	U	$-5f^36d7s^2$	$^5L_6$	4	$^4I_{\frac{9}{2}}$	......	....			5,915.40
93	Np									
94	Pu									
95	Am									
96	Cm									
97	Bk									
98	Cf									

a Data taken from current literature. Use has been made of Moore, "Atomic Energy Levels," vols. I–III, and "Smithsonian Physical Tables," 9th ed.
b Normal state of ion—3d^4.
c Normal state of ion—3d^5.
d Normal state of ion—3d^9.
e Normal state of ion—5d^2.
f Normal state of ion—4f^26s.
g Normal state of ion—4f^36s.
h Structure of closed shells [Xe]4$f^{14}5d^{10}$.
i N. F. Ionov and M. A. Mittsev, *Zhur. Eksp. i Teoret. Fiz.* **38**, 1350 (1960).

† From "American Institute of Physics Handbook," 2d ed.   Copyright 1963.
McGraw-Hill Book Company.   Used by permission.